# Subnuclear Structure of Matter: Achievements and Challenges



EPJ Web of Conferences Volume 134 (2017)

Bonn, Germany
6 – 9 June 2016

**Editors:**

**F. Klein**
**U.G. Meibner**
**U. Thoma**

ISBN: 978-1-5108-3606-8

**Printed from e-media with permission by:**

Curran Associates, Inc.
57 Morehouse Lane
Red Hook, NY 12571

**Some format issues inherent in the e-media version may also appear in this print version.**



Printed by Curran Associates, Inc. (2017)

For additional information, please contact EDP Sciences – Web of Conferences at the address below.

EDP Sciences – Web of Conferences
17, Avenue du Hoggar
Parc d'Activité de Courtabœuf
BP 112
F-91944 Les Ulis Cedex A
France

Phone: +33 (0) 1 69 18 75 75
Fax: +33 (0) 1 69 28 84 91

contact-edps@webofconferences.org

**Additional copies of this publication are available from:**

Curran Associates, Inc.
57 Morehouse Lane
Red Hook, NY 12571 USA
Phone: 845-758-0400
Fax: 845-758-2633
Email: curran@proceedings.com
Web: www.proceedings.com

# TABLE OF CONTENTS

edp sciences

EPJ Web of Conferences

# Statement of Peer review

In submitting conference proceedings to EPJ Web of Conferences, I certify to the Publisher that I adhere to the **Policy on Publishing Integrity** of the journal in order to safeguard good scientific practice in publishing.

1. All articles have been subjected to peer review administered by the proceedings editors.
2. Reviews have been conducted by expert referees, who have been requested to provide unbiased and constructive comments aimed, whenever possible, at improving the work.
3. Proceedings editors have taken all reasonable steps to ensure the quality of the materials they publish and their decision to accept or reject a paper for publication has been based only on the merits of the work and the relevance to the journal.

Title, date and place of the conference

Title: Subnuclear structure of matter. Achievements and Challenges
Date: June 6 - June 9, 2016
Location: Bonn, Germany

Proceedings editor(s):

F. Klein, U.-G. Meißner, U. Thoma

Date and editor's signature

17, avenue du Hoggar - PA de Courtabœuf – BP 112 - 91944 Les Ulis Cedex A (France)
Tel. : 33 (0)1 69 18 75 75 - Fax : 33(0)1 69 07 45 17 – www.edpsciences.org

DOI: 10.1051/epjconf/201713400001

**SFB/TR 16**

**"Subnuclear Structure of Matter"**

# Preface

The symposium "Subnuclear Structure of Matter: Achievements and Challenges" took place in Bonn (Germany) from June 6th to June 9th, 2016. It was intended to summarize all the work that has been done in the framework of the Collaborative Research Center SFB/TR 16 "Subnuclear Structure of Matter" of the Universities Bonn, Bochum and Gießen, funded for 12 years (three funding periods) by the Deutsche Forschungsgemeinschaft (DFG). The research in this CRC was centered around the baryon spectrum made of light quarks, combining photoproduction experiments at the electron accelerator ELSA at Bonn with state-of-the-art theoretical approaches as well as detector, accelerator and polarized target developments. In essence, the research performed within the SFB/TR 16 contributed significantly to our understanding of the strong interactions in the non-perturbative regime of QCD.

The symposium consisted of invited talks, that fell into two groups. On the one hand, there were reports of the various projects funded in the SFB/TR 16, mostly delivered by one of the project leaders. The corresponding contributions are given in these proceedings, together with the summaries of projects funded in earlier periods only. On the other hand, external speakers were invited to present various topics related to this research and to elucidate the role of the SFB/TR 16 in advancing our understanding of the respective topic. These were Volker Burkert, David Richards, Kurt Aulenbacher, Reinhard Schumacher, Barry Holstein, Kenta Itahashi, Achim Denig, Christian Lang, Andreas Thomas, and Michael Döring. We are grateful to these colleagues for making this effort and in particular to Volker Burkert, Barry Holstein and Michael Döring for providing a write-up to these proceedings.

All this work would not have been possible without the substantial funding from the DFG to the SFB/TR 16 and also due to the sizeable support from the Universities. In particular, we are grateful to Bonn University for the on-going major funding to support the local infrastructure, especially ELSA. This has not only laid the foundation for all the high-class physics results reported here, but also helped to educate a large number of students over the years who made very essential contributions to the physics results reported in these proceedings.

F. Klein, U.-G. Meißner, U. Thoma

# Spectrum and Structure of Excited Baryons with CLAS

*Volker D.* Burkert[1,*], for the CLAS collaboration

[1] *Thomas Jefferson National Accelerator Facility*
*12000 Jefferson Avenue, Newport News, Virginia, USA*

**Abstract.** In this contribution I discuss recent results in light quark baryon spectroscopy involving CLAS data and higher level analysis results from the partial wave analysis by the Bonn-Gatchina group. New baryon states were discovered largely based on the open strangeness production channels $\gamma p \to K^+\Lambda$ and $\gamma p \to K^+\Sigma^0$. The data illustrate the great potential of the kaon-hyperon channel in the discovery of higher mass baryon resonances in s-channel production. Other channels with discovery potential, such as $\gamma p \to p\omega$ and $\gamma p \to \phi p$ are also discussed. In the second part I will demonstrate on data the sensitivity of meson electroproduction to expose the active degrees of freedom underlying resonance transitions as a function of the probed distance scale. For several of the prominent excited states in the lower mass range the short distance behavior is described by a core of three dressed-quarks with running quark mass, and meson-baryon contributions make up significant parts of the excitation strength at large distances. Finally, I give an outlook of baryon resonance physics at the 12 GeV CEBAF electron accelerator.

## 1 Foreword

In the first physics talk at this symposium Ulrike Thoma asked me to present the research on excited baryon states with the CLAS detector at Jefferson Lab. Before I come to this subject I like to point out how pleased I am to participate in and to contribute to the celebration of the CRC 16 program, whose impact on baryon spectroscopy is well known in the broader hadron community. Many of the results I show in my presentation would not have been possible without the CRC 16 project and its phenomenological component, which is generally known as the Bonn-Gatchina (BnGa) coupled-channel analysis framework. When I refer to newly discovered baryon states they will have data from CLAS as a major input, but the high level analysis includes other data sets as well and the results of this analysis is the responsibility of the CRC 16 project, which by having strong experimental and analysis components is making lasting contributions to our understanding of the nucleon spectrum.

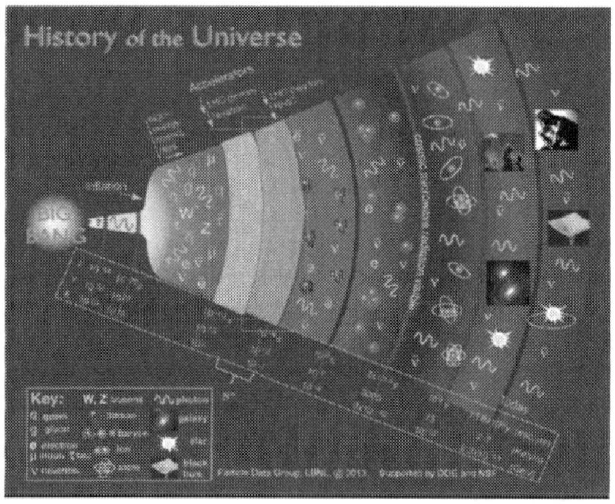


**Figure 1.** The evolution of the universe as pictured by the LBNL Particle Data Group, 2013. The area highlighted in yellow is where the quark-gluon plasma is prevalent. The green area indicates the cross over to the hadronic phase that is driven by the presence of excited baryons. Electron accelerators CEBAF, ELSA and MAMI have the energy to access this region (color highlights added by author.)

## 2 Introduction

Dramatic events occurred in the evolution of the microsecond old universe that had tremendous implication for the further development of the universe to the state it is in today. As the universe expanded and cooled sufficiently into the GeV range (see Fig. 1), the transition occurred from the phase of free (unconfined) quarks and gluons, to the hadron phase with quarks and gluons confined in volumes of $\approx$ 1 fm$^3$, i.e. protons, neutrons, and other baryons. In course of this process, elementary, nearly massless quarks acquire dynamical mass due to the coupling to the dressed gluons, and the fact that chiral symmetry is broken dynamically [1]. This transition is not a simple first order phase transition, but a "cross over" between two phases

*Talk presented at the CRC-16 Symposium, Bonn University, June 6-9, 2016

DOI: 10.1051/epjconf/201713401001

of strongly interacting matter, which is moderated by the excitation of baryon resonances starting from the highest mass states and ending with the low mass resonances. Figure 2 shows a generic QCD phase diagram indicating the region of the cross over from the de-confined region to the confined region of hadrons.

A quantitative understanding of this transition requires more excited baryons of all flavors, than have currently been included in the "Review of Particle Properties (RPP)". The full set of states predicted by quark models with $SU(6)$ symmetry and by Lattice QCD are needed [2, 3] to describe the evolution process. The presence of the full complement of excited baryons, the acquisition of dynamical mass by light quarks, and the transition from unconfined quarks to confinement are intricately related and are at the core of the problems we are trying to solve in hadron physics today. We do have all the tools at our disposal to study the individual excited states in relative isolation, and to probe the quark mass versus the momentum or distance scale, and to search for so far undiscovered baryon states.

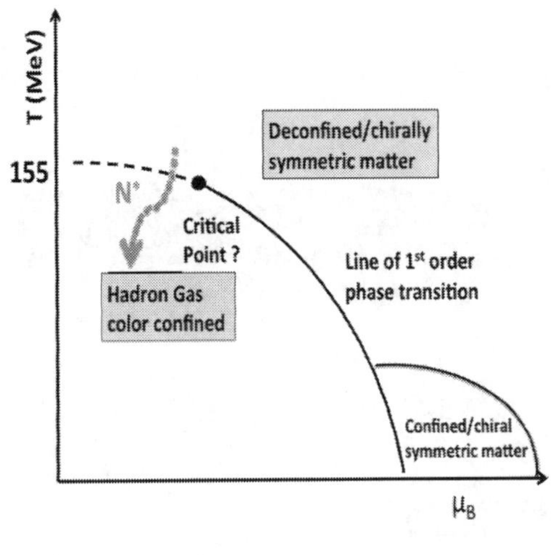


**Figure 2.** Generic QCD phase diagram showing the de-confined quark-gluon plasma phase (yellow shade) and the hadron gas phase (green shade). The thick green line indicates a possible path from the de-confined to the confined phase due to the presence of excited baryons bypassing the line of first order phase transition.

Accounting for the excitation spectrum of protons and nucleons and accessing many facets of strong interaction mechanisms underlying the generation of excited baryons and their intrinsic structures, is one of the most important and certainly the most challenging task in hadron physics. It has been the focus of the CLAS N* program at Jefferson Lab and likely will remain so with its extensions towards higher energies with CLAS12.

**Figure 3.** The CLAS detector in an open maintenance position exposing the large drift chambers covering nearly the full polar angle range. The time-of-flight scintillator bars and the forward electromagnetic calorimeters are also visible.

## 3 Establishing the light quark baryon spectrum

Obtaining an accurate account of the full nucleon resonance spectrum is the basis for making progress in our understanding of strong QCD as it relates to light quark sector. We may learn from atomic spectroscopy of the hydrogen atom whose series of sharp energy levels could be explained within the Bohr model. However, deviations from this model eventually led to the QED, which is applicable to all atoms. In the same vein, we need precise measurements of the nucleon excitation spectrum to test our best models. The symmetric quark model provides a description of the lower mass spectrum in terms of isospin and spin-parity quantum numbers, but masses are off, and there are many states predicted within the $SU(6)\otimes O(3)$ symmetry that are missing from the observed spectrum. Although we have already the correct theory, we cannot really test it on the nucleon spectrum, because the full spectrum is not known, and the theory is currently not in a position to predict more than what the quark model already has done. Our task is therefore two-fold: 1) to establish the experimental nucleon spectrum, and 2) to develop strong QCD to be able to reliably explain it in detail, including masses and hadronic and electromagnetic couplings. The experimental part has been the goal of the $N^*$ program with CLAS detector (see Fig. 3) and with other facilities, especially CB-ELSA. Significant progress has been made in recent years that also included development of multi-channel partial wave analysis frameworks. Much of recent progress came as a result of precise data, including measurement of polarization observables collected in the strangeness channel, which I will discuss in the following section.

DOI: 10.1051/epjconf/201713401001

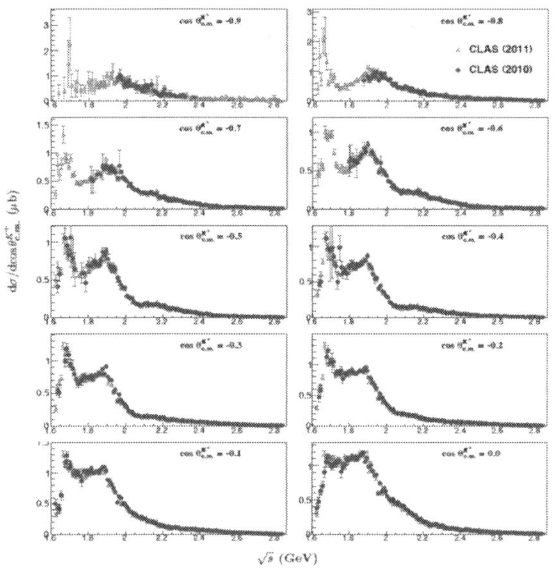


**Figure 4.** CLAS cross section data on $\gamma p \rightarrow K^+\Lambda$ differential cross section in the backward polar angle range. There are 3 structure visible that indicate resonance excitations, at 1.7, 1.9, and 2.2 GeV. The blue full circles are based on the topology $K^+p\pi^-$, the red open triangles are based on topology $K^+p$ or $K^+\pi^-$, which extended coverage towards lower W at backward angles and allows better access to the resonant structure near threshold.

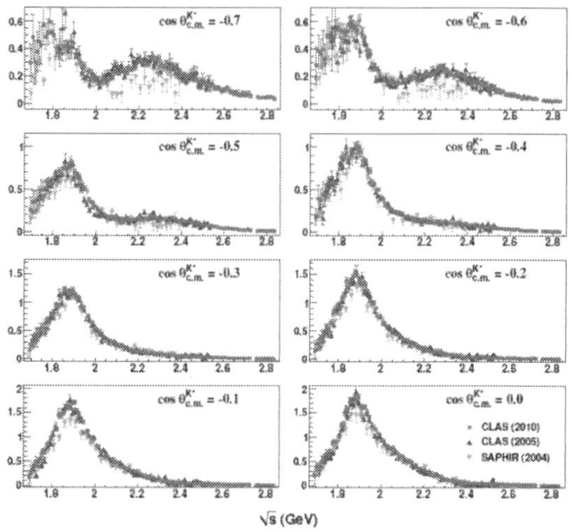


**Figure 5.** Invariant mass dependence of the $\gamma p \rightarrow K^+\Sigma^\circ$ differential cross section in the backward polar angle range. At the most backward angles there are significant discrepancies between the older SAPHIR data and both CLAS data sets, in particular in the mass range from 2.1 to 2.4 GeV, and at backward angles.

### 3.1 Hyperon photoproduction

Here one focus has recently been on precision measurements of the $\gamma p \rightarrow K^+\Lambda$ and $\gamma p \rightarrow K^+\Sigma^\circ$ differential cross section [4, 5]; and using polarized photon beams, with circular or linear polarization [4–8], several polarization observables can be measured by analyzing the weak decay of the recoil $\Lambda \rightarrow p\pi^-$, and $\Sigma^\circ \rightarrow \gamma\Lambda$. Samples of the data are shown in Fig. 4 and Fig. 5. It is well known that the energy-dependence of a partial-wave amplitude for one particular channel is influenced by other reaction channels due to unitarity constraints. To fully describe the energy-dependence of an amplitude one has to include other reaction channels in a coupled-channel approach. Such analyses have been developed by the Bonn-Gatchina group [9], at EBAC [10], by the Argonne-Osaka group [11], and the Jülich/GWU group [12].

The Bonn-Gatchina group has claimed a set of eight states that are either newly discovered or have significantly improved evidence for their existence. These states entered in recent editions of the Review of Particle Properties [13]. Figure 7 shows the nucleon and $\Delta$ spectrum with the new additions highlighted.

Figure 6 shows the nucleon resonances observed in the Bonn-Gatchina multi-channel partial wave analysis using the hyperon photoproduction data from CLAS and other data sets. Figure 7 shows how the new states fit into the previously observed nucleon and $\Delta$ states for masses up to 2.4 GeV.

| State $N(mass)J^P$ | PDG pre 2012 | PDG 2014/2016 | $K\Lambda$ | $K\Sigma$ | $N\gamma$ |
|---|---|---|---|---|---|
| $N(1710)1/2^+$ | *** | **** | **** | ** | **** |
| $N(1880)1/2^+$ | | ** | ** | | ** |
| $N(1895)1/2^-$ | | ** | ** | * | ** |
| $N(1900)3/2^+$ | ** | *** | *** | ** | *** |
| $N(1875)3/2^-$ | | *** | *** | ** | *** |
| $N(2150)3/2^-$ | | ** | ** | | ** |
| $N(2000)5/2^+$ | * | ** | ** | * | ** |
| $N(2060)5/2^-$ | | ** | | ** | ** |

**Figure 6.** Recently discovered nucleon resonances with their star ratings in the Review of Particle Properties, 2014/16. The states have been observed in the Bonn-Gatchina multi-channel partial wave analysis of photo-produced $K^+\Lambda$ and $K^+\Sigma^\circ$ final states.

New data on $K^+\Lambda$ production using a linearly polarized photon beam and measuring the $\Lambda$ recoil polarization along the $\Lambda$ momentum [14] are shown in Fig 8. These data show strong sensitivity to excited baryon states, in-

DOI: 10.1051/epjconf/201713401001

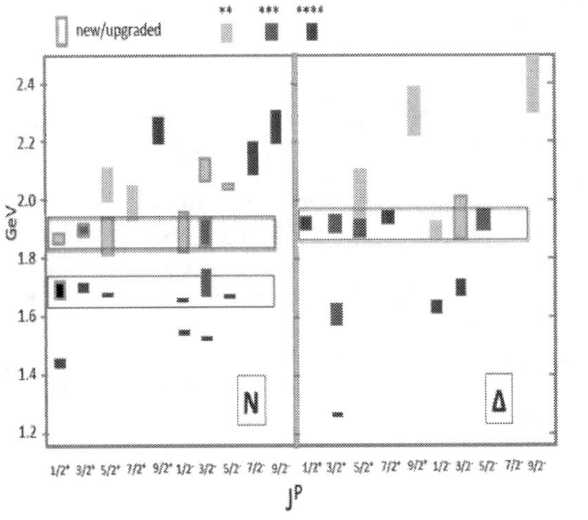


**Figure 7.** The nucleon and Δ spectrum for masses up to 2.4 GeV. The green frames indicate states that have newly been found in the BnGa analysis or have been upgraded in their star rating in the RPP [13]. The red and blue boxes highlight near mass degenerate states with different spin and parity. The states at 1.9 GeV are all from the latest additions to RPP. They seem to repeat a pattern of states observed near 1.7 GeV. A similar pattern is observed in the Δ sector at 1.9 GeV, with a new addition at $J^P = \frac{3}{2}^-$ filling in a void of in the previously observed pattern.

cluding possible new states, but they have not been included in previous full multi-channel partial wave analyses.

### 3.2 Vector meson photoproduction

A large volume of precision photoproduction data have been taken with CLAS on $\gamma p \to p\omega$ covering the mass range from threshold to 2.8 GeV [15]. The final state has been measured in its dominant decay channel $\omega \to \pi^+\pi^-\pi^\circ$. Using neutral-charge vector mesons in partial wave analyses is more involved compared to pseudoscalar mesons as the spin $J = 1$ meson coupling with the spin 1/2 proton results in more amplitudes defining the process. The advantage, however, is that the process can only proceed through isopsin-1/2 nucleon resonances, which simplifies the analysis. There are also large t-channel and diffractive processes that give large background contributions. As of this writing this data set has not yet been included in a multi-channel partial wave analysis.

In spite of these complications, the omega data have been employed in a single channel event-based analysis and fit in a 2-pole K-matrix procedure [16]. Figure 9 demonstrates the sensitivity of this channel to $N(2000)\frac{5}{2}^+$, a 2-star candidate state whose existence is strongly supported by this analysis.

The process $\gamma p \to p\phi$ has been measured in its dominant final state $K\bar{K}$ with both charge channels $\phi \to K^\circ_s K^\circ_l$ [17], and $\phi \to K^+K^-$ [18]. Figure 10 shows the

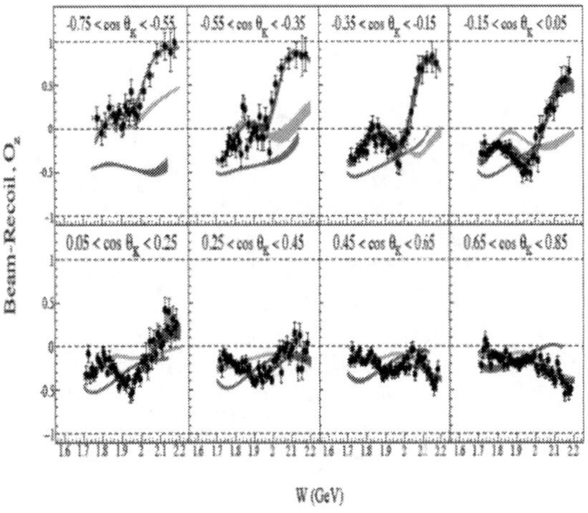


**Figure 8.** Double polarization observable $O_z$ for different $K^+$ polar angle bins. The bands are projections by ANL-Osaka (red), BnGa 2014 (green), and BnGa 2016 (blue). The latter included the new data in the fit. The large discrepancies at $W > 2$GeV indicate possible high mass resonance strength that was not included in earlier fits.

cross section in the backward hemisphere. There are small enhancements near $\sqrt{s} = 2.25$ GeV, indicating resonance structure that is however overwhelmed by the increasing background contributions at more forward angles.

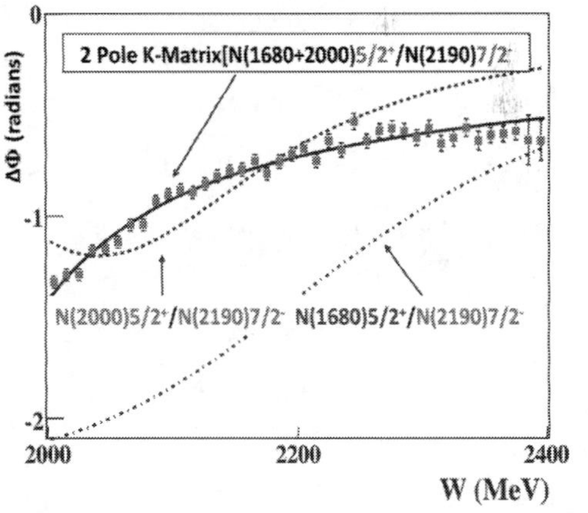


**Figure 9.** The phase motion of $\gamma p \to p\omega$ with the $\omega$ fully reconstructed from its decay $\omega \to \pi^+\pi^-\pi^\circ$ and the differential cross section and (unpolarized) spin-density matrix elements determined from the data.

There is a strong bump near $\sqrt{s} = 2.2$ GeV seen in Fig. 11 at the forward most angle bin. The bump is present in both charge channels and does not appear to be due to a genuine baryon resonance as it is not present at larger an-

DOI: 10.1051/epjconf/201713401001

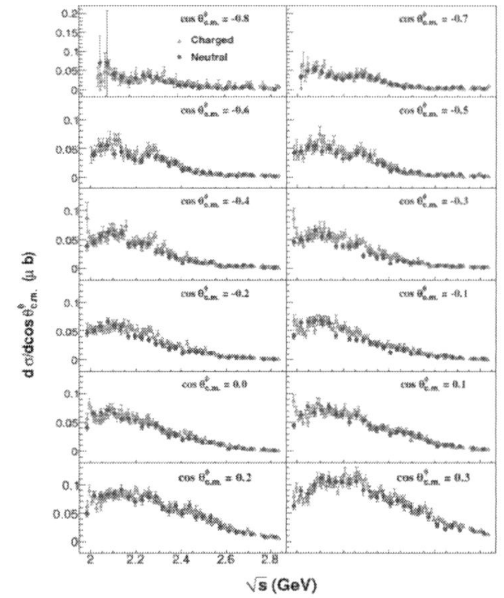


**Figure 10.** Differential cross section of $\phi$ photoproduction, shown in the backward hemisphere. The open red circles are from the charged final state, the solid blue points from the neutral charge final state. There are enhancements in both charge channels at the most backward angle bins near $\sqrt{s} = 2.25$ GeV.

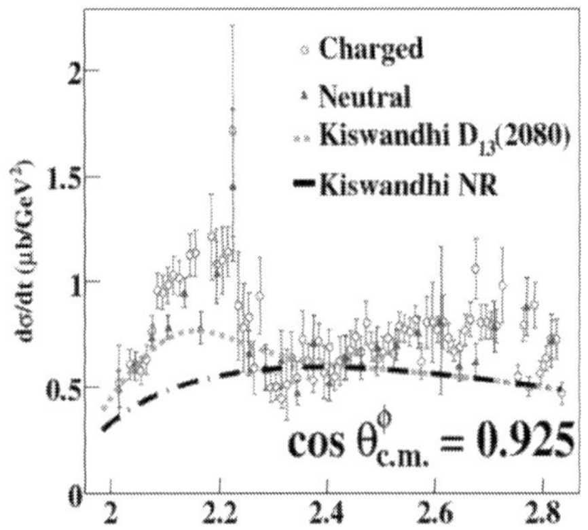


**Figure 11.** Differential cross section $\gamma p \rightarrow \phi p$ in the forward most angle bin.

gles and does not have a partial wave content that would be typical for a baryon resonance with specific spin and parity. There are model speculations [19, 20] that the structure might be due to an diquark-antitriquark in a $(su)(\bar{s}ud)$ configuration, which could explain the strong forward bump correlated with much smaller enhancements at backward angles. A reaction model and partial wave analysis will be needed to come to a more definitive interpretation of the observed cross section behavior.

## 4 Electroexcitation of light-quark baryon resonances

As high precision photoproduction processes have been successfully used to search for new baryon states, and will continue in searches for new states at higher masses, we also need to exploit electron scattering to probe the interior of excited states and understand their intrinsic spatial and spin structure. Meson electroproduction has revealed intriguing new information regarding the active degrees of freedom underlying the structure of the excited states and their scale dependences. At short distances the resonance electrocouplings are sensitive to the momentum dependence of the light quark masses as predicted in DSE calculations starting from the QCD Lagrangian [22]. They have also been incorporated in the $Q^2$ dependence of the quark masses in the LF RQM [30]. At larger distances the role of meson-baryon contributions and of meson cloud effects become relevant.

### 4.1 The N$\Delta(1232)$ transition

One of the important insights emerging from extensive experimental and theoretical research is clear evidence that resonances are not excited from quark transitions alone, but there can be significant contributions from meson-baryon interactions as well, and that these two processes contribute to the excitation of the same state. This evidence has been obtained in part through the observation that the quark transition processes often do not have sufficient strength to explain fully the measured resonance transition amplitudes at the real photon point. The best studied example is the $\Delta(1232)\frac{3}{2}^+$ resonance, which, when excited electromagnetically, is dominantly due to a magnetic dipole transition from the nucleon ground state, but only about 2/3 of the transverse helicity amplitudes at the photon point can be explained by the $q^3$ content of the state. In contrast, at $Q^2 \geq 3$ GeV$^2$ the quark contribution is nearly exhausting the measured strength as can be seen in Fig. 12.

At low $Q^2$ a satisfactory description of this transition has been achieved in hadronic models that include pion-cloud contributions and also in dynamical reaction models, where the missing strength has been attributed to dynamical meson-baryon interaction in the final state. A recent calculation within the light front relativistic quark model (LF RQM) [29] with the $q^3$ contributions normalized to the high $Q^2$ behavior, finds the meson-cloud contributions to set in at $Q^2 \leq 3.0$ GeV$^2$ as shown in Fig. 12. The excitation of this and other states using electron scattering should be highly sensitive to the different Fock components in the wave function of these excited states as it is expected that they have different excitation strengths when probed at large and at short distance scales, i.e. with virtual photons at low and high $Q^2$. At high $Q^2$ we expect

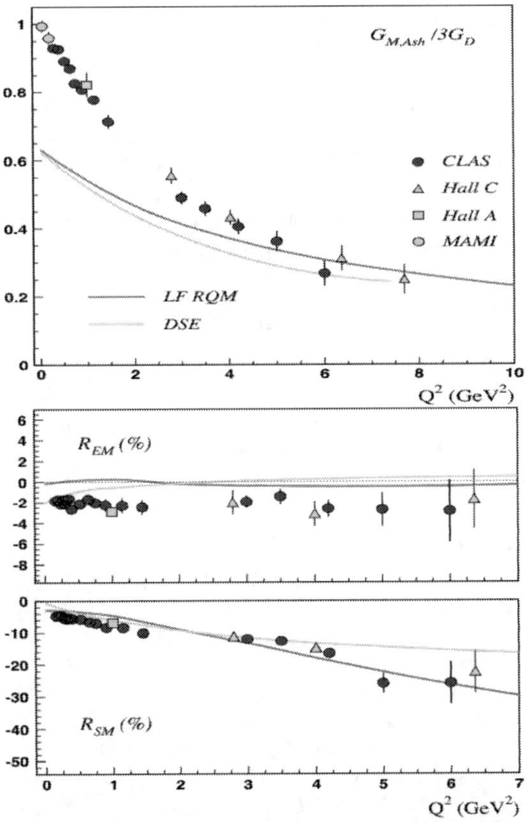


**Figure 12.** The $N\Delta$ magnetic transition form factor (top) and the electric and scalar quadrupole ratios $R_{EM}$ and $R_{SM}$ from CLAS (blue circles), Hall C (triangles) and Hall A (square) experiments from $ep \to ep\pi^0$. The green circles are from photoproduction data. The green line is the result of the first principles DSE calculation [22]. The red line is the prediction of the LF RQM [23] with running quark mass and configuration mixing [29]. The model is normalized to the data at $Q^2 > 4$ GeV$^2$ where the 3-quark core contributions dominate.

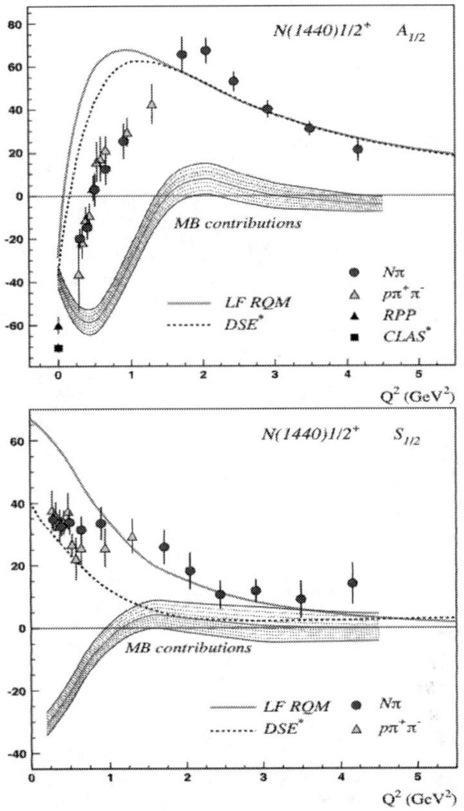


**Figure 13.** The electrocoupling amplitudes $A_{1/2}(Q^2)$ and $S_{1/2}(Q^2)$ for the Roper resonance $N(1440)\frac{1}{2}^+$ (for units see Fig. 12) . Electroproduction data are from CLAS [24–27]. The dashed curve is from the DSE/QCD calculation [28] after renormalization of the quark core contribution at high $Q^2$. The solid curve is from the LF RQM with momentum-dependent quark mass [29, 30]. The shaded band represents the inferred meson-baryon contributions. Although there is some model-dependence concerning the precise forms of these contributions, the marked similarity between those inferred via the LF RQM and those determined in a parameter-free DSE analysis [31] indicates that a quantitative understanding of these effects is near.

the $q^3$ components to be the only surviving contributions, while the higher Fock states may have large, even dominant strength at low $Q^2$.

### 4.2 Solving the Roper puzzle

It is well known that the Roper $N(1440)\frac{1}{2}^+$ state presented the biggest puzzle of the prominent nucleon resonances and for decades defied explanations within models. The non-relativistic constituent quark model has it as the first radial excitation of the nucleon ground state. However, its physical mass is about 300 MeV lower than what is predicted. The most recent LQCD projections have the state even 1 GeV above the nucleon ground state, i.e. near 1.95GeV [32]. The electromagnetic transition amplitude extracted from pion photo production data is large and negative, while the non-relativistc constituent quark model (nrCQM) predicts a large and positive amplitude. Furthermore, the early electroproduction results showed a rapid disappearance of its excitation strength at $Q^2 \leq 0.5$GeV$^2$, while the model predicted a strong rise in magnitude. These apparent discrepancies led to attempts at alternate interpretations of the state, e.g. as the lowest gluonic excitation of the nucleon [33] and as dominantly $N\rho$ [34] or $N\sigma$ [35] molecules.

Recent development of the dynamically coupled channel model by the EBAC group, has led to a possible resolution of the discrepancy in the mass value, by including resonance couplings to inelastic decay channels in their calculations [37]. The inelastic channels cause the dressed Roper pole to move by over 350 MeV from its bare value of 1.736 GeV to 1.365 GeV, i.e. close to where it is found experimentally.

Measurements of meson electroproduction data in a large range of $Q^2$ [25, 40–42] provided the basis for anal-

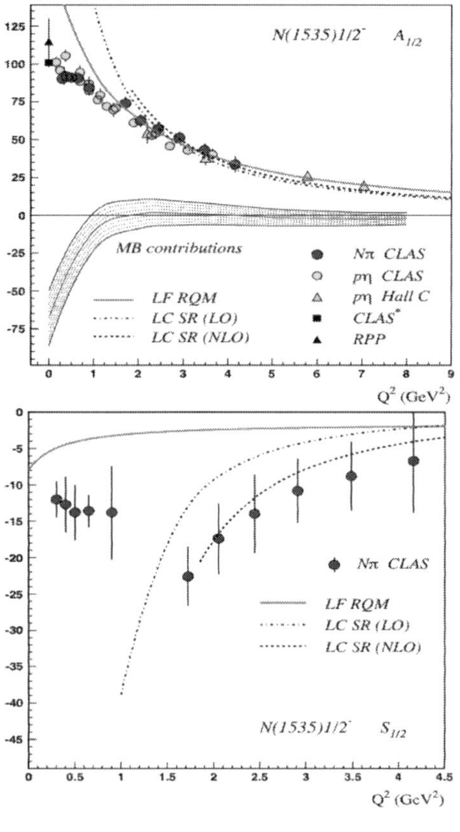


**Figure 14.** The transverse and scalar amplitudes for the $N(1535)\frac{1}{2}^-$ determined in $ep \to ep\eta$ (open symbols) and in $ep \to eN\pi$ (full circles). The shaded bands indicates the size of the meson-baryon contributions evaluated by subtracting LF RQM projection from smoothened data points. Curves represent LF RQM (solid) and LC SR (dashed-dotted).

yses of the electrocoupling amplitudes of the $pN(1440)\frac{1}{2}^+$ transition. The electrocoupling amplitudes are shown in Fig. 13. The LF RQM predicts the correct sign of the transverse amplitude at $Q^2 = 0$ and a sign change at small $Q^2$. The behavior at low $Q^2$ is described well when the $q^3$ component in the wave function is complemented by meson-baryon contributions, e.g. $N\rho$ [34] or $N\sigma$ [35], and also in effective field theories [36] employing pions, $\rho$ mesons, the nucleon and the Roper $N(1440)\frac{1}{2}^+$ as effective degrees of freedom. The high-$Q^2$ behavior is well reproduced in the QCD/DSE approach and the LF RQM which include momentum-dependent quark masses, in QCD/DSE [28] due to the full incorporation of the momentum-dependent dressed quark mass in QCD and modeled in LF RQM [30]. The latter is parameterized as $Q^2$-dependent mass in the quark wave function.

### 4.3 $N(1535)\frac{1}{2}^-$, parity partner of the nucleon

The parity partner of the ground state nucleon lies 600 MeV above the mass of the nucleon. The shift is thought to be due to the breaking of chiral symmetry in the excitation of nucleon resonances. Figure 14 shows the $A_{1/2}$ and $S_{1/2}$ amplitudes. The former is well described by the LF RQM [30] and the LC SR (NLO) [39] evaluation for $Q^2 \geq 1.0\text{GeV}^2$ and $Q^2 \geq 2\text{GeV}^2$, respectively. The scalar amplitude $S_{1/2}$ departs from the LF RQM predictions significantly, it is, however, well described by the LC SR (NLO) calculation at $Q^2 \geq 1.5\text{GeV}^2$. This result points at a promising approach of relating the resonance electrocouplings to calculations from first principles of QCD. The state has also been discussed as having large strangeness components [46], an assertion that might account for the discrepancy in the scalar amplitude with the data at low $Q^2$, although no specific predictions for the $S_{1/2}$ amplitude are available that include such contributions.

### 4.4 Light-cone charge transition densities

The electrocoupling amplitudes of the $N(1440)\frac{1}{2}^+$ and $N(1535)\frac{1}{2}^-$ states exhibit quite different $Q^2$ dependences. To see in what way this reflects the spatial structure of the two states we use the prescription in Ref. [45] to perform a Fourier transform of the resonance transition form factors by extrapolating the range in $Q^2$, where electrocouplings are known, to infinity. Obviously, this involves some model assumptions, for which we assume dimensional scaling behavior of the amplitudes. For the $N(1535)\frac{1}{2}^-$ we have data up to $Q^2 = 8$ GeV$^2$, and the extrapolation is relatively safe, less so for $N(1440\frac{1}{2}^+$ where the data end at $Q^2 \approx 5$ GeV$^2$. Asymptotically, the amplitudes of both states must have the same power behavior of $A_{1/2} \propto 1/Q^3$. The light-cone transition charge densities are shown in Fig. 15 for the unpolarized transition ($\rho_0$) and for the transition from transversely polarized protons ($\rho_T$), respectively. The charge density $\rho_0$ of the proton-Roper transition shows clearly a softer center and a more extended "cloud". In the polarized case, the charge center of $\rho_T$ moves farther to positive $b_y$ values, generating a larger electric dipole behavior than is the case for $N(1535)\frac{1}{2}^-$. The latter shows a stronger central peak for $\rho_T$, that remains fixed at $b_y = 0$. The results indicate a stronger *hadronic* component for the Roper resonance than for the $N(1535)\frac{1}{2}^-$ that appears more *quark like.* Note that for ease of comparison both scales, dimensions and color codes are the same so that the two resonances can be compared directly.

### 4.5 Spin structure of the $\gamma pN(1520)\frac{3}{2}^-$ transition

The transition $\gamma pN(1520)\frac{3}{2}^+$ was early on predicted in non-relativistic quark models (nr QM) to exhibit a radical change of its helicity structure from dominantly $A_{3/2}$ at $Q^2 = 0$ to dominant $A_{1/2}$ with increasing $Q^2$. The verification of this prediction was an early triumph of the quark model and demonstrated that the quark model is not just applicable to baryon spectroscopy but also to the study of the internal structure of baryon states. Figure 16 shows all three electrocoupling amplitudes. $A_{3/2}$ is clearly dominant over $A_{1/2}$ at the photon point, while at $Q^2 > 0.5\text{GeV}^2$ $A_{1/2}$ is larger than $A_{3/2}$, and at $Q^2 > 2\text{GeV}^2$ $A_{1/2}$ is clearly the dominant amplitude.

### 4.6 Baryon states in the 1.7 GeV mass range

Differential electroproduction cross sections of $ep \rightarrow e\pi^+n$ have recently been published in the mass range from 1.6 to 2.0 GeV [43]. The so-called third resonance region is the domain of several nearly mass degenerate states with masses near 1.7 GeV. Several states, e.g. $N(1675)\frac{5}{2}^-$ and $N(1650)\frac{1}{2}^-$ belong to the $[70,1^-]_1$ supermultiplet, while the $N(1680)\frac{5}{2}^+$ quark state is assigned to $[56,2^+]_2$. Depending on the multiplet assignment, we may expect quite different strengths and $Q^2$ dependences of the quark components. For example, the quark structure of the $N(1675)\frac{5}{2}^-$ leads to a suppressed 3-quark transition amplitude from the proton, i.e. $A^q_{1/2} = A^q_{3/2} = 0$. This suppression of the quark components has been verified for the $N(1675)\frac{5}{2}^-$ in explicit quark model calculations. We employed this suppression to directly access the non-quark components [47]. The results are shown in Fig. 17. The constituent quark model predictions are from Ref. [49]. Shown predictions for the meson-baryon (MB) contributions are absolute values of the results from the dynamical coupled-channel model (DCCM) [10].They are in qualitative agreement with the amplitudes extracted from experimental data, i.e. have considerable coupling through the $A_{1/2}$ amplitude and much smaller $A_{3/2}$ amplitude at $Q^2 \geq 1.8$ GeV$^2$.

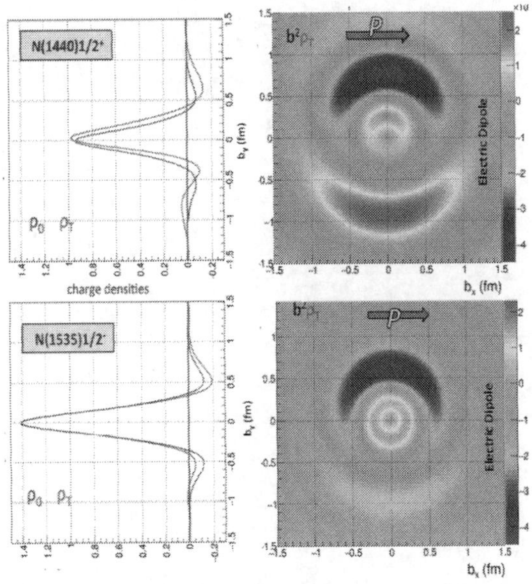


**Figure 15.** The light-cone charge transition densities for the Roper resonance (top panels) and the $N(1535)$ state. The left panels show projections on the transverse impact parameter for unpolarized protons (blue lines) and for protons polarized along the x-axis (red line). For better visibility the densities have been multiplied with $b^2$, which explains the zero in the center $b_{x,y} = 0$ and the enhancements at large $b_{x,y}$.

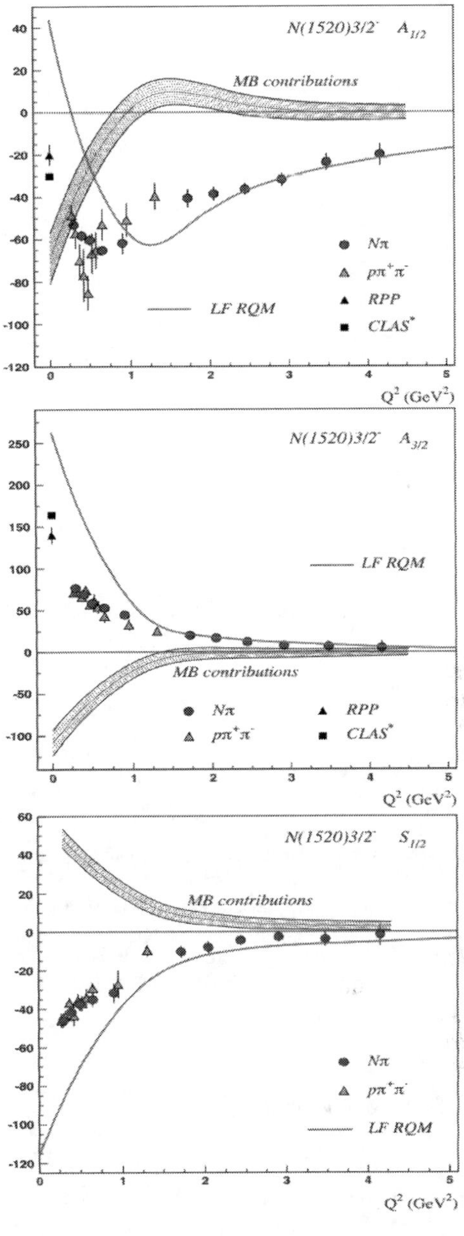


**Figure 16.** The transverse and scalar amplitudes for the $N(1520)\frac{1}{2}^-$ determined from $ep \rightarrow eN\pi$ and $ep \rightarrow ep\pi^+\pi^-$ (same units as in Fig. 12). Electroproduction data are from CLAS [24–27]. The solid curve is the LF RQM calculations as in Fig. 13.

Figure 18 shows the results for the $N(1680)\frac{5}{2}^+$ resonant state [43]. There is a rapid drop with $Q^2$ of the $A_{3/2}$ amplitude, which dominates at $Q^2 = 0$, while the $A_{1/2}$ amplitude, which at $Q^2 = 0$ makes a minor contribution, becomes the leading amplitude at larger $Q^2$. This change of the helicity structure is expected, but it is less rapid than predicted by quark models. Further studies are required to come to more definitive conclusions on such behavior.

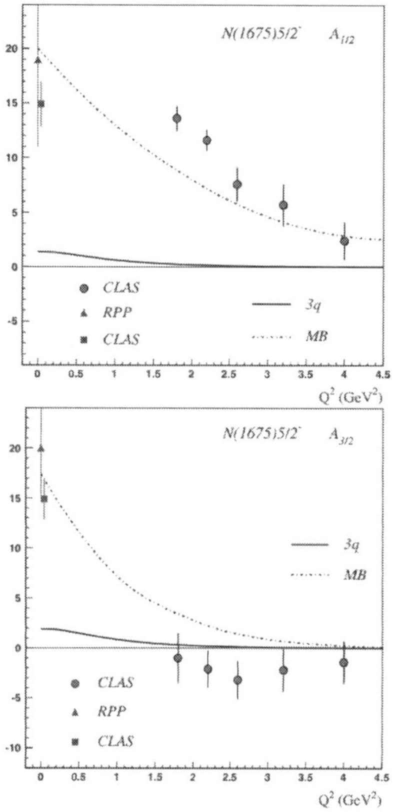


**Figure 17.** The transverse amplitudes $A_{1/2}$ (top) and $A_{3/2}$ (bottom) of the $N(1675)\frac{5}{2}^+$ have been determined in $ep \rightarrow e\pi^+n$ (same units as in Fig. 12). $A_{1/2}$ has a significantly non-zero magnitude dominated by meson-baryon contributions, while within the systematic uncertainties $A_{3/2}$ is consistent with zero at $Q^2 > 1.7\text{GeV}^2$. Quark contributions are very small in both cases. Data from Ref. [43, 47].

## 5 Summary & Outlook

In section 2 we referred to the dramatic transitions that occurred in the microsecond old universe. The energy available at the CEBAF electron accelerator as well as at ELSA and MAMI are well matched to study the details of this transition in searching for "missing" baryon states and in probing the momentum-dependent light-quark mass in the $Q$-dependence of the resonance transition amplitudes. Both of these provide insight into the phenomena of confinement, whose manifestation in strong QCD is still an open problem.

The $N^*$ program pursued by the CLAS collaboration has made very significant contributions towards improving our understanding of strong QCD by charting the kinematic landscape of the excited light-quark baryon states, which led to revealing of many new excited states. The search for new baryon states continues as much of the data the CLAS detector produced in the nucleon resonance region is still to be analyzed, so does the equally successful research on electromagnetic transition form factors of

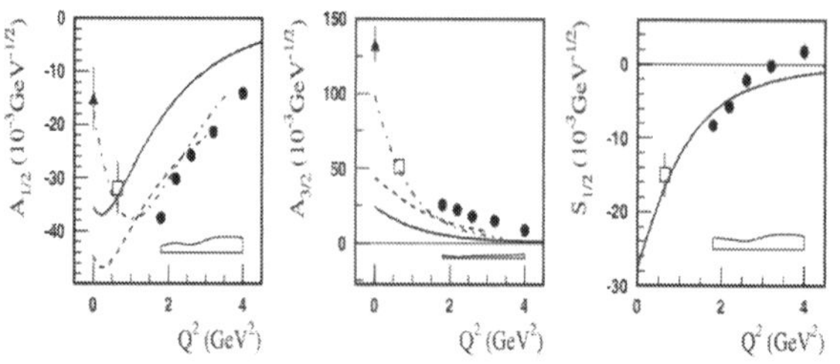


**Figure 18.** The transverse amplitudes and the scalar amplitude of the $N(1680)\frac{5}{2}^-$ have been determined in $ep \rightarrow e\pi^+n$ (data from Ref. [43]). The curves correspond to quark model calculations: dahed: [48], solid: [49], and dashed-dotted: [50]. The open boxes at the bottom indicate the model dependencies.

many excited states. The studies also led to advances in the theoretical exploration of strong QCD through the Dyson-Schwinger Equation based computations, through Light Cone Sum Rule calculations guided by LQCD, and through advances in Light Front RQM.

The multi-channel analysis of the Bonn-Gatchina group using the open strangeness channels $\gamma p \rightarrow K^+\Lambda$ has revealed a series of new nucleon states in the mass range of 1.9 to 2.2 GeV. It is in this mass range where a novel kind of matter is predicted to emerge [38], the hybrid baryons, or gluonic excitations, where the glue plays an active role in the excitation of baryon resonances. This new kind of matter is not distinct from ordinary quark matter in terms of their $J^P$ quantum numbers, but due to the different intrinsic structure, the new states should exhibit $Q^2$ dependences of the electrocouplings that are different [44] from ordinary quark excitations or meson-baryon excitations.

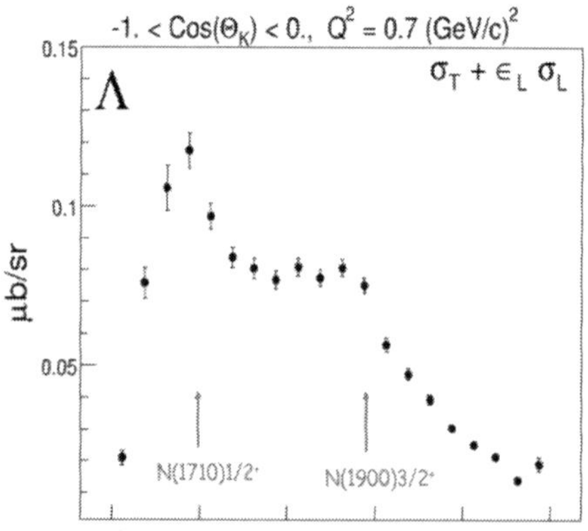


**Figure 19.** Integrated cross section in the backward hemisphere of $ep \rightarrow eK^+\Lambda$. The shown resonances are known to couple significantly to $K\Lambda$. Other states in the same mass ranges may also contribute to the enhancements.

One of the goals of the new program to study the electrocouplings of $N^*$ states at high energies is to find this new baryon matter. Figure 19 shows that in the $K^+\Lambda$ channel several of the higher mass states at 1.7 GeV and at 1.9 GeV are clearly visible in the large angle integrated cross section, and their electrocouplings may be studied. This requires data of much higher statistics than are currently available. Several new experiments have been approved to extend the $N^*$ program to higher masses and into the virtual photon dimension employing the higher energy available with the Jefferson Lab energy upgrade to 12 GeV. The $N^*$ program at the higher energies will make

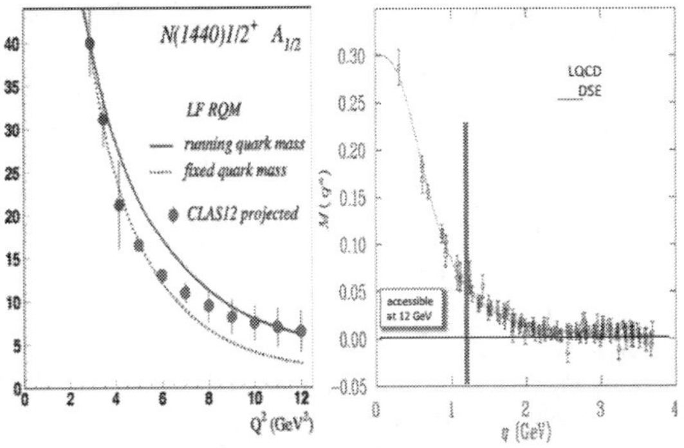


**Figure 20.** Right panel: The quark mass versus momentum transfer. The region left to the red line is accessible at $Q^2 <$ 12GeV$^2$, where the quark mass has dropped to about $\approx$ 50 MeV. Left panel: The two blue lines show the estimated effect of the running quark mass for the Roper resonance as computed in the LF RQM. The blue points are from measurement at beam energy of 6 GeV [25], while the red points are projections for 11 GeV. At the highest $Q^2$ the projected value of $A_{1/2}$ for the running quark mass is twice as high as its value for the fixed quark mass (same units as in Fig. 12).

use of the upgraded CLAS12 detector system that is currently being finalized. A cad drawing is shown in Fig. 21. The higher energy allows for the exploration of a larger range in the momentum-dependent quark mass, down to an effective dynamical mass of 50 MeV. This is shown in Fig. 20. For the 3-quark system the value $q$ of the momentum transfer to a single quark is approximately given as $q = \sqrt{Q^2}/3$. These studies will address the most challenging open problems in hadron physics on the nature of the dynamical hadron masses, the emergence of quark-gluon confinement from QCD, and its connection to dynamically generated symmetry breaking as the source of nearly all mass in the visible part of the universe.

## Acknowledgments

I am grateful to Inna Aznauryan, Ralf Gothe, Kijun Park and Viktor Mokeev who have contributed in many ways to the results discussed in this talk. I also like to thank F.-X. Girod for providing the revealing graphs of the light-cone transition charge densities, Eugene Pasyuk for a careful reading of the manuscript, and Craig Roberts for clarifying the DSE results on the resonance transition form factors.

**Figure 21.** The CLAS12 detector is projected for completion in 2017. It is based on a Torus magnet covering the forward angle range of 5° to 35°, and a 5T Solenoid magnet in the central angle range of 35° to 125°. New features of CLAS12 are improved particle id with FTOF for charged particle ID, improved Cherenkov counter for electron identification, and for $\pi$ and K separation.

## References

[1] I. C. Cloet and C. D. Roberts, Prog. Part. Nucl. Phys. **77**, 1 (2014)
[2] A. Bazavov *et al.*, Phys. Rev. Lett. **113**, no. 7, 072001 (2014)
[3] A. Bazavov *et al.*, Phys. Lett. B **737**, 210 (2014)
[4] M. E. McCracken *et al.* [CLAS Collaboration], Phys. Rev. C **81**, 025201 (2010)
[5] R. K. Bradford *et al.* [CLAS Collaboration], Phys. Rev. C **75**, 035205 (2007)
[6] R. Bradford *et al.* [CLAS Collaboration], Phys. Rev. C **73**, 035202 (2006)
[7] J. W. C. McNabb *et al.* [CLAS Collaboration], Phys. Rev. C **69**, 042201 (2004)
[8] B. Dey *et al.* [CLAS Collaboration], Phys. Rev. C **82**, 025202 (2010)
[9] A. Anisovich, R. Beck, E. Klempt, V. Nikonov, A. Sarantsev and U. Thoma, Eur. Phys. J. A **48**, 15 (2012)
[10] B. Julia-Diaz, T.-S. H. Lee, A. Matsuyama, T. Sato and L. C. Smith, Phys. Rev. C **77**, 045205 (2008)
[11] H. Kamano, S. X. Nakamura, T.-S. H. Lee and T. Sato, Phys. Rev. C **88**, no. 3, 035209 (2013)
[12] D. Rönchen *et al.*, Eur. Phys. J. A **50**, no. 6, 101 (2014)
[13] K. A. Olive *et al.* [Particle Data Group], Chin. Phys. C **38**, 090001 (2014).
[14] C. A. Paterson *et al.* [CLAS Collaboration], Phys. Rev. C **93**, no. 6, 065201 (2016)

[15] M. Williams *et al.* [CLAS Collaboration], Phys. Rev. C **80**, 065208 (2009)

[16] M. Williams *et al.* [CLAS Collaboration], Phys. Rev. C **80**, 065209 (2009)

[17] H. Seraydaryan *et al.* [CLAS Collaboration], Phys. Rev. C **89**, no. 5, 055206 (2014)

[18] B. Dey *et al.* [CLAS Collaboration], Phys. Rev. C **89**, no. 5, 055208 (2014)

[19] R. F. Lebed, Phys. Rev. D **92**, no. 11, 114006 (2015)

[20] R. F. Lebed, Phys. Rev. D **92**, no. 11, 114030 (2015)

[21] I. G. Aznauryan and V. D. Burkert, Prog. Part. Nucl. Phys. **67**, 1 (2012)

[22] J. Segovia, I. C. Cloet, C. D. Roberts and S. M. Schmidt, Few Body Syst. **55**, 1185 (2014)

[23] I. G. Aznauryan and V. D. Burkert, Phys. Rev. C **92**, no. 3, 035211 (2015)

[24] I. G. Aznauryan *et al.* [CLAS Collaboration], Phys. Rev. C **78**, 045209 (2008)

[25] I. G. Aznauryan *et al.* [CLAS Collaboration], Phys. Rev. C **80**, 055203 (2009)

[26] V. I. Mokeev *et al.* [CLAS Collaboration], Phys. Rev. C **86**, 035203 (2012)

[27] V. I. Mokeev *et al.*, Phys. Rev. C **93**, no. 2, 025206 (2016)

[28] J. Segovia et al., Phys. Rev. Lett. **115**, no. 17, 171801 (2015)

[29] I. G. Aznauryan and V. D. Burkert, arXiv:1603.06692 [hep-ph].

[30] I. G. Aznauryan and V. D. Burkert, Phys. Rev. C **85**, 055202 (2012)

[31] C. D. Roberts and J. Segovia, arXiv:1603.02722 [nucl-th].

[32] R. G. Edwards, J. J. Dudek, D. G. Richards and S. J. Wallace, Phys. Rev. D **84**, 074508 (2011)

[33] T. Barnes and F. E. Close, Phys. Lett. B **123**, 89 (1983).

[34] F. Cano and P. Gonzalez, Phys. Lett. B **431**, 270 (1998)

[35] I. T. Obukhovsky, A. Faessler, D. K. Fedorov, T. Gutsche and V. E. Lyubovitskij, Phys. Rev. D **84**, 014004 (2011)

[36] T. Bauer, S. Scherer and L. Tiator, Phys. Rev. C **90**, no. 1, 015201 (2014)

[37] N. Suzuki et al., Phys. Rev. Lett. **104**, 042302 (2010)

[38] J. J. Dudek and R. G. Edwards, Phys. Rev. D **85**, 054016 (2012)

[39] I. V. Anikin, V. M. Braun and N. Offen, Phys. Rev. D **92**, no. 1, 014018 (2015)

[40] K. Park *et al.* [CLAS Collaboration], Phys. Rev. C **77**, 015208 (2008)

[41] G. V. Fedotov *et al.* [CLAS Collaboration], Phys. Rev. C **79**, 015204 (2009)

[42] H. Egiyan *et al.* [CLAS Collaboration], Phys. Rev. C **73**, 025204 (2006)

[43] K. Park *et al.* [CLAS Collaboration], Phys. Rev. C **91**, 045203 (2015)

[44] Z. p. Li, V. Burkert and Z. j. Li, Phys. Rev. D **46**, 70 (1992).

[45] L. Tiator and M. Vanderhaeghen, Phys. Lett. B **672**, 344 (2009)

[46] B. S. Zou, Nucl. Phys. A **790**, 110 (2007)

[47] I. G. Aznauryan and V. D. Burkert, Phys. Rev. C **92**, no. 1, 015203 (2015)

[48] D. Merten, U. Loring, B. Metsch and H. Petry, Eur. Phys. J. A **18**, 193 (2003).

[49] E. Santopinto and M. M. Giannini, Phys. Rev. C **86**, 065202 (2012)

[50] F. E. Close and Z. P. Li, Phys. Rev. D **42**, 2194 (1990).

# Towards Precision Spectroscopy of Baryonic Resonances

*Michael* Döring[1,2,⋆], *Maxim* Mai[1,3,4], and *Deborah* Rönchen[3,4]

[1] *George Washington University, Washington, DC 20052, USA*
[2] *Thomas Jefferson Accelerator Facility, Newport News, VA 23606, USA*
[3] *Helmholtz-Institut für Strahlen- und Kernphysik der Universität Bonn, Nußallee 14-16, 53115 Bonn, Germany*
[4] *Bethe Center for Theoretical Physics, Universität Bonn, 53115 Bonn, Germany*

**Abstract.** Recent progress in baryon spectroscopy is reviewed. In a common effort, various groups have analyzed a set of new high-precision polarization observables from ELSA. The Jülich-Bonn group has finalized the analysis of pion-induced meson-baryon production, the photoproduction of pions and eta mesons, and (almost) the $K\Lambda$ final state. As data become preciser, statistical aspects in the analysis of excited baryons become increasingly relevant and several advances in this direction are proposed.

## 1 Introduction

Increasingly accurate data is produced from photoproduction experiments, allowing for much improved amplitude analyses. Double polarization observables are being released that allow to reduce ambiguities and uncertainties of multipole solutions. In addition, more and more final states like $\eta N$ and $KY$ are measured that be simultaneously analyzed using multi-channel partial-wave analysis techniques. As a result, resonance properties are determined to greater precision allowing for improved comparison with theory.

Yet, the multipoles extracted by different groups differ from each other because different parameterizations are used, the analyzed data bases are different, and the quality of the fits ($\chi^2$ or likelihood) is different. On top of this, the different groups treat systematic uncertainties differently. Given the spectacular experimental progress in recent years, it becomes, therefore, desirable to address these problems.

To check whether the data themselves can help to reduce discrepancies in different multipole solutions, the SAID, MAID, Bonn-Gatchina and Jülich-Bonn groups teamed up to compare solutions before and after the inclusion of new high-precision $\pi^0$ photoproduction data produced at ELSA [1–4]. The quality of the fits to the new data is similar as shown in Ref. [5].

Comparing the multipoles, the visual inspection already indicates that solutions came closer to each other. We have also quantified this by calculating the pairwise and overall variances between the SAID, Bonn-Gatchina, and Jülich-Bonn solutions before and after the inclusion of the new data. That variance, at a given energy, is defined as the sum over the squared differences in all multipoles up to an angular momentum of the final state of $L = 4$. In Fig. 1

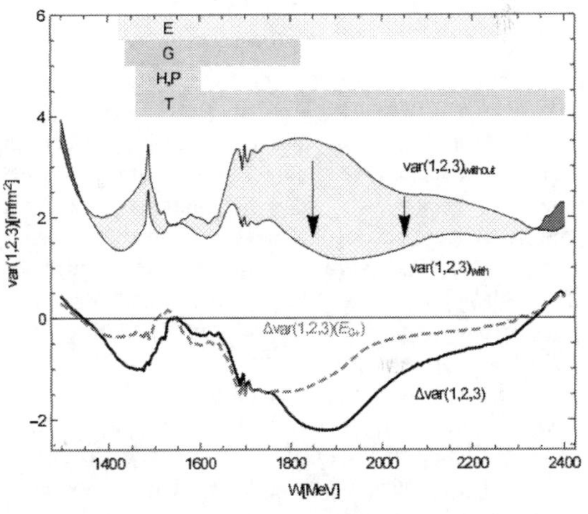


**Figure 1.** The variance of all three PWAs summed over all $\gamma p \to \pi^0 p$ multipoles up to $L = 4$ [5]. The range covered by the new double polarization observables [1–4] are indicated by shaded areas. Over the largest part of the energy range the new data have enforced an improvement of the overall consistency. The improvement is displayed as light green area and, separately as difference of the variance. The contribution to the improvement from the $E_{0+}$ wave is shown as the dashed curve. Ranges with an overall deterioration are marked in red.

the the overall variance is shown as a function of total energy. Indeed, in the energy region covered by the new high precision data, discrepancies are substantially reduced as indicated with the arrows in the upper part of the figure. The solid line in the lower part shows the change, while the dashed line indicates the change from the $E_{0+}$ multipole alone. At lower energies, the solutions come closer to each other even outside the window covered by the new

⋆e-mail: doring@gwu.edu

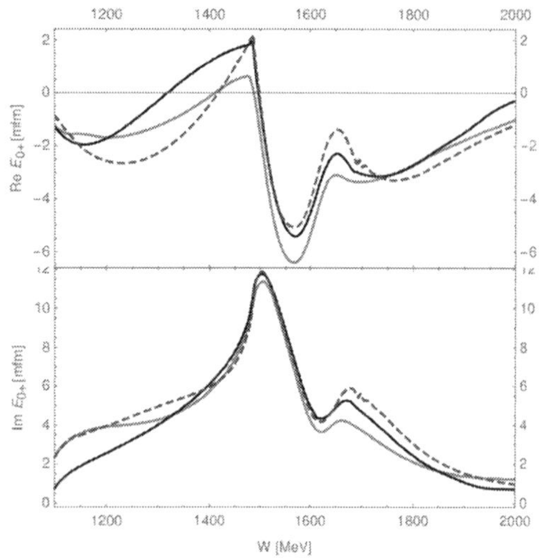


**Figure 2.** The $E_{0+}$ multipole after inclusion of new $\pi^0 p$ observables [1–4]. Black solid line: BnGa, blue dashed lines: JüBo, red solid lines: SAID.

data, while at very low and high energies solutions show slightly larger discrepancies (red areas).

This is a clear indication that indeed high precision data help to determine multipoles more precisely and remove discrepancies between the solutions of different groups. A large part of the reduction in variance originates from the $E_{0+}$ multipole. In Fig. 2 this multipole, after inclusion of the new data in the fits, is shown. It is apparent that despite the qualitative agreement, the details of the multipoles are still different. Most visible is the difference in the description of the cusp in Re $E_{0+}$ at the $\eta N$ threshold. In conclusion, while new high-precision data significantly reduce discrepancies among different partial-wave analysis groups, the details show that it is too early to state perfect matching.

The analysis of photoproduction data could also benefit from improved data on pion-induced reactions, as recently achieved for elastic pion-nucleon scattering [6]. In Ref. [7] the data on the reactions $\pi N \to \eta N$, $KY$ were discussed and analyzed. For example, the measured observables $d\sigma/d\Omega$, $P$, and $\beta$ for the reaction $\pi N \to K\Lambda$ provide, in principle, a complete experiment up to a discrete ambiguity. However, the data are in many cases conflicting, in general not very precise, or even known to be problematic as in $\pi N \to \eta N$ at higher energies. Re-measurements of these reactions would be desirable and provide complementary information on the baryon spectrum. Physics opportunities with hadron beams have been discussed in Ref. [8].

## 2 Analysis of $\eta$ and $K\Lambda$ photoproduction

The Jülich-Bonn group (JüBo) has performed comprehensive analyses of pion photoproduction [9], $\eta$ photoproduction [10] and $K\Lambda$ photoproduction [11]. In several occasions, the group has analyzed new experimental data that were then published together with the analyses [12, 13]. The amplitude used for these analysis is based on a field-theoretical framework using chiral interactions. For a description of the approach, see, e.g., Ref. [7].

As an example for the current progress, analyzed observables in kaon photoproduction at selected energies are shown in Fig. 3. The two curves correspond to solutions with a new $P13(1900)$ resonance (red solid lines) and without (blue dash-dotted lines). A final assessment whether the improvement of the fit through this state is significant will follow soon. The Bonn-Gatchina, Giessen, and other analyses include this state [14, 15].

Another interesting observable is shown in Fig. 4. The $E$ observable in the reaction $\gamma p \to \eta p$ has been recently measured by the CLAS collaboration [13] for the first time ever. The different lines correspond to two fits to differently binned data. The open circles show the fit result if the finite bin width is corrected for. There is a clear structure slightly below $W = 1.7$ GeV. However, in the analysis it was found that the structure can be explained as a conventional effect from the interference of a few multipoles, and no new narrow structure is needed for the description of the data [13].

## 3 Developments for future analyses

### 3.1 Statistical meaning of $\pi N$ partial waves

A notorious problem in multi-channel analyses is to statistically quantify the significance of, e.g., a new resonance. In order to build on the progress made in the earlier $\pi N$ elastic analyses, multi-channel analyses have usually fitted $\pi N$ amplitudes, derived from previous studies [16], together with measured data on other reactions. The fitted amplitude pseudo-data have either been taken from single-energy analyses (SE) or energy-dependent (ED) fits covering the resonance region. The SE analysis amplitudes, derived from fits to narrow energy bins of data, have associated errors which have been used in the multi-channel fits, or enlarged when these fits have become problematic. The smoother ED amplitudes have also been taken at discrete energies, typically with subjective errors not derived from the fit to data.

There are several problems associated with fits to amplitude pseudo-data, which were recently addressed in Ref. [17]. The most obvious of these is the fact that the goodness of fit to these sets of amplitudes cannot be translated into a quality of fit to the underlying dataset. In addition, uncertainties on the SE amplitudes [16] do not account for correlated errors, which can be substantial in some cases.

In baryon spectroscopy, based on multi-reaction analysis, this has unwanted side effects. First, a statistical analysis of fit results is difficult if one of the input channels is not given by data providing a meaningful $\chi^2$. Second, as a consequence, the significance of resonance signals, detected in such multi-reaction fits, is difficult to quantify.

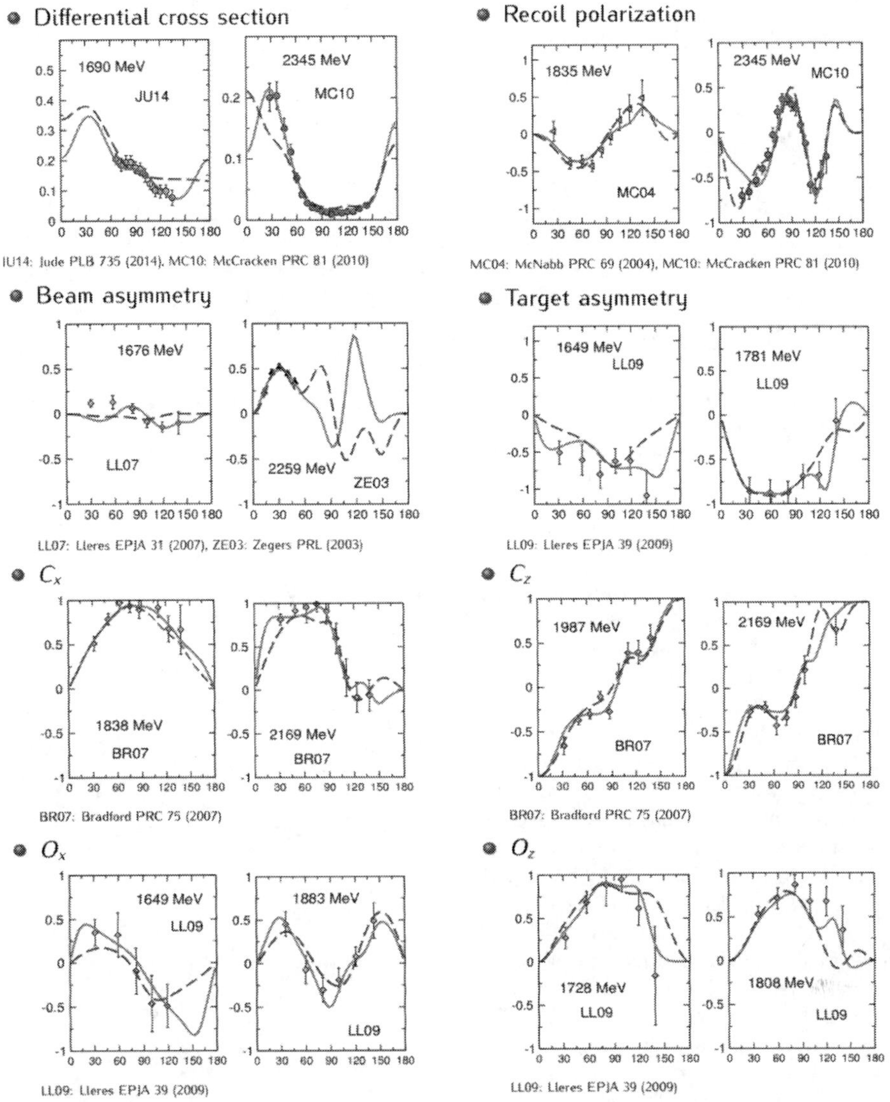


**Figure 3.** Selected results for the preliminary analysis of kaon photoproduction. The red lines show the best fit with a new $P13(1900)$ resonance, the blue dashed lines without.

In Ref. [17], the covariance matrices of the SE solutions have been calculated, plus additional quantities needed to perform correlated $\chi^2$ fits, overcoming the mentioned problem. In that way, other groups can continue using the partial wave solutions of the SAID group for their multi-channel fits, and, at the same time, the $\chi^2$ returned in such fits is very close to the one as if $\pi N \to \pi N$ data were fitted directly.

### 3.2 Model selection techniques

As discussed, discrepancies among different partial-wave analyses are reduced once high-precision polarization data are included in the analyses. Second, to better judge the uncertainties in a given analysis, the inclusion of elastic pion-nucleon scattering can be improved as discussed in the previous section. Yet, to obtain a reliable spectrum of excited baryons and their properties, it is necessary to make a selection of how many resonance states should be included in the model amplitude. In particular, for broad resonances this is not clear at all because, maybe, a very similar data description could be obtained by a background term instead of a resonance term. In the SAID analysis [16] the selection process is data driven, because resonances are generated only if required by data and no manual intervention is needed. In any case, it would be desirable to aim at solutions that, with a minimum of resonances, describe the data well. Blindfolding the model selection with the bias of minimal resonance content is therefore a way to exclude solutions that are over-populated

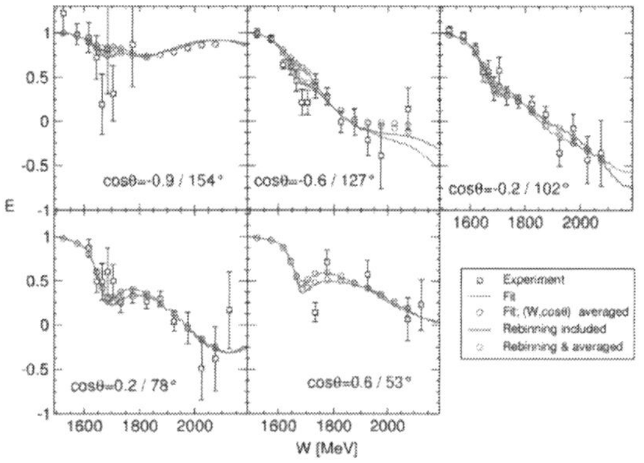


**Figure 4.** The $E$ observable in $\gamma p \rightarrow \eta p$ as measured by the CLAS collaboration [13] for the first time. The different lines correspond to two fits to differently binned data. The open circles show the fit result if the finite bin width is corrected for.

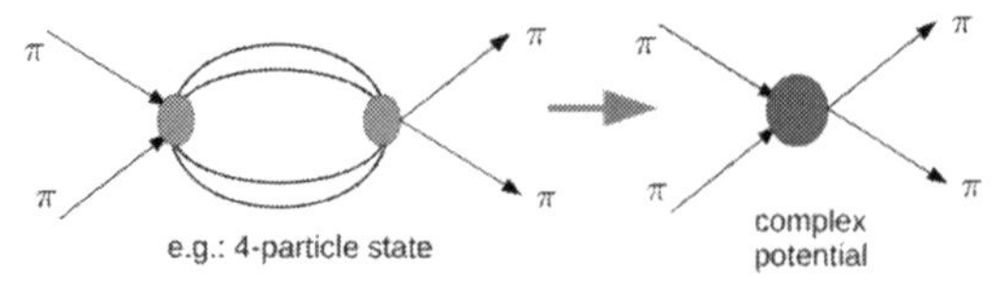


**Figure 5.** Optical potential. Formally, multi-particle and multi-channel scattering problems can be rewritten in terms of one channel and an optical potential that encodes the complexity of the scattering problem. In the figure, a four-pion state is absorbed into the (complex) optical potential.

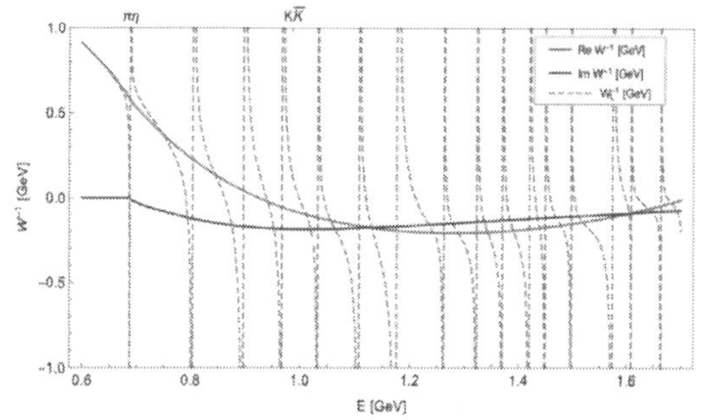


**Figure 6.** The real-valued optical potential for $\pi\eta$, $K\bar{K}$ scattering in the finite volume (gray dashed line). In the infinite volume, the optical potential (red line) develops also an imaginary part (blue) line.

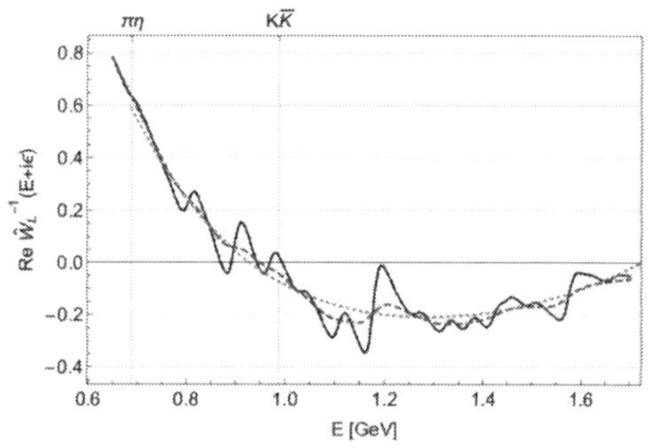


**Figure 7.** Real part of the optical potential in the finite volume, at complex energies $E \rightarrow E + i\epsilon$ for $\epsilon = 20$ MeV (black solid line) and $\epsilon = 50$ MeV (blue dashed line). The infinite-volume optical potential is shown with the red dotted line.

with resonances. Model selection can be performed by penalizing the occurrence of resonances through additional terms in the $\chi^2$. A very successful method is provided by the so-called LASSO (least absolute shrinkage and selection operator) [18, 19]. However, the size of the penalty is a-priori not determined - obviously, too large of a penalty will remove all resonances from the amplitude while too small of a penalty will lead to an overpopulated spectrum. Therefore, the LASSO penalty parameter needs to determined by an independent method, in the context of data analysis from cross validation or information theory [19].

Based on an existing model of the $\bar{K}N \rightarrow K\Xi$ reaction [20], we are performing a test of principle using the LASSO. Indeed, by starting with a largely over-populated resonance amplitude, the LASSO finds exactly those resonances significant that were singled out in Ref. [20] by hand in extensive tests of resonance combinations. A publication is in preparation.

Yet, a caveat tied to the discussed techniques remains: Many of the data in pion- and photon-induced reactions are contradictory or systematic errors are underestimated. To achieve progress in baryon spectroscopy in the discussed directions, it is there necessary to re-assess the data base.

### 3.3 Complex systems in the finite volume

It is, in any case, interesting to note that the discussed model selection techniques can also be used in the analysis of data from lattice QCD calculations. In Ref. [21] the problem of extrapolating lattice QCD results, obtained in a small volume, to the infinite volume is discussed. In particular, a quantity known from nuclear physics, the optical potential, is proposed to serve for this step. Fig. 5 shows an example. To demonstrate the principle, in a numerical simulation we have evaluated the optical potential in the finite volume using a known amplitude for coupled-channel $\pi\eta$, $K\bar{K}$ scattering. The real-valued potential is shown in Fig. 6 with the gray dashed line. It is characterized by finite-volume artefacts in form of multiple poles. In contrast, the infinite-volume optical potential, using the same underlying physical process, is smooth up to threshold openings and has an imaginary part as the figure shows. This demonstrates that the limit $L \rightarrow \infty$ (where $L$ is the side length of the finite cubic volume) is not straightforward to take. Instead, the limit is well-defined for small imaginary parts $E \rightarrow E + i\epsilon$. For such finite $\epsilon$, Fig. 7 shows the real part of the optical potential (the imaginary part shows a similar behavior). At complex energies, the poles on the real axis appear as oscillations. Considering the infinite-volume optical potential at the same time (red dotted line in the figure) it becomes clear that the $L \to \infty$ extrapolation can be obtained by smoothing the oscillations originating from the finite volume.

The problem is now reduced to smoothing a highly oscillatory curve with a low-order polynomial or a different suitable parameterization. A way to achieve this is, again, provided by the LASSO method. In this context, the second derivative of the curve is penalized such that smooth functions are preferred. In combination with cross validation, it was shown in Ref. [21] that it is indeed possible to reconstruct the infinite-volume optical potential from lattice data obtained in the finite volume.

Future applications of this method become increasingly relevant, because finite-volume artefacts get large at smaller quark masses close to the physical point. Second, lattice simulations are now at the verge of moving beyond the simplest hadronic systems addressing multi-channel and multi-particle states. The optical potential can absorb the corresponding degrees of freedom and represents, thus, a promising technique in the analysis of future lattice QCD simulations.

## 4 Summary

Progress in the analysis of pion and photon-induced reactions has been made in the last years. Using a coupled-channel approach based on field theory, the Jülich-Bonn group extracted amplitudes and resonance properties from the world data of pion and eta photoproduction. The fit to $K\Lambda$ data is finished.

To get closer to definite answers regarding the spectrum of excited baryons, new methods were proposed addressing mainly the statistical aspect of analysis, that will become increasingly relevant given recent and future high-precision data. One aspects concerns the need to perform entirely data-driven analyses, for which the necessary information within the SAID approach has been calculated and made available. Model selection techniques can be used to derive amplitudes with a minimal resonance content and minimize the danger of false-positive resonance signals. These techniques are applicable not only in baryon spectroscopy but also in the analysis of lattice QCD data, in particular through the use of the optical potential in the infinite-volume extrapolation of complex hadronic systems that have become into the reach of ab-initio QCD calculations.

**Acknowledgements:** The authors gratefully acknowledge the computing time granted on the supercomputer JURECA at Jülich Supercomputing Centre (JSC). This work is also partially supported by the DFG (Deutsche Forschungsgemeinschaft) and NSFC through the Sino-German CRC 110 ("Symmetries and the Emergence of Structure in QCD"). M.M. is supported by a DFG research fellowship (MA 7156/1-1). M.D. is supported by the National Science Foundation (PIF grant No. PHY 1415459 and CAREER grant No. 1452055) and by the U.S. Department of Energy, Office of Science, Office of Nuclear Physics under contract DE-AC05-06OR23177.

## References

[1] J. Hartmann *et al.* [CBELSA/TAPS Collaboration], Phys. Lett. B **748** (2015) 212.

[2] A. Thiel *et al.*, arXiv:1604.02922 [nucl-ex].

[3] M. Gottschall *et al.* [CBELSA/TAPS Collaboration], Phys. Rev. Lett. **112** (2014), 012003.

[4] A. Thiel *et al.*, Phys. Rev. Lett. **109** (2012) 102001.

[5] R. Beck *et al.*, arXiv:1604.05704 [nucl-th], accepted for publication in Eur. Phys. J. A.

[6] A. Gridnev *et al.* [EPECUR Collaboration], Phys. Rev. C **93** (2016), 062201.

[7] D. Rönchen *et al.*, Eur. Phys. J. A **49** (2013) 44.

[8] W. J. Briscoe, M. Döring, H. Haberzettl, D. M. Manley, M. Naruki, I. I. Strakovsky and E. S. Swanson, Eur. Phys. J. A **51** (2015), 129.

[9] D. Rönchen *et al.*, Eur. Phys. J. A **50** (2014), 101 Erratum: [Eur. Phys. J. A **51** (2015), 63].

[10] D. Rönchen, M. Döring, H. Haberzettl, J. Haidenbauer, U.-G. Meißner and K. Nakayama, Eur. Phys. J. A **51** (2015), 70.

[11] D. Rönchen *et al.*, publication in preparation.

[12] S. Strauch *et al.* [CLAS Collaboration], Phys. Lett. B **750** (2015), 53.

[13] I. Senderovich *et al.* [CLAS Collaboration], Phys. Lett. B **755** (2016), 64.

[14] E. Gutz *et al.* [CBELSA/TAPS Collaboration], Eur. Phys. J. A **50** (2014), 74.

[15] V. Shklyar, H. Lenske and U. Mosel, Phys. Rev. C **87** (2013), 015201.

[16] R. L. Workman, R. A. Arndt, W. J. Briscoe, M. W. Paris and I. I. Strakovsky, Phys. Rev. C **86** (2012), 035202.

[17] M. Döring, J. Revier, D. Rönchen and R. L. Workman, Phys. Rev. C **93** (2016), 065205.

[18] B. Guegan, J. Hardin, J. Stevens and M. Williams, JINST **10** (2015), P09002.

[19] *The Elements of Statistical Learning: Data Mining, Inference, and Prediction*, T. Hasti, R. Tibshirani, J. Friedman, Springer 2009, second edition.

[20] B. C. Jackson, Y. Oh, H. Haberzettl and K. Nakayama, Phys. Rev. C **91** (2015), 065208.

[21] D. Agadjanov, M. Döring, M. Mai, U.-G. Meißner and A. Rusetsky, JHEP **1606** (2016) 043.

# Connecting Compton and Gravitational Compton Scattering

Barry R. Holstein[1]

[1] Department of Physics-LGRT, University of Massachusetts, Amherst, MA 01003, USA

**Abstract.** The study of Compton scattering—$S + \gamma \to S + \gamma$—at MAMI and elsewhere has led to a relatively successful understanding of proton structure via its polarizabilities. The recent observation of gravitational radiation observed by LIGO has raised the need for a parallel understanding of gravitational Compton scattering—$S + g \to S + g$—and we show here how it can be obtained from ordinary Compton scattering by use of the double copy theorem.

## 1 Introduction

The recent observation at LIGO of gravitational radiation has verified the existence of gravitons and has emphasized the importance of studying processes involving their interactions [1]. In the case of the electromagnetic interaction, the Compton scattering process—$S + \gamma \to S + \gamma$—has been utilized for many years as a probe of particle structure by measuring the response of a scattering system to external electromagnetic fields, which are characterized by so-called polarizabilities [2]. The use of causality allows another type of probe by the use of various sum rules. An example of this kind is the well-known GDH (Gerasimov-Drell-Hearn) sum rule, which relates an integral over experimental photoabsorption cross-sections to the anomalous magnetic moment of the target [3],[4]. It is of interest then to study the related process of gravitational Compton scattering—$S + g \to S + g$. In this case the evaluation of such amplitudes is simplified enormously by use of the double copy theorem, which asserts that such gravitational Compton amplitudes can be expressed in terms of the product of simple Compton scattering amplitudes and an elementary kinematic factor [5],[6]. In the case of the gravitational interaction, angular momentum conservation demands that there can exist no anomalous gravitomagnetic moment and it is then interesting to ask if there is a corresponding gravitational GDH sum rule.

It is the purpose of this note to study such issues. In section 2 then, we review the subject of Compton scattering and its connection with polarizabilities. We show how such quantities have been calculated by use of chiral perturbation theory and confront the theoretical and experimental values. In section 3 we examine the parallel gravitational Compton scattering reaction and show how factorization allows an enormously simplified evaluation of such amplitudes. The validity of sum rules is also studied. In section 4 we show how the gravitational Compton amplitudes can be used together with unitarity to provide an elementary evaluation of higher order gravitational scattering and our results are summarized in a brief concluding section.

## 2 Compton Scattering

The Compton scattering process has long been studied both experimentally and theoretically [7]. On the theoretical side Low's theorem [8] asserts that the Compton scattering amplitude is given up to $O(\omega^2)$ by the simple Born amplitude plus a possible anomalous magnetic moment term [9] where here the Born amplitude arises from the sum of lowest order contributions from the three diagrams shown in Figure 1.

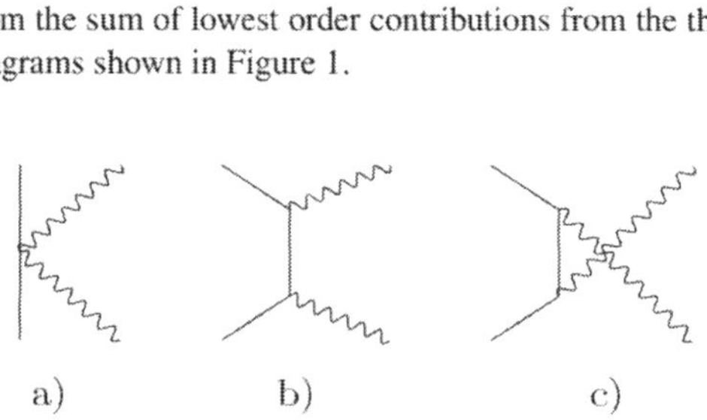


**Figure 1.** Shown are the a) seagull, b) Born, and c) Cross-Born diagrams contributing to Compton scattering. Here the solid lines represent massive scalars while the wiggly lines are photons.

In the case of a spinless particle, the Compton amplitude is found from the minimal gauged Klein-Gordon Lagrangian

$$\mathcal{L}(S = 0) = (D_\mu\phi)^\dagger D^\mu\phi - m^2\phi^2 - \frac{1}{4}F_{\mu\nu}F^{\mu\nu} \qquad (1)$$

where $D_\mu = \partial_\mu + ieA_\mu$ is the covariant derivative and $F_{\mu\nu} = \partial_\mu A_\nu - \partial_\nu A_\mu$ is the electromagnetic field tensor. The contributions from the three (individually non-gauge-invariant) diagrams are

$$\begin{aligned}
\mathrm{Amp}_C^{1a}(S=0) &= -2e^2\epsilon_f^*\cdot\epsilon_i \\
\mathrm{Amp}_C^{1b}(S=0) &= \frac{2e^2\epsilon_f^*\cdot p_f\epsilon_i\cdot p_i}{p_i\cdot k_i} \\
\mathrm{Amp}_C^{1c}(S=0) &= -\frac{2e^2\epsilon_f^*\cdot p_i\epsilon_i\cdot p_f}{p_i\cdot k_f}
\end{aligned} \tag{2}$$

and the resultant Born amplitude

$$\begin{aligned}
&\mathrm{Amp}_C^{Born}(S=0)\sum_{i=a}^{c}=\mathrm{Amp}_C^{1i}(S=0) \\
&= 2e^2\left[\frac{\epsilon_f^*\cdot p_f\epsilon_i\cdot p_i}{p_i\cdot k_i}-\frac{\epsilon_f^*\cdot p_i\epsilon_i\cdot p_f}{p_i\cdot k_f}-\epsilon_f^*\cdot\epsilon_i\right]
\end{aligned} \tag{3}$$

is verified to be gauge-invariant. Low's theorem in this case reads [9]

$$\mathrm{Amp}_C^{tot}(S=0)=\mathrm{Amp}_C^{Born}(S=0)+\mathrm{Amp}_C^{NB}(S=0) \tag{4}$$

where $\mathrm{Amp}_C^{NB}(S=0)=O(\omega^2)$ and possible higher order corrections can be found from an effective Lagrangian

$$\mathcal{L}_{eff}=C_1F_{\mu\nu}F^{\mu\nu}\phi^\dagger\phi+C_2F_{\mu\lambda}F^{\lambda}{}_{\nu}\partial^\mu\phi^\dagger\partial^\nu\phi+\ldots \tag{5}$$

where the ellipsis indicates higher derivative contributions. This result is conventionally written as

$$H_{eff}=\frac{(\boldsymbol{p}-e\boldsymbol{A})^2}{2m}-\frac{1}{2}4\pi\alpha_E\boldsymbol{E}^2-\frac{1}{2}4\pi\beta_M\boldsymbol{B}^2+\ldots \tag{6}$$

where $\alpha_E, \beta_M$ are the electric, magnetic polarizabilities and measure the size of the induced electric, magnetic dipole moments when the system is placed in an external electric or magnetic field via [2]

$$\boldsymbol{d}_E=4\pi\alpha_E\boldsymbol{E}\quad\text{and}\quad\boldsymbol{d}_M=4\pi\beta_M\boldsymbol{B} \tag{7}$$

Since Eq. (6) implies a Compton scattering cross section

$$\begin{aligned}
\frac{d\sigma_C(S=0)}{d\Omega} &= \frac{\alpha^2}{2m^2}\frac{\omega_f^2}{\omega_i^2}\Big\{1+\cos^2\theta \\
&- \frac{m\omega_f\omega_i}{\alpha}\Big[(\alpha_E+\beta_M)(1+\cos\theta)^2 \\
&+ (\alpha_E-\beta_M)(1-\cos\theta)^2\Big]+\ldots\Big\}
\end{aligned} \tag{8}$$

where $\alpha=e^2/4\pi$ is the fine structure constant, we see that the polarizabilities can be measured via Compton scattering. In a series of measurements over the years we now have reliable and precise values for the proton poarizabilities, as listed by the Particle Data Group [10]

$$\alpha_E^p=(11.2\pm0.4)\times10^{-4}\,\mathrm{fm}^3 \tag{9}$$

and

$$\beta_M^p=(2.5\pm0.4)\times10^{-4}\,\mathrm{fm}^3 \tag{10}$$

and these numbers contain useful information about proton structure. Thus, since the hydrogen atom electric polarizability is of order the atomic volume, while the proton electric polarizabiity is of order $10^{-3}$ of the proton volume [11], we learn that the proton is a much stiffer object than the atom. Likewise, since the $\Delta$ pole makes a strongly paramagnetic contribution to $\beta_M$ [12]—$\beta_M^{\Delta-pole}\sim 10\times10^{-4}\,\mathrm{fm}^3$—the measured value given in Eq. (10) indicates a significant diamagnetic contribution, which may arise from the pion cloud.

Causality can be used to write fixed-angle dispersion relations [13]. Thus in the forward direction, we have

$$\mathrm{Amp}_C^{tot}(S=0)\xrightarrow{\theta\to0}\boldsymbol{\epsilon}_f^*\cdot\boldsymbol{\epsilon}_i f(\omega) \tag{11}$$

with

$$f(\omega)=f(-\omega)=-\frac{e^2}{4\pi m}+(\alpha_E+\beta_M)\omega^2+\ldots \tag{12}$$

The amplitude $f(\omega)$ obeys a once-subtracted dispersion relation

$$f(\omega)=-\frac{e^2}{4\pi m}+\frac{\omega^2}{2\pi^2}\int_0^\infty\frac{d\omega'\sigma_{tot}(\omega')}{\omega'^2-\omega^2} \tag{13}$$

where $\sigma_{tot}(\omega')$ is the photoabsorption cross section and, expanding both sides of Eq. (13) in $\omega^2$, we find the Baldin sum rule [14]

$$\alpha_E+\beta_M=\frac{1}{2\pi^2}\int_0^\infty\frac{d\omega'\sigma_{tot}(\omega')}{\omega'^2} \tag{14}$$

In the case of spin-$\frac{1}{2}$, the Born term becomes

$$\begin{aligned}
&\mathrm{Amp}_C^{Born}(S=\tfrac{1}{2})=e^2\bar{u}(p_f) \\
\times&\left[\frac{\left(\not{\epsilon}_f^*+\frac{i\kappa}{2m}\sigma_{\mu\nu}\epsilon_f^{*\mu}k_f^\nu\right)(\not{p}_i+\not{k}_i+m)\left(\not{\epsilon}_i-\frac{i\kappa}{2m}\sigma_{\mu\nu}\epsilon_i^\mu k_i^\nu\right)}{2p_i\cdot k_i}\right. \\
-&\left.\frac{\left(\not{\epsilon}_i-\frac{i\kappa}{2m}\sigma_{\mu\nu}\epsilon_i^\mu k_i^\nu\right)(\not{p}_f-\not{k}_i+m)\left(\not{\epsilon}_f^*+\frac{i\kappa}{2m}\sigma_{\mu\nu}\epsilon_f^{*\mu}k_f^\nu\right)}{2p_i\cdot k_f}\right] \\
\times&\ u(p_i)
\end{aligned} \tag{15}$$

where $\kappa$ is the anomalous magnetic dipole moment. Low's theorem, as before, reads

$$\mathrm{Amp}_C^{tot}(S=\tfrac{1}{2})=\mathrm{Amp}_C^{Born}(S=\tfrac{1}{2})+\mathrm{Amp}_C^{NB}(S=\tfrac{1}{2}) \tag{16}$$

with $\mathrm{Amp}_C^{NB}=O(\omega^2)$, but now the forward scattering amplitude has the form

$$\mathrm{Amp}_C^{tot}(S=\tfrac{1}{2})\xrightarrow{\theta\to0}\chi_f^\dagger\chi_i\boldsymbol{\epsilon}_f^*\cdot\boldsymbol{\epsilon}_i f(\omega)+i\chi_f^\dagger\boldsymbol{\sigma}\chi_i\cdot\boldsymbol{\epsilon}_f^*\times\boldsymbol{\epsilon}_i g(\omega) \tag{17}$$

with

$$f(\omega)=f(-\omega)=-\frac{e^2}{4\pi m}+O(\omega^2) \tag{18}$$

and

$$g(\omega)=-g(-\omega)=\frac{e^2\kappa^2\omega}{2\pi m}+O(\omega^3) \tag{19}$$

Causality implies that we still have the Baldin sum rule, which now reads

$$\alpha_E+\beta_M=\frac{1}{2\pi^2}\int_0^\infty\frac{d\omega'(\sigma_{\frac{3}{2}}(\omega')+\omega_{\frac{1}{2}}(\omega'))}{\omega'^2} \tag{20}$$

 DOI: 10.1051/epjconf/201713401003

where $\sigma_{\frac{3}{2}}(\omega')$, $\sigma_{\frac{1}{2}}(\omega')$ are the photoabsorption cross sections for parallel, antiparallel photon and taget helicities, and we also have a sum rule for $g(\omega)$ which, since it involves the spin-flip amplitude, is *unsubtracted*

$$\frac{\pi^2 e^2 \kappa^2}{2m^2} = \int_0^\infty \frac{d\omega'(\sigma_{\frac{3}{2}}(\omega') - \sigma_{\frac{1}{2}}(\omega'))}{\omega'} \tag{21}$$

Eq. (21) is the GDH (Gerasimov-Drell-Hearn) sum rule [3],[4] and is well-satisfied in the case of the proton [7]

$$LHS = 205\,\mu\text{b} \quad \text{vs.} \quad RHS = 212 \pm 15\,\mu\text{b} \tag{22}$$

In addition to the leading order polarizabilities $\alpha_E, \beta_M$, one can probe for additional structure by defining higher order quantities [15]

$$\begin{aligned} H_{eff} &= \frac{(\boldsymbol{p} - e\boldsymbol{A})^2}{2m} - \frac{1}{2}4\pi\alpha \boldsymbol{E}^2 - \frac{1}{2}4\pi\beta_M \boldsymbol{B}^2 \\ &- \frac{1}{2}4\pi\left(\gamma_{E1E1}\boldsymbol{\sigma}\cdot(\boldsymbol{E}\times\dot{\boldsymbol{E}} + \gamma_{M1M1}\boldsymbol{\sigma}\cdot(\boldsymbol{B}\times\dot{\boldsymbol{B}}\right. \\ &- \left.2\gamma_{M1E2}\sigma_i B_j E_{ij} + 2\gamma_{E1M2}\sigma_i E_j B_{jk}\right) \\ &- \frac{1}{2}4\pi\left(\alpha_{E1\nu}\dot{\boldsymbol{E}}^2 + \beta_{M1\nu}\dot{\boldsymbol{B}}^2\right) \\ &- \frac{1}{12}4\pi\left(\alpha_{E2}E_{ij}^2 + \beta_{M2}B_{ij}^2\right) + \ldots \end{aligned} \tag{23}$$

where $E_{ij} = \frac{1}{2}(\nabla_i E_j + \nabla_j E_i)$ and $B_{ij} = \frac{1}{2}(\nabla_i B_j + \nabla_j B_i)$ are field gradients. These definitions lead to additional sum rules

$$\begin{aligned} &\alpha_{E1\nu} + \beta_{M1\nu} + \frac{1}{12}(\alpha_{E2} + \beta_{M2}) \\ &= \frac{1}{2\pi^2}\int_0^\infty \frac{d\omega'(\sigma_{\frac{3}{2}}(\omega') + \sigma_{\frac{1}{2}}(\omega'))}{\omega'^4} \\ &\gamma_{E1E1} + \gamma_{M1M1} + \gamma_{E2M1} + \gamma_{M2E1} \\ &= \frac{1}{4\pi^2}\int_0^\infty \frac{d\omega'(\sigma_{\frac{3}{2}}(\omega') - \sigma_{\frac{1}{2}}(\omega'))}{\omega'^3} \end{aligned} \tag{24}$$

and

$$\begin{aligned} &\gamma_{E1E1\nu} + \gamma_{M1M1\nu} + \gamma_{M2E1\nu} + \gamma_{M2E1\nu} \\ &+ \gamma_{E2E2} + \gamma_{M2M2} + \frac{8}{5}\gamma_{M3E2} + \frac{8}{5}\gamma_{E3M2} \\ &= \frac{1}{4\pi^2}\int_0^\infty \frac{d\omega'(\sigma_{\frac{3}{2}}(\omega') - \sigma_{\frac{1}{2}}(\omega'))}{\omega'^5} \end{aligned} \tag{25}$$

On the theoretical side, there have been various theoretical approaches used to predict polarizabilities, but recently much work has been done using chiral perturbation theory [16], using both the heavy baryon [17] and covariant baryon versions [18]. In the case of $\alpha_E, \beta_M$ we have the results shown in Table 1. The values of the spin polarizabilities recently measured by the A2 collaboration at MAMI are compared with theoretical predictions in Table 2. Finally, some calculated values of higher order polarizabilities are shown in Table 3 The rough general agreement here between the experimental and theoretical numbers for the polarizabilities suggests the important role played by the pion cloud.

| | HBχpt | Bχpt | PDG [10] |
|---|---|---|---|
| $\alpha_E$ | $10.7 \pm 0.5$ | $10.6 \pm 0.5$ | $11.2 \pm 0.4$ |
| $\beta_M$ | $3.2 \mp 0.5$ | $3.2 \mp 0.5$ | $2.5 \mp 0.4$ |

**Table 1.** Shown are theoretical and experimental values of the electric and magnetic polarizabilities compared with the values quoted by the Particle Data Group. All numbers are in units of $10^{-4}$ fm$^3$.

| | DR [19] | HBχpt | Bχpt | Expt. [20] |
|---|---|---|---|---|
| $\gamma_{E1E1}$ | -5.6 | $-1.1 \pm 1.8$ | $-3.3 \pm 0.8$ | $-3.5 \pm 1.2$ |
| $\gamma_{M1M1}$ | 3.8 | $2.2 \pm 0.6$ | $2.9 \pm 1.5$ | $3.2 \pm 0.9$ |
| $\gamma_{E1M2}$ | -0.7 | $-0.4 \pm 0.4$ | $0.2 \pm 0.2$ | $-0.7 \pm 1.2$ |
| $\gamma_{E2M1}$ | 2.9 | $1.9 \pm 0.4$ | $1.1 \pm 0.4$ | $2.0 \pm 0.3$ |

**Table 2.** Shown are theoretical and experimental values of the spin polarizabilities. All numbers are in units of $10^{-4}$ fm$^4$.

| | HBχpt(NNLO) |
|---|---|
| $\alpha_{E2}$ | $17.3 \pm 3.9$ |
| $\beta_{M2}$ | $-15.5 \pm 3.5$ |
| $\alpha_{E1\nu}$ | $-1.3 \pm 0.4$ |
| $\beta_{M1\nu}$ | $7.1 \pm 2.5$ |

**Table 3.** Shown are theoretical values of certain higher order electric and magnetic polarizabilities. All numbers are in units of $10^{-4}$ fm$^5$.

## 3 Gravitational Compton Scattering

When considering gravitational Compton scattering, Low's theorem can again be applied [9] but now any corrections to the simple Born amplitudes must be of order $O(\omega^4)$, where here the Born amplitude arises from the sum of the four diagrams shown in Figure 2.

The transition from electromagnetic to gravitational interactions is at lowest order given by the substitution [21]

$$\mathcal{L}_{em} = -e j_\mu A^\mu \quad \longrightarrow \quad \mathcal{L}_{grav} = -\frac{\kappa}{2}T_{\mu\nu}h^{\mu\nu} \tag{26}$$

Here, in the electromagnetic case, $j_\mu = \frac{1}{e}\frac{\delta\mathcal{L}}{\delta A^\mu}$ is the electromagnetic current, $\alpha = e^2/4\pi$ is the fine structure constant, and $A^\mu$ is the vector potential, while for the gravitational interaction $T_{\mu\nu} = -\frac{2}{\sqrt{-\det g}}\frac{\delta\mathcal{L}_{mat}}{\delta g_{\mu\nu}}$ is the energy-momentum tensor, $G = \kappa^2/32\pi$ is the Cavendish constant, and $g_{\mu\nu} = \eta_{\mu\nu} + h_{\mu\nu}$ is the metric tensor. Writing the spin-0

DOI: 10.1051/epjconf/201713401003

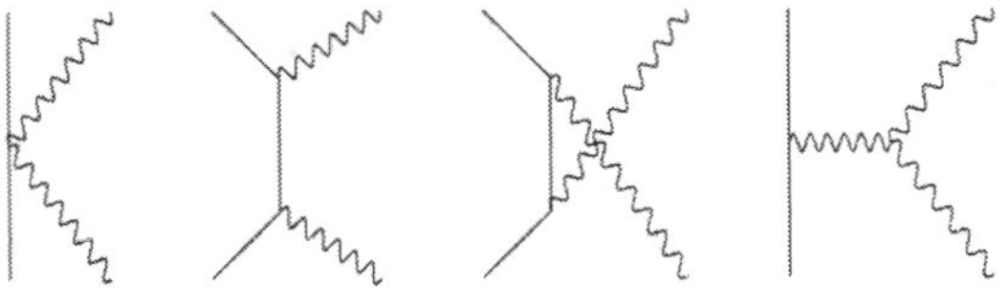

a) b) c) d)

**Figure 2.** Shown are the a) seagull, b) Born, c) Cross-Born, and d) graviton pole diagrams contributing the gravitational Compton scattering. Here the solid lines represent massive spinless particles while the wiggly lines are gravitons.

matter Lagrangian as

$$\mathcal{L}_{mat} = g_{\mu\nu}(D^\mu\phi)^\dagger D^\nu\phi - m^2\phi^\dagger\phi \tag{27}$$

we find the electromagnetic current

$$j_\mu = \phi^\dagger\partial_\mu\phi - (\partial\phi)^\dagger\phi \tag{28}$$

and energy-momentum tensor

$$T_{\mu\nu} = (\partial_\mu\phi)^\dagger\partial_\nu\phi + (\partial_\nu\phi)^\dagger\partial_\mu\phi - \frac{1}{2}g_{\mu\nu}\mathcal{L} \tag{29}$$

We have then the general matrix element forms [21]

$$\langle p_f|j_\mu(x)|p_i\rangle = e^{-iq\cdot x}G(q^2)2P_\mu \tag{30}$$

and

$$\begin{aligned}\langle p_f|T_{\mu\nu}(x)|p_i\rangle &= e^{-iq\cdot x}\Big[2P_\mu P_\nu F_1(q^2)\\ &+ (q_\mu q_\nu - \eta_{\mu\nu}q^2)F_2(q^2)\Big]\end{aligned} \tag{31}$$

where $q = p_i - p_f$ is the momentum transfer and $P_\mu = \frac{1}{2}(p_f + p_i)_\mu$ is the mean momentum. We have the constraints $G(q^2 = 0) = 1$ from current conservation and $F_1(q^2 = 0) = 1$ by energy-momentum conservation.

The contributions to gravitational Compton scattering from the seagull, direct-Born and cross-Born diagrams is then easily found and is only slightly more complex than the corresponding electromagnetic forms:

$$\begin{aligned}\mathrm{Amp}^{2a}_{GC}(S=0) &= \kappa^2\Big[\epsilon_f^*\cdot\epsilon_i(\epsilon_i\cdot p_i\epsilon_f^*\cdot p_f\\ &+ \epsilon_i\cdot p_f\epsilon_f^*\cdot p_i) - \frac{1}{2}k_i\cdot k_f(\epsilon_f^*\cdot\epsilon_i)^2\Big]\\ \mathrm{Amp}^{2b}_{GC}(S=0) &= 2\kappa^2\frac{(\epsilon_i\cdot p_i)^2(\epsilon_f^*\cdot p_f)^2}{p_i\cdot k_i}\\ \mathrm{Amp}^{2c}_{GC}(S=0) &= -2\kappa^2\frac{(\epsilon_f^*\cdot p_i)^2(\epsilon_i\cdot p_f)^2}{p_i\cdot k_f}\end{aligned} \tag{32}$$

However, the sum of these three diagrams is neither complete nor gauge-invariant. We *must* add the contribution from Figure 2d, the graviton pole diagram

$$\begin{aligned}&\mathrm{Amp}^{2d}_{GC}(S=0) = \frac{4\kappa^2}{k_i\cdot k_f}\Big[\epsilon_f^*\cdot p_f\epsilon_f^*\cdot p_i(\epsilon_i\cdot(p_i-p_f))^2\\ +\ &\epsilon_i\cdot p_i\epsilon_i\cdot p_f(\epsilon_f^*\cdot(p_i+p_f))^2\\ +\ &\epsilon_i\cdot(p_i-p_f)\epsilon_f^*\cdot(p_f-p_i)(\epsilon_f^*\cdot p_f\epsilon_i\cdot p_i+\epsilon_f^*\cdot p_i\epsilon_i\cdot p_f)\\ -\ &\epsilon_f^*\cdot\epsilon_i\Big(\epsilon_i\cdot(p_i-p_f)\epsilon_f^*\cdot(p_f-p_i)(p_i\cdot p_f-m^2)\\ +\ &k_i\cdot k_f(\epsilon_f^*\cdot p_f\epsilon_i\cdot p_i+\epsilon_f^*\cdot p_i\epsilon_i\cdot p_f)\\ +\ &\epsilon_i\cdot(p_i-p_f)(\epsilon_f^*\cdot p_fp_i\cdot k_f+\epsilon_f^*\cdot p_ip_f\cdot k_f)\\ +\ &\epsilon_f^*\cdot(p_f-p_i)(\epsilon_i\cdot p_ip_f\cdot k_i+\epsilon_i\cdot p_fp_i\cdot k_i)\Big)\\ +\ &(\epsilon_f^*\cdot\epsilon_i)^2\Big(p_i\cdot k_ip_f\cdot k_i+p_i\cdot k_fp_f\cdot k_f\\ -\ &\frac{1}{2}(p_i\cdot k_ip_f\cdot k_f+p_i\cdot k_fp_f\cdot k_i)\\ +\ &\frac{3}{2}k_i\cdot k_f(p_i\cdot p_f-m^2)^2\Big)\Big]\end{aligned} \tag{33}$$

where the complex form given here arises from the graviton propagator

$$D_{\alpha\beta;\gamma\delta} = \frac{iP_{\alpha\beta;\gamma\delta}}{q^2+i\epsilon}\quad\text{with}\quad P_{\alpha\beta;\gamma\delta}\Big[I_{\alpha\beta;\gamma\delta} - \frac{1}{2}\eta_{\alpha\beta}\eta_{\gamma\delta}\Big] \tag{34}$$

and the triple graviton vertex, which is a consequence of the nonlinearity of general relativity [22]

$$\begin{aligned}&\tau^{\mu\nu}_{\alpha\beta,\gamma\delta}(k,q) = \frac{i\kappa}{2}\Big\{(I_{\alpha\beta,\gamma\delta} - \frac{1}{2}\eta_{\alpha\beta}\eta_{\gamma\delta})\\ \times\ &\Big[k^\mu k^\nu + (k-q)^\mu(k-q)^\nu + q^\mu q^\nu - \frac{3}{2}\eta^{\mu\nu}q^2\Big]\\ +\ &2q_\lambda q_\sigma\Big[I^{\lambda\sigma,}{}_{\alpha\beta}I^{\mu\nu,}{}_{\gamma\delta} + I^{\lambda\sigma,}{}_{\gamma\delta}I^{\mu\nu,}{}_{\alpha\beta}\\ -\ &I^{\lambda\mu,}{}_{\alpha\beta}I^{\sigma\nu,}{}_{\gamma\delta} - I^{\sigma\nu,}{}_{\alpha\beta}I^{\lambda\mu,}{}_{\gamma\delta}\Big]\\ +\ &[q_\lambda q^\mu(\eta_{\alpha\beta}I^{\lambda\nu,}{}_{\gamma\delta} + \eta_{\gamma\delta}I^{\lambda\nu,}{}_{\alpha\beta})\\ +\ &q_\lambda q^\nu(\eta_{\alpha\beta}I^{\lambda\mu,}{}_{\gamma\delta} + \eta_{\gamma\delta}I^{\lambda\mu,}{}_{\alpha\beta})\\ -\ &q^2(\eta_{\alpha\beta}I^{\mu\nu,}{}_{\gamma\delta} + \eta_{\gamma\delta}I^{\mu\nu,}{}_{\alpha\beta})\\ -\ &\eta^{\mu\nu}q^\lambda q^\sigma(\eta_{\alpha\beta}I_{\gamma\delta,\lambda\sigma} + \eta_{\gamma\delta}I_{\alpha\beta,\lambda\sigma})]\\ +\ &[2q^\lambda(I^{\sigma\nu,}{}_{\alpha\beta}I_{\gamma\delta,\lambda\sigma}(k-q)^\mu + I^{\sigma\mu,}{}_{\alpha\beta}I_{\gamma\delta,\lambda\sigma}(k-q)^\nu\\ -\ &I^{\sigma\nu,}{}_{\gamma\delta}I_{\alpha\beta,\lambda\sigma}k^\mu - I^{\sigma\mu,}{}_{\gamma\delta}I_{\alpha\beta,\lambda\sigma}k^\nu)\\ +\ &q^2(I^{\sigma\mu,}{}_{\alpha\beta}I_{\gamma\delta,\sigma}{}^\nu + I_{\alpha\beta,\sigma}{}^\nu I^{\sigma\mu,}{}_{\gamma\delta})\\ +\ &\eta^{\mu\nu}q^\lambda q_\sigma(I_{\alpha\beta,\lambda\rho}I^{\rho\sigma,}{}_{\gamma\delta} + I_{\gamma\delta,\lambda\rho}I^{\rho\sigma,}{}_{\alpha\beta})]\\ +\ &[(k^2 + (k-q)^2)\Big(I^{\sigma\mu,}{}_{\alpha\beta}I_{\gamma\delta,\sigma}{}^\nu + I^{\sigma\nu,}{}_{\alpha\beta}I_{\gamma\delta,\sigma}{}^\mu\\ -\ &\frac{1}{2}\eta^{\mu\nu}P_{\alpha\beta,\gamma\delta}\Big)\\ -\ &(k^2\eta_{\gamma\delta}I^{\mu\nu,}{}_{\alpha\beta} + (k-q)^2\eta_{\alpha\beta}I^{\mu\nu,}{}_{\gamma\delta})]\Big\}\end{aligned} \tag{35}$$

where

$$I_{\alpha\beta;\gamma\delta} = \frac{1}{2}(\eta_{\alpha\gamma}\eta_{\beta\delta} + \eta_{\alpha\delta}\eta_{\beta\gamma})$$

Adding the four diagrams together we find the final gauge-invariant form. Obviously the overall structure of the result is complex and difficult to understand. However, there occurs a miraculous simplification, which goes under the name of the "double copy theorem", which asserts that the gravitational Compton amplitude can be factorized via [5],[6].

$$\text{Amp}_{CG}^{tot}(S=0)=\sum_{i=a}^{d}\text{Amp}_{GC}^{2i}(S=0)=F\cdot(\text{Amp}_C^{tot}(S=0))^2 \tag{36}$$

where $\text{Amp}_C(S=0)$ is the spin-0 Compton amplitude given in Eq. (3) and

$$F=\frac{\kappa^2}{8e^4}\frac{p_i\cdot k_i p_i\cdot k_f}{k_i\cdot k_f} \tag{37}$$

is a kinematic factor. We leave it to the reader to verify this identity. Remarkably this factorization also obtains for general spin via

$$\text{Amp}_{CG}^{tot}(S)=\sum_{i=a}^{d}\text{Amp}_{GC}^{2i}(S)=F\cdot\text{Amp}_C^{tot}(S=0)\cdot\text{Amp}_C^{tot}(S) \tag{38}$$

We can demonstrate this explicitly for the simplest case of spin-$\frac{1}{2}$ for which the general form of the energy-momentum tensor matrix element is [21]

$$\begin{aligned}&\langle p_f|T_{\mu\nu}(x)|p_i\rangle=e^{-iq\cdot x}\bar{u}(p_f)\\ \times&\ \Big[\left(P_\mu\gamma_\nu+P_\nu\gamma_\mu\right)F_1(q^2)\\ -&\ \frac{i}{2m}\left(P_\mu\sigma_{\nu\lambda}q^\lambda+P_\nu\sigma_{\mu\lambda}q^\lambda\right)F_2(q^2)\\ +&\ \frac{1}{4m^2}\left(q_\mu q_\nu-\eta_{\mu\nu}q^2\right)F_3(q^2)\Big]u(p_i)\end{aligned} \tag{39}$$

where now $F_1(q^2=0)=1$ from energy-momentum conservation and $F_2(q^2=0)=1$ from angular-momentum conservation. The latter condition requires the vanishing of the anomalous gravitomagnetic moment. Using the lowest order forms, we find for gravitational Compton scattering

$$\begin{aligned}&\text{Amp}_{GC}^{2a}(S=\tfrac{1}{2}))=\kappa^2\bar{u}(p_f)\\ \times&\ \Big[\frac{3}{16}\epsilon_f^*\cdot\epsilon_i(\not\epsilon_i\epsilon_f^*\cdot(p_i+p_f)+\not\epsilon_f^*\epsilon_i\cdot(p_i+p_f))\\ +&\ \frac{i}{8}\epsilon_f^*\cdot\epsilon_i\epsilon^{\rho\sigma\eta\lambda}\gamma_\lambda\gamma_5(\epsilon_{i\eta}\epsilon_{f\sigma}^*k_{f\rho}-\epsilon_{f\eta}^*\epsilon_{i\sigma}k_{i\rho})\Big]u(p_i)\\ &\text{Amp}_{GC}^{2b}(S=\tfrac{1}{2})=\kappa^2\frac{\epsilon_f^*\cdot p_f\epsilon_i\cdot p_i}{8p_i\cdot k_i}\\ \times&\ \bar{u}(p_f)[\not\epsilon_f^*(\not p_i+\not k_i+m)\not\epsilon_i]u(p_i)\\ &\text{Amp}_{GC}^{2c}(S=\tfrac{1}{2})=-\kappa^2\frac{\epsilon_f^*\cdot p_i\epsilon_i\cdot p_f}{8p_i\cdot k_f}\\ \times&\ \bar{u}(p_f)[\not\epsilon_i(\not p_i-\not k_f+m)\not\epsilon_f^*]u(p_i)\\ &\text{Amp}_{GC}^{2d}(S=\tfrac{1}{2})=\frac{\kappa^2}{k_i\cdot k_f}\bar{u}(p_f)\\ \times&\ \Big[(\not\epsilon_i\epsilon_f^*\cdot k_i+\not\epsilon_f^*\epsilon_i\cdot k_f)\\ \times&\ (\epsilon_i\cdot p_i\epsilon_f^*\cdot p_f-\epsilon_i\cdot p_f\epsilon_f^*\cdot p_i)\\ -&\ (\epsilon_f^*\cdot\epsilon_i)\big(k_i\cdot k_f(\not\epsilon_f^*\epsilon_i\cdot k_f+\not\epsilon_i\epsilon_f^*\cdot p_i)\\ +&\ \not k_i(\epsilon_f^*\cdot p_f\epsilon_i\cdot p_i-\epsilon_f^*\cdot p_i\epsilon_i\cdot p_f)\\ +&\ p_i\cdot k_i(\not\epsilon_i\epsilon_f^*\cdot k_i+\not\epsilon_f^*\epsilon_i\cdot k_f)\big)\\ +&\ (\epsilon_f^*\cdot\epsilon_i)^2\not k_i(p_i\cdot k_i-\frac{1}{2}k_i\cdot k_f)\Big]u(p_i)\end{aligned} \tag{40}$$

Again we leave it as an exercise to show that

$$\begin{aligned}\text{Amp}_{CG}^{tot}(S=\tfrac{1}{2})&=\sum_{i=a}^{d}\text{Amp}_{GC}^{2i}(S=\tfrac{1}{2})\\ &=F\cdot\text{Amp}_C^{tot}(S=0)\cdot\text{Amp}_C^{tot}(S=\tfrac{1}{2})\end{aligned} \tag{41}$$

In the case of gravitational Compton scattering, Low's theorem reads [9].

$$\text{Amp}_{GC}(S)=\text{Amp}_{GC}^{Born}(S)+\text{Amp}_{GC}^{NB}(S) \tag{42}$$

where $\text{Amp}_{GC}^{Born}(S)$ is the sum of the four diagrams shown in Figure 2 and $\text{Amp}_{GC}^{NB}(S)=O(\omega^4)$. The origin of the non-Born terms is again associated with polarizability effects but now represents a *quadrupole* polarizability [23]. Recall that in the electromagnetic case the polarizability $\alpha_E,\beta_M$ represents an induced *dipole* moment which is produced in the presence of an external electromagnetic field. In the gravitational case, the polarizability represents the induced *quadrupole* moment which is generated in the presence of an external gravitational field gradient

$$Q_{ij}=\alpha_G R_{0i;0j} \tag{43}$$

where

$$Q_{ij}=\int d^3r\rho(\mathbf{r})\left(\frac{3}{2}r_ir_j-\frac{1}{2}\delta_{ij}\mathbf{r}^2\right) \tag{44}$$

is the quadrupole moment and $R_{\alpha\beta;\gamma\delta}$ is the curvature tensor [23]. The corresponding effective Lagrangian is

$$\mathcal{L}_{eff}=-\frac{1}{2}\alpha_G R_{\alpha\beta;\gamma\delta}R^{\rho\beta;\sigma\delta}(\partial^\alpha\partial^\gamma\phi^\dagger)\partial_\rho\partial_\sigma\phi \tag{45}$$

Then just as the electromagnetic polarizability leads, via two-photon exchange, to the long range Casimir-Polder interaction [24]

$$V_{em}(r)=-\frac{23(\alpha_E^{(1)}\alpha_E^{(2)}+\beta_M^{(1)}\beta^{(2)})-7(\alpha_E^{(1)}\beta_M^{(2)}+\beta_M^{(1)}\alpha_E^{(2)})}{4\pi r^7} \tag{46}$$

the gravitational polarizability leads, via two-graviton exchange, to a long-range interaction [23]

$$V_{grav}(r)=-\frac{3897\alpha_G^{(1)}\alpha_G^{(2)}}{1024\pi^3r^{11}} \tag{47}$$

Because of the parallels between the electromagnetic and gravitational interactions, it is interesting to ask if there is a gravitational analog to the GDH sum rule, which if so would have a form

$$0=\frac{1}{2\pi^2}\int_0^\infty\frac{d\omega'(\sigma_{\frac{5}{2}}(\omega')-\sigma_{\frac{3}{2}}(\omega'))}{\omega'} \tag{48}$$

because of the vanishing of the anomalous gravitomagnetic moment noted above. A 1968 paper by Gross and Jackiw shows that fixed-angle dispersion relations do exist [9], but there is an important wrinkle which asserts that sum rules such as Eq. (48) do not. The important point here can be seen by use of factorization, which asserts that in the case of spinless scattering we have the forward scattering amplitude

$$\text{Amp}_C(S=0)(\theta) \xrightarrow{\theta\to 0} -2e^2 \tag{49}$$

We have then

$$\text{Amp}_{GC}(\theta) \xrightarrow{\theta\to 0} (-2e^2)^2 \cdot \frac{\kappa^2}{8e^4}\frac{1}{1-\cos\theta} = \frac{\kappa^2}{2(1-\cos\theta)} \tag{50}$$

so that the forward-angle gravitational Compton amplitude becomes infinite. Because of this divergence, the optical theorem [11]

$$\text{Im}\, f(\theta) = \frac{\omega}{4\pi}\sigma_{tot} \tag{51}$$

cannot be used. It is interesting to note that in the case of the Compton scattering of a particle with spin

$$\text{Amp}_C(S)(\theta) \xrightarrow{\theta=0} -2e^2\boldsymbol{\epsilon}_i\cdot\boldsymbol{\epsilon}_f^* + i\boldsymbol{\sigma}\cdot\boldsymbol{\epsilon}_f^*\times\boldsymbol{\epsilon}_i\frac{t}{2(s-m^2)} + \ldots \tag{52}$$

We see explicitly then that the spin-flip Compton amplitude in the forward direction vanishes, so that only the *anomalous* magnetic moment contributes. However, in the gravitational case, because the factor $F \xrightarrow{t\to 0} \sim 1/t$ the gravitational Compton spin-flip amplitude is nonvanishing

$$\text{Amp}_{GC}(S)(\theta) \xrightarrow{\theta=0} 8\pi G\Big[(\boldsymbol{\epsilon}_i\cdot\boldsymbol{\epsilon}_f^*)^2\frac{(s-m^2)^2}{t} + i\boldsymbol{\sigma}\cdot\boldsymbol{\epsilon}_f^*\times\boldsymbol{\epsilon}_i\frac{s-m^2}{2} + \ldots\Big] \tag{53}$$

However, the cross section still diverges so that there exists no gravitational analog of the GDH sum rule.

## 4 Gravitational Scattering

In the previous section we showed how the factorization implied by the double copy theorem could be used to evaluate the gravitational scattering reaction in terms of relatively simple Compton scattering amplitudes. In this section we show how this result can be employed to evaluate the one loop corrections to electromagnetic or gravitational scattering. At lowest order, of course, the results of single-photon or single-graviton exchange are well-known and lead to the Coulomb and Newtonian interactions when Fourier-transformed. That is, in the case of photon exchange between systems having charge $Q_1e$ and $Q_2e$, we have, in the nonrelativistic limit,

$$\begin{aligned}\mathcal{M}_{em}(q) &= \frac{Q_1eQ_2e}{\sqrt{16E_1E_2E_3E_4}}(p_1+p_2)^\mu\frac{\eta_{\mu\nu}}{q^2}(p_3+p_4)^\nu \\ &\xrightarrow{NR} \frac{Q_1Q_2e^2}{\boldsymbol{q}^2}\end{aligned} \tag{54}$$

so

$$V_{em}^{(0)}(r) = \int\frac{d^3q}{(2\pi)^3}e^{-i\boldsymbol{q}\cdot\boldsymbol{r}}\mathcal{M}_\gamma(q) = \frac{Q_1Q_2\alpha}{|\boldsymbol{r}|} \tag{55}$$

while one-graviton exchange between systems having mass $m_1$ and $m_2$ is

$$\begin{aligned}\mathcal{M}_{grav}(q) &= \frac{\frac{\kappa}{2}m_1\frac{\kappa}{2}m_2}{\sqrt{16E_1E_2E_3E_4}}\frac{1}{2}(p_1+p_2)^\mu(p_1+p_2)^\nu \\ &\times \frac{P_{\mu\nu;\alpha\beta}}{q^2}\frac{1}{2}(p_3+p_4)^\alpha(p_3+p_4)^\beta \\ &\xrightarrow{NR} -\frac{\kappa^2m_1m_2}{8\boldsymbol{q}^2}\end{aligned} \tag{56}$$

so

$$V_{grav}^{(0)}(r) = \int\frac{d^3q}{(2\pi)^3}e^{-i\boldsymbol{q}\cdot\boldsymbol{r}}\mathcal{M}_{grav}(q) = -\frac{Gm_1m_2}{|\boldsymbol{r}|} \tag{57}$$

These well-known classical physics results are augmented in the quantum mechanical case by loop corrections, and the one-loop corrections to these forms can be found by the use of so-called *on-shell* methods. In this technique, one uses unitarity to determine the discontinuity across the two-photon or two graviton t-channel cuts (*cf.* Figure 3) [25].

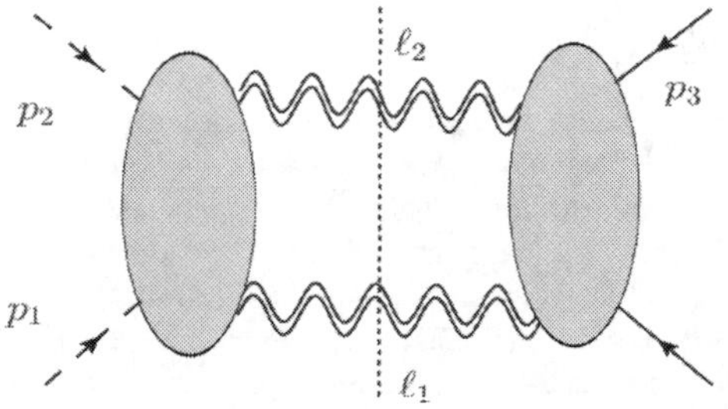


**Figure 3.** The two photon cut for the amplitude between a massive scalar particles. The grey blobs represent tree-level Compton amplitudes.

Thus, in the gravitational case we have the form

$$\begin{aligned}&\text{Disc Amp}_G(p_3,p_4;p_1,p_2)(S) \\ &= \frac{-i}{2!}\frac{1}{\sqrt{16E_1E_2E_3E_4}}\int\frac{d^3\ell_1}{2\ell_1^0(2\pi)^3}\int\frac{d^3\ell_2}{2\ell_2^0(2\pi)^3} \\ &\times (2\pi)^4\delta^4(p_1+p_2-\ell_1-\ell_2)\text{Amp}_{GC}^{*\rho\sigma;\xi\zeta}(p_3p_4;\ell_1\ell_2)(S) \\ &\times P_{\rho\sigma;\alpha\beta}P_{\xi\zeta;\gamma\delta}\text{Amp}_{GC}^{\alpha\beta,\gamma\delta}(\ell_1\ell_2;p_1p_2)(S)\end{aligned} \tag{58}$$

Using factorization, this discontinuity calculation can be written in terms of simple Compton amplitudes. In this way the discontinuity of one-loop corrections to the gravitational scattering of spinless systems can be straightforwardly found to be

$$\begin{aligned}&\text{Disc Amp}_G(p_3p_4;p_1p_2)(S=0) = -i\frac{\kappa^4m_Am_B}{128\pi} \\ &\times \left[\frac{6\pi(m_A+m_B)}{\sqrt{-q^2}} + \frac{41}{5} + 4\pi i\frac{m_Am_Bm_r}{p_0q^2} + \ldots\right]\end{aligned} \tag{59}$$

The scattering amplitude itself then is

$$\begin{aligned}\mathrm{Amp}_G(p_3p_4;p_1p_2)(S=0) &= -\frac{\kappa^4 m_A m_B}{1024\pi^2}\\ \times &\left[-\frac{6\pi(m_A+m_B)}{\sqrt{-q^2}}+\frac{41}{5}\log -q^2\right.\\ + &\left. 4\pi i\frac{m_A m_B m_r}{p_0 q^2}\log -q^2+\ldots\right]\end{aligned} \tag{60}$$

where $m_r = m_A m_B/(m_A+m_B)$ is the reduced mass and $p_0$ is the magnitude of the center of mass frame momentum. Eq. (60) agrees with the result calculated by Bjerrum-Bohr et al. and by Kirilin and Khriplovich by much more laborious Feynman diagram methods [27],[26]. The imaginary piece of the scattering amplitude arises from the second Born approximation, which is subtracted when defining the second order potential

$$\begin{aligned}V^{(1)}_{grav}(r) &= \int\frac{d^3q}{(2\pi)^3}e^{-iq\cdot r}\left(\mathcal{M}_{grav}(q)-\mathrm{Amp}^{(2)}_{Born}(q)\right)\\ &= -\frac{3G^2m_Am_B(m_A+m_B)}{r^2}-\frac{41G^2m_Am_B\hbar}{10\pi r^3}\end{aligned} \tag{61}$$

The simplification here comes from a reordering of summation and integration. That is, in the conventional procedure one evaluates individual and gauge-dependent Feynman diagrams (including ghost loops) by performing four-dimensional integrations. The various diagrams are then summed to produce the final gauge-invariant result. In the on-shell procedure the summation is performed first, producing gauge-invariant Compton amplitudes, which are then used as a component of a two-dimensional integration. Also, because the amplitudes are on-shell, no ghost contributions are required. In this way what was originally a two-*year* calculation becomes a two-*day* evaluation.

## 5 Conclusions

Above we have discussed how Compton scattering—$S+\gamma\to S+\gamma$—has been used to probe the structure of a system with spin-S by measurement of its response when placed in external electromagnetic fields. The resulting deformations are described in terms of polarizabilities [2], which in the case of the proton have been the subject of experimental study for many years at various laboratories [7]. Most recently experiments by the A2 collaboration at MAMI utilizing polarization have measured not only the size of the leading electric and magnetic polarizabilities $\alpha_E$ and $\beta_M$ but also of the higher order spin polarizabilities $\gamma_{E1E1}$, $\gamma_{M1M1}$, $\gamma_{E2M1}$, and $\gamma_{M2E1}$ [20]. On the theoretical side these polarizabilities have been calculated via various methods, most recently by the means of chiral perturbation theory [17],[18]. The *chi*pt estimates are in basic agreement with the experimental results, which indicates the importance of the pion cloud in studying such quantities.

The recent LIGO measurement of gravitational waves and the confirmation thereby of the existence of gravitons [1] emphasizes the need to study the parallel gravitaional Compton scattering reaction—$S+g\to S+g$—and this reaction was discussed in turn. The standard way to calculate such reactions, by the use of Feynman diagrams, was shown to be enormously simplified by the use of te recently discovered double-copy theorem [5], which asserts that the desired gravitational Compton amplitudes can be written in terms of a product of ordinary Compton amplitudes together with a simple kinematic factor. This procedure was then shown to allow, via the use of unitarity and on-shell methods, a relatively simple evaluation of the one-loop corrections to gravitational scattering by evaluation of the discontinuity over the two-graviton t-channel cut.

## Acknowledgement

It is a pleasure to acknowledge the warm hospitality of the Bonn physics group.

## References

[1] B.P. Abbott et al., Phys. Rev. Lett. **116**, 061102 (2016).
[2] B.R. Holstein and S. Scherer, Ann. Rev. Nucl. Part. Sci. **64**, 51-81 (2014).
[3] S. Gerasimov, Yad. Fiz. **2**, 598 (1965).
[4] S.D. Drell and A.C. Hearn, Phys. Rev. Lett. **16**, 908 (1966).
[5] See, *e.g.*, Z. Bern, Living Rev. Relativity **5**, 5 (2002).
[6] S.Y. Choi et al., Phys. Rev. **D51**, 2751 (1995).
[7] See, *e.g.* F. Hagelstein et al., Prog. Part. Nucl. Phys. **88**, 29 (2016).
[8] F.E. Low, Phys. Rev. **96**, 1428 (1954).
[9] D.J. Gross and R. Jackiw, Phys. Rev. **166**, 1287 (1968).
[10] K.A. Olive et al., Chin. Phys. **C38**, 090001 (2014).
[11] See, *e.g.*, E. Merzbacher, *Quantum Mechanics*, Wiley, New York (1972).
[12] N.C. Mukhopadhyay et al., Phys. Rev. **D47**, (1993).
[13] D. Drechsel et al., Phys. Rept. **378**, 99 (2003).
[14] A.M. Baldin, Nucl. Phys. **18**, 310 (1960); L.I. Lapidus, Sov. Phys. JETP **16**, 964 (1963).
[15] B.R. Holstein et al., Phys. Rev. **C61**, 034316 (2000).
[16] J. Gasser, M.E. Sainio, and A. Svarc, Nucl. Phys. **B307** 779 (1988).
[17] V. Bernard, N. Kaiser, and U.-G. Meißner, Int. J. Mod. Phys. **E193** (1995).
[18] V. Lensky and V. Pascalutsa, Eur. Phys. J. **C65**, 195 (2010).
[19] B. Pasquini, D. Drechsel, and M. Vanderhaeghen, Phys. Rev. **C76**, 015203 (2007).
[20] P.P. Martel et al., Phys. Rev. Lett. **114**, 112501 (2015).
[21] See *e.g.*, J.F. Donoghue and B.R. Holstein, J. Phys. **G42**, 103102 (2015).

[22] J.F. Donoghue, Phys. Rev. **D50**, 3874 (1994).
[23] L.H. Ford et al., Phys. Rev. Lett. **116** 151301 (2016).
[24] H.B.G. Casimir and D. Polder, Phys. Rev. **73**, 366 (1948).
[25] N.E.J. Bjerrum-Bohr, J.F. Donoghue, and P. Vanhove, JHEP **01**, 111 (2014).
[26] I.B. Khriplovich and G.G. Kirilin, Econf C0306234, 1361 (2003).
[27] N.E.J. Bjerrum-Bohr, J.F. Donoghue, and B.R. Holstein, Phys. Rev. **D67**, 084033 (2003).

# Spectroscopy of baryon resonances

*Reinhard* Beck[1,⋆] and *Ulrike* Thoma[1,⋆⋆]

[1] *Helmholtz–Institut für Strahlen– und Kernphysik, Universität Bonn, 53115 Bonn, Germany*

**Abstract.** Within project A.1 of the SFB/TR16 "Subnuclear Structure of Matter", a large amount of data on photoproduction reactions has been accumulated at the Bonn Electron Stretcher Accelerator ELSA with the CBELSA/TAPS detector and was analysed in detail. In particular, data have been taken with unpolarized or with linearly or circularly polarized photons and with unpolarized or with longitudinally or transversely polarized protons. Photoproduction off neutrons was studied to determine the helicity amplitudes for the excitation of resonances off neutrons. In a partial wave analysis of the data, new resonances have been found and the properties of new and of known resonances have been determined, including the measurement of partial widths of so far unmeasured decay modes.

## 1 Introduction

The discovery of the $\Omega^-$ baryon with strangeness $S = -3$ at the predicted mass in the reaction $K^-p \rightarrow (\Xi^0\pi^-)K^+K_s$ [1] established SU(3) symmetry and the conjecture that mesons and baryons are composed of constituent quarks [2, 3] and paved the path to an understanding of the particle zoo. Like the $\Delta^{++}(1232)$ and the $\Delta^-(1232)$, the $\Omega^-$ was interpreted as bound state containing three identical quarks $(sss)$ while the two $\Delta(1232)$ were interpreted as $(uuu)$ and $(ddd)$ resonances. The spin and parity of these three baryons was $J^P = 3/2^+$; with a quark spin $s = 1/2$, the three quark spins had to be aligned to yield a total quark spin $S = 3/2$. As ground states, these baryons had to have no intrinsic orbital angular momentum. However, with three identical quarks in a relative $S$-wave, this configuration was in conflict with the well-established Pauli principle which forbids two fermions with identical quantum numbers.

The solution was to introduce a further quantum number, color, and to postulate that three quarks with different colors (and a colored quark and an antiquark with anticolor) form a color singlet, and that the color wavefunction is antisymmetric with respect to the exchange of two colors [4, 5]. The concept of color proved to be extremely important. From the request of local gauge invariance, a new theory of strong interactions, Quantum Chromo Dynamics or QCD, was deduced [6, 7] which assigned to quarks a new triple-valued charge called color. The fact that free quarks were never observed was understood by the hypothesis that color is confined [8], even though their mutual interactions were supposed to be weak at large momentum transfers or at small distances [9–11]. Even the baryon resonances seemed to be adequately described: Isgur and Karl demonstrated in their seminal papers that the hyperfine interaction between quarks suggested in QCD can, without further free parameters, explain the size and pattern of the splittings and the mixing angles observed experimentally in nonstrange, negative parity baryons [12]. The masses and mixing angles of the positive-parity baryons were also reasonably well reproduced in their QCD-inspired model [13]. In short: meson and baryon spectroscopy seemed to be well understood, and QCD was considered to be the established theory of strong interactions, with no mysteries and nothing more to learn. A survey of results on QCD from high-energy experiments can be found in an article by S. Bethke, G. Dissertori, and G. P. Salam [15] in the Review of Particle Properties RPP [16].

A few years before Isgur and Karl calculated the baryon spectrum, a family of new meson resonances had been discovered [17, 18] with the $J/\psi$ as ground state. The interest of the particle physics community turned first to glueballs [19] and hybrids [20], then to the weak-interaction bosons [21, 22], to the Higgs particle [23], and to tests of the Standard Model (see, e.g. [24]).

It was realized only later that important physics was left behind unexplored. New theoretical concepts were formulated which needed to be confronted with experimental data but, on the experimental side, the evidence for a large number of nucleon and $\Delta$ resonances was challenged when new and precise data were included in the analysis. From a theoretical point of view, it was realized that low-energy approximations of QCD [25] lead to the Gell-Mann-Oakes-Renner relation [26] which assigns a mass of a few MeV to light (current) quarks while quark models assume a (constituent) quark mass of about 350 MeV. How is the mass generated? The mass gap between current and constituent quarks was interpreted by spontaneous breaking of the chiral symmetry expected for

⋆e-mail: beck@hiskp.uni-bonn.de
⋆⋆e-mail: thoma@hiskp.uni-bonn.de

nearly massless quarks [27, 28]. An important consequence is the large mass gap between chiral partners: the masses of the nucleon, with spin-parity $J^P = 1/2^+$, and its chiral partner $N(1535)1/2^-$, with spin-parity $J^P = 1/2^-$ and mass $M = 1535\,\text{MeV}$, differ by about 600 MeV. But it is not understood how QCD generates the structure of hadrons and how symmetries influence their structures and their dynamics.

An important development occurred in the realm of low-energy QCD [29]. QCD functions can be expanded in powers of the external momenta and of the quark masses. The Ward identities of chiral symmetry determine the expansion in terms of a few coupling constants (LECs) of a unique effective low-energy Lagrangian. Chiral Dynamics has become an rapidly expanding field (see [30] for a review), and is one of the cornerstones of the SFB/TR16 (see contributions of projects A.6, A.7, B.3, B.6, C.2, C.9 to these proceedings) and the CRC110 (projects A.2, A.3, A.4, A.5, B.6, B.7, B.8). Precise data on photoproduction of baryon resonances are an important ingredient for further advances of the field.

On the experimental side, in 1994 there were 17 $N^*$ and 13 $\Delta^*$ resonances below a mass of 2200 MeV [31]. These resonances were all deduced from an analysis of data on $\pi N$ elastic and charge exchange scattering [32, 33]. These numbers are surprisingly low and, even worse, a large fraction of these resonances were not observed in the a later analysis based on a larger data sample [34] reporting 11 $N^*$ and 9 $\Delta^*$ only. The latter analysis included high precision data from the meson factories (see [34] for a list of data). This analysis predicted the spin rotation parameters [35–37] and the backward asymmetry [38] in the elastic pion-proton scattering from ITEP/PNPI very precisely while the predictions from [32, 33] show clear discrepancies with the data. Hence the number of trustworthy resonances was reduced to 11 $N^*$ and 9 $\Delta^*$ only.

At this point, we mention a basic problem in the analysis of data on $\pi N$ elastic and charge exchange scattering. The process is governed by two amplitudes. One of them represents a nucleon spin-flip in the scattering process, the other one a non-flip process. The two complex amplitudes are written in terms of two magnitudes and two phases. Two constraints come from experiment: the differential cross section $d\sigma/d\Omega$ and the asymmetry due to a target polarization $P$. Refs. [32, 33] make extensive use of dispersion relations to relate the ratio of the real to imaginary part of the scattering amplitude. In [34], an iterative procedure is used which includes energy-dependent fits to guide the solution. Thus there is an intrinsic under-determination of the scattering amplitude.

In the case of photoproduction, there are four complex amplitudes, nucleon spin-flip and non-flip, photon and nucleon spin aligned or anti-aligned. Thus at least seven (in fact eight) appropriate observables need to be determined in single pseudoscalar meson photoproduction, and one overall phase still remains open [60]. The measurement of a sufficient number of observables is the central theme of the efforts to measure the excitation spectrum of the nucleon in photoproduction. In general, if the target and beam polarization can be controlled and the recoil polarization can be determined in an experiment, even 16 observables become accessible overdetermining the amplitudes.

At a first glance it seems that $\pi N$ elastic and charge exchange scattering or pion-induced inelastic reactions might be the better choice to study the nucleon excitation spectrum since fewer observables need to be measured. However, the larger number of contributing amplitudes results in a larger number of interfering amplitudes, and the interference between large and small amplitudes increases the sensitivity to tiny effects. This is a great advantage of photoproduction experiments. A second advantage is that resonances with very small coupling to $\pi N$ but "normal" $\gamma N$ couplings may become experimentally accessible. The de-coupling of resonances from $\pi N$ was made responsible for the non-observation of the many resonances predicted by quark models.

## 2 What has been reached?

Before discussing in detail the outcome of the experiments carried out within the A.1 project of the SFB/TR16, we would like to mention here the main achievements. In 2011, we performed a first systematic search for new baryon resonances [39–42]. A very important result was that the resonances proposed by the groups led by Höhler [32] and Cutkosky [33] were confirmed in the Bonn-Gatchina (BnGa) partial wave analysis (see project A.2 "Partial Wave Analysis"). This analysis included $\pi N$ scattering data but the results were largely based on the new photoproduction data taken within the SFB/TR16 and at other places. Thirteen resonances were observed which were not seen in the most recent analysis based on $\pi N$ elastic scattering data [34]. There were six new entries in the

**Table 1.** New resonances found in the BnGa-PWA, resonances for which the star rating was improved, and resonances confirmed by the BnGa-PWA but not seen in the GWU-Arndt'06-analysis.

| | RPP 2010 | our analyses | RPP 2012 | GWU'06 |
|---|---|---|---|---|
| N(1700)3/2⁻ | *** | *** | *** | no evidence |
| N(1710)1/2⁺ | *** | *** | *** | no evidence |
| N(1860)5/2⁺ | | * | ** | |
| N(1875)3/2⁻ | | *** | *** | no evidence |
| N(1880)1/2⁺ | | ** | ** | no evidence |
| N(1895)1/2⁻ | | ** | ** | no evidence |
| N(1900)3/2⁺ | ** | *** | *** | no evidence |
| N(1990)7/2⁺ | ** | ** | ** | no evidence |
| N(2000)5/2⁺ | ** | ** | ** | no evidence |
| N(2060)5/2⁻ | | *** | ** | no evidence |
| N(2150)3/2⁻ | | ** | ** | no evidence |
| Δ(1900)1/2⁻ | * | * | ** | no evidence |
| Δ(1920)3/2⁺ | *** | *** | *** | no evidence |
| Δ(1940)3/2⁻ | * | ** | ** | no evidence |

 DOI: 10.1051/epjconf/201713402001

**Table 2.** Properties of the $N(1900)3/2^+$ resonance from the BnGa multichannel partial wave analysis. Along with the name of the resonance, the star rating of the Particle Data Group [16] is given. BnGa suggested an upgrade to four stars, marked by a ★. The helicity couplings $A_{1/2,3/2}$ are given in GeV$^{-\frac{1}{2}}$. The electromagnetic transition amplitudes $\gamma p$, $\gamma p^{1/2}$, $\gamma p^{3/2}$ are defined as elements of the transition matrix and dimensionless. $\pi N \to \pi N$ stands for the elastic pole residue, $2(\pi N \to X)/\Gamma$ for inelastic pole residues. They are normalized by a factor $2/\Gamma$ with $\Gamma = \Gamma_{\text{pole}}$.

| **$N(1900)3/2^+$** | | | ***★ |
|---|---|---|---|
| $N(1900)3/2^+$ pole parameters | | | |
| $M_{pole}$ | 1910±30 | $\Gamma_{pole}$ | 280±50 |
| $A^{1/2}$ | 0.026±0.014 | Phase | (60±35)° |
| $A^{3/2}$ | -(0.070±0.030) | Phase | (70±50)° |
| $N(1900)3/2^+$ transition residues | | | phase |
| $\pi N \to \pi N$ | | 4±2 (MeV) | -(10±40)° |
| $2(\pi N \to \Delta(1232)\pi_{L=1})/\Gamma$ | | 0.07±0.04 | -(65±30)° |
| $2(\pi N \to \Delta(1232)\pi_{L=3})/\Gamma$ | | 0.10±0.05 | (80±30)° |
| $2(\pi N \to N\sigma)/\Gamma$ | | 0.03±0.02 | -(110±35)° |
| $2(\pi N \to N(1520)\pi)/\Gamma$ | | 0.07±0.04 | -(105±35)° |
| $N(1900)3/2^+$ photocouplings | | | phase |
| $(\gamma p)^{1/2} \to \Delta(1232)\pi_{L=1}$ | | 7±4 $10^{-3}$ | (15±30)° |
| $(\gamma p)^{1/2} \to \Delta(1232)\pi_{L=3}$ | | 12±6 $10^{-3}$ | (160±30)° |
| $(\gamma p)^{1/2} \to N\sigma$ | | 3±1.5 $10^{-3}$ | -(30±30)° |
| $(\gamma p)^{1/2} \to N(1520)\pi$ | | 8±4 $10^{-3}$ | -(15±35)° |
| $(\gamma p)^{3/2} \to \Delta(1232)\pi_{L=1}$ | | 14±7 $10^{-3}$ | -(145±30)° |
| $(\gamma p)^{3/2} \to \Delta(1232)\pi_{L=3}$ | | 24±10 $10^{-3}$ | (15±30)° |
| $(\gamma p)^{3/2} \to N\sigma$ | | 7+4 $10^{-3}$ | (160+35)° |
| $(\gamma p)^{3/2} \to N(1520)\pi$ | | 16±8 $10^{-3}$ | -(170±25)° |
| $\gamma p \to \Delta(1232)\pi_{L=1}$ | $E_{1+}$ | 7±4 $10^{-3}$ | -(150±30)° |
| $\gamma p \to \Delta(1232)\pi_{L=3}$ | $E_{1+}$ | 13±7 $10^{-3}$ | (5±25)° |
| $\gamma p \to N\sigma$ | $E_{1+}$ | 4±2 $10^{-3}$ | (155±30)° |
| $\gamma p \to N(1520)\pi$ | $E_{1+}$ | 8±5 $10^{-3}$ | (170±25)° |
| $\gamma p \to \Delta(1232)\pi_{L=1}$ | $M_{1+}$ | 10±6 $10^{-3}$ | (50±30)° |
| $\gamma p \to \Delta(1232)\pi_{L=3}$ | $M_{1+}$ | 18±9 $10^{-3}$ | -(155±30)° |
| $\gamma p \to N\sigma$ | $M_{1+}$ | 5±3 $10^{-3}$ | -(10±25)° |
| $\gamma p \to N(1520)\pi$ | $M_{1+}$ | 12±8 $10^{-3}$ | (10±25)° |
| $N_{3/2^+}(1900)$ Breit-Wigner parameters | | | |
| $M_{BW}$ | 1910±30 | $\Gamma_{BW}$ | 270±50 |
| Br($\pi N$) | 3±2% | Br($N\sigma$) | 4±3% |
| Br($\Delta(1232)\pi_{L=1}$) | 17±8% | Br($\Delta(1232)\pi_{L=3}$) | 33±12% |
| Br($N(1520)\pi$) | 15±8% | | |
| $A^{1/2}_{BW}$ | 0.024±0.014 | $A^{3/2}_{BW}$ | -0.067±0.030 |

Review of Particle Properties, and two resonances were upgraded in their star-rating because of the BnGa analysis. This success underlines the impact photoproduction experiments can make in the study of baryon resonances. Table 1 lists the impact of the new photoproduction data on the number of observed resonances. At present, we perform a new systematic search for *missing resonances* based on the data taken up to now. The fits are performed with five further resonances not yet included in the Review of Particle Properties. It is unclear at present, how significant these are, if some resonances will be upgraded due to the new study, and whether further resonances will be required to achieve a good fit.

A further important step forward are new informations about the properties of baryon resonances. In the recent years, we included the reactions $\gamma p \to p\pi^0\eta$ [43–48] and $\gamma p \to p\pi^0\pi^0$ [49–57] into the BnGa data base. As a result, more than 500 resonance properties derived from these analyses are listed in the 2016 edition of the Review of Particle Properties [58]. Table 2 gives the properties of $N(1900)3/2^+$ as an example. The $\pi N$ branching ratio is rather small: this is the reason why it was not seen in [32–34]. The sum of the observed branching ratios as given in table 2 is 72±14%. Measured are also branching ratios for decays into $N\eta$ (10±4%), $\Lambda K$ (13±5%), and $\Sigma K$ (6±3% [39]), and into $N\omega$ (15±8% [59]). The total sum is already more than, but compatible with, 100%, leaving little room for decays into $N\rho$. These are presently under study. This example shows the strength of a coupled channel analysis: the sum of all decay branching ratios of a trustworthy resonance must be compatible with or at least close to 1.

Not all the quantities listed in Table 2 are physically significant. Nevertheless they demonstrate that our knowledge on the spectrum of nucleon excitations has increased very substantially.

## 3 The CBELSA/TAPS experiment at ELSA

### 3.1 Observables

Fig. 1 shows the total cross section for $\gamma p \to p\pi^0$, together with the Breit-Wigner shapes of the contributing resonances based on RPP values for masses, widths, $N\pi$ decay branching ratios and helicity amplitudes. There are clear peaks which can be assigned to photoproduction of $\Delta(1232)3/2^+$, $N(1520)3/2^-$, and $N(1680)5/2^+$. But there are more resonances which are not seen as obvious peaks: baryon resonances are broad and overlap and, obviously, small resonance contributions are difficult to extract from the data. The resonances appear in different partial waves, and these need to be separated and identified. As mentioned in the introduction, photoproduction is governed by

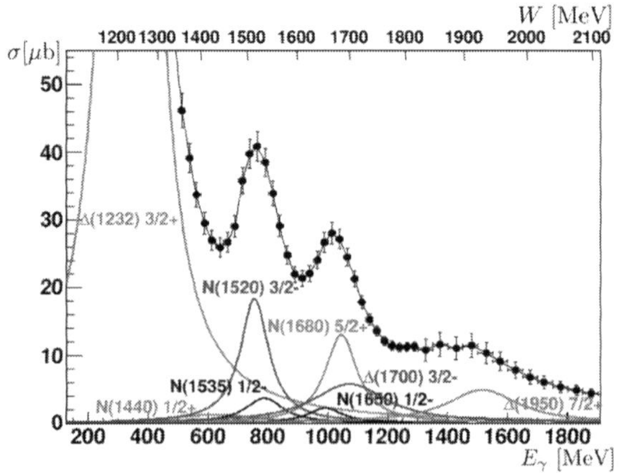


**Figure 1.** $\gamma p \to p\pi^0$ total cross section (CBELSA/TAPS-data), also indicated are the pure Breit-Wigner shapes of the different resonances contributing to the reaction. The phase space dependence is not included here and the couplings are taken from the RPP [16]. $\pi N$-P-, -D-, and -F-wave resonances are shown in green, blue and red, respectively.

**Table 3.** Observables in single pseudoscalar meson photoproduction. Measurements of the observables marked in blue have been performed within the SFB/TR16 for various final states with neutral mesons.

| $\gamma$ | | Target | | | Recoil | | | Target–Recoil | | | |
|---|---|---|---|---|---|---|---|---|---|---|---|
| | | | | | $x'$ | $y'$ | $z'$ | $x'$ | $x'$ | $z'$ | $z'$ |
| | | $x$ | $y$ | $z$ | | | | $x$ | $z$ | $x$ | $z$ |
| unpol. | $\sigma$ | 0 | $T$ | 0 | 0 | $P$ | 0 | $T_{x'}$ | $-L_{x'}$ | $T_{z'}$ | $L_{z'}$ |
| linear | $-\Sigma$ | $H$ | $-P$ | $-G$ | $O_{x'}$ | $-T$ | $O_{z'}$ | $-L_{z'}$ | $T_{z'}$ | $-L_{x'}$ | $-T_{x'}$ |
| circ. | 0 | $F$ | 0 | $-E$ | $C_{x'}$ | 0 | $C_{z'}$ | 0 | 0 | 0 | 0 |

four complex amplitudes; for an energy-independent reconstruction of the amplitudes, at least eight carefully chosen observables need to be measured [60] even though still one phase per interval in energy and angle remains undetermined.

Alternatively, energy-dependent fits can be made like Bonn-Gatchina (A.2), Jülich-Bonn (B.3), or Gießen (B.7), or multipole analyses in which the angular dependence is exploited in the first step. Truncated multipole analyses require measurements of a smaller number of observables [61]; a first attempt in [62] was successful.

Table 3 shows the 16 observables which are experimentally measurable. These polarization observables, or at least 8 carefully chosen ones [60], need to be measured over the full solid angle with high precision. This aim defines the requirements the photon beam, the target and the detector to study light-baryon resonances should meet: The electron accelerator needs to deliver unpolarized or polarized electrons with a large duty cycle from which an intense well collimated photon beam with no, linear, or circular polarization can be produced. Of course the energy of the photon beam should span an energy range which allows to cover the excitation spectrum. A target, with protons or "neutrons", should be polarizable in longitudinal (along the photon beam) or transverse direction. No polarization is required to measure the differential cross section $d\sigma/d\Omega$. The beam asymmetry $\Sigma$ is given by the difference of two perpendicular settings of the linear photon polarization plane. When the target is polarized perpendicular to the beam direction, $T$ can be determined. If we do not consider the difficult measurement of the recoil polarization, all other observables require then both, photon and target to be polarized. An ideal detector would measure charged and neutral (i.e. photons) particles with high precision and complete solid angle coverage.

For two-meson production or the production of mesons with non-zero spin, the number of observables to be measured is significantly higher. Additional observables become accessible using a polarized beam and/or target. Following [63] at least 15 carefully chosen observables need to be measured for double pseudoscalar meson photoproduction. Additional observables can be deduced with the same kind of polarization settings as in the case of photoproduction of one pseudoscalar meson. However, to achieve a complete experiment in double-pseudoscalar meson photoproduction also triple-polarization observables need to be measured. The latter are not needed in the case of single-pseudoscaler meson photoproduction.

In Bonn, the ELectron Stretcher Accelerator ELSA [64] with its possibility to accelerate polarized electrons [66] (see contribution of project D.2 to these proceedings) and its goniometry to produce linear photon polarization [67] fulfills the requirements discussed. Unpolarized as well as circularly and linearly polarized photon beams are available for the experiments. In addition, the Bonn frozen spin polarized target [68] makes polarized protons and "neutrons" available (see contribution of project D.1 to these proceedings). Using the CBELSA/TAPS experiment, the polarization observables marked in blue in Table 3 have been determined within project A.1 for various different final states. Depending on their cross sections, additional statistics is often still desirable.

### 3.2 Detector requirements

The *missing resonances* were expected to decouple from the $\pi N$-channel [69]. Obviously, it is difficult or even impossible to observe baryon resonances with very small or nearly vanishing $\pi N$ coupling in $\pi N$ elastic and charge exchange scattering, or in photoproduction of single pions. However, baryon resonances have to decay and may couple with rather different strengths to different final states. The study of various final states, including the multimeson photoproduction, is necessary to gain a complete picture of the spectrum and the properties of nucleon excitations.

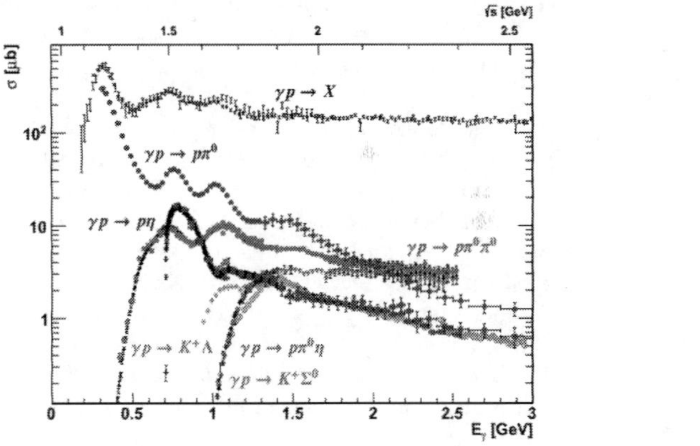


**Figure 2.** Total photon absorption cross section as well as total cross sections for different final states: $\gamma p \to X$ [58], $\gamma p \to \pi^0 p$ [70–72], $\gamma p \to p\eta$ [73–79], $\gamma p \to K^+\Lambda$ [80, 81], $\gamma p \to K^+\Sigma^0$ [80, 81], $\gamma p \to p\pi^0\eta$ [45, 46, 48], $\gamma p \to p\pi^0\pi^0$ [49–54, 57].

Figure 2 shows the inclusive total cross section for $\gamma p \to X$ [58], and the contributions from different final states. At low energies, the formation of $\Delta(1232)3/2^+$ dominates photoproduction. Over a wide range, the inclusive total cross section adopts a value of about 200 $\mu$b. The $\gamma p \to p\pi^0$ cross section falls off rapidly with energy; the peaks have been assigned to $\Delta(1232)3/2^+$, and dominantly to $N(1520)3/2^-$, and $N(1680)5/2^+$ (of course also other resonances with smaller cross section contribute to the second and third resonance region, see also Fig. 1). The

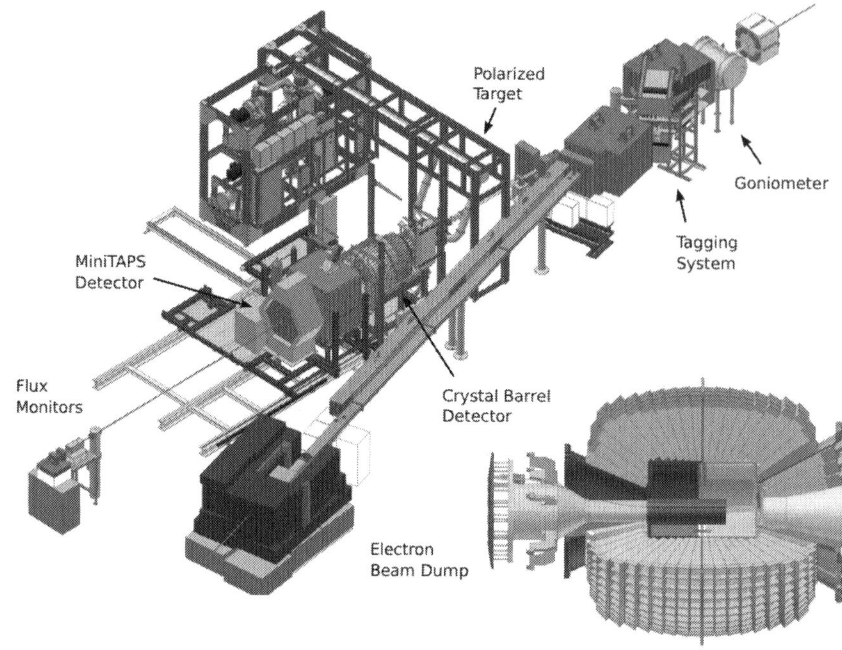


**Figure 3.** Experimental Setup of the CBELSA/TAPS Experiment.

small additional enhancement at $E_\gamma = 1.5\,\text{GeV}$ is mostly due to $\Delta(1950)7/2^+$. Above the threshold of the reaction, the $\gamma p \to p\eta$ cross section is dominated by formation of $N(1535)1/2^-$. The two reactions with open strangeness in the final state shown ($\gamma p \to K^+\Lambda$ and $\gamma p \to K^+\Sigma$) have cross sections which are, at intermediate energies, in size similar to the $\gamma p \to p\eta$ cross section. The reactions $\gamma p \to \pi^0\pi^0 p$ and $\gamma p \to \pi^0\eta p$ gain importance with increasing energy. A comparison of the different total cross sections by eye already indicates that different resonances couple with rather different strengths to the different final states. This clearly indicates the importance of the analysis of different final states to gain a complete understanding of the spectrum. The large gap between the inclusive total cross sections and the cross sections for exclusive final states is mainly due to reactions like $\gamma p \to \pi^+\pi^- p$, $\pi^+\pi^0 n$, and $3\pi N$.

The ideal detector would hence measure the momenta of charged and neutral final-state particles and identify photons and neutrons, and charged pions, kaons, and protons. In practice, present-day detectors are either very well suited for the measurements of charged or neutral decay products. The CBELSA/TAPS detector (section 3.3) is ideally suited to measure neutral mesons decaying into photons. This is advantageous since background amplitudes like charged pion t-channel exchange or Kroll-Rudermann-terms do not contribute. Hence the fractional contribution of the resonant amplitudes is larger than for reactions with charged mesons in the final state. Project B.1 of this SFB added further final states such as e.g. $K^0\Sigma^+$ [84]. The BGO-OD-experiment provides the option to measure charged particles in forward direction with very good resolution. For further details on the BGO-OD-experiment and its capabilities to measure charged and neutral particles, see the contribution of project B.1 to these proceedings.

The CLAS detector at JLab is very well suited to measure charged particles, making data on additional final states such as e.g. $K^+\Lambda$ and $K^+\Sigma^0$ or $p\pi^+\pi^-$-photoproduction available [82, 83]. For $K^+\Lambda$ and $K^+\Sigma^0$ photoproduction, CLAS (Jlab) has e.g. performed not only all measurements needed to reach a complete experiment, but all 16 polarization observables have been measured. In this case the self-analyzing power of the hyperon-decay gives directly access to the polarization of the baryon in the final state, which is of course of great advantage.

A combined analysis of these data sets in a multi-channel PWA (see project A.2 "Partial Wave Analysis") is of great advantage compared to the analysis of individual channels to extract the resonances from the data. All data constrain the solution substantially. A similar argument holds for the photoproduction off the proton and neutron. While the isospin-3/2-$\Delta$-resonances are produced with the same strength off the proton and the neutron, this is not true for the isospin-1/2-$N^*$-resonances. Some $N^*$-resonances might be strongly produced off the neutron but not off the proton or vice versa. Therefore resonances, maybe even resonances with specific symmetries in their wave function or nature, may escape detection if only proton data are analyzed. While first neutron data have been taken within this SFB, neutron data are still scarce. A combined PWA of proton and neutron data will further constrain the different isospin contributions.

### 3.3 The CBELSA/TAPS experiment at ELSA

Fig. 3 shows an overview of the CBELSA/TAPS experiment. The experiment is located at the electron accelerator ELSA [64]. ELSA provides either unpolarized or longitudinally polarized electrons with energies up to 3.5 GeV (for further details see contribution of project D.2 to these proceedings). The electrons hit a radiator target placed inside of a goniometer. Here photons are produced via bremsstrahlung. The electrons which underwent the bremsstrahlung process are momentum analyzed in the tagging system. They are deflected in a magnetic field and detected in a ladder of organic scintillator bars and fibres. The bremsstrahlung photons hit an unpolarized or polarized target (see contribution of project D.1 "Polarized Target" to these proceedings), which is surrounded by a three-layer scintillation fiber detector (513 fibers) used for the identification of charged particles and by an electromagnetic calorimeter consisting of 1230 CsI(Tl)-crystals (part of the former Crystal Barrel calorimeter used at the Low Energy Antiprotion Ring LEAR at CERN [85]). In the forward direction below polar angles of 30°, two further calorimeters, the Forward Plug (90 CsI(Tl)-crystals) and the forward TAPS-wall (216 $BaF_2$ crystals) are placed. Plastic scintillators in front of the crystals allow charged particle identification. Inbetween the Crystal Barrel and the TAPS detector, a $CO_2$ Cherenkov detector is placed to suppress electromagnetic background produced in the target.

For the data discussed within these proceedings the signals from the forward calorimeters, the fiber detector and the tagging system were used as first-level trigger. The Crystal Barrel calorimeter was only included in the second-level trigger identifying the number of clusters in the calorimeter. Presently the calorimeter readout is upgraded to provide in future also timing and first level trigger capabilities of the calorimeter as discussed in detail in the contribution of project D.3 to these proceedings.

### 3.4 Polarized photon beams

The bremsstrahlungs process produces an unpolarized photon beam with approximately a $1/E_\gamma$ intensity distribution. Linear photon polarization is obtained when electrons are scattered off a diamond radiator. Different settings of the diamond radiator are required to maximize the polarization in the desired energy region (see Fig. 4).

The bremstrahlungs-photons are polarized circularly when they are produced by longitudinally polarized electrons. For data with circularly polarized photons, a Møller target is used as bremsstrahlung target to monitor the electron polarization via Møller scattering in parallel to the data taking. The electron polarization was typically 64%; the polarization transfer decreases with decreasing photon energy according to [65]:

$$P_\odot = \frac{4x - x^2}{4 - 4x + 3x^2} \qquad \text{with } x = \frac{E_\gamma}{E_{el}}. \qquad (1)$$

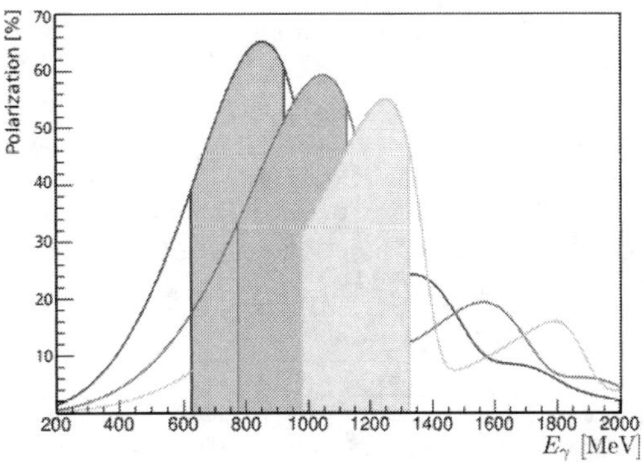


**Figure 4.** Three different settings of the coherent edge in the 2008 data taking period, the shaded areas mark the energy regions used in this analysis (from [86, 87]).

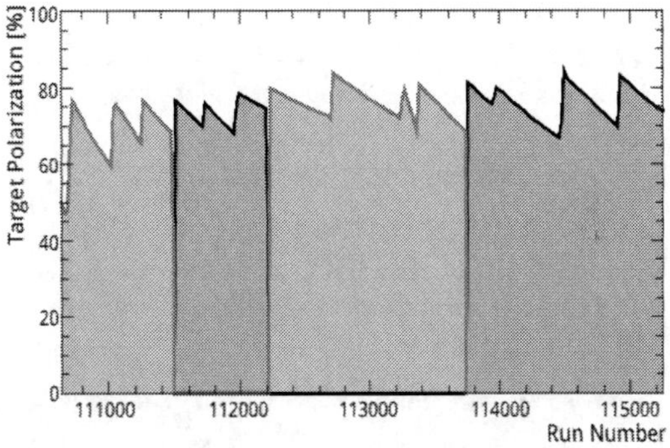


**Figure 5.** The target polarization during the 2008 beamtime of 40 days with positive (red) and negative (black) polarization (from [87]).

### 3.5 The polarized target

Double polarization experiments have been performed using the Bonn polarized frozen spin butanol target (project D.1) reaching typical mean polarizations of about 70-75% during the measurements.

Since the target ($C_4H_9OH$) consists not only of hydrogen but is diluted by the carbon and oxygen, measurements using a carbon foam target have been performed to determine the so-called dilution factor ($d$). The carbon foam target was included in the cryostat as otherwise the butanol. To reach similar conditions the target had approximately the same target area density and dimensions as the unpolarizable components in the butanol target. Measurements using a liquid hydrogen target were used for further systematic studies. The dilution factor depends on the kinematic variables (e.g. $E_\gamma$, $\theta$ for two-particle final states). One example for an analysis using the C-foam-target to determine the dilution factor is shown below (Fig. 6). A corresponding dilution factor

$$d(E_\gamma, \theta) = \frac{N_{\text{butanol}} - N_{\text{carbon}}}{N_{\text{butanol}}} \qquad (2)$$

needs to be determined for each specific analysis. The obtained values of course depend on the cuts. The more restrictive the cuts are, the higher becomes the dilution factor. For forward going pions a smaller fraction of protons is detected in the calorimeter. The cuts are less restrictive, therefore more events stemming from the bound nucleons remain and the dilution factor decreases (Fig. 6).

## 4 $\pi^0$, $\eta$, and $\eta'$ photoproduction

### 4.1 Differential cross sections and the beam asymmetry $\Sigma$

Differential cross sections for the reactions $\gamma p \to \pi^0 p$ [70, 71] and $\gamma p \to \eta p$ [73, 74] have been measured at the beginning of the SFB, followed by additional measurements of $\gamma p \to \pi^0 p$ [72] and $\gamma p \to \eta p$ and $\gamma p \to \eta' p$ [77] covering the full solid angle also in forward direction. Recently, differential cross sections and spin density matrix elements were determined for single $\omega$ meson photoproduction off the proton [88]. The reaction $\gamma p \to \pi^0 \omega p$ was shown to be dominated by $\Delta(1232)\omega$ production [89]. Measurements of the beam asymmetry were carried out for various final states: $p\pi^0$ [90, 91], $p\eta$ [67], $p\omega$ [92], and $K_s^0\Sigma^+$ [84], (see also contribution of project B.1 to these proceedings).

### 4.2 $\gamma p \to \pi^0 p$: The double polarization observable $G$

Within the SFB measurements with a linearly polarized photon beam and a longitudinally polarized target have been performed. The linearly polarized photons stem from coherent bremstrahlung of electrons with an energy of

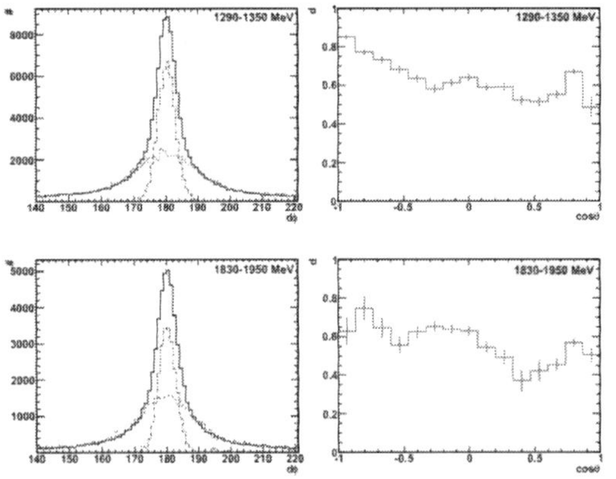


**Figure 6.** Determination of the dilution factor; as an example the analysis of $\gamma p \to p\pi^0$ using a longitudinally polarized target and a circularly polarized beam was chosen. Left: Coplanarity distribution used to determine the fraction of events in the butanol data produced off bound nucleons in carbon (oxygen), black: butanol data, red: carbon data, blue: resulting hydrogen spectrum. The coplanarity spectrum of the carbon data was adjusted to the butanol data outside the central peak area where the free protons contribute. The resulting scaling factor is then used to determine the dilution factor (right) depending on the angle [101].

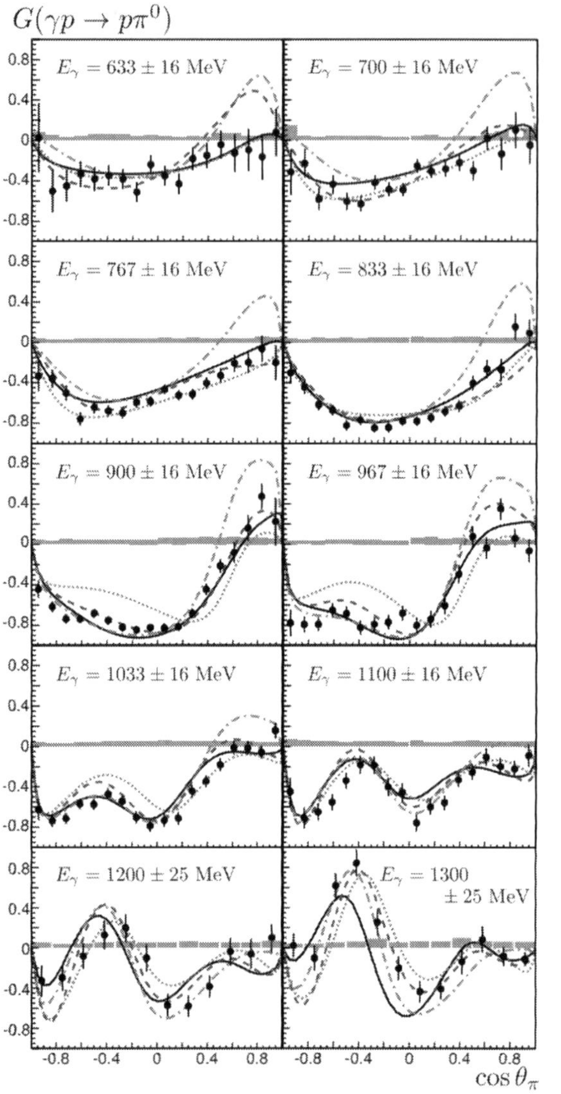


**Figure 7.** The double-polarization observable $G$ for every second measured photon energy bin [87] (black dots), compared to the PWA predictions: MAID 2007 [93] (green dotted line), SAID CM12 [94] (red dashed-dotted line), JüBo 2013-01 [95] (blue dashed line), and BnGa 2011-02 [42] (black solid line).

3.2 GeV off an aligned diamond. The corresponding cross section can be written as:

$$\frac{d\sigma}{d\Omega}(\phi) = \frac{d\sigma_0}{d\Omega} \cdot \left(1 - p_\gamma^{\text{lin}} \Sigma \cos(2\phi) + P_\gamma^{\text{lin}} p_z G \sin(2\phi)\right) \quad (3)$$

where $\frac{d\sigma_0}{d\Omega}$ is the unpolarized cross section, $p_\gamma^{\text{lin}}$ the degree of linear polarization, $p_z$ the polarization of the target and $\phi$ the angle between the polarization and the production plane. Data have been taken with three settings for the linear polarization with coherent edges at 950 MeV, 1150 MeV, 1350 MeV, the maximum linear polarization reached was 65% at 860 MeV, 59% at 1050 MeV, and 55% at 1270 MeV (see Fig. 4).

The final results for our new G data [87] are presented in Fig. 7. It shows the double-polarization observable $G$ as a function of $\cos\theta_\pi$ for photon energies from $E_\gamma = 617$ MeV to $E_\gamma = 1325$ MeV. For every second energy bin, the data are shown in black with their statistical errors; the systematic errors are given as gray histogram on the 0-axis. Symmetry principles enforce $G = 0$ for $\cos\theta_\pi = \pm 1$. Data at exactly these points do not exist but the most forward or backward data points are compatible with a PWA-curve vanishing at $\cos\theta_\pi = \pm 1$. At low photon energies, for $E_\gamma < 900$ MeV, the values of $G$ as functions of $\cos\theta_\pi$ all show negative values, with a single minimum at negative values of $\cos\theta$. With increasing photon energy, the distributions become more complicated, with up to three local minima and two local maxima.

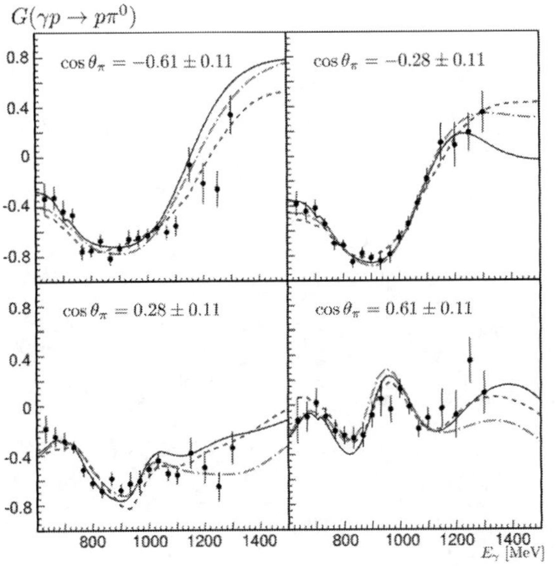


**Figure 8.** The energy dependence of $G$ [87] for four different angles, compared to different solutions of the PWAs: BnGa 2014-02 [48] (black solid line), SAID [97, 99] (red dashed-dotted line), and JüBo [98, 99] (blue dashed line).

The new data for G are compared in Fig. 7 with predictions from different PWA models: with BnGa 2011-02 [42], MAID 2007 [93], SAID CM12 [94], and JüBo 2013-01 [95]. The comparison reveals that at lower energies, the MAID and BnGa analyses can describe the data well, while for the SAID results, a deviation at $\cos\theta_\pi \geq 0.4$ becomes apparent. The predictions of JüBo show a similar disagreement as SAID for the two lowest photon energy bins at $E_\gamma = 633$ and 667 MeV. The differences between the predictions of the MAID model and the data in the medium energy region were discussed in [86] and traced back to the multipoles $E_{0+}$ and $E_{2-}$. The largest deviations between the different models can be observed in the higher energy bins ($E_\gamma > 1150$ MeV). These differences most likely occur since resonance contributions in the fourth resonance region are not well known.

The new $G$ data were included in the BnGa, JüBo, and the SAID analyses and new fits were performed. The new fits, BnGa 2014-02 [48] (black solid line), SAID [97] (red dashed-dotted line), and JüBo [98] (blue dashed line), reproduce the data reasonably well [87]. For convenience, we show the data in Fig. 8 for four slices in $\cos\theta$ as a function of the photon energy. All PWA fits can describe the new data very well at lower photon energies, above $E_\gamma > 1150$ MeV the fit results start to diverge. Here more precise data for $G$ are needed to constrain the PWA solutions.

The impact of the new data can be best seen when the multipoles $E_{0+}$, $E_{2-}$ and $M_{2-}$ exciting the dominantly contributing resonances are compared; a longer paper with a comparison of all leading multipoles has been published recently [99]. The values of the double-polarization observable $G$ reported in this subsection have been used together with the data on the observables $E$ [100, 101], and $T$, $P$, and $H$ [102, 103]. As an example, the real and imaginary parts of the $E_{0+}$ multipole derived from these new fits to the data are shown in Fig. 9 and compared with the $E_{0+}$ multipole derived from the older fits. For the imaginary part of the multipole, the spread of the solutions reduces considerably: the new data with double-polarization observables $(G, E, T, P, H)$ have a decisive influence on the resulting multipoles. For the real part, some reduction in the spread is observed even though much less pronounced. Certainly, more precise polarization data and further analyses are both required before the remaining discrepancies are resolved.

These new data had a decisive impact on the $\pi$-photoproduction multipoles. When these data and other new data were included in the partial-wave analyses Bonn-Gatchina and SAID and in the dynamical coupled-channel approach Jülich-Bonn, the mutual agreement was improved considerably [99].

### 4.3 $\gamma p \to \pi^0 p$: The double polarization observable $E$

The helicity asymmetry $E$ requires circularly polarized photons and longitudinally polarized protons. It can be determined as

$$E = \frac{N_{1/2} - N_{3/2}}{N_{1/2} + N_{3/2}} \cdot \frac{1}{d} \cdot \frac{1}{p_\odot p_T} \tag{4}$$

where $N_{1/2}$ and $N_{3/2}$ are the number of events observed with photon and target polarization in opposite or parallel directions, normalized to the same number of incident photons. $d$ is the dilution factor (see Fig. 6), $p_\odot$, and $p_T$ are the polarization degrees of the circularly polarized photon beam and the target, respectively.

Figure 10 (left) shows the total cross section for $\gamma p \to \pi^0 p$ [100] in comparison to some PWA fits. The total cross section is well described by all three models. Also shown are the contributions from anti-parallel ($\sigma_{1/2}$) and parallel ($\sigma_{3/2}$) spin settings of photon and proton. It becomes directly obvious by eye, that the resonant contributions to both cross sections are rather different. $\sigma_{3/2}$ shows a clear peaking structure not only in the second but also in the

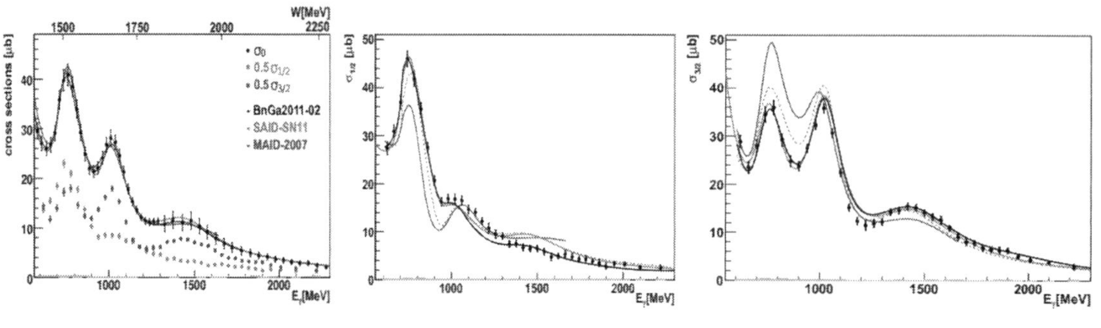


**Figure 10.** Left: The total cross section $\sigma_0$ [71] plotted together with its two helicity components as a function of $E_\gamma$. The error bars give the statistical errors, the systematic errors are shown as a dark gray band. $\sigma_{1/2}$ (center) and $\sigma_{3/2}$ (right) for $\gamma p \to p\pi^0$, PWA predictions: black solid curve: BnGa2011-02, blue solid curve: MAID, red solid curve: CM12, red dashed curve: SN11. For further details see [100, 101].

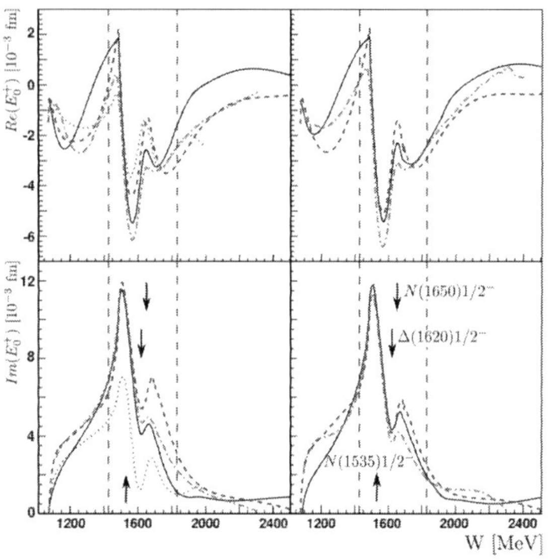


**Figure 9.** The real and imaginary part of the multipole $E_{0+}$, before the new data were included (left) and after (right), determined by the BnGa PWA (2011-02 [42] resp. 2014-2 [48]: black solid line), the SAID (CM12 [94] resp. new fit [97]: red dashed-dotted line), the JüBo model (2015-B [95] resp. new fit [98]: blue dashed line) and the MAID model (green dotted line) [93]. The dashed lines mark the region covered by the observable $G$. Additionally, the positions of the resonances $N(1535)1/2^-$, $N(1650)1/2^-$ and $\Delta(1620)1/2^-$, which contribute to the $E_{0+}$ multipole, are marked with arrows.

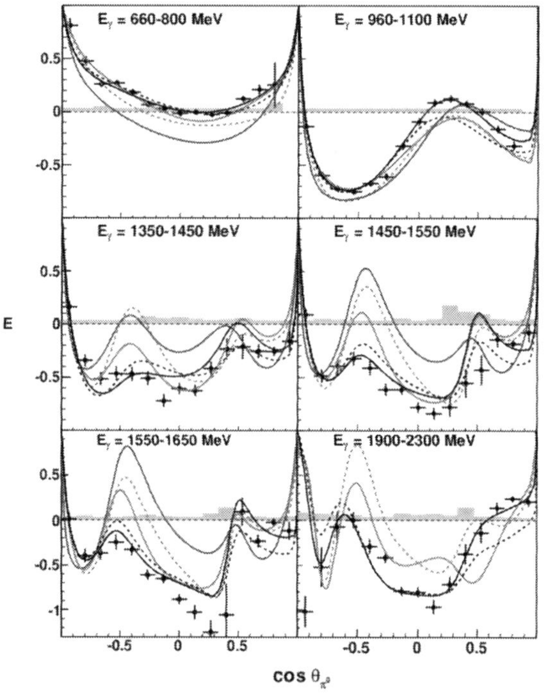


**Figure 11.** The helicity asymmetry $E$ as a function of $\cos\theta_\pi$ for selected $E_\gamma$ bins. PWA predictions: black dashed curve: BnGa2011-02, blue solid curve: MAID, red solid curve: CM12, red dashed curve: SN11. Fit to the data points (BnGa2011-02E): black solid curve. Figures taken from [100].

third and fourth resonance region, $\sigma_{1/2}$ shows only a peak in the second resonance region while only a shoulder in the third and no peaking structure at all is observed in the fourth resonance region. Significant differences are seen when the partial wave analysis predictions are compared to $\sigma_{1/2}$ and $\sigma_{3/2}$ separately (see Fig. 10, center and right). Even at low energies where one might have expected that everything is well understood based on the existing photoproduction and $\pi N$-elastic scattering data large discrepancies occur. They become even more obvious if the predictions are compared to the double-polarization observable $E$ as function of $\cos(\theta^{\pi^0}_{\text{CMS}})$ (see Fig. 11). None of the partial wave analyses predicted the data precisely. A BnGa refit included these data and gave reasonable consistency between data and fit curve.

The data were included in the multi-channel BnGa-PWA and were the basis for a precise determination of the helicity amplitudes of several baryon resonances. Together with our data on $G$, $T$, $P$ and $H$, these data led to a determination of the $N(1520)3/2^-$ helicity amplitudes

 DOI: 10.1051/epjconf/201713402001

from an energy-independent multipole analysis [102] and to a significantly better determination of the contributing multipole amplitudes (see Fig. 18 in the contribution of project A.2 "Partial Wave Analysis" to these proceedings). Averaged over all multipoles and energies the error bands of the multipoles reduced by a factor 2.25. This demonstrates the value of the new polarization data taken with the CBELSA/TAPS-experiment at ELSA with polarized photon beams and polarized targets in an impressive way.

### 4.4 $\gamma p \to \pi^0 p$: The polarization observables $T$, $P$, $H$

The observables $T$, $P$, and $H$ can be measured simultaneously when data with a transversely polarized target and a linearly polarized photon beam are taken. The measurements [102, 103] were performed using two orthogonal settings of the polarization plane (∥ and ⊥). The polarization reached its maximum of $p_\gamma = 65\%$ at 850 MeV and dropped down to 40% at 700 MeV. The mean target-proton polarization reached was $p_T \approx 75\%$. Data were taken with two opposite settings of the target polarization direction (↓ and ↑). In that case, the azimuthal distribution of events is given by

$$\begin{aligned}\frac{N(\phi)}{N_0} = 1 &- p_\gamma \Sigma_{\text{eff}} \cos(2\phi) + d\, p_T T \sin(\phi - \alpha) \\ &- d\, p_T\, p_\gamma P \cos(2\phi) \sin(\phi - \alpha) \\ &+ d\, p_T\, p_\gamma H \sin(2\phi) \cos(\phi - \alpha), \end{aligned} \quad (5)$$

where $\alpha$ is the azimuthal angle between the target polarization vector and the photon polarization plane. $T$, $P$, and $H$ are determined, for each $(E_\gamma, \cos\theta_{\pi^0})$ bin, by fits to the measured azimuthal distribution of events. $T$ is determined from a fit to the azimuthal yield asymmetry:

$$\Delta N(\phi)_{\text{T}} = \frac{1}{d \cdot \text{p}_{\text{T}}} \cdot \frac{N_\uparrow - N_\downarrow}{N_\uparrow + N_\downarrow} = T \cdot \sin(\beta - \phi) \quad (6)$$

A typical example for such a fit is shown in Fig. 13, left panel. $P$ and $H$ are extracted from data where not only the target polarization is changed but also the photon polarization plane from ∥ to ⊥, using the equation:

$$\Delta N(\phi)_{\text{BT}} = \frac{1}{d \cdot \text{p}_\gamma \text{p}_{\text{T}}} \cdot \frac{(N_{\perp\uparrow} - N_{\perp\downarrow}) - (N_{\parallel\uparrow} - N_{\parallel\downarrow})}{(N_{\perp\uparrow} + N_{\perp\downarrow}) + (N_{\parallel\uparrow} + N_{\parallel\downarrow})}$$

$$= P \sin(\beta - \phi) \cos(2(\alpha - \phi)) - H \cos(\beta - \phi) \sin(2(\alpha - \phi)) \quad (7)$$

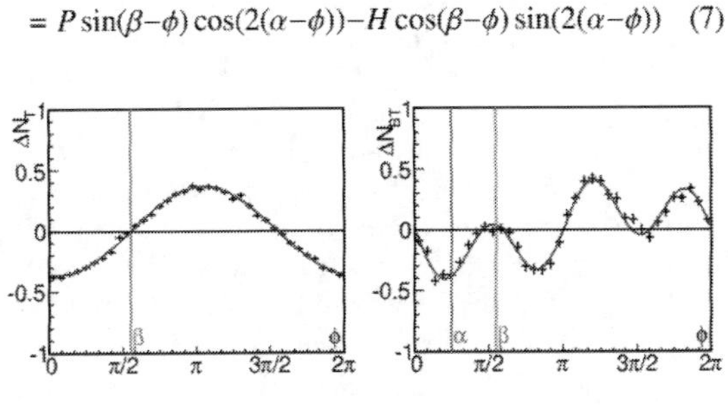


**Figure 13.** Event yield asymmetry as a function of $\phi$, left: $\Delta N(\phi)_{\text{T}}$, right: $\Delta N(\phi)_{\text{BT}}$, fitted by the functions given in eqs. (6) and (7), respectively. Shown is the energy bin $W = 1.524$–$1.542$ GeV as example [103].

The observables $P$ and $H$ are determined by a fit to the $\Delta N(\phi)_{\text{BT}}$ distributions (see Fig. 13, right as example). The polarization observables $T$, $P$, and $H$ are shown in Fig. refpic:tph for selected bins. The lower energy data (up to W = 1620 MeV) were used as basis for an energy-independent trucated partial wave analysis in sliced energy bins (see the discussion in project A.2 "Partial Wave Analysis" and in [102]). In addition to the data on $T$, $P$, and $H$, differential cross section data (e.g. [71]), data on $\Sigma$ [90, 91, 96], $G$ [86] and on $E$ [100] were included for this reaction, making in total seven observables available for the PWA. It should be mentioned that this was the first time that the available data allowed for an energy-independent reconstruction of multipoles in the energy range covering the second resonance region of $N(1520)3/2^-$ and $N(1535)1/2^-$. Based on this analysis, the helicity amplitudes of $N(1520)3/2^-$ were deduced with minimal model assumptions [102]. The result is inconsistent at the level of more than $2\sigma$ with older (model-dependent) determinations and supports those of the BnGa PWA.

### 4.5 $\gamma p \to p\pi^0$: An expansion into partial waves

It is desirable to have simple methods that help to infer some of the physics contained in newly measured polarization data, without having to do a full energy-dependent partial wave analysis (see project A.2 "Partial Wave Analysis"). The ansatz described here is capable of giving a good estimate of the maximal angular momentum content $\text{L}_{\text{max}}$ needed to describe a given data set in any kind of partial wave analysis. More details on the method, as well as its application to a wide array of polarization data, will be published in a forthcoming paper [105].

The model-independent decomposition of the full $T$-matrix of pseudoscalar meson photoproduction into, for instance, four CGLN [107] spin-amplitudes $F_i$ and the expansion of these amplitudes into electric ($E_{\ell\pm}$) and magnetic ($M_{\ell\pm}$) multipoles is well-known. In single-meson photoproduction, 16 polarization observables (including the unpolarized cross section) are measurable (see Table 3). In case the multipole expansion is combined with a consistent definition of the 16 polarization observables in terms of CGLN-amplitudes [108], one arrives at a polynomial parametrization for the angular distributions of the profile functions. Here for example, the parametrization is given for the profile function $\check{T}$ of the target asymmetry $T$

$$\begin{aligned}\check{T}(W,\theta) &= T \cdot \left(\frac{d\sigma}{d\Omega}\right)_0 \\ &= \rho \sum_{n=1}^{2\text{L}_{\text{max}}} (a_{\text{L}_{\text{max}}})_n^{\check{T}} (W)\, P_n^1(\cos\theta)\,. \end{aligned} \quad (8)$$

The phase-space factor $\rho := q/k$ was defined, with $k$ and $q$ being the moduli of the CMS-three-momenta of the initial photon and final state meson. Finally, $P_n^1$ are the associated Legendre polynomials [109], $P_n^m$ with $m = 1$. It has to be noted, that the upper index $m$ depends on the parametrized observable. For the profile functions $\check{E}$ and $\check{G}$, for instance,

 DOI: 10.1051/epjconf/201713402001

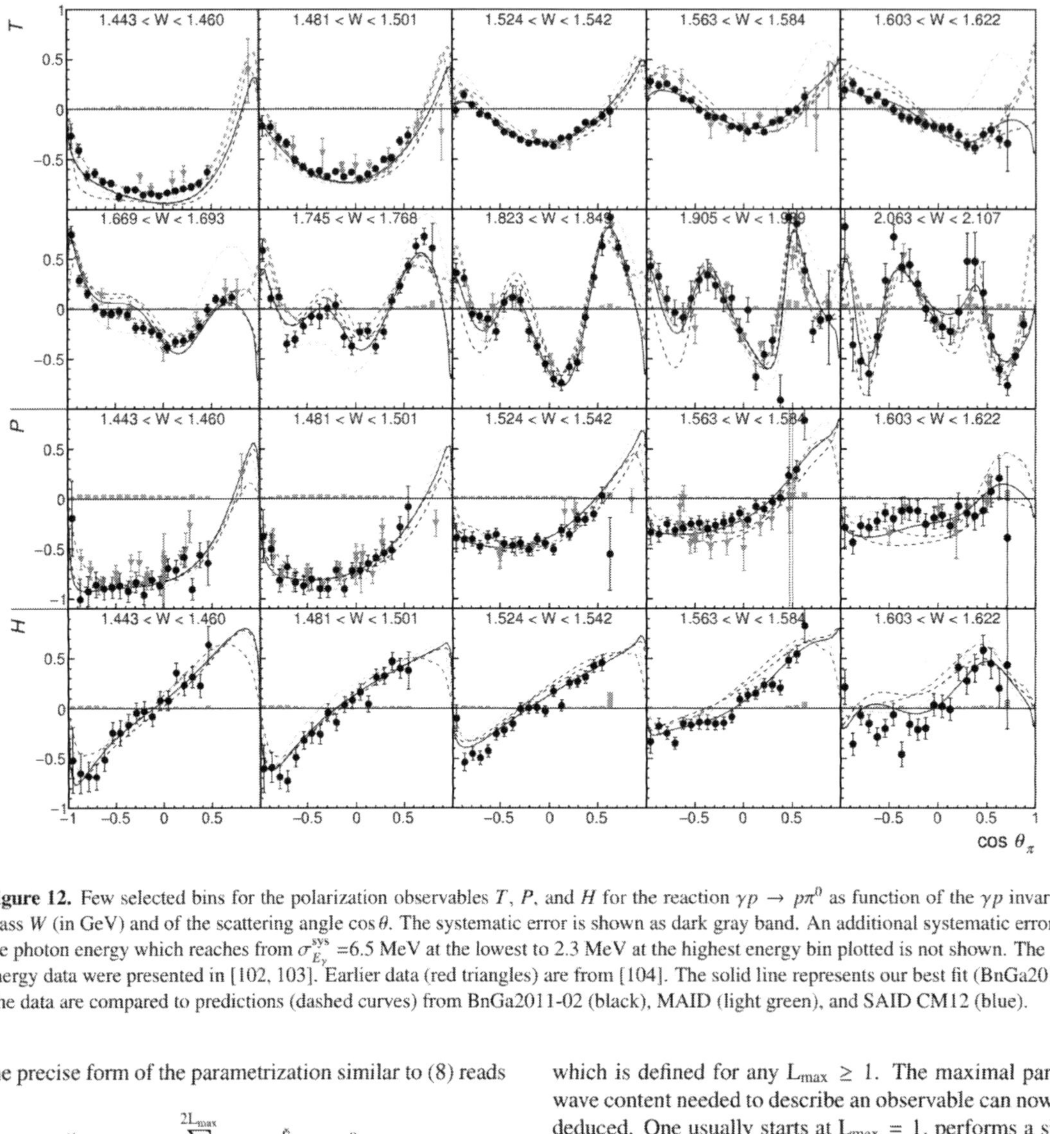


**Figure 12.** Few selected bins for the polarization observables $T$, $P$, and $H$ for the reaction $\gamma p \to p\pi^0$ as function of the $\gamma p$ invariant mass $W$ (in GeV) and of the scattering angle $\cos\theta$. The systematic error is shown as dark gray band. An additional systematic error on the photon energy which reaches from $\sigma^{sys}_{E_\gamma}$ =6.5 MeV at the lowest to 2.3 MeV at the highest energy bin plotted is not shown. The low energy data were presented in [102, 103]. Earlier data (red triangles) are from [104]. The solid line represents our best fit (BnGa2014). The data are compared to predictions (dashed curves) from BnGa2011-02 (black), MAID (light green), and SAID CM12 (blue).

the precise form of the parametrization similar to (8) reads

$$\check{E}(W,\theta) = \rho \sum_{n=0}^{2\mathrm{L}_{\max}} \left(a_{\mathrm{L}_{\max}}\right)_n^{\check{E}}(W)\, P_n^0(\cos\theta), \qquad (9)$$

$$\check{G}(W,\theta) = \rho \sum_{n=2}^{2\mathrm{L}_{\max}} \left(a_{\mathrm{L}_{\max}}\right)_n^{\check{G}}(W)\, P_n^2(\cos\theta). \qquad (10)$$

The Legendre coefficients $a_{\mathrm{L}_{\max}}$ are themselves bilinear Hermitian forms in the multipoles. In case of $\check{T}$, one can express the Legendre coefficients as

$$\left(a_{\mathrm{L}_{\max}}\right)_n^{\check{T}}(W) = \left\langle \mathcal{M}_{\mathrm{L}_{\max}}(W) \middle| C_n^{\check{T}} \middle| \mathcal{M}_{\mathrm{L}_{\max}}(W) \right\rangle, \qquad (11)$$

defined by certain (calculable) matrices $C_n^{\check{T}}$ of dimension $4\,\mathrm{L}_{\max} \times 4\,\mathrm{L}_{\max}$ and the multipoles $\mathcal{M}_{\mathrm{L}_{\max}}$ [105]. The multipoles are here written as the $4\mathrm{L}_{\max}$-dimensional complex vector

$$\left|\mathcal{M}_{\mathrm{L}_{\max}}\right\rangle = \left[E_{0+}, E_{1+}, M_{1+}, M_{1-}, E_{2+}, E_{2-}, M_{2+}, M_{2-}, \ldots, E_{\mathrm{L}_{\max}+}, E_{\mathrm{L}_{\max}-}, M_{\mathrm{L}_{\max}+}, M_{\mathrm{L}_{\max}-}\right]^T, \qquad (12)$$

which is defined for any $\mathrm{L}_{\max} \geq 1$. The maximal partial wave content needed to describe an observable can now be deduced. One usually starts at $\mathrm{L}_{\max} = 1$, performs a simple $\chi^2$-test for a fit to the data with a parametrization such as (8). The angular momentum $\mathrm{L}_{\max}$ is increased in case the test is not passed, i.e. until $\chi^2/\mathrm{ndf} \simeq 1$ is fulfilled. This procedure is continued up to the lowest order $\mathrm{L}_{\max}$ where the $\chi^2$-test suggests a good fit. Then, this method directly returns an estimate for the maximal angular momentum content needed. Interesting phenomena can be seen drawing the $\chi^2/\mathrm{n}_{\mathrm{df}}$ of each fitted order $\mathrm{L}_{\max} \geq 1$ against the center-of-mass energy $W$. Such plots are shown in Figs. 14, 15, and 16 for $G$ [86, 87], $E$ [101], and $T$ [102, 103]. For specific truncation orders, bumps can be observed in the $\chi^2$-plots within certain energy regions. These bumps can be cautiously attributed to well-known resonances (see Fig. 1). For more details, see reference [105].

The evaluation of the matrices defining bilinears such as (11), which are calculable, opens the door to the comparison of the fitted Legendre coefficients to energy dependent partial wave analysis models. In this case here, the Bonn-Gatchina solution BnGa 2014-02 was chosen

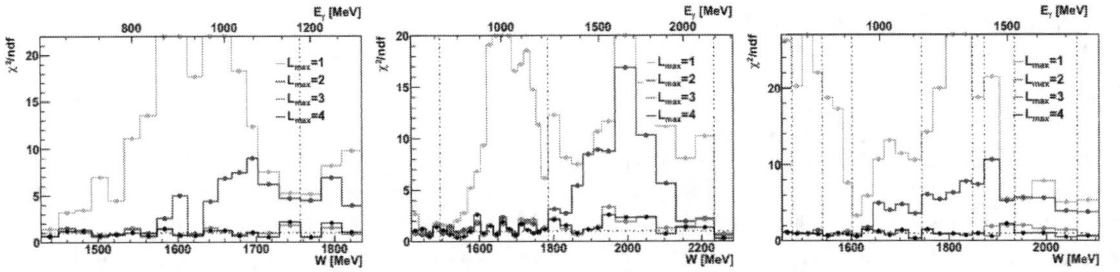


**Figure 14.** The $\chi^2/n_{df}$ values of fits with associated Legendre polynomials according to eq. (8) to the recent new double polarization observable $\check{G}$ [86, 87] (left) $\check{E}$ [100, 101] (center), and $\check{T}$ [102, 103] (right). Only statistical errors were used in the fits. The partial wave expansion was truncated at $L_{max} = 1 \ldots 4$. The $\chi^2/n_{df}$ values are plotted for fits with different $L_{max}$ as a function of the center of mass energy W. Results are preliminary and will be published in [105].

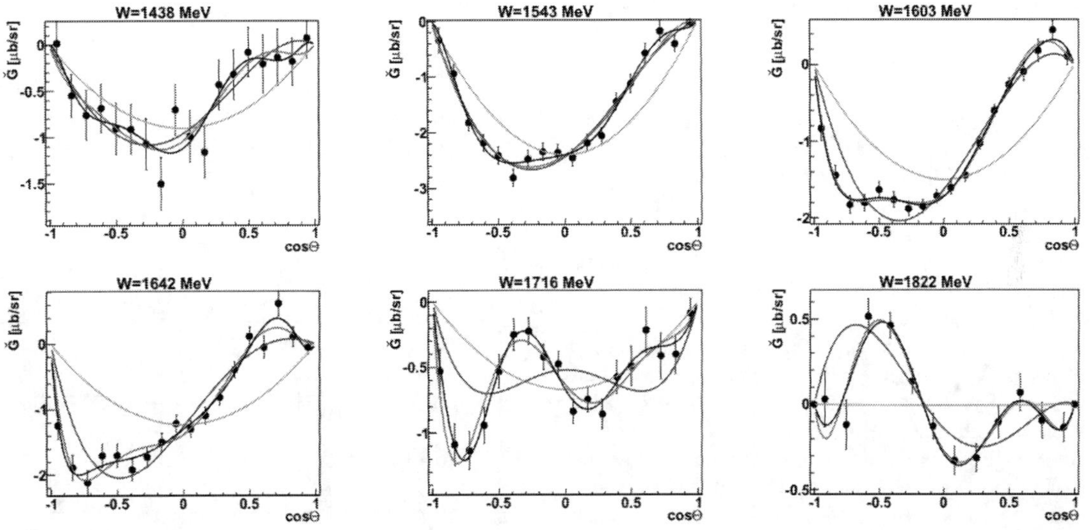


**Figure 15.** 6 out of 19 selected angular distributions of $\check{G}$ (black points) with the different $L_{max}$ fits (solid lines) in the range from $W = 1438$ MeV to 1822 MeV. Colors are the same as in Fig. 14.

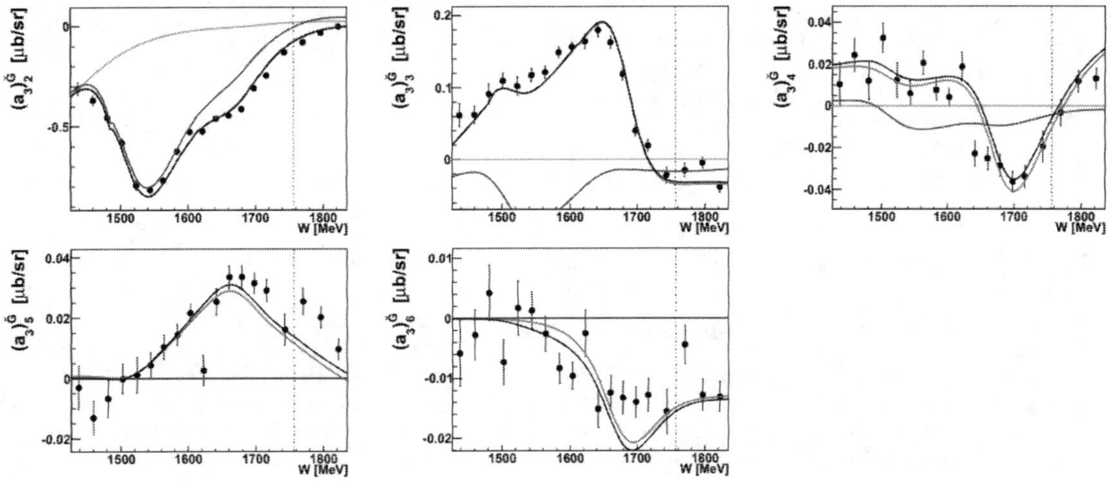


**Figure 16.** Comparison of the fit coefficients for $L_{max} = 3$ (black points), $(a_3)^{\check{G}}_{2\ldots6}$ (see eq. 10), with the BnGa2014-02 solution truncated at different $L_{max}$ (solid lines). Colors are the same as in Fig. 14.

(see project A.2). Then, the $a_{L_{max}}$ evaluated from model-multipoles up to a certain order can be plotted in comparison to the fitted values. This has been done for the $G$-observable in Figure 16. Then, the Legendre coefficients can help us to learn about the influence of the contributing multipoles, at least in the context of the employed model.

The $\chi^2$-plot in Fig. 14 suggests a good description using $D$-waves for energies below $W = 1600$ MeV for $T$, $E$, and $G$. For higher energies, all the $\chi^2$-plots consistently enforce at least $F$-waves, since $\chi^2/n_{df}$ becomes larger than 1 for the $L_{max} = 2$-fit. This behavior is well-reflected in the coefficient plot for $(a_3)_2^{\check{G}}$. However, the $F$-waves from Bonn-Gatchina are needed to reproduce $(a_3)_3^{\check{G}}$ for all energies and therefore in particular also in the lower energy region. This is an example of an interference phenomenon. There can be interferences between dominant lower and even small higher partial waves, taking place in the lower Legendre coefficients, to which the simple method described here is insensitive.

An important lesson learned from model comparisons such as these is the fact that the data needs to be precise enough to be able to detect also the interferences of large resonant multipoles with small background partial waves. Quite generally, Legendre coefficients are well suited for model comparisons, since here effects can become apparent which are practically invisible if one compares just the angular distributions of the data to a model. Apart from the above-mentioned model-comparison, the investigation of Legendre coefficients is also useful for the comparison of different data sets. In particular, the size and influence of different systematic effects in the data can become more visible. As an outlook, in the case the data shall be processed further in a truncated partial wave analysis (TPWA), one can get a first good idea about the cutoff $L_{max}$ from in this kind of analysis.

## 4.6 $\gamma p \to \eta p$: The polarization observables $E$, $G$, $T$, $P$, $H$

New single and double polarization observables were also determined for $\gamma p \to p\eta$ [110]. The observables $E$ [111], $G$ [112], $T, P$, and $H$ [113] have been measured with the CBELSA/TAPS experiment. Data on $T$ and $F$ [114] were obtained in Mainz, and data on $E$ have recently been published by CLAS [115]. The latter data are consistent with our CBELSA/TAPS-data but are not yet included in the BnGa-PWA.

A few bins of the newly measured double polarization observables are shown in Figs. 17-19 for $E$, $T$, $P$, $H$, and $G$, together with the predictions from different PWAs. None of the predictions comes even close to the data, even though they do describe all the so far existing $\eta$-photoproduction data ($d\sigma/d\Omega$, $\Sigma$) reasonably well. This clearly shows that the amplitudes for $\eta$-photoproduction were even worse constrained by existing data than for $\pi^0$-photoproduction. Of course this was to be expected since $\pi N$-elastic scattering provides already significant information on the $\pi N$ coupling of resonances, while data of comparable accuracy is missing for the $\eta N$-final state. Data on $\pi N \to \eta N$ are scarce.

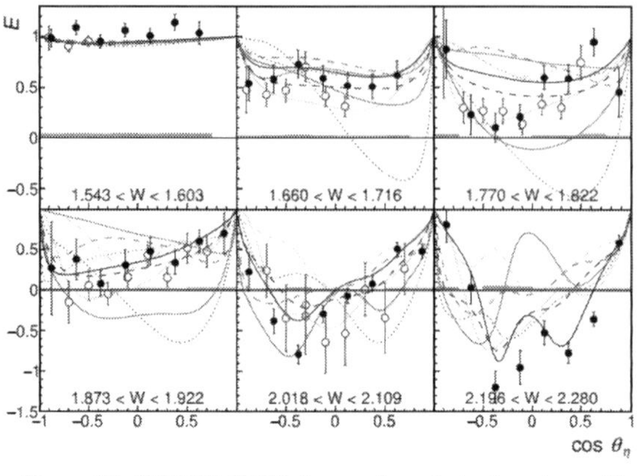


**Figure 17.** CBELSA/TAPS data on the polarization observable $E$ (only selected bins shown) for $\gamma p \to p\eta$ (black) plotted together with the respective data from CLAS [115] (blue, open circles) and the predictions from MAID (green), SAID (blue, dotted), JüBo2015 (2015-3) (cyan, dashed (solid)), and BnGa2011-01(02) (red, solid (dashed)). New BnGa-Fits are shown in black: with (solid) and without (dashed) a new $5/2^-$-state at 2200 MeV included.

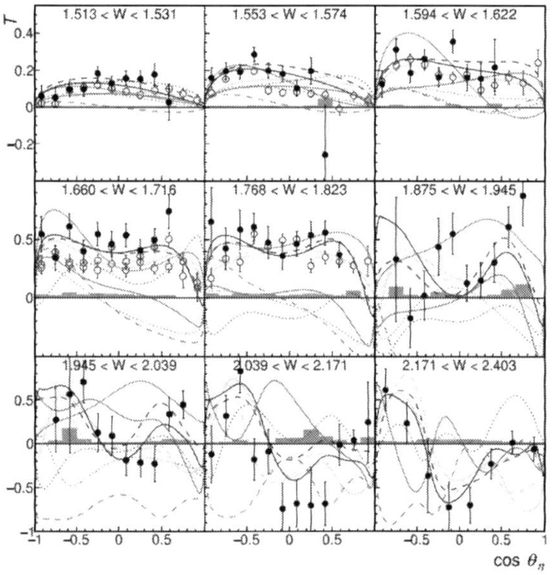


**Figure 18.** CBELSA/TAPS data on the polarization observable $T$ (only selected bins shown) for $\gamma p \to p\eta$ (black) plotted together with the respective data from MAMI [114] (blue, open circles). For an explanation of the curves see Fig. 17.

The new polarization data on $E$, $T$, $P$, $H$, $G$, $F$ (except those from [115]) were included into the BnGa-PWA. Some of the results are presented in Table 4. Further results on other resonances are given in [110]. The $N\eta$-branching ratio found in this new PWA reduces the large and controversially discussed difference in $N\eta$-branching ratios of the $N(1535)1/2^-$ and the $N(1650)1/2^-$ significantly. The unexpectedly large $N\eta$-branching ratio difference observed before had led to a large number of very different explanations. Isgur and Karl [12] explained the observed difference in the $p\eta$-coupling by mixing of

 DOI: 10.1051/epjconf/201713402001

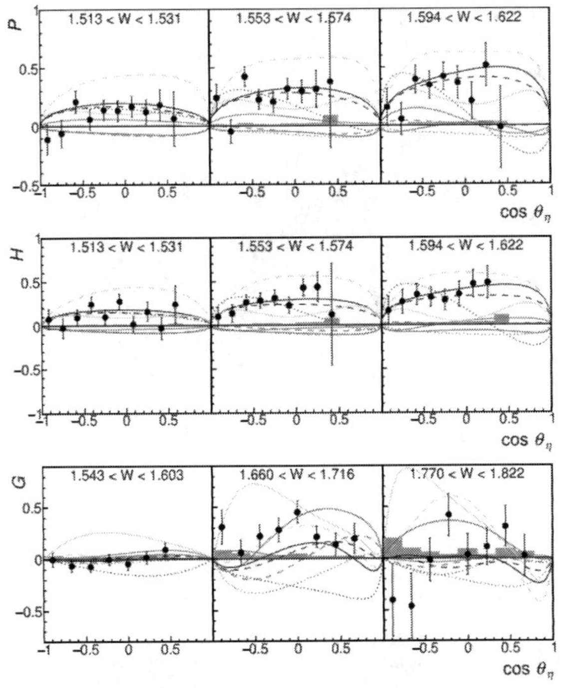


**Figure 19.** CBELSA/TAPS data on the polarization observable $P$ (top), $H$ (middle), and $G$ (bottom) for $\gamma p \to p\eta$ (only selected bins shown). For an explanation of the curves see Fig.17.

the two quark-model states with defined total quark spin $S = 1/2$ and $S = 3/2$; Kaiser, Siegel and Weise [116] and Mai, Bruns, and Meißner [117] generated $N(1535)1/2^-$ dynamically (see also project B.3 "Chiral dynamics with (non)strange quarks"). In [116] it is interpreted it as quasi-bound $K\Sigma$-$K\Lambda$-state decaying strongly into $N\eta$ due to coupled channel effects; Zou [118] assigned a large pentaquark component to the $N(1535)1/2^-$ wave function.

The new result seems to be at variance with these explanations. The branching ratios given in Table 4 are derived based on the pole position properties of the resonances (for more details see contribution of project A.2 to these proceedings and [110]). One interesting observation to note, is that the $K$-matrix-coupling to $p\eta$ at the lower mass $K$-matrix pole is large while the one at the higher mass (close to 1650 MeV) is almost negligible.

**Table 4.** Branching ratios BR for $N^* \to N\eta$ decays and the photon helicity amplitudes $A_{1/2}$, $A_{3/2}$ for 4 of the contributing nucleon resonances, derived from their properties at the pole position. The helicity amplitudes are given in units of GeV$^{-1/2}$ (taken from [110]). Small numbers below the BRs or below the helicity amplitudes give either the RPP [58] estimate or the entries reported in [42].

| Res. $A_{1/2}$ | BR($N^* \to N\eta$) $A_{3/2}$ | Res. $A_{1/2}$ | BR($N^* \to N\eta$) $A_{3/2}$ |
|---|---|---|---|
| $N(1535)$ | 0.42±0.04 | $N(1710)$ | 0.25±0.09 |
| $1/2^-$ | 0.42±0.10 | $1/2^+$ | 0.10 - 0.30 |
| 0.093±0.009 | - | 0.040±0.020 | - |
| 0.115±0.015 | - | 0.035±0.012 | - |
| $N(1650)$ | 0.32±0.04 | $N(1720)$ | 0.03±0.02 |
| $1/2^-$ | 0.05 - 0.15 | $3/2^+$ | 0.021±0.014 |
| 0.032±0.006 | - | 0.115±0.045 | 0.135±0.040 |
| 0.045±0.010 | - | 0.100±0.020 | 0.150±0.030 |

An improved description of the new data on $T$ and $E$ is reached if a further high mass $N(2200)5/2^-$ is included in the fits (see Figs. 17 and 18). Since no convincing evidence for the state was observed in the other reactions one must consider this new resonance with some caution. Of course one would like to see the results of the BnGa-PWA confirmed by other analyses including the new data. Such an analysis is for instance planned within the JüBo approach.

## 5 Photoproduction of meson pairs

### 5.1 Differential cross section and measurements with linearly polarized beam

Data on the reactions $\gamma p \to p\pi^0\pi^0$ and $\gamma p \to p\pi^0\eta$ were acquired in different run periods with unpolarized (985,000 $p\pi^0\pi^0$ and 145,000 $p\pi^0\eta$ events) and with linearly polarized photons (620,000 / 105,500 events). Figure 20 shows two Dalitz plots for two-pion production, for $1500 < E_\gamma < 1700$ MeV and for $1800 < E_\gamma < 2200$ MeV. Prominent structures in the distributions can be seen: at the lower photon energy, $\Delta(1232)$ and $N(1520)3/2^-$ are produced as intermediate isobar in a sequential decay chain. At the higher photon energy, $N(1680)5/2^+$ shows up in addition. In the reaction $\gamma p \to p\pi^0\eta$, $N(1535)1/2^-$ is seen above the $p\eta$ threshold (see Fig. 21). With increasing photon energy, the $\Delta(1232)\eta$ isobar can be observed as well as contributions from $pa_0(980)$.

### 5.2 $\vec{\gamma}p \to p\pi^0\eta$: First measurement of the polarization observables $I^s$ and $I^c$ in photoproduction of two mesons

Using linear photon polarization, new observables show up, called $I^s$ and $I^c$, which have been determined for the first time in [47]. Before [47] two-meson photoproduction using linearly polarized photons has been treated in a quasi two-body approach only resulting in the extraction of the beam asymmetry $\Sigma$ [45, 51, 119] known from single-meson photoproduction. In fact, however, three-body final states like $p\pi^0\eta$ or $p\pi^0\pi^0$ yield additional degrees of freedom, reflected in a different set of polarization observables [63], which can be accessed in a full three-body approach. The cross-section for the photoproduction of pairs of pseudoscalar mesons, disregarding target- and recoil polarization, is given by:

$$\frac{d\sigma}{d\Omega} = \frac{d\sigma_0}{d\Omega}\left[1 + \delta_\odot I^\odot(\phi^*) + \delta_l\left(I^c(\phi^*)\cos 2\phi + I^s(\phi^*)\sin 2\phi\right)\right].$$

Here $\frac{d\sigma_0}{d\Omega}$ denotes the unpolarized cross-section and $\delta_\odot$ ($\delta_l$) the degree of circular (linear) polarization. The polarization observables $I^s$, $I^c$, and $I^\odot$ emerge as the amplitudes of the respective modulations of the azimuthal distributions of the final state particles (Fig. 22, right). These occur if the non-coplanar kinematics of the reaction is considered by taking into account the angle $\phi^*$ between the reaction plane (spanned by the incoming photon and one recoiling particle) and the decay plane (comprising all three final

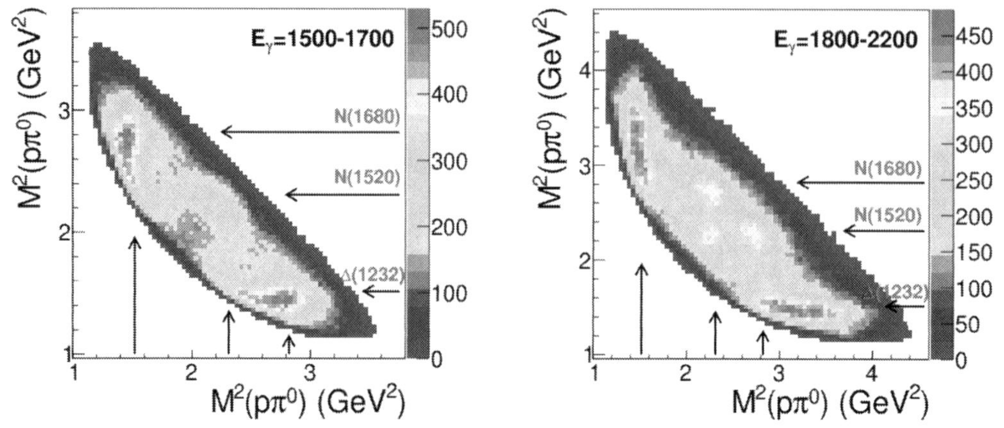


**Figure 20.** Dalitz plots for 1500 MeV < E$_\gamma$ < 1700 MeV (left) and 1800 MeV < E$_\gamma$ < 2200 MeV. The arrows indicate the positions of $\Delta(1232)$, $N(1520)3/2^-$, and $N(1680)5/2^+$.

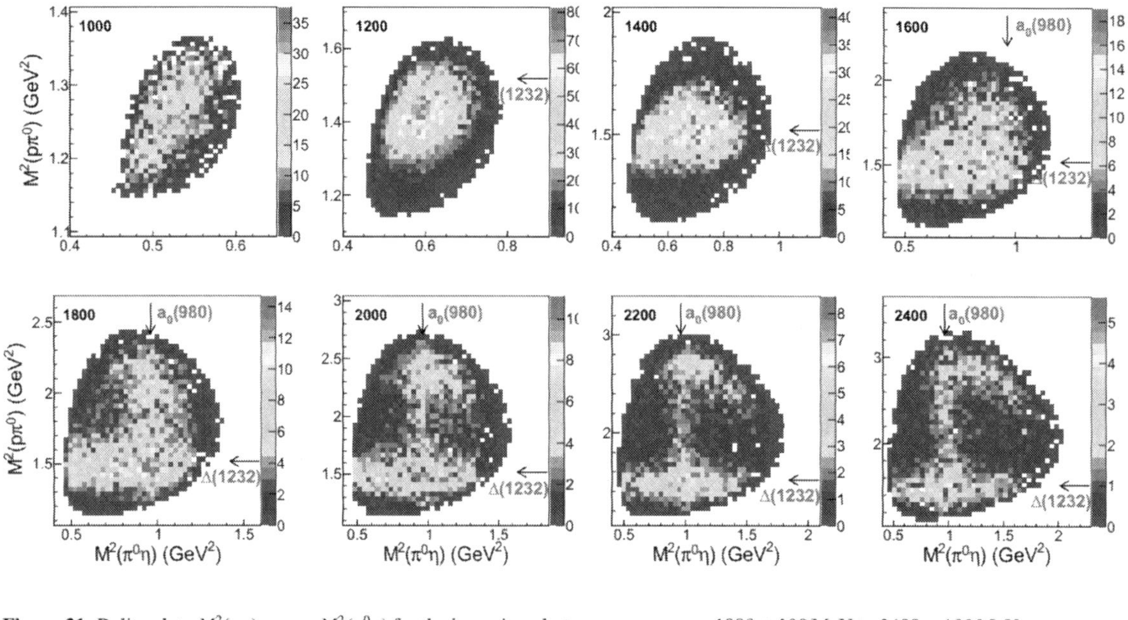


**Figure 21.** Dalitz plots $M^2(p\eta)$ versus $M^2(\pi^0\eta)$ for the incoming photon energy ranges 1000 ± 100 MeV to 2400 ± 100 MeV.

state particles) (Fig. 22, left). If intergrated over $\phi^*$, $I^s$ and $I^{\odot}$ vanish, while $I^c$ transfers into the beam asymmetry $\Sigma$, well known for two-particle final states.

The full three-body approach with linearly polarized photons was applied for the first time for the reaction $\gamma p \to p\pi^0\eta$ [47, 48]. This reaction allows us to study high-lying $\Delta$-resonances via their isospin-selective $\Delta(1232)\eta$ decay. $\Delta(1700)3/2^-$ and $\Delta(1940)3/2^-$ were thus identified. The latter observation led to an upgrade of the $\Delta(1940)3/2^-$ resonance from a one-star to a two-star resonance in the RPP [14]. In addition, the couplings of resonances to $pa_0(980)$ were investigated. Evidence for the corresponding intermediate states is seen in Fig. 23, top. Also the $N(1535)1/2^-\pi$ intermediate state was observed.

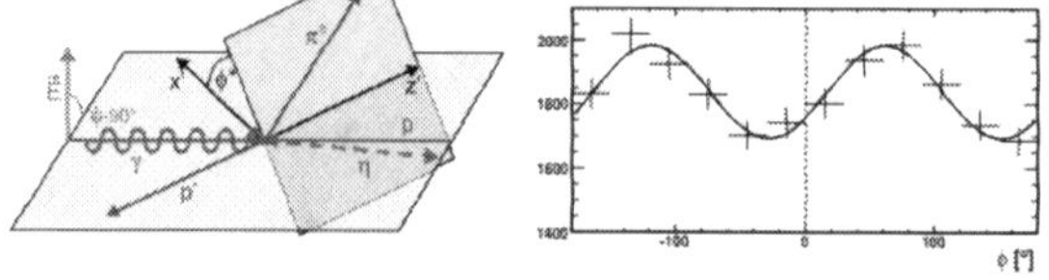

**Figure 22.** Left: Definition of the relevant angles in a three-body final state. Right: Example for a measured $\phi$-distribution for a limited $\phi^*$-range ($\gamma p \to p\pi^0\pi^0$, 970 MeV< $E_\gamma$ < 1200 MeV, $\phi^*$ : 18 − 36°). Here, both the cos2$\phi$- as well as the sin2$\phi$-modulation (as a phase-shift) of the cross-section are visible.

Examples for the extracted asymmetries $I^s$ and $I^c$ are shown in Fig. 23 (bottom). The data on $I^s$ and $I^c$

are compared to three theoretical models, to an early BnGa-PWA [43, 44], to the analysis by Fix et al. [120, 121], neglecting $pa_0$-couplings and assuming $\Delta(1232)\eta$ dominance in most resonance decays, and Döring et al. [122], describing the reaction by the excitation of the $\Delta(1700)3/2^-$ resonance, which in this model is dynamically generated. Close to threshold all models are able to reproduce the observables reasonably well. In the higher-energy region, shown in Fig. 23 (bottom), only the BnGa-PWA, including $\Delta(1232)\eta$, $p\,a_0(980)$ and $N(1535)1/2^-\pi$ decays (as well as excited isospin $I=1/2$ states) gives a reasonable description of the data. This demonstrates the high sensitivity of the observables $I^s$ and $I^c$ to the contributing resonances and their decays.

### 5.3 $\vec{\gamma}p \to p\pi^0\pi^0$: $I^s$ and $I^c$ in $2\pi^0$-photoproduction

Also in the photoproduction of two neutral pions, the observables $I^s$ and $I^c$ exhibit a high sensitivity to the reaction dynamics. Fig. 24 (top) shows the Dalitz plot for the reaction exhibiting various intermediate resonant structures.

In the analysis of our data on $\gamma p \to p\pi^0\pi^0$ [53, 54] with no photon polarization two equally good descriptions of the data were obtained within the BnGa-PWA. One solution yielded $S$-wave dominance in the decay $\Delta(1700)3/2^- \to \Delta\pi$, one preferred $D$-wave. Both solutions are shown in Fig. 24, bottom, as dashed and solid curves, respectively. The sensitivity of these observables to the two decay modes is clearly visible. Additionally, a prediction by Arenhövel et al. [123] is shown (red line), assuming $N(1520)3/2^-$ (and not $\Delta(1700)3/2^-$) to be the dominant isobar to the reaction, and assuming $\Delta\pi$ to be the dominant decay mode of resonances. These assumptions are distinctively dismissed by the data.

The importance of $N(1520)3/2^-$ as an intermediate resonance in cascade decays of higher-lying resonances is demonstrated in Fig. 25. The Dalitz plot is shown for $1300 < E_\gamma < 1650$ MeV which covers the invariant mass range from 1820 to 2000 MeV. For this mass range, $I^s$ and $I^c$ have been extracted from the data for two different regions, I and II, of the Dalitz plot, as denoted by the lines. In the region where $\Delta(1232)3/2^+$ dominates the intermediate state, the $I^s$ and $I^c$ show little to no amplitude (see the upper row of the four subfigures in Fig. 25). In region II where the cascade $N^* \to N(1520)\pi^0 \to p\pi^0\pi^0$ makes a significant contribution, $I^s$ and $I^c$ show a significant structure (see the lower row of the four subfigures in Fig. 25). The latter four distributions can be fitted with very simple assumptions. A resonance in one initial partial wave with defined $J^P$ is assumed to decay via

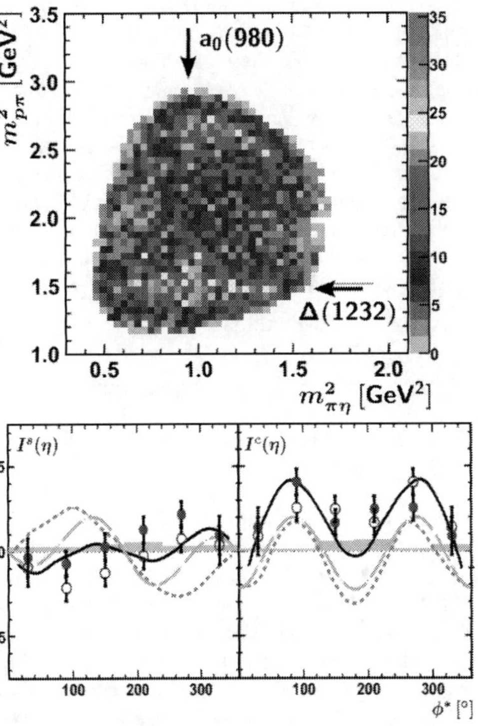


**Figure 23.** Top: $m^2(p\pi^0)$-$m^2(\pi^0\eta)$ Dalitz plot for $E_\gamma = 2150\pm50$ MeV. Marked are the signatures of the $\Delta(1232)3/2^+$ and the scalar meson $a_0(980)$, additionally, the signal for the $N(1535)1/2^-$ is visible in the diagonal. Bottom: Exemplary distributions for the observables $I^s$ and $I^c$ (W=1946±48 MeV), extracted from the $\phi$-distributions of the final state $\eta$-meson. Solid (black) curve: BnGa-PWA [43, 44], long-dashed (green) curve: Fix et al. [120, 121], short-dashed (red) curve: Döring et al. [122].

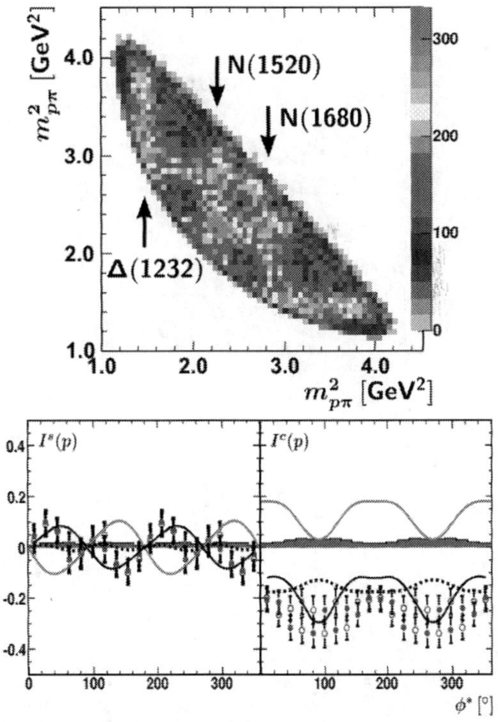


**Figure 24.** Top: Dalitz plot for the reaction $\gamma p \to p\pi^0\pi^0$, exhibiting a rich spectrum of intermediate resonance structures ($E_\gamma$= 2000-2100 MeV). Bottom: Exemplary distributions for the observables $I^s$ and $I^c$ ($E_\gamma$=970-1200 MeV), extracted from the $\phi$-distributions of the final state proton. Solid (black) curve: BnGa-PWA [53, 54] with $D$-wave dominance $\Delta(1700)3/2^- \to \Delta\pi$, dashed: BnGa-PWA with $S$-wave dominance $\Delta(1700)3/2^- \to \Delta\pi$, solid, (red): Arenhövel et al. [123].

a cascade $N^* \to N^+(1520)\pi^0 \to p\pi^0\pi^0$. It turns out that fits with $J^P = 1/2^+$, $5/2^+$ and $7/2^+$ give a bad description while a fit with $J^P = 3/2^+$ yields a good description of the data. In the W≈ 1900 MeV region, the cascade $N^* \to N(1520)\pi^0 \to p\pi^0\pi^0$ is dominated by the $J^P = 3/2^+$ partial wave (see Fig. 6 in the contribution of project A.2 to these proceedings and [56] for a detailed discussion).

These observations made directly from the data are confirmed by the BnGa multi-channel partial wave analysis (project A.2) which includes a large number of pion and photo-induced reactions in addition to the data presented here. The result of a mass scan in which a single Breit-Wigner amplitude with $J^P = 3/2^+$ is added above $N(1720)3/2^+$ to the set of amplitudes is shown in Fig. 26. The mass of the added $J^P = 3/2^+$-state is fixed while all other parameters are fitted freely. The $\chi^2$-difference, normalized to the minimal $\chi^2$ of the fit, shows a clear minimum in the respective mass range. The same is true if the $\chi^2$-change only due to the $\gamma p \to p\pi^0\pi^0$-reaction is looked at.

The BnGa-PWA provides additionally indications for the existence of a second resonance around 1900 MeV in the $3/2^+$ wave beyond the known $N(1900)3/2^+$ [56] even though the fit to the data with $N(1900)3/2^+$ only is also acceptable.

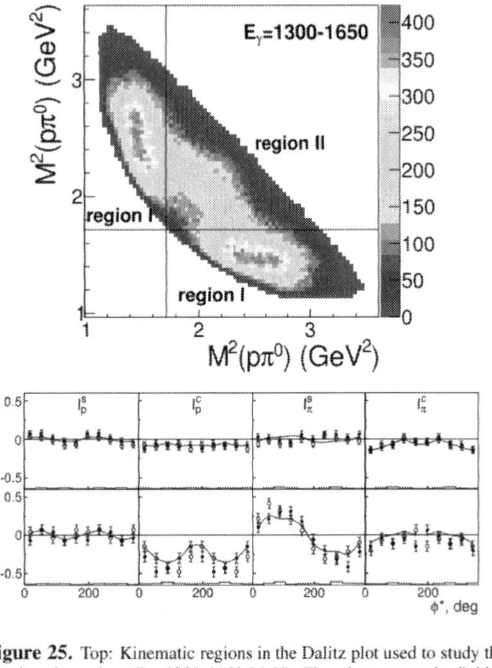


**Figure 25.** Top: Kinematic regions in the Dalitz plot used to study the reaction dynamics ($E_\gamma$=1300-1650 MeV). The phase space is divided into two regions I and II. In region II the $\Delta(1232)$ contribution is suppressed and the $N(1520)3/2^-$ contribution enhanced. Bottom: $I^s_p$, $I^c_p$, $I^s_\pi$, and $I^c_\pi$ as functions of $\phi^*$ for $1300 < E_\gamma < 1650$ MeV. The subfigures on the first (second) row show the distributions for events in region I (II) of the Dalitz plot. Dots: $I^s$, $I^c$; open circles: derived from symmetry; yellow band: syst. uncertainties. The solid curve represents a two-parameter fit assuming $J^P = 3/2^+$ as initial partial wave (Figures taken from [56]).

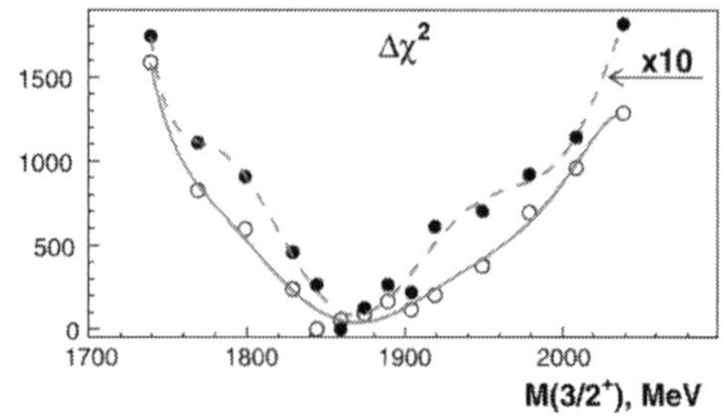


**Figure 26.** Mass scan for a resonance with $J^P = 3/2^+$. Shown is the $\chi^2$-change of the BnGa fit as a function of the imposed mass. In the mass scan, a Breit–Wigner amplitude with $J^P = 3/2^+$ is added above $N(1720)3/2^+$; its mass is fixed while all other parameters are fitted freely. The curves are included to guide the eye. The open circles give the total change in $\chi^2$, normalized to the minimal $\chi^2$. The black circles represent the $\chi^2$-contribution from $p\pi^0\pi^0$; it is multiplied by a factor 10 for better visibility. The $\gamma p \to p\pi^0\pi^0$ data are fitted by an event-based likelihood fit; the likelihood converted into a pseudo-$\chi^2$ by $\Delta(lnL) = 2\Delta\chi^2$. (Figure taken from [57]).

The partial wave analysis (within the Bonn-Gatchina approach) of the multi-meson data yields a large number of branching ratios of resonances into $\Delta\pi$, $\Delta\eta$, $N(\pi\pi)_S$, $N(1440)1/2^+$, $N(1520)3/2^-$, $N(1535)1/2^-$, and $N(1680)5/2^+$. Many of these branching ratios have been determined for the first time. Decays of $\Delta(1910)1/2^+$, $\Delta(1920)3/2^+$, $\Delta(1905)5/2^+$, $\Delta(1950)7/2^+$ into $\Delta(1232)\pi$ and of the corresponding spin-parity series in the nucleon sector, $N(1880)1/2^+$, $N(1900)3/2^+$, $N(2000)5/2^+$, and $N(1990)7/2^+$ are observed. For the nucleon resonances, these decay modes were reported for the first time. Further new decay modes proceed via $N(1520)3/2^-\pi$, $N(1535)3/2^-\pi$, $N(1680)5/2^+\pi$, and $N\sigma$. The latter decay modes are observed in the decay of $N^*$ resonances, in $\Delta^*$ decays, however, at most weakly. In [55, 56] it is argued that these decay modes provide evidence for a 3-quark nature of $N^*$ resonances rather than a quark-diquark structure (for further explanations see also the contribution of project A.2 to these proceedings).

# 6 Photoproduction off neutrons

## 6.1 Study of a narrow peak in the $n\eta$ mass distribution

A narrow dip-bump structure at 1.67 GeV was observed in photoproduction of $\eta$ mesons off neutrons [124]. It is discussed in more detail in the contribution off project A.2 "Partial Wave Analysis" to these proceedings. The peaking structure was observed in the $n\eta$ invariant mass distribution but not seen in the cross section on the proton. The structure was suggested to signal the existence of a relatively narrow ($M \approx 1.68$ GeV, $\Gamma \leq 30$ MeV) baryon state.

The CBELSA/TAPS collaboration studied the reaction $\gamma d \to \eta\, n\, p_{\text{spectator}}$ and confirmed the structure [125] in the $n\,\eta$ invariant mass while no similar structure was observed in the $n\,\eta'$ mass distribution [126]. The structure became even more pronounced when the recoiling neutron was detected and effects from Fermi smearing were avoided [127]. In 2012, the structure was listed in the RPP as one-star resonance. At Mainz, the structure was studied with high statistics and again confirmed [128, 129]. The peak is discussed controversially in the literature, and different interpretations were given [130–136]. So far, the most recent and most precise data [128, 129] were only fitted in [136]. The latter fit gave a nearly perfect description of the data assuming that the structure emerges from the interference of $N(1535)1/2^-$ and $N(1650)1/2^-$ within the $1/2^-$-partial wave. The fit deteriorated when in the fit a resonance was imposed with properties as suggested in the experimental studies. In 2016, the PDG decided to remove $N(1685)$ from the listings.

### 6.2 Helicity amplitudes off neutrons

Data on photoproduction off neutrons provide for the respective $\gamma n$-helicity amplitudes. To obtain sufficient information, all existing data on $\gamma n$ were included in the BnGa partial wave analysis [135]. The fit was further constrained by the data photoproduction off protons. Recently, CLAS reported new and very precise data on $\gamma n \to K^0\Lambda$; the new data allowed for a new determination of the helicity amplitudes for photoproduction of neutrons [137].

## 7 Recent results

In this section we present results which are based on data taken during the SFB which are not yet published or submitted for publication and which partly are still preliminary.

### 7.1 $\gamma p \to p\pi^0\pi^0$: The polarization observables $T, P, H$

The data taken with transversely polarized target (see Section 4.4 and 4.6) were also analyzed to investigate the reaction $\gamma p \to \pi^0\pi^0 p \to 4\gamma p$ [139]. The four photons were combined pair-wise to form the pions. The resulting invariant mass spectrum is shown in Fig. 27 where the mass of one photon pair is plotted against the other. Peaks due to the double pion final state as well as the $\pi^0\eta$ final state are clearly visible. The shown spectrum still contains combinatorical background which can be eliminated nearly completely by employing a kinematic fit using the masses of the final state particles as constraints. The background contamination in the final data set is in the order of 1.5%. The dilution factor has been determined again by analyzing the C-foam-target data as described in Section 3.5.

Taking the full 3-body-kinematics of the $p\pi^0\pi^0$ final state into account additional polarization observables, not present for two-body final states, arise. Considering only linearly polarized photons and transversely polarized target protons, the differential cross section is given by [63]

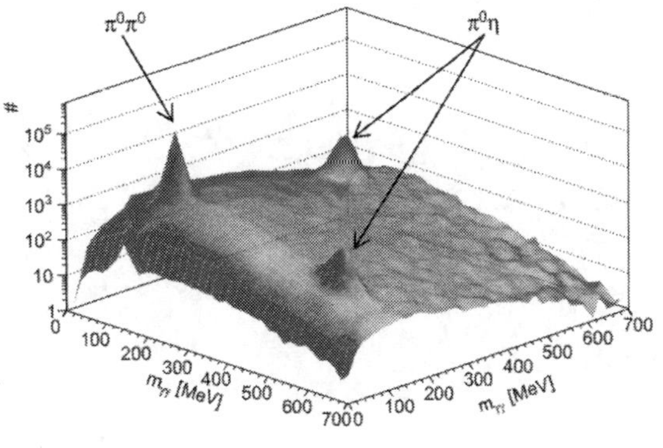


**Figure 27.** Invariant mass of one photon pair plottet versus the invariant mass of the other. Combinatorical background (which can be eliminated) is still included here.

$$\frac{d\sigma}{d\Omega} = \frac{d\sigma_0}{d\Omega} \cdot \Big\{ \big[ 1 + \Lambda_x P_x + \Lambda_y P_y \big] + \delta_\ell \cdot \sin(2\phi) \cdot \big[ I^s + \Lambda_x P_x^s + \Lambda_y P_y^s \big] + \delta_\ell \cdot \cos(2\phi) \cdot \big[ I^c + \Lambda_x P_x^c + \Lambda_y P_y^c \big] \Big\} \quad (13)$$

with the degree of linear polarization $\delta_\ell$ and the target polarization degree parallel (perpendicular) to the reaction plane $\Lambda_x$ ($\Lambda_y$). After integration over $\phi^*$, the polarization observables $P_y, P_y^c$ and $P_x^s$ correspond to the observables $T, P$, and $H$ in photoproduction of single pseudoscalar mesons (Fig. 22, left).

As an example, the target asymmetry $P_y$ is shown in Fig. 28 as a function of the angle $\phi^*$. Here only every second energy bin is shown. The PWA predictions from the BnGa group show significant discrepancies to the data: The data shows a much more pronounced $\phi^*$-dependence than the BnGa'2014-predictions. Results for the double polarization observables $P$ and $H$ are shown in Fig. 29 as functions of the invariant mass of the two pions. The PWA predictions completely fail to describe the data indicating again that double pion photoproduction is not yet fully understood.

Unlike the single meson photoproduction where there are only two kinematic variables (e.g. $E_\gamma, \cos\vartheta$), a three body final state like $p\pi^0\pi^0$ has a five-dimensional phase space. For a complete experiment the polarization observables need to be determined at every (5-dimensional) kinematic point. Therefore, it is important to determine the observables multi-dimensional. Fig. 30 shows the target asymmetry $P_y$ for one energy bin simultaneously binned in $\cos\vartheta, m_{p\pi^0}$ and $\phi^*$. The observable clearly shows quite a different behavior in different kinematic regions. Obviously the integrated target asymmetry shown in Fig. 28 carries only a part of the full information. In Fig. 28, all data shown in Fig. 30 are contained in the third energy bin of Fig. 28 (and the two bins adjacent in energy which are not shown). The data ist presently included in the PWA

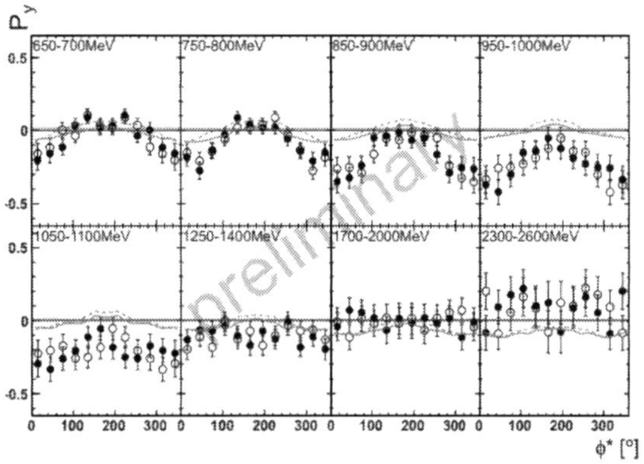


**Figure 28.** Target asymmetry $P_y$ as a function of $\phi^*$ (black dots). The open symbols are calculated from the symmetry property $P_y(2\pi - \phi^*) = P_y(\phi^*)$. Red solid line: predictions from BnGa'2014-02, red dashed line: BnGa'2014-01.

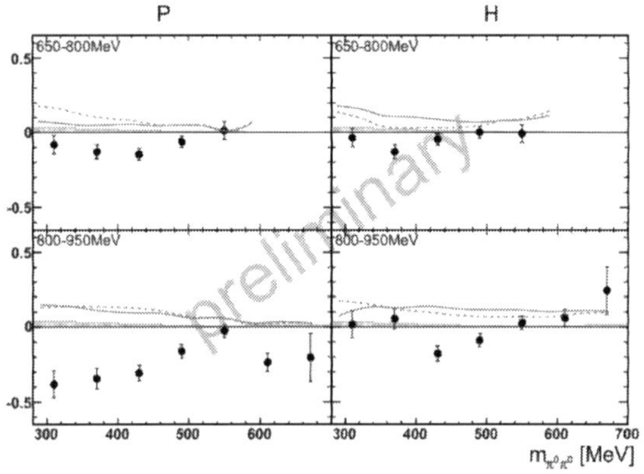


**Figure 29.** Double polarization observables $P$ and $H$ as a function of the double pion invariant mass. Predictions from BnGa'2014-02 (red solid line) and BnGa'2014-01 (red dashed) fail to describe the data.

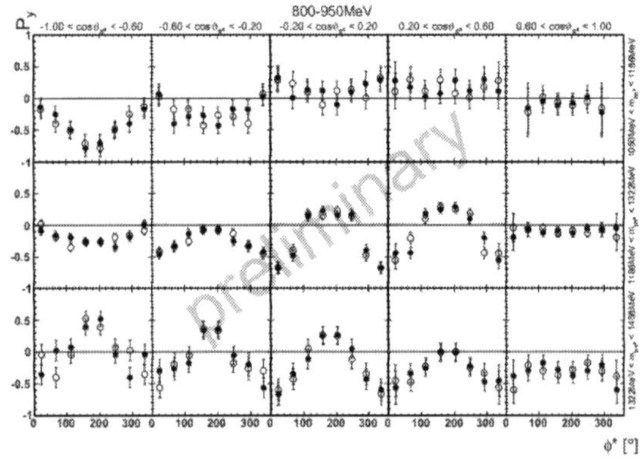


**Figure 30.** Four-dimensional determination of the target asymmetry $P_y$: Shown is the energy bin 800-950 MeV, within one row only $\cos\vartheta$ is varied, within one column only $m_{p\pi^0}$ and $\phi^*$ is on the $x$-axis of the individual histograms. The open symbols are calculated from the symmetry property $P_y(2\pi - \phi^*) = P_y(\phi^*)$.

and one can expect that it will help to constrain the $p\ 2\pi^0$-decays of $N^*$ and $\Delta^*$-resonances more precisely.

### 7.2 $\gamma p \to p\pi^0\pi^0$: The beam asymmetry $\Sigma$ at high energies

In the fall of 2013, a large-statistics data set with a linearly polarized photon beam impinging on an unpolarized liquid hydrogen ($LH_2$) target was taken. The coherent edge was placed at remarkably high beam energies of 1750 MeV resp. 1850 MeV. This allows for studies of the beam asymmetry $\Sigma$ (or of $I^c$ and $I^s$ in the case of two pseudoscalar mesons in the final state) without the problems stemming from the inherent carbon dilution of the polarized target. The analysis of the reaction $\gamma p \to \pi^0\pi^0 p$ is currently ongoing [138]. Due to the five-dimensional reaction phase space, it will especially benefit from the large amount of data that has been recorded. Within a preliminary reaction selection, $3.6 \times 10^6$ $2\pi^0$ events were reconstructed.

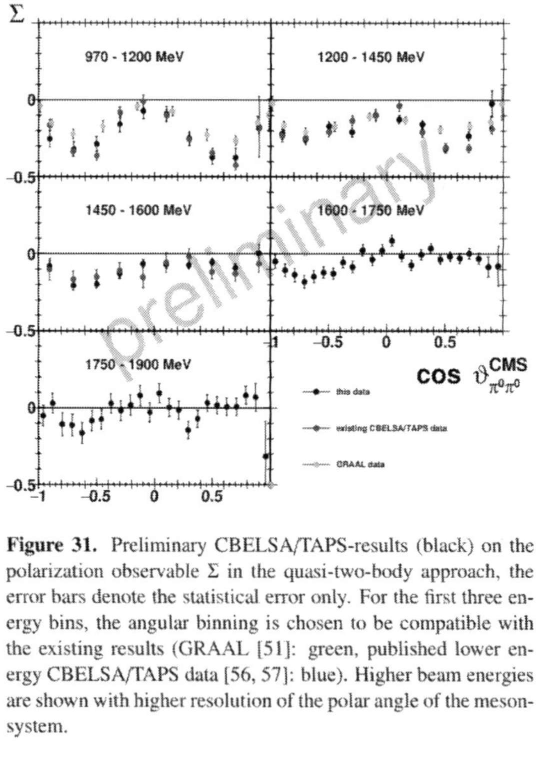


**Figure 31.** Preliminary CBELSA/TAPS-results (black) on the polarization observable $\Sigma$ in the quasi-two-body approach, the error bars denote the statistical error only. For the first three energy bins, the angular binning is chosen to be compatible with the existing results (GRAAL [51]: green, published lower energy CBELSA/TAPS data [56, 57]: blue). Higher beam energies are shown with higher resolution of the polar angle of the meson-system.

A first view on the data is shown in Fig. 31. For lower beam energies, the data can be compared to existing results from the GRAAL [51] and the CBELSA/TAPS collaboration [56, 57]. Compared with the GRAAL data, the beam asymmetry is found with a slightly larger amplitude, but in good agreement with the CBELSA/TAPS data. Possibly, the differences are caused by a difference in the acceptance coverage of the two detector systems. Note that at higher energies in the region of higher polarization, $\Sigma$ is shown with a finer binning.

### 7.3 $\gamma p \rightarrow p\pi^0\pi^0$: Preliminary results on the double polarization observables $E$

The double polarization observable $E$ is currently also extracted for two neutral pions in the final state [138]. Compared to single meson photoproduction, the required polarization settings (circularly polarized beam with polarization $\delta_\odot$ and longitudinally polarized target with polarization $\Lambda_z$) give rise to further polarization observables. Following [63] the relevant part of the cross-section reads:

$$\frac{d\sigma}{d\Omega} = \frac{d\sigma_0}{d\Omega} \cdot \left[1 + \delta_\odot I^\odot + \Lambda_z P_z + \Lambda_z \delta_\odot P_z^\odot\right], \quad (14)$$

where $P_z^\odot$ corresponds to $-E$ in single meson photoproduction if integrated over $\phi^*$.

Changing the sign of the beam and the target polarization, the data set can be divided into four groups; the polarization observables are then accessible as pairwise asymmetries between the subsets. However, the single polarization observables are still under investigation ($P_z$) or should be measured with a liquid-hydrogen target ($I^\odot$) to avoid the data being diluted with reactions off bound protons. In the following, preliminary results for the polarization observable $E$ will be shown.

Events in the detector with one charged and four neutral hits (considered as the decay photons of the two pions) are further selected by means of orthogonal cuts. Cut boundaries are fine-tuned to match the experimental conditions at different kinematic regions and account for the target field direction, as the longitudinal magnetic holding field influences the trajectory of the proton in azimuthal direction. So far, the data still includes combinatorial background. The carbon contribution still remaining after the kinematic cuts is eliminated by subtracting the carbon yield, measured separately and scaled correctly, from each of the subsets mentioned above.

Overall, approximately $5 \times 10^5$ events in the dataset fulfill the cut conditions. Thus, the five-dimensional reaction phase space is explored by various (generally 2-dimensional) projections. Figure 32 shows the helicity asymmetry as a function of the beam energy and the polar angle of the vector sum of the meson system. Instead of the proton, one pion can be considered as the recoiling particle, which is used for the projections of figure 33. Compared to the fully-predictive partial wave analysis solutions, discrepancies are still occurent, most visible at medium or highest beam energies. Within a three-body approach, the resulting observable can be tested to be invariant under parity conservation, which translates into the constraint $E(\phi^*) = E(2\pi - \phi^*)$ and appears to be in good agreement with the data. After the data analysis is finalised the data will be included in the BnGa-PWA.

## 8 Summary and Outlook

One of the most important topics in baryon spectroscopy is to identify the relevant degrees of freedom and the effective forces which determine the spectrum and properties of baryon resonances. Here significant progress has been

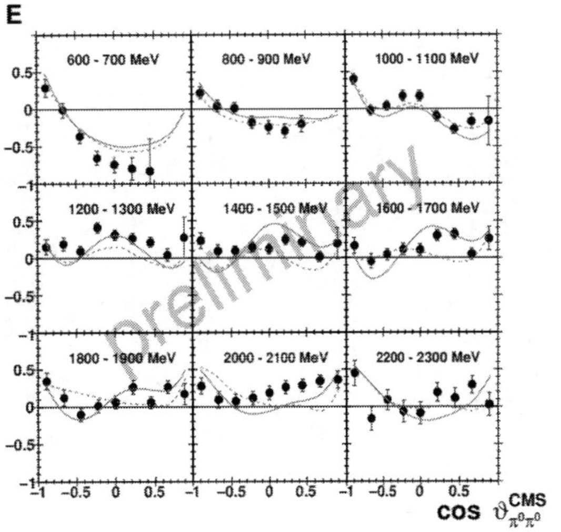


**Figure 32.** Preliminary results on the polarization observable $E$ in the two-body approach. The error bars denote the statistical error. Only every second energy bin is shown. Results are shown along with predictions of the Bonn-Gatchina partial wave analysis group (— BnGa 2014-02 --- BnGa 2014-01)

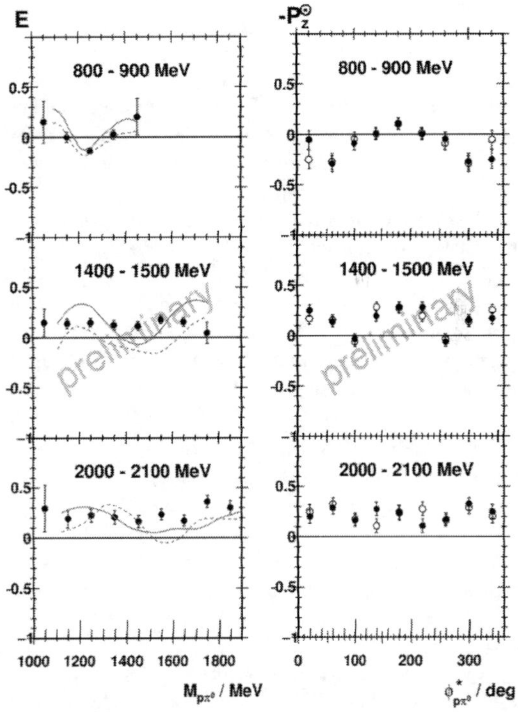


**Figure 33.** Preliminary results on the polarization observable $E$ in two-body and three-body (right) approach, the error bars denote the statistical error. Only selected energy bins are shown. On the left hand side, the data are compared to the predictions of the Bonn-Gatchina partial wave analysis group (— BnGa 2014-02, --- BnGa 2014-01). Within a three-body approach, the resulting observable can be tested to be invariant under parity conservation, which translates into the constraint $E(\phi^*) = E(2\pi - \phi^*)$. The open symbols show this symmetrization of the data.

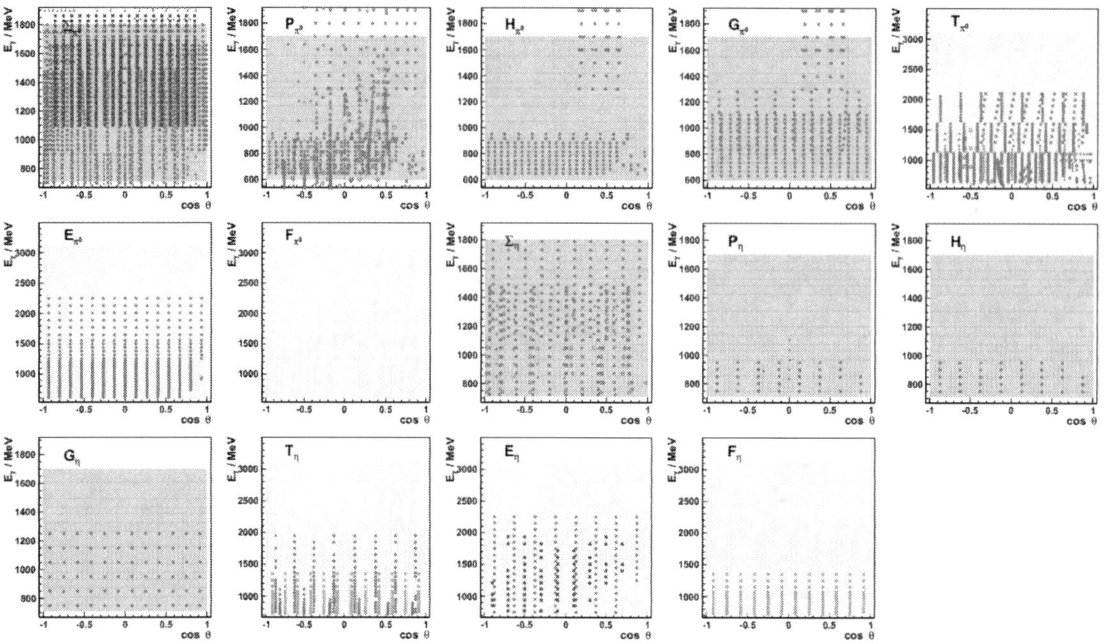


**Figure 34.** Gray: Existing data (SAID-data base), red: new published data from the CBELSA/TAPS-experiment, blue: new CLAS-data, green: new MAMI data. Green (blue area): energy regime accessible at ELSA with circularly (linearly) polarised photons. Of course the target polarization $T$ can be determined independently from the beam-polarization.

made. Based on the new high statistics unpolarized and polarized data taken at ELSA and elsewhere, new baryon resonances could be identified and known or less well known states could be confirmed or have in part been re-established. The number of the now observed resonances excludes e.g. quasi-stable diquarks as part of the dynamics of excited baryons. In addition, based on the new photoproduction data, properties of the baryon resonances such as their partial decays widths in different final states could be determined (see contribution of the project A.2 "Partial Wave Analysis" to these proceedings).

Of course not all of the resonances found in the data are considered to be *established* ($\geq$ 3-star) resonances yet. Further and higher statistics data is needed. Higher statistics is especially needed at higher energies and for channels with lower cross section. The need for the measurement of additional polarization observables relates to areas in which so far no, or only very restriced data has been taken. Fig. 34 shows for two reactions $\gamma p \to p\pi^0$ and $\gamma p \to p\eta$ the polarization observables which have been investigated. Data in the so far not covered regions (Figs. 34), which include especially the regime of higher energies/masses are clearly of large interest. Respective measurements which provide of course not only access to $\pi^0$- and $\eta$-photoproduction but also to additional single and multi-meson final states are planned at ELSA. Here it needs to be stressed, that especially also reactions with more than a single pseudoscalar meson in the final state can be expected to provide new and interesting results on the presently least understood higher mass baryon spectrum. These measurements will extend our knowledge of the baryon spectrum into the regime of higher masses.

Another interesting and so far not well investigated area is photoproduction of the neutron: baryon resonances may strongly couple to the neutron but they may only weakly or not at all couple to the proton. Therefore resonances, maybe even resonances of specific symmetries or nature, may escape detection if only proton data is analyzed. This bias can be overcome if also photoproduction data off the neutron, which is presently still scarce, is taken and investigated. Another not well investigated degree of freedom in the experiments is the isospin dependence. The electromagnetic transition operator can be split into an isovector and an isoscalar part, giving rise to three independent matrix elements [140]: $A^{IS}=\langle\frac{1}{2},\pm\frac{1}{2}|\hat{S}|\frac{1}{2},\pm\frac{1}{2}\rangle$, $\mp A^{IV}=\langle\frac{1}{2},\pm\frac{1}{2}|\hat{V}|\frac{1}{2},\pm\frac{1}{2}\rangle$, $A^{V3}=\langle\frac{3}{2},\pm\frac{1}{2}|\hat{V}|\frac{1}{2},\pm\frac{1}{2}\rangle$, using the notation $\langle I_f, I_{f3}|\hat{O}|I_i, I_{i3}\rangle$. Photoproduction of isovector mesons, such as the pions, involve all three matrix elements, while in the case of isoscalar meson photoproduction only two, $A^{IS}$ and $A^{IV}$, contribute. To study the isospin structure of the amplitudes, photoproduction experiments on the neutron are unavoidable. This becomes directly obvious in the isoscalar meson production. It is hence important to study photoproduction off neutrons with resonable precision compared to what was achieved for protons. Respective measurements will be performed at ELSA.

The work reported here would not have been possible without the enduring enthusiasm of many bachelor, master, PhD students, and post-docs as well as the continuous technical support from staff members of the participating Universities. The support of the ELSA staff members and the polarized target group is gratefully recognized. The results discussed here have been obtained within the CBELSA/TAPS collaboration. At the beginning of the SFB/TR16, Christian Weinheimer was the designated A.1 project leader but accepted a professorship at the University of Münster shortly after the start of the SFB/TR16. Rainer Novotny was project leader of A.1 during the first funding period focussing on the hardware developments. With the second funding period project A.1 was split into the project presented here and the instrumentation project D.3. The instrumentation results including the ones obtained during the first funding period have been summarized within project D.3. We would like to thank E. Klempt, F. Afzal, J. Hartmann, P. Mahlberg, H. van Pee, T. Seifen, A. Thiel, and Y. Wunderlich for contributions to this report. We appreciate the support of the Schweizerischer Nationalfonds, the US National Science Foundation, and the Russian Foundation for Basic Research, in particular the generous support from the Deutsche Forschungsgemeinschaft (SFB/TR16).

## References

[1] V. E. Barnes *et al.*, "Observation of a Hyperon with Strangeness -3," Phys. Rev. Lett. **12**, 204 (1964).

[2] M. Gell-Mann, "A Schematic Model Of Baryons And Mesons," Phys. Lett. **8**, 214 (1964).

[3] G. Zweig, "An SU(3) Model For Strong Interaction Symmetry And Its Breaking.," published in 'Developments in the Quark Theory of Hadrons'. Volume 1. Edited by D. Lichtenberg and S. Rosen. Nonantum, Mass., Hadronic Press, 1980. pp. 22-101.

[4] O. W. Greenberg, "Spin and Unitary Spin Independence in a Paraquark Model of Baryons and Mesons," Phys. Rev. Lett. **13**, 598 (1964).

[5] M. Y. Han and Y. Nambu, "Three Triplet Model with Double SU(3) Symmetry," Phys. Rev. **139**, B1006 (1965).

[6] S. Weinberg, "Nonabelian Gauge Theories Of The Strong Interactions," Phys. Rev. Lett. **31**, 494 (1973).

[7] H. Fritzsch, M. Gell-Mann, H. Leutwyler, "Advantages of the Color Octet Gluon Picture," Phys. Lett. **B47**, 365 (1973).

[8] K. G. Wilson, "Confinement of Quarks," Phys. Rev. **D10**, 2445 (1974).

[9] S. Coleman and D. J. Gross, "Price of Asymptotic Freedom," Phys. Rev. Lett. **31**, 851 (1973).

[10] D. J. Gross and F. Wilczek, "Ultraviolet Behavior of Nonabelian Gauge Theories," Phys. Rev. Lett. **30**, 1343 (1973).

[11] H. D. Politzer, "Reliable Perturbative Results for Strong Interactions?," Phys. Rev. Lett. **30**, 1346 (1973).

[12] N. Isgur and G. Karl, "Hyperfine Interactions in Negative Parity Baryons," Phys. Lett. B **72**, 109 (1977).

[13] N. Isgur and G. Karl, "P Wave Baryons in the Quark Model," Phys. Rev. D **18**, 4187 (1978).

[14] J. Beringer et al. (Particle Data Group), Phys. Rev. **D86**, 010001 (2012).

[15] S. Bethke, G. Dissertori and G. P. Salam, "Quantum Chromo Dynamics", in: [16]

[16] K. A. Olive, "Review of Particle Physics," Chin. Phys. C **38**, 090001 (2014).

[17] J. J. Aubert *et al.*, "Experimental Observation of a Heavy Particle J," Phys. Rev. Lett. **33**, 1404 (1974).

[18] J. E. Augustin *et al.*, "Discovery of a Narrow Resonance in $e^+e^-$ Annihilation," Phys. Rev. Lett. **33**, 1406 (1974).

[19] D. Robson, "A Basic Guide for the Glueball Spotter," Nucl. Phys. B **130**, 328 (1977).

[20] M. S. Chanowitz and S. R. Sharpe, "Hybrids: Mixed States of Quarks and Gluons," Nucl. Phys. B **222**, 211 (1983), Erratum: [Nucl. Phys. B **228**, 588 (1983)].

[21] G. Arnison *et al.* [UA1 Collaboration], "Experimental Observation of Isolated Large Transverse Energy Electrons with Associated Missing Energy at $s^{1/2}$ = 540 GeV," Phys. Lett. B **122**, 103 (1983).

[22] D. Decamp *et al.* [ALEPH Collaboration], "Determination of the Number of Light Neutrino Species," Phys. Lett. B **231**, 519 (1989).

[23] G. Aad *et al.* [ATLAS Collaboration], "Observation of a New Particle in the Search for the Standard Model Higgs Boson with the ATLAS Detector at the LHC," Phys. Lett. B **716**, 1 (2012).

[24] A. Freitas, "Understanding Electroweak Physics in the Standard Model and Beyond," Nucl. Part. Phys. Proc. **273-275**, 21 (2016).

[25] S. Weinberg, "Phenomenological Lagrangians," Physica A **96**, 327 (1979).

[26] M. Gell-Mann, R.J. Oakes and B. Renner, "Behavior of Current Divergences under SU(3) x SU(3)," Phys. Rev. **175**, 2195 (1968).

[27] Y. Nambu, "Quasi-particles and Gauge Invariance in the Theory of Superconductivity," Phys. Rev. **117**, 648 (1960).

[28] J. Goldstone, "Field Theories With Superconductor Solutions," Nuovo Cim. **19**, 154 (1961).

[29] J. Gasser and H. Leutwyler, "Chiral Perturbation Theory to One Loop," Annals Phys. **158**, 142 (1984).

[30] V. Bernard and U.-G. Meißner, "Chiral Perturbation Theory," Ann. Rev. Nucl. Part. Sci. **57**, 33 (2007).

[31] L. Montanet *et al.* [Particle Data Group Collaboration], "Review of Particle Properties. Particle Data Group," Phys. Rev. D **50**, 1173 (1994).

[32] G. Höhler, F. Kaiser, R. Koch and E. Pietarinen, "Handbook of Pion Nucleon Scattering", Phys. Daten **12N1**, 1 (1979).

[33] R. E. Cutkosky, C. P. Forsyth, J. B. Babcock, R. L. Kelly and R. E. Hendrick, "Pion - Nucleon Partial Wave Analysis", 4th Int. Conf. on Baryon Resonances, Toronto, Canada, Jul 14-16, 1980. Published in Baryon 1980:19 (QCD161:C45:1980).

[34] R. A. Arndt, W. J. Briscoe, I. I. Strakovsky and R. L. Workman, "Extended Partial-wave Analysis of $\pi N$ Scattering Data", Phys. Rev. C **74**, 045205 (2006).

[35] I. G. Alekseev *et al.*, "Influence of Spin-rotation Measurements on Partial-wave Analyses of Elastic Pion

Nucleon Scattering," Phys. Rev. C **55**, 2049 (1997).

[36] I. G. Alekseev *et al.*, "Measurements of Spin Rotation Parameter A in Pion Proton Elastic Scattering at 1.62 GeV/c," Phys. Lett. B **485**, 32 (2000).

[37] I. G. Alekseev *et al.*, "Measurement of the Spin Rotation parameter A in the Elastic Pion Proton Scattering at 1.43 GeV/c," Eur. Phys. J. C **45**, 383 (2006).

[38] I. G. Alekseev *et al.*, "Backward Asymmetry Measurements in the Elastic Pion-proton Scattering at Resonance Energies," Eur. Phys. J. A **39**, 163 (2009).

[39] A. V. Anisovich, E. Klempt, V. A. Nikonov, A. V. Sarantsev and U. Thoma, "Nucleon Resonances in the Fourth Resonance Region," Eur. Phys. J. A **47**, 153 (2011).

[40] A. V. Anisovich, E. Klempt, V. A. Nikonov, A. V. Sarantsev, H. Schmieden and U. Thoma, "Evidence for a Negative-parity Spin-doublet of Nucleon Resonances at 1.88 GeV," Phys. Lett. B **711**, 162 (2012).

[41] A. V. Anisovich, E. Klempt, V. A. Nikonov, A. V. Sarantsev and U. Thoma, "Evidence for a Spin-quartet of Nucleon Resonances at 2 GeV," Phys. Lett. B **711**, 167 (2012).

[42] A. V. Anisovich, R. Beck, E. Klempt, V. A. Nikonov, A. V. Sarantsev and U. Thoma, "Properties of Baryon Resonances from a Multichannel Partial Wave Analysis," Eur. Phys. J. A **48**, 15 (2012).

[43] I. Horn *et al.* [CB-ELSA Collaboration], "Evidence for a Parity Doublet $\Delta(1920)P33$ and $\Delta(1940)D33$ from $\gamma p \to p\pi^0\eta$," Phys. Rev. Lett. **101**, 202002 (2008).

[44] I. Horn *et al.* [CB-ELSA Collaboration], "Study of the Reaction $\gamma p \to p\pi^0\eta$," Eur. Phys. J. A **38**, 173 (2008).

[45] J. Ajaka *et al.*, "Simultaneous Photoproduction of $\eta$ and $\pi^0$ Mesons on the Proton," Phys. Rev. Lett. **100**, 052003 (2008).

[46] V. L. Kashevarov *et al.* [Crystal Ball at MAMI and TAPS and A2 Collaborations], "Photoproduction of $\pi^0\eta$ on Protons and the $\Delta(1700)D_{33}$ Resonance," Eur. Phys. J. A **42**, 141 (2009).

[47] E. Gutz *et al.* [CBELSA and TAPS Collaborations], "Photoproduction of Meson Pairs: First Measurement of the Polarization Observable $I^s$," Phys. Lett. B **687**, 11 (2010).

[48] E. Gutz *et al.* [CBELSA/TAPS Collaboration], "High Statistics Study of the Reaction $\gamma p \to p\pi^0\eta$," Eur. Phys. J. A **50**, 74 (2014).

[49] F. Zehr *et al.*, "Photoproduction of $\pi^0\pi^0$ and $\pi^0\pi^\pm$ Pairs off the Proton from Threshold to the Second Resonance Region," Eur. Phys. J. A **48**, 98 (2012).

[50] V. L. Kashevarov *et al.*, "Study of the $\gamma p \to \pi^0\pi^0 p$ Reaction with the Crystal Ball/TAPS at the Mainz Microton," Phys. Rev. C **85**, 064610 (2012).

[51] Y. Assafiri *et al.*, "Double $\pi^0$ Photoproduction on the Proton at GRAAL," Phys. Rev. Lett. **90**, 222001 (2003).

[52] A. V. Sarantsev *et al.* [CB-ELSA and A2-TAPS Collaborations], "New Results on the Roper Resonance and the $P_{11}$ Partial Wave", Phys. Lett. B **659**, 94 (2008).

[53] U. Thoma, M. Fuchs *et al.* [CB-ELSA Collaboration], "$N^*$ and $\Delta^*$-decays into $N\pi^0\pi^0$", Phys. Lett. B **659**, 87 (2008).

[54] A. V. Sarantsev *et al.*, "New Results on the Roper Resonance and the $P_{11}$ Partial Wave," Phys. Lett. B **659**, 94 (2008).

[55] A. Thiel *et al.* [CBELSA/TAPS Collaboration], "Three-body Nature of $N^*$ and $\Delta^*$ Resonances from Sequential Decay Chains," Phys. Rev. Lett. **114**, 091803 (2015).

[56] V. Sokhoyan *et al.* [CBELSA/TAPS Collaboration], "Data on $I^s$ and $I^c$ in $\vec{\gamma} p \to p\pi^0\pi^0$ Reveal Cascade Decays of N(1900) via N(1520)$\pi$," Phys. Lett. B **746**, 127 (2015).

[57] V. Sokhoyan *et al.* [CBELSA/TAPS Collaboration], "High-statistics Study of the Reaction $\gamma p \to p\ 2\pi^0$," Eur. Phys. J. A **51**, 95 (2015), Erratum: [Eur. Phys. J. A **51**, 187 (2015)].

[58] K. A. Olive, "Review of Particle Physics," Chin. Phys. C **40**, 100001 (2016).

[59] I. Denisenko *et al.*, "$N^*$ Decays to $N\omega$ from New Data on $\gamma p \to \omega p$," Phys. Lett. B **755**, 97 (2016).

[60] W. T. Chiang and F. Tabakin, "Completeness Rules for Spin Observables in Pseudoscalar Meson Photoproduction," Phys. Rev. C **55**, 2054 (1997).

[61] Y. Wunderlich, R. Beck and L. Tiator, "The complete-experiment Problem of Photoproduction of Pseudoscalar Mesons in a Truncated Partial-wave Analysis," Phys. Rev. C **89**, 055203 (2014).

[62] A. V. Anisovich *et al.*, "Energy-independent PWA of the Reaction $\gamma p \to K^+\Lambda$," Eur. Phys. J. A **50**, 129 (2014).

[63] W. Roberts and T. Oed, "Polarization Observables for Two-pion Production off the Nucleon" Phys. Rev. C **71**, 055201 (2005).

[64] W. Hillert, "The Bonn Electron Stretcher Accelerator ELSA: Past and Future," Eur. Phys. J. A **28S1**, 139 (2006).

[65] H. Olsen, L. C. Maximon, "Photon and Electron Polarization in High-Energy Bremsstrahlung and Pair Production with Screening," Phys. Rev. **114**, 887 (1959).

[66] W. Hillert, M. Gowin and B. Neff, "A new injector for polarized electrons at ELSA," Symposium on the Gerasimov-Drell-Hearn Sum Rule and the Nucleon Spin Structure in the Resonance Region (GDH 2000), 14-17 Jun 2000. Mainz, Germany, Proceedings. Edited by D. Drechsel and L. Tiator. Singapore, World Scientific, 2001.

[67] D. Elsner *et al.* [CBELSA and TAPS Collaborations], "Measurement of the beam asymmetry in $\eta$-photoproduction off the proton," Eur. Phys. J. A **33**, 147 (2007).

[68] H. Dutz *et al.*, "The New Bonn frozen spin target for experiments with real photons," Nucl. Instrum. Meth. A **340**, 272 (1994).

[69] R. Koniuk, N. Isgur, "Where have all the Resonances gone? An Analysis of Baryon Couplings in a Quark Model with Chromodynamics", Phys. Rev. Lett. **44**,

845 (1980).

[70] O. Bartholomy, V. Crede, H. van Pee *et al.* [CB-ELSA Collaboration], "Neutral Pion Photoproduction off Protons in the Energy Range 0.3 GeV < $E_\gamma$ < 3 GeV," Phys. Rev. Lett. **94**, 012003 (2005).

[71] H. van Pee, O. Bartholomy *et al.* [CB-ELSA Collaboration], "Photoproduction of $\pi^0$-mesons off Protons from the $\Delta(1232)$ region to $E_\gamma = 3$ GeV," Eur. Phys. J. A **31**, 61 (2007).

[72] V. Crede *et al.* [CBELSA/TAPS Collaboration], "Photoproduction of Neutral Pions off Protons", Phys. Rev. C **84**, 055203 (2011).

[73] V. Crede, O. Bartholomy *et al.* [CB-ELSA Collaboration], "Photoproduction of $\eta$ Mesons off Protons for 0.75 GeV < $E_\gamma$ < 3 GeV," Phys. Rev. Lett. **94**, 012004 (2005).

[74] O. Bartholomy *et al.* [CB-ELSA Collaboration], "Photoproduction of $\eta$-mesons off Protons," Eur. Phys. J. A **33**, 133 (2007).

[75] F. Renard *et al.* [GRAAL Collaboration], "Differential Cross-section Measurement of $\eta$ Photoproduction on the Proton from Threshold to 1100 MeV," Phys. Lett. B **528**, 215 (2002).

[76] M. Dugger *et al.* [CLAS Collaboration], "Eta Photoproduction on the Proton for Photon Energies from 0.75 GeV to 1.95 GeV," Phys. Rev. Lett. **89**, 222002 (2002), Erratum: [Phys. Rev. Lett. **89**, 249904 (2002)].

[77] V. Crede *et al.* [CBELSA/TAPS Collaboration], "Photoproduction of $\eta$ and $\eta'$ mesons off protons", Phys. Rev. C **80**, 055202 (2009).

[78] E. F. McNicoll *et al.* [Crystal Ball at MAMI Collaboration], "Study of the $\gamma p \to \eta p$ Reaction with the Crystal Ball Detector at the Mainz Microtron (MAMI-C)," Phys. Rev. C **82**, 035208 (2010), Erratum: [Phys. Rev. C **84**, 029901 (2011)].

[79] B. Krusche *et al.*, "New Threshold Photoproduction of $\eta$ Mesons off the Proton," Phys. Rev. Lett. **74**, 3736 (1995).

[80] R. Bradford *et al.* [CLAS Collaboration], "Differential Cross Sections for $\gamma + p \to K^+ + Y$ for $\Lambda$ and $\Sigma^0$ Hperons," Phys. Rev. C **73**, 035202 (2006).

[81] A.V. Anisovich, V. Kleber, E. Klempt, V.A. Nikonov, A.V. Sarantsev and U. Thoma, "Baryon Resonances and Polarization Transfer in Hyperon Photoproduction," Eur. Phys. J. A **34**, 243 (2007).

[82] Evgeny Golovach, Viktor Mokeev, and the CLAS Collaboration, "CLAS data on $\gamma p \to \pi^+\pi^- p$", private communication to BnGa.

[83] Volker Crede and the CLAS Collaboration, "CLAS polarization data on $\gamma p \to \pi^+\pi^- p$", private communication to BnGa.

[84] R. Ewald *et al.* [CBELSA/TAPS Collaboration], "Measurement of Polarization Observables in $K^0_s\Sigma^+$ Photoproduction off the Proton," Phys. Lett. B **738**, 268 (2014).

[85] E. Aker *et al.* [Crystal Barrel Collaboration], "The Crystal Barrel Spectrometer at LEAR," Nucl. Instrum. Meth. A **321**, 69 (1992).

[86] A. Thiel *et al.*, "Well-established Nucleon Resonances Revisited by Double-polarization Measurements," Phys. Rev. Lett. **109**, 102001 (2012).

[87] A. Thiel *et al.* [CBELSA/TAPS Collaboration], "Double-polarization Observable G in Neutral-pion Photoproduction off the Proton," arXiv:1604.02922 [nucl-ex].

[88] A. Wilson *et al.* [CBELSA/TAPS Collaboration], "Photoproduction of $\omega$ mesons off the proton," Phys. Lett. B **749**, 407 (2015).

[89] J. Junkersfeld *et al.* [CB-ELSA Collaboration], "Photoproduction of $\pi^0\omega$ off Protons for $E_\gamma < 3$ GeV," Eur. Phys. J. A **31**, 365 (2007).

[90] D. Elsner *et al.*, "Linearly Polarized Photon Beams at ELSA and Measurement of the Beam Asymmetry in $\pi^0$-photoproduction off the Proton," Eur. Phys. J. A **39**, 373 (2009).

[91] N. Sparks *et al.* [CBELSA/TAPS Collaboration], "Measurement of the Beam Asymmetry $\Sigma$ in the Forward Direction for $\vec{\gamma} p \to p\pi^0$ Photoproduction", Phys. Rev. C **81**, 065210 (2010).

[92] F. Klein *et al.* [CBELSA/TAPS Collaboration], "Beam Asymmetry in Near Threshold $\omega$ Photoproduction off the Proton," Phys. Rev. D **78**, 117101 (2008).

[93] M. Hilt, B. C. Lehnhart, S. Scherer and L. Tiator, "Pion Photo- and Electroproduction in Relativistic Baryon Chiral Perturbation Theory and the Chiral MAID Interface," Phys. Rev. C **88**, 055207 (2013).

[94] R. L. Workman, M. W. Paris, W. J. Briscoe and I. I. .Strakovsky, "Unified Chew-Mandelstam SAID Analysis of Pion Photoproduction Data," Phys. Rev. C **86**, 015202 (2012).

[95] D. Rönchen *et al.*, "Photocouplings at the Pole from Pion Photoproduction," Eur. Phys. J. A **50**, 101 (2014) Erratum: [Eur. Phys. J. A **51**, 63 (2015)].

[96] O. Bartalini *et al.*, "Measurement of $\pi^0$ Photoproduction on the Proton from 550 MeV to 1500 MeV at GRAAL," Eur. Phys. J. A **26**, 399 (2005).

[97] R. L. Workman, private communication.

[98] D. Roenchen, private communication.

[99] A. V. Anisovich *et al.*, "The Impact of New Polarization Data from Bonn, Mainz and Jefferson Laboratory on $\gamma p \to \pi N$ Multipoles," Eur. Phys. J. A **52**, 284 (2016).

[100] M. Gottschall *et al.* [CBELSA/TAPS Collaboration], "First Measurement of the Helicity Asymmetry for $\gamma p \to p\pi^0$ in the Resonance Region," Phys. Rev. Lett. **112**, 012003 (2014).

[101] M. Gottschall *et al.* [CBELSA/TAPS Collaboration], "Double Polarization Observable E for the Reaction $\gamma p \to \pi^0 p$", in preparation.

[102] J. Hartmann *et al.*, "The $N(1520)3/2^-$ Helicity Amplitudes from an Energy-independent Multipole Analysis Based on new Polarization Data on Photoproduction of Neutral Pions," Phys. Rev. Lett. **113**, 062001 (2014).

[103] J. Hartmann *et al.* [CBELSA/TAPS Collaboration], "The Polarization Observables T , P , and H and their Impact on $\gamma p \to p\pi^0$ Multipoles," Phys. Lett. B **748**,

212 (2015).

[104] V.G. Gorbenko *et al.*, Pisma Zh. Eksp. Teor. Fiz. **19**, 659 (1974), V.G. Gorbenko *et al.*, Yad. Fiz. **27**, 1204 (1978), A.A. Belyaev *et al.*, Nucl. Phys. B **213**, 201 (1983), J.O. Maloy, Ph.D. Thesis, 1961. 141pp, V.G. Gorbenko *et al.*, Pisma Zh. Eksp. Teor. Fiz. **22**, 393 (1975), P.J. Bussey *et al.*, Nucl. Phys. B **154**, 492 (1979), S. Kato *et al.*, Nucl. Phys. B **168**, 1 (1980), A.S. Bratashevsky *et al.*, Nucl. Phys. B **166**, 525 (1980), A.S. Bratashevsky *et al.*, Ukr. Fiz. Zh. (Russ.Ed.) **31**, 1306 (1986)

[105] Y. Wunderlich, F. Afzal, A. Thiel, and R. Beck, "Determining the dominant partial wave contributions from angular distributions of single- and double-polarization observables in pseudoscalar meson photoproduction," arXiv:1611.01031 [physics.data-an].

[106] PhD thesis Yannik Wunderlich, Bonn University, in prep.

[107] G.F. Chew, M.L. Goldberger, F.E. Low, Y. Nambu, "Relativistic Dispersion Relation Approach to Photomeson Production," Phys. Rev. **106**, 1345 (1957).

[108] A. M. Sandorfi, S. Hoblit, H. Kamano and T.-S. H. Lee, "Determining Pseudoscalar Meson Photoproduction Amplitudes from Complete Experiments," J. Phys. G **38**, 053001 (2011).

[109] M. Abramowitz and I.A. Stegun, Handbook of Mathematical Functions, Dover Publishing (1972).

[110] J. Müller, J. Hartmann, M. Grüner *et al.* [CBELSA/TAPS Collaboration], "New Data on $\vec{\gamma}\vec{p} \to \eta p$ with Polarized Photons and Protons and their Implications for $N^* \to N\eta$ Decays", subm. to Phys. Lett. B.

[111] PhD thesis Jonas Müller, Bonn University, in prep.

[112] PhD thesis Marcus Grüner, Bonn University, 2016

[113] PhD thesis Jan Hartmann, Bonn University, in prep.

[114] C. S. Akondi *et al.*, "Measurement of the Transverse Target and Beam-target Asymmetries in $\eta$ Meson Photoproduction at MAMI," Phys. Rev. Lett. **113**, 102001 (2014).

[115] I. Senderovich *et al.* [CLAS Collaboration], "First Measurement of the Helicity Asymmetry E in $\eta$ Photoproduction on the Proton," Phys. Lett. B **775**, 64 (2016).

[116] N. Kaiser, P. B. Siegel and W. Weise, "Chiral Dynamics and the S11 (1535) Nucleon Resonance," Phys. Lett. B **362**, 23 (1995).

[117] M. Mai, P. C. Bruns and U.-G. Meißner, "Pion Photoproduction off the Proton in a Gauge-invariant Chiral Unitary Framework," Phys. Rev. D **86**, 094033 (2012).

[118] B. S. Zou, "On the Nature of the Lowest $1/2^-$ Baryon Nonet and Decuplet," Eur. Phys. J. A **35**, 325 (2008).

[119] E. Gutz *et al.* [CBELSA/TAPS Collaboration], "Measurement of the Beam Asymmetry Σ in $\pi^0\eta$ Production off the Proton with the CBELSA/TAPS Experiment", Eur. Phys. J. A **35**, 291 (2008).

[120] A. Fix, V. L. Kashevarov, A. Lee and M. Ostrick, "Isobar Model Analysis of $\pi^0 - \eta$ Photoproduction on Protons," Phys. Rev. C **82**, 035207 (2010).

[121] A. Fix and H. Arenhövel, "Polarization Observables in $\pi^0\eta$-photoproduction on the Proton," Phys. Rev. C **83**, 015503 (2011).

[122] M. Döring, E. Oset and U.-G. Meißner, "Evaluation of the Polarization Observables $I^S$ and $I^C$ in the Reaction $\gamma p \to \pi^0\eta p$," Eur. Phys. J. A **46**, 315 (2010).

[123] H. Arenhovel, W. Leidemann, E. L. Tomusiak, "General Survey of Polarization Observables in Deuteron Electrodisintegration," Eur. Phys. J. A **23**, 147 (2005).

[124] V. Kuznetsov *et al.*, "Evidence for a Narrow Structure at $W \approx 1.68$ GeV in $\eta$ Photoproduction off the Neutron," Phys. Lett. B **647**, 23 (2007).

[125] I. Jaegle *et al.* [CBELSA/TAPS Collaboration], "Quasi-free Photoproduction of $\eta$-mesons of the Neutron", Phys. Rev. Lett. **100**, 252002 (2008).

[126] I. Jaegle *et al.* [CBELSA/TAPS Collaboration], "Photoproduction of $\eta'$-mesons off the Deuteron", Eur. Phys. J. A **47** 11 (2011).

[127] I. Jaegle *et al.* [CBELSA/TAPS Collaboration], "Quasi-free Photoproduction of $\eta$-mesons off the Deuteron", Eur. Phys. J. A **47** 89 (2011).

[128] D. Werthmüller *et al.* [A2 Collaboration], "Narrow Structure in the Excitation Function of $\eta$ Photoproduction off the Neutron," Phys. Rev. Lett. **111**, 232001 (2013).

[129] D. Werthmüller *et al.* [A2 Collaboration], "Quasi-free Photoproduction of $\eta$ Mesons off Protons and Neutrons," Phys. Rev. C **90**, 015205 (2014).

[130] K. S. Choi, S. i. Nam, A. Hosaka and H. C. Kim, "A New $N^*(1675)$ Resonance in the $\gamma N \to \eta N$ reaction," Phys. Lett. B **636**, 253 (2006).

[131] V. Shklyar, H. Lenske and U. Mosel, "Eta-photoproduction in the Resonance Energy Region," Phys. Lett. B **650**, 172 (2007).

[132] R. Shyam and O. Scholten, "Photoproduction of $\eta$ Meson within a Coupled-channels K-matrix Approach," Phys. Rev. C **78**, 065201 (2008).

[133] A. V. Anisovich, I. Jaegle, E. Klempt, B. Krusche, V. A. Nikonov, A. V. Sarantsev and U. Thoma, "Photoproduction of $\eta$ mesons off Neutrons from a Deuteron Target," Eur. Phys. J. A **41**, 13 (2009).

[134] M. Doring and K. Nakayama, "On the Cross Section Ratio $\sigma(n)/\sigma(p)$ in $\eta$ Photoproduction," Phys. Lett. B **683**, 145 (2010).

[135] A. V. Anisovich, E. Klempt, V. Kuznetsov, V. A. Nikonov, M. V. Polyakov, A. V. Sarantsev and U. Thoma, "Study of the Narrow Structure at 1685 MeV in $\gamma p \to \eta p$," Phys. Lett. B **719**, 89 (2013).

[136] A. V. Anisovich, E. Klempt, B. Krusche, V. A. Nikonov, A. V. Sarantsev, U. Thoma and D. Werthmüller, "Interference Phenomena in the $J^P = 1/2^-$ Wave in $\eta$ Photoproduction," Eur. Phys. J. A **51**, 72 (2015).

[137] A.V. Anisovich, V. Burkert, N. Compton, K. Hicks, E. Klempt, V.A. Nikonov, A.V. Sarantsev, and U. Thoma, "Neutron helicity amplitudes", in preparation.

[138] PhD thesis Philipp Mahlberg, Bonn University, in preparation.

[139] PhD thesis Tobias Seifen, Bonn University, in preparation.

[140] K. M. Watson, "The Hypothesis of Charge Independence for Nuclear Phenomena," Phys. Rev. **85**, 852 (1952).

# Partial wave analysis

*Eberhard* Klempt[1,*], *Andrey V.* Sarantsev[1,2,**], and *Ulrike* Thoma[1,***]

[1] *Helmholtz–Institut für Strahlen– und Kernphysik, Universität Bonn, 53115 Bonn, Germany*
[2] *National Research Centre "Kurchatov Institute", Petersburg Nuclear Physics Institute, Gatchina, Russia*

**Abstract.** The project A.2 of the SFB/TR16 "Subnuclear Structure of Matter" aimed at a combined multichannel partial wave analysis of all or of most data on pion and photo-induced reactions for invariant masses from the Δ(1232) region to 2300 MeV. In this review the most important results are presented. It is shown that the data and their partial wave analysis have improved our understanding of the dynamics of excited light-quark baryons significantly.

## 1 Introduction

"Why baryons?" asked Nathan Isgur [1] at NSTAR2000, and gave three answers: Because nucleons are the stuff of which our world is made, because they are the simplest system in which the non-abelian character of QCD is manifest, and because their complexity - compared to mesons - may reveal physics hidden from us in mesons. And he predicted that baryon spectroscopy will be one of the most interesting and fruitful areas for at least thirty years. Recent reviews of baryon spectroscopy by Klempt and Richard [2] and Crede and Roberts [3] underline the interest in the field.

Isgur's high expectations were based on severe discrepancies between the experimental findings and the spectrum of nucleon excitations predicted in quark models (e.g. [4–9]) or in recent calculations within lattice QCD [10] (even though these calculations still use quark masses corresponding to a large pion mass, $m_\pi$ =396 MeV).

In 2000, the knowledge of the excitation spectrum of the nucleon stemmed essentially from data on $\pi N$ elastic scattering [11–13]. In quark models, the predicted spectrum is organized in excitation shells, and a shell structure is clearly seen in the experimental spectrum. However, the resonance spectrum which emerged exhibited two severe problems:

**A)** The number of states expected for a system with three constituent quarks exceeded by far the number of observed states.

The authors of, e.g., ref. [6] list 66 $N^*$ and 35 $\Delta^*$ resonances predicted to have masses below 2.5 GeV. However, the Review of Particle Properties of 2000 [14] quotes only 19 $N^*$ and 19 $\Delta^*$ in this mass range. And, even worse, Arndt and his collaborators [15] found only 8 $N^*$ and 10 $\Delta^*$-states. This latter analysis included high precision data from the meson factories at LAMPF, PSI, and TRIUMF. A strong argument in favor of the Arndt *et al.* analysis was the correct prediction of spin rotation parameters [16–18] and of the backward asymmetry [19] in the elastic pion-proton scattering from ITEP/PNPI; the predictions from [11, 12] show clear discrepancies with the data.

Clearly, not all 101 resonances expected below 2500 MeV need to be found to reach a good unterstanding of the spectrum. In the first excitation shell 5 negative-parity $N^*$'s and 2 $\Delta^*$'s are predicted and experimentally established. In the second shell, 14 positive-parity $N^*$'s and 8 $\Delta^*$'s are predicted; their calculated masses fall below 2100 MeV. Based on [11, 12] (or [15]), only 8 (or 4) positive-parity $N^*$'s and 6 (or 5) $\Delta^*$'s are actually observed. Only 4 (instead of 14) positive-parity $N^*$'s are *safe*: this small number casts into doubt if models based on constituent quarks with some residual interactions are the right degrees of freedom to describe the nucleon excitation spectrum.

The problem of missing baryon resonances is aggravated by the prediction of additional states, hybrid baryons, in which the gluonic string mediating the interaction between the quarks is itself excited. Hybrid baryons carry the same quantum numbers as $qqq$ baryons, $qqq$ and hybrid configurations can mix and are thus difficult to establish. In a quark model [20], the baryon hybrid spectrum (with no mixing with $qqq$ baryons) is calculated to intrude the spectrum of baryon resonances at 2 GeV and above. In lattice calculations [21], most baryons with a dominant hybrid content show up above 2500 MeV, except for $J^P = 1/2^+$ where $N(1440)1/2^+$ and $N(1710)1/2^+$ are predicted to acquire a significant hybrid component.

A well-known solution of the problem of the *missing resonances* is the conjecture that baryon excitations could be driven by the dynamics of a quark and a (quasi-stable) diquark. There is a long discussion on the nature and rel-

*e-mail: klempt@hiskp.uni-bonn.de
**e-mail: andsar@hiskp.uni-bonn.de
***e-mail: thoma@hiskp.uni-bonn.de

evance of the diquark concept (see e.g. [22–25]). This hypothesis decreases the number of expected resonances very significantly and helps to solve the problem of the *missing baryon resonances*. Santopinto, e.g., calculated the $N^*$ and $\Delta^*$ excitation spectrum [26] with the assumption that the baryon is made up from a point-like diquark and a quark. The results match data perfectly, provided $N^*$- and $\Delta^*$-resonances are omitted that had an one- or two-star ranking only in PDG2004 [27].

Alternatively, the *missing resonances* may have escaped detection due to their small coupling to the $\pi N$ channel [28]. The question arizes: are there baryon resonances with small $\pi N$ couplings which can be produced by electromagnetic excitation and which decay into other final states?

A further point of concern is the observation that

**B)** the pattern of observed states is neither compatible with quark model calculations nor with present lattice calculations.

In particular the low mass of the Roper resonance $N(1440)1/2^+$ has led to vivid discussions in the literature. Inconsistencies in different data sets led Morsch and Zupranski [29] to conjecture that the 1400 MeV region might even house two resonances, a hypothesis which we ruled out in [30]. Most papers on the Roper resonance address the question if it is a "normal" $qqq$ resonance, a low-mass hybrid baryon [31, 32], or if it is dynamically generated from meson-baryon interactions [33]. If the Roper resonance would be exotic, $N(1710)1/2^+$ would be the true first radial excitation of the nucleon. However, electroexcitation of the Roper resonance provides strong evidence for this state as a predominantly radial excitation of a 3-quark ground state [34]. Note, that any *exotic* interpretation of the Roper resonance should explain the low masses of the corresponding states in the $\Lambda$, $\Sigma$, and $\Xi$ spectra. Likewise, the $\Lambda(1405)1/2^-$ resonance has attracted intense interest since Dalitz proposed it to be generated dynamically [35]. This topic is treated in project B.3.

The generation of baryon resonances from the interaction of mesons and baryons is a highly attractive approach to understand the dynamics of baryon resonances. At present, however, the number of expected resonances is unknown, and the relation between quark model states and dynamically generated resonances is not understood. It is hence important to measure the decay modes of all major decay modes.

At higher energies, the masses of $\Delta(1900)1/2^-$, $\Delta(1940)3/2^-$, $\Delta(1930)5/2^-$ are incompatible with quark model predictions (e.g. [6]). They belong to the third excitation shell, and their masses should fall into the 2100 - 2300 MeV range. Instead, they are approximately mass degenerate with the positive-parity $\Delta(1910)1/2^+$, $\Delta(1920)3/2^+$, $\Delta(1905)5/2^+$, $\Delta(1950)5/2^+$. Such parity doublets occur frequently in hadron spectra [36]; Fig. 1 shows parity doublets in the $N^*$ spectrum. Parity doublets are observed undoubtedly. Similar observations in the meson [37] and baryon [38] sector have led to the idea that chiral symmetry might be effectively restored in highly

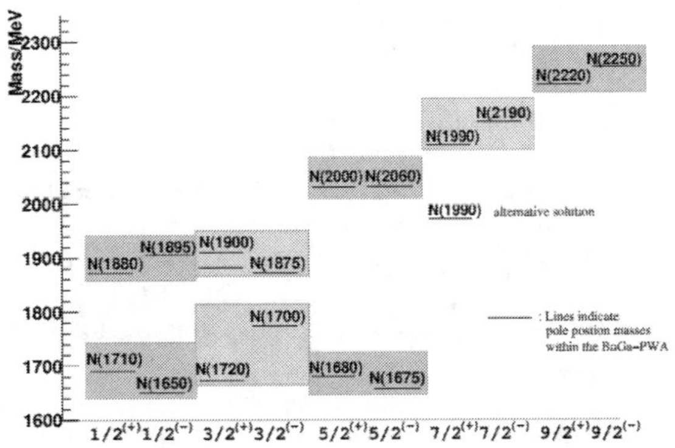


**Figure 1.** Examples of nucleon resonances organized as parity doublets. The lines indicate the masses of BnGa pole position of the respective states. For the $7/2^+$-state two alternative solutions exist: one close to the $N(2190)7/2^-$, the other one at a lower mass, both are shown. The pole masses of $N(1720)3/2^+$ and $N(1700)3/2^-$ are found 110 MeV apart, all other parity doublets indicated have a significantly smaller mass splitting.

excited hadrons. If this would indeed be the case chiral partners were expected to exist for all high-mass states. Independently of the idea of chiral symmetry restoration, the question whether or not parity partners exist for all high-mass states is an interesting question to be answered experimentally.

Photoproduction of baryon resonances is hence studied to find answers to the following questions:

1. Is the number of observed resonances compatible with diquark models?
2. Is their evidence for three-quark dynamics in excited baryon states?
3. What is the dynamical reason for the existence of parity doublets?
4. Which resonances are seen in the third excitation shell?
5. Can baryon resonances be assigned to specific $SU(6) \otimes O(3)$ multiplets?

Based on the new data taken during the SFB/TR16 funding period, further questions arose:

6. Is photoproduction of $\omega$ mesons governed by diffraction, or play resonances a significant role?
7. Is the narrow peak seen in the reaction $\gamma n \to \eta n$ a resonance?
8. What is the relation between quark-model resonances and dynamically generated resonances?
9. Do the new photoproduction data with polarized photon beams and polarized targets lead to consistent multipoles when analyzed with different PWA methods?
10. Can photoproduction multipoles be determined in slices of fixed masses?

Photoproduction data were taken within this SFB/TR16 (Projects A.1 "Spectroscopy of baryon resonances", B.1 "Protoproduction of mesons"). These data and data from other laboratories provide important information used in the comprehensive analysis of photo- and pion-induced reactions performed within the Bonn-Gatchina partial wave analysis (BnGa-PWA) (project A.2 within the SFB/TR16) presented here. The importance of the recent single and double polarization observables becoming available from ELSA, JLab, and MAMI cannot be overestimated. They provide very important constraints for the PWA.

## 2 The Bonn-Gatchina PWA, data base and method

As composite systems, nucleons exhibit a rich spectrum of radial and orbital excitations. The excitation levels of the nucleon are extremely short-lived and decay in a variety of different decay modes. Many states are predicted which overlap and are very difficult to resolve. Hence the aim to identify all $N^*$ and $\Delta^*$ resonances contributing to photoproduction and to measure all important decay modes requires a simultaneous fit to a large body of data in a multi-channel partial wave analysis. The following data were included in the analysis:

1. Data on the photoproduction off protons and off neutrons obtained at ELSA and at other laboratories. This includes data on single as well as on double meson photoproduction. The latter data are fitted in terms of an event-based maximum likelihood fit to take all the correlations between the variables the event depends on properly into account.
2. Real and imaginary part of the partial wave amplitudes for $\pi N$ elastic scattering either from Höhler's group or from Arndt's group.
3. Pion-induced inelastic reactions.

The latter data (2) and (3) ensure consistency with the $\pi N$-scattering data and allow the extraction of the $\pi N$ coupling strength of baryon resonances. Fits to photoproduction data alone would provide only access to the product of the photon helicity amplitudes $A_{1/2}$, $A_{3/2}$ and the square root of the decay branching ratio into the specific final state, $A_{1/2\,or\,3/2}\cdot\sqrt{BR_{final\,i}}$ with $BR_{final\,i} = \Gamma_{final\,i}/\Gamma_{tot}$ (with $\Gamma_{final\,i}$ being the partial decay width into the final state i and $\Gamma_{tot}$ the total width of the resonance).

The following reactions are presently included in the BnGa-PWA. (Most data can be found via the web page of the GWU PWA group [39]):

| Photoproduction | $\pi N$-scattering |
|---|---|
| $\gamma p \to \pi^0 p,\ \pi^+ n$ | $\pi^- p \to \pi^- p$ (elastic)-partial waves |
| $\gamma p \to \eta p,\ \eta' p$ | $\pi^- p \to \eta n$ |
| $\gamma p \to \omega p,\ K^{*+}\Lambda$ | |
| $\gamma p \to K^+\Lambda$ | $\pi^- p \to K^0\Lambda,\ K^+\Sigma^-,\ K^0\Sigma^0,\ K^-\Sigma^+$ |
| $\gamma p \to K^{+/0}\Sigma^{0/+}$ | $\pi^+ p \to K^+\Sigma^+$ |
| $\gamma p \to p\pi^0\pi^0,\ p\pi^+\pi^-$ | $\pi^- p \to n\pi^0\pi^0,\ n\pi^+\pi^-$ |
| $\gamma p \to p\pi^0\eta$ | |
| $\gamma n \to \pi^0 n,\ \pi^- p$ | |
| $\gamma n \to \eta n$ | |

This data base has been extended continuously over the last 12 years. In particular the new polarization data were included in the fits whenever they became available. Very recently also (so far unpublished) data on photoproduction of two charged pions (CLAS) were added.

A multi-channel analysis of a large body of data has a large advantage; in contrast to an analysis of $\pi N \to \pi N$ elastic scattering alone, the simultaneously fitted photoproduction data provide access to resonances which might have escaped detection in $\pi N$ scattering due to their insufficient $\pi N$-coupling. At the same time the photoproduction data constrain the elastic amplitudes by fixing the structure of the otherwise unknown inelasticities. Polarization observables measured in photoproduction reactions provide sensitive information on the interferences of different partial waves. This helps to extract even small resonance contributions from the data. Simultaneous fits to different single and double polarization observables exploit different interference effects and are hence very sensitive to small amplitudes.

The fit method has been documented in a series of papers [40–43]. Here we just mention that the amplitudes for pion- and photo-induced reactions are represented by a sum of K-matrix partial wave amplitudes and by background amplitudes describing $u$- and $t$-channel exchanges. The corresponding exchange amplitudes are written in the form of the exchange of Regge trajectories. The K-matrix amplitudes contain poles and simple functions of $s$ to model additional background. Methods for the decomposition of the background amplitudes from $t$- and $u$-channel exchanges into partial waves have been developed and implemented. These allow the direct comparison of our partial waves (or multipoles) with those of other analyses as discussed, e.g., in section 3.9.

Due to the so far still incomplete data base, different solutions are found which describe the data with similar quality. We always quote errors which cover all solutions or we discuss different classes of solutions in which the number and the properties of some resonances may differ [44–49].

It should also be mentioned, that within the BnGa-approach, we do not attempt to provide a microscopical description of the background starting from a Lagrangian like it is done by other groups like Argonne-Osaka [50], Jülich-Bonn [51–56], Dubna-Mainz-Taipeh [57], EBAC-Jlab [58], Valencia [59], or within the Gießen Coupled Channel analysis [60, 61] (see also contribution of project B.7 "Coupled channels approach to photo-meson production on the nucleon" to these proceedings). These models have the advantage that the number of free parameters for the background is minimized while problems might occur if the dynamical model is incomplete and the fit is constrained to it.

A purely phenomenological model as the BnGa-PWA satisfying basic requirements is hence a meaningful tool to access properties of resonances. The method employed here satisfies unitarity and the amplitudes are analytic functions in the complex energy plane. The method has recently been improved by introducing a $N/D$-based method in our fitting [62]. Studies on baryons indicate that the effects on the amplitudes are mostly small, and that changes in the resonance parameters are covered by the errors we have quoted in our analyses based on the K-matrix formalism.

## 3 Results

### 3.1 Can we rule out diquark models?

The Particle Data Group (PDG) summarizes biannually the status of particle physics. Their Review of Particle Physics (RPP) contains a listing of known baryon resonances and their properties. Before the 2007 issue, the information on nucleon and $\Delta$ excitations came nearly entirely from the groups led by Höhler [11], Cutkosky [12], and Arndt [15] analyzing data on $\pi N$ elastic and charge exchange scattering, with a few additions from Manley [63]. However, the GWU group reduced continuously the number of observed resonances required in their fits [13, 15, 64]. Before 2006, data on photoproduction added information on the helicity couplings of resonances listed in the RPP, but the resonances were introduced into the respective analyses with fixed masses and widths. Early results of the BnGa group using data on photoproduction from the CB-ELSA experiment [65–68] were not yet included in the RPP [44, 69] or not yet used to define particle properties [30, 70]

The BnGa group was the first one which combined pion and photoproduced reactions in a coupled channel fit, fitted masses and widths freely *and* searched systematically for new resonances [46–49]. In this way, the baryon resonances dismissed in the Arndt *et al.* analysis [15] could be re-established [46]. In [49], we have reported the observation of 21 $N^*$ and of 10 $\Delta^*$ resonances and determined their masses, width, and a number of decay modes. Six nucleon resonances were *new* in the sense that they had no own PDG entry before (even though earlier experiments may also have found evidence for their existence). These results entered the RPP2012 [71].

Table 1 lists the resonances found in the BnGa analysis which were *new* or for which the star rating was increased.

**Table 1.** New states found by the BnGa PWA, which have been introduced in the new PDG'2012, or states in which the BnGa PWA improved the star rating of the resonance. "no evidence" in the GWU Arndt'06 analysis indicates states given in PDG'2010 which were not confirmed in the GWU Arndt'06-analysis.

| | RPP 2010 | our analyses | RPP 2012 | GWU'06 |
|---|---|---|---|---|
| $N(1860)5/2^+$ | | * | ** | |
| $N(1875)3/2^-$ | | *** | *** | |
| $N(1880)1/2^+$ | | ** | ** | |
| $N(1895)1/2^-$ | | ** | ** | |
| $N(1900)3/2^+$ | ** | *** | *** | no evidence |
| $N(2060)5/2^-$ | | *** | ** | |
| $N(2150)3/2^-$ | | ** | ** | |
| $\Delta(1940)3/2^-$ | * | ** | ** | no evidence |

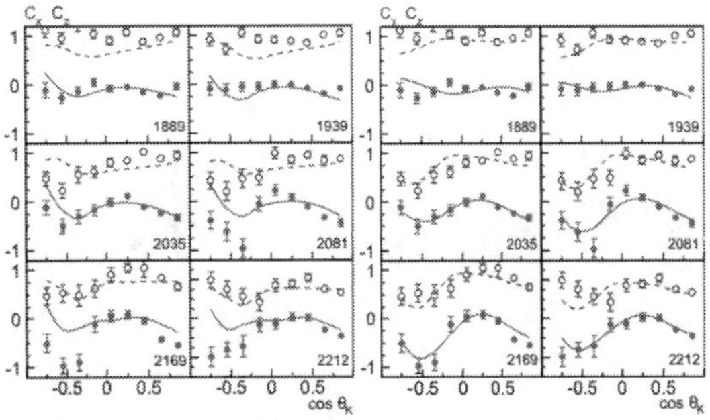


**Figure 2.** Double polarization observables $C_x$ (red points) and $C_z$ (open blue circles) for $\gamma p \to K^+\Lambda$ [72]. The solid ($C_x$) and dashed ($C_z$) curves are our result obtained without (left panel) and with the $N(1900)3/2^+$ state (right panel) included in the fit.

When comparing RPP'2000 with RPP'2010, we notice that not a single new baryon resonance entered the RPP during these 10 years. With the new high-quality photoproduction data, including single and double polarization observables, the situation has now changed dramatically.

The evidence for one of the states, $N(1900)3/2^+$, can be seen in Fig. 2. The figure shows the CLAS beam-recoil polarization data ($C_x$, $C_z$) for the reaction $\gamma p \to K^+\Lambda$ (selected bins only) [72] in comparison to the best fit obtained without and with $N(1900)3/2^+$ included in the PWA [73]. The improvement due to the $N(1900)3/2^+$ is clearly visible.

The number of observed states is now definitely larger than expected from diquark models and supports symmetric quark models in which three constituent quarks participate in the dynamics of excited states.

### 3.2 Baryon cascade decays reveal three-quark dynamics

Three-body decays of baryon resonances give access to sequential decays of baryon resonances via intermediate baryon or meson resonances. The photoproduction reactions $\gamma p \to \pi^0\pi^0 p$ [74] and $\gamma p \to \pi^0\eta p$ [75, 76] were included in the PWA using an event-based maximum likelihood fit to properly consider all the correlations the reaction depends on in the given 5-dimensional phase space. The data were taken at ELSA with the CBELSA/TAPS experiment within project A.1. Data from GRAAL [77] and MAMI [78] on $\gamma p \to \pi^0\pi^0 p$ were included in the analysis, the MAMI data on $\gamma p \to \pi^0\eta p$ [79] were added to the data base later.

A first example can be seen in Fig. 3. In the reaction $\gamma p \to \pi^0\eta p$ and for a mass range of $2000 \pm 100$ MeV masses, $\Delta^*$ decays into $\Delta(1232)\eta$ are observed as well as $N^*$ or $\Delta^*$ decays into $N a_0(980)$. Contributions from the isobar $N(1535)1/2^-\pi$ emerge clearly, in particular in the $2200\pm 100$ MeV mass range. A partial wave analysis gives the contributing resonances and their decay frequencies into these isobars [76]. The most significant contribution stems from a $\Delta(1232)\eta$ threshold enhancement and is assigned to $\Delta(1700)3/2^-$ production. But also $\Delta(1940)3/2^-$ decays significantly into the $\Delta(1232)\eta$ isobar: because of

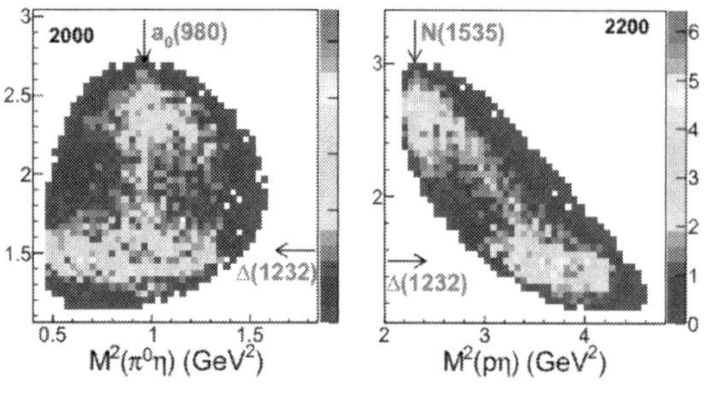


**Figure 3.** Dalitz plots (taken from [76]) for the reaction $\gamma p \to \pi^0\eta p$ for the mass range $2000 \pm 100$ MeV and $2200 \pm 100$ MeV (CBELSA/TAPS-data). Evidence is seen for the intermediate states $\Delta(1232)\eta$, $p\,a_0(980)$, and $N(1535)1/2^-\pi$.

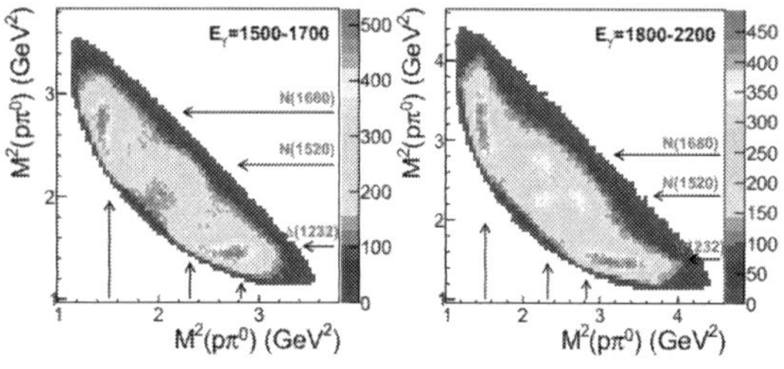


**Figure 4.** Dalitz plots for the reaction $\gamma p \to p\pi^0\pi^0$ (CBELSA/TAPS) [74]. Shown are examples from two energy bins. the pictures are taken from [74], which also provides firther details).

this observation, the PDG upgraded the $\Delta(1940)3/2^-$ resonance from one star to two stars.

Fig. 4 shows two Dalitz plots on $\gamma p \to \pi^0\pi^0 p$. Several isobars are seen. Higher-mass baryon resonances seemingly prefer to decay into other high-mass resonances with a low decay momentum, instead of choosing a high-momentum decay into a low-mass meson and a nucleon: Apparently, the intrinsic wave functions of resonances do not contain large components with high momenta. QCD likes to invest into mass and not only into momentum.

The analysis of the data allows us to extract branching ratios for the decays of resonances into $\Delta\pi$, $\Delta\eta$, $N(\pi\pi)_S$, $N(1440)1/2^+\pi$, $N(1520)3/2^-\pi$, $N(1535)1/2^-\pi$, and $N(1680)5/2^+\pi$. In total, more than 400 properties of

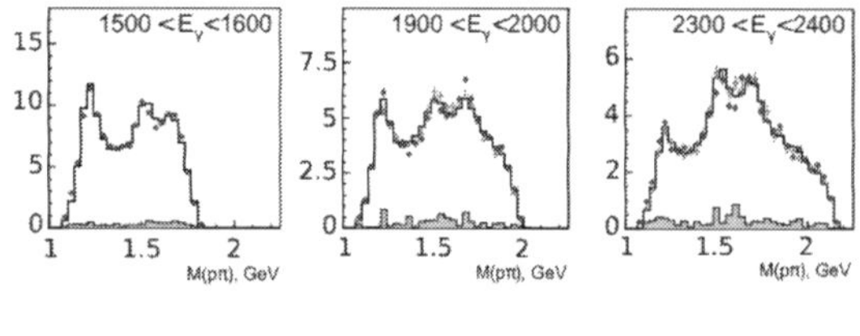


**Figure 5.** $d\sigma/dM(p\pi^0)$ for $\gamma p \to p\pi^0\pi^0$: examples for three $p\pi^0$-invariant mass bins. The blue an red data points show two different CBELSA/TAPS data sets. The systematic uncertainty of the data is shown as grey band. In addition, there is a normalization uncertainty of 10% [74]. The result of the BnGa-PWA is shown as black curve.

baryon resonances were determined, many of them were obtained for the first time. They will be listed in the 2016 edition of the RPP. Fig. 5 shows invariant masses in comparison to the BnGa fit.

In the 1900 MeV region, for $1300 < E_\gamma < 1650$ MeV, the cascade $N^* \to N(1520)\pi^0 \to p\pi^0\pi^0$ is dominated by the $J^P = 3/2^+$ partial wave. This can be seen by plotting the variables $I_s, I_c$ defined by

$$\frac{d\sigma}{d\Omega} = \left(\frac{d\sigma}{d\Omega}\right)_0 \{1 + P_l[I^s \sin(2\phi) + I^c \cos(2\phi)]\}. \quad (1)$$

$I_s, I_c$ carry information on the spin density matrix of the $N(1520)3/2^-$ isobar. The variables can be plotted as functions of $\phi^*$ [74, 80] and are shown repeatedly in Fig. 6. Superimposed are two-parameter fits assuming different partial waves in the initial state. The coupling strength and the fractional contribution of the higher partial-wave contribution are varied in the fits. Only one fit gives a reasonable fit to the data; it assumes $J^P = 3/2^+$ in the initial state. This is an important result. It is shown that in this cascade, one partial wave is dominant. The result is fully model independent and has been achieved without any partial wave analysis just calculating the respective angular distributions!

The decays of the $N^*$ and $\Delta^*$-resonances reveal evidence for the three-particle dynamics of excited nucleon resonances. $\Delta(1910)1/2^+$, $\Delta(1920)3/2^+$, $\Delta(1905)5/2^+$, and $\Delta(1950)7/2^+$ show rather small de-

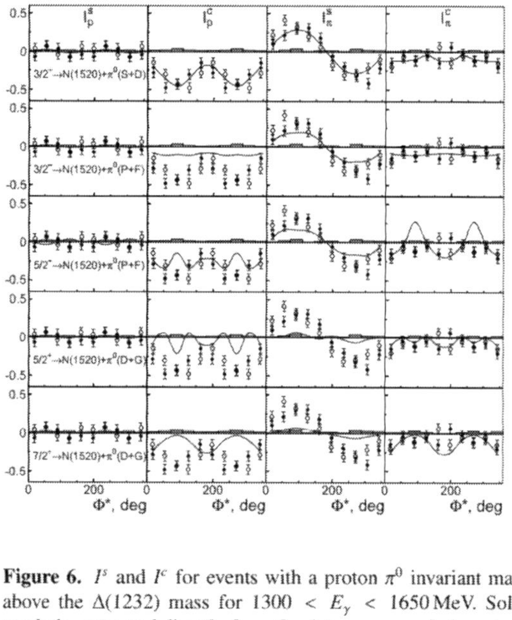


**Figure 6.** $I^s$ and $I^c$ for events with a proton $\pi^0$ invariant mass above the $\Delta(1232)$ mass for $1300 < E_\gamma < 1650$ MeV. Solid symbols: extracted directly from the data, open symbols: mirror points due to symmetry. The grey areas represent the systematic errors. The data are shown repeatedly and compared to simulated distributions for the reaction chains $(J^P) \to N(1520)3/2^- + \pi^0$. Distributions for $(1/2^\pm) \to N(1520)3/2^- + \pi^0$ vanish and are not shown. In the fits, the systematic errors are added quadratically to the statistical errors. Picture taken from [80] which provides further details.

cay fractions into the intermediate orbitally ($\ell = 1$) excited baryon resonances $N(1520)3/2^-$, $N(1535)1/2^-$, $N(1680)5/2^+$ and a pion. At the same time $N(1880)1/2^+$, $N(1900)3/2^+$, $N(2000)5/2^+$, and $N(1990)7/2^+$ decay into $N(1520)3/2^-\pi$, $N(1535)1/2^-\pi$, $N(1680)5/2^+\pi$, $N\sigma$ with a significant fraction. Since the individual branching ratios have large error bars [81]: the data were re-fitted using two assumptions:

1. Decays of the four $\Delta^*$'s into $N(1440)\pi$, $N(1520)\pi$, $N(1535)\pi$, and $N(1680)\pi$ were forbidden. This has little effect on the fit and the $\chi^2$ deteriorated by 692 units. We consider this to be at the border of becoming statistically significant. Of course, these decays are not forbidden but obviously, the branching ratios for these decay modes from the four $\Delta^*$'s are small.
2. If decay modes into orbitally excited states were forbidden for the four $N^*$ resonances, the $\chi^2$ change became 3880 units and the fit quality deteriorated visibly. The four $N^*$ resonances decay via orbitally excited intermediate states with a significant decay fraction.

Why are the decays of the four $N^*$ resonances into orbitally excited intermediate resonances so frequent ($\approx 23\%$), and why are these decay modes suppressed for the four $\Delta^*$ resonances [81]?

These phenomena can be understood inspecting their wave functions. If one neglects mixing, the four positive-parity $\Delta^*$-resonances and $N^*$-resonances could both be assigned to a spin-quartet with intrinsic orbital and spin angular momenta $L = 2$ and $S = 3/2$. The four positive-parity $\Delta^*$ resonances have a spin and a flavor wave function which is symmetric with respect to the exchange of any pair of quarks. The color wave function is completely antisymmetric, hence the spatial wave function $\phi_{n,l}$ must be symmetric as well:

$$S = \frac{1}{\sqrt{2}}\{\left[\phi_{0s}(\vec{\rho}) \times \phi_{0d}(\vec{\lambda})\right] + \left[\phi_{0d}(\vec{\rho}) \times \phi_{0s}(\vec{\lambda})\right]\}^{(L=2)}, \quad (2)$$

with $\rho$ and $\lambda$ being the oscillators which can be excited in the 3-quark-system. For the nucleon states a mixed symmetry spatial wave function is required: For $L = 2$ the wave functions have equal admixtures of

$$\mathcal{M}_S = \frac{1}{\sqrt{2}}\{\left[\phi_{0s}(\vec{\rho}) \times \phi_{0d}(\vec{\lambda})\right] - \left[\phi_{0d}(\vec{\rho}) \times \phi_{0s}(\vec{\lambda})\right]\}^{(L=2)} \quad (3)$$

$$\mathcal{M}_{\mathcal{A}} = \left[\phi_{0p}(\vec{\rho}) \times \phi_{0p}(\vec{\lambda})\right]^{(L=2)}, \quad (4)$$

and both parts need to be present to fulfill the Pauli principle. The part $\mathcal{M}_{\mathcal{A}}$ describes a component in which the $\rho$ and the $\lambda$ oscillator are both excited simultaneously. We now assume that the $\mathcal{M}_{\mathcal{A}}$-part can only de-excite in a two-step process: In the first step one of the intrinsic oscillators is de-excited, and the intermediate state still carries an intrinsic excitation, an intermediate resonance is created. The latter excitation requires a second decay, from the intermediate resonance to the ground state. The large contribution of cascade decays to decays of $N^*$ in the fourth resonance region and their smallness in $\Delta^*$ decays provide evidence that two independent oscillators are excited in high-mass baryon resonances.

**Table 2.** Branching ratios BR for $N^* \to N\eta$ decays and the photon helicity amplitudes $A_{1/2}$, $A_{3/2}$ for four of the contributing nucleon resonances (taken from [82]). The helicity amplitudes are given in units of GeV$^{-1/2}$. Small numbers below the BRs or below the helicity amplitudes give either the RPP [87] estimate or the entries reported in [49].

| Res. $A_{1/2}$ | BR($N^* \to N\eta$) $A_{3/2}$ | Res. $A_{1/2}$ | BR($N^* \to N\eta$) $A_{3/2}$ |
|---|---|---|---|
| $N(1535)$ | 0.42±0.04 | $N(1710)$ | 0.25±0.09 |
| $1/2^-$ | 0.42±0.10 | $1/2^+$ | 0.10 - 0.30 |
| 0.093±0.009 | - | 0.040±0.020 | - |
| 0.115±0.015 | - | 0.035±0.012 | - |
| $N(1650)$ | 0.32±0.04 | $N(1720)$ | 0.03±0.02 |
| $1/2^-$ | 0.05 - 0.15 | $3/2^+$ | 0.021±0.014 |
| 0.032±0.006 | - | 0.115±0.045 | 0.135±0.040 |
| 0.045±0.010 | - | 0.100±0.020 | 0.150±0.030 |

## 3.3 Negative-parity resonances

### 3.3.1 Negative-parity resonances below 1800 MeV

The five negative-parity nucleon and the two $\Delta$ excitations in the first excitation shell are well established and their properties well defined. However, this does not exclude surprises when new data become available.

New single and double polarization data for the $\gamma p \to p\eta$ final state were reported recently. The observables $E, G, T, P$, and $H$ have been measured with the CBELSA/TAPS experiment ([82], see project A.1) at ELSA. A few bins of the newly measured double polarization observables $T$ and $E$ are shown in Fig. 7. Data on $T$ and $F$ [83] were obtained in Mainz and data on $E$ have been published recently by CLAS [84]. The latter data are consistent with our CBELSA/TAPS-data but were not yet included in our fits.

The predictions for the newly measured observables $E$ and $T$ from the different PWAs are also shown in Fig. 7. None of the predictions comes close to the data, even though they do describe reasonably well all $\eta$-photoproduction data ($d\sigma/d\Omega$, $\Sigma$) which existed before. This clearly shows that $\eta$-photoproduction was less constrained by existing data than $\pi$-photoproduction. Of course, this was to be expected since $\pi N$-elastic scattering does already provide significant information on the $\pi N$ coupling of resonances, while comparable data are missing for the $\eta N$-final state: data on $\pi N \to \eta N$ are scare.

Table 2 presents the $N^* \to N\eta$ branching ratios of four resonances used in the fit. There are a few interesting observations:

While the $N(1535)1/2^- \to N\eta$ branching ratio is found to be $0.42 \pm 0.04$ (most recent PDG value: $(0.42 \pm 0.10)$ GeV$^{-1/2}$), there is a very significant change in the $N(1650)1/2^- \to N\eta$ branching ratio observed. It changes from $0.05 - 0.15$ (RPP) to 0.32±0.04 in our fit. The $N(1710)1/2^+ \to N\eta$ branching ratio settles at 0.27±0.09, in the higher half of its previous range $0.10 - 0.30$, while $N(1720)3/2^+$ contributes little [82]. These results clearly show the power of polarization observables to constrain PWAs; an earlier PWA [65, 85] indicated a large $N(1720)3/2^+$ contributions.

The large $N(1650)1/2^- \to N\eta$ branching ratio found is surprising, given the previously large and puzzling dif-

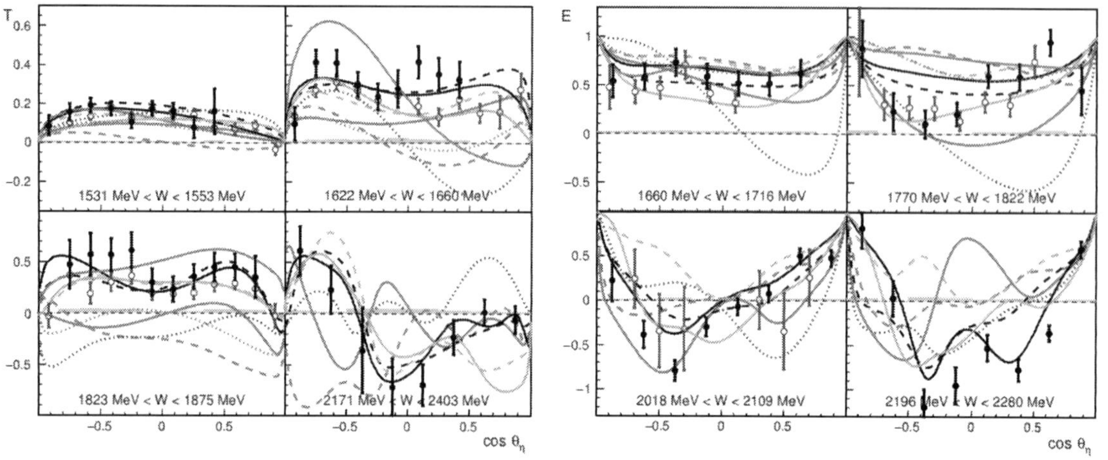


**Figure 7.** The polarization observable $T$ (left) and $E$ (right) as function of $\cos\theta_\eta$ for a few selected W-bins [82] (CBELSA/TAPS-data (black), from [82]). Very recent data from CLAS [84] are shown for comparison as blue open points (due to different binning, the energies differ by up to half of the bin size). The curves represent different models. Black: BnGa fit with (solid) and without (dashed) $N(2200)5/2^-$; red solid (dashed): BnGa2011-01 (-02); green: MAID, dark blue (dotted): SAID (GE09); magenta (dotted): Gießen , light blue: JüBo 2015 (dashed) and JüBo 2015-3 [84] (solid).

ference in the $N\eta$ branching ratios of the $N(1535)1/2^-$/ $N(1650)1/2^-$ nucleon-resonance pair which was extensively discussed in literature (see [86]). In the standard quark model, this has been taken as evidence for a large mixing of SU(6)×O(3) states (see Review on Quark Models in the RPP2014 [87]). A further intriguing result is the inversion of the relative sign of the electromagnetic couplings of the $N(1535)1/2^-$ and the $N(1650)1/2^-$ state for photoproduction off the proton and the neutron [88].

These results refer to properties of the resonances derived at the pole position. To calculate branching ratios (or Breit-Wigner parameters) from the multi-channel multi-pole K-matrix parameterization we first calculated the couplings of the resonance to the different channels as residues at the pole position. The Breit-Wigner/Flatte couplings were then introduced to be proportional to these couplings with one common factor $f$. The factor $f$ as well as the Breit-Wigner mass are then adjusted until they match exactly the pole position obtained from the full amplitude resulting from the K-matrix-fit to the data. Then, the helicity Breit-Wigner/Flatte couplings were determined making again sure that they reproduce the helicity residues of the full amplitude at the pole position. The resulting branching ratios for four of the contributing resonances are given in Table 2.

It is interesting to note that the K-matrix couplings maintain a small $N(1650)1/2^- \to N\eta$ coupling in the fits and do not show the inversion of the relative sign of $N(1535)1/2^-$ and $N(1650)1/2^-$ electromagnetic couplings.

### 3.3.2 Negative-parity resonances above 1800 MeV

We continue the discussion of fits to the double polarization data on $\gamma p \to \eta p$. The fit did not reproduce well the high-mass range; hence we tried to improve the description by adding one further resonance with spin-parity $J^P = 1/2^\pm, \cdots, 9/2^\pm$. The best result was achieved

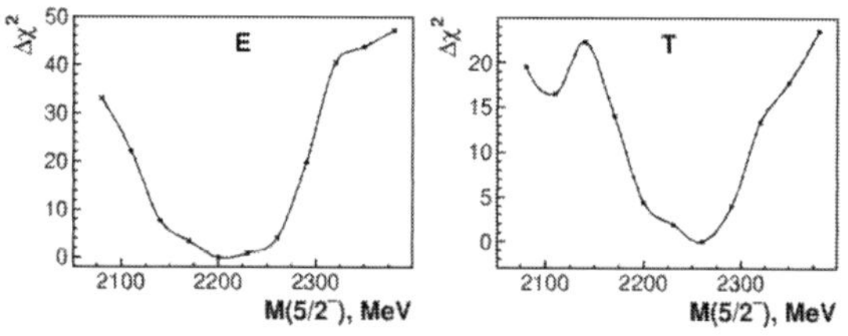


**Figure 8.** Mass scan for a resonance with $J^P = 5/2^-$. Shown is the $\chi^2$-change for $E$ and $T$ of the fit as a function of the imposed mass. In the mass scan, starting from the best fit solution, a Breit-Wigner amplitude is added with its mass fixed while all other parameters are fitted freely. The curves are included to guide the eye. The figure is taken from [82].

by introducing a resonance with $J^P = 5/2^-$, $M = (2200 \pm 50)$ MeV, $\Gamma = (260 \pm 50)$ MeV. Figure 8 shows the $\chi^2$-minimum for the two observables $E$ and $T$ if a scan of the $5/2^-$-resonance mass is performed. In the scan, the mass of the $5/2^-$-resonance was set to prefixed values while all other parameters were fitted freely. Significant minima are observed. However, no convincing evidence was observed in other reactions. Hence one must consider the

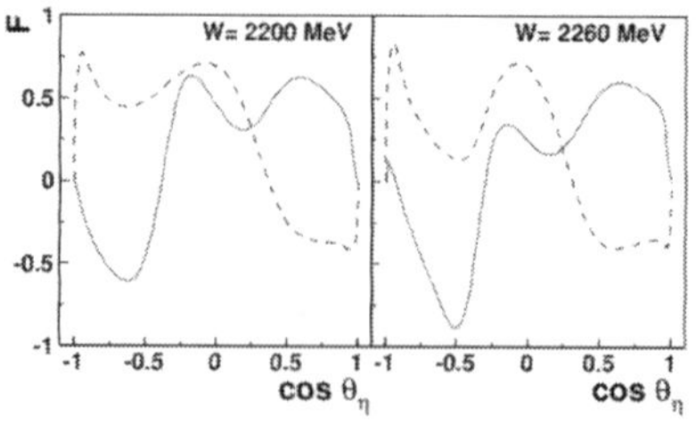


**Figure 9.** Predictions for the double polarization observable $F$ for $\gamma p \to p\eta$: with $N(2200)5/2^-$: red solid curve, without $N(2200)5/2^-$: blue dashed curve.

DOI: 10.1051/epjconf/201713402002

**Table 3.** Masses and widths of selected negative-parity resonances. The mass values are from [49]. The second column gives the PDG [96] star rating for resonances known in 2010. †: in 2010, this entry summarized observations ranging from 1900 to 2240 MeV and was removed from the RPP in 2012.

| | | | | |
|---|---|---|---|---|
| $N(1895)1/2^-$: | new | $M_{BW} = 1895 \pm 15$ | $\Gamma_{BW} = 90^{+30}_{-15}$ | [MeV ] |
| $N(1875)3/2^-$: | new | $M_{BW} = 1880 \pm 20$ | $\Gamma_{BW} = 200 \pm 25$ | [MeV ] |
| $N(2150)3/2^-$: | new | $M_{BW} = 2150 \pm 60$ | $\Gamma_{BW} = 330 \pm 45$ | [MeV ] |
| $N(2060)5/2^-$: | new | $M_{BW} = 2060 \pm 15$ | $\Gamma_{BW} = 375 \pm 25$ | [MeV ] |
| $N(2200)5/2^-$: | † | $M_{BW} = 2200 \pm 50$ | $\Gamma_{BW} = 260 \pm 50$ | [MeV ] |
| $N(2190)7/2^-$: | 4* | $M_{BW} = 2180 \pm 20$ | $\Gamma_{BW} = 335 \pm 40$ | [MeV ] |
| $N(2250)9/2^-$: | 4* | $M_{BW} = 2200 \pm 40$ | $\Gamma_{BW} = 480 \pm 60$ | [MeV ] |

evidence for this new resonance with some caution. Based on the PWA-solutions with and without the $N(2200)5/2^-$-resonance, the double polarization observable $F$ has been predicted. Fig. 9 shows the sensitivity of this observable. Measuring F would definitely help to settle the question whether or not a $5/2^-$-resonance exists in a mass range around 2200 MeV. Of course also improved statistics on the already measured observables can be expected to lead to an improved sensitivity.

New high-mass negative-parity resonances were reported in [90]. $N(1875)3/2^-$ proved to be particularly sensitive to the reaction $\gamma p \to \Sigma^+ K^0_S$ [91–93] studied in project B.1. Also, the data on the beam asymmetry on $\pi^0$ and $\eta$ have to be mentioned [94, 95]. Evidence for the new resonances can be seen in the mass scans shown in Fig. 10. The masses of high-mass negative-parity resonances as obtained within the BnGa multichannel PWA are collected in Table 3.

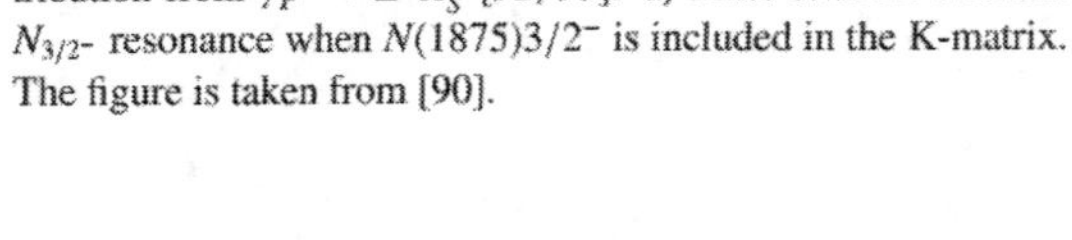


**Figure 10.** a) Mass scan for a $N_{1/2^-}$ resonance; change of the total $\chi^2$ of the fit as a function of the assumed mass; b,c) Mass scan for a $N_{3/2^-}$ resonance; $\chi^2$ of the fit as a function of the assumed mass for an assumed width of 100 MeV. b) total $\chi^2$, c) $\chi^2$ contribution from $\gamma p \to \Sigma^+ K^0_S$ [92, 93]. d) Mass scan for a fourth $N_{3/2^-}$ resonance when $N(1875)3/2^-$ is included in the K-matrix. The figure is taken from [90].

## 3.4 Assignment to SU(6)xO(3)-multiplets

An assignment of resonances to SU(6)xO(3) multiplets is discussed by Klempt and Metsch in [97]. Figure 11 shows the resonances which have at least 2 stars in the RPP and their assignment to SU(6)xO(3) multiplets. Resonances where the RPP'2014 states "Latest GWU analysis [15] finds no evidence for the resonance" are indicated by blue / light blue crosses. All states marked by blue, red, or magenta boxes have been identified by the BnGa-PWA.

### *3.4.1 Negative parity resonances*

The first excitation shell ($1\hbar\omega$) is (since long) complete. The high-mass negative-parity resonances are collected in Table 4. If the three $\Delta^*$ resonances $\Delta(1900)1/2^-$, $\Delta(1940)3/2^-$, $\Delta(1930)5/2^-$ are interpreted as a spin triplet, they have a fully symmetric spin and flavor wave function; hence their spatial wave function must be symmetric, too. Thus, they belong to a 56 representation, and there must be a doublet of negative-parity nucleon resonances close in mass. These are identified with $N(1895)1/2^-$ and $N(1875)3/2^-$. These five resonances can be assigned to the third excitation shell with $L_{3q} = 1$. They have one unit of radial excitation $N = N_\rho + N_\lambda = 1$).

There are five negative-parity nucleon resonances $N(2150)3/2^-$; $N(2200)5/2^-$; $N(2060)5/2^-$; $N(2190)7/2^-$; $N(2250)9/2^-$ in the mass range from 2050 to 2250 MeV. It is tempting to organize these resonances in the same way as those in the first excitation shell, as quartet with an intrinsic quark spin $S = 3/2$, and a $S = 1/2$ doublet. In this interpretation, one resonance with $J^P = 7/2^-$ is missing. We have added to Table 4 $\Delta(2223)5/2^-$ [15], a very wide resonance which did not enter the RPP. It would combine with $\Delta(2200)7/2^-$ to the expected spin doublet.

### *3.4.2 Positive parity resonances*

Well known are the four $\Delta^*$-resonances, $\Delta(1910)1/2^+$, $\Delta(1920)3/2^+$, $\Delta(1905)5/2^+$, and $\Delta(1905)5/2^+$, as well as the doublet of $N^*$-states, $N(1720)3/2^+$ and $N(1680)5/2^+$ which are except the 3-star $\Delta(1920)3/2^+$ all 4-star resonances. In addition to these states, the BnGa analysis gives evidence for the existence of four to six positive-parity resonances $N(1880)1/2^+$, $N(1900)3/2^+$ (with a $N(1960)3/2^+$ companion, see below), $N(1860)5/2^+$ (weak evidence from BnGa but confirmed by other analyses), $N(2000)5/2^+$, and $N(1990)7/2^+$ [49]. $N(1880)1/2^+$ was also reported in [63, 98]. The evidence for the $N(1900)3/2^+$ has been confirmed by the analyses of other groups, see [63, 98–100], and by the Gießen group [60, 101] within project B.7. In [102], $N(1900)3/2^+$ was tentatively interpreted as a member of a spin quartet, an identification which is challenged by later results (see below). The other spin partners $N(2000)5/2^+$ and $N(1990)7/2^+$

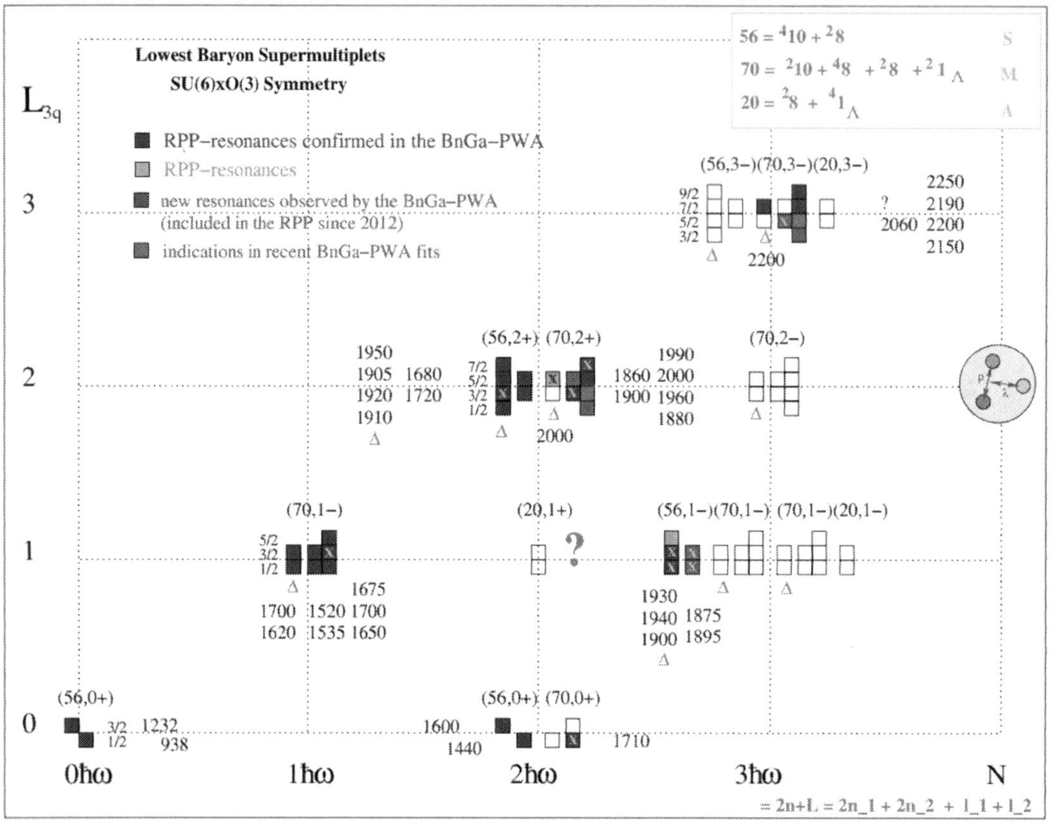


**Figure 11.** $N^*$- and $\Delta^*$-resonances assigned to SU(6)xO(3)-multiplets. The assignments are not unambiguous and, likely, mixing occurs. Only resonances which have at least a 2-star-assignment in the RPP2014 are shown. The numbers given at the left, right or below the multiplets correspond to the masses of the states. Above the multiplets, the SU(6) multiplet is given together with the total orbital angular momentum $L$ between the quarks and the parity of the multiplet ($D, L^P_N$, where $D$ is the SU(6) dimensionality, $L$ is the intrinsic orbital angular momentum, $P$ the parity and N the shell number of the harmonic oscillator). The light blue or blue crosses indicate the states where the RPP2014 states "Latest GWU analysis [15] finds no evidence for the resonance".

**Table 4.** Nucleon and Δ resonances assigned to the third excitation shell. The masses of nucleon resonances are from our work, most Δ resonances from [96], one from [15]. $N(2190)7/2^-$ could belong to the $S = 3/2$ quartet or to a $S = 1/2$ doublet. $N(2190)7/2^-$ may both consist of two unresolved resonances, one belonging to the quartet, the other one to the doublet.

| | | | | | |
|---|---|---|---|---|---|
| $L = 1, N = 1$ | S=$\frac{1}{2}$ | $N(1895)1/2^-$ | $N(1875)3/2^-$ | | |
| | S=$\frac{3}{2}$ | $\Delta(1900)1/2^-$ | $\Delta(1940)3/2^-$ | $\Delta(1930)5/2^-$ | |
| $L = 3, N = 0$ | S=$\frac{3}{2}$ | $N(2150)3/2^-$ | $N(2220)5/2^-$ | $N(2190)7/2^-$ | $N(2250)9/2^-$ |
| | S=$\frac{1}{2}$ | | $N(2060)5/2^-$ | | |
| | S=$\frac{1}{2}$ | | $\Delta(2223)5/2^-$ | $\Delta(2200)7/2^-$ | |

were already known since long, even though reported mass values scattered wildly.

In the $3/2^+$-sector, a much better fit is achieved in the BnGa analysis when a second $N\,3/2^+$ resonance is assumed to exist in addition to the $N(1900)3/2^+$-resonance. Its mass is not well defined, the most likely value is 1960 MeV. Clearly, quark models have ample possibilities to house two $N\,3/2^+$ resonance at this mass. Expected is, e.g., a 70-plet which contains a quark-spin quartet and a doublet. These six expected states could be the six resonances seen in the BnGa PWA.

The assignment to a quark-spin quartet or a doublet is possible considering the electric couplings. The photocouplings of baryon resonances are mostly given as helicity amplitudes $A_{1/2}$, $A_{3/2}$, and for excitations off protons, we gave these quantities only [103]. In a fit to data on photoproduction off neutrons, we evaluated also the electric and magnetic multipole amplitudes [104]. In many cases,

**Table 5.** The ratio of the electric and magnetic multipole amplitudes $E/M$ for resonances in the fourth resonance region. For resonances with $J^P = 1/2^{\pm}$ $E/M$ ratio cannot be determined since they are excited by magnetic quadrupole or electric dipole amplitudes only.

| | $N(1960)3/2^+$ | $N(2000)5/2^+$ | $N(1990)7/2^+$ |
|---|---|---|---|
| $E/M$ | 0.15±0.12 | $0.4^{+0.5}_{-0.4}$ | 0.23±0.17 |
| | $N(1900)3/2^+$ | $N(1860)5/2^+$ | |
| $E/M$ | 3.2±1.9 | undefined | |

the ratio is not well defined. Often, the electric multipole amplitude is larger than the magnetic one: $E > M$ holds for many resonances. But there are a few cases where the magnetic multipole prevails over the electric one. A well known example is, of course, $\Delta(1232)$ where a spin flip is required to go from the nucleon to the $\Delta(1232)$ resonance. The $E/M$ ratio is also small for $N(1960)3/2^+$, for $N(1990)7/2^+$, and, perhaps, for $N(2000)5/2^+$. Hence we conjecture that these three resonances belong to a spin-quartet while $N(1900)3/2^+$ belongs to a spin-doublet. For $N(1860)5/2^+$, the $E/M$ ratio cannot yet be defined.

These findings are supported by Montagne and Stancu [105] who analyze the nucleon excitation spectrum in an $1/N_c$ expansion scheme. They expect a spin quartet in the SU(6) multiplet $^4N[70, 2^+]$ with masses

| | | | |
|---|---|---|---|
| 2080±39 | 2042±41 | 1955±32 | 1878±34 |
| $N(1990)7/2^+$ | $N(2000)5/2^+$ | $N(xxx)3/2^+$ | $N(1880)1/2^+$. |

The calculated mass values are used to identify candidates which are observed experimentally. These are listed in the second line. For the $3/2^+$ state, Montagne and Stancu predict 1955 MeV, and we identify this missing resonance with $N(1960)3/2^+$.

Likewise, the SU(6) multiplet $^2N[70, 2^+]$ is expected with masses

| | |
|---|---|
| 1959±29 | 1902±22 |
| $N(1860)5/2^+$ | $N(1900)3/2^+$ |

and the identification in [105] with known states is again listed in the second line.

In the second excitation shell, a large fraction of the predicted resonances has been found. Missing are a second $\Delta^*$ with spin-parity $J^P = 3/2^+$ at about 1900 - 2000 MeV, the doublet partner of $\Delta^*(2000)5/2^+$ (2-star-state, not confirmed by the BnGa-PWA) and the $3/2^+$ companion of $N(1710)1/2^+$. Also the one-star $\Delta(1750)1/2^+$ is not shown, the evidence for its existence is poor only.

Summarizing the discussion of the multiplet structure, we can conclude that the number of resonances is well predicted by the quark model based on SU(6)xO(3)-symmetry. Of course significant deficiencies are present if quark model predictions of masses or decay widths are compared to the experimental findings. A few entries are missing in the second excitation shell but the resonances found so far provide a reasonable coverage of the expected spectrum. Empty are, however, the entries for the $(20, 1^+)$ multiplet. Possible reasons for their absence will be discussed in the Outlook (Section 4).

### 3.5 What is the dynamical reason for the existence of parity doublets?

Quark models use constituent light-quark masses of about 350 MeV to reproduce the masses of ground-state baryons. However, low-energy approximations of QCD [106] lead to the Gell-Mann-Oakes-Renner relation [107] which assigns a mass of a few MeV to light (current) quarks. The mass gap between current and constituent quarks is interpreted by spontaneous breaking of the chiral symmetry expected for nearly massless quarks [108, 109]. With no spontaneous symmetry breaking, the masses of positive and negative-parity hadrons would be degenerate in mass when they have the same total angular momentum: the nucleon and its chiral partner $N(1535)1/2^-$ would have the same mass.

In the higher mass region, parity doublets are observed and often, nucleon and $\Delta$ resonances of given spin and parity form a quartet of almost mass-degenerate states. (One particular case is studied in [110, 111]). This observation has led to the conjecture that chiral symmetry might be restored effectively in highly excited baryons [38]. The mass generation mechanism in excited hadrons is, according to Glozman [112], very different compared to the mechanism in the lower-mass states. In the latter states, the mass is supposed to be driven by chiral symmetry breaking in the vacuum. For highly excited states, the quark condensate is believed to be almost irrelevant and the mass of resonances within a parity doublet could have a chirally symmetric origin.

This idea is best tested in baryons. In meson spectroscopy, resonances falling onto a leading Regge trajectory (with $J = L + S$) have no mass-degenerate parity partner (see Fig. 55 in the review of Klempt and Zaitsev [113]). However, in $\bar{p}p$ formation of, e.g., $f_4(2050)$, $L = 3$ is required; formation of its (unobserved or non-existing) mass-degenerate parity partner $\eta_4$ requires $L = 4$. The non-observation of a mass-degenerate $\eta_4$ might be due to a suppression by the angular-momentum barrier. In $\pi N$ scattering, $\Delta(1950)7/2^+$ formation requires $L = 3$ and formation of its $7/2^-$ parity partner $L = 4$.

This is different in photoproduction of baryons. In photoproduction, a $7/2^+$ resonance needs an electric or magnetic $E_4^+$ / $M_4^+$ amplitude (with $L = 4$ between photon and nucleon); a $7/2^-$ resonance needs a $E_3^-$ or $M_3^-$ amplitude. Photoproduction hence provides the best and possibly the only chance to find a decisive support or an experimental argument against the hypothesized restoration of chiral symmetry.

Slightly above 1900 MeV, four positive-parity and three negative-parity $\Delta^*$ resonances have been observed:

| | | | |
|---|---|---|---|
| $\Delta(1910)1/2^+$ | $\Delta(1920)3/2^+$ | $\Delta(1905)5/2^+$ | $\Delta(1950)7/2^+$ |
| $\Delta(1900)1/2^-$ | $\Delta(1940)3/2^-$ | $\Delta(1930)5/2^-$ | ? |

New photoproduction data on $T$, $P$, $H$ [115, 116] $E$ [117], and $G$ [81, 118] for $\gamma p \to \pi^0 p$ taken within the A.1 project at Bonn and data on $E$ for $\gamma p \to \pi^+ n$ [119] were included in the multi-channel BnGa-PWA and a large number of fits were performed. The fit required contributions from a $\Delta\, 7/2^+$ and a $\Delta\, 7/2^-$ resonance. Figure 12

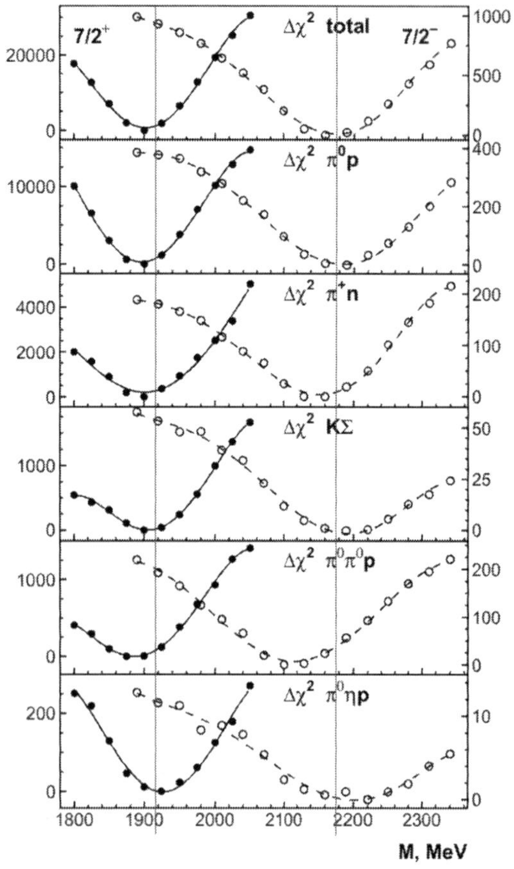


**Figure 12.** The increase in pseudo-$\chi^2$ of the fit to a large body of pion- and photo-produced reactions when the mass of $\Delta(1950)7/2^+$ (solid points) or $\Delta(2200)7/2^-$ (open circles) is scanned. The scale on the left (right) abscissa refers to the $7/2^+$ ($7/2^-$) partial wave. The curves are to guide the eye [121]. The red line indicate the mass as obtained by the fit as given in Table 6.

**Table 6.** Properties of $\Delta(1950)7/2^+$ and $\Delta(2200)7/2^-$: Breit-Wigner masses, widths, helicity amplitudes and branching ratios. The RPP estimates are given as small numbers. The helicity couplings $A_{1/2}$, $A_{3/2}$ are given in units of $10^{-3}$ GeV$^{-1/2}$ (see [121]).

| $\Delta(1950)7/2^+$ **** | | $\Delta(2200)7/2^-$ * | |
|---|---|---|---|
| $M$=1917±4<br>1915 - 1950 | Γ=251±8 MeV<br>235 - 335 | $M$=2176±40 | Γ=210±70 MeV |
| $A_{1/2}$=-67±5<br>-76±12 | $A_{3/2}$=-94±4<br>-97±10 | $A_{1/2}$=60±20 | $A_{3/2}$=20±8 |
| $N\pi$ 46±2%<br>35 - 45 | $\Sigma K$ 0.6±0.2%<br>seen | $N\pi$ 3.5±1.5% | $\Sigma K$ 4±3% |
| $\Delta\pi$ 5±3%<br>20 - 30 | $\Delta\pi$ -<br>seen | $\Delta\pi$ 70±15% | $\Delta\pi$ 15±10% |
| $\Delta\eta$ 0.3±0.3% | $\Delta\eta$ - | $\Delta\eta$ ~ 1% | $\Delta\eta$ - |

shows a scan in the $J^P = 7/2^+$ and $J^P = 7/2^-$ partial waves. In the mass scans, the $J^P = 7/2^+$ and $J^P = 7/2^-$ partial waves are described by multichannel Breit-Wigner amplitudes rather than by K-matrix amplitudes. Hence the optimal parameters in the scan for mass and width can differ from the optimal values from the main fit. Figure 12 (top) shows the change of the resulting pseudo-$\chi^2$ as a function of the imposed mass of the $J^P = 7/2^+$ or the $J^P = 7/2^-$ resonance. The total pseudo-$\chi^2$ has clear minima at a mass of about 1900 MeV for $J^P = 7/2^+$ and somewhat below 2200 MeV for $J^P = 7/2^-$. The scans demonstrate clearly that the masses of $\Delta(1950)7/2^+$ and $\Delta(2200)7/2^-$ are different. Both resonances are seen with compatible masses in $\pi N$, $K\Sigma$, $\pi^0\pi^0 p$, and even in $\pi^0\eta p$. The properties of the resonances as derived in these fits are listed in Table 6. A $7/2^-$-state mass degenerate with $\Delta(1950)7/2^+$ has been found in none of the reactions. The $\Delta(2200)7/2^-$ observed confirms a 1-star-state given in the

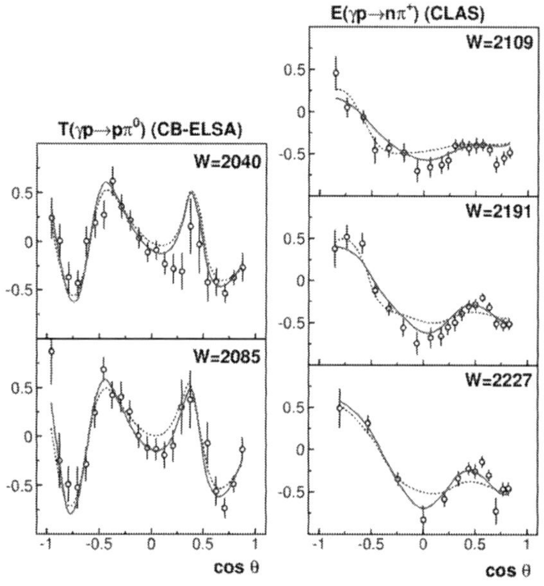


**Figure 13.** Selected data and fits. Data: Target asymmetry T [115, 116]; helicity dependence $E$ [119]. Best fit: solid curve, fit without $\Delta(2200)7/2^-$: dotted curve. The fit deteriorated by $\Delta\chi^2 = 597$ when $\Delta(2200)7/2^-$ was removed.

RPP. The PDG decided to upgrade $\Delta(2200)7/2^-$ to a 3* resonance [120].

The resonances with $J^P = 1/2^\pm, 3/2^\pm, 5/2^\pm$ are nearly *mass-degenerate*. The ideas of Glozman require necessarily a $\Delta\,7/2^-$ resonance at about the same mass [114]. For such a resonance, there is no evidence at all!

Fig. 13 shows examples for data bins in comparison to the best fit achieved with or without including a negative parity $J = 7/2$, $\Delta(2200)$-state. The data description clearly improves if the $7/2^-$-state is included in the fit. Further examples for polarization observables and more analysis details can be found in [121].

What is then the reason for the often observed parity doublets? Klempt argues that the dynamical reason for the parity doublets lies in the string nature of the quark-quark

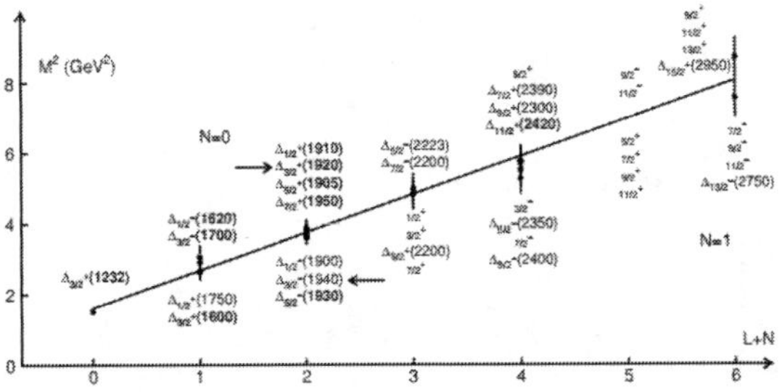


**Figure 14.** Mass of Δ resonances as a function of the leading intrinsic orbital angular momentum $L$ and the radial excitation quantum number $N$ (corresponding to $n_\rho + n_\lambda$ in quark models). Resonances assigned to $N = 0$ and $N = 1$ are listed above or below the trajectory. The solid line represent a linear trajectory. Figure taken from [126]

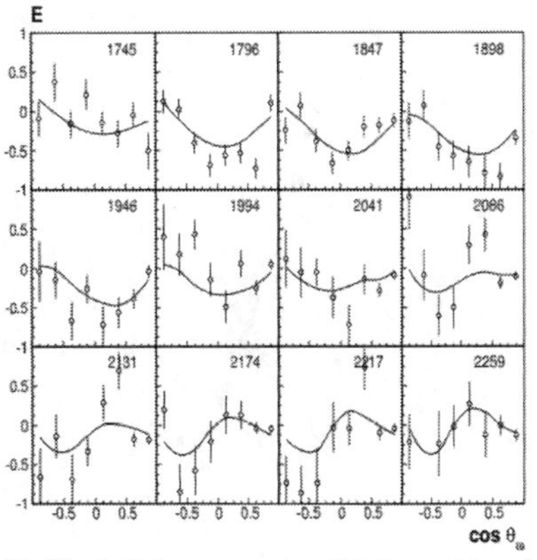


**Figure 15.** The helicity asymmetry $E$ in bins of the photon energy, taken from [137].

interaction. His empirical mass formula [122] was derived in AdS/QCD [123], and was shown to be, compared to quark models requiring a much larger number of parameters, at least twice as precise in calculating baryon masses, using just two parameters [124]. In Ref. [125], Klempt pointed out that the $\Delta\, 7/2^-$ mass is decisive for the interpretation. In string models, the mass increases with $L + N$ instead of the harmonic oscillator which suggests $L + 2N$ ($\vec{L} = \vec{l}_\rho + l_\lambda$, $N = n_\rho + n_\lambda$). Evidence for the string like behavior was presented in [126] from which we take Fig. 14 to demonstrate the $L + N$ dependence of baryon masses. The future will tell us whether the next-generation lattice calculations performed at lower pion masses will provide results closer to the experimental findings: will there be parity doublets for most resonances, except for those on the leading Regge trajectories?

### 3.6 Photoproduction of $\omega$ mesons

At high energies, real or virtual photons interact with protons by diffraction [127]. The photon converts into a vector meson ($\rho^0$, $\phi$, $\omega$) and scatters off the proton by the exchange of Pomerons carrying the quantum numbers of the vacuum [128, 129]. Close to the $\omega$ production threshold, $N^*$ resonances are likely to contribute to the reaction [130, 131]. The Gießen group pioneered the PWAs of the reactions $\pi^- p \to \omega n$ and $\gamma p \to \omega p$ [60], see project B.7.

The CLAS collaboration reported a high-statistics study of $\omega$ photoproduction [133, 134] and claimed evidence for several $N^*$s decaying into $N\omega$. The A2-collaboration at MAMI reported differential cross sections from threshold to $E_\gamma = 1.4$ GeV with a 15 MeV binning and full angular coverage [135]. No resonant contributions were discussed. The CBELSA/TAPS collaboration measured $d\sigma/d\Omega$ and spin density matrix elements [136], the beam asymmetry $\Sigma$ [132], the helicity difference $E$, and the correlation $G$ between linear photon and longitudinal target polarization [137]. Photoproduction of $\omega$ mesons off neutrons was studied in [138].

**Table 7.** Branching ratios (B.R. in %) for $N^* \to N\omega$ decays from the BnGa PWA, see [62]. Small numbers were reported in [60]. The $\delta(\chi^2)$ values give the change in $\chi^2$ when the $N\omega$ decay mode is excluded.

| Resonance | B.R. | $\delta(\chi^2)$ | Resonance | B.R. | $\delta(\chi^2)$ |
|---|---|---|---|---|---|
| $N(1700)3/2^-$ | 22±12 | 100 | $N(1900)3/2^+$ | 15±8<br>13±9 | 70 |
| $N(1710)1/2^+$ | 2±2<br>8±5 | 26 | $N(2000)5/2^+$ | 18±8<br>1±1 | 42 |
| $N(1720)3/2^+$ | 26±14 | 105 | $N(2060)5/2^-$ | 4±3 | 37 |
| $N(1875)3/2^-$ | 13±7<br>20±4 | 98 | $N(2100)1/2^+$ | 15±10 | 78 |
| $N(1880)1/2^+$ | 20±8 | 33 | $N(2150)3/2^-$ | 12±8 | 99 |
| $N(1895)1/2^-$ | 28±12 | 100 | $N(2190)7/2^-$ | 14±6 | 131 |

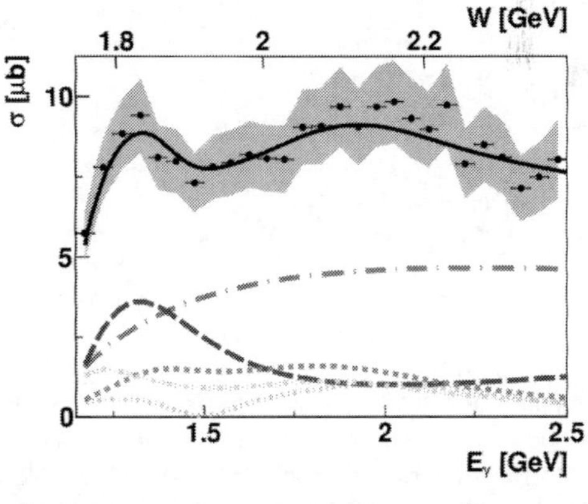


**Figure 16.** Total cross section for $\gamma p \to p\omega$ from the CBELSA/TAPS experiment (•) as a function of the initial photon ($E_\gamma$) and center-of-mass (W) energy, see [62, 136]. The error bars represent the statistical, the grey band the systematic uncertainty. The Bonn-Gatchina fit is represented by a solid black line. The largest contributions are Pomeron-exchange (red), resonant production of the $J^P = 3/2^+$ partial wave (blue), $3/2^-$ partial wave (purple), $5/2^+$ partial wave (blue), and $1/2^-$ partial wave (green).

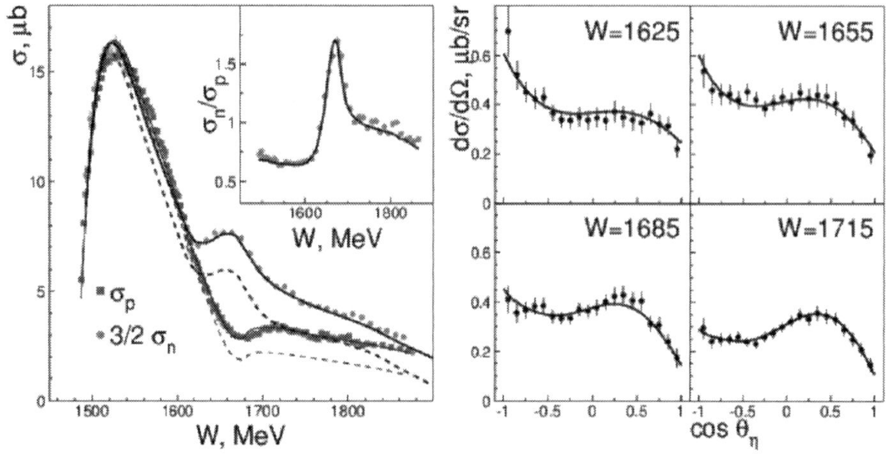


**Figure 17.** Left: The total cross section for $\gamma n \to \eta n$ (multiplied by 3/2), $\gamma p \to \eta p$, and their ratio (as inset) [150] . The solid curves represent our fit [88] folded with the experimental resolution (thick $\eta n$, thin $\eta p$), the dashed curves the contributions from the $S_{11}$ waves. Right: Selected differential cross section bins for $\gamma n \to \eta n$ in the region of the narrow structure.

The data were included in the BnGa data base. Figure 15 shows the helicity difference $E$ with the BnGa fit. The PWA identified twelve $N^* \to N\omega$ branching ratios (see Table 7); most of these were determined for the first time. The total cross section, shown in Fig. 16, is obtained by integration of the differential cross sections. Fig. 16 also shows the dominant partial wave contributions.

Only $N^*$ contribute to $\gamma p \to \omega p$. $\Delta^*$ decays can be studied in the process $\gamma p \to \pi^0 \omega p$ to which the isobar $\Delta(1232)\omega$ makes the most significant contribution [139]. Similarly, the reaction $\gamma p \to K^*\Lambda$ has no $\Delta^*$ contributions [140]. Instead, the BnGa PWA finds contributions from $N(1895)1/2^-$, $N(1880)1/2^+$, $N(2100)1/2^+$, and two or three higher-mass $N^*$'s with ill-defined quantum numbers. $N^*$ and $\Delta^*$ both contribute to $\gamma p \to K^*\Sigma$ [141] but a PWA has not been attempted.

### 3.7 Is the narrow peak seen in the reaction $\gamma n \to \eta n$ a resonance?

The "discovery" in 2003 of $\Theta^+(1540)$, an exotic baryon with positive strangeness stimulated further studies [142–145] and the search for its non-strange partner. Indeed, a narrow structure was observed in the reaction $\gamma n \to \eta n$ studied at GRAAL [146]. Its properties, mass of 1685 MeV, its narrow width, the production off neutrons but not off protons, and its preference to decay into $\eta n$ would be consistent with predictions for the nonstrange member of a multiplet of pentaquarks with spin-parity $J^P = 1/2^+$ [142]. The observation attracted considerable interest, and the structure was listed in the RPP as one–star resonance $N(1685)$ [71].

The evidence for the existence of $\Theta^+(1540)$ faded away in a number of precision experiments (even though further evidence is reported in several more recent experiments, see [147] for a recent review. In contrast to $\Theta^+(1540)$, the peak at about 1685 MeV became increasingly significant with better experiments at ELSA and MAMI [148–150], its existence is beyond doubt. But the interpretation as a resonance is unclear.

The Gießen group explained the narrow peak in the $\eta$ photoproduction on the neutron in terms of coupled-channel effects due to $N(1650)1/2^-$ and $N(1710)1/2^+$ [151]; Shyam and Scholten interprete the peak by interference effects between the $N(1650)1/2^-$, $N(1710)1/2^+$, and $N(1720)3/2^+$ resonances [152] within a coupled-hannel analysis; Döring and Nakayama ascribe the peak to effects from strangeness threshold openings [153]. The Bonn-Gatchina group [154, 155] and Zhong and Zhao [156] demonstrated that the narrow peak can be explained naturally by interference effects in the $J^P = 1/2^-$ wave, even though - according to [154, 155] - solutions with a narrow resonance were not yet excluded by the data.

Fig. 17 shows the total cross section for $\gamma p \to \eta p$, for $\gamma n \to \eta n$ and - as insert - the ratio $\sigma_n/\sigma_p$ from the most recent experiment [150]. The new data exceed the earlier data both in quality and in statistics. In the new experiment, the hit proton or neutron was detected, and hence the $N\eta$ invariant mass was reconstructed without smearing due to the Fermi motion. These experimental achievements greatly enhanced the visibility of the narrow structure. The contributions from $\eta$ photoproduction off neutrons were determined by two different methods; the differences were used to estimate the systematic error.

In [150], the narrow structure was tentatively interpreted as a narrow nucleon resonance at $W = (1670 \pm 5)$ MeV mass and $\Gamma = (30 \pm 5)$ MeV width. The product coupling of the hypothesized resonance given by its helicity amplitude $A^n_{1/2}$ and the branching ratio for its neutron-$\eta$ decay $b_\eta$ was determined to $\sqrt{b_\eta}A^n_{1/2} = (12.3 \pm 0.8)\,10^{-3}\,\mathrm{GeV}^{-1/2}$.

The data were fitted within the Bonn-Gatchina partial-wave analysis [88]. Masses, widths, decay couplings of resonances were all frozen by a fit to $\pi N$ elastic and inelastic reactions and to data on photoproduction off protons (see [49, 159] for the data included). Here, only the helicity amplitudes for the photoproduction of nucleon resonances off neutrons and contributions from $t$- and $u$-channel exchange were used as free parameters. To constrain the data on $\gamma n \to \eta n$ further, we included GRAAL data on the beam asymmetry for this reaction [157] and data on $\gamma n \to \pi N$ [158]. These additional data were described well and have no impact on the conclusions; hence we retain from a more detailed discussion of those data.

The new and precise data are well described by the interference within the $S_{11}$ wave and are now definitely incompatible with the existence of a nucleon resonance with the reported properties. If a narrow resonance with the properties reported in [150] is imposed in the fit, the $\chi^2$ of the fit deteriorates very significantly. The reason for the peak structure in $\eta n$ and dip structure in $\eta p$ lies in the opposite relative sign of the helicity amplitudes for the two resonances $N(1535)S_{11}$ and $N(1650)S_{11}$. The narrow structure observed in photoproduction of $\eta$ mesons off neutrons does not support the existence of a nucleon resonance with exotic properties. We mention that data on $\eta'$ production off neutrons [89] do not show a similar structure.

### 3.8 What is the relation between quark-model resonances and dynamically generated resonances?

A significant number of nucleon resonances can be generated dynamically from the interaction of their decay products. In coupled-channel dynamical approaches, some baryon resonances need to be put in "by hand", others are generated by the unitarization of the background amplitude. Some of the dynamically generated poles can be identified with quark-model states. But there are poles which do not correspond to quark mode states, and there are resonances which are not (yet?) dynamically generated. The relation between quark-model resonances and dynamically generated resonances remains an unsolved problem in hadron spectroscopy.

### 3.9 Photoproduction multipoles from different PWA groups

Amplitudes are more sensitive when different models are to be compared. In [160] we have reported transition amplitudes to $\Lambda K$, $\Sigma K$, and $N\eta$ for pion and photo-produced reactions. Fig. 18 shows, as example, the BnGa-$E_{0^+}$ and the $E_{2^-}$ multipoles for $\pi^0$ photoproduction before (2011, red) and after (2014, blue, for further multipoles see [115, 116]) the inclusion of the new double polarization data on $E, G, T, P$, and $H$.

The error bands for the BnGa2011 solutions (red) were derived from the ($1\sigma$) spread of 12 different solutions with different assumptions on the ingredients: the number of poles in the $J^P = 3/2^+$ wave was 3 or 4, in the $J^P = 5/2^+$ wave 2 or 3, the $N(1700)3/2^-$ width of the pole converged to a wide (~600 MeV) or a narrow (~250 MeV) value, a K-matrix formalism was used or, alternatively, a $N/D$-based parametrization [62]. The $\chi^2$ ranged from its minimum $\chi^2_{\rm min}$ to $\chi^2_{\rm min}$ + 800. (Note that the absolute $\chi^2$ value is meaningless since part of the data are multiparticle final states and fitted in an event-based likelihood fit. The log likelihood value is then converted into a pseudo-$\chi^2$ [49].)

In the new fits including the new polarization data, the fits started from the same solutions, then all parameters were re-optimized. All fits converged, but 6 fits resulted in a $\chi^2$ larger than the new $\chi^2$-minimum ($\chi^2_{\rm new\,min}$) by 1000 units or more. These fits, mostly those with only 3 poles in the $I = 1/2$, $J^P = 3/2^+$ wave, were then removed from the error analysis. The resulting error bands for all remaining solutions within the same $\chi^2$-boundaries as for the old solutions are also shown in Fig. 18.

The new error bands (blue) are significantly smaller than the previous ones. Averaged over all multipoles and energies, the errors are reduced by a factor of 2.25.

Fig. 19 shows the new double-polarization data from JLab and Bonn with fits from MAID , SAID, JüBo, and BnGa, for fits which were made before the data were available (left) and with the new data included (right). Large deviations are observed even at lower photon energies, where one might have expected that $\pi^0$-photoproduction is well understood. After a refit, all PWAs are capable to describe the new data.

In Fig. 20 (left) the $E_{0^+}$-multipoles from MAID, SAID, JüBo, and BnGa are shown (for further multipoles see [161]). It is striking that even at low energies, large differences between all four solutions are seen. Obviously, the data that existed prior to the double polarization mea-

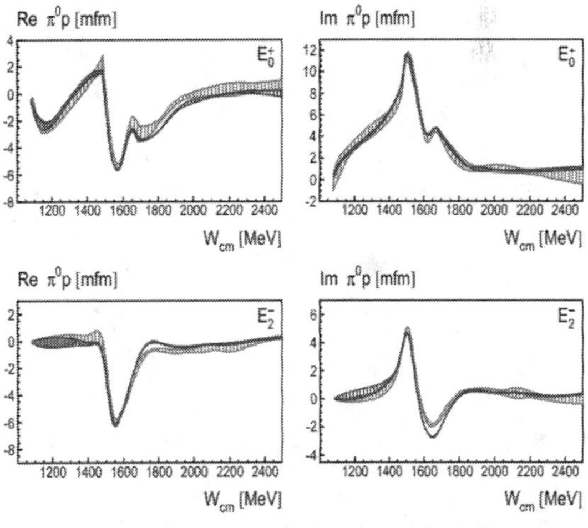


**Figure 18.** Multipole decomposition of the $\gamma p \to p\pi^0$ transition amplitudes, real part (left), imaginary part (right), showing two examples the $E_0^+$ and the $E_2^-$-multipole [115, 116]. The red shaded areas give the range from a variety of different fits derived from solution BnGa2011-01 and BnGa2011-02 [49]. The blue shaded area represents the range of solutions when the new data are included in the fit (BnGa2014).

surements (e.g. the $p\pi^0$-double polarization data discussed in project A.1) did not constrain the solution to a sufficient accuracy. For a further comparison of PWA-predictions to the new polarization data see also the contribution of project A.1 to these proceedings.

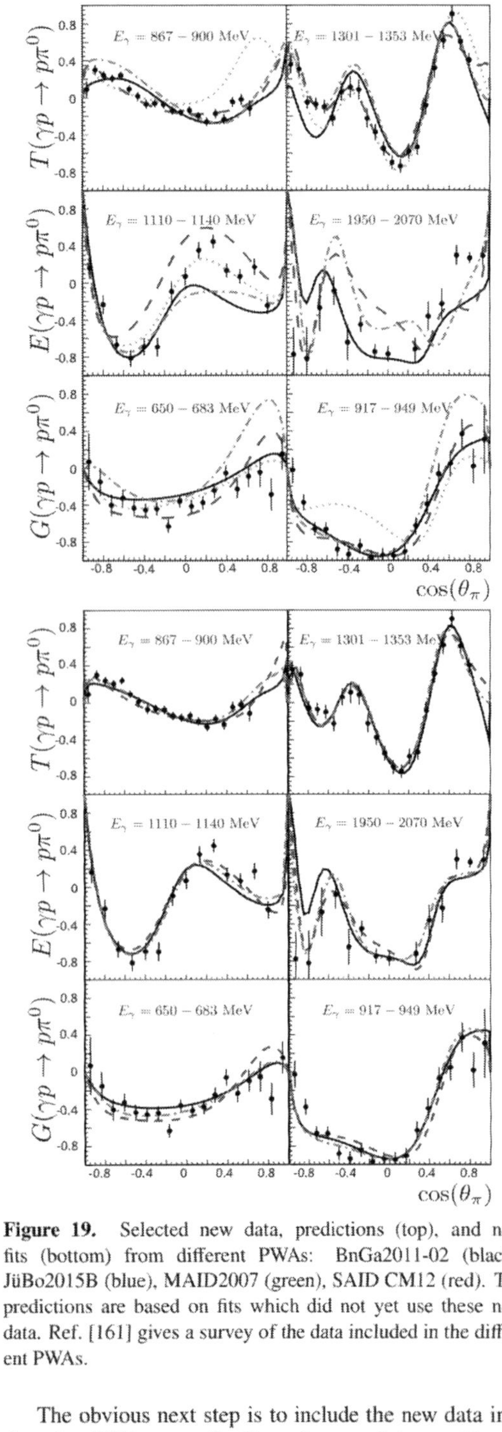


**Figure 19.** Selected new data, predictions (top), and new fits (bottom) from different PWAs: BnGa2011-02 (black), JüBo2015B (blue), MAID2007 (green), SAID CM12 (red). The predictions are based on fits which did not yet use these new data. Ref. [161] gives a survey of the data included in the different PWAs.

The obvious next step is to include the new data into the other PWAs as well. Since the new data provide additional constraints, a better consistency of the multipoles can be expected. The improvement in the data description is shown in Fig. 19 (bottom). The improvement in consistency of the multipoles is shown in Fig. 20 using the $E_0^+$-multipole as example. For a complete picture of the multipoles see [161]. Fig. 21 shows the improvement in the overall consistency of the BnGa, SAID and JüBo-analyses considering all $\gamma p \to p\pi^0$ multipoles up to L = 4. It becomes clearly visible that the new data enforces the multipoles to approach each other, the variance is reduced by about a factor of two.

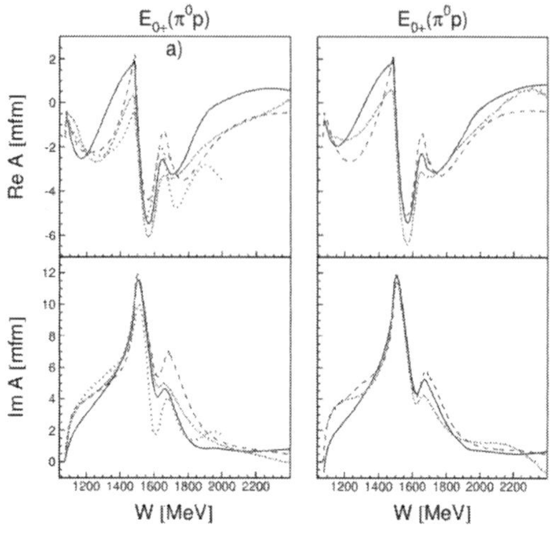


**Figure 20.** Real (top) and imaginary (bottom) part of the $E_0^+$-multipole before (left) and after (right) including new data, MAID (green), SAID (red), JüBo (blue), and BnGa (black).

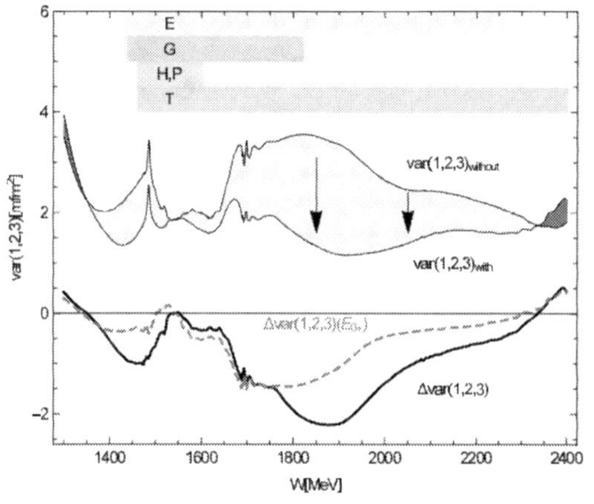


**Figure 21.** The variance of the BnGa, SAID and JüBo-analyses summed over all $\gamma p \to p\pi^0$ multipoles up to $L = 4$. The range covered by the new double polarization observables are indicated by shaded areas. Over the largest part of the energy range the new data have enforced an improvement of the overall consistency. The improvement is displayed as light green area and, separately as difference of the variance. The contribution to the improvement from the $E_0^+$ wave is shown as the dashed curve. Ranges with an overall deterioration are marked in red. For details see [161]

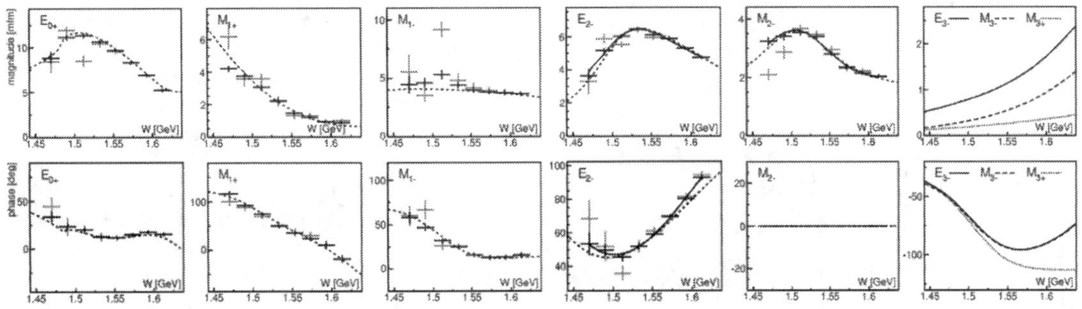


**Figure 22.** Magnitude (upper row) and phase (lower row) of multipoles derived from a fit to data in slices of energy. Grey (red) crosses show the results of an unbiased fit, black crosses represent a fit with a penalty function (see text). The dashed lines show the new energy-dependent Bonn-Gatchina fit (BnGa2014), the solid lines represent a Breit-Wigner plus background fit to the black crosses for $E_{2}$- and $M_{2}$-. The largest contributions from higher waves (from BnGa2014) are shown as well. The $E_{3}$- and $M_{3}$- multipoles excite the close-by resonance $N(1680)5/2^{+}$, $M_{3}$- excites $\Delta(1950)7/2^{+}$. Figure taken from [115].

**Table 8.** The $N(1520)\,3/2^{-}$ helicity amplitudes (in $\mathrm{GeV}^{-1/2}$) as determined in [115].

| $N(1520)\,3/2^{-}$ | BnGa [115] | CM12 [167] | SN11 [168] | BnGa [49] | PDG [96] |
|---|---|---|---|---|---|
| $A_{1/2}$ | -0.022<br>±0.009 | -0.019<br>±0.002 | -0.016<br>±0.002 | -0.022<br>±0.004 | -0.024<br>±0.009 |
| $A_{3/2}$ | 0.118<br>±0.020 | 0.141<br>±0.002 | 0.156<br>±0.002 | 0.131<br>±0.010 | 0.166<br>±0.005 |

### 3.10 Can photoproduction multipoles be determined in slices of fixed masses?

For a long time, the ultimate aim of PWAs of photo-induced reactions was to construct the four CGLN amplitudes from data binned in photon energy (or invariant $\gamma N$ mass) and the meson production angle $\theta$ [162]. It was shown that eight well chosen experiments are required (and sufficient) to determine these amplitudes [163–165], at least when these observables are measured with sufficiently (i.e. extremely) high precision. Ideally, one would expand the constructed amplitude at each mass value in $\cos\theta$-dependent multipoles; the results are then grouped according to their multipole structure and then given as functions of invariant mass. These energy-independent multipoles can than be fitted with dynamical reaction models to separate backgrounds and poles in a given multipole, and to determine the pole properties. However, even with extremely precise data, there is one overall phase per every bin in mass and meson production angle which is difficult to deal with. (One could, e.g., take one partial wave from an energy-dependent fit and determine the other phases relative to this wave.) A more fruitful path is a truncated partial wave analysis [166].

The measurement of single and double-polarization observables for the reaction $\gamma p \to p\pi^{0}$ were used to reconstruct multipoles with $L = 0$, 1 and 2 [115]. Multipoles with $L \geq 3$ were fixed to those from the BnGa energy-dependent partial wave analysis while magnitudes and phases for $E_{0^+}$, $E_{1^+}$, $E_{2^+}$, $E_{2^-}$, $M_{1^+}$, $M_{2^+}$, and $M_{1^-}$ were left free (see Fig. 22). The helicity amplitudes of $N(1520)\,3/2^{-}$ were thus deduced with minimal model assumptions. The results are collected in Table 8. They are partly inconsistent with older (model-dependent) determinations and support those of the energy-dependent BnGa PWA.

Production of $\Lambda$ hyperons is particularly fruitful to construct multipoles from the data since their decay reveals their degree of polarization. In a first step, we analyzed the reaction $\pi^{-}p \to \Lambda K^{0}$ [169] and determined low-lying multipoles. Ambiguities and large fluctuations showed up. Consistent results for the $S_{11}$, $P_{11}$, and

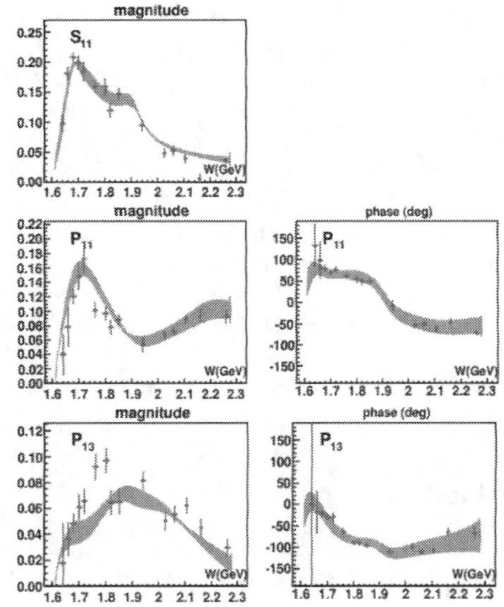


**Figure 23.** Fit with free $S$ and $P$-waves and $D, F$, and $G$-waves fixed to the energy-dependent solution BnGa2011-02 for $\gamma p \to K^{+}\Lambda$, see [170]

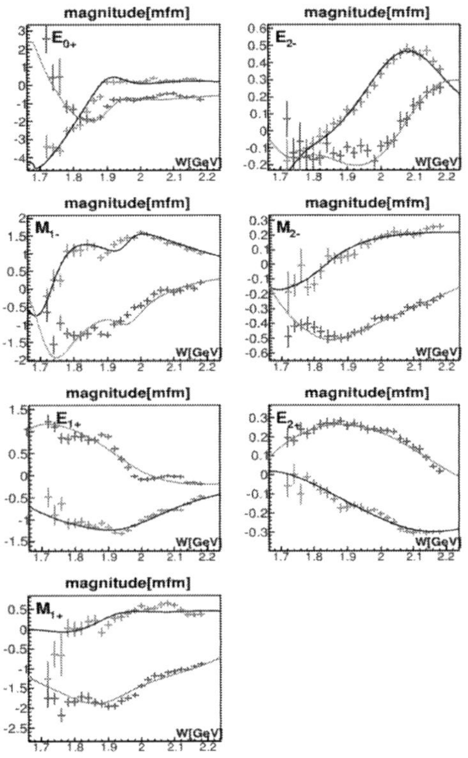


**Figure 24.** Preliminary results: real (red) and imaginary (blue) part of multipoles derived from a fit to data in slices of energy. The left column of subfigures is obtained without any constraints, for the multipoles in the right column, a penalty function is used. The lines show the energy-dependent Bonn-Gatchina fit. The results shown here were not included in the fit.

$P_{13}$ were obtained when the higher multipoles were fixed to the BnGa PWA solution (see Fig. 23). A first attempt to determine photoproduction multipoles for $\gamma p \to K^+\Lambda$ was made in [170]. However, new precise data became available recently [171]. They allowed for a much more precise reconstruction of the multipoles. In the new analysis [172], the four largest multipoles $E_{0+}$, $M_{1-}$, $E_{1+}$, and $M_{1+}$ were left free completely, the next three multipoles $E_{2-}$, $M_{2-}$, and $E_{2+}$ were "persuaded" by a penalty function to stay close to the BnGa energy-dependent fit (see Fig. 24). The multipoles are presently fitted applying the Laurent-Pittarinen expansion, a technique developed at Zagreb [173–175]. The preliminary results confirm with a model-independent method the existence of $N(1895)1/2^-$, $N(1880)1/2^+$, and $N(1900)3/2^+$ whose existence must now be considered as *established*; $N(1875)3/2^-$ is strongly supported.

## 4 Outlook

One of the most important topics in baryon spectroscopy is to identify the effective degrees of freedom and respective forces which determine the spectrum and properties of baryon resonances. This includes answering the question if the *missing resonances* are proof of a quark-diquark structure of baryon excitations. The answer to the latter question is mostly settled thanks to the efforts to collect a large amount of high-quality data (differential cross sections and single and double polarization observables) for a large variety of different reactions. The numerosity of observed resonances excludes quasi-stable diquarks as part of the dynamics of excited baryons.

Some of resonances used to argue in this paper are not yet considered to be *established* ($\geq$ 3-star-resonances). However, further data are still in the pipe line or will be taken in future; higher statistic data and data on new polarisation observables will be obtained. Hopefully, the few still missing resonances (below $\sim$2.2 GeV) will be discovered in these forthcoming data. When the data on $\gamma p \to \pi^+\pi^- p$ are included in the PWA data base, practically all possible decay modes are included, and the sum of all branching ratios has to be close to 1. When this condition is not fulfilled for a particular resonance, then that resonance should be considered with caution.

One class of baryon resonances has, however, so far escaped discovery: the lowest mass 20plet in Fig. 11 remains empty. A resonance X in the 20plet has an totally antisymmetric spin-flavor wave function. In quark models, these states have spacial wave functions in which both oscillators are excited simultaneously $\ell_\rho = \ell_\lambda = 1$ and $|\vec{L}_{tot}| = |\vec{\ell}_\rho + \vec{\ell}_\lambda| = 1$.

High-mass nucleon resonances can have wave functions with a component in which both oscillators are excited. In section 3.2 we have argued that this component decays into an excited state plus meson, and not into the ground state plus meson. This observation implies that the resonance X in the lowest-mass 20plet cannot be excited from the ground state. Possibly, X can be seen only in a double cascade. In a first step, a high-mass resonance R is excited, in a second step, the high-mass resonance de-excites into a member of the 20plet, which then decays in a two-step cascade into ground state plus two mesons. This scenario demands a partial wave analysis of four-body final states. The decay sequences might be

$$\gamma p \to R \to X + \pi \to N^*\pi + \pi \to N\pi\pi + \pi\,. \qquad (5)$$

The main problem is likely the combinatorial background.

An alternative approach is to search for the 20plet in the spectrum of $\Lambda$ excitations. The 20plet reduces to a spin-1/2 octet and a spin-3/2 singlet. One might search for a cascade

$$\begin{aligned} \gamma p \to R \quad &\to \quad \Lambda(xxx)J^P + K^+ \qquad (6)\\ &\to \quad N^*K^0 + K^+ \to N\pi K^0 + K^+\,. \end{aligned}$$

Expected are three states with $J^P = 1/2^+, 3/2^+, 5/2^+$, and a $J^P = 1/2^+, 3/2^+$ doublet. Their preferential decay modes should be two-step cascades. The lowest mass resonance belongs to a spin-1/2 octet and is predicted to have a mass of 1747 MeV [7]. Cascades like in (6) are energetically disfavored and thus, $\Lambda(1747)1/2^+$ might be narrow.

Of course, one has to keep in mind that our argument given in section 3.2, that wave function components where both oscillators are excited simultaneously decay into an

DOI: 10.1051/epjconf/201713402002

excited state plus meson and not into the ground state plus meson was deduced from the observation of different $p2\pi$-decay modes of $N^*$- and $\Delta^*$-resonances which individually still have large error bars. New high statistics $\gamma p \to p\pi^+\pi^-$ data, as well as new $\gamma p \to p2\pi^0$ data taken with polarized beam and polarized target are presently included in our data base. These data will provide further contraints to more firmly determine these decay modes.

The number of resonances could, of course, also be higher than quark models predict. Theories which generate baryons from meson-baryon interactions have made great success in reproducing precisely properties of some baryon resonances. So far they fail, however, to generate the full spectrum, and the relation between quark-model states and dynamically generated resonances is still unclear. Intriguing are poles like $N(1750)1/2^+$ [54] which are dynamically generated and not seen in other PWAs. Is this a genuine resonance? Then it should be seen as well in phenomenological coupled-channel analyses. More precise data will provide the potential for new discoveries, a more precise determination of the photoproduction multipoles and a better definition of branching ratios.

$N^*$-resonances may strongly couple to the neutron but they may only weekly or not at all couple to the proton. Therefore resonances, maybe even resonances of specific symmetries or nature, may escape detection if only proton data is analyzed. This bias can be overcome if also photoproduction data off the neutron, which is presently still scarce, is investigated. A model which predicts e.g. resonances with a large photocoupling for production off neutrons and a small coupling for production off protons is the chiral soliton model [142, 143].
One of the last missing degrees of freedom in the experiments is the isospin dependence since the electromagnetic interaction does not conserve isospin. The electromagnetic transition operator can be split into an isovector and an isoscalar part, giving rise to three independent matrix elements [176]. Photoproduction of isovector mesons, such as the pions, involve all three matrix elements, while in the case of isoscalar meson photoproduction only two contribute. To study the isospin structure of the amplitudes, photoproduction experiments on the neutron are unavoidable. This becomes directly obvious in the isoscalar meson production. It is hence important to study photoproduction off neutrons with the same precision as achieved for protons.

Topical will also be the search for baryon resonances with a large hybrid content. Dudek and Edwards calculated the hybrid contribution to the wave functions of baryon resonances and found that are are a few distinctive resonances in which the gluon field is excited [21]. Ordinary baryons and hybrid baryons show a different $Q^2$ dependence in electroproduction. In quark model calculations, the transverse helicity amplitude for a hybrid state decreases much faster than that for a radially excited state; in addition, the hybrid longitudinal helicity amplitude vanishes [31, 32]. Identification of hybrid contributions in high-mass baryon resonances is certainly a most challenging and rewarding task for future experimental and theoretical studies. Electroproduction of baryon resonances provides the necessary information to disentangle the $qqq$ component in the baryonic wave function, the meson-baryon cloud contributions and a potential hybrid component. The latter two contributions are expected to contribute especially at low $Q^2$.

Summarizing, we have seen that a number of important questions in baryon spectroscopy has found an answer. At the same time, new questions have emerged. To answer these questions, more precise data are required. Here, data on polarization observables will play a most significant role.

We appreciate the constructive atmosphere within the SFB/TR16 and would like to thank all SFB members, professors, post-docs, students, technicians, and administrative assistents for the fruitful cooperation. In particular, we would like to mention the contributions of V. Nikonov and A. Anisovich to this project. The work reported here would not have been possible without the support from the Deutsche Forschungsgemeinschaft within the SFB/TR16.

## References

[1] N. Isgur, "Why $N*$'s are important," nucl-th/0007008, JLAB-THY-00-20; in: NSTAR2000: Excited Nucleons and Hadronic Structure, Feb. 16-19, 2000, Newport News, Virginia. Proceedings: N. Isgur, V.D. Burkert *et al.* (eds.), Singapore, World Scientific, 2001.

[2] E. Klempt and J.M. Richard, "Baryon spectroscopy," Rev. Mod. Phys. **82**, 1095 (2010).

[3] V. Crede and W. Roberts, "Progress towards understanding baryon resonances," Rept. Prog. Phys. **76**, 076301 (2013).

[4] L.Y. Glozman, W. Plessas, K. Varga and R.F. Wagenbrunn, "Unified description of light and strange baryon spectra," Phys. Rev. D **58**, 094030 (1998).

[5] S. Capstick and W. Roberts, "Quark models of baryon masses and decays", Prog. Part. Nucl. Phys. **45**, S241 (2000).

[6] U.Löring, B.C. Metsch and H.R. Petry, "The light baryon spectrum in a relativistic quark model with instanton-induced quark forces: The non-strange baryon spectrum and ground-states", Eur. Phys. J. A **10**, 395 (2001).

[7] U. Löring, B.C. Metsch and H.R. Petry, "The light baryon spectrum in a relativistic quark model with instanton induced quark forces: The strange baryon spectrum", Eur. Phys. J. A **10**, 447 (2001).

[8] M. Ronniger and B.C. Metsch, "Effects of a spin-flavour dependent interaction on the baryon mass spectrum," Eur. Phys. J. A **47**, 162 (2011).

[9] E. Santopinto and J. Ferretti, "Strange and nonstrange baryon spectra in the relativistic interacting quark-diquark model with a Gürsey and Radicati-inspired exchange interaction," Phys. Rev. C **92**, 025202 (2015).

[10] R.G. Edwards, J.J. Dudek, D.G. Richards and S.J. Wallace, "Excited state baryon spectroscopy from lattice QCD", Phys. Rev. D **84**, 074508 (2011).

[11] G. Höhler, F. Kaiser, R. Koch and E. Pietarinen, "Handbook Of Pion Nucleon Scattering", Published by Fachinform. Zentr. Karlsruhe 1979, 440 P. (Physics Data, No.12-1 (1979)).

[12] R.E. Cutkosky, C.P. Forsyth, J.B. Babcock, R.L. Kelly and R.E. Hendrick, "Pion - Nucleon Partial Wave Analysis", 4th Int. Conf. on Baryon Resonances, Toronto, Canada, Jul 14-16, 1980. Published in Baryon 1980:19 (QCD161:C45:1980).

[13] R.A. Arndt, Z.-J. Li, L.D. Roper, R.L. Workman and J.M. Ford, "Pion - nucleon partial wave analysis to 2 GeV", Phys. Rev. D **43**, 2131 (1991).

[14] D.E. Groom *et al.* [Particle Data Group Collaboration], "Review of Particle Physics. Particle Data Group," Eur. Phys. J. C **15**, 1 (2000).

[15] R.A. Arndt, W.J. Briscoe, I.I. Strakovsky and R.L. Workman, "Extended partial-wave analysis of $\pi N$ scattering data", Phys. Rev. C **74**, 045205 (2006).

[16] I.G. Alekseev *et al.*, "Influence of spin-rotation measurements on partial-wave analyses of elastic pion nucleon scattering", Phys. Rev. C **55**, 2049 (1997).

[17] I.G. Alekseev *et al.*, "Measurements of spin rotation parameter A in pion proton elastic scattering at 1.62-GeV/c", Phys. Lett. B **485**, 32 (2000).

[18] I.G. Alekseev *et al.*, "Measurement of the spin rotation parameter A in the elastic pion proton scattering at 1.43-GeV/c", Eur. Phys. J. C **45**, 383 (2006).

[19] I.G. Alekseev *et al.*, "Backward asymmetry measurements in the elastic pion-proton scattering at resonance energies", Eur. Phys. J. A **39**, 163 (2009).

[20] S. Capstick and P.R. Page, "Hybrid and conventional baryons in the flux tube model," Phys. Rev. C **66**, 065204 (2002).

[21] J.J. Dudek and R.G. Edwards, "Hybrid baryons in QCD," Phys. Rev. D **85**, 054016 (2012).

[22] M. Anselmino, E. Predazzi, S. Ekelin, S. Fredriksson and D.B. Lichtenberg, "Diquarks," Rev. Mod. Phys. **65**, 1199 (1993).

[23] M. Kirchbach, M. Moshinsky and Y. F. Smirnov, "Baryons in O(4) and vibron model," Phys. Rev. D **64**, 114005 (2001).

[24] R.L. Jaffe and F. Wilczek, "Diquarks and exotic spectroscopy," Phys. Rev. Lett. **91**, 232003 (2003).

[25] R.L. Jaffe, "Exotica," Phys. Rept. **409**, 1 (2005).

[26] E. Santopinto, "An interacting quark-diquark model of baryons," Phys. Rev. C **72**, 022201 (2005).

[27] S. Eidelman *et al.* [Particle Data Group Collaboration], "Review of Particle Physics. Particle Data Group," Phys. Lett. B **592**, 1 (2004).

[28] R. Koniuk, N. Isgur, "Where have all the resonances gone? An analysis of baryon couplings in a quark model with chromodynamics", Phys. Rev. Lett. **44**, 845 (1980).

[29] H.P. Morsch and P. Zupranski, "Structure of the $P_{11}(1440)$ resonance from $\alpha p$ and $\pi N$ scattering," Phys. Rev. C **61**, 024002 (2000).

[30] A. V. Sarantsev *et al.*, "New results on the Roper resonance and the $P_{11}$ partial wave," Phys. Lett. B **659**, 94 (2008).

[31] Z. P. Li, "Photoproduction signatures of hybrid baryons: an application of the quark model with gluonic degrees of freedom," Phys. Rev. D **44**, 2841 (1991).

[32] Z.P. Li, V. Burkert and Z.J. Li, "Electroproduction of the Roper resonance as a hybrid state," Phys. Rev. D **46**, 70 (1992).

[33] S. Schneider, S. Krewald and U.-G. Meissner, "The reaction $\pi N \rightarrow \pi\pi N$ in a meson-exchange approach," Eur. Phys. J. A **28**, 107 (2006).

[34] I.G. Aznauryan *et al.* [CLAS Collaboration], "Electroexcitation of nucleon resonances from CLAS data on single pion electroproduction," Phys. Rev. C **80**, 055203 (2009) .

[35] R.H. Dalitz, "K-nucleon bound-state interpretation of the 1385 MeV $\pi - \Lambda$ resonance," Phys. Rev. Lett. **6**, 239 (1961).

[36] L.Y. Glozman, "Restoration of chiral and U(1)A symmetries in excited hadrons," Phys. Rept. **444**, 1 (2007).

[37] L.Y. Glozman, "Chiral multiplets of excited mesons," Phys. Lett. B **587**, 69 (2004).

[38] L.Y. Glozman, "Parity doublets and chiral symmetry restoration in baryon spectrum," Phys. Lett. B **475**, 329 (2000).

[39] W.J. Briscoe *et al.*, gwdac.phys. gwu.edu/.

[40] A. Anisovich, E. Klempt, A. Sarantsev and U. Thoma, "Partial wave decomposition of pion and photoproduction amplitudes", Eur. Phys. J. A **24**, 111 (2005).

[41] E. Klempt, A.V. Anisovich, V.A. Nikonov, A.V. Sarantsev and U. Thoma, "Phase motion of baryon resonances", Eur. Phys. J. A **29**, 307 (2006).

[42] A.V. Anisovich and A.V. Sarantsev, "Partial decay widths of baryons in the spin-momentum operator expansion method", Eur. Phys. J. A **30**, 427 (2006).

[43] A.V. Anisovich, V.V. Anisovich, E. Klempt, V.A. Nikonov and A.V. Sarantsev, "Baryon-baryon and baryon-antibaryon interaction amplitudes in the spin-momentum operator expansion method," Eur. Phys. J. A **34**, 129 (2007).

[44] A.V. Anisovich, A. Sarantsev, O. Bartholomy, E. Klempt, V.A. Nikonov and U. Thoma, "Photoproduction of baryons decaying into $N\pi$ and $N\eta$", Eur. Phys. J. A **25**, 427 (2005).

[45] A.V. Anisovich, V. Kleber, E. Klempt, V.A. Nikonov, A.V. Sarantsev and U. Thoma, "Baryon resonances and polarization transfer in hyperon photoproduction", Eur. Phys. J. A **34**, 423 (2007).

[46] A.V. Anisovich, E. Klempt, V.A. Nikonov, M.A. Matveev, A.V. Sarantsev and U. Thoma, "Photoproduction of pions and properties of baryon resonances from a Bonn-Gatchina partial wave analysis," Eur. Phys. J. A **44**, 203 (2010).

[47] A.V. Anisovich, E. Klempt, V.A. Nikonov, A.V. Sarantsev and U. Thoma, "P-wave excited baryons from pion- and photo-induced hyperon

production", Eur. Phys. J. A **47**, 27 (2011).

[48] A.V. Anisovich, E.Klempt, V.A. Nikonov, A.V. Sarantsev and U. Thoma, "Nucleon resonances in the fourth resonance region", Eur. Phys. J. A **47**, 153 (2011).

[49] A. V. Anisovich, R. Beck, E. Klempt, V. A. Nikonov, A. V. Sarantsev and U. Thoma, "Properties of baryon resonances from a multichannel partial wave analysis", Eur. Phys. J. A **48**, 15 (2012).

[50] T. Sato and T.-S.H. Lee, "Dynamical Models of the Excitations of Nucleon Resonances", J. Phys. G G **36**, 073001 (2009).

[51] M. Döring, C. Hanhart, F. Huang, S. Krewald and U.-G. Meißner, "The Role of the background in the extraction of resonance contributions from meson-baryon scattering", Phys. Lett. B **681**, 26 (2009).

[52] M. Döring, C. Hanhart, F. Huang, S. Krewald and U.-G. Meißner, "Analytic properties of the scattering amplitude and resonances parameters in a meson exchange model", Nucl. Phys. A **829**, 170 (2009).

[53] M. Döring, C. Hanhart, F. Huang, S. Krewald, U.-G. Meissner and D. Rönchen, "The reaction $\pi^+ p \to K^+\Sigma^+$ in a unitary coupled-channels model," Nucl. Phys. A **851**, 58 (2011).

[54] D. Rönchen *et al.*, "Coupled-channel dynamics in the reactions $\pi N \to \pi N, \eta N, K\Lambda, K\Sigma$," Eur. Phys. J. A **49**, 44 (2013).

[55] D. Rönchen *et al.*, "Photocouplings at the Pole from Pion Photoproduction," Eur. Phys. J. A **50**, 101 (2014) Erratum: [Eur. Phys. J. A **51**, 63 (2015)].

[56] D. Rönchen, M. Döring, H. Haberzettl, J. Haidenbauer, U.-G. Meißner and K. Nakayama, "Eta photoproduction in a combined analysis of pion- and photon-induced reactions," Eur. Phys. J. A **51**, 70 (2015).

[57] G.Y. Chen, S.S. Kamalov, S.N. Yang, D. Drechsel and L. Tiator, "Nucleon resonances in $\pi N$ scattering up to energies $s^{1/2} < 2.0$-GeV", Phys. Rev. C **76**, 035206 (2007).

[58] A. Matsuyama, T. Sato and T. -S. H. Lee, "Dynamical coupled-channel model of meson production reactions in the nucleon resonance region", Phys. Rept. **439**, 193 (2007).

[59] S. Sarkar, E. Oset and M.J. Vicente Vacas, "Baryonic resonances from baryon decuplet-meson octet interaction", Nucl. Phys. A **750**, 294 (2005). [Erratum-ibid. A **780** (2006) 78].

[60] G. Penner and U. Mosel, "Vector meson production and nucleon resonance analysis in a coupled channel approach for energies $m_N < s^{1/2} < 2$-GeV. 1. Pion induced results and hadronic parameters", Phys. Rev. C **66**, 055211 (2002).

[61] G. Penner and U. Mosel, "Vector meson production and nucleon resonance analysis in a coupled channel approach for energies $m_N < s^{1/2} < 2$-GeV. 2. Photon induced results," Phys. Rev. C **66**, 055212 (2002).

[62] I. Denisenko *et al.*, "$N^*$ decays to $N\omega$ from new data on $\gamma p \to \omega p$," Phys. Lett. B **755**, 97 (2016).

[63] D.M. Manley and E.M. Saleski, "Multichannel resonance parametrization of $\pi N$ scattering amplitudes", Phys. Rev. D **45**, 4002 (1992).

[64] R.A. Arndt, W.J. Briscoe, I.I. Strakovsky, R.L. Workman and M.M. Pavan, "Dispersion relation constrained partial wave analysis of $\pi N$ elastic and $\pi N \to \eta N$ scattering data: The Baryon spectrum", Phys. Rev. C **69**, 035213 (2004).

[65] V. Crede *et al.* [CB-ELSA Collaboration], "Photoproduction of $\eta$ mesons off protons for $0.75\,\mathrm{GeV} < E_\gamma < 3\,\mathrm{GeV}$," Phys. Rev. Lett. **94**, 012004 (2005).

[66] H. van Pee *et al.* [CB-ELSA Collaboration], "Photoproduction of $\pi^0$-mesons off protons from the $\Delta(1232)$ region to $E_\gamma = 3\,\mathrm{GeV}$," Eur. Phys. J. A **31**, 61 (2007).

[67] O. Bartholomy *et al.* [CB-ELSA Collaboration], "Neutral pion photoproduction off protons in the energy range $0.3\,\mathrm{GeV} < E_\gamma < 3\,\mathrm{GeV}$," Phys. Rev. Lett. **94**, 012003 (2005).

[68] O. Bartholomy *et al.* [CB-ELSA Collaboration], "Photoproduction of $\eta$-mesons off protons," Eur. Phys. J. A **33**, 133 (2007).

[69] A.V. Sarantsev, V.A. Nikonov, A.V. Anisovich, E. Klempt and U. Thoma, "Decays of baryon resonances into $\Lambda K^+, \Sigma^0 K^+$ and $\Sigma^+ K^0$," Eur. Phys. J. A **25**, 441 (2005).

[70] U. Thoma *et al.*, "$N^*$ and $\Delta^*$ decays into $N\pi^0\pi^0$," Phys. Lett. B **659**, 87 (2008).

[71] J. Beringer *et al.* [Particle Data Group Collaboration], "Review of Particle Physics (RPP)," Phys. Rev. D **86**, 010001 (2012).

[72] R. K. Bradford *et al.* [CLAS Collaboration], "First measurement of beam-recoil observables C(x) and C(z) in hyperon photoproduction," Phys. Rev. C **75**, 035205 (2007).

[73] V.A. Nikonov, A.V. Anisovich, E.Klempt, A.V. Sarantsev and U. Thoma, "Further evidence for N(1900) P(13) from photoproduction of hyperons," Phys. Lett. B **662**, 245 (2008).

[74] V. Sokhoyan *et al.* [CBELSA/TAPS Collaboration], "High-statistics study of the reaction $\gamma p \to p\, 2\pi^0$," Eur. Phys. J. A **51**, 95 (2015); Erratum: [Eur. Phys. J. A **51**, no. 12, 187 (2015)].

[75] E. Gutz *et al.* [CBELSA and TAPS Collaborations], "Photoproduction of meson pairs: First measurement of the polarization observable $I^{**}s$," Phys. Lett. B **687**, 11 (2010).

[76] E. Gutz *et al.* [CBELSA/TAPS Collaboration], "High statistics study of the reaction $\gamma p \to p\pi^0\eta$," Eur. Phys. J. A **50**, 74 (2014).

[77] Y. Assafiri *et al.*, "Double $\pi^0$ photoproduction on the proton at GRAAL," Phys. Rev. Lett. **90**, 222001 (2003).

[78] V.L. Kashevarov *et al.*, "Study of the $\gamma p \to \pi^0\pi^0 p$ reaction with the Crystal Ball/TAPS at the Mainz," Phys. Rev. C **85**, 064610 (2012).

[79] J.R.M. Annand *et al.* [A2 Collaboration], "First measurement of target and beam-target asymmetries in the $\gamma p \to \pi^0\eta p$ reaction," Phys. Rev. C **91**, 055208 (2015).

[80] V. Sokhoyan *et al.* [CBELSA/TAPS Collaboration], "Data on $I^s$ and $I^c$ in $\vec{\gamma}p \to p\pi^0\pi^0$ reveal cascade decays of $N(1900)$ via $N(1520)\pi$," Phys. Lett. B **746**, 127 (2015).

[81] A. Thiel *et al.* [CBELSA/TAPS Collaboration], "Three-body nature of $N^*$ and $\Delta^*$ resonances from sequential decay chains," Phys. Rev. Lett. **114**, 091803 (2015).

[82] J. Müller *et al.* [CBELSA/TAPS Collaboration], "New data on $\vec{\gamma}\vec{p} \to \eta p$ with polarized photons and protons and their implications for $N^* \to N\eta$ decays", submitted to Physics Letters B.

[83] C.S. Akondi *et al.*, "Measurement of the transverse target and beam-target asymmetries in $\eta$ meson photoproduction at MAMI," Phys. Rev. Lett. **113**, 102001 (2014).

[84] I. Senderovich *et al.* [CLAS Collaboration], "First measurement of the helicity asymmetry $E$ in $\eta$ photoproduction on the proton," Phys. Lett. B **755**, 64 (2016).

[85] V. Crede *et al.* [CBELSA/TAPS Collaboration], "Photoproduction of $\eta$ and $\eta'$ mesons off protons," Phys. Rev. C **80**, 055202 (2009).

[86] B. Krusche *et al.*, Prog. Part. Nucl. Phys. **51**, 399 (2003).

[87] K.A. Olive *et al.* [Particle Data Group Collaboration], "Review of Particle Physics," Chin. Phys. C **38**, 090001 (2014).

[88] A.V. Anisovich, E. Klempt, B. Krusche, V.A. Nikonov, A.V. Sarantsev, U. Thoma and D. Werthmüller, "Interference phenomena in the $J^P = 1/2^-$ wave in $\eta$ photoproduction," Eur. Phys. J. A **51**, 72 (2015).

[89] I. Jaegle *et al.* [CBELSA/TAPS Collaboration], "Photoproduction of $\eta'$-mesons off the deuteron," Eur. Phys. J. A **47**, 11 (2011).

[90] A.V. Anisovich, E. Klempt, V.A. Nikonov, A.V. Sarantsev, H. Schmieden and U. Thoma, "Evidence for a negative-parity spin-doublet of nucleon resonances at 1.88 GeV," Phys. Lett. B **711**, 162 (2012).

[91] R. Castelijns *et al.* [CBELSA/TAPS Collaboration], "Nucleon resonance decay by the $K^0\Sigma^+$ channel," Eur. Phys. J. A **35**, 39 (2008).

[92] R. Ewald *et al.* [CBELSA/TAPS Collaboration], "Anomaly in the $K^0_S\Sigma^+$ photoproduction cross section off the proton at the $K^*$ threshold," Phys. Lett. B **713**, 180 (2012).

[93] R. Ewald *et al.* [CBELSA/TAPS Collaboration], "Measurement of polarisation observables in $K^0_s\Sigma^+$ photoproduction off the proton," Phys. Lett. B **738**, 268 (2014).

[94] D. Elsner *et al.* [CBELSA and TAPS Collaborations], "Measurement of the beam asymmetry in $\eta$-photoproduction off the proton," Eur. Phys. J. A **33**, 147 (2007).

[95] N. Sparks *et al.* [CBELSA/TAPS Collaboration], "Measurement of the Beam Asymmetry $\Sigma$ in the Forward Direction for $\vec{\gamma}p \to p\pi^0$ Photoproduction," Phys. Rev. C **81**, 065210 (2010).

[96] K. Nakamura *et al.* [Particle Data Group Collaboration], "Review of particle physics," J. Phys. G **37**, 075021 (2010).

[97] E. Klempt and B.C. Metsch, "Multiplet classification of light-quark baryons," Eur. Phys. J. A **48**, 127 (2012).

[98] M. Shrestha and D.M. Manley, "Multichannel parametrization of $\pi N$ scattering amplitudes and extraction of resonance parameters," Phys. Rev. C **86**, 055203 (2012).

[99] O. V. Maxwell, "Electromagnetic production of kaons from protons, and baryon electromagnetic form factors," Phys. Rev. C **85**, 034611 (2012).

[100] T. Mart and M.J. Kholili, "Origin of the second peak in the cross section of the $K^+\Lambda$ photoproduction," Phys. Rev. C **86**, 022201 (2012)

[101] V. Shklyar, H. Lenske and U. Mosel, "$\eta$-meson production in the resonance-energy region," Phys. Rev. C **87**, 015201 (2013).

[102] A.V. Anisovich, E. Klempt, V.A. Nikonov, A.V. Sarantsev and U. Thoma, "Evidence for a spin-quartet of nucleon resonances at 2 GeV," Phys. Lett. B **711**, 167 (2012).

[103] A.V. Anisovich, V. Burkert, E. Klempt, V.A. Nikonov, A.V. Sarantsev and U. Thoma, "Helicity amplitudes for photoexcitation of nucleon resonances off neutrons," Eur. Phys. J. A **49**, 67 (2013).

[104] A.V. Anisovich *et al.*, "Neutron helicity amplitudes", in preparation.

[105] N. Matagne and F. Stancu, "Updated $1/N_c$ expansion analysis of $[\mathbf{56}, \mathbf{2}^+]$ and $[\mathbf{70}, \ell^+]$ baryon multiplets," Phys. Rev. D **93**, 096004 (2016).

[106] S. Weinberg, "Phenomenological Lagrangians," Physica A **96**, 327 (1979).

[107] M. Gell-Mann, R. J. Oakes and B. Renner, "Behavior of current divergences under SU(3) x SU(3)," Phys. Rev. **175**, 2195 (1968).

[108] Y. Nambu, "Quasiparticles and Gauge Invariance in the Theory of Superconductivity," Phys. Rev. **117**, 648 (1960).

[109] J. Goldstone, "Field theories with superconductor solutions," Nuovo Cim. **19**, 154 (1961).

[110] I. Horn *et al.* [CB-ELSA Collaboration], "Evidence for a parity doublet $\Delta(1920)P_{33}$ and $\Delta(1940)D_{33}$ from $\gamma p \to p\pi^0\eta$," Phys. Rev. Lett. **101**, 202002 (2008).

[111] I. Horn *et al.* [CB-ELSA Collaboration], "Study of the reaction $\gamma p \to p\pi^0\eta$," Eur. Phys. J. A **38**, 173 (2008).

[112] L. Y. Glozman and A. V. Nefediev, "Chiral restoration in excited nucleons versus SU(6)," Nucl. Phys. A **807**, 38 (2008).

[113] E. Klempt and A. Zaitsev, "Glueballs, Hybrids, Multiquarks. Experimental facts versus QCD inspired concepts," Phys. Rept. **454**, 1 (2007).

[114] L.Y. Glozman, private communication to E.K.

[115] J. Hartmann *et al.*, "The $N(1520)3/2^+$ helicity amplitudes from an energy-independent multipole analysis based on new polarization data on photoproduction of

neutral pions," Phys. Rev. Lett. **113**, 062001 (2014).

[116] J. Hartmann *et al.* [CBELSA/TAPS Collaboration], "The polarization observables T, P, and H and their impact on $\gamma p \to p\pi^0$ multipoles," Phys. Lett. B **748**, 212 (2015).

[117] M. Gottschall *et al.* [CBELSA/TAPS Collaboration], "First measurement of the helicity asymmetry for $\gamma p \to p\pi^0$ in the resonance region," Phys. Rev. Lett. **112**, 012003 (2014).

[118] A. Thiel *et al.*, "Double-polarization observable G in neutral-pion photoproduction off the proton," submitted to EPJA, arXiv:1604.02922 [nucl-ex].

[119] S. Strauch *et al.* [CLAS Collaboration], "First Measurement of the Polarization Observable E in the $\vec{p}(\vec{\gamma}, \pi^+)n$ Reaction up to 2.25 GeV," Phys. Lett. B **750**, 53 (2015).

[120] E. Klempt, private communication.

[121] A. V. Anisovich, V. Burkert, E. Klempt, V. A. Nikonov, E. Pasyuk, A. V. Sarantsev, S. Strauch and U. Thoma, "Existence of $\Delta(2200)7/2^-$ precludes chiral symmetry restoration at high mass," arXiv:1503.05774 [nucl-ex].

[122] E. Klempt, "A Mass formula for baryon resonances," Phys. Rev. C **66**, 058201 (2002).

[123] H. Forkel and E. Klempt, "Diquark correlations in baryon spectroscopy and holographic QCD," Phys. Lett. B **679**, 77 (2009).

[124] E. Klempt, "Nucleon Excitations," Chin. Phys. C **34**, 1241 (2010).

[125] E. Klempt, "Do parity doublets in the baryon spectrum reflect restoration of chiral symmetry?," Phys. Lett. B **559**, 144 (2003).

[126] E. Klempt, "Delta resonances, Quark models, chiral symmetry and AdS/QCD," Eur. Phys. J. A **38**, 187 (2008).

[127] G. Wolf, "Review of High Energy Diffraction in Real and Virtual Photon Proton scattering at HERA," Rept. Prog. Phys. **73**, 116202 (2010).

[128] A. Donnachie and P. V. Landshoff, "Exclusive vector meson production at HERA," Phys. Lett. B **348**, 213 (1995).

[129] A. Donnachie and P. V. Landshoff, "Exclusive vector photoproduction: Confirmation of Regge theory," Phys. Lett. B **478**, 146 (2000).

[130] J. Barth *et al.*, "Low-energy of photoproduciton of $\omega$-mesons," Eur. Phys. J. A **18**, 117 (2003).

[131] J. Ajaka *et al.* [GRAAL Collaboration], "Evidence for nucleon-resonance excitation in $\omega$-meson photoproduction," Phys. Rev. Lett. **96**, 132003 (2006).

[132] F. Klein *et al.* [CBELSA/TAPS Collaboration], "Beam asymmetry in near threshold $\omega$ photoproduction off the proton," Phys. Rev. D **78**, 117101 (2008).

[133] M. Williams *et al.* [CLAS Collaboration], "Differential cross sections and spin density matrix elements for the reaction $\gamma p \to p\omega$," Phys. Rev. C **80**, 065208 (2009).

[134] M. Williams *et al.* [CLAS Collaboration], "Partial wave analysis of the reaction $\gamma p \to p\omega$ and the search for nucleon resonances," Phys. Rev. C **80**, 065209 (2009).

[135] I.I. Strakovsky *et al.*, "Photoproduction of the $\omega$ meson on the proton near threshold," Phys. Rev. C **91**, 045207 (2015) .

[136] A. Wilson *et al.* [CBELSA/TAPS Collaboration], "Photoproduction of $\omega$ mesons off the proton," Phys. Lett. B **749**, 407 (2015).

[137] H. Eberhardt *et al.*, "Measurement of double polarisation asymmetries in $\omega$-photoproduction," Phys. Lett. B **750**, 453 (2015).

[138] F. Dietz *et al.* [CBELSA/TAPS Collaboration], "Photoproduction of $\omega$ mesons off protons and neutrons," Eur. Phys. J. A **51**, 6 (2015).

[139] J. Junkersfeld *et al.* [CB-ELSA Collaboration], "Photoproduction of $\pi^0\omega$ off protons for $E_\gamma \leq 3\,\text{GeV}$," Eur. Phys. J. A **31**, 365 (2007).

[140] A.V. Anisovich *et al.* [CLAS Collaboration], "Differential cross sections and polarization observables from the CLAS $K^*$ photoproduction and search for new $N^*$ states", in collaboration review.

[141] M. Nanova *et al.* [CBELSA/TAPS Collaboration], "$K^0\pi^0\Sigma^+$ and $K^{*0}\Sigma^+$ photoproduction off the proton," Eur. Phys. J. A **35**, 333 (2008).

[142] D. Diakonov, V. Petrov and M.V. Polyakov, "Exotic anti-decuplet of baryons: Prediction from chiral solitons," Z. Phys. A **359**, 305 (1997).

[143] M.V. Polyakov and A. Rathke, "On photoexcitation of baryon anti-decuplet," Eur. Phys. J. A **18**, 691 (2003).

[144] V. Kuznetsov and M.V. Polyakov, "New Narrow Nucleon $N^*(1685)$," JETP Lett. **88**, 347 (2008).

[145] R.A. Arndt, Y.I. Azimov, M.V. Polyakov, I.I. Strakovsky and R.L. Workman, "Nonstrange and other unitarity partners of the exotic $\Theta^+$ baryon," Phys. Rev. C **69**, 035208 (2004).

[146] V. Kuznetsov *et al.* [GRAAL Collaboration], "Evidence for a narrow structure at W 1.68-GeV in $\eta$ photoproduction on the neutron," Phys. Lett. B **647**, 23 (2007).

[147] T. Liu, Y. Mao and B.Q. Ma, "Present status on experimental search for pentaquarks," Int. J. Mod. Phys. A **29**, 1430020 (2014).

[148] I. Jaegle *et al.* [CBELSA and TAPS Collaborations], "Quasi-free photoproduction of $\eta$-mesons of the neutron," Phys. Rev. Lett. **100**, 252002 (2008).

[149] I. Jaegle *et al.*, "Quasi-free photoproduction of $\eta$-mesons off the deuteron," Eur. Phys. J. A **47**, 89 (2011).

[150] D. Werthmüller *et al.* [A2 Collaboration], "Narrow Structure in the Excitation Function of $\eta$ Photoproduction off the Neutron," Phys. Rev. Lett. **111**, 232001 (2013).

[151] V. Shklyar, H. Lenske and U. Mosel, "$\eta$-photoproduction in the resonance energy region," Phys. Lett. B **650**, 172 (2007).

[152] R. Shyam and O. Scholten, "Photoproduction of $\eta$ meson within a coupled-channels K-matrix approach," Phys. Rev. C **78**, 065201 (2008).

[153] M. Döring and K. Nakayama, "On the cross section ratio $\sigma(n)/\sigma(p)$ in $\eta$ photoproduction," Phys. Lett. B **683**, 145 (2010).

[154] A.V. Anisovich, I. Jaegle, E. Klempt, B. Krusche, V.A. Nikonov, A.V. Sarantsev and U. Thoma, "Photoproduction of $\eta$ mesons off neutrons from a deuteron target," Eur. Phys. J. A **41**, 13 (2009).

[155] A. V. Anisovich, E. Klempt, V. Kuznetsov, V. A. Nikonov, M. V. Polyakov, A. V. Sarantsev and U. Thoma, "Study of the narrow structure at 1685 MeV in $\gamma p \to \eta p$," Phys. Lett. B **719**, 89 (2013).

[156] X.H. Zhong and Q. Zhao, "$\eta$ photoproduction on the quasi-free nucleons in the chiral quark model," Phys. Rev. C **84**, 045207 (2011).

[157] A. Fantini *et al.*, "First measurement of the $\Sigma$ beam asymmetry in $\eta$ photoproduction on the neutron," Phys. Rev. C **78**, 015203 (2008).

[158] R. Di Salvo *et al.*, "Measurement of $\Sigma$ beam asymmetry in $\pi^0$ photoproduction off the neutron in the second and third resonances region," Eur. Phys. J. A **42**, 151 (2009).

[159] A.V. Anisovich, E. Klempt, V.A. Nikonov, A.V. Sarantsev and U. Thoma, "Sign ambiguity in the K$\Sigma$ channel," Eur. Phys. J. A **49**, 158 (2013).

[160] A.V. Anisovich, R. Beck, E. Klempt, V.A. Nikonov, A.V. Sarantsev and U. Thoma, "Pion- and photo-induced transition amplitudes to $\Lambda K$, $\Sigma K$, and $N\eta$," Eur. Phys. J. A **48**, 88 (2012).

[161] A.V. Anisovich *et al.*, "The Impact of New Polarization Data from Bonn, Mainz and Jefferson Laboratory on $\gamma p \to \pi N$ Multipoles," Eur. Phys. J. A **52**, 284 (2016).

[162] G.F. Chew, M.L. Goldberger, F.E. Low and Y. Nambu, "Relativistic dispersion relation approach to photomeson production," Phys. Rev. **106**, 1345 (1957). [163]

[163] I.S. Barker, A. Donnachie and J K. Storrow, "Complete Experiments in Pseudoscalar Photoproduction," Nucl. Phys. B **95**, 347 (1975).

[164] W.T. Chiang and F. Tabakin, "Completeness rules for spin observables in pseudoscalar meson photoproduction," Phys. Rev. C **55**, 2054 (1997).

[165] A. M. Sandorfi, S. Hoblit, H. Kamano and T.-S. H. Lee, "Determining pseudoscalar meson photoproduction amplitudes from complete experiments," J. Phys. G **38**, 053001 (2011).

[166] Y. Wunderlich, R. Beck and L. Tiator, "The complete-experiment problem of photoproduction of pseudoscalar mesons in a truncated partial-wave analysis," Phys. Rev. C **89**, 055203 (2014).

[167] R.L. Workman, M.W. Paris, W.J. Briscoe and I.I. Strakovsky, "Unified Chew-Mandelstam SAID analysis of pion photoproduction data," Phys. Rev. C **86**, 015202 (2012).

[168] R.L. Workman, W.J. Briscoe, M.W. Paris and I.I. Strakovsky, "Updated SAID analysis of pion photoproduction data," Phys. Rev. C **85**, 025201 (2012).

[169] A.V. Anisovich, R. Beck, E. Klempt, V.A. Nikonov, A.V. Sarantsev, U. Thoma and Y. Wunderlich, "Study of ambiguities in $\pi^- p \to \Lambda K^0$ scattering amplitudes," Eur. Phys. J. A **49**, 121 (2013).

[170] A.V. Anisovich *et al.*, "Energy-independent PWA of the reaction $\gamma p \to K^+\Lambda$," Eur. Phys. J. A **50**, 129 (2014).

[171] C. A. Paterson *et al.* [CLAS Collaboration], "Photoproduction of $\Lambda$ and $\Sigma^0$ hyperons using linearly polarized photons," Phys. Rev. C **93**, 065201 (2016).

[172] A.V. Anisovich *et al.*, "First (nearly) model-independent confirmation of resonances in the fourth resonance region", in preparation.

[173] A. Svarc, M. Hadzimehmedovic, H. Osmanovic, J. Stahov, L. Tiator and R.L. Workman, "Introducing the Pietarinen expansion method into the single-channel pole extraction problem," Phys. Rev. C **88**, 035206 (2013).

[174] A. Svarc, M. Hadzimehmedovic, H. Osmanovic, J. Stahov, L. Tiator and R.L. Workman, "Pole positions and residues from pion photoproduction using the Laurent-Pietarinen expansion method," Phys. Rev. C **89**, 065208 (2014).

[175] A. Svarc, M. Hadzimehmedovic, H. Osmanovic, J. Stahov, L. Tiator and R.L. Workman, "Generalization of the model-independent Laurent–Pietarinen single-channel pole-extraction formalism to multiple channels," Phys. Lett. B **755**, 452 (2016).

[176] K. M. Watson, "The Hypothesis of Charge Independence for Nuclear Phenomena," Phys. Rev. **85**, 852 (1952).

EPJ Web of Conferences **134**, 02003 (2017) DOI: 10.1051/epjconf/201713402003
*Subnuclear Structure of Matter: Achievements and Challenges*

# Magnetic Moments of Excited Baryons

*Volker* Metag[1,*]

[1] *II. Physikalisches Institut, Universität Gießen, Heinrich Buff Ring 16, 35392 Gießen*

**Abstract.** In project **A.3**, the reaction $\gamma p \rightarrow \pi^0\gamma' p$ has been studied using the TAPS photon spectrometer in the energy range $\sqrt{s}$= 1221-1331 MeV. Energy tagged photon beams have been produced with the Glasgow tagging spectrometer from electron beams provided by the MAMI-B accelerator. Angle and energy differential cross sections have been measured and compared to theoretical calculations. This comparison allows the magnetic moment of the $\Delta^+$ isobar to be extracted for the first time to $\mu_{\Delta^+}$ = [2.7$^{+1.3}_{-1.0}$(stat)±1.5(syst)±3(theo)] $\mu_N$. In an extension of the A3 project to the meson sector, the time-like transition form factor of the $\eta$ meson has been measured with the Crystal Ball/TAPS detector system at MAMI-C.

## 1 Introduction

The main goal of the Transregio TR16 is to explore the excitation energy spectrum of the nucleon in order to unravel its complex structure. Apart from determining the mass, width and quantum number of excited states, information on the polarizability, magnetic moments and form factors provide further insight in the structure of the nucleon. For baryon ground states these properties have been measured with great precision. Corresponding information on excited states has so far been lacking. Because of its prominent position in the excitation energy spectrum, the properties of the $\Delta(1232)$ resonance are of particular importance.

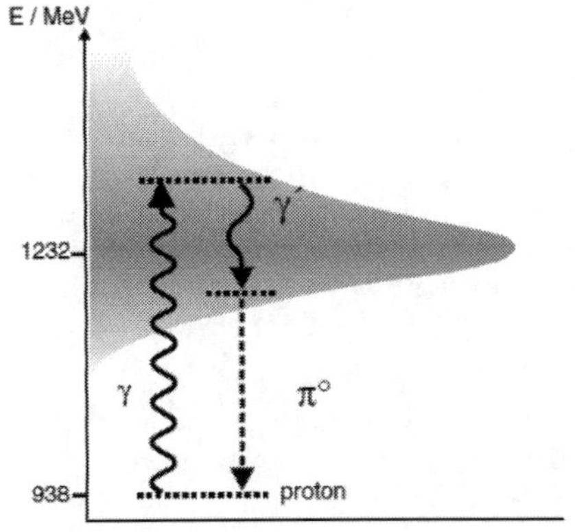


**Figure 1.** Scheme for studying the static electromagnetic properties of the $\Delta(1232)$ isobar. The $\gamma'$ transition carries the information on the magnetic moment of the $\Delta^+$.

Following pioneering experiments determining the magnetic moment of the $\Delta^{++}$ resonance [1, 2], a first attempt has been made within the A3 project to measure the magnetic moment of the $\Delta^+$ resonance. The idea of the experiment is sketched in Fig. 1. The broad $\Delta^+$ resonance ($\Gamma$ = 120 MeV) is populated in the high mass regime via photon excitation, followed by a $\gamma$ transition within the resonance and a subsequent $\pi^0$ decay of the $\Delta^+$ resonance to the nucleon ground state. Spin and parity conservation require that - to lowest multipole order - this $\gamma'$ transition is of magnetic dipole (M1) type. The next allowed multipole is electric quadrupole (E2), but the E2/M1ratio of the $\Delta$ - N transition has been measured to be small (0.025) [3], indicating a small quadrupole deformation of the $\Delta^+$ resonance. The magnetic octupole transition (M3) is suppressed by two additional powers of the photon energy. Thus, the measurement of the $\gamma p \rightarrow \pi^0\gamma' p$ reaction provides access to $\mu_{\Delta^+}$. The sensitivity of the $\Delta^+ \rightarrow \Delta^+\gamma'$ amplitude to the magnetic moment $\mu_{\Delta^+}$ has been investigated theoretically by Machavariani et al. [4, 5] and Drechsel [6]. The same $\pi^0\gamma' p$ final state can, however, also result from bremsstrahlung radiation of the $\Delta^+$ or the proton. For a comparison with the experiment, these channels have to be included in a calculation as in [7]. Fig. 2 shows the graph of interest (a2) and all other graphs contributing to the $\gamma p \rightarrow \pi^0\gamma' p$ reaction. Our experimental data will be compared to these extended calculations by Drechsel and Vanderhaegen [7].

## 2 The experiment

The reaction $\gamma p \rightarrow \pi^0\gamma' p$ was measured at the electron accelerator MAMI-B, using the Glasgow Tagged photon facility [8] and the photon spectrometer TAPS [9, 10]. A quasi monochromatic photon beam was produced in the energy range of 205-820 MeV via bremsstrahlung tagging. The 510 hexagonal $BaF_2$ TAPS modules were arranged in

*e-mail: Volker.Metag@exp2.physik.uni.giessen.de

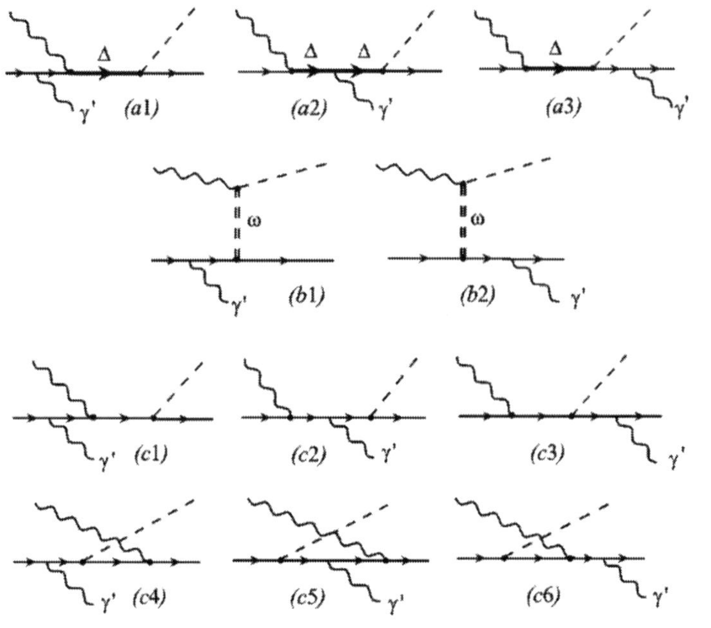


**Figure 2.** Diagrams considered in the calculation of the $\gamma p \rightarrow \pi^0\gamma' p$ reaction in the $\Delta(1232)$ energy region [7].

a horizontal plane around the liquid hydrogen target in 6 blocks of 62 detectors and a forward wall of 138 detectors. The setup covered about 40% of the full solid angle. Further details of the experimental setup and the running conditions can be found in [11].

## 3 Data analysis

The $\gamma p \rightarrow \pi^0\gamma' p$ was measured exclusively since the 4-momentum vectors of all final state particles were determined. The $\pi^0$ mesons were detected via their two photon decay and identified via an invariant mass analysis. The third photon in the final state was assigned to the $\gamma'$ transition. The protons were identified by time-of-flight, exploiting the excellent time resolution of TAPS, and by the energy deposited in the $BaF_2$ crystals. The main experimental background was the $2\pi^0$ channel where one of the 4 decay photons escaped undetected. This background could be removed by a cut in the $\pi^0 p$ missing mass spectrum shown in Fig. 3. For the $\gamma p \rightarrow \pi^0\gamma' p$ reaction the missing mass is 0, while $2\pi^0$ events with one decay photon lost were centered around a missing mass of 0.02 GeV$^2$.

## 4 Results

The measured angle and energy differential cross sections are shown in Fig. 4. For the higher incident photon energies a clear deviation from a $1/E_\gamma$ bremsstrahlung distribution is observed, indicating contributions from $\gamma'$ transitions within the $\Delta^+$ resonance. The data are compared with calculations by Drechsel and Vanderhaegen [7] for

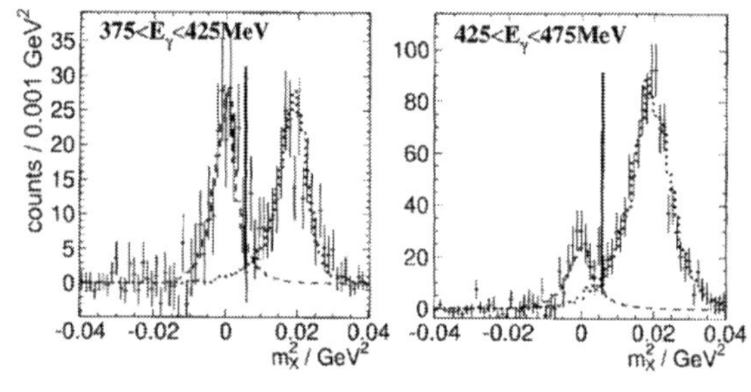


**Figure 3.** Missing mass of the $\pi^0 p$ system for two incident photon energy bins. The peak near 0.02 GeV$^2$ originates from $2\pi^0$ production and is cut away. The peak at 0 GeV$^2$ corresponds to the $\pi^0\gamma'$ $p$ exit channel. The dashed and dotted curves show the corresponding line shapes simulated with GEANT3.

different values of $\kappa_{\Delta^+}$, related to the magnetic moment of the $\Delta^+$ resonance by

$$\mu_{\Delta^+} = (1 + \kappa_{\Delta^+})\frac{m_N}{m_\Delta}\mu_N. \qquad (1)$$

The calculations follow the overall shape of the experimental distributions but overestimate the data for the highest excitation energy. A model independent determination of the $\gamma p \rightarrow \pi^0\gamma' p$ cross section is possible in the soft photon limit, which relates the $\pi^0\gamma' p$ production to the

 DOI: 10.1051/epjconf/201713402003

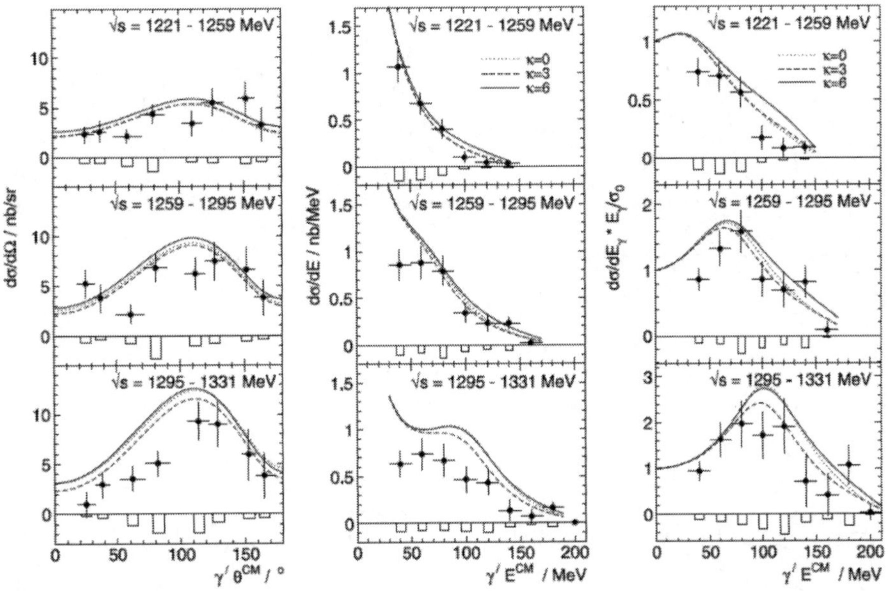


**Figure 4.** Differential cross sections for three excitation energy bins $\sqrt{s}$ in the center of mass system. The curves represent calculations [7] for $\kappa_{\Delta^+} = 0$, 3, and 6. Bars indicate the systematic errors. (Left) Angular distribution of the $\gamma$' photon. (Middle) Energy distribution of the $\gamma$' photon. (Right) Energy distribution divided by the soft photon limit $\frac{\sigma_0}{E_\gamma}$ for data and calculations.

$\pi^0 p$ production in the limit of vanishing photon energy $E_{\gamma'}$. The ratio of both cross sections approaches unity for small transition energies $E_{\gamma'}$ (see Fig. 4). The cross section ratios are less sensitive to experimental and theoretical uncertainties and are thus used to extract $\mu_{\Delta^+}$. A combined maximum likelihood analysis of all three cross section ratios in Fig. 4 (right column) gives a magnetic moment of the $\Delta^+$ resonance of $\mu_{\Delta^+} = [2.7^{+1.3}_{-1.0}(\text{stat})\pm1.5(\text{syst})\pm3(\text{theo})]\ \mu_N$; the goodness of the fit is $\chi^2/f = 1.8$ ($f = 21$). The first error represents the statistical uncertainty, and the second one reflects the systematic errors of the experiment indicated in Fig. 4. The systematic uncertainty of the model calculation is estimated to be $\approx \pm3\ \mu_N$. This determination of $\mu_{\Delta^+}$ is unfortunately not precise enough to allow a detailed test of different baryon structure calculations.

The above analysis is simplified in so far as the magnetic moment of the $\Delta$ - as of any other unstable state - is a complex quantity. Treating the $\Delta$ as a resonance (not as a stable particle), an explicit prescription has been developed in project A6 how to extract the $\Delta N\gamma*$ transition form factor from lattice data [12].

In continuation of the A3 project a follow-up experiment was performed at MAMI-B employing the Crystal Ball/TAPS detector system with almost full coverage of the solid angle. The improved experimental conditions led to 60 times higher statistics. The results of these measurements were published in [13]. Despite of the much better data set an improved value of the magnetic moment $\mu_{\Delta^+}$ could not be extracted because the available calculations did not reproduce in detail the measured high statistics $\gamma$' angle and energy distributions.

## 5 The time-like transition form factor of the $\eta$ meson

In an extension of project A3 to the meson sector the time-like transition form factor of the $\eta$ meson has been measured using the Crystal Ball/TAPS detector system at MAMI-C. Electromagnetic transition form factors are an important tool for studying the intrinsic structure of hadrons. The decay of light pseudoscalar mesons $\pi^0, \eta$, and $\eta'$ as well as vector mesons $\rho, \omega$ and $\phi$ allow a test of the vector meson dominance (VMD) conjecture where the coupling of a photon to a hadron is mediated by a virtual vector meson. For point-like particles the decay rates can be exactly calculated within Quantum Electrodynamics (QED). Information on the intrinsic structure of mesons is encoded in the multiplicative transition form factor. The transition form factors can be studied for momentum transfers not accessible in annihilation or scattering experiments. Furthermore, new theoretical efforts aiming at a systematic description of light pseudoscalar and vector mesons and their couplings to electromagnetism [14] can be tested. An overview over earlier work in this field can be found in [15].

In this experiment the $\eta$ Dalitz decay $\eta \rightarrow \gamma\gamma* \rightarrow \gamma\ e^+ e^-$ with a branching ratio of $6 \cdot 10^{-3}$ has been investigated. The reconstruction of such a weak decay channel is a major experimental challenge and could only succeed

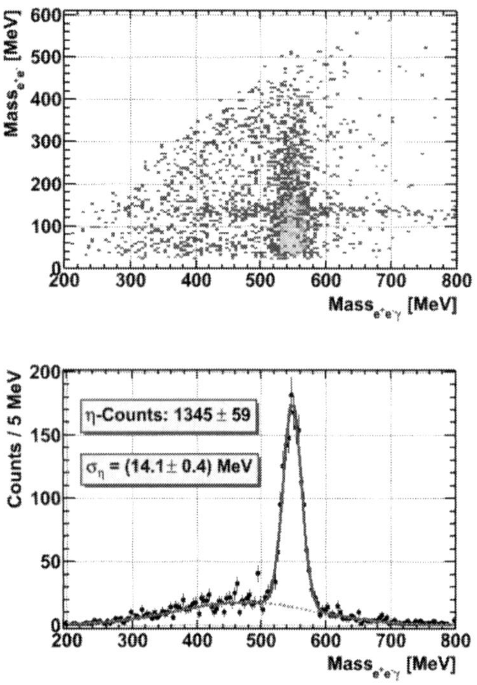


**Figure 5.** Distribution of reconstructed $e^+e^-\gamma$ candidate events after applying the selection cuts. The lower figure shows the projection onto the $e^+e^-\gamma$ invariant mass axis [16].

by using the almost complete $4\pi$ coverage of the combined Crystal Ball and TAPS detector system at MAMI-C. The disadvantage of the detector system is that it does not provide a magnetic field for charged particle tracking. Only the energy and angle of registered particles is available and the information whether they are charged or neutral. The sign of the charge cannot be determined.

The identification of this weak decay channel required a series of cuts described in detail in [16]. These cuts reduced the reconstruction efficiency of the $\gamma p \to p\eta \to p\gamma e^+e^-$ reaction to about 5%. For events that have survived these cuts the correlation between the $e^+e^-$ invariant mass and the $e^+e^-\gamma$ invariant mass as well as the projection on the $e^+e^-\gamma$ invariant mass axis are shown in Fig. 5. A total of 1345 $\eta \to \gamma e^+e^-$ events residing on a small background could be reconstructed. Fitting the $\gamma e^+e^-$ mass spectra for different slices in $\mathrm{Mass}_{e^+e^-}$, the $\mathrm{Mass}_{e^+e^-}$ distribution shown in Fig. 6 is obtained. The mass distribution is compared to the QED prediction for a point-like meson. Dividing the measured $\mathrm{Mass}_{e^+e^-}$ distribution by the QED prediction, the time-like transition form factor of the $\eta$ meson is obtained as shown in Fig. 7. Fitting the form factor using the standard VMD parametrization

$$F(m^2) = \frac{1}{1-(m/\Lambda)^2}, \qquad (2)$$

a value of $\Lambda^{-2}$ = (1.92 ± 0.35(stat)±0.13(syst) ) $\mathrm{GeV}^{-2}$ is obtained corresponding to $\Lambda$ = (720 ± 60(stat)±25

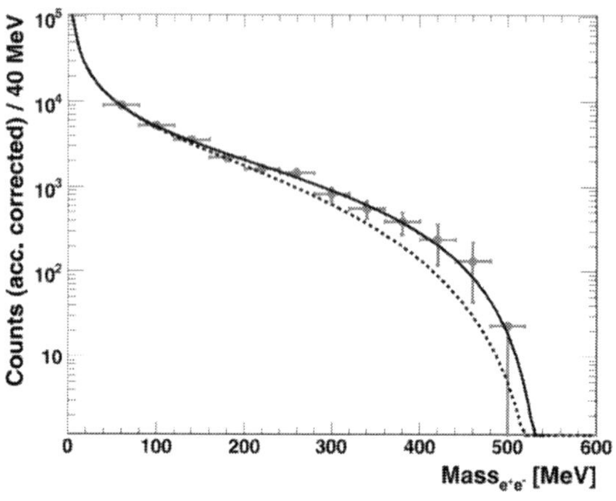


**Figure 6.** Invariant $e^+e^-$ mass distribution of reconstructed events after acceptance correction [16].The errors given are purely statistical.The solid curve is a fit within the VMD-model, using a monopole form factor (Eq. 2). The dotted curve is the QED prediction.

(syst)) MeV. The systematic error is estimated from a comparison of results obtained for different cut settings. The statistics is improved by an order of magnitude compared to the most recent form factor measurement in the $e^+e^-$ channel [17]. The data are in good agreement with the high statistics NA60 data, obtained in the $\mu^+\mu^-$ channel, however, without a full reconstruction of the $\eta$ meson, and with recent theoretical calculations [19]. After establishing in this pioneering experiment that weak

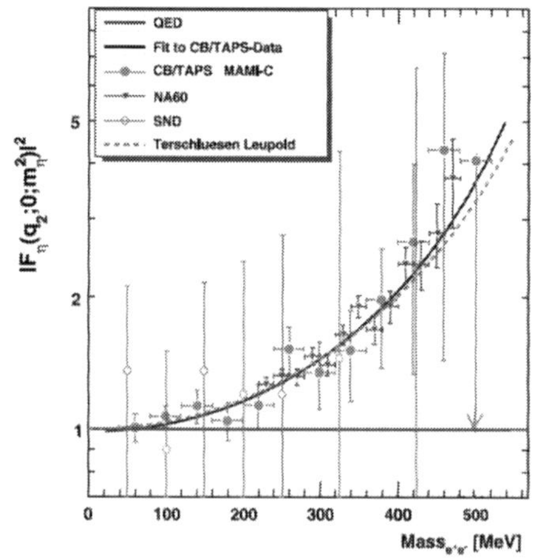


**Figure 7.** Dalitz transition form factor. The red circles are the data of this work and the black curve is a fit to these data [16]. The green (open) circles show the result of the SND experiment [17]. The blue (inverted) triangles represent the results obtained by the NA60 collaboration in the $\mu^+\mu^-$ channel [18]. The green (dashed) curve is a calculation performed by [19].

decay channels like the $\eta$ Dalitz decay can be identified in an exclusive analysis using a highly segmented $4\pi$

calorimeter without magnetic field, a series of measurements has been performed within the A2 collaboration at MAMI. An improved measurement of the $\eta$ Dalitz decay has been reported in [20]. Including all runs on a liquid hydrogen target and applying kinematic fitting routines a large data set has been accumulated for further improving the $\eta$ form factor determination and to extend the analysis to the $\omega \rightarrow \pi^0 e^+ e^-$ Dalitz decay [21]. On the theoretical side, improved calculations based on dispersion theory have been performed within project B3 for the $\eta, \omega$ and $\phi$ transition form factors [22, 23].

I would like to thank all the colleagues in the A2 collaboration at MAMI for their support and contributions to the results described in this article. The results are based on the theses of the strongly engaged PhD-students M. Kotulla and H. Berghäuser. Illuminating discussions with D. Drechsel, V. Pascalutsa and M. Vanderhagen on the theoretical interpretation of the data are highly appreciated. The work reported here would not have been possible without the support from the Deutsche Forschungsgemeinschaft within the SFB/TR16.

## References

[1] B. Nefkens *et al.*, Phys. Rev. D **18**, 3911 (1978).
[2] A. Bosshard *et al.*, Phys. Rev. D **44**, 930 (1991).
[3] R. Beck *et al.*, Phys. Rev. Lett. **78**, 606 (1997).
[4] A. Machavariani, A. Faessler and A. Buchmann, Nucl. Phys. A **646**, 231 (1999).
[5] A. Machavariani, A. Faessler and A. Buchmann, Nucl. Phys. A **686**, 601 (2002).
[6] D. Drechsel, Phys. Lett. B **484** , 236 (2000).
[7] D. Drechsel and M. Vanderhaegen, Phys. Rev. C **64**, 065202 (2001).
[8] S. Hall *et al.*, Nucl. Inst. Methods Phys. Res. A **368**, 698 (1996).
[9] R. Novotny, *et al.*, IEEE Trans. Nucl. Sci **38**, 392 (1991).
[10] A. R. Gabler *et al.*, Nucl. Instr. Methods Phys. Res. A **346**, 168 (1994).
[11] M. Kotulla *et al.*, Phys. Rev. Lett. **89**, 278001 (2002).
[12] A. Agadjanov *et al.*, Nucl. Phys. B **886**, 1199 (2014).
[13] S. Schuman *et al.*, Eur. Phys. J. A **43**, 269 (2010).
[14] M. F. M. Lutz and S. Leupold, Nucl. Phys. A **813**, 96 (2008).
[15] L. G. Landsberg, Phys. Rep. **128**, 301 (1985).
[16] H. Berghäuser *et al.*, Phys. Lett. B **701**, 562 (2011).
[17] M. N. Achasov *et al.*, Phys. Lett. B **504**, 275 (2001).
[18] R. Arnaldi *et al.*, Phys. Lett. B **677**, 260 (2009).
[19] C. Terschlüsen, Diploma thesis, Univ. of Giessen, 2010.
[20] P. Aguar-Bartolome *et al.*, Phys. Rev. C **89**, 044608 (2014).
[21] S. Prakhov, private communication (2016).
[22] C. W. Xiao *et al.*, arXiv:1509.02194.
[23] S. P. Schneider *et al.*, Phys. Rev. D **86** 054013 (2012).

EPJ Web of Conferences **134**, 02004 (2017) DOI: 10.1051/epjconf/201713402004
*Subnuclear Structure of Matter: Achievements and Challenges*

# Neutron anomaly in $\gamma N \to \eta N$ and the flavour SU(3) symmetry

*Maxim V.* Polyakov[1,*] and *Klaus* Goeke[1,**]

[1] *Institute for Theoretical Physics II, Ruhr-University Bochum, 44801 Bochum, Germany*

**Abstract.** We study the implications of the flavour SU(3) symmetry for various interpretations (existence of a narrow anti-decuplet resonance, interference of known resonances and a cusp effect) of the neutron anomaly in $\gamma N \to \eta N$ cross section. We show that the explanation of the neutron anomaly due to interference of known N(1535) and N(1650) resonances implies that N(1650) resonance should have a huge coupling to $\phi$-meson – at least 5 times larger than the corresponding $\rho^0$ coupling. In terms of quark degrees of freedom that implies that the well-known N(1650) resonance must be a "cryptoexotic pentaquark"– its wave function should contain predominantly an $s\bar{s}$ component, implying that the N(1650) resonance is dominantly a pentaquark. The explanation of the neutron anomaly as a cusp effect implies very strong violation of the flavour SU(3) symmetry.

## 1 Introduction

The discovery of the neutron anomaly[1] in the $\gamma N \to \eta N$ cross section was reported in Ref. [1], in this paper the GRAAL data on the photon scattering off the deuteron were analysed. Presently three other collaborations ( LNS [2],CBELSA/TAPS[3], and A2 [4]) confirmed the neutron anomaly beyond any doubts. For an illustration of the neutron anomaly in $\gamma N \to \eta N$ we show on Fig. 1 the most recent results of the A2 collaboration [4].

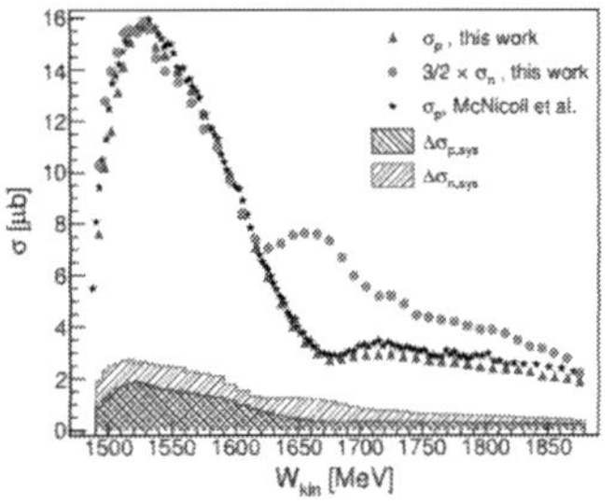


**Figure 1.** Figure from Ref. [4]. Total cross sections as a function of the final-state invariant mass $m(\eta N)$: Blue triangles: proton data. Red circles: neutron data scaled by $3/2$. Black stars: free proton data from MAMI-C [6]. Hatched areas: total systematic uncertainties of proton (blue) and neutron (red) data.

Furthermore the neutron anomaly at the same invariant mass of $W \sim 1680$ MeV was also observed in the Compton scattering [5].

In our view the observation of the neutron anomaly is the most striking discovery in the field of the nucleon resonances spectroscopy during the last decade. It is important to figure out the physics nature of the phenomenon. In the present contribution we study the implications of the flavour SU(3) symmetry for various explanations of the neutron anomaly.

## 2 Flavour SU(3) decomposition of the $\gamma N \to \eta N$ amplitude

In the SU(3) symmetry limit the amplitudes $A(\gamma p \to \eta p)$ and $A(\gamma n \to \eta n)$ can be decomposed through the amplitudes corresponding to the irreducible representations of the SU(3) group in the s-channel. The photon and nucleon belong to the octet representation of the SU(3) group, therefore the possible representations in the s-channel are those which appear in the product $\mathbf{8} \times \mathbf{8} = \mathbf{1} + \mathbf{8}_{\mathrm{F}} + \mathbf{8}_{\mathrm{D}} + \mathbf{10} + \overline{\mathbf{10}} + \mathbf{27}$. Obviously, the $\mathbf{1}$ and $\mathbf{10}$ representations do not enter the decomposition of the $\gamma N \to \eta N$ amplitude. The SU(3) decomposition for the amplitudes has the following form:

$$\begin{aligned} A(\gamma p \to \eta p) &= \frac{1}{3} A_{\mathrm{D}}^{(8)} + A_{\mathrm{F}}^{(8)} + A^{(27)}, \\ A(\gamma n \to \eta n) &= \frac{2}{3} A_{\mathrm{D}}^{(8)} + A^{(\overline{10})} + \frac{1}{2} A^{(27)}. \end{aligned} \quad (1)$$

We see that the anti-decuplet amplitude $A^{(\overline{10})}$ do not enter the $\gamma p$ channel, whereas the antisymmetric octet amplitude $A_{\mathrm{F}}^{(8)}$ do not enter the $\gamma n$ channel.

In order to describe the phenomenon of the neutron anomaly one needs that the amplitude $A(\gamma n \to \eta n)$ is very different (larger size and more rapid energy dependence) from $A(\gamma p \to \eta p)$ on a narrow invariant energy interval (several tens of MeV) around $W \sim 1680$ MeV. The decomposition (1) offers three possibilities to arrange such difference (ordered according to the *Prinzip der Denkökonomie*):

* e-mail: maxim.polyakov@tp2.rub.de
** deceased

[1] Existence of the narrow ($\Gamma \sim$10-40 MeV) peak in the $\gamma n \to \eta n$ cross section around 1680 MeV and its absence in the $\gamma p \to \eta p$ process

(I) the anti-decuplet amplitude $A^{(\overline{10})}$ has large size and rapid energy dependence on a narrow energy interval around 1680 MeV,

(II) there is a conspiracy and a fine tuning among the SU(3) amplitudes $A_{\rm F}^{(8)}$, $A_{\rm D}^{(8)}$ and $A^{(27)}$ on that narrow energy interval,

(III) extraordinarily strong violation of the SU(3) symmetry on that narrow energy interval.

We emphasise that the option (II) can explain the neutron anomaly only in the $\eta$-photoproduction. In other channels, e.g. the Compton scattering [5], the assumed conspiracy and fine tuning are destroyed due to different from (1) SU(3) decomposition of the Compton amplitude. The anti-decuplet amplitude $A^{(\overline{10})}$ enters only the $\gamma n$ channel independently of the final state. Therefore the option (I) predicts the neutron anomaly for the Compton scattering as well.

Usually the approximate flavour SU(3) symmetry works pretty well. As a rule its predictions are satisfied with an accuracy of about 30% or better, see e.g. a review [7]. A very large violation of the SU(3) symmetry would be a serious challenge to our common wisdom about hadron dynamics. It seems that a possible realisation of the option (III) is provided by Ref. [8]. In this paper the neutron anomaly was explained by the threshold effect due to $K\Lambda$ and $K\Sigma$ intermediate states. It was argued in Ref. [8] that the intermediate $K^+\Sigma^-$ state in the $\gamma n$ channel produces the cusp effect at $W \sim 1685$ MeV which can explain the peak in that channel. In order to suppress the corresponding peak in the $\gamma p$ channel the authors of Ref. [8] fitted their model parameter in such a way that the cusp due to $K^+\Lambda$ intermediate state cancels the cusp effect due to $K^+\Sigma^0$ state (a kind of fine tuning). Again, the explanation of the neutron neutron anomaly of Ref. [8] is not universal, *i.e.* it works only for $\eta$-photoproduction and fails for Compton scattering, the same as for the option (II).

In the following sections we analyse the physics realisations of the two first possibilities discussed above.

## 3 (I) Dominance of the anti-decuplet amplitude

The simplest physics realisation of the option (I) is an existence of a narrow anti-decuplet of baryons. The existence of such narrow exotic baryon multiplet was predicted in Ref. [9]. Main properties of N* from the anti-deculpet which were predicted theoretically in years 1997-2004 (before the discovery of the neutron anomaly) are the following:

- quantum numbers are $P_{11}$ ($J^P = \frac{1}{2}^+$, isospin=$\frac{1}{2}$) [9],
- narrow width of $\Gamma \leq 40$ MeV [9–11],
- mass of $M \sim 1650 - 1720$ MeV [10–12],
- strong suppression of the proton photocoupling relative to the neutron one [13] ,
- the $\pi N$ coupling is suppressed, N* prefers to decay into $\eta N$, $K\Lambda$ and $\pi\Delta$ [9–11].

It seems that the nucleon resonance with such properties can explain concisely the neutron anomaly in both $\eta$-photoproduction and the Compton scattering.

Detailed account for predictions and evidences for narrow anti-decuplet nucleon was presented at length previously in the literature (see e.g. [14, 19]). Not to dwell on this once again, we just give the Table 1 which summarises extracted properties of the putative anti-decuplet nucleon resonance.

**Table 1.** Our estimate of properties of the putative narrow antidecuplet N* extracted from the data.

| observable | extracted value |
|---|---|
| mass (MeV) | 1680 ± 15 |
| $\Gamma_{\rm tot}$ (MeV) | $\leq 40$ |
| $\Gamma_{\pi N}$ (MeV) | $\leq 0.5$ |
| $\sqrt{{\rm Br}_{\eta N}}A^n_{1/2}$ ($10^{-3}$ GeV$^{-1/2}$) | 12-18 |
| $\sqrt{{\rm Br}_{\eta N}}A^p_{1/2}$ ($10^{-3}$ GeV$^{-1/2}$) | 1-3 |

## 4 (II) Conspiracy and fine tuning among non-exotic SU(3) amplitudes

A physics realisation of the option (II) was suggested in Refs. [20–22] by the Bonn-Gatchina group (BnGa). In these papers the neutron anomaly was explained by the interference effect of well-known wide $S_{11}$ resonances N(1535) and N(1650). In order to arrange a narrow structure in the neutron channel the photocouplings of these two resonances should be fine tuned. In particular, the proton and neutron photocouplings of N(1650) must have the same sign. To describe the most recent and the most precise data of the A2 collaboration on the neutron anomaly [4] BnGa obtained the following ratio of the proton to neutron photocouplings [22]:

$$R_{pn} \equiv \frac{A^p_{1/2}(1650)}{A^n_{1/2}(1650)} = 1.74 \pm 0.66 \,. \qquad (2)$$

Employing the flavour SU(3) symmetry one can express the ratio of the $F_V$ and $D_V$ octet vector couplings in terms of the ratio $R_{pn}$:

$$\frac{F_V}{D_V} = -\frac{1}{3}\left(2R_{pn}+1\right) = -1.50 \pm 0.44 \,. \qquad (3)$$

The resulting from the analysis [22] $F_V$ to $D_V$ ratio is negative and larger than 1 in the absolute value. To our best knowledge such values of $F_V/D_V$ have been never obtained in any model of baryon resonances (variants of quark model, MIT bag model, soliton models, etc). Let us see what are physics implications of such unusual values of the $F_V/D_V$ ratio.

The flavour SU(3) symmetry allows to express various flavour combinations of the vector current couplings in terms of $F_V/D_V$-ratio (and hence in terms of $R_{pn}$ (2) owing Eq. (3)). One can easily derive the following relations for various vector couplings of N(1650) (valid also for any octet nucleon resonance $N'$):

$$R_\omega \equiv \frac{g_{\omega NN'}}{g_{\rho^0 NN'}} = \frac{R_{pn}+1}{R_{pn}-1} + \sqrt{\frac{2}{3}}\, r_0, \qquad (4)$$

$$R_\phi \equiv \frac{g_{\phi NN'}}{g_{\rho^0 NN'}} = -\sqrt{2}\,\frac{R_{pn}+1}{R_{pn}-1} + \sqrt{\frac{1}{3}}\, r_0. \qquad (5)$$

Here $r_0$ is the ratio of the flavour singlet $((\bar{u}\gamma_\mu u + \bar{d}\gamma_\mu d + \bar{s}\gamma_\mu s)/\sqrt{3})$ vector current coupling to the isovector $((\bar{u}\gamma_\mu u - \bar{d}\gamma_\mu d)/\sqrt{2})$ that. The value of $r_0$ is not fixed by the SU(3) symmetry, however with help of Eqs. (4,5) we can express $\phi$-meson coupling $R_\phi$ in terms of the $\omega$-meson coupling $R_\omega$ and the proton to neutron ratio of the photocoupling $R_{pn}$:

$$R_\phi = \frac{1}{\sqrt{2}}\left(R_\omega - 3\,\frac{R_{pn}+1}{R_{pn}-1}\right). \qquad (6)$$

Additionally, from Eqs. (4,5) one can easily derive the following inequality:

$$R_\omega^2 + R_\phi^2 \geq 3\left(\frac{R_{pn}+1}{R_{pn}-1}\right)^2. \qquad (7)$$

If we take the BnGa value (2) for $R_{pn}$ we obtain from (7):

$$R_\omega^2 + R_\phi^2 \geq 27. \qquad (8)$$

One sees that in the scenario of Refs. [20–22] the $\omega$- and $\phi$-meson couplings of N(1650) can not be small simultaneously[2].

Experimentally [24] the decay N(1650) $\to \rho N$ is seen and sizable, however the decay N(1650) $\to \omega N$ is not seen. If we conservatively assume that the yield of $\omega$ mesons does not exceed factor of four relative to the yield of $\rho$ mesons, *i.e* $R_\omega^2 \leq 4$ [3] then from Eq. (6) with BnGa value for the $p/n$ ratio of N(1650) photocouplings (2) we obtain:

$$|R_\phi| \geq 6. \qquad (9)$$

We see that the explanation of the neutron anomaly by the interference of known N(1535) and N(1650) resonances advocated in [20–22] implies that the $\phi$-meson (almost pure $s\bar{s}$ state) coupling to N(1650)→N transition should be huge. In terms of quark degrees of freedom it means that N(1650) has a large admixture of $s\bar{s}$ component, *i.e.* in the scenario of Refs. [20–22] N(1650) is dominantly "cryptoexotic pentaquark" . It turns out that the "conventional" interpretation of the neutron anomaly by the interference of known resonances [20–22] metamorphose into unconventional physics picture of N(1650).

[2]Note that if we take N(1650) photocouplings from the SAID analysis [23], than $F_V/D_V = 0.6 \pm 0.6$ (range of values typical for all models of baryon resonances) and $R_\omega^2 + R_\phi^2 \geq 0.2$ (small values of $\omega$ and $\phi$ couplings are not excluded).

[3]If one combines the result of Ref. [33] for $g_{\rho NN'}$ with the result of Ref. [34] for $g_{\omega NN'}$ one obtains $R_\omega^2 \simeq 1$

## 5 (III) Very large violation of the flavour SU(3) symmetry

In Ref. [8] the neutron anomaly was explained[4] by the threshold effect due to $K\Lambda$ and $K\Sigma$ intermediate states. We shall see that this explanation implies an extraordinarily large violation of the SU(3) symmetry. Usually the approximate flavour SU(3) symmetry works pretty well. As a rule its predictions are satisfied with an accuracy of about 30% or better, see e.g. a review [7]. A very large violation of the SU(3) symmetry is a serious challenge to our common wisdom about hadron dynamics.

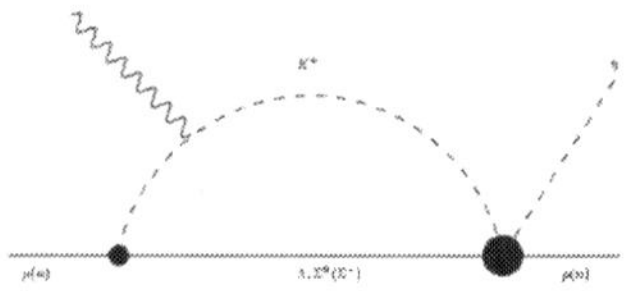

**Figure 2.** The diagram from [8] which contributes to the cusp at $KY$ threshold.

The neutron anomaly was described in Ref. [8] by the diagram shown on Fig. 2. It was argued in Ref. [8] that the intermediate $K^+\Sigma^-$ state in the $\gamma n$ channel produces the cusp effect at $W \sim 1685$ MeV which can explain the peak in that channel. In order to suppress the corresponding peak in the $\gamma p$ channel the authors of Ref. [8] fitted their model parameter in such a way that the cusp due to $K^+\Lambda$ intermediate state cancels the cusp effect due to $K^+\Sigma^0$ state (a kind of fine tuning). Let us see how the assumed cancelation can be reconciled with SU(3) chiral dynamics.

The cusp contribution of the diagram on Fig. 2 to $\gamma N \to \eta N$ can be written as:

$$\begin{aligned} A_{\text{cusp}}(\gamma p \to \eta p) &= D\left(A_\Sigma - \sqrt{\frac{1}{3}}A_\Lambda\right) \\ &- F\left(A_\Sigma + \sqrt{3}A_\Lambda\right), \\ A_{\text{cusp}}(\gamma n \to \eta n) &= 2(D-F)A_\Sigma. \end{aligned} \qquad (10)$$

Here $A_\Sigma$ and $A_\Lambda$ are (up to a common factor) $A(K^+\Sigma^0 \to \eta p)$ and $(K^+\Lambda \to \eta p)$ amplitudes at the corresponding $KY$ thresholds (right blob on Fig. 2). In Eqs. (10,10) we expressed $g_{KYN}$ coupling constants (left blob on Fig. 2) in terms of $F$ and $D$ pseudoscalar ground state octet couplings. To arrange the cancelation of the cusp effects suggested in Ref. [8] on sees from Eq. (10) that the amplitudes $A_\Sigma$ and $A_\Lambda$ should be related by:

$$\frac{A_\Lambda}{A_\Sigma} \simeq \sqrt{3}\frac{D-F}{D+3F} = \sqrt{3}\cdot 0.16. \qquad (11)$$

Here the numerical value is obtained by taking the experimental value of $F = 0.46$, $D = 0.8$ [26]. The $A_\Lambda/A_\Sigma$ ratio is predicted by the SU(3) chiral dynamics (see e.g [27]) as:

[4]More precisely the authors were able to describe only the ratio of the neutron to proton cross section having a dozen of free adjustable parameters.

$$\frac{A_\Lambda}{A_\Sigma} = \sqrt{3}\left(1+O\left(\frac{m_K}{M_N}\right)\right), \qquad (12)$$

where the symmetry breaking corrections $O(m_K/M_N)$ numerically can be rather sizable. The estimates of these corrections in Ref. [28] give $A_\Lambda/A_\Sigma = \sqrt{3}\cdot 0.8$, previous estimates in [29] give $A_\Lambda/A_\Sigma = \sqrt{3}\cdot(0.6 \div 0.7)$. We see that the result (11), which is needed for the explanation of the neutron anomaly by the cusp effect as in Ref. [8], is in a disagreement with the SU(3) chiral dynamics prediction (12), even if we take into account the $O(m_K/M_N)$ corrections of Refs. [28, 29]. We note, however, that the most recent analysis of the anti-kaon – baryon scattering [36–38] revealed that the NLO chiral corrections can be very large . It would be very interesting to see what is the resulting ratio in Eq. (12).

## 6 Conclusions

In summary, we analysed the implication of the flavour SU(3) symmetry for explanations of the neutron anomaly in the $\gamma N \to \eta N$ cross section. The SU(3) symmetry suggests three classes of scenarios: (A) dominance of the anti-decuplet channel at narrow energy interval (B) fine tuning of parameters of known wide resonances to arrange very specific interference pattern ([20–22]) or (C) fine tuning of parameters to arrange strong cusp effects in the neutron channel and their strong cancellation in the proton channel ([8]) .

The first two scenarios need exotic nucleon resonances – this can be either (A) a narrow anti-decuplet of baryons, or (B) well know N(1650) resonance with very large $s\bar{s}$ component, *i.e.* the well-known N(1650) resonance must be a "cryptoexotic pentaquark". In the scenario (C) it seems that one tunes parameters in such unnatural way that the flavour $SU(3)$ symmetry is very strongly violated. Also the scenario (C) assumes a fine tuning of a dozen of free adjustable parameters.

We stressed that the scenarious (B,C) (in contrast to the first scenario) can explain the neutron anomaly only in the $\gamma N \to \eta N$ process and it fails to explain the neutron anomaly in the Compton scattering. It seems that the simplest, universal (for both $\eta$-photoproduction and the Compton scattering) and concise way to explain the neutron anomaly is the existence of a narrow anti-decuplet of baryons. The most recent data [39] on double polarization observables for $\gamma n \to \eta n$ reaction strongly support our scenario (A) for explanation of the neutron anomaly. In that reference the authors came to the conclusion that "...the angular dependence to different model predictions favors a scenario with a contribution from a narrow $P_{11}$ resonance."

## 7 Epilogue

The project **A.4** was funded over the first two periods of the SFB/TR 16. The PIs were Klaus Goeke and Maxim Polyakov and the project was called "Multiquark states of baryons". Apart from the neutron anomaly that was discussed in detail above, we briefly summarize here other pertinent work performed within this project.

### 7.1 $\Theta^+$: puzzle of its existence

In 1997 the Diakonov, Petrov and Polyakov predicted in the Chiral Quark Soliton Model ($\chi$QSM) exotic baryons that have a particularly small decay width (<15 MeV) compared to usual resonances (about 100 MeV). Six years after the prediction the first positive indication of a baryon with positive strangeness (i.e. $\Theta^+$) was provided by the LEPS collaboration at a mass of 1540 MeV. Experiments from DIANA, CLAS, SAPHIR and others found also positive evidence for the $\Theta^+$. However in the last years several experiments were performed without evidence of $\Theta^+$, such that presently the majority of physicists in the hadronic community doubts the existence of $\Theta^+$. The main weakness is that one does not know the production mechanism of these exotic particles, if they exist. Therefore theoretical studies are needed and suggested and hence performed in the present project [40–42]. The studies are all performed assuming the predicted quantum numbers of $\Theta^+$, the above mentioned energy, and a width below 1 MeV [43, 44].

### 7.2 Possibility to identify $\Theta^+$ baryon from interference

In this work [45], we suggest to search for the narrow $\Theta^+$ resonance in a non-standard way, exploiting the interference of the small $\Theta^+$ production amplitude with the large production amplitude of a known resonance, yielding the same final state. Although the interference idea is very general, we apply it primarily to the CLAS experiment whose impressive amount of data can be used to look for the $\Theta^+$ resonance in interference with the large $\phi$ photoproduction. Since the final state in both cases is the same, the two amplitudes *must* interfere unless forbidden kinematically. A simple account for kinematics shows that the two amplitudes interfere in the CLAS experiment.

The $\Theta^+$ production amplitude *squared* has been estimated in Ref. [46] with the tiny result for the cross section in the sub-nanobarn range – too small to be observable even with the large CLAS statistics. However, the interference cross section is *linear* in the $\Theta^+$ coupling and hence can be substantially larger. In the CLAS experiment studying the $\gamma p \to K^0\bar{K}^0 p$ reaction, the $K_S$ (decaying into $\pi^+\pi^-$) and $p$ have been detected. The second kaon was reconstructed from the missing mass of all detected particles. A large portion of events were due to the production of the $\phi$ meson decaying into $K_L K_S$. These events have been rejected in this analysis, however they are exactly what are needed now. The reanalysis of the CLAS data using the $K_L K_S$ pairs from the decay of $\phi$-meson the was performed in Ref. [47]. The observation of a narrow peak structure at ~ 1.54 GeV with a Gaussian width of ~ 6 MeV in the missing mass of $K_S$ was reported. The observed structure may be due to the interference between a strange (or anti-strange) baryon resonance in the $pK_L$ system and the $\phi(K_S K_L)$ photoproduction leading to the same final state. The statistical significance of the observed excess of events estimated as the log likelihood ratio of the resonant signal+background hypothesis and the $\phi$-production based background-only hypothesis corresponds to $5.3\sigma$.

The method is directly applicable also to the $\bar{K}^0 K^+ n$ reactions at relatively low energies, where one should look for the $\Theta - \Lambda(1520)$ interference.

### 7.3 $K^*$-couplings for the anti-decuplet and photo-production of the $\Theta^+$ on the nucleon

Current approaches to the theoretical description of $\Theta$-production are often based on $K$ and $K^*$ meson or reggeon exchanges. The vertex for the $KN\Theta$-coupling may be considered as known, if one assumes the spin-parity and width of $\Theta^+$ to be known (the corresponding form factor is still a problem, of course). Contrary to this, properties of the $K^*$ exchange are totally unknown. However, they may be essential, *e.g.*, for comparisons of $\Theta^+$-photoproduction off the proton and/or the neutron. In the works [43, 44, 46] we estimate the coupling of the $K^*$ vector meson to the $N \to \Theta^+$ transition employing unitary symmetry, vector meson dominance, and results from the GRAAL Collaboration for $\eta$ photo-production off the neutron. Our small numerical value for the coupling constant is consistent with the non-observation of the $\Theta^+$ in recent CLAS searches for its photo-production. We also estimate the $K^*$-coupling for the $N \to \Sigma^*$ excitation, with $\Sigma^*$ being the $\Sigma$-like anti-decuplet partner of the $\Theta^+$-baryon. As described in our paper ref. [43] we obtain the following estimates:

$$\begin{aligned} |f_2(K^{*0}\, p\, \Theta^+)| = |f_2(K^{*+}\, n\, \Theta^+)| &= \sqrt{6}\, |f_2(\rho^0\, n\, n^*)| \\ &= (1.10 - 3.14)\,, \\ |f_2(K^{*-}\, p\, \Sigma^{*0})| = (0.45 - 1.28)\,, \quad & |f_2(\overline{K}^{*0}\, p\, \Sigma^{*+})| \\ &= (0.64 - 1.81)\,. \end{aligned}$$

Incidentally, the value of 1.1 for $|f_2(K^{*+}\, n\, \Theta^+)| = |f_2(K^{*0}\, p\, \Theta^+)|$ was used earlier in Ref. [46] to estimate the production cross section of $\Theta^+$ in photo-reactions. With a $\Theta^+$-width of 1 MeV and the above value of $f_2(K^*\, N\, \Theta)$, calculations [46] find small cross sections $\sigma_{tot}(\gamma p \to \overline{K}^0 \Theta^+) < 0.22$ nb and $\sigma_{tot}(\gamma n \to K^- \Theta^+) < 1$ nb, which are below the limits given recently by the CLAS Collaboration.

Note also the difference between proton and neutron targets. To clarify the meaning of our numerical values for $f_2$, we consider in more detail the photo-production $\gamma + p \to \overline{K}^0 + \Theta^+$ with $K^*$-exchange as the main contribution. We can compare experimental limits for this reaction, obtained by the CLAS Collaboration, and the theoretical calculations in our model [46]. Our analysis shows that the CLAS analysis very likely is not sensitive enough to reveal the $\Theta^+$ (if it exists).

### 7.4 SU(3) systematization of baryons

We studied the spectrum of all baryons with the mass less than approximately 2000-2200 MeV using the Gell-Mann–Okubo mass formulas and SU(3)-symmetric predictions for two-body hadronic decays [48–50]. We successfully placed almost all known baryons in twenty-one SU(3) multiplets and, thus, confirmed the prediction that the approximate SU(3) symmetry works remarkably well.

Our results are summarized in ref.[48]. In fact there are predictions of 17 new particles, which are absent in the Review of Particle Physics. Among the predicted baryons, the most remarkable is the $\Lambda$ hyperon with $J^P = 3/2^-$, the mass around 1850 MeV, the total width $\approx$ 130 MeV, significant branching into the $\Sigma\pi$ and $\Sigma(1385)\pi$ states and a very small coupling to the $N\overline{K}$ state [51]. Our analysis gives a model-independent confirmation of the constituent quark model prediction that there should exist a new $\Lambda$ baryon with the mass between 1775 MeV and 1880 MeV, which almost decouples from the $N\overline{K}$ state.

Methods of the approximate flavor SU(3) symmetry of the strong interactions are also applicable to the antidecuplet. Our analysis clearly demonstrates that, if the antidecuplet exists, it must mix with some other SU(3) multiplet [48–50]. We considered a scenario that the antidecuplet mixes with a non-exotic octet. This enables us to accommodate in a simple way all presently available experimental information on the antidecuplet decays and to predict the unmeasured $\overline{\mathbf{10}}$ decays. Ref. [48] summarizes our SU(3) predictions for the antidecuplet. Assuming $\Gamma_{\Theta^+} \leq 1$ MeV we obtain fkor all partial widths of the anti-decuplet of less than 2.5 MeV with the only exception of $\Gamma_{N_{\overline{10}} \to \Delta\pi} = (2.6 - 15.6)$MeV and $\Gamma_{\Sigma_{\overline{10}} \to \Sigma(1385)\pi} = (0.33 - 1.96)$MeV.

### 7.5 Baryons from the chiral quark-soliton picture

During the second funding period our work was also concentrated on studies of hadrons which emerge in the chiral quark-soliton picture of the baryons. Firstly, we worked out in Ref. [52] in detail the properties of possible anti-decuplet baryons. We have proposed a scenario in which the Roper octet can mix with the putative anti-decuplet of exotic baryons and predicted the properties of its cryptoexotic states. In our view, this is the best what one can do in the framework of the chiral-quark soliton model. It was found that the cryptoexotic anti-decuplet nucleon state must have a much larger photocoupling to the neutron than the proton. Also we found that the coupling of anti-decuplet $N^*$ to the $\pi N$ channel is strongly suppressed. As the byproduct of our studies of the ground state baryons in the chiral quark-soliton model, we developed a method to calculate the electromagnetic mass differences in the octet and decuplet of baryons [53]. In this work we showed that the electromagnetic mass differences in the octet and the decuplet are related to each other due to the hedgehog symmetry of the model. These relations are in a good agreement with the experimental data.

We appreciate very much fruitful collaboration with M. Amarian, Ya. I. Azimov, T. Boika, D.I Diakonov, V. Kuznetsov,

V. Petrov, M. Praszałowicz, and I. Strakovsky. We acknowledge important contributions from V. Guzey during the first funding period. The work reported here would not have been possible without the support from the Deutsche Forschungsgemeinschaft within the SFB/TR16.

## References


[1] V. Kuznetsov [GRAAL Collaboration], "$\eta$ photoproduction off the neutron at GRAAL: Evidence for a resonant structure at W = 1.67-GeV," arXiv:hep-ex/0409032,
V. Kuznetsov *et al.*, Phys. Lett. B **647**, 23 (2007) [arXiv:hep-ex/0606065].

[2] F. Miyahara *et al.*, Prog. Theor. Phys. Suppl. **168**, 90 (2007).

[3] I. Jaegle *et al.* [CBELSA Collaboration and TAPS Collaboration], Phys. Rev. Lett. **100** (2008) 252002 [arXiv:0804.4841 [nucl-ex]],
I. Jaegle, B. Krusche, A. V. Anisovich, J. C. S. Bacelar, B. Bantes, O. Bartholomy, D. E. Bayadilov and R. Beck *et al.*, Eur. Phys. J. A **47** (2011) 89 [arXiv:1107.2046 [nucl-ex]].

[4] D. Werthmuller *et al.* [A2 Collaboration], Phys. Rev. Lett. **111** (2013) 23, 232001 [arXiv:1311.2781 [nucl-ex]],
L. Witthauer *et al.* [A2 Collaboration], Eur. Phys. J. A **49** (2013) 154 [arXiv:1312.1571 [nucl-ex]],
D. Werthmuller *et al.* [ A2 Collaboration], Phys. Rev. C **90** (2014) 015205 [arXiv:1407.6974 [nucl-ex]].

[5] V. Kuznetsov *et al.*, Phys. Rev. C **83** (2011) 022201(R) [arXiv:1003.4585 [hep-ex]].

[6] E. F. McNicoll *et al.*, Phys. Rev. C **82** (2010) 035208.

[7] N. P. Samios, M. Goldberg and B. T. Meadows, Rev. Mod. Phys. **46** (1974) 49.

[8] M. Doring and K. Nakayama, Phys. Lett. B **683**, 145 (2010) [arXiv:0909.3538 [nucl-th]].

[9] D. Diakonov, V. Petrov and M. V. Polyakov, Z. Phys. A **359** (1997) 305 [arXiv:hep-ph/9703373].

[10] R. A. Arndt *et al.*, Phys. Rev. C **69**, 035208 (2004) [arXiv:nucl-th/0312126].

[11] J. R. Ellis, M. Karliner and M. Praszalowicz, JHEP **0405** (2004) 002; M. Praszalowicz, Acta Phys. Polon. B **35** (2004) 1625 [arXiv:hep-ph/0402038].

[12] D. Diakonov and V. Petrov, Phys. Rev. D **69** (2004) 094011 [arXiv:hep-ph/0310212].

[13] M. V. Polyakov and A. Rathke, Eur. Phys. J. A **18**, 691 (2003) [arXiv:hep-ph/0303138].

[14] V. Kuznetsov *et al.*, Acta Phys. Polon. B **39**, 1949 (2008) [arXiv:0807.2316 [hep-ex]];

[15] V. Kuznetsov and M. V. Polyakov, JETP Lett. **88**, 347 (2008) [arXiv:0807.3217 [hep-ph]].

[16] V. Kuznetsov, M. V. Polyakov and M. Thurmann, arXiv:1102.5209 [hep-ph].

[17] A. V. Anisovich, E. Klempt, V. Kuznetsov, V. A. Nikonov, M. V. Polyakov, A. V. Sarantsev and U. Thoma, Phys. Lett. B **719** (2013) 89 [arXiv:1108.3010].

[18] Y. I. Azimov, *et al.*, Eur. Phys. J. A **25**, 325 (2005) [arXiv:hep-ph/0506236].

[19] K. Goeke, M. V. Polyakov and M. Praszalowicz, Acta Phys. Polon. B **42** (2011) 61 [arXiv:0912.0469 [hep-ph]],
M. Praszalowicz, Acta Phys. Polon. Supp. **3** (2010) 917 [arXiv:1005.1007 [hep-ph]].

[20] A. V. Anisovich *et al.*, Eur. Phys. J. A **41**, 13 (2009) [arXiv:0809.3340 [hep-ph]].

[21] A. V. Anisovich, V. Burkert, E. Klempt, V. A. Nikonov, A. V. Sarantsev and U. Thoma, Eur. Phys. J. A **49** (2013) 67 [arXiv:1304.2177 [nucl-ex]].

[22] A. V. Anisovich, E. Klempt, V. A. Nikonov, A. V. Sarantsev and U. Thoma, arXiv:1402.7164 [nucl-ex].

[23] W. Chen, H. Gao, W. J. Briscoe, D. Dutta, A. E. Kudryavtsev, M. Mirazita, M. W. Paris and P. Rossi *et al.*, Phys. Rev. C **86** (2012) 015206 [arXiv:1203.4412 [hep-ph]],
R. L. Workman, M. W. Paris, W. J. Briscoe and I. I. Strakovsky, Phys. Rev. C **86** (2012) 015202 [arXiv:1202.0845 [hep-ph]].

[24] J. Beringer *et al.* [Particle Data Group Collaboration], Phys. Rev. D **86** (2012) 010001.

[25] V. Guzey and M. V. Polyakov, "SU(3) systematization of baryons," hep-ph/0512355.

[26] F. E. Close and R. G. Roberts, Phys. Lett. B **316** (1993) 165 [hep-ph/9306289],
B. Borasoy, Phys. Rev. D **59** (1999) 054021 [hep-ph/9811411].

[27] N. Kaiser, T. Waas and W. Weise, Nucl. Phys. A **612** (1997) 297 [hep-ph/9607459],
B. Borasoy, E. Marco and S. Wetzel, Phys. Rev. C **66** (2002) 055208 [hep-ph/0212256].

[28] Y. Ikeda, T. Hyodo and W. Weise, Nucl. Phys. A **881** (2012) 98 [arXiv:1201.6549 [nucl-th]].

[29] B. Borasoy, R. Nissler and W. Weise, Eur. Phys. J. A **25** (2005) 79 [hep-ph/0505239].

[30] A. Fix, L. Tiator and M. V. Polyakov, Eur. Phys. J. A **32**, 311 (2007) [arXiv:nucl-th/0702034].

[31] V. Shklyar, H. Lenske and U. Mosel, Phys. Lett. B **650**, 172 (2007) [arXiv:nucl-th/0611036].

[32] M. Praszalowicz, arXiv:1005.1007 [hep-ph].

[33] A. M. Gasparyan, J. Haidenbauer, C. Hanhart and J. Speth, Phys. Rev. C **68** (2003) 045207 [nucl-th/0307072].

[34] V. Shklyar, H. Lenske, U. Mosel and G. Penner, Phys. Rev. C **71** (2005) 055206 [Erratum-ibid. C **72** (2005) 019903] [nucl-th/0412029].

[35] T. M. Knasel *et al.*, Phys. Rev. **D11** (1975) 1-13; B. Nelson *et al.*, Phys. Rev. Lett. **31** (1973) 901-904.

[36] M. Mai and U. G. Meissner, Nucl. Phys. A **900** (2013) 51 doi:10.1016/j.nuclphysa.2013.01.032 [arXiv:1202.2030 [nucl-th]].

[37] Z. H. Guo and J. A. Oller, Phys. Rev. C **87** (2013) no.3, 035202 doi:10.1103/PhysRevC.87.035202 [arXiv:1210.3485 [hep-ph]].

[38] A. Cieply, M. Mai, U. G. Meißner and J. Smejkal, Nucl. Phys. A **954** (2016) 17 doi:10.1016/j.nuclphysa.2016.04.031 [arXiv:1603.02531 [hep-ph]].

[39] L. Witthauer *et al.*, Phys. Rev. Lett. **117** (2016) no.13, 132502. doi:10.1103/PhysRevLett.117.132502

[40] K. Goeke, H. C. Kim, M. Praszalowicz and G. S. Yang, Prog. Part. Nucl. Phys. **55** (2005) 350 [arXiv:hep-ph/0411195].

[41] K. Goeke, H. C. Kim and M. Praszalowicz, Europhys. News **36** (2005) 151.

[42] M. Praszalowicz and K. Goeke, Acta Phys. Polon. B **36**, 2255 (2005) [arXiv:hep-ph/0506041].

[43] Y. Azimov, V. Kuznetsov, M. V. Polyakov and I. Strakovsky, Phys. Rev. D **75**, 054014 (2007) [arXiv:hep-ph/0611238].

[44] Y. Azimov, V. Kuznetsov, M. V. Polyakov and I. Strakovsky, Eur. Phys. J. A **25**, 325 (2005) [arXiv:hep-ph/0506236].

[45] M. Amarian, D. Diakonov and M. V. Polyakov, arXiv:hep-ph/0612150.

[46] H. Kwee, M. Guidal, M. V. Polyakov and M. Vanderhaeghen, Phys. Rev. D **72**, 054012 (2005) [arXiv:hep-ph/0507180].

[47] M. J. Amaryan *et al.*, Phys. Rev. C **85** (2012) 035209 doi:10.1103/PhysRevC.85.035209, 10.1103/PhysRevC.85.049901 [arXiv:1110.3325 [hep-ex]].

[48] V. Guzey and M. V. Polyakov, arXiv:hep-ph/0512355.

[49] V. Guzey, AIP Conf. Proc. **775**, 3 (2005).

[50] V. Guzey and M. V. Polyakov, Annalen Phys. **13**, 673 (2004).

[51] V. Guzey, arXiv:hep-ph/0512276.

[52] K. Goeke, M. V. Polyakov and M. Praszalowicz, Acta Phys. Polon. B **42**, 61 (2011)

[53] G.-S. Yang, H.-C. Kim and M. V. Polyakov, Phys. Lett. B **695**, 214 (2011)

EPJ Web of Conferences **134**, 02005 (2017)
*Subnuclear Structure of Matter: Achievements and Challenges*
DOI: 10.1051/epjconf/201713402005

# Quark model description of baryon properties

*Bernard Ch. Metsch*[1,*] and *Herbert-R. Petry*

[1] *Helmholtz Institut für Strahlen- und Kernphysik, Nußallee 14-16, D 53115 Bonn, Germany*

**Abstract.** The goal of this project was a systematic calculation, consistent with the mass spectrum, of electroweak properties of baryons, as well as of the hadronic two-body decays of baryon resonances. The basis of the calculations was a relativistically covariant quark model for mesons and baryons, which allows for the calculation of (momentum dependent) hadronic form factors and coupling vertices in the relevant photon and hadron induced reactions and simultaneously describes the hadronic mass spectrum up to 3 GeV.

## 1 Introduction

We developed a constituent quark model for baryons and mesons within a field-theoretical framework starting from the homogeneous Bethe-Salpeter for bound hadronic states. This model thus fully respects Poincaré covariance, mandatory to describe deeply bound states and/or scattering processes at larger momentum transfer.

Before the start of the first funding period of the SFB/TR16 this model, which describes the quark dynamics through an instantaneous confinement potential with a suitable Dirac structure, rising linearly with interquark distances and a spin-flavour dependent interaction motivated by instanton effects to describe the major spin-dependent mass splittings, had been applied to a calculation of the spectrum of light flavoured mesons and baryons up to masses of about 3 GeV, see e.g. [1–3] as well as some selected electroweak observables such as electromagnetic form factors, charge radii and magnetic moments, see e.g. [4, 5].

## 2 Earlier achievements

During the first funding period the following issues have been investigated, also see [6]:

*Electromagnetic properties:*

The calculation of some momentum transfer dependent helicity amplitudes as they occur in electro- or photo-induced reactions on the basis of the Salpeter amplitudes determined by the mass spectrum is (in lowest order in the Mandelstam formalism) parameter free. These have been calculated systematically both for non-strange baryons [7] and for hyperons, see [8]. For non-strange photo-excitation it was found, that sizeable transition amplitudes are to be expected only for the lowest states of each total spin. Higher excited states can therefore not be directly reached in photo-production experiments. To exemplify the quality of the prediction a comparison of calculated and experimental helicity amplitudes at a finite momentum transfer of $Q^2 = 0.65\,\mathrm{GeV}^2$ is shown in Table 1 .

**Table 1.** Calculated helicity amplitudes (in units of $10^{-3}\,\mathrm{GeV}^{-\frac{1}{2}}$) of proton resonances at $Q^2 = 0.65\,\mathrm{GeV}^2$ compared to data from a combined analysis of single- and double-pion electroproduction.

| | Calc. | | | Aznauryan *et al.*[9] | | |
|---|---|---|---|---|---|---|
| $B^*$ | $A_p^{1/2}$ | $A_p^{3/2}$ | $S_p^{1/2}$ | $A_p^{1/2}$ | $A_p^{3/2}$ | $S_p^{1/2}$ |
| $P_{11}(1440)$ | 62 | | 17 | $21 \pm 4$ | | $33 \pm 6$ |
| $D_{13}(1520)$ | −55 | 19 | −20 | $-65 \pm 4$ | $62 \pm 5$ | $-35 \pm 4$ |
| $S_{31}(1620)$ | 2 | | −14 | $16 \pm 4$ | | $-28 \pm 3$ |
| $S_{11}(1650)$ | 1 | | −3 | $43 \pm 7$ | | $-6 \pm 3$ |
| $F_{15}(1680)$ | −51 | 25 | −28 | $-32 \pm 5$ | $51 \pm 4$ | $-15 \pm 3$ |
| $D_{33}(1700)$ | 49 | 41 | −15 | $44 \pm 4$ | $36 \pm 4$ | $-7 \pm 4$ |
| $D_{13}(1700)$ | −19 | −7 | −6 | $-21 \pm 2$ | $10 \pm 1$ | 0 |
| $P_{13}(1720)$ | 66 | −16 | −10 | $55 \pm 3$ | $-68 \pm 4$ | 0 |

Although helicity amplitudes calculated for hyperons are not directly accessible in experiments, they play a key role in some contributions to strangeness photo- and electroproduction. As for the non-strange baryons it was found that many amplitudes are larger in magnitude at a finite momentum transfer than at the photon point. It was concluded in [8] that interesting isospin dependencies occur when comparing the decay of hyperons in $\Lambda$- with those in $\Sigma$-resonances. Furthermore the calculated photonic decay width for $\Lambda(1405) \to \Lambda(1116) + \gamma$ exceeds the experimental value by at least an order of magnitude and thus lends further support to the conclusion that $\Lambda(1405)$ can not merely be described as a dominant three-quark excited state. In contrast the electromagnetic decay of the lowest negative parity $\Sigma(1620)$ resonance can be well accounted for by the present quark model calculation.

*e-mail: metsch@hiskp.uni-bonn.de

In a field theoretical context static electromagnetic observables, such as electric charge radii and magnetic moments are usually extracted from elastic form factors, *i.e.* by extrapolation to vanishing momentum transfer of matrix elements of appropriate current operators, a procedure which partially suffers from numerical inaccuracies. In a novel approach [10, 11] developed in the first funding period this circuitous route was circumvented and such observables can, very similar to the procedure in non-relativistic quantum mechanics, be computed directly (and numerically more accurately) as expectation values of operators with respect to the Salpeter amplitudes obtained from the calculation of the mass spectrum in the baryon rest frame. The expressions, derived essentially from linear response, are readily interpreted physically, possess an obvious non-relativistic limit and allow for a discrimination of *e.g.* the spin- and orbital angular momentum contribution to the magnetic moments. Some results are compared to experimental data in Table 2, where also the results obtained by the Graz group [16], which use Dirac's point form formulation of relativistic quantum mechanics to calculate the relevant current matrix elements. Although at first face this is a completely different method the results are strikingly similar; in both methods the importance of relativistic covariance is paramount. The new method also allows for a prediction of magnetic moments, see also Table 2, of excited baryons, a research topic of some interest within the SFB/TR16, also see project B.4.

**Table 2.** Comparison of calculated (BS) and experimental magnetic moments (in units of $\mu_N$) of some selected baryon resonances. Experimental data from [12] and [*a*] [13], [*b*] [14], [*c*] [15]. The column labeled PFSA contains the results obtained in [16].

| B | BS | Exp. | PFSA |
|---|---|---|---|
| p | 2.77 | 2.793 | 2.70 |
| n | −1.71 | −1.913 | −1.70 |
| $\Lambda$ | −0.61 | $-0.613 \pm 0.004$ | −0.65 |
| $\Sigma^+$ | 2.51 | $2.458 \pm 0.01$ | 2.35 |
| $\Sigma^0$ | 0.75 | – | 0.72 |
| $\Sigma^-$ | −1.02 | $-1.16 \pm 0.025$ | −0.92 |
| $\Xi^0$ | −1.33 | $-1.25 \pm 0.014$ | −1.24 |
| $\Xi^-$ | −0.56 | $-0.6507 \pm 0.0025$ | −0.68 |
| $\Delta^{++}$ | 4.14 | 3.7 − 7.5 | 4.17 |
| $\Delta^+$ | 2.07 | $2.7^{+1.0}_{-1.3} \pm 1.5 \pm 3$[a] | 2.08 |
| $\Delta^0$ | 0 | – | 0 |
| $\Delta^-$ | −2.07 | – | −2.08 |
| $\Sigma^{*+}$ | 2.51 | – | 2.07 |
| $\Sigma^{*0}$ | 0.27 | – | 0.09 |
| $\Sigma^{*-}$ | −1.97 | – | −1.89 |
| $\Xi^{*0}$ | 0.59 | – | 0.18 |
| $\Xi^{*-}$ | −1.83 | – | −1.73 |
| $\Omega^-$ | −1.66 | $2.02 \pm 0.05$[b,c] | −1.59 |
| $P^{11}(1440)^+$ | 1.55 | – | – |
| $D^{13}(1520)^+$ | 1.44 | – | – |
| $S^{11}(1535)^+$ | 0.37 | – | – |
| $S^{11}(1650)^+$ | 1.85 | – | – |

*Baryonic semi-leptonic decays:*

As an additional (again parameter free) application and therefore genuine test of our relativistic quark model, we calculated weak decays of the ground state decuplet and octet states with light flavours and also the semi-leptonic decay of *e.g.* $\Lambda_c$, see also below. The computed observables [17, 18] include helicity amplitudes, axial-vector to vector coupling ratios, weak-magnetism ratios, differential and total decay rates of electronic and muonic semi-leptonic baryon decays. We have compared our theoretical predictions with all available experimental data and in general found a good agreement. We have given many predictions, as an example see Fig. 1, of observables for which currently no experimental results exist. There clearly exists an experimental deficit regarding differential decay rates of semi-leptonic baryon decays. An experimental verification of the predicted axial-vector to vector coupling ratio $g_A/g_V = -0.95$ for the decay $\Lambda_c^+ \to \Lambda^0 e^+ \bar{\nu}_e$ would be very important in this respect.

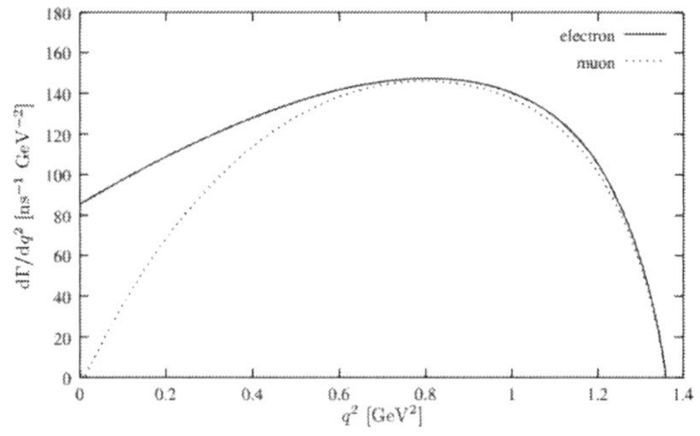


**Figure 1.** Calculated differential decay width for the semi-leptonic decay $\Lambda_c^+ \to \Lambda^0 \ell^+ \bar{\nu}_\ell$, for $\ell = e$ (solid line) and $\ell = \mu$ (dotted line).

*Strong decays:*

Like the electroweak amplitudes discussed above, strong two-body decays of baryon resonances can in lowest order in the context of the Mandelstam formalism be calculated parameter free. Because of the neglect of important hadronic final state interactions in this approximation, a quantitatively accurate reproduction of experimental decay widths is hardly to be expected, and, in fact, the calculated values are in general too small. Nevertheless, in a systematic theoretical study of roughly 1700 strong two-body decays of light-flavoured baryon excitations into a pseudoscalar octet meson and a (possibly excited) baryon final state [18], as performed during the first funding period, it was found that the theoretically investigated strong decays can be classified into three categories: those (about 15-20%) for which the calculated partial decay width is relatively large and numerically a substantial fraction of

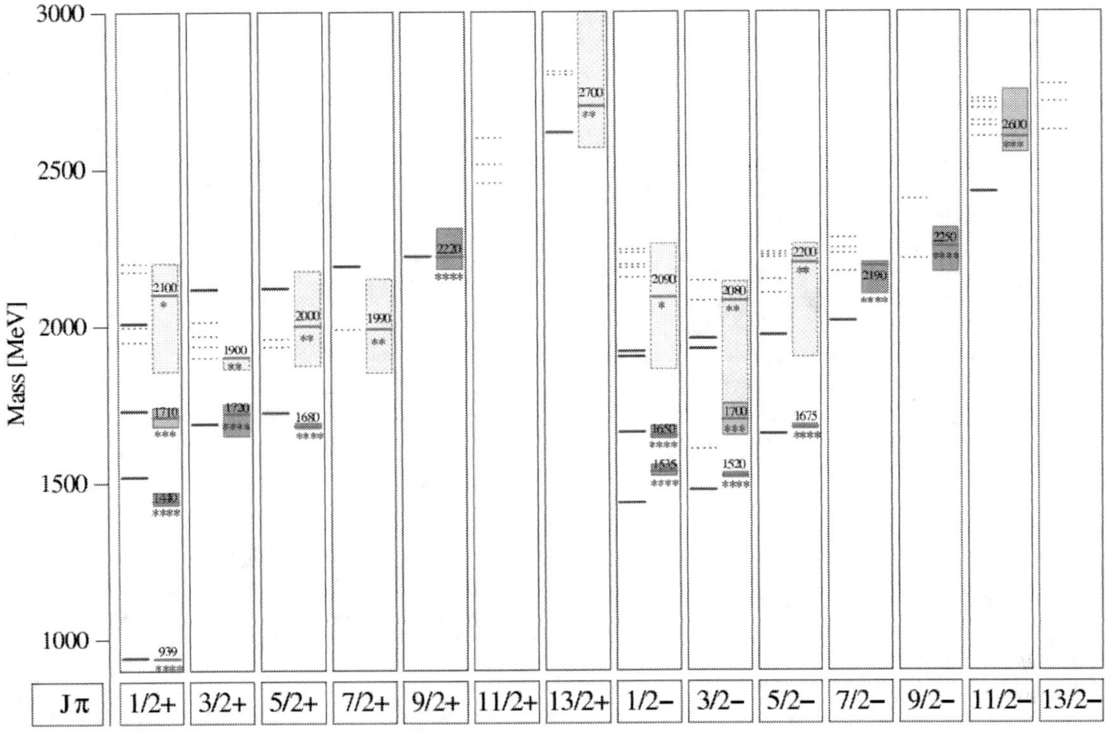


**Figure 2.** The calculated nucleon resonance spectrum (left in each column) compared to the experimental masses (right in each column) from [12]. Dotted lines indicate that the calculated $\pi N$-decay width of the resonance vanishes, a solid line indicates those states for which the calculated width is finite and for most states at least in qualitative agreement with the experimental value.

the experimental value, those for which the calculated width is small but clearly non-zero and finally those with (virtually) vanishing decay width. The resonances with decays belonging to the first and second categories are indeed those which have been observed experimentally in $\pi$- or $K$-nucleon scattering, see *e.g.* Fig. 2, explaining why the majority of the excited states could not be detected in such reactions. The results can be summarised as follows:

- The present calculation does offer a solution to the problem of the missing resonances: They simply do not couple to $N\pi$ or $N\bar{K}$.
- Most of the observed resonances indeed have a non-vanishing theoretical coupling to $N\pi$ or $N\bar{K}$.
- The results allow for unique assignments between theoretical and experimental states. These assignments are not possible based on the mass spectra alone.
- For $N$- and $\Lambda$-resonances $N\pi$ and $N\bar{K}$ are the prefered decay channels. For each set of resonances about 60% of the couplings vanish and about 25% calculated decay widths are larger than 5 MeV.
- Most $\Delta$- and $\Sigma$-resonances do not decay into $N\pi$- or $N\bar{K}$, respectively. For each set of resonances about 80% of the couplings vanish and about 5% of the decay widths are larger than 5 MeV. For these decuplet state some decays are theoretically prefered: $\Delta \to \Delta\eta$, $\Sigma \to \Sigma\pi$ and $\Sigma \to \Delta\bar{K}$.
- Clear selection rules with respect to an intrinsic spin-flip ($\frac{3}{2} \to \frac{1}{2}$) were observed: Such a spin-flip transition is prefered when $\Lambda$ is a final state, is not relevant when a $\Delta$ is involved somewhere and is avoided in all the other cases.
- In general even the non-vanishing decay widths tend to be too small compared to experiment. Note however that the decay widths for the missing resonances are very often smaller by even several orders of magnitude.
- There is always a resonance observed in experiment, if the theoretical coupling is non-vanishing to $N\pi$ or $N\bar{K}$, respectively.
- In total about 60% of the couplings to ground states and 90% of those to excited states vanish, but there are many allowed quasi-two-body decays for almost every N- or $\Delta$-resonance, including the missing resonances, indicating the possibility to detect these states in quasi-two-body decay cascades.

The calculations could be extended to even more missing states and to strong two-body decays involving vector mesons. Instantons also lead to an effective three-body potential, which, because of its colour dependence is spectroscopically irrelevant but in principle leads to a highly selective three-body contribution to strong decays involving (pseudo)scalar mesons, in particular $\eta$ and $\eta'$. It indeed proved to have a strong impact on the numerical results for particular strong two-body decays of mesons. Unfortunately, the numerical implementation of 't Hooft's three-body interaction for baryonic decays turned out to be numerically much more involved than it is for mesonic decays where the amplitude essentially factorises. This subject was therefore not further pursued.

In view of the results discussed so far we concluded that this relativistic constituent quark model accounts for the majority of the salient features in the baryon spectrum, such as linear Regge trajectories, the major spin dependent mass splittings, in particular the occurrence of spin-parity doublets and offers a parameter-free quantitative description of many electroweak observables, such as static moments, helicity amplitudes and semi-leptonic decay observables. Furthermore, as discussed above, the lowest order calculation of strong meson-baryon decay amplitudes, at least qualitatively explains why many excited baryon states have not been observed in meson-baryon scattering experiments so far.

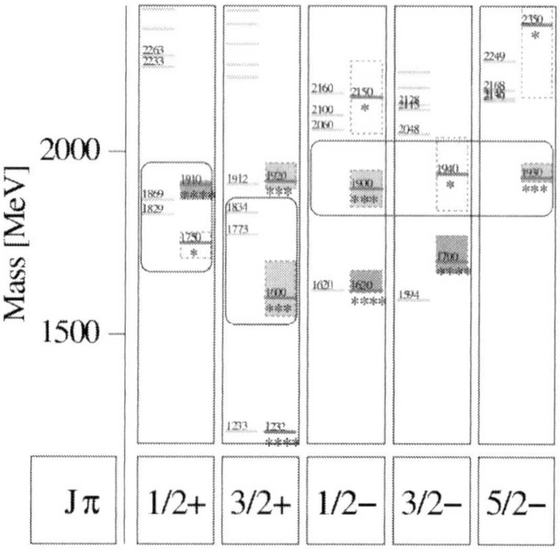


**Figure 3.** Discrepancies between experimental (right part of each column, from [20]) and calculated (left part of each column, as in [2]) positions of the lowest Δ-resonances. Shaded boxes indicate the uncertainty in the resonance position.

## 3 Deficiences and improvements

Although this in our opinion validated the calculations performed so far, some unresolved issues remained:

1. The quantitative description of strong two-body decay amplitudes is clearly deficient, probably mainly because final state interactions and thus coupled channel effects are completely ignored. An attempt to improve upon this turned out to be extremely difficult. In the instantaneous approximation it is possible to determine the strong two-body decay amplitudes merely on the mass shell. Even with an adequate parametrisation of non-resonant amplitudes as an ingredient in a simplified coupled channel calculation (essentially in a $K$-matrix approach) the resulting decay widths hardly changed[19].

2. In the mass spectrum of negative parity Δ-resonances there was significant experimental evidence for the existence of three resonances in the mass range 1900 – 1940 MeV, of which the conspicuous low position could not be accounted for by the calculation as described above, see Fig. 3. Moreover, also the calculated position of the first excited positive parity Δ resonances was much too high. An improvement of the description was obtained in the manner described in [22].

### 3.1 Effects of an additional spin and flavour dependent interaction

Among the various possibilities we persued the simple *Ansatz* to phenomenologically introduce an additional spin and flavour dependent interaction, more or less inspired by the results obtained by the Graz-Group with a version of a so-called Goldstone-boson-exhange interaction, see *e.g.* [21]. Here, we decided to parameterise the new spin-flavour-dependent interaction purely phenomenologically as a local potential in configuration space, its simple form given by

$$V(\vec{x}) = \sum_{a=1}^{8} g_a^2 \left[\lambda^a \gamma_5 \otimes \lambda^a \gamma_5\right] v_{\lambda_a}(|\vec{x}|)\ ,$$

where the Gell-Mann matrices $\lambda^a$ determine the flavour-, $(\gamma_5 \cdot \gamma_5)$ determines the spin- and the simple gaussian

$$v_\lambda(r) := \frac{1}{\lambda^3\, \pi^{\frac{3}{2}}} \exp\left(-\frac{r^2}{\lambda^2}\right)$$

determines the interquark-distance- dependence.

After an adjustment of the parameters, for the values, see [22] (and for a slightly modified set of parameter values, see [23]) indeed an improved description of the Δ mass spectrum was found:, see *e.g.* Fig. 4, Remarkably enough, in this manner, not only it was possible to account for negative-parity Δ-resonances slightly below 2 GeV, but also a much more accurate description of the $\Delta(\frac{3}{2}^+; 1660)$ and the postive-parity $\Delta(\frac{1}{2}^+)$-resonances was obtained. Furthermore, with this additional spin-flavour dependent interaction the position of the Roper-resonance below the first negative parity nucleon resonances is now accurately accounted for, see Fig. 5. In addition the mass-splittings of these negative-parity resonances are now better reproduced. For similar improvements in the spectra of strange baryons we refer to [22].

### 3.2 Impact on electromagnetic observables

The phenomenological introduction of the additional spin-flavour depependent interaction kernel increased the number of parameters of seven in the original model (Model $\mathcal{A}$,

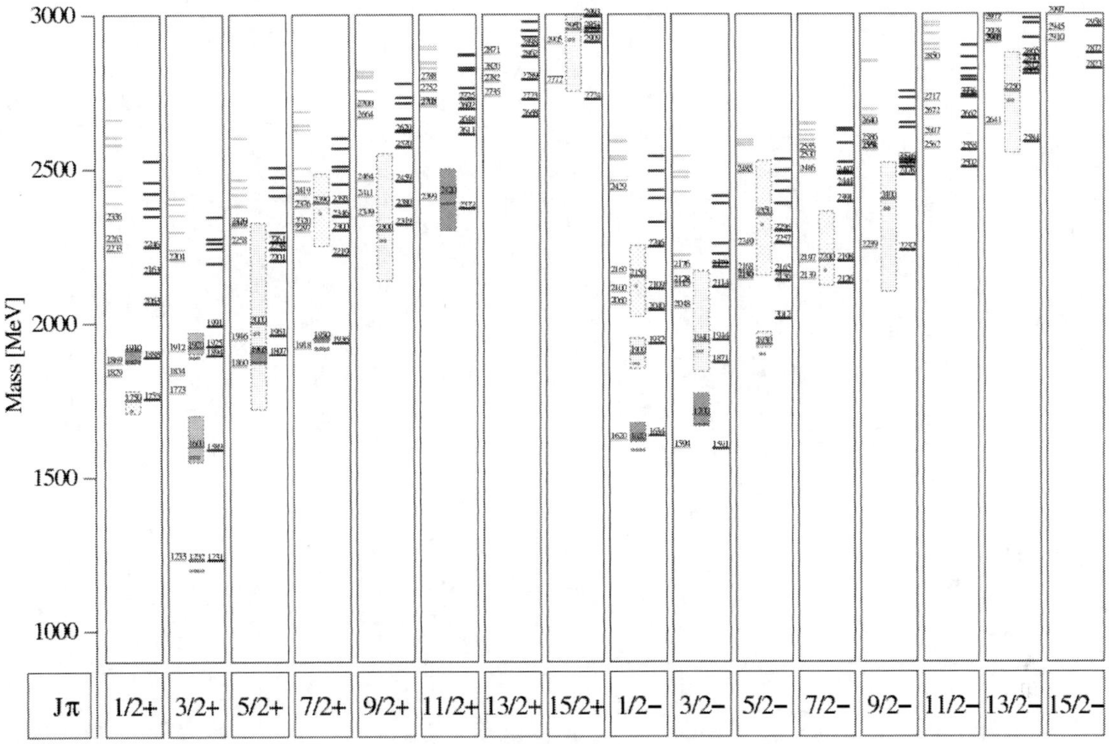


**Figure 4.** Comparison of the Δ-spectrum calculated with the additional phenomenological spin-flavour dependent interaction (right side of each column), see [22], with experimental data from the Particle Data Group [20] (central in each column) and with the results from the model as discussed in [2] (left side in each column). Shaded boxes indicate the uncertainty in the resonance position.

see [2, 3]) to ten in the new version (Model *C*, see [23]), which we considered to be still acceptable in view of the multitude of baryon masses described accurately in this manner. As before, after fixing the parameters by a fit to the mass spectra, the calculation of electromagnetic properties in lowest order is a genuine prediction. In Fig. 6 we display the electric form factor of the proton, divided by the dipole-shape

$$G_D(Q^2) = \frac{1}{(1 + Q^2/M_V^2)^2}, \quad (1)$$

with $M_V^2 = 0.71$ GeV$^2$ (see [24, 27]) is shown for two versions of parameters (see Table 1 in [23]) in comparison to experimental data.

It is found that with the new model describes the data much better than the original model, see [5]. This applies also to the electric neutron from factor, see Fig. 7; this small quantity is very sensitive to parameter changes and the deviation between both curves could be interpreted as an estimate of the uncertainty in the model prediction. It demonstrates, however, that it is indeed possible to account for the momentum transfer dependence and the position of the maximum rather accurately. Also for the magnetic form factors displayed in Fig. 8 and 9 improvements are observed: this concerns in particular the description of the magnetic proton form factor at higher momentum transfers $Q^2 \gtrsim 1$ GeV$^2$. The momentum transfer dependence (up to $Q^2 \approx 6$ GeV$^2$) of a multitude of transverse and longitudinal helicity amplitudes has been presented and discussed in [23], both in the original model and in

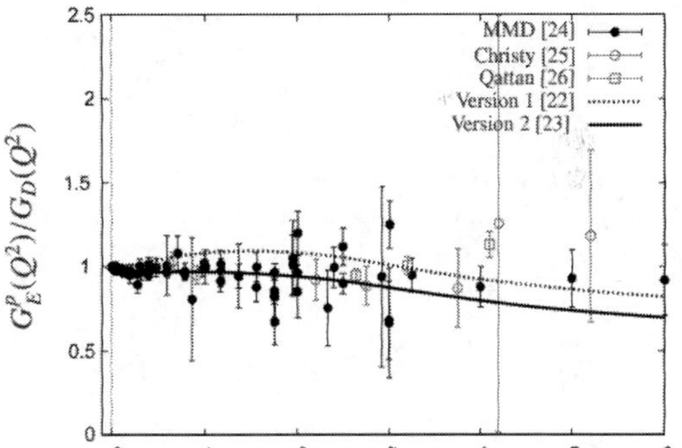


**Figure 6.** The electric form factor of the proton divided by the dipole form $G_D(Q^2)$, Eq. (1). MMD-Data are taken from Mergell *et al.* [24], supplemented by data from Christy *et al.* [25] and Qattan *et al.* [26]. The solid line represents the results with the model parameters listed in [23] while the dashed line those listed in [22]. Red data points are taken from polarisation experiments and black points are obtained by Rosenbluth separation.

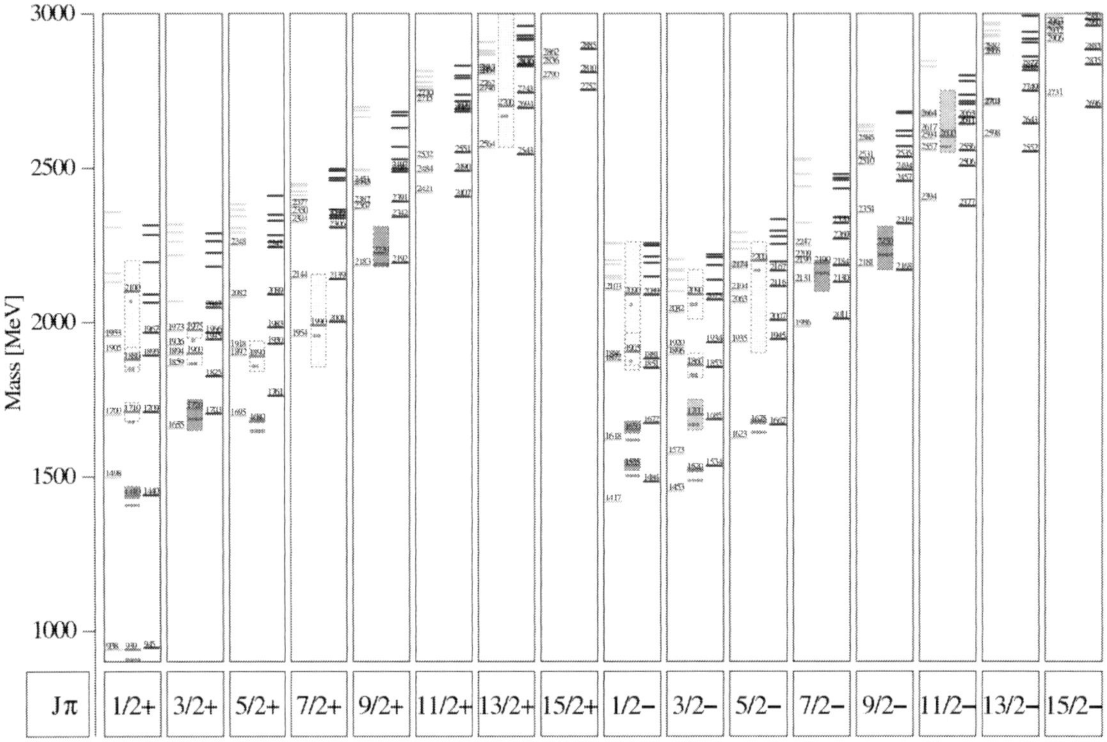


**Figure 5.** Comparison of the $N$-spectrum calculated with the additional phenomenological spin-flavour dependent interaction (right side of each column), see [22], with experimental data from the Particle Data Group [20] (central in each column) and with the results from the model as discussed in [2] (left side in each column). Shaded boxes indicate the uncertainty in the resonance position.

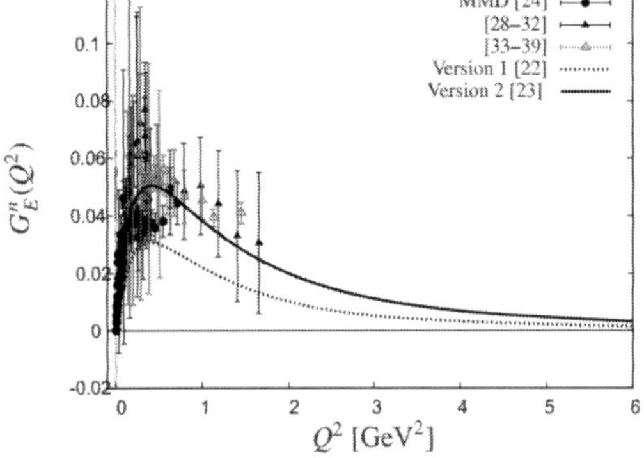


**Figure 7.** The electric form factor of the neutron. MMD-Data are taken from the compilation of Mergell *et al.* [24]. The solid line represents the results with the model parameters listed in [23] while the dashed line those listed in [22]. Red data points are taken from polarisation experiments and black points are obtained by Rosenbluth separation.

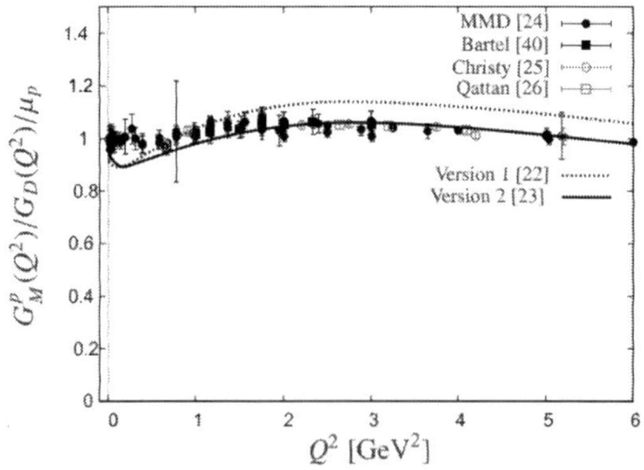


**Figure 8.** The magnetic form factor of the proton divided by the dipole form $G_D(Q^2)$, Eq. (1) and the magnetic moment of the proton $\mu_p = 2.793\,\mu_N$ . MMD-Data are taken from the compilation of Mergell *et al.* [24]. Additionally, polarisation experiments are marked in red. The black marked data points are obtained by Rosenbluth separation.

the model with the additional spin-flavour dependent interaction. Here also results on photocouplings for a few dozen $N \to N^*$- and $N \to \Delta^*$- transitions can be found.

### 3.3 Summary

The same calculational framework as presented by Merten *et al.* [5] was used for the computation of current-matrix elements on the basis of Salpeter-amplitudes obtained in

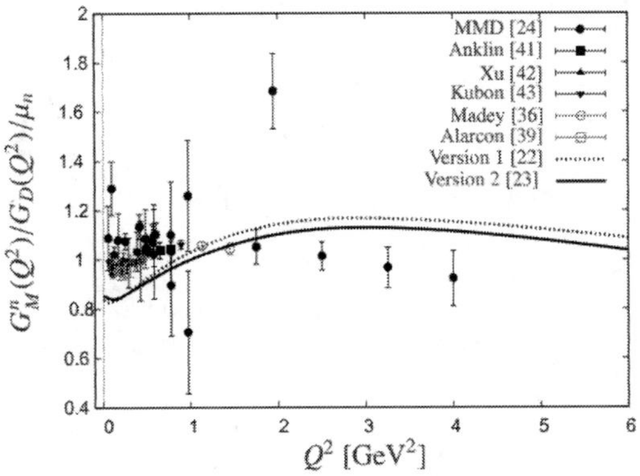


**Figure 9.** The magnetic form factor of the neutron divided by the dipole form $G_D(Q^2)$, Eq. (1) and the experimental value of the magnetic moment of the neutron $\mu_n = -1.913\,\mu_N$. Note that the calcultated value of the neutron magnetic moment is in fact too small, see also Table 2, where it was calulated with an alternative method. MMD-Data are taken from the compilation by Mergell *et al.* [24] and from more recent results from MAMI [41, 43]. Additionally, polarisation experiments are marked by red data points. The black marked points are obtained by Rosenbluth separation.

a calculation of the baryon mass spectra, see [22], where, in addition to confinement and a spin-flavour dependent interaction motivated by instanton effects as used in an older model version [2, 3], an additional phenomenological short ranged spin-flavour dependent interaction was introduced in order to improve in particular upon the description of some excited negative parity $\Delta$ resonances at $\approx$ 1.9 GeV. In some significant details a superior description of the mass spectrum of all baryon resonances at the expense of only three extra phenomenological parameters was thus obtained. It was found that the description of ground state electromagnetic form factors was also significantly improved. A systematic investigation of electromagnetic transition form factors in the form of momentum transfer dependent transverse and longitudinal helicity amplitudes revealed that both in the original and in the novel model the transverse $A^{\frac{1}{2}}$-helicity amplitudes are significantly better described than the $A^{\frac{3}{2}}$-amplitudes.

Stimulating discussions with E. Klempt, A.V. Sarantsev and U. Thoma within the framework of the DFG supported Collaborative Research Centre SFB/TR16 as well as contributions by Christian Haupt, Sascha Migura, Michael Ronniger and Daniela Tolentino-Zang are gratefully acknowledged.

## References

[1] U. Löring, K. Kretzschmar, B.Ch. Metsch, H.R. Petry, Eur. Phys. J. A **10** (2001) 309-346.

[2] U. Löring, B.Ch. Metsch, H.R. Petry, Eur. Phys. J. A **10** (2001) 395-446.

[3] U. Löring, B.Ch. Metsch, H.R. Petry, Eur. Phys. J. A **10** (2001) 447-486.

[4] U. Löring, K. Kretzschmar, D. Merten, B. Metsch, H.-R. Petry, Eur. Phys. J. A **14** (2002) 477-489.

[5] D. Merten, U. Löring, B. Metsch, H.-R. Petry, Eur. Phys. J. A **18** (2003) 193-195.

[6] B.C. Metsch, Eur. Phys. J. A **35** (2008) 275. doi:10.1140/epja/i2007-10557-4.

[7] S. Kreuzer, "Ein massenabhängiges Confinement Potential für das Bethe-Salpeter Modell", Diplomarbeit, Universität Bonn, August 2006;

[8] T. Van Cauteren, J. Ryckebusch, B. Metsch and H. R. Petry, Eur. Phys. J. A **26** (2005) 339. [arXiv:nucl-th/0509047].

[9] I. G. Aznauryan, V. D. Burkert, G. V. Fedotov, B. S. Ishkhanov, Phys. Rev. C **72** (2005) 045201d [arXiv:hep-ph/0508057].

[10] C. Haupt, B. Metsch and H. R. Petry, Eur. Phys. J. A **28** (2006) 213 [arXiv:hep-ph/0602151].

[11] C. Haupt, "Electromagnetic properties of Baryons," Dissertation, Universität Bonn, April 2006.

[12] Particle Data Group, W.-M. Yao *et al.*, J. Phys. G **33** (2006) 1.

[13] M. Kotulla *et al.*, Phys. Rev. Lett. **89** (2002) 272001 [arXiv:nucl-ex/0210040].

[14] N. B. Wallace *et al.*, Phys. Rev. Lett. **74** (1995) 3732.

[15] H. T. Diehl *et al.*, Phys. Rev. Lett. **67** (1991) 804.

[16] K. Berger *et al.*, Phys. Rev. D **70** (2004) 094027 [arXiv:nucl-th/0407009].

[17] S. Migura, D. Merten, B. Metsch and H. R. Petry, Eur. Phys. J. A **28** (2006) 55. [arXiv:hep-ph/0602152]

[18] S. Migura, "Weak and strong baryon decays in a constituent quark model", Dissertation, Universität Bonn, Juni 2006.

[19] D. M. Tolentino Zang, "Final state interactions in strong two-body baryon decays", Dissertation, Universität Bonn, July 2014.

[20] C. Amsler *et al.* (Particle Data Group), Phys. Lett. B **667**, (2008) 1.

[21] T. Melde, W. Plessas, and B. Sengl, Phys. Rev. D **77**, 114002 (2008).

[22] M. Ronniger and B. C. Metsch, Eur. Phys. J. A **47** (2011) 162 doi:10.1140/epja/i2011-11162-8 [arXiv:1111.3835 [hep-ph]].

[23] M. Ronniger and B. C. Metsch, Eur. Phys. J. A **49** (2013) 8 doi:10.1140/epja/i2013-13008-9 [arXiv:1207.2640 [hep-ph]].

[24] P. Mergell, U.-G. Meißner and D. Drechsel, Nucl. Phys. A **596**, 367 (1996).

[25] M.E. Christy *et al.*, Phys. Rev. C **70**, 015206 (2004).

[26] I. A. Qattan *et al.*, Phys. Rev. Lett. **94**, 142301 (2005).

[27] A. Bodek, S. Avvakumov, R. Bradford, and H. Budd, J. Phys. Conf. Ser. **110**, 082004 (2008).

[28] T. Eden *et al.*, Phys. Rev. C **50**, 1749 (1994).

[29] C. Herberg *et al.*, Eur. Phys. J. A **5**, 131 (1999).

[30] M. Ostrick *et al.*, Phys. Rev. Lett. **83**, 276 (1999).

[31] I. Passchier *et al.*, Phys. Rev. Lett. **82**, 4988 (1999).

[32] R. Schiavilla *et al.*, Phys. Rev. C **64**, 041002 (2001).

[33] D. Rohe *et al.*, Phys. Rev. Lett. **83**, 21 (1999).

[34] J. Golak *et al.*, Phys. Rev. C **63**, 034006 (2001).

[35] H. Zhu *et al.*, Phys. Rev. Lett. **87**, 081801 (2001).
[36] R. Madey *et al.*, Phys. Rev. Lett. **91**, 122002 (2003).
[37] G. Warren *et al.*, Phys. Rev. Lett. **92**, 042301 (2004).
[38] D. I. Glazier *et al.*, Eur. Phys. J. A **24**, 101 (2005).
[39] R. Alarcon *et al.*, Eur. Phys. J. A **31**, 588 (2007).
[40] T. Bartel *et al.*, Nucl. Phys. B **58**, 469 (1973).
[41] H. Anklin *et al.*, Phys. Lett. B **428**, 248 (1998).
[42] W. Xu *et al.*, Phys. Rev. Lett. **85** (2000).
[43] G. Kubon *et al.*, Phys. Lett. B **524**, 26 (2002).

# Baryon resonances in a finite volume

*Ulf-G.* Meißner[1,2,*] and *Akaki* Rusetsky[1,**]

[1] *Helmholtz-Institut für Strahlen- und Kernphysik (Theorie) and Bethe Center for Theoretical Physics, Universität Bonn, D-53115 Bonn, Germany*
[2] *Institute for Advanced Simulation (IAS-4), Institut für Kernphysik (IKP-3) and Jülich Center for Hadron Physics, Forschungszentrum Jülich, D-52425 Jülich, Germany*

**Abstract.** We review the main results obtained in the project **A.6**, which dealt with hadron-hadron scattering and resonances in a finite volume. These results provide tools for the extraction of the resonance characteristics from lattice QCD simulations. In particular, in this review we focus on the evaluation of the matrix elements of the weak and electromagnetic currents between the states containing at least one resonance, as well as on the description of the three-particle states in a finite volume. We also discuss insights obtained using unitarized chiral perturbation theory in a finite volume.

## 1 Introduction

The project **A.6** has emerged from the project **C.3** "Spin structure of the nucleon (theory)," and covers the last two funding periods of the SFB/TR 16 (2008-2016). The necessity to start this project was justified by the intensified efforts to study the properties of resonances by using lattice QCD. It should be pointed out that Monte Carlo simulations in lattice QCD constitute only a part of these efforts. In addition, one needs a framework, where the "raw" results of the simulations should be interpreted in terms of the observable hadronic characteristics, which can be directly compared to experiments. Without possessing tools for such an interpretation, the simulations alone might turn out to be useless.

In order to illustrate this statement, note first that a physical stable particle corresponds to an eigenstate of the finite-volume lattice Hamiltonian (let us remind the reader that the lattice simulations are always performed in a finite volume). Further, finite-volume artifacts in the measured values of the energies are usually small, being exponentially suppressed at large volumes. Thus, one may directly extract the masses of the stable particles from the results of simulations. On the other hand, resonances correspond to the poles of the $S$-matrix at complex energies on unphysical Riemann sheets, and not to the isolated eigenvalues of the Hamiltonian. Consequently, measuring the individual energy levels on the lattice does not immediately provide information about the location of a resonance pole. In order to extract the latter, one should first use a formula which relates the energy, measured in a finite volume, to the value of the infinite-volume phase at the same energy (this relation goes under the name of the Lüscher formula [1]). In the next step, using the extracted phase shifts at the different energies, one performs an analytic continuation into the complex plane and determines the position of the pole.

The above example clearly demonstrates the need of a framework, which should be used for the analysis of the lattice data. In case of more complicated systems (coupled-channel scattering, calculation of the matrix elements with resonances, many-particle systems), more elaborate tools are needed. In our study of these systems, we have extensively used non-relativistic effective field theories (EFT) in a finite volume, which enable one to systematically relate the measured quantities in a finite and in the infinite volume. The Lüscher formula, mentioned above, directly follows from this approach in the simple case of the elastic scattering. We briefly review the EFT approach in Sect. 2.

During 2008-2016, the investigations carried out within the project **A.6**, contributed significantly to the major developments in the field. In particular, we studied the extraction of the mass and width of the $\Delta$-resonance from the lattice data, performing the finite-volume calculations in Chiral Perturbation Theory (ChPT) at order $\epsilon^4$ within the Small Scale Expansion (SSE) [2]. The group-theoretical aspects of the Lüscher approach, including the full analysis in the moving frames for the particles with both the integer and half-integer spin were addressed in Refs. [3, 4]. In Refs. [5, 6], using the non-relativistic EFT in a finite volume, we have generalized the Lüscher approach to the case of the coupled channels and discussed the extraction of the $\bar{K}N$ scattering lengths from the data, as well as the application in the scalar meson sector. A further development of these ideas are contained in Refs. [7–10], where the finite-volume generalization of the Unitary ChPT (UChPT) has been introduced. The parameters of

* e-mail: meissner@hiskp.uni-bonn.de
** e-mail: rusetsky@hiskp.uni-bonn.de

UChPT are fitted to the measured energy levels in a finite volume. At the next stage, UChPT in the infinite volume *with the same set of the parameters* is used to reproduce the $S$-matrix elements and determine the position of its poles in the complex plane. Fitting the synthetic data, it was demonstrated that the method is very robust and works even if the data have sizable statistical errors. The main insights from this framework are summarized in Sec. 5.

We would like to stress that using an effective field theory approach in a finite volume has led to the *two completely new developments* in the field. The first one is the extraction of the resonance matrix elements [11–14] – by this, we mean the matrix elements of currents sandwiched by the states containing at least one resonance (we shall give a precise definition of this quantity in Sect. 3). Examples are given by the form factors of the $\Delta$ and $\rho$, the $\Delta N\gamma^*$ transition form factor, as well as the semileptonic form factors in the $B \to K^*\ell^+\ell^-$ decays. The second development deals with the study of the three-particle systems in a finite volume [15–20]. These investigations, which are of ultimate importance for the analysis of the inelastic resonances (e.g., the Roper resonance) as well as the data obtained in nuclear lattice simulations, will be addressed in Sect. 6 in more detail.

Finally, it should be pointed out that there have been very interesting developments in using the non-relativistic EFT in the infinite volume. In particular, it has proven to be very useful in study of the three-particle decays [21–23], $K_{\ell 4}$ decays [24], as well as in study of hadronic atoms [25, 26]. Further developments include the study of the various aspects of the chiral expansion in the infinite volume [27–31], as well as the formulation of the alternative method for the extraction of a resonance mass and width on the lattice [32].

Disclaimer: Note that the present work is the report on the activities carried out within the project **A.6** during the period 2008-2016. It is impossible to give all references relevant for the subjects covered by the research in eight years. Below, we shall mainly cite the work done within **A.6**, including a few more references whenever necessary. No claim is made on having a full list of references. We also point towards the connection to other projects.

## 2 Non-relativistic effective theories

In order to extract the hadron characteristics in the scattering sector from lattice data, the size of the lattice, $L$, should obey certain constraints. Namely, $L$ must be much larger than $R$, a typical scale of the hadronic interactions. Usually, $R$ is defined by the mass of the lightest particle in the system, which in the case of QCD is the pion mass. Consequently, one should work in the limit $M_\pi L \gg 1$. In the case of boxes of such size, the wave function of a system in the vicinity of the walls of the box has already its asymptotic form which, for elastic scattering, is determined solely by the scattering phase shift. Consequently, the discrete energy spectrum in a finite volume, which is obtained by imposing (periodic) boundary conditions on the wave function at the walls, is determined solely by the phase shift at the same energy. This statement conveys the main message contained in the Lüscher equation [1] which, in the S-wave and in the absence of the partial-wave mixing, takes the simple form

$$\begin{aligned} \tan\delta_0(p) &= -\tan\phi(q)\,, \\ \tan\phi(q) &= -\frac{\pi^{3/2}q}{Z_{00}(1;q^2)}\,, \quad q=\frac{pL}{2\pi}\,. \end{aligned} \tag{1}$$

Here, $\delta_0(p)$ is the S-wave elastic scattering phase and $p$ denotes the magnitude of the relative three-momentum in the center-of-mass (CM) frame. Further, $Z_{00}(1,q^2)$ stands for the Lüscher zeta-function

$$Z_{00}(1;q^2)=\frac{1}{\sqrt{4\pi}}\lim_{s\to 1}\sum_{\mathbf{n}\in\mathbb{Z}^3}\frac{1}{(\mathbf{n}^2-q^2)^s}\,. \tag{2}$$

The same result can be easily obtained within the EFT approach, which has the advantage that it can be easily generalized to treat more complicated systems. Below we closely follow the line of the original paper [33]. Since the characteristic momenta on the lattice $p \sim L^{-1}$, the condition $M_\pi L \gg 1$ implies $p \ll M_\pi$. Systems with such small momenta can be described by the so-called non-relativistic EFT, which contains only particles (antiparticles are consistently included in the couplings of the Lagrangian). In other words, the particle number is a conserved quantity in the non-relativistic EFT, since the particle number non-conserving processes contain the large energy scale $M_\pi$.

In the theory which conserves the particle number, the structure of the Feynman diagrams is remarkably simple. Namely, in the infinite volume, the two-particle scattering amplitude $T$ is given by the sum of the bubble diagrams shown in Fig. 1 and, therefore, obeys the Lippmann-Schwinger equation

$$T = V + VG_0T\,. \tag{3}$$

Here, $V$ denotes the sum of all two-particle irreducible diagrams (the potential) and $G_0$ stands for the free two-particle Green function.

Unlike Ref. [33], we retain the relativistic dispersion law $w(\mathbf{p}) = \sqrt{m^2+\mathbf{p}^2}$ for all particles, even if one is working in the non-relativistic theory. In Refs. [23, 34] one can find a very detailed discussion of the covariant non-relativistic approach. The one-particle propagator is given by

$$D(p)=\frac{1}{2w(\mathbf{p})}\,\frac{1}{w(\mathbf{p})-p_0-i0}\,. \tag{4}$$

As seen, this propagator contains only the particle pole. The advantage of this approach is that the kinematic relations hold automatically and should not be adjusted order by order. For example, the thresholds are always at the right place from the beginning. The price for this is a more complicated structure of the corresponding Feynman integrals. In particular, as shown in Refs. [23, 34], if one works out these integrals just by using dimensional regularization, the result does not obey the counting rules. In order to observe these rules, one has to apply the so-called threshold expansion [35], which boils down to the following prescription:

**Figure 1.** The sum of the bubble diagrams in the two-particle $T$-matrix in the non-relativistic EFT. The small filled square denotes the sum of all two-particle irreducible diagrams (the potential).

- Expand all integrands in the Feynman integrals in three-momenta;
- Perform the integration in the dimensional regularization order by order and sum up the result again.

As shown in Refs. [23, 34], after applying the threshold expansion, the Lippmann-Schwinger equation (3) turns into an algebraic equation. Indeed, the two-particle loop in the CM frame after the threshold expansion gives

$$\int \frac{d^Dk}{(2\pi)^D i}\,\frac{1}{2w_1(\mathbf{k})2w_2(\mathbf{k})}\,\frac{1}{w_1(\mathbf{k})-k_0-i0} \times \frac{1}{w_2(\mathbf{k})-E+k_0-i0} = \frac{ip}{8\pi E}\,. \quad (5)$$

Here, $E$ denotes the total energy of two particles in the CM frame $E = w_1(\mathbf{p})+w_2(\mathbf{p})$. Choosing a proper normalization and performing the partial-wave expansion, Eq. (3) in the S-wave can be rewritten as

$$T(p,p) = V(p,p) + ipV(p,p)T(p,p)\,. \quad (6)$$

Consequently, in this scheme, $pV(p,p) = \tan\delta_0(p)$ denotes the on-shell $K$-matrix. It should be pointed out that the threshold expansion automatically "projects" all quantities, entering the Lippmann-Schwinger equation, onto the mass shell, since the off-shell effects lead to an regular integrand and vanish in this scheme. This constitutes a major advantage with respect to the relativistic approach, where, in some cases, checking the explicit cancellation of the off-shell effects is a rather non-trivial matter.

Next, we place the system in a three-dimensional box with a side length $L$ and calculate the finite-volume analog of the scattering $T$-matrix. Note that the momenta in a box are quantized:

$$\mathbf{p} = \frac{2\pi}{L}\mathbf{n}, \quad \text{where } \mathbf{n}\in\mathbb{Z}^3\,. \quad (7)$$

Consequently, in all diagrams, the loop integrals will be replaced by discrete sums. On the other hand, the quantity $V$, which encodes the short-range physics related to the couplings of the effective Lagrangian, is not affected by the finite-volume artifacts up to exponentially suppressed terms $\propto \exp(-M_\pi L)$. The two-particle loop in a finite volume is given by

$$\frac{1}{L^3}\sum_{\mathbf{k}} \frac{1}{2w_1(\mathbf{k})2w_2(\mathbf{k})}\,\frac{1}{w_1(\mathbf{k})-k_0-i0} \times \frac{1}{w_2(\mathbf{k})-E+k_0-i0} = \frac{1}{4\pi^{3/2}EL}Z_{00}(1;q^2)\,, \quad (8)$$

and, if the partial-wave mixing is neglected, the finite-volume counterpart of Eq. (3) reads

$$T_L(p,p) = V(p,p) + \frac{2}{\sqrt{\pi}L}Z_{00}(1;q^2)V(p,p)T_L(p,p)\,. \quad (9)$$

In a finite volume, the quantity $T_L(p,p)$ possesses an infinite tower of poles, corresponding to the discrete eigenvalues of the full Hamiltonian. It is straightforward to ensure that the position of these poles is given exactly by Eq. (1). Thus, the Lüscher equation is reproduced in the EFT approach.

The above derivation clearly demonstrates the path, which is followed by the EFT approach in more complicated cases. The basis is given by the effective Lagrangian which is the same in a finite and in the infinite volume, up to exponentially suppressed corrections. This stems from the fact that the Lagrangian encodes the short-distance physics which is not affected by the infrared effects due to the presence of the box. Only the Feynman rules are modified, namely, the integrals are replaced by sums everywhere. The fact that the Lagrangian is the same enables one to relate the results of calculation in a finite and in the infinite volume and thus to establish the bridge between the observables we are looking for.

In the next sections, we shall consider three examples of such calculations. Namely, the resonance matrix elements and the three-particle states will be treated in detail, as well as the extraction of the resonance pole positions with the use of the UChPT.

## 3 Resonance matrix elements

As already mentioned in the Introduction, stable particles correspond to isolated energy levels of the Hamiltonian in a finite volume. Consequently, matrix elements of the weak and electromagnetic currents, evaluated between these eigenvalues, yield the pertinent form factors in the infinite-volume limit. The situation with resonances, which correspond to complex poles in the $S$-matrix, is different. It can be shown that the pertinent matrix elements, measured in a finite volume, do not have a well-defined infinite-volume limit. Hence, the following conceptual issues should be addressed:

1. Strictly speaking, one does not directly observe a resonance in the experiment, only the final products of its decay are detected. Is it possible to avoid the notion of a resonance whatsoever and to directly relate the decay amplitude to the results of the lattice measurements?
2. How does one define the resonance matrix element in the infinite volume and how does one measure it?
3. How does one extract this matrix element from the lattice data?
4. Do the procedures described in 1. and 3. agree in the limit of an infinitely narrow resonance?

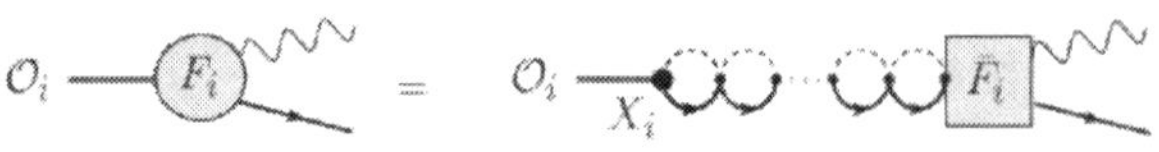


**Figure 2.** Matrix element in the EFT, exemplified on the decay $\Delta \to N\gamma^*$. The quantity $\bar{F}_i$ denotes the irreducible vertex. This quantity yields the full matrix element $F_i$ after dressing with any number of $\pi N$ bubbles that describe interactions in the initial state. The solid line with the operator $O_i$ denotes the resonance, here the delta, and solid, dashed, wiggly lines refer to nucleons, pions, and photons, respectively.

The answer to the question 1. is provided by the so-called Lellouch-Lüscher formula [36], which relates the matrix element of a current, measured in a finite volume, to the matrix element of the same current in the infinite volume, evaluated between the scattering states that, in our case, emerge in the decay of a resonance (see also Ref. [37], where the moving frames were considered). In the original paper [36], the (non-resonant) weak decay process $K \to 2\pi$ was considered. In Ref. [38], the approach was extended to the extraction of the timelike pion electromagnetic form factor, where the $\rho$ plays a prominent role. All these results can be most easily obtained within the EFT approach which, in addition, can be straightforwardly generalized to treat more complicated systems [11–14, 39–44].

The answer to question 2. is in fact well known since decades. Consider an arbitrary (local or non-local) scalar operator $O(x)$, which has the internal quantum numbers of a given resonance. The statement that a resonance is present is equivalent to the claim that the two-point function

$$D(p^2) = i\int d^4x\, e^{ipx}\langle 0|TO(x)O^\dagger(0)|0\rangle \qquad (10)$$

has a pole in the complex variable $z = p^2$ on the lower half of the second Riemann sheet at $z = p_R^2$:

$$\lim_{p^2\to p_R^2} D(p^2) = \frac{B_R}{p_R^2 - p^2} + \cdots . \qquad (11)$$

The real and imaginary parts of $p_R^2$ are related to the resonance mass $m_R$ and the width $\Gamma_R$, according to $\mathrm{Re}\, p_R^2 = m_R^2 - \Gamma_R^2/4$, $\mathrm{Im}\, p_R^2 = -m_R\Gamma_R$.

In order to define resonance matrix elements of, e.g., the electromagnetic current $J_\mu$, we consider the following three-point function:

$$V_\mu(p,q) = -\int d^4x\, d^4y\, e^{ipx-iqy}\langle 0|TO(x)J_\mu(0)O^\dagger(y)|0\rangle . \qquad (12)$$

The form factor of a resonance is then *defined* as

$$i(p+q)_\mu F((p-q)^2) \doteq \langle p|J_\mu(0)|q\rangle = \lim_{p^2,q^2\to p_R^2} B_R^{-1/2}(p_R^2 - p^2)V_\mu(p,q)(p_R^2 - q^2)B_R^{-1/2} , \qquad (13)$$

where $B_R$ is the residue at the resonance pole, see Eq. (11). Note that the matrix element displayed in Eq. (13) should be understood as a mere notation: in the spectrum, there exists no isolated resonance state with a definite momentum. Moreover, as it is clear from Eq. (13), this definition of the resonance matrix elements necessarily implies an analytic continuation into the complex plane. We would like to stress that we are not aware of any consistent field-theoretical prescription, where the analytic continuation would not be employed. The generalization to resonances with arbitrary spin is straightforward.

The key question to be addressed here is 3. As demonstrated in Refs. [11–14], this can be done by using the non-relativistic EFT framework and evaluating the matrix elements both in a finite and in the infinite volume. An example of this procedure, corresponding to the decay process $\Delta \to N\gamma^*$, is given in Fig. 2. It is seen that the matrix element $F_i$ we are looking for is given as a product of the two-particle irreducible vertex $\bar{F}_i$ and any number of two-particle bubbles. The difference between the two cases is concentrated exactly in these bubbles, because the two-particle irreducible blocks are exactly the same in a finite and in the infinite volume (the finite-volume effects are exponentially suppressed). Moreover, the analytic continuation in the irreducible blocks can be done by using an analog of the effective-range expansion, because these blocks can be expressed in terms of low-energy polynomials. Finally, one may formulate a unique prescription, which relates the matrix element, measured on the lattice, to the resonance matrix element evaluated at the pole. We do not give the details of this prescription here. The interested reader can find these details in Refs. [11–14]. Moreover, it can be directly shown that, for infinitely narrow resonances, the matrix elements measured on the lattice coincide with the ones extracted by using this procedure.

There is, however, one complication which should also be mentioned. If a resonance appears *both* in the initial and the final states (an example: the electromagnetic form factor of the $\Delta$, see Ref. [12]), the triangle diagram becomes the source of a problem that hinders an analytic continuation to the resonance pole (the so-called finite fixed points emerge [12]). In this case, one may use lattice measurements of the matrix element between two nearby eigenstates in order to cancel the contribution of the triangle diagram.

Finally, note that the analytic continuation is performed in the CM frame, for a fixed momentum transfer three-vector. After analytic continuation, the fourth component of the momentum transfer becomes complex, as well as the virtuality $q^2$. Albeit at present neither the experimental nor the lattice collaborations provide the values of the form factors at complex virtuality, we believe that this is the only consistent way of presenting the results. The form factors defined at real virtuality, in general, are off the pole (the elastic form factors in the Breit frame

DOI: 10.1051/epjconf/201713402006

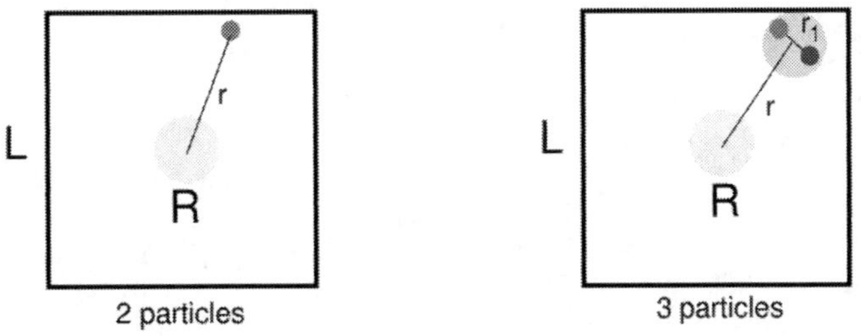


**Figure 3.** Two- and three-particle systems, placed in a finite box. In case of the two-particle system, if the box size $L$ is much larger than the interaction range $R$, the particle close to the box experiences the asymptotic wave function only. Matters are different for the three-particle system, even if $L \gg R$, two particles close to the wall can experience strong interactions.

are an exception). Consequently, the background does not vanish and the result will be process-dependent.

## 4 Three particles in a finite volume

The rapid advance of calculations of the properties of various nuclei on the lattice invokes the necessity of the study of the many-body systems in a finite volume. An obvious candidate to start with is the three-body system, as studied e.g. in project **B.6**. Another justification is provided by the necessity of the proper treatment of the inelastic resonances, e.g., the Roper resonance, whose decay rate into the three-particle final state is sizable (quark mass expansions for the Roper are studied in projects **C.3** and **C.9**).

The *conceptual* difference between the two-particle and three-particle problems in a finite volume is clearly demonstrated in Fig. 3. In the two-particle case, after removing the CM motion, we are left with the effective one-particle problem. If the interaction range is much smaller than the size of the box, the effective particle in the vicinity of the walls is *necessarily* far away from the interaction region, i.e., the wave function nearby the walls can be replaced by the asymptotic wave function. This means that the boundary conditions, which determine the spectrum in a finite volume, contain the asymptotic wave function only – in other words, the spectrum in a finite volume is solely determined by the $S$-matrix elements in the infinite volume.

The situation is different in the case of the three-particle system. Again, after removing the CM motion, we are left with two effective particles. However, in this case, these two particles may be close to the walls and to each other simultaneously, even if the interaction range is much smaller than the size of the box. For this reason, it is not *a priori* clear that the three-body spectrum in a finite volume is solely determined by the pertinent $S$-matrix elements.

The study of the three-body systems in a finite volume is still in its infancy. Below, we list some of the major achievements in the last few years:

- In the pioneering work [15], it has been demonstrated that the finite-volume spectrum is indeed determined solely by the infinite-volume $S$-matrix despite the problem mentioned above.
- In Refs. [17–19], the spectrum of the three-particle system in a finite volume has been studied by solving the EFT equations numerically. The volume-dependence of the energy levels is obtained. For details of the formalism in the infinite volume, see the contribution from project **B.6**.
- In Refs. [45–48], the three-particle quantization condition (an analog of the Lüscher equation in the two-particle systems) has been obtained and explored for various simple systems.
- In Ref. [20], an explicit expression for the energy shift of the three-body bound state in a finite volume in the unitary limit (infinitely large two-body scattering length and vanishing effective range) was obtained. The expression has the following form:

$$\frac{\Delta E}{E} = \text{const}\,|A|^2(\kappa L)^{-3/2}\exp\left(-\frac{2}{\sqrt{3}}\kappa L\right), \qquad (14)$$

  where $A$ stands for the asymptotic normalization coefficient of the three-particle wave function, $E$ is the binding energy, $m$ is the particle mass and $\kappa = \sqrt{mE}$ denotes the three-body binding momentum.

There is still a host of interesting issues which should be addressed. In particular, it would be interesting to see, how the above explicit result can be obtained by using the general three-body quantization condition from Refs. [45–48].

## 5 Insights from UChPT in a finite volume

The extraction of the scalar meson resonance parameters is particularly challenging, not only in Nature but also in a finite volume. Unitarized Chiral Perturbation Theory (UChPT) provides a successful approach that reproduces well the $\pi\pi/\pi\eta$ and $K\bar{K}$ data up to 1200 MeV. In this scheme, the resonances $f_0(600), f_0(980), a_0(980)$ are generated dynamically from the coupled channel interaction, see e.g. Refs. [49–51].

In Ref. [6] we have discussed the extraction of the mass and width of the scalar mesons $f_0(980)$ and $a_0(980)$ from the finite-volume spectrum in lattice QCD. In particular, it is argued that the nature of these states can be studied by invoking twisted boundary conditions, as well as investigating the quark mass dependence of the spectrum. It was further shown that a two-particle threshold can mimic the occurrence of a resonance through the appearance of an avoided level crossing, see Fig. 4.

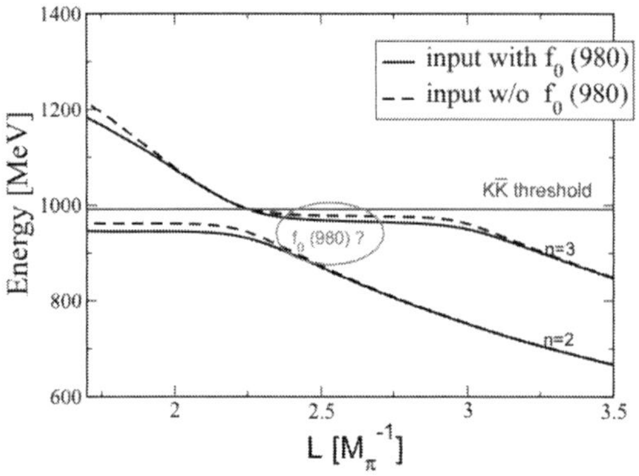


**Figure 4.** Energy levels calculated in finite volume UChPT for the $\pi\pi/K\bar{K}$ system in the vicinity of the $K\bar{K}$ threshold and the $f_0(980)$ resonance (solid lines). Even if one removes the scalar resonance, the avoided level crossing still appears (dashed lines).

In the paper [7], we have discussed the extraction of the parameters of the scalar mesons $f_0(600)$, $f_0(980)$ and $a_0(980)$ resonances from lattice data. In order to facilitate this extraction, we have used UChPT in a finite volume in the fit. Fitting synthetic data, we have demonstrated that the approach works, and the pole position for the above resonances can indeed be extracted by analyzing the volume-dependent energy spectrum in the vicinity of the resonance energy, provided sufficiently many volumes are simulated. The different sources of errors were analyzed in detail. The key point is that the use of the phenomenological input from UChPT stabilizes the fit. It is, thus, very challenging to apply these results to present and forthcoming lattice data in order to extract valuable information concerning hadronic resonances. From the theoretical point of view one should also keep in mind that in field theory there are parts of the kernel of the Bethe-Salpeter equation which are volume dependent, although exponentially suppressed. Quantitative studies of these terms and the extent of the large $L$ suppression would be a very good complement to the work we have carried out in that work.

Dynamical coupled-channel approaches are a widely used tool in hadronic physics that allow to analyze different reactions and partial waves in a consistent way. In such approaches the basic interactions are derived within an effective Lagrangian framework and the resulting pseudopotentials are then unitarized in a coupled-channel scattering equation. In Ref. [8], we have proposed a scheme that allows for a solution of the arising integral equation in discretized momentum space for periodic as well as twisted boundary conditions. This permits to study finite size effects as they appear in lattice QCD simulations. The new formalism, at this stage with a restriction to S-waves, was applied to coupled-channel models for the light scalar mesons, and also for the $\Lambda(1405)$ baryon. Lattice spectra were predicted, see Fig. 5 for the coupled $\bar{K}N$-$\pi\Sigma$ systems that features the $\Lambda(1405)$ resonance.

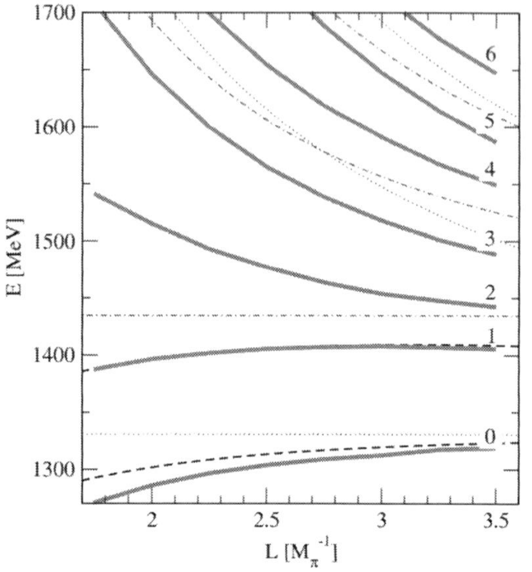


**Figure 5.** Spectrum $E(L)$ of the $I = 0$, $S = -1$, $J^P = \frac{1}{2}^-$ meson-baryon sector [$\Lambda(1405)$] for periodic boundary conditions. Solid lines: results for the $\bar{K}N$ Jülich model of hadron exchange [52]. Dashed lines: For comparison, the first two levels as obtained from the chiral unitary approach of Ref. [53]. Dotted (dash-dotted) lines: free $\pi\Sigma$ ($\bar{K}N$) levels.

In the study [9] we have provided a formulation for the scattering of two particles confined in a finite box with total *nonzero* momentum, adapted to the UChPT framework. This extended previously known techniques for zero momentum, discretizing the energy levels by imposing the boundary conditions in the moving frame. Given a hadronic interaction, we have predicted levels for the first five boosts $\vec{P} = (2\pi/L)(0,0,0)$ to $(0,0,2)$ and attributed these to the subgroups of cubic symmetry. Employing UChPT including NLO terms, we have derived the levels for the mixed-partial wave system with $I = 1/2$, $S = -1$ and $L = 0,1$ [$\kappa(800)$ and $K^*(892)$, respectively] as well as for the scalar sector with $I = 0$, $S = 0$ and $L = 0,2$ where the $\sigma(600)$ resides. We have demonstrated for the $\kappa(800)/K^*(892)$ system that partial wave mixing is a very large effect for realistic box sizes and needs to be taken into account. To disentangle the S-wave from the P- or D-wave, we have derived a set of equations in the one-channel formalism that are shown to be very precise as long as one stays below the inelastic thresholds. Furthermore, we have worked out a scheme in which the hadronic interaction is expanded in energy to allow for the extraction of the infinite volume limit, simultaneously fitting levels for different boosts and at different energies. The model-independent information from the lowest order in the chiral expansion is kept explicitly in this expansion, greatly stabilizing the fit to lattice data. Such statistical analyses can be used for actual lattice data, or, as done here, serve to determine promising lattice setups and the accuracy of lattice data to allow for reliable resonance extraction. We have tested this new method for the example of the $\sigma(600)$. We conclude that using information from moving frames is indeed rewarding since, with only a few different box sizes, phase shifts and resonance parameters of excited mesons can be determined.

Simulating the $\kappa(800)$ on the lattice is a challenging task that starts to become feasible due to the rapid progress in recent-years lattice QCD calculations. As the resonance is broad, special attention to finite-volume effects has to be paid, because no sharp resonance signal as from avoided level crossing can be expected. In the paper [10], we have investigated the finite volume effects in the framework of unitarized chiral perturbation theory using next-to-leading order terms. After a fit to meson-meson partial wave data, lattice levels for $\pi K$ scattering are predicted. In addition, levels are shown for the quantum numbers in which the $\sigma(600)$, $f_0(980)$, $a_0(980)$, $\phi(1020)$, $K^*(892)$, and $\rho(770)$ appear, as well as the repulsive channels. Methods to extract the $\kappa(800)$ signal from the lattice spectrum are presented. Using pseudo-data, we estimate the precision that lattice data should have to allow for a clear-cut extraction of this resonance. To put the results into context, in particular the required high precision on the lattice data, the $\sigma(600)$, the P-wave resonances $K^*(892)$ and $\rho(770)$, and the repulsive $\pi K, \pi\pi$ phases were analyzed as well.

## 6 Conclusions, outlook

The SFB/TR 16 was finished at the end of June 2016. However, the investigations, which were launched within this project, are continued within other projects, part of which are funded by the DFG (e.g. within the CRC 110 "Symmetries and the Emergence of Structure in QCD"). In this respect, we would like to especially mention the study of the electromagnetic and weak matrix elements in a finite volume, as well as the extensive investigations of the three-particle problem.

The work reported here would not have been possible without the support from the Deutsche Forschungsgemeinschaft within the SFB/TR16. We would like to thank all our collaborators, who have contributed to this work: A. Agadjanov, D. Agadjanov, V. Baru, V. Bernard, M. Bissegger, S. Bour, G. Colangelo, M. Döring, E. Epelbaum, A. Fuhrer, J. Gasser, M. Göckeler, F.-K. Guo, C.O. Güllstrom, H.-W. Hammer, J. Haidenbauer, Ch. Hanhart, D. Hoja, R. Horsley, A. Khelashvili, S. König, H. Krebs, S. Kreuzer, B. Kubis, A. Kupsc, M. Lage, D. Lee, F. Llanes-Estrada, V.E. Lyubovitskij, E. Oset, K. Polejaeva, P.E.L. Rakow, G. Rios, G. Schierholz and J.M. Zanotti.

## References

[1] M. Lüscher, Nucl. Phys. B **354**, 531 (1991).

[2] V. Bernard, D. Hoja, U.-G. Meißner and A. Rusetsky, JHEP **0906**,061 (2009).

[3] V. Bernard, M. Lage, U.-G. Meißner and A. Rusetsky, JHEP **0808**, 024 (2008).

[4] M. Göckeler, R. Horsley, M. Lage, U.-G. Meißner, P. E. L. Rakow, A. Rusetsky, G. Schierholz and J. M. Zanotti, Phys. Rev. D **86**, 094513 (2012).

[5] M. Lage, U.-G. Meißner and A. Rusetsky, Phys. Lett. B **681**, 439 (2009).

[6] V. Bernard, M. Lage, U.-G. Meißner and A. Rusetsky, JHEP **1101**, 019 (2011).

[7] M. Döring, U.-G. Meißner, E. Oset and A. Rusetsky, Eur. Phys. J. A **47**, 139 (2011).

[8] M. Döring, J. Haidenbauer, U.-G. Meißner and A. Rusetsky, Eur. Phys. J. A **47**, 163 (2011).

[9] M. Döring, U.-G. Meißner, E. Oset and A. Rusetsky, Eur. Phys. J. A **48**, 114 (2012).

[10] M. Döring and U.-G. Meißner, JHEP **1201**, 009 (2012).

[11] D. Hoja, U.-G. Meißner and A. Rusetsky, JHEP **1004**, 050 (2010).

[12] V. Bernard, D. Hoja, U.-G. Meißner and A. Rusetsky, JHEP **1209**, 023 (2012).

[13] A. Agadjanov, V. Bernard, U.-G. Meißner and A. Rusetsky, Nucl. Phys. B **886**, 1199 (2014).

[14] A. Agadjanov, V. Bernard, U.-G. Meißner and A. Rusetsky, Nucl. Phys. B **910**, 387 (2016).

[15] K. Polejaeva and A. Rusetsky, Eur. Phys. J. A **48**, 67 (2012).

[16] S. Bour, S. Koenig, D. Lee, H.-W. Hammer and U.-G. Meißner, Phys. Rev. D **84**, 091503 (2011).

[17] S. Kreuzer and H.-W. Hammer, Phys. Lett. B **673**, 260 (2009).

[18] S. Kreuzer and H.-W. Hammer, Eur. Phys. J. A **43**, 229 (2010).

[19] S. Kreuzer and H.-W. Hammer, Phys. Lett. B **694**, 424 (2011).

[20] U.-G. Meißner, G. Rios and A. Rusetsky, Phys. Rev. Lett. **114**, 091602 (2015).

[21] M. Bissegger, A. Fuhrer, J. Gasser, B. Kubis and A. Rusetsky, Nucl. Phys. B **806**, 178 (2009).

[22] C.-O. Güllstrom, A. Kupsc and A. Rusetsky, Phys. Rev. C **79**, 028201 (2009).

[23] J. Gasser, B. Kubis and A. Rusetsky, Nucl. Phys. B **850**, 96 (2011).

[24] G. Colangelo, J. Gasser and A. Rusetsky, Eur. Phys. J. C **59**, 777 (2009).

[25] J. Gasser, V. E. Lyubovitskij and A. Rusetsky, Ann. Rev. Nucl. Part. Sci. **59**, 169 (2009).

[26] V. Baru, E. Epelbaum and A. Rusetsky, Eur. Phys. J. A **42**, 111 (2009).

[27] F. K. Guo, C. Hanhart, F. J. Llanes-Estrada and U.-G. Meißner, Phys. Lett. B **678**, 90 (2009).

[28] F.-K. Guo, C. Hanhart, F. J. Llanes-Estrada and U.-G. Meißner, Phys. Lett. B **703**, 510 (2011).

[29] H. Krebs, E. Epelbaum and U.-G. Meißner, Phys. Rev. C **80**, 028201 (2009).

[30] H. Krebs, E. Epelbaum and U.-G. Meißner, Phys. Lett. B **683**, 222 (2010).

[31] A. Agadjanov, D. Agadjanov, A. Khelashvili and A. Rusetsky, Eur. Phys. J. A **49**, 120 (2013).

[32] U.-G. Meißner, K. Polejaeva and A. Rusetsky, Nucl. Phys. B **846**, 1 (2011).

[33] S. R. Beane, P. F. Bedaque, A. Parreno and M. J. Savage, Nucl. Phys. A **747**, 55 (2005).

[34] G. Colangelo, J. Gasser, B. Kubis and A. Rusetsky, Phys. Lett. B **638**, 187 (2006).

[35] M. Beneke and V. A. Smirnov, Nucl. Phys. B **522**, 321 (1998).

[36] L. Lellouch and M. Lüscher, Commun. Math. Phys. **219**, 31 (2001).

[37] C. h. Kim, C. T. Sachrajda, S. R. Sharpe, Nucl. Phys. B **727**, 218 (2005).

[38] H. B. Meyer, Phys. Rev. Lett. **107**, 072002 (2011).

[39] M. T. Hansen and S. R. Sharpe, Phys. Rev. D **86**, 016007 (2012).

[40] R. A. Briceno, M. T. Hansen and A. Walker-Loud, Phys. Rev. D **91**, 034501 (2015).

[41] R. A. Briceno and M. T. Hansen, Phys. Rev. D **92**, 074509 (2015).

[42] R. A. Briceno, J. J. Dudek, R. G. Edwards, C. J. Shultz, C. E. Thomas and D. J. Wilson, Phys. Rev. Lett. **115**, 242001 (2015).

[43] R. A. Briceno and M. T. Hansen, arXiv:1509.08507 [hep-lat].

[44] R. A. Briceno, J. J. Dudek, R. G. Edwards, C. J. Shultz, C. E. Thomas and D. J. Wilson, Phys. Rev. D **93**, 114508 (2016).

[45] M. T. Hansen and S. R. Sharpe, Phys. Rev. D **90**, 116003 (2014).

[46] M. T. Hansen and S. R. Sharpe, Phys. Rev. D **92**, 114509 (2015).

[47] M. T. Hansen and S. R. Sharpe, Phys. Rev. D **93**, 014506 (2016).

[48] M. T. Hansen and S. R. Sharpe, Phys. Rev. D **93**, 096006 (2016).

[49] J. A. Oller, E. Oset, Nucl. Phys. **A620** (1997) 438 [Erratum-ibid. A **652** (1999) 407].

[50] J. R. Pelaez, Phys. Rev. Lett. **92** (2004) 102001.

[51] N. Kaiser, Eur. Phys. J. **A3** (1998) 307.

[52] J. Haidenbauer, G. Krein, U.-G. Meißner and L. Tolos, Eur. Phys. J. A **47** (2011) 18.

[53] E. Oset and A. Ramos, Nucl. Phys. A **635** (1998) 99.

EPJ Web of Conferences **134**, 02007 (2017) DOI: 10.1051/epjconf/201713402007
*Subnuclear Structure of Matter: Achievements and Challenges*

# Baryons with functional methods

*Christian S. Fischer*[1,2,⋆]

[1] *Institut für Theoretische Physik, Justus-Liebig–Universität Giessen, 35392 Giessen, Germany*
[2] *HIC for FAIR Giessen, 35392 Giessen, Germany*

**Abstract.** We summarise recent results on the spectrum of ground-state and excited baryons and their form factors in the framework of functional methods. As an improvement upon similar approaches we explicitly take into account the underlying momentum-dependent dynamics of the quark-gluon interaction that leads to dynamical chiral symmetry breaking. For light octet and decuplet baryons we find a spectrum in very good agreement with experiment, including the level ordering between the positive- and negative-parity nucleon states. Comparing the three-body framework with the quark-diquark approximation, we do not find significant differences in the spectrum for those states that have been calculated in both frameworks. This situation is different in the electromagnetic form factor of the $\Delta$, which may serve to distinguish both pictures by comparison with experiment and lattice QCD.

## 1 Introduction

In these proceedings we summarise our progress on the properties of baryons as obtained from the functional approach (Dyson-Schwinger and Bethe-Salpeter equations) to QCD. All topics discussed in this contribution are covered in much more detail in a very recent review on baryons as covariant three-quark bound states [1].

Understanding the baryon excitation spectrum of QCD and the internal structure of baryons using electromagnetic and other probes is one of the key elements in unravelling the structure of the strong interaction. In the past years, significant experimental progress has been made by the analysis of data from photo- and electroproduction experiments at JLAB, ELSA and MAMI [2–4]. As a result, a number of new radial and orbital excitations of the ground-state baryons have been added to the PDG [5].

On the theoretical side, a number of non-perturbative methods contribute a number of interesting aspects to our understanding of baryons, which are not always in agreement with each other. In general, the quark model still serves as a possible standard by which one may define and distinguish the 'expected' from the 'unexpected'. Its discussion along the years has created a list of standard problems in baryon physics [6, 7]. Amongst these are the problem of missing resonances and, related, the question whether baryons should better be described as quark-diquark rather than three-quark states, thus reducing the number of possible internal excitations. On the other hand, dynamical coupled-channel calculations have suggested that parts of the spectrum may even be generated purely dynamically, i.e. without a 'bare' three-quark or quark-diquark seed [8, 9]. Lattice QCD contributes its own share with fairly small error bars and very good agreement between different lattice groups and approaches for most ground-state masses. Differences, however, occur for the excited states, where a number of technical subtleties still obscure the picture.

In the past years we explored another non-perturbative approach to QCD, namely the functional methods of Dyson-Schwinger and Bethe-Salpeter equations (DSEs and BSEs). They can be used on different levels of sophistication, ranging from NJL-like truncations with contact interactions, quark-diquark models using ansätze for propagators and wave functions as input, rainbow-ladder truncations using effective quark-gluon interactions, to beyond-rainbow-ladder approaches explicitly solving a range of underlying DSEs for the QCD Green's functions; see [1] for an overview. In the work summarised here we have used the rainbow-ladder truncation as a minimal baseline but also explored effects beyond rainbow-ladder. In addition, we do not rely on the widely used quark-diquark approximation but provided first solutions for the masses of ground and selected excited states as well as baryon form factors from the three-body Faddeev equation. We are therefore in a position to be able to systematically compare both approaches using the same underlying assumptions for the quark-gluon interaction.

Our results for the light baryon spectrum will be discussed in Sec. 3. Corresponding results for form factors will be the topic of Sec. 4, with a focus on the $\Delta$ electromagnetic form factors. First, however, we outline the framework in the next section.

⋆e-mail: christian.fischer@physik.uni-giessen.de

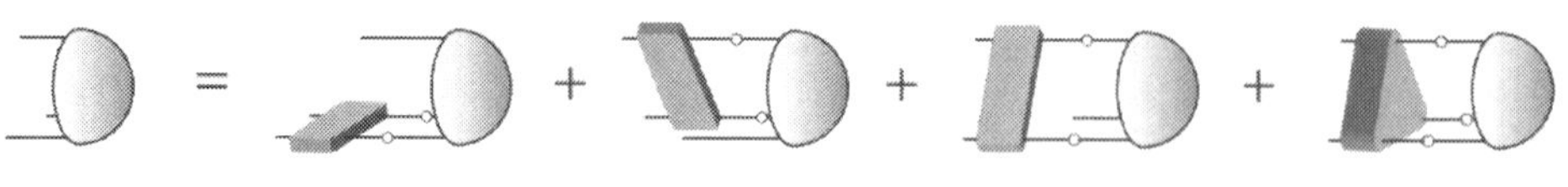

**Figure 1.** Three-quark Faddeev equation.

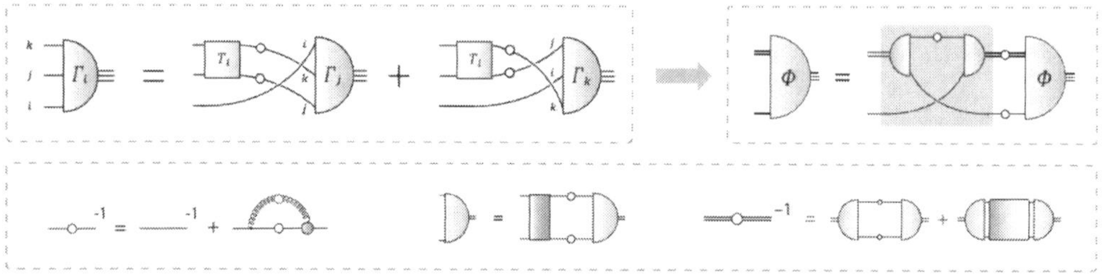

**Figure 2.** Simplification of the Faddeev equation in Fig. 1 (*top left*) to the quark-diquark Bethe-Salpeter equation (*top right*). The bottom panel shows the ingredients that enter in the equation and are calculated beforehand: the quark propagator, diquark Bethe-Salpeter amplitudes and diquark propagators.

## 2 The functional approach

In functional frameworks the masses and wave functions of baryons are extracted from their gauge-invariant poles in the (gauge-dependent) quark six-point Green function. This procedure is related to the corresponding one in lattice QCD, see Ref. [1] for more detailed explanations. An exact equation that can be derived in this approach is the covariant three-body Faddeev equation in Fig. 1. Its ingredients are the fully dressed quark propagator (solid line with open circle) as well as the quark two-body and irreducible three-body interactions. The latter ones have been neglected so far. The equation determines the baryon's Faddeev amplitude (the shaded half-sphere) and the baryon bound-state mass.

A considerable simplification is obtained once the quark-diquark approximation is chosen. To this end, one expands the quark-quark scattering matrix in terms of separable diquark contributions, resulting in the quark-diquark Bethe-Salpeter equation in the top-right corner of Fig. 2. Its main ingredients are again the dressed quark propagator, the diquark propagator (double line with open circle) as well as the diquark Bethe-Salpeter amplitude. In turn, the latter needs to be calculated from the diquark Bethe-Salpeter equation (bottom center diagram of Fig. 2). The dressed quark propagator is determined from its Dyson-Schwinger equation (bottom left diagram of Fig. 2).

An important question in connection with the diquark approximation is how many diquarks to consider. It turns out that the diquarks with smallest mass are the scalar and axialvector ones [10]. While both of these contribute to the nucleon, the $\Delta$ baryon can only be made of an axialvector diquark due to the symmetry properties of the diquark wave functions. Thus, scalar and axialvector diquarks together constitute a minimal set of diquarks necessary to consider. However, as it turns out, the parity partners of the nucleon and the the $\Delta$ cannot be described adequately without including the heavier pseudoscalar and vector diquarks in addition [11]. Thus in the diquark calculation of the spectrum discussed below all four types of diquarks have been taken into account.

A systematic comparison between the three-body and diquark-quark approach is only possible once the same underlying quark-gluon interaction is chosen [12]. This distinguishes a quark-diquark approach based on the dynamics of the underlying QCD from quark-diquark models [13–15]. In our case we choose the effective interaction introduced by Maris and Tandy in Ref. [16]; details are discussed e.g. in [12, 17]. In this rainbow-ladder framework the two-body interaction kernel between two quarks is approximated by the exchange of an effective gluon, which contains the gluon propagator supplemented by those effects of the quark-gluon vertex that can be captured by a function depending on the gluon momentum only.

In the rainbow-ladder approximation one arrives at an interaction entirely of vector-vector type. In the heavy quark sector it is known that such types of interactions are not sufficient to account for the different effects due to spin-spin and spin-orbit dependent forces between the quarks, and hence cannot describe all meson channels sufficiently well. In practice, it turns out that the meson spectrum in the pseudoscalar, vector and some tensor channels are well represented in rainbow-ladder, while others suffer from too much binding [18, 19]. In [12, 20] a simple procedure has been devised to remedy this problem within the quark-diquark approach. Scalar and axialvector diquarks are well represented in rainbow-ladder, whereas the pseudoscalar and vector ones are too light. Reducing the interaction strength in these channels by a common factor gauged by the $\rho - a_1$ splitting remedies this problem.

## 3 Light baryon spectrum

In Fig. 3 we show the resulting spectra from the three-body calculation (left diagram, open boxes), the quark-

DOI: 10.1051/epjconf/201713402007

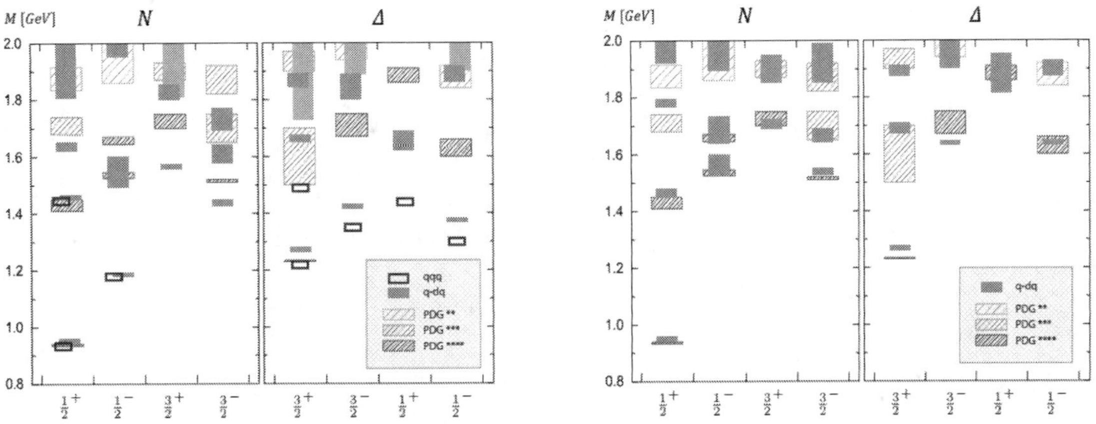


**Figure 3.** *Left:* Nucleon and Δ baryon spectrum for $J^P = 1/2^{\pm}$ and $3/2^{\pm}$ states determined within rainbow-ladder. The three-body results (open boxes) are compared to the quark-diquark spectrum with full diquark content (filled boxes), together with the PDG values and their experimental uncertainties [5]. The widths represent an estimate of the systematic error of our results based on the extrapolated eigenvalue curves of the BSE, see [12] for details. *Right:* Nucleon and Δ spectrum with reduced strength in the pseudoscalar and vector diquark channels; see text for a detailed discussion.

diquark approximation without correction (left diagram, filled boxes) and the quark-diquark approximation with improved pseudoscalar and vector diquarks (right diagram, filled boxes) [12] and compare them with the two-, three- and four-star states given by the PDG [5].

Let us first discuss the nucleon channel. In the left diagram we observe good agreement of the three-body with the quark-diquark approach for the ground-state nucleon as well as the first radially excited state. The mass of the latter agrees well with the Breit-Wigner mass of the Roper shown as a red shaded box[1]. The next two excited states are (numerically) only accessible in the quark-diquark framework and lie in the same ballpark as the PDG's N(1710) and N(1880). In the nucleon channel with negative parity, both approaches still agree very well but the states are generally much too low. Clearly, this reflects the above discussed deficiency of the rainbow-ladder approximation.

A similar picture is observed in the various Δ channels. The quantitative agreement between the quark-diquark and the three-body approach is somewhat less pronounced than in the nucleon case but still satisfied on a semi-quantitative level. Taken at face value, this could mean that interactions beyond the diquark approximation could be more important for Δ states than for baryons of nucleon type. This hypothesis needs to be tested in the future. Again, the (s-wave) Δ-channel delivers results for the ground and first radially excited state in agreement with experiment, while its parity partners and the other two channels are poorly represented.

The picture changes when the deficiencies of the rainbow-ladder approximation are accounted for by adjusting the binding in the pseudoscalar and vector diquark channels. As explained above, this is done by introducing one parameter that is adjusted by the $\rho - a_1$ splitting; everything else is unchanged. The influence of this adjustment depends on the quantum numbers of the baryon due to different weights of the corresponding diquark contributions as compared to the scalar and axialvector one. It is therefore highly non-trivial that the states arrange themselves in the one-to-one correspondence with the experimental ones as observed in the right diagram of Fig. 3. Without further corrections, the level ordering between the positive and negative nucleon parity sector comes out correctly on a quantitative level. Not only the first radial excitations but also the second and third ones are close to experimental states. We therefore arrive at a consistent and quantitative description of the light baryon spectrum below 2 GeV.

Of course this does not mean that beyond rainbow-ladder effects, like pion-cloud and coupled-channel effects, or further non-Abelian corrections on the gluon level are absent [21]. Our results merely suggest that these effects are opposite in sign and largely cancel each other in the nucleon and Δ channels, whereas in the other channels there is a net effect that can be absorbed by the extra parameter for the pseudoscalar and vector diquarks that we introduced. In fact, in the meson sector there are strong indications that cancellations between non-Abelian corrections [22] and pion cloud effects [23] occur. Both effects have already been explored for baryons on an exploratory level [24, 25] and work will continue in this direction.

Also very encouraging are the results for the strange baryon sector reported for the three-body framework in [29]. Here the masses of ground-state octet and decuplet baryons have been reproduced on the few-percent level, while the calculation of excited states still needs to be performed.

[1] The mass evolution of the Roper with varying pion mass is discussed in [12] and compared with results from lattice QCD.

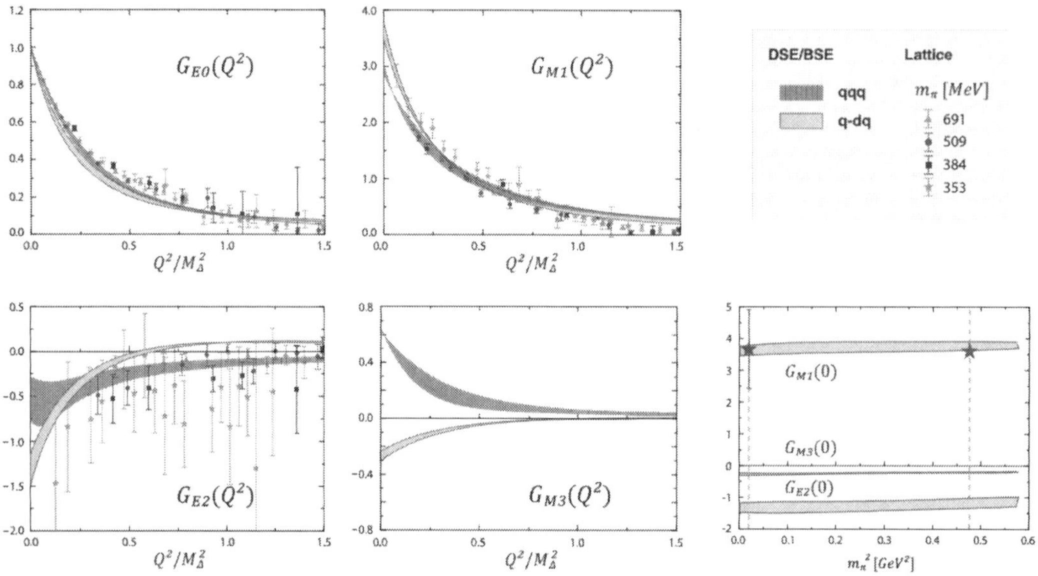


**Figure 4.** Electric monopole, magnetic dipole, electric quadrupole and magnetic octupole form factors for the $\Delta^+$ baryon. We compare lattice data [26] with the Dyson-Schwinger three-quark [27] and quark-diquark calculations [28]. The bands for the DSE results reflect the systematic uncertainty. The plot on the bottom right shows the current-mass evolution of the static quantities from the quark-diquark calculation [28], where stars denote the experimental magnetic moments of the $\Delta^+$ and $\Omega^-$ baryon. (The value $m_\pi^2 \approx 0.47$ where the magnetic moment of the all-strange $\Omega^-$ baryon is plotted corresponds to a pion mass evaluated with constituent strange quarks.)

## 4 Selected form factors: electromagnetic $\Delta$ form factor

Form factors in the functional approach are calculated by coupling external currents to equations for Green functions. In the three-body approach, electromagnetic form factors [30], axial form factors [31] and hyperon form factors [32] have been calculated and reviewed in [1]. Here we discuss aspects of the electromagnetic form factors for the $\Delta$, which have been determined in the three-body approach [27] as well as in the quark-diquark approximation [28]. In Fig. 4 we display corresponding results for the electric monopole and quadrupole form factor as well as the magnetic dipole and octupole form factor. We compare them with results from lattice QCD [26] using different values for the up/down quark masses corresponding to different pion masses.

It is interesting to compare the three-body with the quark-diquark results. For the electric monopole form factor both approaches agree very well. The limit of vanishing photon momentum, $Q^2 \to 0$, is constrained by charge conservation, which is guaranteed automatically in our framework due to gauge invariance. For all $Q^2$ values the agreement with lattice results is furthermore on a quantitative level. The electric quadrupole form factor, however, is more sensitive to the internal structure of the $\Delta$ baryon, as can be seen in the bottom left diagram of Fig. 4. Here the results from the three-body and the quark-diquark approach display a different momentum behaviour and even a different sign at large momenta. The same signature is found for the magnetic form factors. While the dipole results agree with each other (apart from a small region at very low $Q^2$), the octupole component has a different sign for all $Q^2$. Thus, high precision lattice results are clearly able to distinguish between them. Given the presently available set of data for the electric quadrupole component, there may be a slight tendency towards the three-body results. However, the accuracy is not good enough to make a definite statement. Thus, it would be very interesting to repeat the lattice calculation with today's technical and numerical resources.

While experimental results for the $\Delta$ form factors are not available, static quantities like its magnetic moment can be extracted. These are shown in the bottom right plot of Fig. 4. Compared are the bands for the quark-diquark results with the experimental data for the $\Delta^+$ and the $\Omega^-$ baryon. The agreement with the central value of the experimental error bar is very good. Note that the magnetic moments are real, since in the rainbow-ladder approximation used in the present approach decay channels are not included, see the review Ref. [1] for a more detailed discussion of this aspect. As discussed before, the magnetic moment of the $\Delta^+$ extracted from the magnetic dipole form factor shows a small discrepancy between the three-body and the quark-diquark results. Thus also more precise experimental results have the potential to discriminate between the two.

In general, we confirm the $\Delta$ baryon's deformation as encoded in its electric quadrupole and magnetic octupole form factors. Non-relativistically, a negative sign for the electric quadrupole moment $G_{E2}(0)$ indicates an oblate charge distribution for the $\Delta$ in the Breit frame [26]. From the measurement of the $\gamma N \to \Delta$ transition one can infer the value $G_{E2}(0) = -1.87(8)$ in the large-$N_C$ limit [26, 33]; comparable values are predicted by a range of constituent-quark models [34]. These numbers are in the ballpark of the lattice and DSE results displayed in Fig. 4.

## 5 Conclusions

Baryon spectroscopy, baryon structure and baryon dynamics remain very active and lively fields. Dedicated experiments at many facilities such as Jefferson Lab, BESIII, ELSA, J-PARC, LHCb, MAMI and the future PANDA/FAIR experiment are contributing and will continue to contribute to our understanding of the baryonic world. On the theory side, the past years have seen continuous advances in connecting the underlying quark and gluon interactions of QCD with baryon phenomenology.

In functional approaches, the challenges are to capitalize on continuous improvements to the truncation/approximation schemes. While the aim is not to enter the realm of high precision physics on the sub-percent level, much can be achieved in the way of understanding the underlying mechanisms that generate the observed phenomena. A prime example of this is the question of three-quark vs. a quark-diquark structure of baryons. At face value, the results reported in this contribution suggest that irreducible three-body forces inside baryons may be subdominant compared to two-body interactions. If this is the case, the quark-diquark approximation should be fine. Indeed, accounting for well-known deficiencies in the rainbow-ladder truncation of the quark-quark interaction, we found a spectrum of ground and excited states of light baryons in one-to-one agreement with experiment. This highly non-trivial result is very encouraging and suggests further extensions of our work to the strange baryon sector and quantum numbers beyond $J = \frac{1}{2}$ and $J = \frac{3}{2}$.

On the other hand, we have seen that form factors may serve to distinguish between a three-body and a quark-diquark picture of baryons via static quantities at low momenta and the dynamical behaviour at larger momenta. While the former can be tested by more precise experiments, the latter can be accessed by lattice QCD.

**Acknowledgement**

The work reported here would not have been possible without the support from the Deutsche Forschungsgemeinschaft within the SFB/TR16. I thank the members of SFB/TR16 for many fruitful discussions and the spokesperson Ulrike Thoma for the excellent leadership. I am furthermore very grateful to Gernot Eichmann, Helios Sanchis-Alepuz and Richard Williams for the successful and pleasant collaboration on the topics summarized in this proceedings contribution within the framework of the SFB/TR16. In addition I would like to thank Reinhard Alkofer (together with the above mentioned gentlemen) for many discussions and the collaboration in the production of an extensive review article [1] on the topics of the SFB/TR16.

## References

[1] G. Eichmann, H. Sanchis-Alepuz, R. Williams, R. Alkofer, C.S. Fischer, Prog. Part. Nucl. Phys. **91**, 1 (2016), 1606.09602

[2] I.G. Aznauryan et al. (CLAS), Phys. Rev. **C80**, 055203 (2009), 0909.2349

[3] I.G. Aznauryan et al., Int. J. Mod. Phys. **E22**, 1330015 (2013), 1212.4891

[4] A.V. Sarantsev et al., Phys. Lett. **B659**, 94 (2008), 0707.3591

[5] K.A. Olive et al. (Particle Data Group), Chin. Phys. **C38**, 090001 (2014)

[6] S. Capstick, W. Roberts, Prog. Part. Nucl. Phys. **45**, S241 (2000), nucl-th/0008028

[7] E. Klempt, J.M. Richard, Rev. Mod. Phys. **82**, 1095 (2010), 0901.2055

[8] M. Doring, C. Hanhart, F. Huang, S. Krewald, U.G. Meißner, Nucl. Phys. **A829**, 170 (2009), 0903.4337

[9] P.C. Bruns, M. Mai, U.G. Meißner, Phys. Lett. **B697**, 254 (2011), 1012.2233

[10] P. Maris, Few Body Syst. **32**, 41 (2002), nucl-th/0204020

[11] G. Eichmann, *Progress in the calculation of nucleon transition form factors* (2016), 1602.03462

[12] G. Eichmann, C.S. Fischer, H. Sanchis-Alepuz (2016), 1607.05748

[13] M. Oettel, G. Hellstern, R. Alkofer, H. Reinhardt, Phys. Rev. **C58**, 2459 (1998), nucl-th/9805054

[14] M. Oettel, R. Alkofer, L. von Smekal, Eur. Phys. J. **A8**, 553 (2000), nucl-th/0006082

[15] J. Segovia, B. El-Bennich, E. Rojas, I.C. Cloet, C.D. Roberts, S.S. Xu, H.S. Zong, Phys. Rev. Lett. **115**, 171801 (2015), 1504.04386

[16] P. Maris, P.C. Tandy, Phys. Rev. **C60**, 055214 (1999), nucl-th/9905056

[17] G. Eichmann, H. Sanchis-Alepuz, R. Williams, R. Alkofer, C.S. Fischer (2016), 1606.09602

[18] T. Hilger, C. Popovici, M. Gomez-Rocha, A. Krassnigg, Phys. Rev. **D91**, 034013 (2015), 1409.3205

[19] C.S. Fischer, S. Kubrak, R. Williams, Eur. Phys. J. **A51**, 10 (2015), 1409.5076

[20] H.L.L. Roberts, L. Chang, I.C. Cloet, C.D. Roberts, Few Body Syst. **51**, 1 (2011), 1101.4244

[21] R. Williams, C.S. Fischer, W. Heupel, Phys. Rev. **D93**, 034026 (2016), 1512.00455

[22] C.S. Fischer, R. Williams, Phys. Rev. Lett. **103**, 122001 (2009), 0905.2291

[23] C.S. Fischer, R. Williams, Phys. Rev. **D78**, 074006 (2008), 0808.3372

[24] H. Sanchis-Alepuz, C.S. Fischer, S. Kubrak, Phys. Lett. **B733**, 151 (2014), 1401.3183

[25] H. Sanchis-Alepuz, R. Williams, Phys. Lett. **B749**, 592 (2015), 1504.07776

[26] C. Alexandrou, T. Korzec, G. Koutsou, C. Lorce, J.W. Negele, V. Pascalutsa, A. Tsapalis, M. Vanderhaeghen, Nucl. Phys. **A825**, 115 (2009), 0901.3457

[27] H. Sanchis-Alepuz, R. Williams, R. Alkofer, Phys. Rev. **D87**, 096015 (2013), 1302.6048

[28] D. Nicmorus, G. Eichmann, R. Alkofer, Phys. Rev. **D82**, 114017 (2010), 1008.3184

[29] H. Sanchis-Alepuz, C.S. Fischer, Phys. Rev. **D90**, 096001 (2014), 1408.5577

[30] G. Eichmann, Phys. Rev. **D84**, 014014 (2011), 1104.4505

[31] G. Eichmann, C. Fischer, Eur. Phys. J. **A48**, 9 (2012), 1111.2614

[32] H. Sanchis-Alepuz, C.S. Fischer (2015), 1512.00833

[33] A.J. Buchmann, J.A. Hester, R.F. Lebed, Phys. Rev. **D66**, 056002 (2002), hep-ph/0205108

[34] G. Ramalho, M.T. Pena, F. Gross, Phys. Lett. **B678**, 355 (2009), 0902.4212

EPJ Web of Conferences **134**, 03001 (2017) DOI: 10.1051/epjconf/201713403001
*Subnuclear Structure of Matter: Achievements and Challenges*

# Photoproduction of Mesons

*Hartmut* Schmieden[1,*] and *Friedrich* Klein[1,**]

[1] *Physikalisches Institut, Universität Bonn*

**Abstract.** B.1 is one of the experimental projects within the CRC 16. It aims at the systematic investigation of the photoproduction of mesons off nucleons in order to understand reaction mechanisms and the relevant degrees of freedom in resonance formation. Of particular interest is the photoproduction of mesons heavier than the pion and resonances involving hidden or open strangeness. Essential hardware contributions have been made to the experimental programme of the CRC 16 through tagging systems, and photon-beam polarisation and polarimetry. A new experiment has been set up within the framework of the BGO-OD collaboration. This combines a forward magnetic spectrometer with a central BGO calorimeter with charged particle recognition and identification. The BGO-OD experiment enables reconstruction of complex final states composed of both charged and neutral particles, complementary to the existing CBELSA/TAPS calorimeter which is optimised for multi-photon final states. Selected results of the 12-year CRC period are presented from both experiments.

## 1 Introduction

The goal of CRC 16 project B.1 is the systematic investigation of the photoproduction of mesons off nucleons. Meson production is tightly connected with the nucleon excitation spectrum. Those gross features seem understood in terms of the quark sub-structure, but important aspects still remain puzzling.

In relation to constituent quark models and also recent results in Lattice-QCD [1], a large fraction of the expected excited states, predominantly at higher masses, remains "missing" in experiments [2]. Results in the double-strange sector of $\Xi$-hyperon states even seem to exacerbate the situation: Despite the expected improved separability of resonances due to smaller widths in the 2s-sector, almost no excited states at all are observed in photoproduction experiments [3].

Not only the higher states, also the lowest excitations in the non-strange and 1s-sector exhibit peculiar patterns, which are not described in genuine 3-quark models:
– Parity is reversed between the two lowest lying $N^*$ excitations, $P_{11}(1440)$ with $J^P = 1/2^+$ and $S_{11}(1535)$ with $J^P = 1/2^-$; naturally the lowest state above the ground state would be expected to have negative parity, and the state above this to have positive parity.
– Mass is reversed between the two lowest lying negative parity states in the strange and non-strange sector, i.e. the $\Lambda(1405)$ and the $S_{11}(1535)$; naturally the state containing the s-quark would be expected heavier.[1]

These, and other patterns opened speculations regarding the inner structure of the states and the effective degrees of freedom, which are not necessarily just quarks. Due to the closeness of the chiral symmetry breaking scale to the nucleon mass/size scale, the associated Goldstone bosons also enter as effective "elementary" objects [4, 5], enabling interactions of the light mesons directly with quarks, and the formation of meson-baryon[2] type objects. Corresponding models indeed appear more successful than genuine three constituent quark models in some aspects, for example, determining the parity ordering of the lowest nucleon excitations [5]. Meson-baryon dynamics seem to play an important role for higher baryon excitations too [6–14]. In addition to the interaction of pseudoscalar mesons with baryons, vector mesons should also contribute to resonance formation [15–17]. In this case, degenerate states of $J^P = 1/2^-, 3/2^-$ are expected, in particular in the mass region around 2 GeV [18].

Meson-baryon dynamics may be expected to be particularly important at thresholds, for example through (sub-threshold) rescattering effects. Many unresolved structures seem associated with such production or decay thresholds, as is illustrated in Fig. 1. The above mentioned $\Lambda(1405)$ and $S_{11}(1535)$ lie directly at the $\bar{K}N$ and $\eta p$ thresholds, respectively. Around 1680 MeV at the closely spaced $\bar{K}\Sigma$ and $\omega p$ thresholds, a narrow peak structure is observed in $\eta$-photoproduction off the neutron [19–21]. This seems associated with a dip structure off the proton at the same energy, in particular at forward $\eta$ angles [22], and a pronounced structure in the photon beam asymmetry off the proton at central angles [23]. In the $\phi p$ photoproduction channel, an unexpected narrow peak is observed in the vicinity of the almost degenerate $K^*\Lambda_{g.s.}$ and $K^+\Lambda(1520)$ thresholds [24–26]. And, finally, a dip structure was found

*e-mail: schmieden@physik.uni-bonn.de
**e-mail: klein@physik.uni-bonn.de
[1] despite the special role of the $\Lambda(1405)$ as the $uds$-singlet state
[2] in the meson sector also meson-meson

at the $K^*\Sigma$ threshold in $K_s^0\Sigma^+$ photoproduction off the proton, which is particularly pronounced at forward directions [27]. At the same energy, a structure is suggested in the photon beam asymmetry at central angles [28]. A dip in the cross section can be reproduced in a model based on vector meson-baryon interactions [29]. It appears associated with the formation of a $N^*(2030)$ vectormeson-baryon resonance, possibly in combination with a close-by $N^*(2080)$, and it is predicted to be complemented by a peak structure in $K_s^0\Sigma^0$ photoproduction off the neutron. This issue is not experimentally resolved yet. The phenomenology may be potentially similar to the situation in the $\eta$-channel at 1680 MeV.

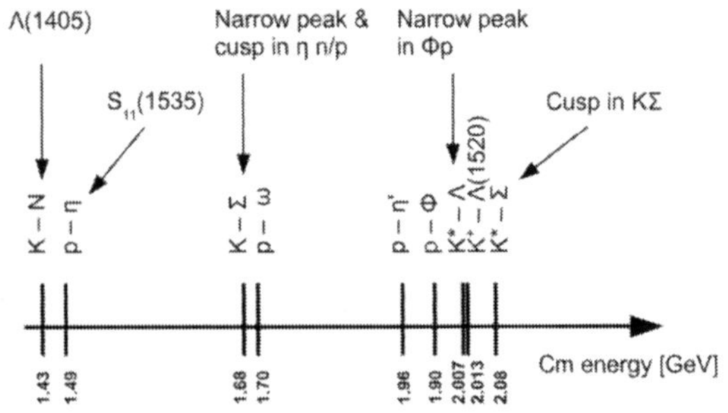


Figure 1: Overview of production/decay thresholds.

Some of the states which persistingly resisted a conventionial three-quark explanation are likely to be "dynamically" generated through meson-baryon interactions, at least to some extent. For the $\Lambda(1405)$ the important role of meson-baryon dynamics was recently supported by a result from LQCD [30]. Approaching realistic quark masses in the calculations, the $\Lambda(1405)$ seems to adopt a pronounced meson-baryon structure.

To better understand these issues, non-pionic channels are of particular interest. For the ELSA experiments this means $\eta$, associated $K$-hyperon and vectormeson photoproduction. The main goal is to disentangle the reaction mechanisms, in order to understand the excitations or structures observed. This is complementing projects preferentially pursuing spectroscopy of states. The results provide qualitative input and quantitative contraints for Partial Wave Analyses, for example, with respect to $s$ versus $t$-channel contributions. In turn, the feedback from spectroscopy and PWA aids the interpretation of the data, and is also in tight cooperation with theory.

An essential aspect of the experiments are polarisation degrees of freedom. For the CBELSA/TAPS experiment, which is run in combination with the Bonn Frozen Spin polarised target, linearly polarised tagged photon beams and the necessary electron polarimetry for circularly polarised beam were provided. This is described in the next section.

The CBELSA/TAPS calorimeter setup, optimised for neutral final states which involve multiple photons, is complemented by the new BGO-OD experiment. Combining a large aperture magnetic spectrometer in forward directions with a central BGO ball, BGO-OD is ideal for reconstructing complex final states including both charged and neutral particles. This makes it especially suited for the strangeness and vectormeson channels addressed. Setup and commissioning results are described in section 3. Some selected results are then discussed in section 4.

## 2 Tagging and Polarimetry at the CBELSA/TAPS Experiment

The photon tagging system is an essential feature of the CBELSA/TAPS experiment. This includes the positioning system to align a diamond crystal to produce linearly polarised coherent bremsstrahlung, and a Møller polarimeter for in-situ measurements of the longitudinal electron beam polarisation. The latter provides the basis to determine the (energy-dependent) degree of circular polarisation of the photon beam.

### 2.1 Photon Tagging and Linear Polarisation

Energy tagging of bremsstrahlung photon beams is a common method in meson photoproduction experiments. The principle is to measure the momentum of the post-bremsstrahlung electrons using a magnetic spectrometer, and associate it with the detected hadronic reaction through time coincidences.

The electron beam extracted from the ELSA accelerator [31] impinges on one of a set of four amorphous copper radiators of thickness 15 – 300 $\mu$m (corresponding to 1.05 – 21 ‰ radiation lengths $X_0$), a diamond crystal for coherent bremsstrahlung of thickness 500 $\mu$m (4.1 ‰ $X_0$), or a magnetisable ferromagnetic foil[3] of effective thickness 58.5 $\mu$m (3.6 ‰ $X_0$) for in-situ Møller measurements when using a polarised electron beam. The radiators are accurately positioned by means of a commercial goniometer system.[4]

A schematic overview of the tagging system is shown in Fig. 2. The post-bremsstrahlung electrons are horizon-

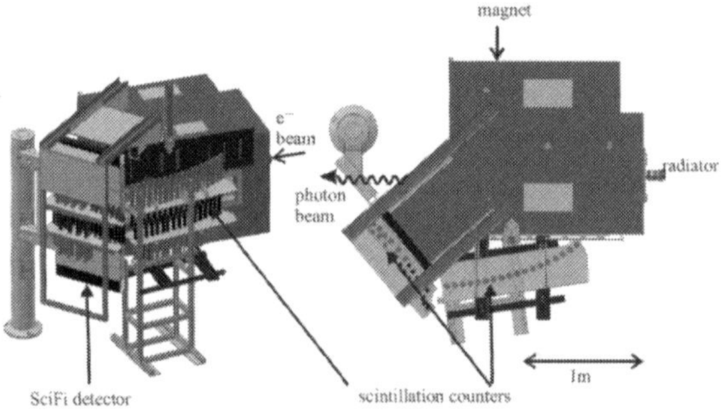


Figure 2: Schematics of the tagging system, 3-dimensional view (left) and top-view (right).

[3]material "Vacoflux", an alloy of 49 % iron, 49 % cobalt, and 2 % vanadium

[4]Newport Corporation *www.newport.com*

tally bent in the field of a *Brown-Boveri* MC type dipole magnet. Maximum field strength is 2 T at a current of 1500 A. Nominal settings are such that electrons of primary beam energy are deflected by 9 degree into the exit beamline and the electron beam dump. The momentum of the post-bremsstrahlung electrons is analysed through a two-part hodoscope of 96 half overlapping plastic scintillators. The backward part is placed in the focal plane of the magnet, against which the forward part had to be tilted due to space limitations. All scintillators are 1 cm thick, with the height and width varying from 6 – 13 cm and 1.4 – 5 cm, respectively. The total tagged photon energy range is from 21 % – 87.1 % $E_0$. Resolutions are between 0.1 % in electron energy for the lowest energy electrons and 0.4 % for the highest energies, which translates into a photon energy resolution of $\Delta E_\gamma / E_\gamma = 0.1...6.0\,\%$. The scintillator hodoscope is augmented by a 480 channel scintillating fibre (SciFi) detector. Arranged in two layers the 2 mm diameter fibres cover the high electron range of the tilted hodoscope part (cf. Fig. 2). In this range the photon energy resolution is significantly improved to $\Delta E_\gamma / E_\gamma < 2.9\,\%$. Setup, calibration and performance of the tagging system are described in detail in Ref. [32].

Linearly polarised photon beams are generated by coherent bremsstrahlung off the diamond radiator. Precise alignment through the goniometer system enables the setting of the energy range of maximum polarisation (the "coherent peak") and the orientation of the polarisation plane. The energy-dependent degree of polarisation can be determined from the energy spectrum of post-bremsstrahlung electrons. Methods and results are described in Ref. [33].

## 2.2 Circular Polarisation and Møller Polarimetry

Circularly polarised photon beams are obtained using a longitudinally polarised electron beam. The photon polarisation $P_\gamma(E_\gamma)$ is energy dependent, with the polarisation transfer from an electron beam of polarisation $P_e$ given by [34]

$$P_\gamma / P_e = x \frac{3 + (1 - x)}{3 + 3(1 - x)^2 - 2(1 - x)}, \qquad (1)$$

where $x = E_\gamma / E_e$ is the fractional energy transfer, which is measured with the tagging system. The determination of $P_\gamma$ requires the additional measurement of the degree of electron polarisation. This is done using a Møller polarimeter integrated into the tagging system.

In Møller scattering, a count rate asymmetry $A_M = (N^{++} - N^{+-})/(N^{++} + N^{+-})$ is obtained between parallel $(++)$ and antiparallel $(+-)$ spin orientations of the electrons which translates into

$$A_M = a_{zz} P_z^B P_z^T, \qquad (2)$$

where $P_z^B$ and $P_z^T$ are the longitudinal beam and target polarisations, and $a_{zz}$ is the asymmetry coefficient, which follows from the Møller cross section and can be calculated in QED. For highly relativistic electrons it is energy independent. It does, however, depend on the Møller scattering angle with a maximum of $|a_{zz}| = 7/9$ for the symmetric case ($\theta^{cm} = 90$ degree) where, in the laboratory frame, both outgoing electrons have the same scattering angle and the same momentum $p'_e = p_0/2$, i.e. half the incoming beam momentum, $p_0$.

Due to the electron spin precession in magnetic (bending) fields, it is important to measure the polarisation of the electron beam directly at the nominal bremsstrahlung radiator position. This is easily achieved by simultaneously using the bremsstrahlung radiator as the Møller target. To determine the asymmetry according to Eq. 2, the target/radiator electrons need to be polarised. This is achieved by using one of the above mentioned ferromagnetic Vacoflux foils as the bremsstrahlung radiator. It is placed inside a small solenoid producing a magnetic field of approximately 0.008 mT, sufficient to magnetise the foil into saturation. The magnetisation of the foil is measured by inductive methods, and the electron polarisation determined using the gyromagnetic ratio. For Vacoflux the latter is known with a precision of $\simeq 0.7\%$.

The polarisation is orientated within the foil plane.[5] Therefore, target polarisation in exactly $z$ – i.e. beam – direction can not be realised. Instead, the foil is slightly tilted against the beam direction, in our case by 20 degree. The (unwanted) small sensitivity induced by this for transverse components of the beam polarisation is compensated for by alternate tilt settings of +20 and −20 degree, which sign-flip and thus cancel the measured transverse components in the asymmetry while leaving the longitudinal one unaffected.

The polarimeter tags Møller scattering events through coincident detection of the two outgoing electrons in a pair of lead glass detectors of thickness 6.4 cm. Lead plates of 1 cm thickness in front of the detectors ensure the shower maximum to lie inside the detector material for the symmetric scattering case at the magic[6] beam energy of 2.4 GeV. To minimise systematic errors through the Levchuk effect [35], symmetric scattering is selected through momentum. The detectors are accordingly arranged after the tagger dipole, and behind the tagger scintillator strip and fibre hododcopes. The plane for Møller pairs to be detected is chosen perpendicular to the bending plane of the tagger magnet. This allows to place the detectors far enough from the magnet's mid plane, thus avoiding the high rate exposition through post-bremsstrahlung electrons.

The polarimeter works fast enough to be used for accelerator diagnostics. At nominal tagger rates of a few MHz a relative statistical error of $(\delta P_e / P_e)_{stat} \simeq 2\%$ is achieved after about 15 minutes, the precision of which can be increased with longer data taking periods. The overall limitation is always the total systematic error of $(\delta P_e / P_e)_{syst} \simeq 1.5\%$ which is dominated by the precision of the target foil polarisation. A detailed description of beam polarimetry for the CBELSA/TAPS experiment can be found in Ref. [36].

[5]To magnetise a thin foil perpendicular to its plane requires a much larger external field exceeding the saturation field of $B_{sat} \simeq 2.25$ T, and would require, for example, superconducting magnets to produce it.

[6]At this energy the electron spin orientation is exactly longitudinal at the radiator position

## 3 BGO-OD

The BGO-OD experiment emerged from the B.1 project of a magnetic spectrometer, originally intended to be used in combination with Crystal Barrel to detect forward going charged particles with high resolution over a large acceptance. The main device is a 94 t large-aperture dipole magnet[7], the "open dipole" (OD). A major part of the former GRAAL group joined the spectrometer project, which led to the formation of the international BGO-OD collaboration in March 2010 with currently 62 members of institutions from Basel University, Bonn University (HISKP and PI), Edinburgh University, INFN Frascati, PNPI Gatchina, Gießen University, Kharkov University, Messina University and INFN, Moscow University, Pavia University and INFN, Rome "Tor Vergata" and INFN, and Torino University and INFN. In particular the Italian groups contribute very significantly to the BGO-OD experiment, most importantly through the 480 channel BGO ball which was previously used at the ESRF Grenoble and is now installed with upgraded electronics at ELSA.

The photon tagging system at BGO-OD works upon the same principles as the system in operation at the CBELSA/TAPS experiment (see section 2.1). Linearly polarised photon beams are available via coherent bremsstrahlung using a diamond radiator [37]. To determine the accuracy of extracting the degree of polarisation, beam asymmetries were determined for both $p\eta$ and $p\pi^0$ final states. The data agreed well with previous data sets and model fits (see [37] for details).

Circularly polarised photon beams are produced using longitudinally polarised electron beams from ELSA. A Møller polarimeter is used to determine the degree of circular polarisation [38].

The BGO ball encloses the cryogenic target[8] over the angular range from 25°–155°, complemented by detectors for charge recognition and tracking[9] (see Fig. 3). Forward directions are covered by the OD spectrometer (see Fig. 4). The magnet is sandwiched by tracking detectors and complemented by a time-of-flight (TOF) spectrometer. The small acceptance gap between the BGO ball and forward spectrometer is covered with a segmented plastic scintillator detector, labelled SciRi in Fig. 3. This acceptance gap will also be covered by a special MRPC[10] (labelled Daisy in Fig. 3), which is presently under construction at Rome. Generally, the BGO-OD setup is ideally suited for unpolarised target experiments with complex final states including both charged and neutral mesons, hence ideal to pursue associated strangeness and vector meson production.

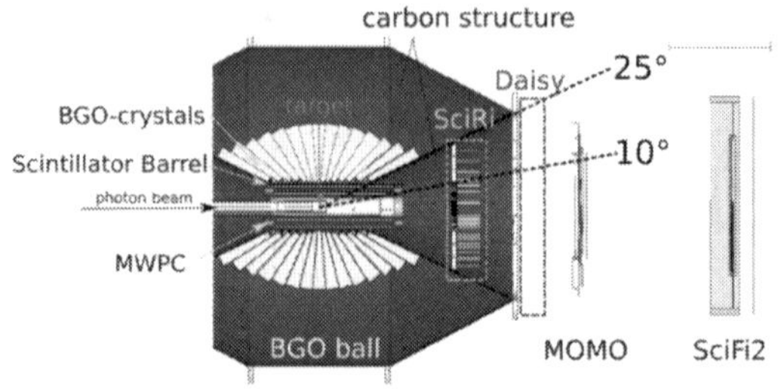


Figure 3: A slice view through the BGO ball.

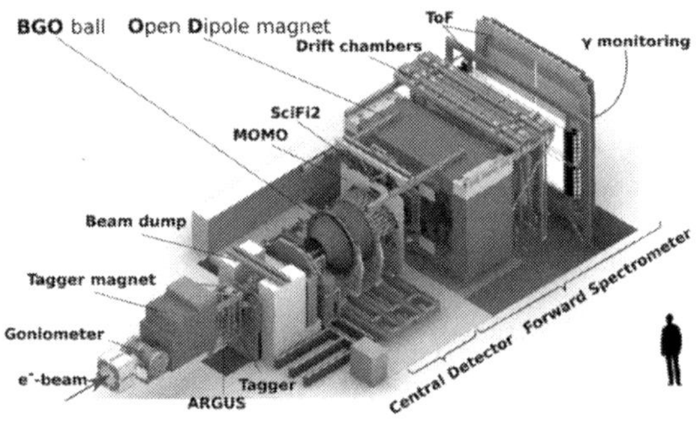


Figure 4: A schematic of the BGO-OD experiment.

The commissing of BGO-OD was succesfully finished in 2015. Additionally to commissioning tests, the first data of approximately 25 days with linearly polarised photon beams was taken during 2015. A small subset of this data is shown in Figs. 5 and 6. Fig. 5 shows the reconstruction of neutral mesons in the BGO ball. With timing cuts applied to ensure a coincidence between the tagged beam photons and events recorded in the BGO, the mass recoiling from two photons in the final state is plotted against the invariant mass of the two photons. Peaks corresponding to the $p\pi^0$ and $p\eta$ final state are visible. The two photon invariant mass for a selection of events where the the mass recoiling from the mesons is consistent with the proton mass is shown in Fig. 5(b).

Fig. 6 shows the same subset of data for particle trajectories reconstructed in the forward spectrometer. Fig. 6(a) shows, for postively charged particles, $\beta$ (calculated from the time of flight walls), plotted against momentum (calculated from the extent of curvature of the particle trajectory in the magnetic field of the open dipole). Enhancement of events for $\pi^+$ and protons are seen to agree well with the expected loci denoted by the red lines. Of these events, a subset was selected where the forward going particle is likely to be a proton. The mass recoiling from the proton in the final state is plotted in Fig. 6(b). Peaks consistent with $\eta$, $\omega$ and $\eta'$ mesons are visible without further analysis.

[7] We thank DESY for providing the magnet on permanent loan basis.
[8] Provided by Messina university and INFN.
[9] A cylindric MWPC is being built by the Pavia group.
[10] Multi resistive plate chamber

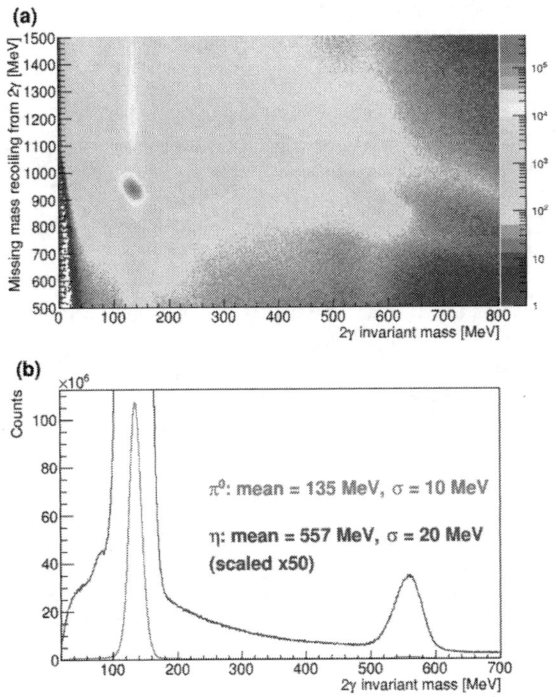


Figure 5: Neutral meson reconstruction in the BGO. (a) Mass recoiling from a two photon system versus the two photon invariant mass. (b) Two photon invariant mass after a selection cut where the mass recoiling from the two photon system is within $2\sigma$ of the proton mass. The mean and sigma values are from a Gaussian fit to the $\pi^0$ and $\eta$ invariant mass peaks (red and blue respectively). To observe the $\eta$ mass peak, the data is replotted and scaled by a factor of 50.

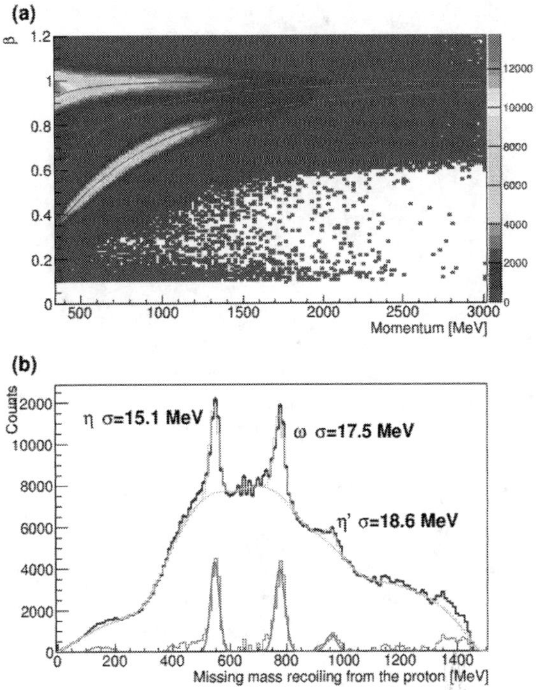


Figure 6: Positively charged particle reconstruction in the forward spectrometer. (a) Particle $\beta$ versus momentum. Red lines indicate the expected loci of $\pi^+$, $K^+$ and protons. (b) The mass recoiling from forward going protons (thick dark blue line). The green line was used to estimate background and extract the peaks at the $\eta$, $\omega$ and $\eta'$ masses (red lines). The peaks were subsequently fitted with a Gaussian function (thick red curves), the widths of which are labelled inset.

Additionally to commissioning tests and the liquid hydrogen target data that was taken, approximately two days of data with a carbon target, and two days of data with a liquid deuterium target were taken during 2015.

## 4 Selected Results

As motivated in section 1, a particular interest is the photoproduction of mesons beyond the pion, and to disentangle reaction dynamics at reaction thresholds. The CBELSA/TAPS detector essentially comprises a highly segmented $4\pi$ electromagnetic calorimeter, optimised for multi-photon final states. Therefore, neutral mesons were detected in their neutral decay modes, e.g. $\eta \to \gamma\gamma$ or $\to 3\pi^0 \to 6\gamma$, $\omega \to \pi^0\gamma \to 3\gamma$, $K^0_s \to \pi^0\pi^0 \to 4\gamma$.

Polarisation observables play an important role in understanding the photoproduction mechanism. The most basic one is the beam asymmetry, $\Sigma$, which is accessible through linearly polarised photon beams, and provides seed for a whole class of double polarisation observables which are sensitive to aspects of spin-orbit interactions involved [39].

The new BGO-OD detector opens a window to more complex final states of mixed neutral and charged particles. First preliminary results are presented including $K^+Y^{(*)}$ with subsequent decay of the (excited) hyperons into different channels. Charged mesons not only can be detected in the magnetic forward spectrometer, but also in the central BGO calorimeter, using the inner scintillator barrel and MWPC as vetoes (section 3). This includes $K^+$ detection in the BGO using the time delayed weak decay as explained in section 4.4.1.

A few personal-choice selected results of the last 12 year CRC/TR-16 funding are presented in the following.

### 4.1 Beam asymmetry in $\eta p$ photoproduction

As described in Ref. [40], the beam asymmetry was measured in the reaction $\vec{\gamma} p \to \eta p$ using linearly polarised tagged photon beams from coherent bremsstrahlung, which is described in section 2.1. By accurate alignment

DOI: 10.1051/epjconf/201713403001

of the diamond crystal the coherent peak was set to the energy range $E_\gamma = 800 - 1400$ MeV. A maximum polarisation of $P_\gamma = 49\%$ was obtained at $E_\gamma = 1305$ MeV. The $\eta$ was detected in both dominant decay modes, $\eta \to 2\gamma$ and $\eta \to 3\pi^0$, and the beam asymmetry extracted from the azimuthal modulation of the cross section [41],

$$\frac{d\sigma}{d\Omega} = \left(\frac{d\sigma}{d\Omega}\right)_0 \left[1 - P_\gamma \Sigma \cos 2\Phi\right]. \qquad (3)$$

Large asymmetries of up to 80% were extracted over a cm angular range of $\Theta_\eta^{cm} = 50 - 150$ degree. It was shown that with both the eta-MAID model and the Bonn-Gatchina partial wave parametrisation the measurement could be described equally well, although their relative resonance contributions differed substantially. This is illustrated in Fig. 7, and was mainly traced back to the interplay of $P_{11}(1710)$ and $P_{13}(1720)$, and $D_{13}(1520)$ and $D_{15}(1675)$ states. These results clearly demontrated the need to measure further double-polarisation observables, in order to resolve such an unsatisfactory situation.

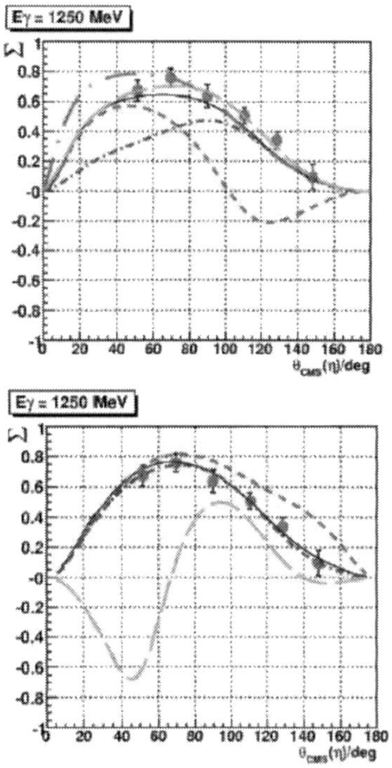


Figure 7: Impact of different resonance contributions to the beam asymmetry in $\eta - MAID$ (left) and the Bonn-Gatchina PWA (right), both plotted at a photon beam energy of 1250 MeV. Full lines represent full calculations, broken curves the effect of "turning off" individual resonances: $P_{13}(1720)$ (long dashed), $P_{11}(1710)$ (long dash-dotted), $D_{13}(1520)$ (short dashed), and $D_{15}(1675)$ (short dash-dotted). The figure is taken from Ref.[40].

### 4.2 Measurements of polarisation observables and spin-density matrix elements in $\omega$-meson photoproduction

$\omega$ photoproduction is an ideal channel to explore the lesser known third resonance region due to the higher production threshold compared to pionic channels. Furthermore, the isoscaler nature of the $\omega$ ($I = 0$) allows only $N^*$ states to contribute to the reaction mechanism, with no interference from $\Delta$ resonance states.

For vector meson photoproduction, at least 23 polarisation observables are required to completely determine the photoproduction amplitudes and relative phases, which is considerably more than the eight required for pseudoscalar meson photoproduction. Despite this, the measurement of only a few polarisation observables provides important insights into the reaction mechanism. The data presented here are from three papers. Klein et al. [42] measured photon beam asymmetries, $\Sigma$, and pion asymmetries, $\Sigma_\pi$, for $\omega$ photoproduction close to threshold. Eberhardt et al. measured the beam-target helicity asymmetries, $E$, $G$ and $G_\pi$ from threshold [43], and Wilson et al. the differential cross sections and spin-density matrix elements [44].

At higher energies, $\omega$ photoproduction is dominated by diffractive scattering, via the exchange of a Pomeron, as shown in Fig. 8, top. $\pi$ exchange, as depicted in Fig. 8, centre, should also be expected due to the sizable observed decay mode $\omega \to \pi^0\gamma$ ($\Gamma = 8.3\%$). The energy and angular dependence of the differential cross section near threshold however cannot be described with these mechanisms alone. Previous beam asymmetries from GRAAL [45], supported the significance of $s$-channel resonances near threshold (Fig. 8, bottom). A study of $\omega$ photoproduction at CLAS [46, 47], via the decay $\omega \to \pi^+\pi^-\pi^0$ was made after the paper described here by Klein et al. [42], but before the paper by Eberhardt et al. [43]. Strong evidence for several $N^*$ states were found, including $N(1680)5/2^+$, $N(1700)3/2^+$ and $N(2190)7/2^+$.

The measurement of polarisation observables is particularly sensitive to the role of $s$-channel resonances near threshold. The beam asymmetry, $\Sigma$, is expected to be close to zero for pure $t$-channel mechanisms, but large with $s$-channel contributions. This can be explained by the manner of the photon coupling: If the photon couples to the vector meson, and with no angular momentum transferred in the $t$-channel (either for Pomeron or pion exchange), no beam asymmetry should be observed. Conversely, the observable $\Sigma_\pi$, which is a measurement of the extent of the asymmetry in the decay $\omega \to \pi^0\gamma$, is predicted to be +0.5 or -0.5 for pure Pomeron and pion exchange respectively. From existing partial wave analyses, significant contributions from $s$-channel mechanisms should render $\Sigma_\pi$ close to zero. The simultaneous measurement of $\Sigma$ and $\Sigma_\pi$ as was made in [42] will therefore constrain the role of the three mechanisms close to the $\omega$ threshold.

The measurements of $E$, $G$ and $G_\pi$ that were made in [43] also act to understand the photoproduction mechanism close to threshold. For Pomeron or pion exchange, $E$ should be zero. $E$ may be non-zero for a mixture of the two exchange mechanisms. For $s$-channel contributions, $E$

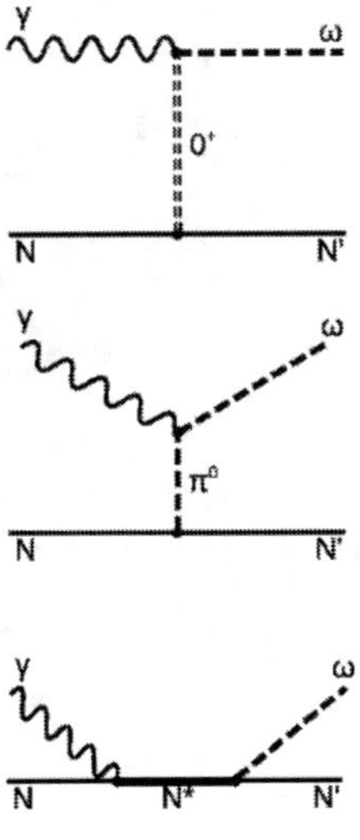


Figure 8: $\omega$ production via $t$-channel $0^+$ (Pomeron) exchange (top), $t$-channel $\pi^0$ exchange (middle) and $s$-channel intermediate resonance (bottom).

will be the projection of the intermediate state spin upon the beam axis.

Based on the unpolarised cross section and the photon beam asymmetry, spin-density matrix elements were determined for the reaction $\gamma p \to p\omega$ using the radiative decay $\omega \to \pi^0\gamma$ [44]. The cross sections cover the full angular range, yielding shape of the angular distribution and unpolarised spin-density matrix elements in agreement with previous data. In the energy range $E_\gamma = 1150 - 1650$ MeV polarised spin-density matrix elements were extracted for the first time using the radiative channel. The data were included in the Bonn-Gatchina partial wave analysis.

The measurements with unpolarised and linearly polarised photon beams [42–44] used an electron energy of 3.2 GeV from ELSA [31]. The electrons were incident upon a diamond radiater, 500 $\mu$m thick, and aligned to produce linearly polarised photons via the coherent bremsstrahlung process. For the measurement of $E$ presented in [43], a longitudinally polarised electron beam with an energy of 2.4 GeV was also used to produce a circularly polarised photon beam. The electrons were incident upon a Møller radiator to produce bremsstrahlung photons and simultaneously determine the degree of polarisation (see section 2.2).

The $\omega \to \pi^0\gamma$ channel was observed using the Crystal Barrel [48] and TAPS [49, 50] detector system. Surrounding the target was a three layer scintillating fibre detector [51] for charged particle identification and reaction vertex determination. For the measurement of $\Sigma$ presented in [42], a liquid hydrogen target was used. For the measurements of $E$, $G$ and $G_\pi$ presented in [43], a 2 cm long, longitudinally polarised butanol target was used [52].

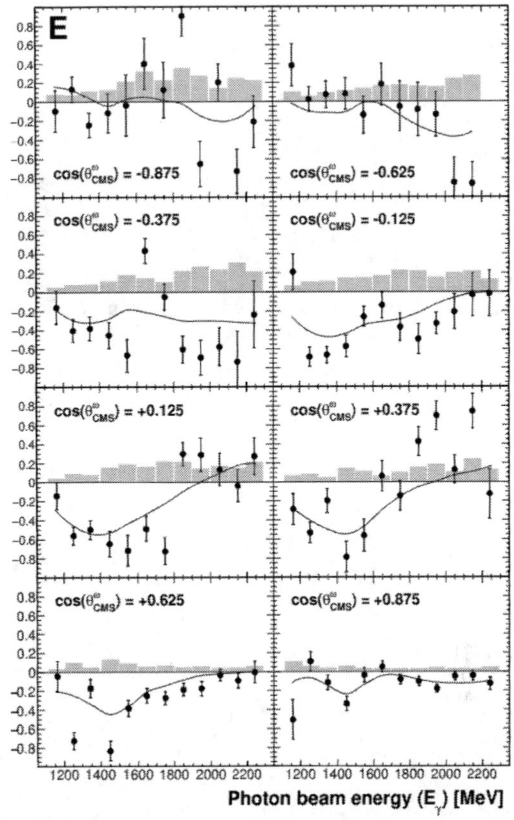


Figure 9: $p\omega$ Beam-target-helicity asymmetry, $E$, as a function of photon beam energy. Systematic errors are on the abscissa. The solid line is the result of the Bonn-Gatchina PWA when including this data. Figure taken from [43].

Fig. 9 shows the $E$ measurement from the analysis of Eberhardt et al. [43]. The overlayed line is the solution from a Bonn Gatchina partial wave analysis, using a large body of data, including this data set. The fit found significant contributions from Pomeron exchange, particularly at higher energies, but only a small contribution from pion exchange. Intermediate $s$-channel resonances were found to be important, with significant contributions from a $J^P = 3/2^+$ and $J^P = 5/2^+$ states below 1.9 GeV, and an improvement with the additional inclusion of a $J^P = 3/2^-$. Table 1 [53] details the changes the new data sets have in constraining the resonance contribution in $\omega$ photoproduction.

Fig. 10 shows the $\Sigma$ and $\Sigma_\pi$ measurements of Klein et al. [42]. It was noted that there is a dependence of $\Sigma$ on the centre of mass polar angle. Given the extent of the variation away from zero, this was considered as evidence of $s$-channel mechanisms playing a significant role. This is supported by the $\Sigma_\pi$ measurement, which remains close to zero. A further comparison to Bonn-Gatchina PWA, which did not include this data set, showed that $s$-channel

DOI: 10.1051/epjconf/201713403001

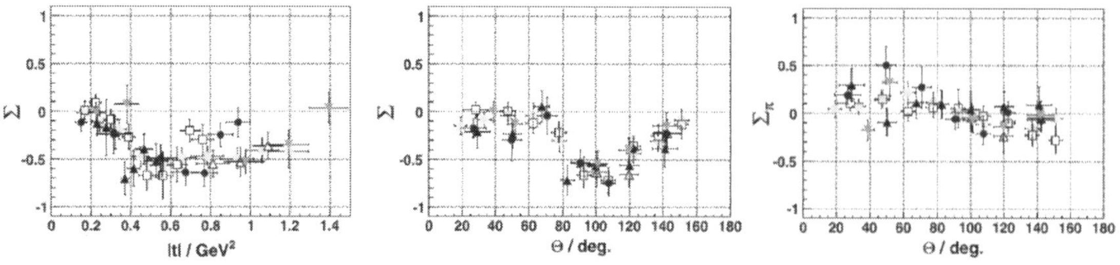


Figure 10: Photon beam asymmetries and pion asymmetries for beam energies 1108-1200 MeV (black triangles), 1200-1300 MeV (open squares), 1300-1400 MeV (black circles), 1400–1500 MeV (open triangles), and 1500–1700 MeV (grey triangles). Figure taken from [42].

resonances are required to have a satisfactory fit to the data.

| Resonance | B.R. | $\delta(\chi^2)$ | Resonance | B.R. | $\delta(\chi^2)$ |
|---|---|---|---|---|---|
| $N(1700)3/2^-$ | $22 \pm 12$ | 100 | $N(1900)3/2^+$ | $15 \pm 8$ | 70 |
| | | | | $13 \pm 9$ | |
| $N(1710)1/2^+$ | $2 \pm 2$ | 26 | $N(2000)5/2^+$ | $18 \pm 8$ | 42 |
| | $8 \pm 5$ | | | $1 \pm 1$ | |
| $N(1720)3/2^+$ | $26 \pm 14$ | 105 | $N(2060)5/2^-$ | $4 \pm 3$ | 37 |
| $N(1875)3/2^-$ | $13 \pm 7$ | 98 | $N(2100)1/2^+$ | $15 \pm 10$ | 78 |
| | $20 \pm 4$ | | | | |
| $N(1880)1/2^+$ | $20 \pm 8$ | 33 | $N(2150)3/2^-$ | $12 \pm 8$ | 99 |
| $N(1895)1/2^-$ | $28 \pm 12$ | 100 | $N(2190)7/2^-$ | $14 \pm 6$ | 131 |

Table 1: Branching ratios for $N^*$ contributions to $N\omega$. $\delta\chi^2$ is the change in $\chi^2$ with the inclusion of the new data. Taken from [53].

There is a change of sign of $E$ at 1650 MeV at $\cos\theta^{\omega}_{CMS} = 0.125$, and a peak like structure at -0.375 and -0.125. This is interesting as it is the region of the cusp-like structure in the $K^0_s\Sigma^+$ cross section over the $K^*$ threshold [27], which is described in the following subsection.

### 4.3 Differential cross section and polarisation observable measurements for $K^0_s\Sigma^+$

In the strange quark sector, $K^0_s$ photoproduction is an ideal tool to understand the effective degrees of freedom that are required to model excitation spectra. Compared to charged $K^+$ photoproduction, few measurements of $K^0_s$ photoproduction have been made. This channel benefits from the lack of the $t$-channel kaon exchange mechanism, as the photon cannot couple to the vanishing charge of the neutral $K^0$. On the contrary, $K^0_s$ photoproduction is sensitive to $t$-channel $K^*$ exchange. Detailed measurements of $K^0_s$ photoproduction over the $K^*$ thresholds therefore allow an understanding of the role of $t$-channel exchange, and potentially, $K^*$-hyperon dynamically generated interactions which may contribute to the $K^0_s$ photoproduction process.

The data presented here are from differential cross section measurements of $\gamma p \to K^0_s\Sigma^+$ from threshold to a photon beam energy of 2250 MeV [27], and the first recoil polarisation and beam asymmetry measurements made for this reaction channel [28].

Both experiments were performed at the Crystal Barrel [48] and TAPS [49, 50] detector system. An electron beam energy of 3.2 GeV was extracted from ELSA [31]. This was incident upon a 500 $\mu$m diamond radiator, which was orientated to produce linearly polarisated photons via the coherent bremsstrahlung process, with peaks in the photon energy spectrum at 1305, 1515 and 1814 MeV. The photon beam was incident upon a 5.3 cm long liquid hydrogen at the centre of the Crystal Barrel. The absolute photon flux was determined by a fast, total absorbing lead tungstate crystal detector behind the TAPS detector.

Simulated $\gamma p \to K^0_s\Sigma^+$ data was used to determine the detection efficiency when identifying the reaction channel via the decays $K^0_s \to 2\pi^0$ and $\Sigma^+ \to \pi^0 p$. This was found to be flat across the entire polar angle range, with an efficiency of approximately 11% near threshold, and reducing to 2% in the highest energy bin of 2150 to 2250 MeV.

Fig. 11 shows the differential cross section, for both with and without the requirement of the charged cluster from the proton detection. There is good agreement between the data sets. Previous data from Crystal Barrel [54] and SAPHIR [55] are also shown, which agree reasonably well with this new data.

Near threshold, the cross section is approximately flat, with respect to the centre of mass polar angle. This is expected for a dominantly $s$-wave production mechanism. With increasing beam energy, the differential cross section becomes more forward peaked, indicative of $t$-channel mechanisms becoming more dominant. This continues up to a beam energy of 1750 MeV. In the energy bin 1750 to 1850 MeV, the cross section sharply drops at forward angles by a factor of approximately four. At higher energies, the cross section remains flat at approximately 0.02 $\mu$b/sr. The cusp-like structure is so pronounced that it is visible in the total integrated cross section shown in Fig. 12.

DOI: 10.1051/epjconf/201713403001

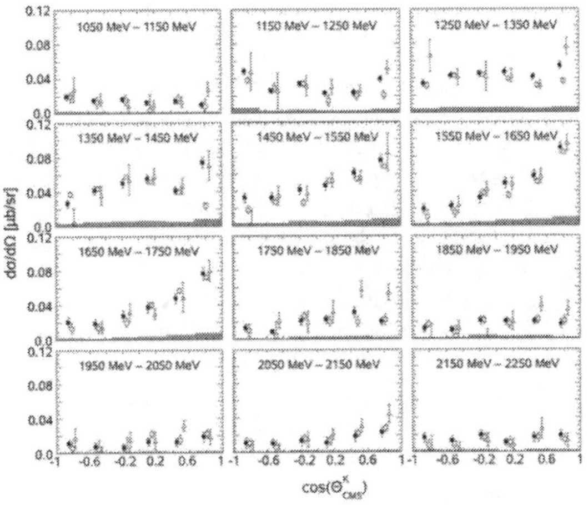

Figure 11: $\gamma p \to K^0_s \Sigma^+$ differential cross sections. Filled squares are this data (grey bars on the abscissa are the systematic errors), open squares are the previous data from the Crystal Barrel [54], and the open triangles are the previous data from SAPHIR [55]. Figure taken from [27].

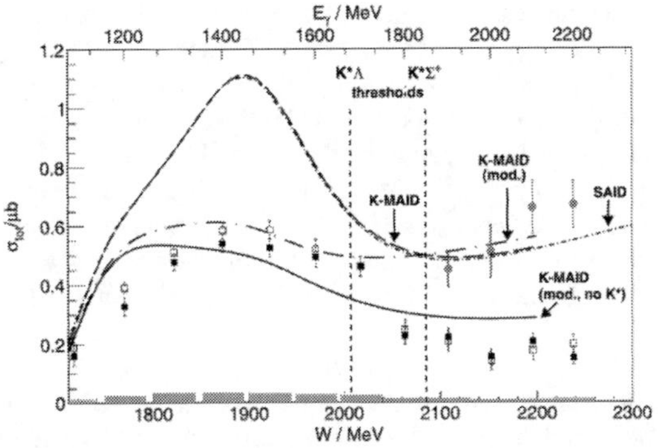

Figure 12: Total $\gamma p \to K^0_s \Sigma^+$ cross sections. Filled squares are this data (grey bars on the abscissa are the systematic errors), open squares are the previous data from the Crystal Barrel [54]. Grey circles are the addition of this data to $K^{*0}\Sigma^+$ cross section data of Nanova et al. [58]. The SAID [56] and modified Kaon Maid (K-MAID) [57] solutions which are labelled inset are described in the text. Figure taken from [27].

The exhibited behaviour suggests there is a $t$-channel mechanism that contributes below 1750 MeV, but not above this energy, leaving only $s$-channel contributions as the dominant mechanism. To test this hypothesis, the SAID [56] solution, and modified versions of the Kaon-MAID calculation [57] are overlayed in fig 12. Using the standard parameterisation, neither SAID nor Kaon-MAID can fit the data below a beam energy of 1700 MeV. An improvement can be made however by manually adjusting the couplings of the $S_{31}(1900)$ state to $G_1 = 0.3$ and $G_3 = 0.3$, and a reduction of the Born couplings to 0.7. Fig. 12 shows this adjusted fit, and an additional fit without the $K^*$ exchange term. The inclusion of the $K^*$ exchange term improves the description of the data below the $K^*$ threshold, but the omission of this term gives an improved description above this energy. The grey data points show this data summed with the $K^{*0}\Sigma^+$ total cross section data of Nanova et al. [58]. The incoherent addition of this data shows an agreement with the modified Kaon-MAID solution when $K^*$ exchange terms are included. This suggests that $K^*$ channels contribute to the $K^0_s\Sigma^+$ cross section below $K^*$ threshold. Above threshold, $K^*$ is produced on-shell and therefore no longer contributes to the $K^0_s\Sigma^+$ cross section.

It is speculated that the mechanism shown in Fig. 13 is responsible for this cusp-like structure. A $K^*$ is produced off-shell, and subsequently rescatters to a $K^0_s$ and $\pi$, where the $\pi$ is absorbed by the hyperon. The coupled channel model by Ramos and Oset, which includes vector meson-baryon interactions, supports this claim [29]. An interesting prediction of this model is a peak like structure in the $K^0_s\Sigma^0$ cross section in photoproduction off the neutron. Measurements of this channel will therefore provide further constraints to meson-baryon interactions and are intended at the BGO-OD experiment.

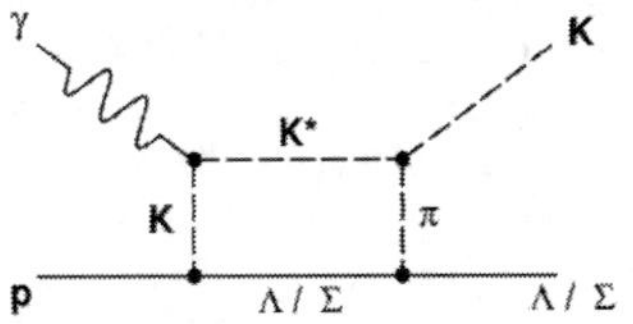



Figure 13: $t$-exchange diagram for $K^0_s$ photoproduction, with an intermediate $K^0_s$ and pion rescattering through sub-threshold $K^*$ decay.

Fig. 14 shows the photon beam asymmetry for $K^0_s\Sigma^+$ photoproduction. In general there is better agreement with the Bonn-Gatchina PWA solution than any of the modified Kaon-MAID parameterisations. An improved variant of the Bonn-Gatchina PWA was included, which used in the fit the $K^0_s\Sigma^+$ data presented here. The new fit suggested a significant $J^P = 3/2^-$ contribution over a large energy range. A larger contribution of $J^P = 3/2^-$ than $J^P = 3/2^+$ gave a better fit to the beam asymmetry data, however the opposite was true for the recoil polarisation data.

At the most forward angle and high energy bin ($\cos\theta^{cm}_K = 0.8$, $E_\gamma = 1550$ MeV), the photon beam asymmetry deviates from the trends described by the Bonn Gatchina PWA, and has a negative value of -0.7. This is an interesting region as it is just below the $K^*$ thresholds and the cusp-like structure that was observed in the differential cross section. Extension of beam asymmetry and recoil polarisation measurements over this energy region will be possible with the BGO-OD experiment.

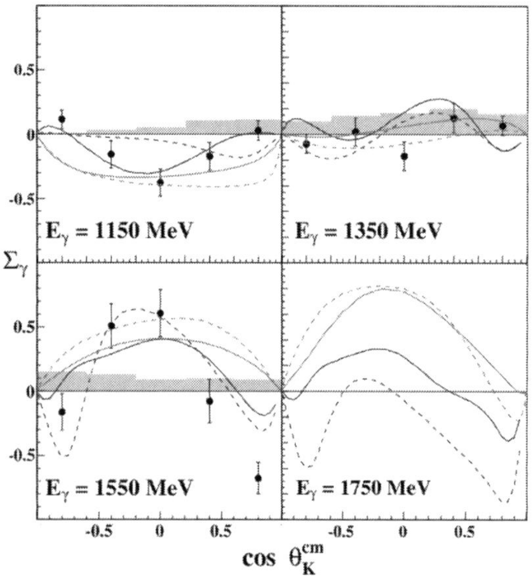


Figure 14: Angular distribution of the photon beam asymmetry $\Sigma_\gamma$ in the three bins of photon energy indicated in the diagrams. The error bars attached to the data points are purely statistical, the systematic errors are indicated by the grey bars on the abscissa. Curves represent the results of the Bonn-Gatchina-PWA [59] solutions, BG2011-02m (black dashed) and BG2011-02 (black solid), and the K-MAID [57] parametrisations, standard (red solid) and modified as described in the text, to study the origin of the cross section anomaly at the $K^*$ threshold (red dashed). The fourth energy bin is added to show the behaviour of the parametrisations across the $K^*$ thresholds where there is no data yet available. Figure taken from [28].

### 4.4 Strangeness photoproduction at the BGO-OD experiment

BGO-OD is ideal for the investigation of reactions dominated by t-channel mechanisms, due to the acceptance at forward angles with high momentum resolution. Neutral particle identification in the central region aids the identification of final states of mixed charge. This unique combination is suited to the investigation of reaction channels with open strangeness and complicated final states.

The preliminary analysis presented in this section demonstrates the BGO-OD capabilities of reconstructing channels with both $K_s^0$ and $K^+$ in the final state.

#### 4.4.1 $K^+$ detection in the BGO ball

$K^+$ reconstruction over the polar angle range of the BGO is crucial to identify associated strangeness final states over a large kinematic range. The techniques presented here are currently being used to extract differential cross section data for $K^+\Lambda$ and $K^+\Sigma^0$ channels. This can be used to help resolve discrepancies in the world data set, in particular at $W \approx 1.9$ GeV, where differences in differential cross section data from SAPHIR [60] and CLAS [61, 62] leave ambiguities to the $s$-channel contributions.

Accurate understanding of $K^+$ detection in the BGO also is crucial for the measurements of channels with higher lying hyperons. It is intended to study, for example, differential cross sections and polarisation observables for $K^+\Lambda(1405)$ and $K^+\Lambda(1520)$. Measurements of $K^+\Lambda(1405)$ at BGO-OD will be complementary to recent measurements made at CLAS [63, 64]. BGO-OD is suited in particular to identify the decay $\Lambda(1405)\rightarrow \pi^0\Sigma^0$ via photon reconstruction in the BGO. This channel discriminates against the mass-overlapping $\Sigma(1385)$, for which it is isospin forbidden. Of course, $K^+$ reconstruction is also important for vector meson studies, for example $\gamma p \rightarrow p\phi$ with the subsequent decay $\phi \rightarrow K^+K^-$.

A technique of $K^+$ detection developed with the Crystal Ball detector [65] was implemented for the BGO ball [66]. It makes use of the weak decay of $K^+$ into the dominant modes $\mu^+\nu_\mu$ and $\pi^+\pi^0$ with time delays characteristic to the $K^+$ lifetime of approximately 12 ns. The BGO time resolution per crystal allows the $K^+$ cluster to be separated into an incident sub-cluster from stopping the $K^+$, and a decay sub-cluster from the subsequent decay. Fig. 15 shows a schematic of the technique and characteristic energy and time spectra. Fig. 15(b) shows the energy of the decay cluster with a peak at 153 MeV, corresponding to the $\mu^+$ energy deposition when the $K^+$ decays via the $\mu^+\nu_\mu$ mode at rest. There is excellent agreement between real and simulated data. Events where the $K^+$ decays pionically are usually lost, since the two sub-clusters cannot be spatially separated. Fig. 15(c) shows the time difference between the incident subcluster from stopping the $K^+$, and the decay subcluster. The real and simulated spectra agree very well, and an exponential fit to the spectra yields a lifetime close to the expected $K^+$ lifetime of 12 ns.

Fig. 16 shows the mass recoiling from the $K^+$ in the final state for both real and simulated data. Peaks at the $\Lambda$, $\Sigma^0$ and $\Lambda(1405)$ or $\Sigma(1385)$ masses are visible. Events from $K^+\Sigma^0$ and $K^+\Lambda$ reaction channels can be separated by identifying the radiative $\Sigma^0 \rightarrow \Lambda\gamma$ decay [66]. A detection efficiency of the decay $\Sigma^0 \rightarrow \Lambda\gamma$ (having already identified the $K^+$) of approximately 50-60% is obtained, with misidentification of $K^+\Lambda$ events of less than 10%.

#### 4.4.2 Kinematic fitting and neural network analysis

Kinematic fitting has been successfully implemented at the BGO-OD experiment, with careful analysis to understand energy and spatial resolution of particle momenta reconstruction. This has been used to reconstruct the reaction $\gamma p \rightarrow K^+\Sigma^0$ over a wide kinematic range [67]. The analysis shown here used the full set of liquid hydrogen target data that was taken during 2015.

The techniques described below are complementary to either detecting the $K^+$ in the forward spectrometer or in the BGO via the weak decay as described in section 4.4.1. The angular acceptance is however higher, and in particular for when the $K^+$ is detected in the BGO, the detection efficiency is greatly increased. These techniques have application in studying higher lying hyperons. For example,

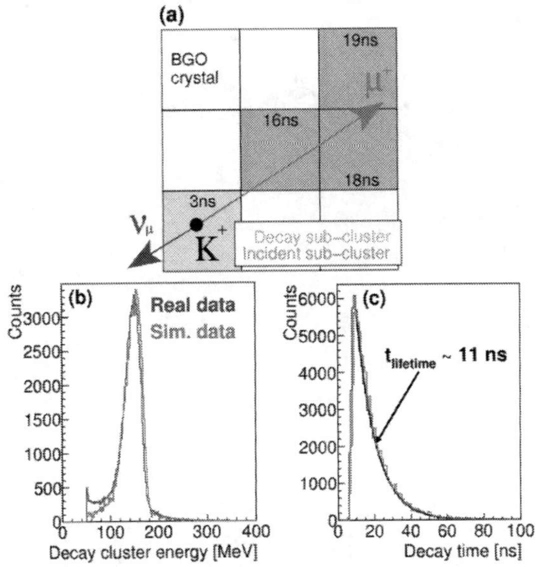


Figure 15: $K^+$ identification in the BGO ball. (a) An initial cluster is separated by time to an incident sub-cluster of crystals (blue) and decay sub-cluster (red). The time inset each crystal indicates the time of the measured energy deposition. (b) The energy of the decay sub-cluster. (c) The time difference between the incident sub-cluster from stopping the $K^+$, and the decay sub-cluster from the $K^+$ decay.

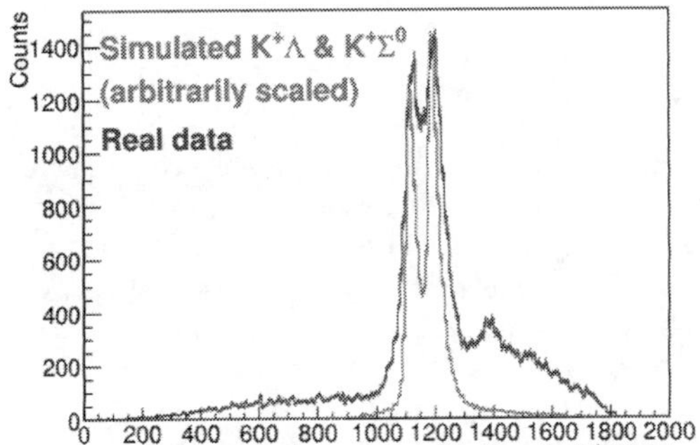


Figure 16: Missing mass recoiling from $K^+$ detected in the BGO. Blue line is real data, red line is simulated $K^+\Lambda$ and $K^+\Sigma^0$ final states, arbitrarily scaled with respect to each other and the real data.

$\gamma p \rightarrow K^+\Lambda(1405)$, via the decay $\Lambda(1405)\rightarrow \pi^0\Sigma^0$, can be studied in a similar manner, but with the additional reconstruction of the $\pi^0$ in the BGO ball.

Events were selected under the assumption that after the decay $\Sigma^0 \rightarrow \gamma\Lambda$, the $\Lambda$ decayed to $p\pi^-$. It was therefore required to identify three charged particles over the

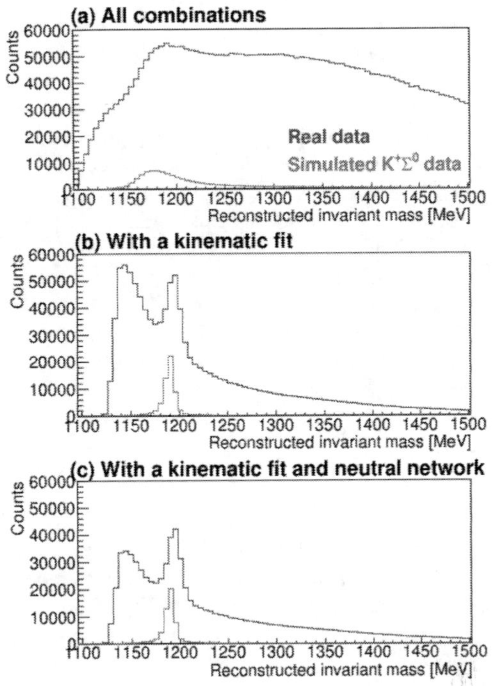


Figure 17: Identification of $K^+\Sigma^0$ via three charged particles in either the BGO, SciRi or the forward spectrometer, and a photon in the BGO [67]. The reconstructed $\Sigma^0$ invariant mass is plotted for real and simulated data (blue and red respectively) for (a) All combinations of selected events, (b) After a kinematic fit and a confidence level cut of 1%, and (c) after the kinematic fit and applying a neutral network to remove background.

range of the forward spectrometer, SciRi and BGO (where a coincidence with the inner plastic scintillator determined the particle charge), and an additional neutral cluster in the BGO. The momenta of the three charged particles were determined by requiring momentum conservation transverse to the beam (in the $xy$ plane). The direction of the three particles is a sufficient constraint to determine their momenta, and energy signals from the detectors were not required. Three solutions per event were made, depending upon the order of assignment of the $K^+$, proton and $\pi^-$ to the three detected charged particles. The beam energy was then used to determine the overall energy conservation for each solution, and the solution with the closest conserved energy was chosen. Fig. 17 shows the reconstructed $\Sigma^0$ invariant mass using this technique for real and simulated data. Fig. 17(a) is the results from the selection process described, with a broad bump at the $\Sigma^0$ mass, but with a large amount of background from misidentified events in the real data. Fig. 17(b) is the same event selection but with a kinematic fit applied. The contraints enforced by the fit were the $\Lambda$ mass and total momentum conservation. A selection cut rejected events with a confidence level below 1%. A narrow peak is now visible at the $\Sigma^0$ mass in real and simulated data. A neural network has been preliminarly implemented to understand the event topology and be able to distinguish true $K^+\Sigma^0$ final states from background. Fig. 17(c) shows the same event selection after the kinematic fit but with an additional selection cut to remove events which the neural network considered to be background. For real data, the integral of the background spectrum reduces, leaving the integral of peak corresponding to $K^+\Sigma^0$ events unchanged. The peak from simulated $K^+\Sigma^0$ remains unchanged.

Fig. 18 shows preliminary $\gamma p \rightarrow K^+\Sigma^0$ differential cross sections using this technique [67]. A complete, solid angle acceptance of $4\pi$ can be accessed with this technique, and with only approximately 25 days of data, it is clear that the statistics are comparable to previous measurements.

#### 4.4.3 $K^+$ at forward angles

Despite the channels $K^+\Lambda$ and $K^+\Sigma^0$ having been extensively studied, there remains a dearth of data at extremely forward $K^+$ centre of mass polar angles. This region is crucial to understand the role of $t$ channel mechanisms which are forward peaking and dominate the reaction process (see for example [70] and reference therein). Accurately knowing the free $\gamma p \rightarrow K^+\Lambda$ production cross section at forward angles is also an important constraint in hypernuclei electroproduction, where there is typically very small $Q^2$ and forward going $K^+$.

The data presented below are preliminary differential cross section for $\gamma p \rightarrow K^+\Lambda$, for $K^+$ travelling at extremely forward angles [38]. The analysis shown is approximately one third of the available liquid hydrogen target data set taken during 2015.

Due to to the comparatively low cross section of final states with associated strangeness, additional cuts are required to observe $K^+$ directly in $\beta$ versus momentum plots for particles detected in the forward spectrometer. The requirement of a $\pi^0$ reconstructed in the BGO and a total BGO energy deposition of less than 250 MeV accentuates $K^+\Lambda$ and $K^+\Sigma^0$ final states where the forward going $K^+$ takes the majority of the available energy. Fig. 19(a) shows the $\beta$ versus momentum plot with these cuts applied, with a visible loci corresponding to $K^+$. Particle mass was calculated from $\beta$ and momentum and shown in Fig. 19(b). Events were selected where this mass was between 420 to 640 MeV, corresponding to a $1\sigma$ of a Gaussian fit to the $K^+$ mass peak. A deliberately narrow selection range was chosen to reduce background, however this range may be extended in future analysis. Fig. 20(a) shows the mass recoiling from the forward going $K^+$, with clear peaks corresponding to $\Lambda$ and $\Sigma^0$. The same technique of identifying the photon from the decay $\Sigma^0 \rightarrow \gamma\Lambda$ that was described in section 4.4.1 was applied, and the $K^+\Sigma^0$ events removed. Fig. 20(b) shows the remaining spectrum, with a clear peak at the $\Lambda$ mass.

A detection efficiency of between 10-20% was determined for $K^+$ detection at forward angles. This is limited by the proportion of $K^+$ which decay on flight and are therfore not identified. With the additional requirement of a

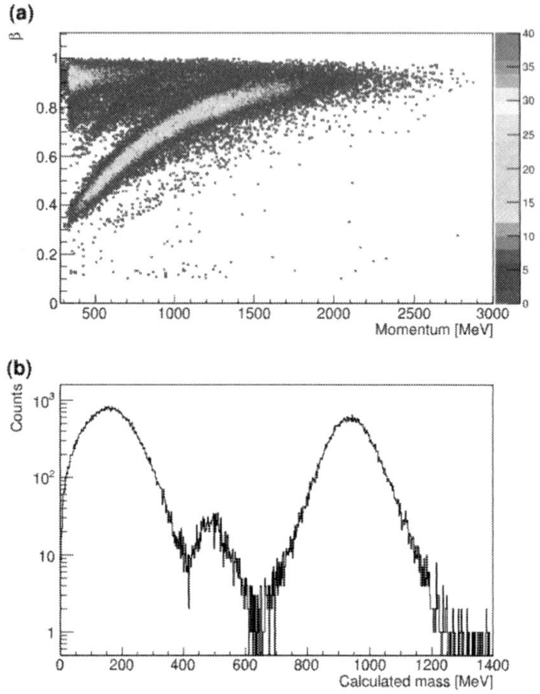


Figure 19: $K^+$ selection in the forward spectrometer. (a) $\beta$ versus momentum for positively charged particles. To accentuate the loci of $K^+$ between the protons and $\pi^+$, a selection cut required an energy sum of less than 250 MeV in the BGO, and the decay $\Sigma \rightarrow \Lambda\gamma$ to be identified, as described in section 4.4.1. (b) Particle mass calculated from $\beta$ and momentum.

$\pi^0$ in the BGO, the detection efficiency reduces to approximately 6%, due to the efficiency of $\pi^0$ reconstruction, and the decay branching ratio of $\Lambda \rightarrow \pi^0 n$ of 35.8%.

Fig. 21 shows preliminary $K^+\Lambda$ differential cross sections at forward angles for two $W$ intervals. The data extends over a larger energy range, however further analysis is required at these intervals to separate signal from background. The data presented here is approximately 1/3 of the available data from 2015 and, nevertheless, the statistics are comparable to the previous data labelled in Fig. 21.

Removing BGO selection cuts ($\pi^0$ selection and energy sum) allows the identification of higher lying hyperon states via $K^+$ detection in the forward spectrometer [66]. Fig. 22(a) shows the mass recoiling from the $K^+$ under these conditions. Peaks corresponding to $\Sigma(1385)$ and $\Lambda(1405)$ (which need further analysis to separate), $\Lambda(1520)$, and potentially higher states up to 1800 MeV are visible. The event topology can be significantly changed by the requirement of additional charged particles and reconstructed $\pi^0$, as shown in Fig. 22(b,c,d). For example, the requirement of a $\pi^0$ and at least another charged particle (besides the $K^+$) accentuates the peak at the $\Sigma(1385)$ and $\Lambda(1405)$ masses, while the requirement of at least

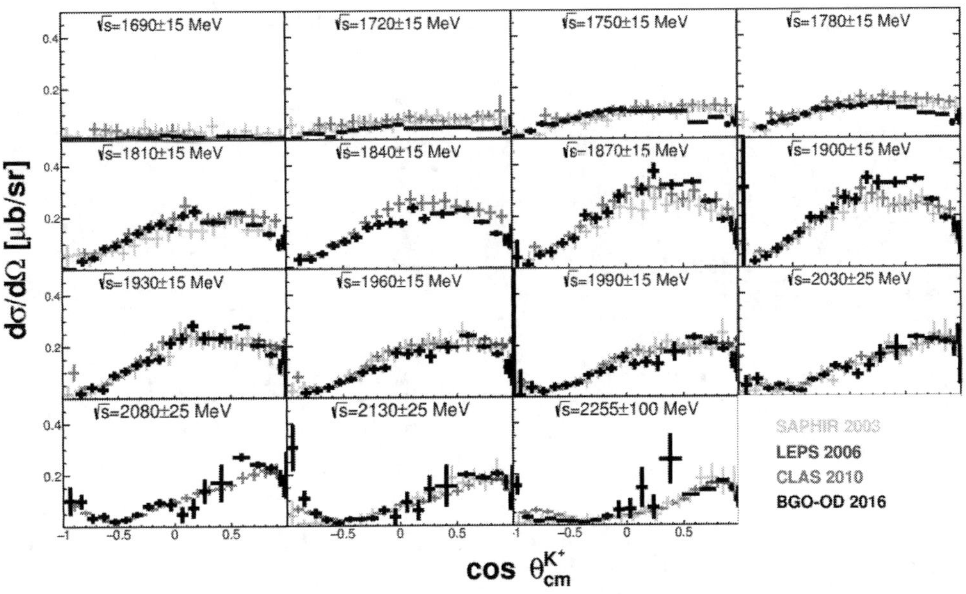


Figure 18: Preliminary $\gamma p \to K^+\Sigma^0$ differential cross section data versus the $K^+$ centre of mass polar angle. Centre of mass energy ranges, $\sqrt{s}$ labelled inset. Black points are this preliminary data [67], green are the SAPHIR data of Glander et al. [60], blue are the LEPS data of Sumihama et al. [68], and red are the CLAS data of Dey et al. [69].

two more charged particles accentuates the peak at the $\Lambda(1520)$ mass. Analysis to isolate particular $K^+Y^*$ states is in progress.

#### 4.4.4 $K_s^0$ detection from a deuterium target

An important measurement to be made with BGO-OD is the $K_s^0\Sigma^0$ differential cross sections over the $K^*$ threshold regions. As was described in section 4.3, the model of Ramos and Oset [29] uses vector meson-baryon interactions to describe $K_s^0\Sigma^+$ data. This model predicts a peak for $K_s^0\Sigma^0$ data at the $K^*$ thresholds. Measurements over this region are therefore crucial in constraing meson - baryon degrees of freedom in the photoproduction mechanism.

Additionally to this measurement, studies of the reaction $\gamma n \to K_s^0 Y$, give important insights and constraints into the reaction mechanism for associated strangeness photoproduction. The photoproduction of charged and neutral kaons are intrinsically related via SU(3) symmetry, leading to relations between hadronic coupling constants. This can be used, for example as shown in reference [71], for better constraints upon isobar models for both charged and neutral kaon photoproduction.

Data from a two day test beam with a liquid deuterium target has been used to demonstrate the $K_s^0$ identification procedure in the pure neutral channel [72]. Analysis with simulated data demonstrated that there is significant background from channels with uncorrelated multiple $\pi^0$ in the final state. A selection cut demanding that no more than five neutral clusters should be identified in the BGO, and no more than two charged particles identified either in the BGO, SciRi or the forward spectrometer was used to suppress this background. After applying timing cuts between neutral clusters and the tagged photon beam, events were selected where the two $\pi^0$ could be reconstructed via pairs of photons with a combined invariant mass between 114 - 162 MeV. With the preliminary energy calibration used, this corresponded to approximately $2\sigma$ of a fitted Gaussian function to the $\pi^0$ mass peak.

Fig. 23(a) shows the $2\pi^0$ invariant mass after selecting events where the mass recoiling from the $2\pi^0$ system is consistent with the $\Sigma^+$ mass (1050 - 1250 MeV). A quadratic function was fitted to the background, with a Gaussian function fitted to the peak corresponding to the $K_s^0$ mass. An additional method was used, where a proton or $\pi^-$ was required in the forward spectrometer. Assuming the decay, $\Lambda \to p\pi^-$, the mass recoiling from the $2\pi^0 p$ or $2\pi^0\pi^-$ system should be the $\pi^-$ or proton masses respectively. Events were selected consistent with this topology, and the resulting $2\pi^0$ invariant mass is shown in Fig. 23(b). The peak corresponding to the $K_s^0$ mass is now more pronounced above the background, albeit with lower statistics.

The same analysis was performed with data using a liquid hydrogen target. No peak structure at the $K_s^0$ mass was observed, however the background looked very similar. This suggests that the peak observed with the deuterium target was from the reactions $\gamma n \to K_s^0\Sigma^0$ and $\gamma n \to K_s^0\Lambda$. Due to limited statistics with the real data set, no serious attempt has been made to separate these two channels. Simulated data has shown however that the technique of identifying the photon from the decay $\Sigma^0 \to \Lambda\gamma$,

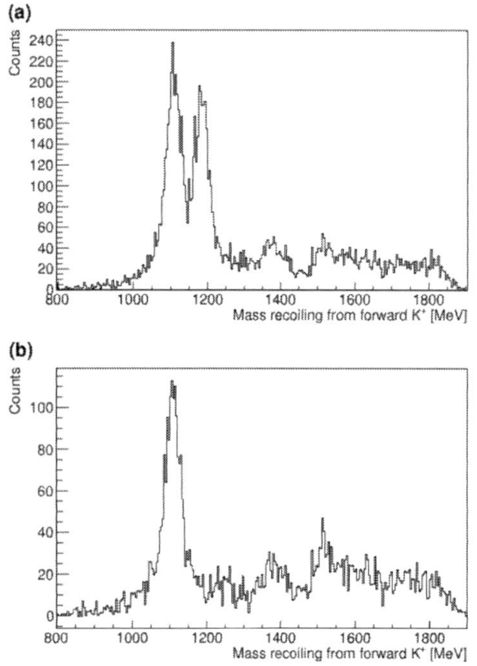


Figure 20: Mass recoiling from forward going $K^+$. (a) To accentuate the peaks corresponding to the $\Lambda$ and $\Sigma^0$ hyperons, a selection cut required an energy sum of less than 250 MeV and for a $\pi^0$ to be identified in the BGO. (b) $K^+\Sigma^0$ events were identified as described in section 4.4.1, and removed, leaving a pronounced peak from the $K^+\Lambda$ final state.

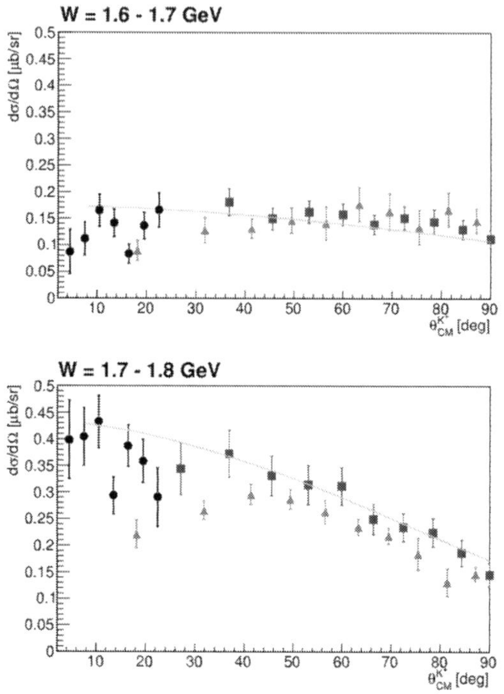


Figure 21: $\gamma p \to K^+\Lambda$ differential cross sections. Black circular data are from this analysis, at a preliminary stage [38]. Blue squares and red triangles are the data of McCracken (CLAS) [62] and Glander (SAPHIR) [60] respectively. The green line is the current parameterisation from Bonn - Gatchina PWA.

as was described for $K^+$ detection in section 4.4.1, will work for these channels as well.

## 5 Summary and Outlook

The goal of project B.1 was the systematic investigation of the photoproduction of mesons off nucleons in order to understand the reaction mechanisms involved, and the relevant degrees of freedom in resonance formation. Of particular interest were mesons heavier than the pion, for example $\eta$, $\omega$ and $K_s^0$, and resonances carrying hidden or open strangeness. Polarisation is an important tool, and essential hardware contributions have been made to the experimental programme of the CRC 16 in this regard. As a requirement for the experiments, photon tagging systems have been designed and set up for the CBELSA/TAPS and BGO-OD experiments. Both include coherent bremsstrahlung radiators for linearly polarised photon beams and Møller polarimeters to determine the energy dependent degree of circular polarisation when a polarised electron beam is used. Using polarised photon beams in conjunction with the Bonn polarised frozen spin target, significant results could be achieved in single and double polarisation experiments which, in combination with the unpolarised cross section measurements, provided essential input for partial wave and other theoretical analyses. Selected highlights are the single and double polarisation measurements in $\omega$-photoproduction which helped to disentangle the reaction mechanism near threshold, and to identify specific $s$-channel resonances involved. In the strange sector, a cusp like structure was observed in $K_s^0\Sigma^+$ production at the $K^*$ threshold which in chiral unitary models seems to be associated with one or two dynamically generated vector meson-baryon states around 2 GeV.

In the frame of the BGO-OD collaboration a new experiment has been set up. It combines a forward magnetic spectrometer with a central BGO calorimeter with charged particle recognition and identification. The BGO-OD experiment enables reconstruction of complex final states composed of both charged and neutral particles, complementary to the existing CBELSA/TAPS calorimeter which is optimised for multi-photon final states. The experiment was commissioned at the end of the CRC period and first physics data were taken. The preliminary results demonstrate the performance of the detector. It opens a new experimental window to investigate hadronic excitation structures, complementary

DOI: 10.1051/epjconf/201713403001

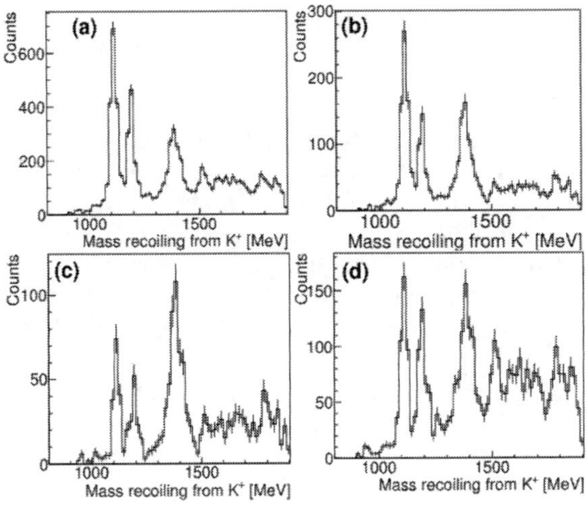


Figure 22: Missing mass recoiling from a forward $K^+$ over a momentum range of 500 - 800 MeV, with different selection criteria: (a) All data. (b) With a $\pi^0$ reconstructed in the BGO. (c) With a $\pi^0$ reconstructed and at least one additional charged particle. (d) With at least two additional charged particles.

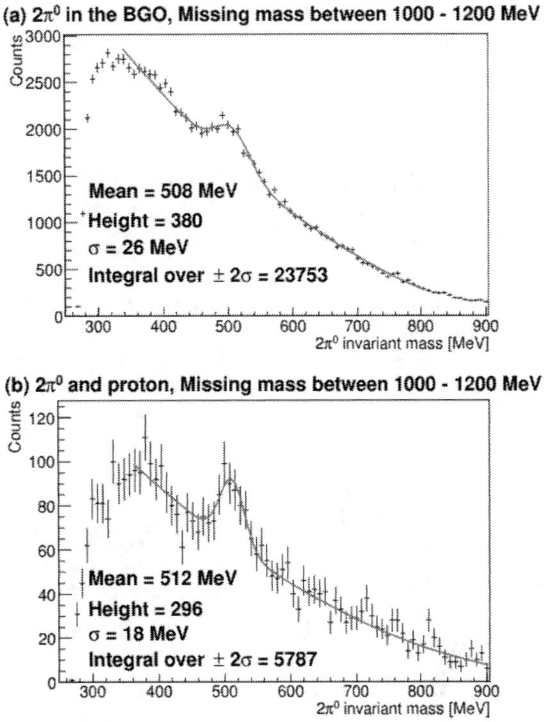


Figure 23: $K_s^0$ detection with a deuterium target (approximately two days of data taking). Selection cuts described in the text enhanced the $\gamma n \to K_s^0\Sigma^+$ signal above background. (a) The invariant mass of $2\pi^0$ after selecting events where the missing mass is consistent with either the $\Lambda$ or $\Sigma^0$ masses. (b) The invariant mass of $2\pi^0$ after selecting events where the missing mass from the $2\pi^0 p$ system is consistent with the $\pi^-$ mass (assuming the decay $\Lambda \to p\pi^-$). For both histograms, a second order polynomial and a Gaussian function describe the background and $K_s^0$ signal respectively. Parameters from the fitted Gaussian function are inset.

to other experiments and facilities.

The results presented by the spokespersons of B.1 are the fruits of work and commitment of numerous people involved in the project over the past 12 years. Many students who received their Bachelor's, Master's, or Diploma degrees significantly contributed to the success of B.1. In particular, the doctoral students carried the main load of technical developments and data analysis. Without their enthusiasm the whole project would have been unthinkable. Equally important have always been the postdoctoral fellows with their dedication to B.1. Particular mention for their contributions deserve Daniel Elsner, Jürgen Hannappel, Thomas C. Jude, and Valentina Vegna. As co-spokesperson during the first 4-year funding period Hartmut Kalinowsky made important contributions in the initial phase of B.1.

For providing excellent beam we thank the whole staff of the ELSA accelerator, in particular the technical director Wolfgang Hillert. The project also benefited from the support of the mechanical and electronics workshops. We thank DESY for providing the BGO-OD spectrometer magnet on permanent loan basis, and the IPN Orsay for loan of the TOF detector of the former GRAAL experiment. Very signifcant contributions to the BGO-OD experiment by Italian Universities and INFN in the frame of the BGO-OD collaboration are gratefully acknowledged, as is the fruitful co-operation with PNPI Gatchina and INR Moscow.

The whole work reported here would not have been possible without the continuous support from the Land Northrine-Westphalia, and from the Deutsche Forschungsgemeinschaft within the SFB/TR (CRC) 16, which we are very grateful for.

## References

[1] R. Edwards et al., Phys. Rev **D 84** (2011) 074508

[2] E. Klempt and J.M. Richard, Rev. Mod. Phys. **82** (2010) 1095

[3] L. Guo et al. (CLAS collaboration), Phys. Rev **C 76** (2007) 025208

[4] A. Manohar and H. Georgi, Nucl. Phys. **B 234** (1984) 189

[5] L.Ya. Glozman and D.O. Riska, Physics Reports **268** (1996) 263

[6] R.H. Dalitz and J.G. McGinley, in *Low and Intermediate Energy Kaon-Nucleon Physics*, ed. by E. Ferrari and G. Violini, Reidel, Boston (1981) 381; R. H. Dalitz, T.C. Wong, and G. Rajasekaran, Phys. Rev. **153** (1967) 1617

[7] P.B. Siegel, and W. Weise, Phys. Rev **C 38** (1988) 2221

[8] N. Kaiser, T. Waas, and W. Weise, Nucl. Phys, **A 612** (1997) 297

[9] C. Garcia-Recio, M.F.M. Lutz, and J. Nieves, Phys. Lett B 582 (2004) 49
[10] M.F.M. Lutz and E.E. Kolomeitsev, Phys. Lett. **B 585** (2004) 243
[11] B. Borasoy, P.C. Bruns, U.-G. Meißner and R. Nißler, Eur. Phys. J. **A 34** (2007) 161
[12] P.C. Bruns, M. Mai, U.-G. Meißner, Phys. Lett. **B 697** (2011) 254
[13] J.A. Oller and U.-G- Meißner, Phys. Lett. **B 500** (2001) 263
[14] B. Borasoy, U.-G. Meißner and R. Nißler, Phys. Rev **C 74** (2006) 055201
[15] P. Gonzalez, E. Oset and J. Vijande, Phys. Rev **C 79** (2009) 025209
[16] S. Sarkar et al., Eur. Phys. J. **A 44** (2010) 431
[17] E. Oset and A. Ramos, Eur. Phys. J. **A 44** (2010) 445
[18] E. Oset et al., AIP Conf.Proc. 1388 (2011) 295, arXiv:1103.0807v1 [nucl-th]
[19] V. Kuznetsov et al., Phys. Lett. **B 647** (2007) 23
[20] I. Jaegle et al., Phys. Rev. Lett. **100** (2008) 252002
[21] I. Jaegle et al., Eur. Phys. J. **A 47** (2011) 89
[22] E.F. McNicoll et al., Phys. Rev **C 82** (2011) 035208; Erratum: Phys. Rev **C 84** (2011) 029901
[23] V. Kuznetsov and M.V. Polyakov, JETP Letters **88** (2008) 347, and V. Kuznetsov et al., Acta. Phys. Polon. **39** (2008) 1949 and references therein
[24] T. Mibe et al., Phys. Rev. Lett. **95** (2005) 182001
[25] B. Dey et al., Phys. Rev **C 89** (2014) 055208
[26] H. Seraydaryan et al., Phys. Rev **C 89** (2014) 055206
[27] R. Ewald et al., Phys. Lett. **B 713** (2012) 180
[28] R. Ewald et al., Phys. Lett. **B 738** (2012) 268
[29] A. Ramos and E. Oset, Phys. Lett. **B 727** (2013) 287
[30] J.M.M. Hall et al., Phys. Rev. Lett. **114** (2015) 132002
[31] W. Hillert, Eur. Phys. J. **A 28**, s01 (2006) 139
[32] K. Fornet-Ponse, Doctoral Thesis, Physikalisches Institut, Bonn (2009)
[33] D. Elsner et al., Eur. Phys. J. **A 39** (2009) 373
[34] H. Olsen and L.C. Maximon, Phys. Rev **114** (1959) 887
[35] L.G. Levchuk, Nucl. Instr. Meth. **A 345** (1994) 496
[36] S. Kammer, Doctoral Thesis, Physikalisches Institut, Bonn (2010)
[37] A. Bella, Doctoral Thesis, Physikalisches Institut, Bonn (2016)
[38] T. Zimmermann, Doctoral Thesis, Physikalisches Institut, Bonn (2016)
[39] H. Schmieden, Chinese Physics **33** (2009) 1146
[40] D. Elsner et al., Eur. Phys. J. **A 33** (2007) 147
[41] G. Knöchlein, D. Drechsel, L. Tiator, Z. Phys. **A 352** (1995) 327
[42] F. Klein et al., Phys. Rev. **D 78** (2008) 11
[43] H. Eberhardt, T.C. Jude, H. Schmieden et al., Phys. Lett. **B 750** (2015) 453
[44] A. Wilson et al., Phys. Lett. **B 749** (2015) 407
[45] J. Ajaka et al., Phys. Rev. Lett. **96** (2006) 132003
[46] M. Williams et al., Phys. Rev. **C 80** (2009) 065208
[47] M. Williams et al., Phys. Rev. **C 80** (2009) 065209
[48] E. Aker et al., Nucl. Instrum. Methods **A 321** (1992) 69
[49] R. Novotny et al., IEEE Trans. Nucl. Sci. **38** (1991) 379
[50] A. R. Gabler et al., Nucl. Instrum. Methods **A 346** (1994) 168
[51] G. Suft et al., Nucl. Instr. Meth. **A 538** (2005) 416
[52] Ch. Bradtke, H. Dutz et al., Nucl. Instr. Meth. **A 436** (1999) 430
[53] I. Denisenko et al., Phys. Lett. **B 755** (2016) 97
[54] R. Castelijns, et al., Eur. Phys. J. **A 35** (2008) 39
[55] R. Lawall, et al., Eur. Phys. J. **A 24** (2005) 275
[56] R.A. Arndt, et al., http://gwdac.phys.gwu.edu
[57] http://www.kph.uni-mainz.de/MAID/ (version 29.3.2007)
[58] M. Nanova, et al., Eur. Phys. J. **A 35** (2008) 333
[59] A. V. Anisovich et al., Eur. Phys. J. **A 47** (2011) 27
[60] K.H. Glander et al., Eur. Phys. J. **A 19** (2004) 251
[61] R. Bradford, R. A. Schumacher et al., Phys. Rev. **C 73** (2006) 035202
[62] M. E. McCracken, M Bellis, C. A. Meyer, M. Williams et al., Phys. Rev. **C 81** (2010) 025201
[63] K. Moriya, R. A. Schumacher et al., Phys Rev **C 87** (2013) 035206
[64] K. Moriya, R. A. Schumacher et al., Phys. Rev. Lett. (2014) 082004
[65] T.C. Jude, D.I. Glazier, D.P. Watts *et al.*, Phys. Lett. **B 735** (2014) 112
[66] T.C. Jude et al., JPS Conf. Proc. **10** (2016) 032002
[67] G. Scheluchin, Doctoral Thesis, Physikalisches Institut, Bonn (in preparation)
[68] M. Sumihama et al., Phys. Rev. **C 73** (2006) 035214
[69] B. Dey et al., Phys. Rev. **C 82** (2010) 025202
[70] D. Skoupil and P. Bydžovský, Phys. Rev. **C 93** (2016) 025204
[71] T. Mart, Phys. Rev. **C 83** (2011) 048203
[72] T.C. Jude et al., priv. comm. (2016)

# Chiral dynamics with (non)strange quarks

*Bastian* Kubis[1,*] and *Ulf-G.* Meißner[1,2,**]

[1] *Universität Bonn, Helmholtz Institut für Strahlen- und Kernphysik, Bethe Center for Theoretical Physics, D-53115 Bonn, Germany*
[2] *Forschungszentrum Jülich, Institute für Kernphysik (IKP-3), Institute for Advanced Simulation (IAS-4) and Jülich Center for Hadron Physics, D-52145 Jülich, Germany*

**Abstract.** We review the results and achievements of the project **B.3**. Topics addressed include pion photoproduction off the proton and off deuterium, three-flavor chiral perturbation theory studies, chiral symmetry tests in Goldstone boson decays, the development of unitarized chiral perturbation theory to next-to-leading order, the two-pole structure of the $\Lambda(1405)$, the dynamical generation of the lowest $S_{11}$ resonances, the theory of hadronic atoms and its application to various systems, precision studies in light-meson decays based on dispersion theory, the Roy–Steiner analysis of pion–nucleon scattering, a high-precision extraction of the elusive pion–nucleon $\sigma$-term, and aspects of chiral dynamics in few-nucleon systems.

## 1 Introduction

The project **B.3** has been funded over the whole period of the SFB/TR 16. In the first funding period, it actually focused on "Meson production off nucleons and light nuclei" (title in the first funding period), spanning chiral perturbation theory (ChPT) calculations in the single-nucleon sector, extensions of three-flavor ChPT to higher energies by unitarization methods, and developments of chiral nuclear effective field theory (EFT) for light nuclei, with an emphasis on meson production reactions. The PIs in this funding period were Ulf Meißner and Akaki Rusetsky. As this project was extremely successful but also very broad, it was decided to move the nuclear aspects into the new project **B.6** "Neutron properties from light nuclei" in funding periods 2 and 3, with the PIs Evgeny Epelbaum and Hans-Werner Hammer (and Ulf Meißner after 2013 when Hans-Werner Hammer moved to Darmstadt). In the second and third funding period, the project focused on precision calculations in meson decays, largely in the framework of dispersion relations, on the further development of unitarized ChPT for meson–baryon interactions, and on Roy–Steiner equations for pion–nucleon scattering. The advances made in these fields will be reported in chronological order, grouped into the three four-year funding periods.

Our contribution is organized as follows: Section 2 summarize the achievements from the first funding period. In the second funding period, Sec. 3 the research was more focused on Goldstone boson decays, Sec. 3.1, and on chiral coupled-channel analysis of meson–baryon scattering and meson production, Sec. 3.2. In the third funding period, Sec. 4, we worked on precision studies in light-meson decays, Sec. 4.1, revisited (anti)kaon–nucleon scattering, see Sec. 4.2, analyzed the $\Lambda(1405)$ in kaon photoproduction, see Sec. 4.3, and, last but not least, performed a detailed Roy–Steiner analysis of pion–nucleon scattering, see Sec. 4.4. In fact, the foundation of the Roy–Steiner studies were already laid out in the second funding period, but we prefer to report these topics in one section. Finally, an **important disclaimer**: This is the review of the project **B.3**, but not a general review of the various fields to which this research contributed. Therefore, only the papers that resulted from this research are cited, with the exception of only a few external papers and of course, in case data were involved, with proper reference to the experimental papers. The links to the other projects are mentioned in passing, but no effort is made to discuss these in any detail.

## 2 First funding period 2004–2008

### 2.1 Pion photoproduction

Neutral pion photoproduction off nucleons is one of the best reactions to test two-flavor baryon chiral dynamics. In Ref. [1] we have analyzed the Fubini–Furlan–Rosetti sum rule (which was derived in 1965 using soft-pion techniques) in the framework of covariant baryon chiral perturbation theory to leading one-loop accuracy and including next-to-leading order (NLO) polynomial contributions. We have shown how the appearing subtraction constants in the invariant amplitudes can be mapped onto certain low-energy constants (LECs) employed in earlier ChPT studies of threshold neutral pion photoproduction off nucleons. In particular, we consider the corrections to the sum rule due to the finite pion mass and show that below the threshold, they agree well with determinations based on fixed-$t$ dispersion relations performed by

*e-mail: kubis@hiskp.uni-bonn.de
**e-mail: meissner@hiskp.uni-bonn.de

 DOI: 10.1051/epjconf/201713403002

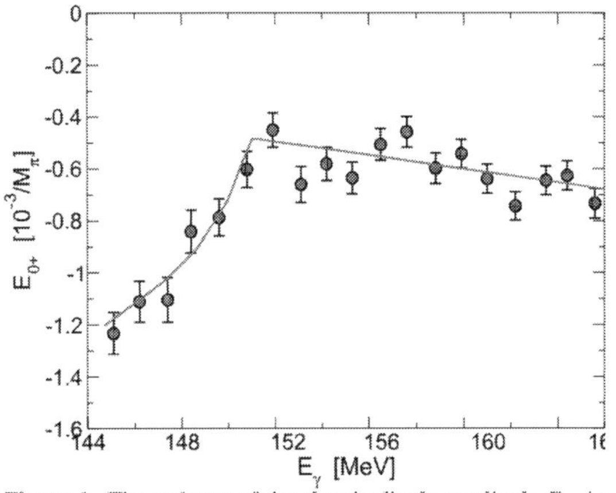


**Figure 1.** The real part of the electric dipole amplitude $E_{0+}$ in the threshold region. The solid line is the theoretical prediction as discussed in the text and the data are from Ref. [2]. The cusp at the opening of the $\pi^+n$ threshold ($E_\gamma = 151.8$ MeV) is clearly visible. Figure taken from Ref. [1].

the Mainz group. We also have shown that this representation allows for a precise determination of certain LECs and we have given the most accurate ChPT prediction for the energy dependence of the electric dipole amplitude $E_{0+}$ in the threshold region, see Fig. 1.

### 2.2 Three-flavor ChPT studies

To make further progress in SU(3) chiral dynamics, we have also studied the application of ChPT in a variety of kaon decays of current interest because high-precision data emerge in particular from NA 48/2 at CERN. Aspects of radiative kaon decays were studied in Refs. [3–5], with particular emphasis on the possibility to extract structure-dependent terms and the role of the axial anomaly. In Ref. [6], an effective field theory framework was formulated that allows one to extract the pion-pion S-wave scattering lengths from the structure of the cusp in the reaction $K \to 3\pi$, originating from the $\pi^0\pi^0$ rescattering in the final state (that was earlier pointed out by Meißner, Müller and Steininger [7] a decade ago). We have also investigated the scalar sector of QCD as tested in $J/\psi$ decays to vector mesons and two pseudoscalars [8] and extracted the large-$N_c$-suppressed low-energy constant $L_4$ and $L_6$ of the fourth-order ChPT Lagrangian. This analysis was extended in Ref. [9] to include isospin-breaking terms in order to investigate the phenomenon of $a_0(980)$–$f_0(980)$ mixing. In Ref. [10], we have updated the analysis of $\omega$-$\phi$ mixing in the framework of chiral perturbation theory with vector mesons. This result was used later in project **C.2** in the investigation of isospin violation in the nucleon form factors [11]. As a basic ingredient for calculations in the meson–baryon sector, we have constructed the complete and minimal effective Lagrangian at one loop, the third-order terms are published in Ref. [12]. At third order, we find 78 independent operators, from which many, however, only contribute in processes with many pions and external fields. This minimal set of operators was later confirmed by the Murcia group [13]. We have also investigated the matching between the two- and three-flavor EFTs for the scalar sector [14]. Furthermore, we have calculated the hyperon vector form factors to one loop based on a covariant formulation of ChPT to study the possibility of extracting $V_{us}$ [15]. We have also written a general review on "Chiral Perturbation Theory" for *Annual Review in Nuclear and Particle Science* [16].

### 2.3 Kaon–nucleon scattering and hadronic atoms

An important building block in the analysis of kaon photo- and electroproduction is (anti)kaon–nucleon scattering. To further constrain the pertinent scattering amplitudes based on coupled-channel unitary extensions of ChPT, we have given the complete expression for the isospin-breaking corrections to the energy levels and the decay widths of kaonic hydrogen up-to-and-including $O(\alpha, m_d - m_u)$ in QCD+QED [17], with $\alpha$ the fine structure constant and $m_q$ the current mass of a quark of flavor $q$. It was demonstrated that although the leading-order corrections emerging due to the unitarity cusp are huge, they can be expressed solely in terms of the $\bar{K}N$ S-wave scattering lengths. We have also found additional large electromagnetic corrections and shown that the most recent DEAR data from kaonic hydrogen [18] are not consistent with most older scattering data (as it was also stressed by the Munich group). This extension of the leading-order Deser formula has become a standard tool in this field. These observations have further been sharpened in Ref. [19], where the strong $K^-p$ scattering length is extracted within chiral SU(3) unitary approaches from a very large variety of fits to low-energy $K^-p$ scattering data. For the first time, we have given theoretical uncertainties in such type of approach, based on a large number of Monte Carlo simulations in the multi-dimensional parameter-space (utilizing a method developed in project **C.2**). Within this scheme, a very good overall agreement with available scattering data is obtained (see Fig. 2) and the resulting scattering length is compared and found to be incompatible with the kaonic hydrogen data from DEAR at the $3\sigma$ level. In Ref. [20], we have discussed the extraction of the S-wave kaon–nucleon scattering lengths $a_0$ and $a_1$ from a combined analysis of existing kaonic hydrogen and synthetic deuterium data within the framework of a low-energy effective field theory. It turns out that with the present DEAR central values for the kaonic hydrogen ground-state energy and width, a solution for $a_0$ and $a_1$ exists only in a restricted domain of input values for the kaon–deuteron scattering length. It is thus very important to measure the kaon–deuteron scattering length, as proposed by the SIDDHARTA collaboration. Finally, we have shown how the nature of the $\Lambda(1520)$ as a dynamically generated resonance can be investigated in proton–proton collisions [21].

### 2.4 Gauge-invariant formulation of coupled-channel dynamics

Chiral unitary approaches have already been utilized successfully by various groups to describe kaon and $\eta$ photo-

DOI: 10.1051/epjconf/201713403002

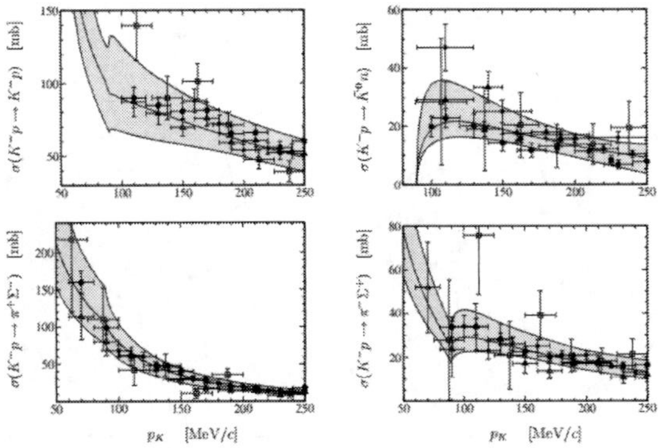


**Figure 2.** Total cross sections for $K^-p$ scattering into various channels. The best fit is represented by the solid line and the shaded area indicates the $1\sigma$ confidence region. Figure taken from Ref. [19].

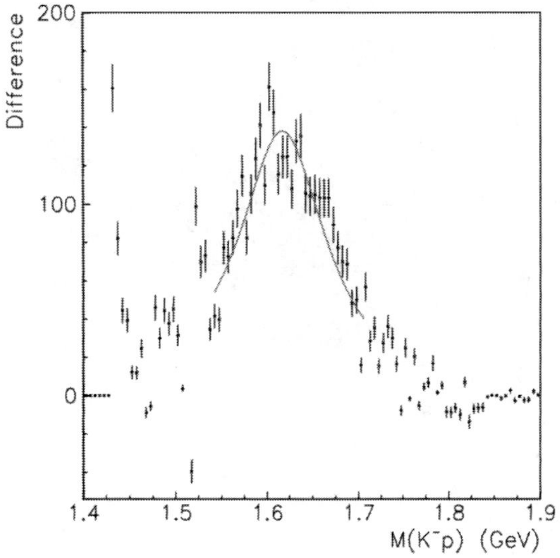


**Figure 3.** The difference between the SAPHIR data for the $K^-p$ invariant mass spectra of $\gamma p \to K^+K^-p$ and our model prediction. The solid line indicates the fit with a relativistic Breit–Wigner function. Figure taken from Ref. [24].

and electroproduction data off the nucleon. In most earlier calculations, the photon only scatters with the incoming baryon to form a meson–baryon pair that then undergoes strong final-state interactions. Such an approach is, however, not gauge invariant. In Ref. [22] we have shown how gauge invariance is obtained for the coupling of a photon to a two-body state described by the solution of the Bethe–Salpeter equation, i.e. the reaction $\gamma\phi B \to \phi B$ (with $\phi$ a Goldstone boson and $B$ a member of the ground state baryon octet). The key observation is that the photon has to couple to all possible initial, intermediate, and final states. The method has been illustrated both for a complex scalar field theory and for interaction kernels derived from chiral effective Lagrangians. This paves the way for a gauge-invariant formulation of meson photoproduction based on chiral coupled-channel dynamics. Subsequently, we have analyzed the so-called "turtle model" of electromagnetic meson production, which as the basic building block contains the full meson–baryon scattering T-matrix and the gauge-invariant subset of meson production vertices with an outgoing kaon. This model is gauge invariant (but not crossing symmetric). It is particularly useful to analyze the existing and upcoming (near threshold) kaon photo- and electroproduction data. As a first step, we have used it to test the various approximations (like the on-shell scheme for the loop integrals or the coupling of the photon to the initial baryon only) made in the literature so far [23].

## 2.5 Phenomenological models of meson photoproduction and OZI analysis

We have also performed some phenomenological studies of kaon and vector meson production based on results from SAPHIR (ELSA) and JLab. In Ref. [24] we studied the contribution of the Drell mechanism driven by $K^+$ and $K^-$ exchange to the reaction $\gamma N \to K\bar{K}N$. Adding a contribution from $K^*$-meson exchange with subsequent excitation of the $\Lambda(1520)$ resonance, we can describe the data of the older LAMP2 experiment. When applying the same model to the more recent SAPHIR data we find an excellent description of the $K^+p$ spectrum and evidence for a hyperon resonance with $M_R = (1617 \pm 2)$ MeV and $\Gamma_R = (117 \pm 4)$ MeV in the $K^-p$ mass distribution, see Fig. 3. We have also studied the violation of the OZI rule in photoproduction [25] and in proton–proton collisions [26]. Furthermore, we have given an explanation of the puzzling $A$-dependence in electromagnetic $\phi$-meson production off nuclei observed at Spring-8 based on a two-step mechanism [27].

## 2.6 Coupled-channel dynamics in $\eta$ and $\eta'$ decays

Many of the technical aspects related to the work on (anti)kaon–nucleon scattering and photoproduction off nucleons can also be addressed in the decays of $\eta$ and $\eta'$ mesons that exhibit strong final-state interactions. Besides, such reactions allow for fine tests of the symmetries of QCD. In Ref. [28] the hadronic decays $\eta, \eta' \to 3\pi$ and $\eta' \to \eta\pi\pi$ were investigated within the framework of a U(3) chiral effective field theory in combination with a relativistic coupled-channel approach. Final-state interactions were included in the S- and P-wave interactions. In Ref. [29] the claim that the light quark mass ratio $(m_d - m_u)/m_s$ can be extracted from the decay width ratio $\Gamma(\eta' \to \pi^0\pi^+\pi^-)/\Gamma(\eta' \to \eta\pi^+\pi^-)$ was critically investigated. In particular, the assumption that the amplitude for $\eta' \to \eta\pi\pi$ is constant over phase space is not correct, as shown by the Dalitz plot in the right panel of Fig. 4. Also, the assumption that the amplitude $A(\eta' \to \pi^0\pi^+\pi^-)$ is entirely given from the amplitude $A(\eta' \to \eta\pi^+\pi^-)$ through $\pi^0 - \eta$ mixing could be shown to be erroneous. For further work on $\eta$ and $\eta'$ decays in this framework, see Ref. [30].

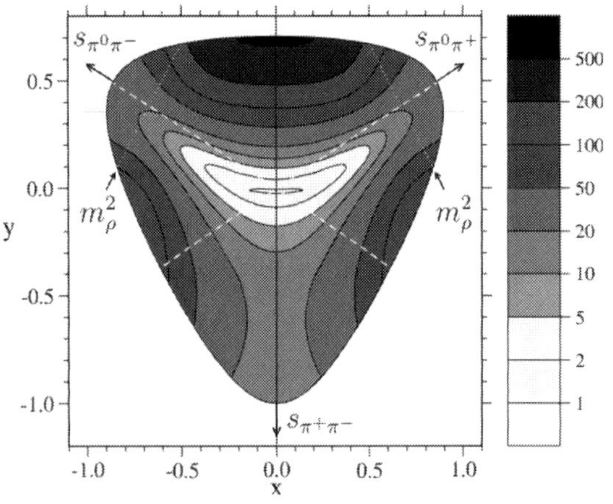


**Figure 4.** Dalitz plot distribution $|A(\eta' \to \pi^0\pi^+\pi^-)|^2$ of the best overall fit including the VES data for $\eta' \to \eta\pi^+\pi^-$. The distribution is normalized to unity at $x = y = 0$. The P-wave contributions to $\pi^0\pi^+$ ($\pi^0\pi^-$) rescattering vanish on the rising (falling) dashed line and the invariant energies associated with the $\rho^\pm(770)$ are indicated by the dotted lines.

## 2.7 Developments in chiral nuclear EFT and light nuclei

To study also meson production off neutrons, it is mandatory to further develop the chiral EFT for few-nucleons systems. Furthermore, few-nucleon systems allow for fine tests of fundamental symmetries and their violation. In Ref. [31] the contributions from the isospin-violating one- and two-pion-exchange potentials to the nucleon–nucleon potential were derived, in particular novel expressions for the subleading charge-symmetry-breaking two-pion exchange potential were given. In Ref. [32] we have classified $A$-nucleon forces according to their isospin dependence and discussed the most general isospin structure of the three-nucleon force. We have derived the leading and subleading isospin-breaking corrections to the three-nucleon force. Furthermore in Ref. [33] we have studied the two-nucleon force at next-to-next-to-leading order in a chiral EFT with explicit $\Delta$ degrees of freedom. Fixing the appearing LECs from a next-to-leading order calculation of pion–nucleon threshold parameters, we find an improved convergence of most peripheral nucleon–nucleon phases compared to the theory with pions and nucleons only. We have also studied the consistency of Weinberg's power counting [34] and further elaborated on the conjectured infrared renormalization group limit cycle in QCD [35, 36]. To extend this program to nuclei beyond $A \geq 5$, we have started a long-term project in nuclear lattice simulations, which amounts to a lattice formulation of the $A$-nucleon Schrödinger equation and its solution via Monte Carlo methods, utilizing the potentials derived in chiral EFT. First results obtained for the $A = 2, 3, 4$ systems already looked promising [37, 38]. Finally, universal properties of few-body systems have been studied in Refs. [39, 40]. In particular, in the investigation of the triton charge form factor [39], the nucleon form factors developed in project **C.2** were used. In a similar fashion, the isospin mixing effects in the nucleon form factors worked out in project **C.2** were used to analyze isospin mixing in $^4$He and its influence on the extraction of the nucleon strange electric form factors [41].

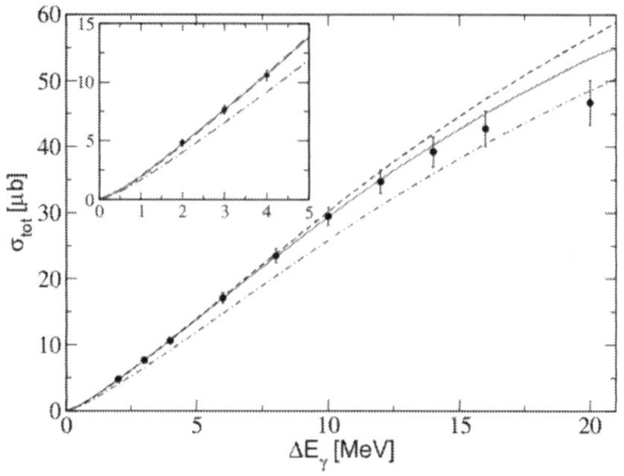


**Figure 5.** Total cross section of the reaction $\gamma d \to \pi^+ nn$ at LO (dashed line), NLO (dash–dotted line) and NNLO (solid line) together with experimental data from MIT-Bates. Figure taken from Ref. [42].

## 2.8 Pion photoproduction off deuterium

In Ref. [42] we have analyzed the reaction $\gamma d \to \pi^+ nn$ up to order $\chi^{5/2}$ in ChPT, where $\chi$ denotes the ratio of the pion to the nucleon mass. Special emphasis was put on the role of nucleon-recoil corrections that are the source of contributions with fractional power in $\chi$. Using the threshold production amplitude for $\gamma p \to \pi^+ n$ calculated in ChPT a long time ago by Bernard, Kaiser, and Meißner as the only input, the total cross section for $\gamma d \to \pi^+ nn$ is described very well, see Fig. 5. A conservative estimate suggests that the theoretical uncertainty for the transition operator amounts to 3% for the computed amplitude near threshold. In Ref. [43] we have discussed the extraction of the neutron–neutron scattering length from experimental spectra on this reaction. We show that for properly chosen kinematics, the theoretical uncertainty of the method can be as low as 0.1 fm.

# 3 Second funding period 2008–2012

## 3.1 Chiral symmetry tests in Goldstone boson decays

One of the highlights of the previous funding period were new developments for precision determinations of the pion–pion scattering lengths. One method exploits the unitarity cusp at the $\pi^+\pi^-$ threshold in the $\pi^0\pi^0$ spectrum in the decay $K^+ \to \pi^0\pi^0\pi^+$, which has been measured with unprecedented precision by the NA48/2 collaboration at CERN; see Fig. 6. To interpret this cusp effect in a model-independent manner and extract the scattering length combination $a_0^0 - a_0^2$ with highest possible precision, we have developed a manifestly covariant version of

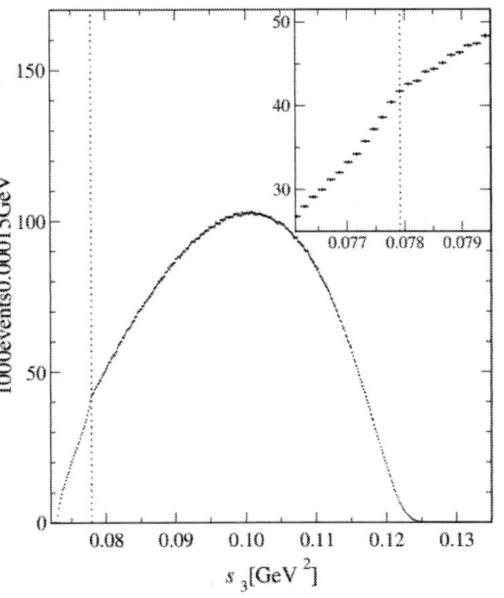


**Figure 6.** Cusp in the decay spectrum $d\Gamma/ds_3$ of the decay $K^\pm \to \pi^0\pi^0\pi^\pm$ as seen by the NA48/2 collaboration. The dotted vertical line marks the position of the $\pi^+\pi^-$ threshold, the insert focuses on the cusp region. Figure taken from Ref. [44], based on data from Ref. [45].

non-relativistic effective field theory (NREFT) [6, 46], extending and partly relying on the expertise on NREFTs in the context of hadronic atoms [47]. An essential ingredient of the most recent experimental analysis [48] was the inclusion of radiative corrections [49], which have an important impact on the analytic structure of the decay amplitude near the cusp. The group's expertise on radiative corrections was subsequently also used for a different class of kaon decays, the flavor-changing neutral current processes $K \to \pi\ell^+\ell^-$ [50]. The theoretical foundations of this theory were recently clarified [51], including thorough investigations of the analytic properties of the decay amplitude at two-loop order. Furthermore, a similar cusp effect has also been predicted and analyzed for the decay $\eta' \to \eta\pi^0\pi^0$ [52].

A complementary determination of $\pi\pi$ scattering lengths is based on $K_{e4}$ decays. It has been pointed out that also here, isospin-breaking corrections play an important role [53], given the tremendous precision of the data collected at NA48/2 [54]. The combination of both kaon-decay-based scattering length extractions has led to the most precise experimental determinations of these fundamental quantities in low-energy strong-interaction physics to date, in excellent agreement with theoretical predictions [55].

Chiral perturbation theory, as a perturbative expansion of observables in the light quark masses, in particular allows to study the pattern of explicit chiral symmetry breaking caused thereby, giving unique access to the masses of the up, down, and strange quarks, which are not otherwise accessible due to their confinement inside hadrons. A particularly important process to constrain the ratios of the light quark masses is the isospin-breaking decay $\eta \to 3\pi$.

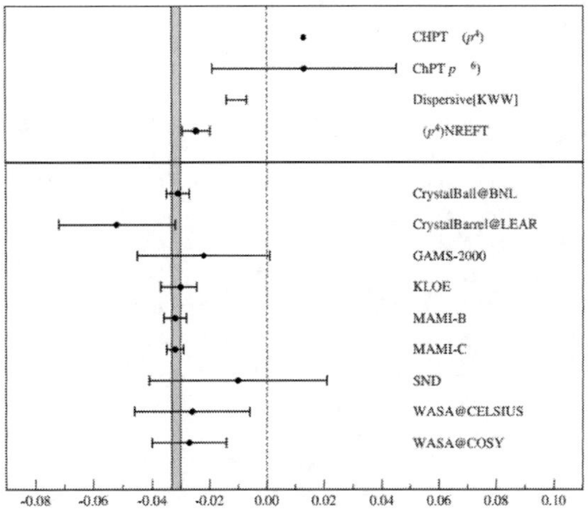


**Figure 7.** Comparison of values for the slope parameter $\alpha$ in $\eta \to 3\pi^0$. Top: theoretical predictions. Bottom: experimental determinations. The gray shaded area is the average from Ref. [63]. Figure taken from [62].

One of its particular advantages is the suppression of electromagnetic (as opposed to $m_u - m_d$) effects; these have been reassessed to higher precision than before [56], but still finding only very small corrections. In order to extract the quark mass ratio from the normalization of the measured $\eta \to 3\pi$ Dalitz plot distribution, the energy dependence of the decay amplitude has to be under control with very high precision—a notoriously difficult task for chiral perturbation theory in particular due to the large pion–pion rescattering effects. One particular issue in the past few years was a discrepancy between the $\eta \to 3\pi^0$ Dalitz plot parameter $\alpha$ as measured with very high precision by various experiments (analyzed in the context of the TR16 project **C.5** [57] and elsewhere [58–60]), and as predicted by chiral perturbation theory [61]. We have made use of the NREFT developed for $K \to 3\pi$ decays to study these corrections [62], giving a theoretical prediction for $\alpha$ compatible with experimental findings, and explaining the failure of chiral perturbation theory for this observable at the same time; see Fig. 7 for an illustration. Also, we have derived a model-independent relation between the Dalitz plot parameters for the charged- and neutral-pion final state that seems to be violated by the most recent (and accurate) $\eta \to \pi^+\pi^-\pi^0$ data [64].

Also in connection with the ongoing $\eta/\eta'$ program in project **C.5** and at various other laboratories, we have predicted the branching ratios of the yet unmeasured anomalous decays of $\eta'$ and $\eta$ into four pions [65]. For the P-wave-dominated final states including charged pions, we predicted branching fractions only slightly below the upper limits at the time; first experimental measurements of these by the BESIII Collaboration found agreement with our predictions within uncertainties [66]. On the other hand, the decays into neutral pions only have to be D-wave driven as long as CP symmetry is conserved, and are therefore more strongly suppressed. CP-violating contributions to these decays based on the QCD $\theta$-term have also been considered (the bounds on which were recently re-studied in the context of the neutron electric dipole moment [67]).

When going to higher energies, the purely perturbative treatment of final-state interactions in meson decays, be it in chiral perturbation theory or the NREFT framework, are doomed to fail, and one has to resort to methods from dispersion theory that resum such rescattering effects to all orders. We have performed such a dispersive analysis for the three-pion decays of the lightest isoscalar vector mesons $\omega$ and $\phi$ [68], allowing for a consistent description of final-state interactions between all three pions. We calculated the Dalitz plot distributions for both decays, comparing to the high-precision experimental determinations of $\phi \to 3\pi$ by KLOE [69] and CMD-2 [70], and predicting the $\omega \to 3\pi$ Dalitz plot about to be analyzed at WASA-at-COSY, KLOE, CLAS, or elsewhere. As a further application of this study, we have used the resulting three-pion decay amplitudes, together with the vector form factor of the pion, as input to predict the vector-meson conversion decays $\omega \to \pi^0\ell^+\ell^-$ and $\phi \to \pi^0\ell^+\ell^-$ [71].

Dispersive methods have also been employed to demonstrate a new, model-independent way to analyze the decays $\eta$, $\eta' \to \pi^+\pi^-\gamma$ [72], which are also driven by the chiral anomaly. This has been applied to experimental data by the WASA-at-COSY [73] and subsequently by the KLOE [74] collaborations.

### 3.2 Chiral coupled-channel analysis of meson–baryon scattering and meson production

Working in a still perturbative setting, we have analyzed meson–baryon scattering lengths in the framework of covariant SU(3) baryon chiral perturbation theory at leading one-loop order [75]. We have computed the complete set of matching relations between the dimension-two low-energy constants in the two- and three-flavor formulations of the theory, and derived new two-flavor low-energy theorems for pion–hyperon and pion–cascade scattering that have already been tested in lattice simulations [76].

As the convergence behavior of three-flavor chiral perturbation theory in the baryon sector is doubtful at best already at threshold, unitarization methods can be used to extend chiral amplitudes to higher energies and investigate properties of resonances [77–79]. This nicely connects to the experimental and partial-wave-analysis resonance studies performed in the projects **A.1**, **A.2**, and **B.1**. We remark, however, that a systematic investigation of such resonance generation has to start by considering the low-lying resonances. In the last funding period, we have made considerable progress within this demanding field:

$S_{11}$ resonances [77]: We have analyzed S-wave pion–nucleon scattering in a unitarized chiral effective Lagrangian including all dimension-two contact terms. We find that both the $S_{11}(1535)$ and the $S_{11}(1650)$ are dynamically generated, see Fig. 8, but the $S_{31}(1620)$ is not. This is a novel result. We have also analyzed the structure of these dynamically generated resonances in terms of their meson–baryon components.

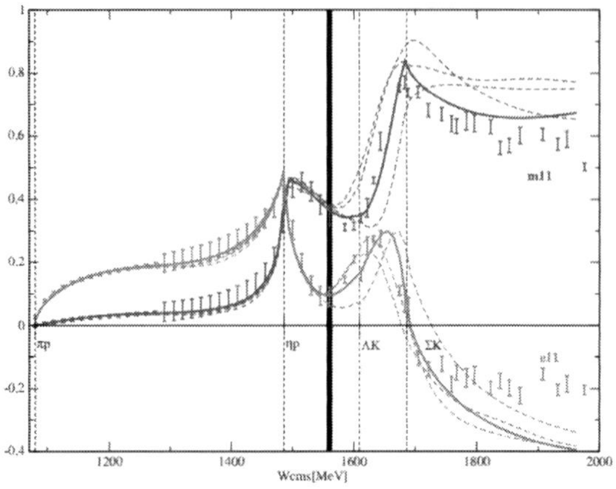


**Figure 8.** Real and imaginary part of the $S_{11}$ partial wave amplitude compared with the SAID-data (WI08-analysis). Full curves correspond to the best fit, the dashed ones to fits with slightly worse $\chi^2_{\text{dof}}$. The bold vertical line limits the region of the fit, where in the non-fit region single energy values are taken from the SAID-data. Figure taken from Ref. [77].

$\eta$ photoproduction [78]: We have analyzed photoproduction of $\eta$ mesons off the proton in a gauge-invariant chiral unitary framework. The interaction kernel for meson–baryon scattering was derived from the leading order chiral effective Lagrangian and iterated in a Bethe–Salpeter equation. The recent precise threshold data from Crystal Ball at MAMI [80] can be described rather well, and the complex pole corresponding to the $S_{11}(1535)$ was extracted. An extension of the kernel to include magnetic couplings was also discussed.

Structure of the $\Lambda(1405)$ [79]: We have performed a combined analysis of antikaon–nucleon scattering cross sections and the recent SIDDHARTA kaonic hydrogen data [81] in the framework of a coupled-channel Bethe–Salpeter approach at next-to-leading order in the chiral expansion of the effective potential. We find a precise description of the antikaon–proton scattering amplitudes and were able to extract accurate values of the scattering lengths, $a_0 = -1.81^{+0.30}_{-0.28} + i0.92^{+0.29}_{-0.23}$ fm, $a_1 = +0.48^{+0.12}_{-0.11} + i0.87^{+0.26}_{-0.20}$ fm. We have also shed new light on the two-pole structure of the $\Lambda(1405)$.

Polarization observables in $\gamma p \to \pi^0\eta p$ [82]: We evaluate the polarization observables $I^S$ and $I^C$ for the reaction $\gamma p \to \pi^0\eta p$, using a chiral unitary framework. These observables have been recently measured for the first time by the CBELSA/TAPS collaboration (see project **A.1**). The theoretical predictions of $I^S$ and $I^C$, given for altogether 18 angle dependent functions, are in good agreement with the measurements; see Fig. 9. Also, the asymmetry $d\Sigma/d\cos(\theta)$ agrees with the data. We show the importance of the $\Delta(1700)D_{33}$ resonance and its S-wave decay into $\eta\Delta(1232)$. The result can be considered as a further confirmation of the dynamical nature of this resonance. At the highest energies, deviations of the predictions from the data start to become noticeable, which leaves room for additional processes and resonances such as a $\Delta(1940)D_{33}$.

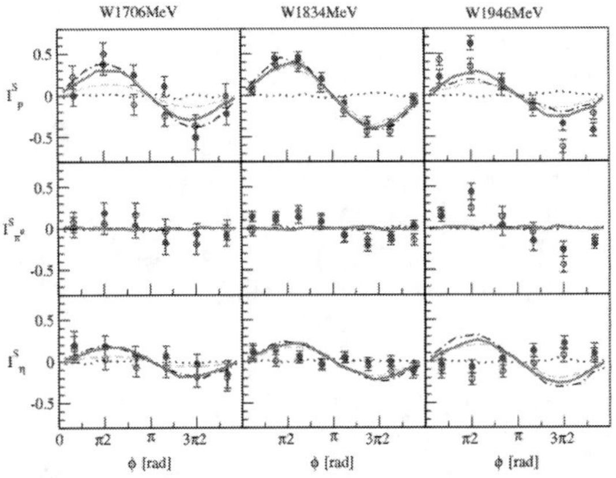

**Figure 9.** Polarization observable $I^S(\phi^*)$ for the three cases of $p$, $\pi^0$, and $\eta$ spectators and for three different energies $W \equiv \sqrt{s}$. The data $I^S(\phi^*)$ are from Ref. [83] (full circles). The empty circles show $-I^S(2\pi - \phi^*)$. (Red) solid lines: Present results, predicted from the model of Refs. [84, 85]. (Black) dotted lines: Without the $\Delta(1700)\eta\Delta$ and $\Delta(1700)K\Sigma(1385)$ couplings predicted from the chiral unitary model. (Green) dashed lines: Without the contributions from the $\pi^0 S_{11}(\eta p)$ final state. (Black) dash-dotted lines: Only contribution from the $\Delta(1700)\eta\Delta$ and $\Delta(1700)K\Sigma(1385)$ tree graphs. Figure taken from Ref. [82].

## 4 Third funding period 2012–2016

### 4.1 Precision studies in light meson decays

In the realm of light-meson physics, most of the work performed in the final funding period has been motivated by hadronic contributions to the anomalous magnetic moment of the muon. In particular, the hadronic virtual corrections denoted by "hadronic light-by-light scattering" (HLbL) have so far only been investigated in model calculations. In view of the upcoming $(g-2)_\mu$ experiments at Fermilab and JPARC, which are expected to improve on the accuracy achieved by the BNL E821 experiment by roughly a factor of four, an improved determination of HLbL contributions with a reliable assessment of the uncertainty is urgently called for. We have therefore suggested to conduct an analysis of at least the lightest, dominant intermediate states using dispersion theory [86–88], thereby reducing the problem to a data-driven driven analysis, much in the same spirit as the relation of hadronic vacuum polarization to $e^+e^- \to$ hadrons cross section measurements. We have therefore investigated several photo–pion and $\eta$, $\eta'$ processes that are important in this respect.

For the reaction $\gamma\gamma \to \pi\pi$, Roy–Steiner equations (see the more detailed discussion in the context of pion–nucleon scattering in Sec. 4.4) have been worked out and solved numerically [89]. The subtraction constants introduced to reduce the influence of high-energy input can be identified with the pion dipole and quadrupole polarizabilities. A sum rule for the isospin $I = 2$ S-wave, together with chiral constraints, produces an improved prediction for the charged-pion quadrupole polarizability $(\alpha_2 - \beta_2)^{\pi^\pm} = (15.3 \pm 3.7) \cdot 10^{-4}\mathrm{fm}^5$. The two-photon coupling of the $\sigma$ resonance was investigated and a correlation with the isospin $I = 0$ pion polarizabilities derived, which, using the chiral-perturbation-theory results for the polarizabilities, led to $\Gamma_{\sigma\gamma\gamma} = (1.7 \pm 0.4)\,\mathrm{keV}$.

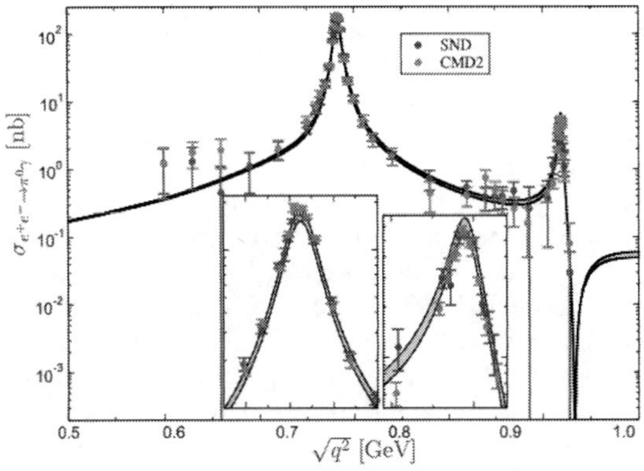

**Figure 10.** $e^+e^- \to \pi^0\gamma$ cross section predicted from $e^+e^- \to 3\pi$, compared to data [92–94]. The inserts zoom in on the $\omega$ and $\phi$ resonance peaks. Figure taken from Ref. [91].

The process $\gamma\pi \to \pi\pi$ is driven by the chiral anomaly, and hence a consequence of one of the most profound predictions of QCD at low energies. In contrast to the decay $\pi^0 \to \gamma\gamma$, similarly anomalous, that has been well tested experimentally, the prediction for $\gamma\pi \to \pi\pi$ has never been verified better than to a level of about 10%. We have therefore suggested a dispersive representation that will allow a much more accurate experimental extraction of the chiral anomaly from Primakoff data as taken by the COMPASS collaboration, as it allows to use not only the (scarce) near-threshold data, but at least the full range in which pion–pion scattering is still elastic to excellent approximation, hence ranging up to energies of about 1 GeV, and therefore including the resonant enhancement by the $\rho$ [90].

Subsequently, we have performed the first dispersive analysis of the $\pi^0$ transition form factor [91], with so far evaluating the singly-virtual one for both timelike and spacelike momenta. For isovector intermediate states, the previously discussed amplitude for $\gamma\pi \to \pi\pi$ serves as a vital input. For the isoscalar part, we have generalized the study of $\omega/\phi \to 3\pi$ [68] to arbitrary virtualities in $e^+e^- \to 3\pi$, fitting the available cross section data. The resulting prediction for $e^+e^- \to \pi^0\gamma$ yields excellent agreement with direct measurements, see Fig. 10. The prediction at (moderate) spacelike momenta will soon be tested by new data from the BESIII collaboration. We can also predict the slope of the singly-virtual transition form factor at vanishing virtuality with unprecedented accuracy, using a sum rule, with the result $a_\pi = (30.7 \pm 0.6) \times 10^{-3}$.

Vector meson conversion decays, such as $\omega \to \pi^0\ell^+\ell^-$ or $\phi \to \eta\ell^+\ell^-$, pose one of the few opportunities to test the *doubly*-virtual transition form factors (of $\pi^0$ and $\eta$, respectively, in the examples given) experimentally with good precision. In particular data on $\omega \to \pi^0\mu^+\mu^-$ for large dimuon invariant masses seems to be in sharp contrast with all theory predictions, as already noted in our own dispersive analysis [71]. As the latter relies on elastic unitarity, we have also investigated general "unitarity bounds" on this form factor, confirming the existence of a disagreement with experimental data near the upper end of the decay region [95]. This disagreement can be further sharpened when using cross section data on $e^+e^- \to \omega\pi^0$ explicitly [96].

The previous dispersive analysis of the $\eta$ (and $\eta'$) singly-virtual transition form factor by the Jülich group [97] builds on the model-independent representation of the decay amplitude $\eta \to \pi\pi\gamma$ [72]. We have improved the latter representation by the inclusion of the leading left-hand-cut contribution, the $a_2(1320)$ tensor resonance (that decays into $\pi\eta$) [98], and shown that some highly interesting aspects of this representation could be tested in a very useful manner by a measurement of the crossed-channel reaction $\gamma\pi \to \pi\eta$—an experiment that is feasible, in a Primakoff reaction, at the COMPASS experiment at CERN. Finally, first steps towards a test of the doubly-virtual $\eta$ transition form factor have been taken, testing the factorization hypothesis in data on $e^+e^- \to \eta\pi\pi$ [99]. A central result is that the left-hand-cut contributions calculated in Ref. [98] are a vital input to allow for (approximate) factorization, which would have to be broken badly without it.

### 4.2 Antikaon–nucleon scattering revisited

We have re-analyzed meson–baryon scattering amplitude in the strangeness $S = -1$ sector in Ref. [100] because of the new SIDDHARTA data and similar NLO work by other groups [101, 102]. The free parameters of our approach, the low-energy constants as well as the regularization scales $\mu$ are adjusted to reproduce *all* known experimental data in the meson–baryon sector. The main bulk of this data consists of the cross sections for the processes $K^-p \to K^-p$, $K^-p \to \bar{K}^0n$, $K^-p \to \pi^0\Lambda$, $K^-p \to \pi^+\Sigma^-$, $K^-p \to \pi^0\Sigma^0$, and $K^-p \to \pi^-\Sigma^+$ for laboratory momentum $P_{\rm lab} < 300\,$MeV. Electromagnetic effects are not included in the analysis and assumed to be negligible at the measured values of $P_{\rm lab}$. Additionally, at the antikaon–nucleon threshold, the decay ratios $\Gamma(K^-p \to \pi^+\Sigma^-)/\Gamma(K^-p \to \pi^-\Sigma^+) = 2.38 \pm 0.04$, $\Gamma(K^-p \to \pi^0\Lambda)/\Gamma(K^-p \to \text{neutral}) = 0.189 \pm 0.015$, and $\Gamma(K^-p \to \pi^\pm\Sigma^\mp)/\Gamma(K^-p \to \text{inelastic}) = 0.664 \pm 0.011$, as well as the energy shift and width of kaonic hydrogen in the 1s state, i.e. $\Delta E - i\Gamma/2 = (283 \pm 42) - i(271 \pm 55)$ eV from the SIDDHARTA experiment at DAΦNE, were considered. The fit procedure was performed in two steps: First, for randomly chosen starting values of the free parameters (in a natural range) the fit was performed to all threshold values and the cross section data at a few momenta $P_{\rm lab} < 300$ MeV. Repeating this procedure several thousand times, we ended with several dozen of parameter sets that describe the data equally well. For each of these sets the amplitudes were analytically continued to the positive and negative complex plane. Thereafter, every unphysical solution, e.g. poles on the first Riemann sheet for ${\rm Im}(W) < 200\,$MeV, was sorted out. The remaining sets were used in the second step as starting point of the fit procedure, including all threshold and cross section data

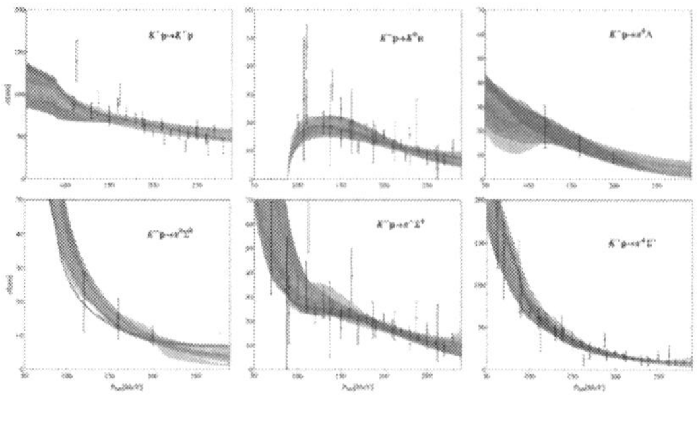

**Figure 11.** Fit results compared to the experimental data of $K^-$-proton scattering. Different colors correspond to the eight best solutions, while the bands represent the $1\sigma$ uncertainty due to errors of the fit parameters. Figure taken from Ref. [100].

points. Eight best solutions were obtained by this procedure, see Table 1, whereas the next best $\chi^2_{\rm d.o.f.}$ are at least one order of magnitude larger. Although the fit results look very promising, see Fig. 11, we would like to point out that there are quite a few free parameters in the model. The latter are assumed to be of natural size, but not restricted otherwise. When continued analytically to the complex $W$ plane, all eight solutions confirm the double pole structure of the $\Lambda(1405)$. Due to the SIDDHARTA data restricting the amplitude close to the the $\bar{K}N$ threshold quite strongly, we observe stability of the position of the narrow pole. The position of the second pole is, on the other hand, less restricted. Therefore, further data are needed to resolve this ambiguity.

We have further presented a comparative analysis of the various chiral unitary SU(3) approaches to (anti)kaon–nucleon scattering [103]. It was demonstrated that the approaches lead to very different predictions for the $K^-p$ amplitude extrapolated to subthreshold energies as well as for the $K^-n$ amplitude. The origin of the poles generated by the models is traced to the so-called zero-coupling limit, in which the inter-channel couplings are switched off. This provides new insights into the pole contents of the various approaches.

### 4.3 The $\Lambda(1405)$ in kaon photoproduction

Recently, very sophisticated measurements of the reaction $\gamma p \to K^+\Sigma\pi$ were performed by the CLAS collaboration at JLAB [104]. There, the invariant mass distribution of all three $\pi\Sigma$ channels was determined in a broad energy range and with high resolution. Finally, from these data the spin–parity analysis of the $\Lambda(1405)$ was performed in Ref. [105]. There, for the first time, the quantum numbers $J^P = 1/2^-$ were deduced from an experimental measurement directly. As a coupled-channel gauge invariant chiral unitary model is very complicated, we have assumed the simplest ansatz for the photoproduction amplitude to analyze the CLAS data [100]

$$\mathcal{M}^j(\tilde{W}, M_{\rm inv}) = \sum_{i=1}^{10} C^i(\tilde{W})\, G^i(M_{\rm inv})\, f_{0+}^{i,j}(M_{\rm inv})\,,$$

DOI: 10.1051/epjconf/201713403002

**Table 1.** Quality of the various fits in the description of the hadronic and the photoproduction data from CLAS. For the precise definition of $\chi^2_{\rm p.p.}$, see Ref. [100].

| Fit # | 1 | 2 | 3 | 4 | 5 | 6 | 7 | 8 |
|---|---|---|---|---|---|---|---|---|
| $\chi^2_{\rm d.o.f.}$ (hadronic data) | 1.35 | 1.14 | 0.99 | 0.96 | 1.06 | 1.02 | 1.15 | 0.90 |
| $\chi^2_{\rm p.p.}$ (CLAS data) | 3.18 | 1.94 | 2.56 | 1.77 | 1.90 | 6.11 | 2.93 | 3.14 |

where $\tilde{W}$ and $M_{\rm inv}$ denote the total energy of the system and the invariant mass of the $\pi\Sigma$ subsystem, respectively. For a specific meson–baryon channel $i$, the energy-dependent (and in general complex valued) constants $C^i(\tilde{W})$ describe the reaction mechanism of $\gamma p \to K^+ M_i B_i$, whereas the final-state interaction is captured by the standard Höhler partial waves $f_{0+}$. For a specific meson–baryon channel $i$, the Green's function is denoted by $G^i(M_{\rm inv})$. The only new parameters of the photoproduction amplitude are the constants $C^i(\tilde{W})$ which, however, are quite numerous (10 for each $\tilde{W}$). These parameters are adjusted to reproduce the invariant mass distribution $d\sigma/dM_{\rm inv}(M_{\rm inv})$ for the final $\pi^+\Sigma^-$, $\pi^0\Sigma^0$, and $\pi^-\Sigma^+$ states and for all 9 measured total energy values $\tilde{W} = 2.0, \ldots, 2.8\,{\rm GeV}$. The achieved quality of the photoproduction fits is listed in the third row of Table 1, where the $\chi^2_{\rm d.o.f.}$ of the hadronic part are stated in the second row. Note that for the comparison of the photoproduction fits the quantity $\chi^2_{\rm d.o.f.}$ is not meaningful due to the large number of generic parameters $C_i(\tilde{W})$. Therefore, we compare the total $\chi^2$ divided by the total number of data points for all three $\pi\Sigma$ final states, denoted by $\chi^2_{\rm p.p.}$. It turns out that even within such a simple and flexible photoproduction amplitude, only the solutions #2, #4, and #5 of the eight hadronic solutions allow for a decent description of the CLAS data. The best solution is indeed #4, which we display in Fig. 12. Incidentally, it also has the lowest $\chi^2_{\rm d.o.f.}$ for the hadronic part. This solution also gives an excellent description of the $\Sigma\pi\pi$ mass distribution from Ref. [106]. With respect to these data, solution #2 is also satisfactory but #5 is not. Therefore, the photoproduction data combined with the scattering and the SIDDHARTA data lead to a sizable reduction in the ambiguity of the second pole of the $\Lambda(1405)$. To be precise, the location of the two poles in these surviving solutions is given by (all energies in MeV)

$$\text{sol. \#2, pole 1}: 1434^{+2}_{-2} - i\,10^{+2}_{-1}, \text{pole 2}: 1330^{+4}_{-5} \;- i\,56^{+17}_{-11},$$

$$\text{sol. \#4, pole 1}: 1429^{+8}_{-7} - i\,12^{+2}_{-3}, \text{pole 2}: 1325^{+15}_{-15} - i\,90^{+12}_{-18}.$$

We conclude that the inclusion of the CLAS data as experimental input can serve as a new important constraint on the antikaon-nucleon scattering amplitude. However, for future studies a theoretically more robust model for the two-meson photoproduction amplitude is required. Finally, note that the two-pole structure of the $\Lambda(1405)$ is now also discussed in the Particle Data Group [107].

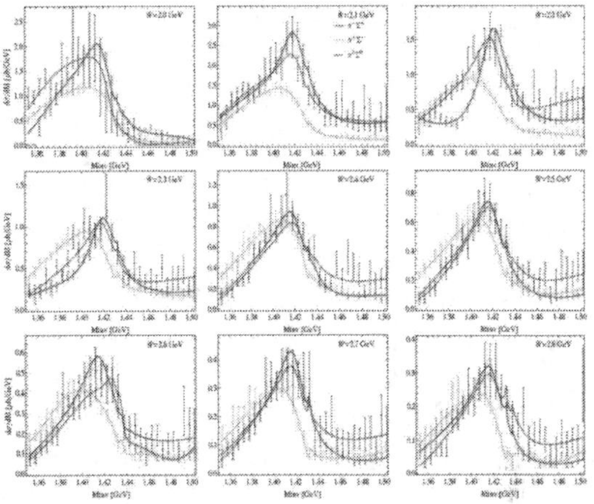

**Figure 12.** Result of the fits to the CLAS data in all three channels $\pi^+\Sigma^-$ (green), $\pi^-\Sigma^+$ (red) and $\pi^0\Sigma^0$ (blue). Correspondingly, green (dashed), red (full) and blue (dotted) lines represent the outcome of the model for the solution #4 in the $\pi^+\Sigma^-$, $\pi^-\Sigma^+$ $\pi^0\Sigma^0$ channels, respectively. Figure taken from Ref. [100].

### 4.4 Pion–nucleon scattering: from scattering lengths to Roy–Steiner equations

In a series of publications [108–111], we have made substantial progress in understanding and determining the (near-)threshold properties of pion–nucleon scattering. In a first step, the complete isospin-breaking corrections, both induced by the light quark-mass difference and electromagnetic effects, were calculated for the pion–nucleon scattering lengths in all physical channels, using covariant baryon chiral perturbation theory [108]. These results were then extended slightly above threshold, so that the energy dependence of the so-called triangle relation (between the three physical channels accessible in charged-pion–proton scattering) could be investigated [109]. Subsequently, we have calculated $\pi^-$ deuteron scattering at threshold to percent-level accuracy, including isospin-violating corrections both in the two- and three-body sector [110, 111]. This was used to extract the isoscalar and isovector pion-nucleon scattering lengths from data on pionic hydrogen [112] and pionic deuterium [113] atoms, leading to the results $a^+ = (7.6 \pm 3.1) \times 10^{-3} M_\pi^{-1}$ and $a^- = (86.1 \pm 0.9) \times 10^{-3} M_\pi^{-1}$, thus indicating a non-vanishing (and positive) $a^+$ at $2\sigma$ significance; see Fig. 13. Finally, we have discussed the validity of the Goldberger–Miyazawa–Oehme sum rule in the presence of isospin vio-

DOI: 10.1051/epjconf/201713403002

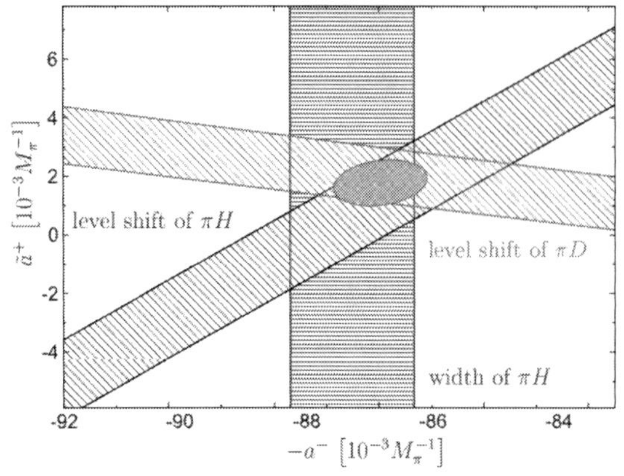


**Figure 13.** Combined constraints on $\tilde{a}^+$ and $a^-$ from data on the width and level shift of $\pi H$, as well as the $\pi D$ level shift. The figure is an updated version of the ones from Refs. [110, 111], taken from Ref. [114], to account for the new value of the $\pi H$ level shift [115].

lation, and used it to determine the charged-pion–nucleon coupling constant, with the result $g_c^2/4\pi = 13.69 \pm 0.20$.

Perhaps the most ambitious single project within **B.3** is the derivation and evaluation of Roy–Steiner equations for pion–nucleon scattering. These are partial-wave dispersion relations, based on hyperbolic dispersion relations and making maximal use of analyticity, unitarity, and crossing symmetry. We have derived a closed system of Roy–Steiner equations [116], where all kernel functions and unitarity relations required for the lowest partial waves have been worked out analytically. In order to suppress the dependence on the high-energy regime, also once- and twice-subtracted versions of these equations were considered, where the subtraction constants were identified with subthreshold parameters. Assuming Mandelstam analyticity we have determined the maximal range of validity of these equations. As a first step towards the solution of the full system, we have cast the equations for the $\pi\pi \to \bar{N}N$ partial waves into the form of a Muskhelishvili–Omnès problem with finite matching point, which were solved numerically in the single-channel approximation. Some consequences for the spectral functions of the nucleon electromagnetic form factors were also discussed. As an application, we have solved a system of coupled integral equations for the $\pi\pi \to \bar{N}N$ and $\bar{K}K \to \bar{N}N$ ($t$-channel) S-waves in order to update the dispersive analysis of the correction $\Delta_\sigma = \sigma(2M_\pi^2) - \sigma_{\pi N}$ in the nucleon scalar form factor, which is needed for the extraction of the pion–nucleon $\sigma$-term from $\pi N$ scattering [117].

The full, self-consistent solution of the Roy–Steiner system of $s$- and $t$-channel partial waves was documented in detail in Ref. [114]. The result of this analysis is a determination of $s$- and $t$-channel S- and P-wave amplitudes at low energies, as well as a consistent set of the subthreshold parameters that act as the subtraction constants of the dispersion relations. S- and P-waves above the matching point of $\sqrt{s} = 1.38\,\text{GeV}$ are taken as input from existing partial-wave analyses, mainly by the GWU/SAID group [118, 119], but also from Karlsruhe–Helsinki [120, 121]; both also provide necessary input for the inelasticities, as well as for all higher partial waves. Finally, the S-wave scattering lengths extracted from hadronic atoms [110, 111] serve as a constraint to the solution.

We regard our work as an update of the rigorous, dispersion-theory-based Karlsruhe–Helsinki analysis, based on modern data input. We have checked throughout that reverting to the input used in that analysis, we reproduce Karlsruhe–Helsinki results, e.g. for the $\sigma$-term, or for the pion–nucleon coupling constant. This is a crucial consistency check of our numerical solution.

One of the most important tasks is a reliable assessment of the errors propagating through the Roy–Steiner solution. We find that for the $s$-channel partial waves at very low energies, the uncertainty is dominated by those in the S-wave scattering lengths (see Fig. 13), while the uncertainties at the matching point dominate there, at "large" energies (for the range of validity of the equations). In the intermediate regime, flat fit minima of the altogether ten subtraction constants that show significant correlations dominate the final error. Resulting $s$-channel S- and P-wave phase shifts including the final error bands are shown in Fig. 14. Similar results for the final $t$-channel solution are given in Ref. [114].

One of the central results is a determination of the pion–nucleon $\sigma$-term [122]. It is given in terms of two subthreshold parameters obtained as solution of the Roy–Steiner equations, through the Cheng–Dashen low-energy theorem, for which we have improved on the isospin-breaking corrections. The result is

$$\sigma_{\pi N} = (59.1 \pm 1.9 \pm 3.0)\,\text{MeV} = (59.1 \pm 3.5)\,\text{MeV}\,, \quad (1)$$

where the first error refers to uncertainties from the Roy–Steiner solution, and the second from those in the low-energy theorem (mainly due to isospin breaking). The fact that the value in Eq. (1) is significantly larger than the "canonical" value of 45 MeV obtained by Gasser, Leutwyler, and Sainio [123], based on the Karlsruhe–Helsinki solution, has caused some controversy; not the least because recent lattice simulations, performed at physical pion masses, seem to confirm the lower value. We have shown that the large $\sigma$-term is a direct consequence of the scattering length input, such that there is actually a conflict between lattice results and hadronic atoms spectroscopy [124]. We have suggested that a lattice determination of the $\pi N$ scattering lengths may be a step towards an understanding of this puzzle.

Finally, we have matched the Roy–Steiner results for the subthreshold parameters of pion–nucleon scattering to chiral perturbation theory up to next-to-next-to-next-to-leading order, to extract the pertinent low-energy constants [125]. The convergence of the chiral series is expected to be best at the subthreshold point, far away from all cuts and threshold singularities, such that the resulting LECs should be more reliable than those extracted from fits in the physical region. Also here, we included a com-

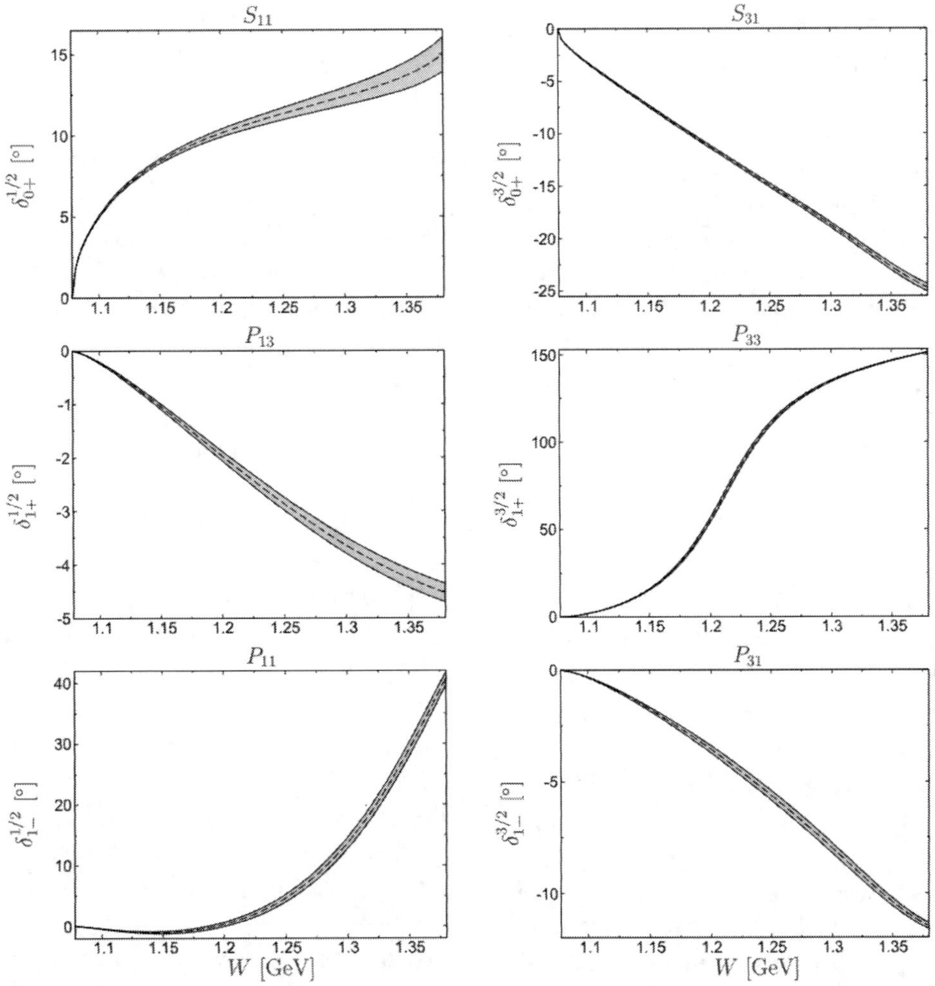


**Figure 14.** Final errors bands for the $\pi N$ phase shifts. The dashed lines indicate the central curves. Figure taken from Ref. [114].

prehensive analysis of systematic uncertainties and correlations. The convergence of the chiral series, studied by investigating the chiral expansion of threshold parameters up to the same order, turns out to be more problematic than expected, an effect which probably can only partly be explained by the role of the Δ resonance. Results for the low-energy constants are also presented in the counting scheme usually applied in chiral nuclear effective field theory, where they serve as crucial input to determine the long-range part of the nucleon–nucleon potential as well as three-nucleon forces.

**Acknowledgements** We thank our collaborators B. Ananthanarayan, Vadim Baru, Véronique Bernard, Moritz Bissegger, Bugra Borasoy, Peter Bruns, Irinel Caprini, Gilberto Colangelo, Ales Cieply, Tobias Dato, Christoph Ditsche, Michael Döring, Evgeny Epelbaum, Matthias Frink, Andreas Fuhrer, Jürg Gasser, Feng-Kun Guo, Hans Werner Hammer, Christoph Hanhart, Johann Haidenbauer, Martin Hoferichter, Hermann Krebs, Siegfried Krewald, Ayse Kucukarslan, Andrzej Kupść, Alexander Kudryavtsev, André Lacour, Timo Lähde, Dean Lee, Vadim Lensky, Randy Lewis, Stefan Leupold, Valery Lyubovitskij, Maxim Mai, Eike Müller, Franz Niecknig, Robin Nißler, Andreas Nogga, Konstantin Ottnad, Eulogio Oset, José Palomar, Nello Paver, José R. Peláez, Daniel Phillips, Lucas Platter, Judith Plenter, Massimiliano Procura, Udit Raha, Luis Roca, Dino Ruić, Jacobo Ruiz de Elvira, Akaki Rusetsky, Dimitrios Sakkas, Rocco Schiavilla, Martin Schmid, Rebekka Schmidt, Sebastian Schneider, Alexander Sibirtsev, Johan Smejkal, Peter Stoffer, Felix Stollenwerk, Anthony Thomas, Michaela Verbeni, Michele Viviani, Andreas Wirzba, and Chu-Wen Xiao. The work reported here would not have been possible without the support from the Deutsche Forschungsgemeinschaft within the SFB/TR16.

## References

[1] V. Bernard, B. Kubis and U.-G. Meißner, Eur. Phys. J. A **25** (2005) 419 [arXiv:nucl-th/0506023].

[2] A. Schmidt *et al.*, Phys. Rev. Lett. **87** (2001) 232501 [Erratum-ibid. **110** (2013) 039903] [nucl-ex/0105010].

[3] J. Gasser, B. Kubis, N. Paver and M. Verbeni, Eur. Phys. J. C **40** (2005) 205 [arXiv:hep-ph/0412130].

[4] E. H. Müller, B. Kubis and U.-G. Meißner, Eur. Phys. J. C **48** (2006) 427 [arXiv:hep-ph/0607151].

[5] B. Kubis, E. H. Müller, J. Gasser and M. Schmid, Eur. Phys. J. C **50** (2007) 557 [arXiv:hep-ph/0611366].

[6] G. Colangelo, J. Gasser, B. Kubis and A. Rusetsky, Phys. Lett. B **638** (2006) 187 [arXiv:hep-ph/0604084].

[7] U.-G. Meißner, G. Müller and S. Steininger, Phys. Lett. B **406** (1997) 154 [Erratum-ibid. B **407** (1997) 454] [hep-ph/9704377].

[8] T. A. Lähde and U.-G. Meißner, Phys. Rev. D **74** (2006) 034021 [arXiv:hep-ph/0606133].

[9] C. Hanhart, B. Kubis and J. R. Peláez, Phys. Rev. D (2007) in print [arXiv:0707.0262 [hep-ph]].

[10] A. Kucukarslan and U.-G. Meißner, Mod. Phys. Lett. A **21** (2006) 1423 [arXiv:hep-ph/0603061].

[11] B. Kubis and R. Lewis, Phys. Rev. C **74** (2006) 015204 [nucl-th/0605006].

[12] M. Frink and U.-G. Meißner, Eur. Phys. J. A **29** (2006) 255 [arXiv:hep-ph/0609256].

[13] J. A. Oller, M. Verbeni and J. Prades, hep-ph/0701096.

[14] M. Frink and U.-G. Meißner, JHEP **0407**, 028 (2004) [arXiv:hep-lat/0404018].

[15] A. Lacour, B. Kubis, and U.-G. Meißner, JHEP **0710** (2007) 083 [arXiv:0708.3957 [hep-ph]].

[16] V. Bernard and U.-G. Meißner, Ann. Rev. Nucl. Part. Sci. **57** (2007) 33 [arXiv:hep-ph/0611231].

[17] U.-G. Meißner, U. Raha and A. Rusetsky, Eur. Phys. J. C **35** (2004) 349 [arXiv:hep-ph/0402261].

[18] G. Beer *et al.* [DEAR Collaboration], Phys. Rev. Lett. **94** (2005) 212302.

[19] B. Borasoy, U.-G. Meißner and R. Nißler, Phys. Rev. C **74** (2006) 055201 [arXiv:hep-ph/0606108].

[20] U.-G. Meißner, U. Raha and A. Rusetsky, Eur. Phys. J. C **47** (2006) 473 [arXiv:nucl-th/0603029].

[21] L. Roca, C. Hanhart, E. Oset and U.-G. Meißner, Eur. Phys. J. A **27** (2006) 373 [arXiv:nucl-th/0602016].

[22] B. Borasoy, P. C. Bruns, U.-G. Meißner and R. Nißler, Phys. Rev. C **72** (2005) 065201 [arXiv:hep-ph/0508307].

[23] B. Borasoy, P. C. Bruns, U.-G. Meißner and R. Nißler, Eur. Phys. J. A **34** (2007) 161 [arXiv:0709.3181 [nucl-th]].

[24] A. Sibirtsev, J. Haidenbauer, S. Krewald, U.-G. Meißner and A. W. Thomas, Eur. Phys. J. A **31** (2007) 221 [arXiv:hep-ph/0509145].

[25] A. Sibirtsev, U.-G. Meißner and A. W. Thomas, Phys. Rev. D **71** (2005) 094011 [arXiv:hep-ph/0503276].

[26] A. Sibirtsev, J. Haidenbauer and U.-G. Meißner, Eur. Phys. J. A **27** (2006) 263 [arXiv:nucl-th/0512055].

[27] A. Sibirtsev, H. W. Hammer, U.-G. Meißner and A. W. Thomas, Eur. Phys. J. A **29** (2006) 209 [arXiv:nucl-th/0606044].

[28] B. Borasoy and R. Nißler, Eur. Phys. J. A **26** (2005) 383 [arXiv:hep-ph/0510384].

[29] B. Borasoy, U.-G. Meißner and R. Nißler, Phys. Lett. B **643** (2006) 41 [arXiv:hep-ph/0609010].

[30] B. Borasoy and E. Lipartia, Phys. Rev. D **71** (2005) 014027 [arXiv:hep-ph/0410141].

[31] E. Epelbaum and U.-G. Meißner, Phys. Rev. C **72** (2005) 044001 [arXiv:nucl-th/0502052].

[32] E. Epelbaum, U.-G. Meißner and J. E. Palomar, Phys. Rev. C **71** (2005) 024001 [arXiv:nucl-th/0407037].

[33] H. Krebs, E. Epelbaum and U.-G. Meißner, Eur. Phys. J. A **32** (2007) 127-137 [arXiv:nucl-th/0703087].

[34] E. Epelbaum and U.-G. Meißner, Few Body Syst. **54** (2013) 2175 [nucl-th/0609037].

[35] E. Epelbaum, H. W. Hammer, U.-G. Meißner and A. Nogga, Eur. Phys. J. C **48** (2006) 169 [arXiv:hep-ph/0602225].

[36] H. W. Hammer, D. R. Phillips and L. Platter, Eur. Phys. J. A **32** (2007) 335 [arXiv:0704.3726 [nucl-th]].

[37] B. Borasoy, H. Krebs, D. Lee and U.-G. Meißner, Nucl. Phys. A **768** (2006) 179 [arXiv:nucl-th/0510047].

[38] B. Borasoy, E. Epelbaum, H. Krebs, D. Lee and U.-G. Meißner, Eur. Phys. J. A **31** (2007) 105 [arXiv:nucl-th/0611087].

[39] L. Platter and H. W. Hammer, Nucl. Phys. A **766** (2006) 132 [arXiv:nucl-th/0509045].

[40] H. W. Hammer and L. Platter, Eur. Phys. J. A **32** (2007) 113 [arXiv:nucl-th/0610105].

[41] M. Viviani *et al.*, Phys Rev. Lett. **99** (2007) 112002 [arXiv:nucl-th/0703051].

[42] V. Lensky, V. Baru, J. Haidenbauer, C. Hanhart, A. E. Kudryavtsev and U.-G. Meißner, Eur. Phys. J. A **26** (2005) 107 [arXiv:nucl-th/0505039].

[43] V. Lensky, V. Baru, E. Epelbaum, C. Hanhart, J. Haidenbauer, A. Kudryavtsev and U.-G. Meißner, Eur. Phys. J. A (2007) in print [arXiv:0704.0443 [nucl-th]].

[44] B. Kubis, EPJ Web Conf. **3** (2010) 01008 [arXiv:0912.3440 [hep-ph]].

[45] J. R. Batley *et al.* [NA48/2 Collaboration], Phys. Lett. B **633** (2006) 173 [arXiv:hep-ex/0511056].

[46] M. Bissegger, A. Fuhrer, J. Gasser, B. Kubis and A. Rusetsky, Phys. Lett. B **659** (2008) 576 [arXiv:0710.4456 [hep-ph]].

[47] J. Gasser, V. E. Lyubovitskij and A. Rusetsky, Phys. Rept. **456** (2008) 167 [arXiv:0711.3522 [hep-ph]].

[48] J. R. Batley *et al.* [Na48/2 Collaboration], Eur. Phys. J. C **64** (2009) 589 [arXiv:0912.2165 [hep-ex]].

[49] M. Bissegger, A. Fuhrer, J. Gasser, B. Kubis and A. Rusetsky, Nucl. Phys. B **806** (2009) 178 [arXiv:0807.0515 [hep-ph]].

[50] B. Kubis and R. Schmidt, Eur. Phys. J. C **70** (2010) 219 [arXiv:1007.1887 [hep-ph]].

[51] J. Gasser, B. Kubis and A. Rusetsky, Nucl. Phys. B **850** (2011) 96 [arXiv:1103.4273 [hep-ph]].

[52] B. Kubis and S. P. Schneider, Eur. Phys. J. C **62** (2009) 511 [arXiv:0904.1320 [hep-ph]].

[53] G. Colangelo, J. Gasser and A. Rusetsky, Eur. Phys. J. C **59** (2009) 777 [arXiv:0811.0775 [hep-ph]].

[54] J. R. Batley *et al.* [NA48-2 Collaboration], Eur. Phys. J. C **70** (2010) 635.

[55] G. Colangelo, J. Gasser and H. Leutwyler, Phys. Lett. B **488** (2000) 261 [hep-ph/0007112].

[56] C. Ditsche, B. Kubis and U.-G. Meißner, Eur. Phys. J. C **60** (2009) 83 [arXiv:0812.0344 [hep-ph]].

[57] M. Unverzagt *et al.* [CB-at-MAMI, TAPS, and A2 Collaborations], Eur. Phys. J. A **39** (2009) 169 [arXiv:0812.3324 [hep-ex]].

[58] C. Adolph *et al.* [WASA-at-COSY Collaboration], Phys. Lett. B **677** (2009) 24 [arXiv:0811.2763 [nucl-ex]].

[59] S. Prakhov *et al.* [Crystal Ball at MAMI and A2 Collaborations], Phys. Rev. C **79** (2009) 035204 [arXiv:0812.1999 [hep-ex]].

[60] F. Ambrosino *et al.* [KLOE Collaboration], Phys. Lett. B **694** (2011) 16 [arXiv:1004.1319 [hep-ex]].

[61] J. Bijnens and K. Ghorbani, JHEP **0711** (2007) 030 [arXiv:0709.0230 [hep-ph]].

[62] S. P. Schneider, B. Kubis and C. Ditsche, JHEP **1102** (2011) 028 [arXiv:1010.3946 [hep-ph]].

[63] K. Nakamura *et al.* [Particle Data Group Collaboration], J. Phys. G **37** (2010) 075021.

[64] F. Ambrosino *et al.* [KLOE Collaboration], JHEP **0805** (2008) 006 [arXiv:0801.2642 [hep-ex]].

[65] F.-K. Guo, B. Kubis and A. Wirzba, Phys. Rev. D **85** (2012) 014014 [arXiv:1111.5949 [hep-ph]].

[66] M. Ablikim *et al.* [BESIII Collaboration], Phys. Rev. Lett. **112** (2014) 251801 [Addendum: ibid. **113** (2014) 039903] [arXiv:1404.0096 [hep-ex]].

[67] K. Ottnad, B. Kubis, U.-G. Meißner and F.-K. Guo, Phys. Lett. B **687** (2010) 42 [arXiv:0911.3981 [hep-ph]].

[68] F. Niecknig, B. Kubis and S. P. Schneider, Eur. Phys. J. C **72** (2012) 2014 [arXiv:1203.2501 [hep-ph]].

[69] A. Aloisio *et al.* [KLOE Collaboration], Phys. Lett. B **561** (2003) 55 [Erratum-ibid. B **609** (2005) 449] [hep-ex/0303016].

[70] R. R. Akhmetshin *et al.*, Phys. Lett. B **642** (2006) 203.

[71] S. P. Schneider, B. Kubis and F. Niecknig, Phys. Rev. D **86** (2012) 054013 [arXiv:1206.3098 [hep-ph]].

[72] F. Stollenwerk, C. Hanhart, A. Kupść, U.-G. Meißner and A. Wirzba, Phys. Lett. B **707** (2012) 184 [arXiv:1108.2419 [nucl-th]].

[73] P. Adlarson *et al.* [WASA-at-COSY Collaboration], Phys. Lett. B **707** (2012) 243 [arXiv:1107.5277 [nucl-ex]].

[74] D. Babusci *et al.* [KLOE Collaboration], Phys. Lett. B **718** (2013) 910 [arXiv:1209.4611 [hep-ex]].

[75] M. Mai, P. C. Bruns, B. Kubis and U.-G. Meißner, Phys. Rev. D **80** (2009) 094006 [arXiv:0905.2810 [hep-ph]].

[76] A. Torok, S. R. Beane, W. Detmold, T. C. Luu, K. Orginos, A. Parreno, M. J. Savage and A. Walker-Loud, Phys. Rev. D **81** (2010) 074506 [arXiv:0907.1913 [hep-lat]].

[77] P. C. Bruns, M. Mai and U.-G. Meißner, Phys. Lett. B **697** (2011) 254 [arXiv:1012.2233 [nucl-th]].

[78] D. Ruić, M. Mai and U.-G. Meißner, Phys. Lett. B **704** (2011) 659 [arXiv:1108.4825 [nucl-th]].

[79] M. Mai and U.-G. Meißner, Nucl. Phys. A **900** (2013) 51 [arXiv:1202.2030 [nucl-th]].

[80] E. F. McNicoll *et al.* [Crystal Ball at MAMI Collaboration], Phys. Rev. C **82** (2010) 035208 [Erratum-ibid.

C **84** (2011) 029901] [arXiv:1007.0777 [nucl-ex]].

[81] M. Bazzi *et al.*, Phys. Lett. B **704** (2011) 113 [arXiv:1105.3090 [nucl-ex]].

[82] M. Döring, E. Oset and U.-G. Meißner, Eur. Phys. J. A **46** (2010) 315 [arXiv:1003.0097 [nucl-th]].

[83] E. Gutz *et al.* [CBELSA/TAPS Collaboration], Phys. Lett. B **687**, 11 (2010) [arXiv:0912.2632 [nucl-ex]].

[84] M. Döring, E. Oset and D. Strottman, Phys. Rev. C **73**, 045209 (2006) [arXiv:nucl-th/0510015].

[85] M. Döring, E. Oset and D. Strottman, Phys. Lett. B **639**, 59 (2006) [arXiv:nucl-th/0602055].

[86] G. Colangelo, M. Hoferichter, M. Procura and P. Stoffer, JHEP **1409** (2014) 091 [arXiv:1402.7081 [hep-ph]].

[87] G. Colangelo, M. Hoferichter, B. Kubis, M. Procura and P. Stoffer, Phys. Lett. B **738** (2014) 6 [arXiv:1408.2517 [hep-ph]].

[88] G. Colangelo, M. Hoferichter, M. Procura and P. Stoffer, JHEP **1509** (2015) 074 [arXiv:1506.01386 [hep-ph]].

[89] M. Hoferichter, D. R. Phillips and C. Schat, Eur. Phys. J. C **71** (2011) 1743 [arXiv:1106.4147 [hep-ph]].

[90] M. Hoferichter, B. Kubis and D. Sakkas, Phys. Rev. D **86** (2012) 116009 [arXiv:1210.6793 [hep-ph]].

[91] M. Hoferichter, B. Kubis, S. Leupold, F. Niecknig and S. P. Schneider, Eur. Phys. J. C **74** (2014) 3180 [arXiv:1410.4691 [hep-ph]].

[92] M. N. Achasov *et al.*, Eur. Phys. J. C **12** (2000) 25.

[93] M. N. Achasov *et al.*, Phys. Lett. B **559** (2003) 171 [hep-ex/0302004].

[94] R. R. Akhmetshin *et al.* [CMD2 Collaboration], Phys. Lett. B **605** (2005) 26 [hep-ex/0409030].

[95] B. Ananthanarayan, I. Caprini and B. Kubis, Eur. Phys. J. C **74** (2014) 3209 [arXiv:1410.6276 [hep-ph]].

[96] I. Caprini, Phys. Rev. D **92** (2015) 014014 [arXiv:1505.05282 [hep-ph]].

[97] C. Hanhart, A. Kupść, U.-G. Meißner, F. Stollenwerk and A. Wirzba, Eur. Phys. J. C **73** (2013) 2668 [Erratum-ibid. C **75** (2015) 242] [arXiv:1307.5654 [hep-ph]].

[98] B. Kubis and J. Plenter, Eur. Phys. J. C **75** (2015) 283 [arXiv:1504.02588 [hep-ph]].

[99] C. W. Xiao, T. Dato, C. Hanhart, B. Kubis, U.-G. Meißner and A. Wirzba, arXiv:1509.02194 [hep-ph].

[100] M. Mai and U.-G. Meißner, Eur. Phys. J. A **51** (2015) 3, 30 [arXiv:1411.7884 [hep-ph]].

[101] Y. Ikeda, T. Hyodo and W. Weise, Phys. Lett. B **706** (2011) 63 [arXiv:1109.3005 [nucl-th]].

[102] Z. H. Guo and J. A. Oller, Phys. Rev. C **87** (2013) 035202 [arXiv:1210.3485 [hep-ph]].

[103] A. Ciepý, M. Mai, U.-G. Meißner and J. Smejkal, Nucl. Phys. A **954** (2016) 17 [arXiv:1603.02531 [hep-ph]].

[104] K. Moriya *et al.* [CLAS Collaboration], Phys. Rev. C **87** (2013) 035206 [arXiv:1301.5000 [nucl-ex]].

[105] K. Moriya *et al.* [CLAS Collaboration], Phys. Rev. Lett. **112** (2014) 082004 [arXiv:1402.2296 [hep-ex]].

[106] R. J. Hemingway, Nucl. Phys. B **253** (1985) 742.

[107] U.-G. Meißner, T. Hyodo, http://pdg.lbl.gov/2015/reviews/rpp2015-rev-lam-1405-pole-struct.pdf.

[108] M. Hoferichter, B. Kubis and U.-G. Meißner, Phys. Lett. B **678** (2009) 65 [arXiv:0903.3890 [hep-ph]].

[109] M. Hoferichter, B. Kubis and U.-G. Meißner, Nucl. Phys. A **833** (2010) 18 [arXiv:0909.4390 [hep-ph]].

[110] V. Baru, C. Hanhart, M. Hoferichter, B. Kubis, A. Nogga and D. R. Phillips, Phys. Lett. B **694** (2011) 473 [arXiv:1003.4444 [nucl-th]].

[111] V. Baru, C. Hanhart, M. Hoferichter, B. Kubis, A. Nogga and D. R. Phillips, Nucl. Phys. A **872** (2011) 69 [arXiv:1107.5509 [nucl-th]].

[112] D. Gotta *et al.*, Lect. Notes Phys. **745** (2008) 165.

[113] T. Strauch *et al.*, Eur. Phys. J. A **47** (2011) 88 [arXiv:1011.2415 [nucl-ex]].

[114] M. Hoferichter, J. Ruiz de Elvira, B. Kubis and U.-G. Meißner, Phys. Rept. **625** (2016) 1 [arXiv:1510.06039 [hep-ph]].

[115] M. Hennebach *et al.*, Eur. Phys. J. A **50** (2014) 190 [arXiv:1406.6525 [nucl-ex]].

[116] C. Ditsche, M. Hoferichter, B. Kubis and U.-G. Meißner, JHEP **1206** (2012) 043 [arXiv:1203.4758 [hep-ph]].

[117] M. Hoferichter, C. Ditsche, B. Kubis and U.-G. Meißner, JHEP **1206** (2012) 063 [arXiv:1204.6251 [hep-ph]].

[118] R. A. Arndt, W. J. Briscoe, I. I. Strakovsky and R. L. Workman, Phys. Rev. C **74** (2006) 045205 [nucl-th/0605082].

[119] R. L. Workman, R. A. Arndt, W. J. Briscoe, M. W. Paris and I. I. Strakovsky, Phys. Rev. C **86** (2012) 035202 [arXiv:1204.2277 [hep-ph]].

[120] R. Koch and E. Pietarinen, Nucl. Phys. A **336** (1980) 331.

[121] G. Höhler, *Pion–Nukleon-Streuung: Methoden und Ergebnisse*, in Landolt-Börnstein, **9b2**, ed. H. Schopper, Springer Verlag, Berlin, 1983.

[122] M. Hoferichter, J. Ruiz de Elvira, B. Kubis and U.-G. Meißner, Phys. Rev. Lett. **115** (2015) 092301 [arXiv:1506.04142 [hep-ph]].

[123] J. Gasser, H. Leutwyler and M. E. Sainio, Phys. Lett. B **253** (1991) 252.

[124] M. Hoferichter, J. Ruiz de Elvira, B. Kubis and U.-G. Meißner, Phys. Lett. B **760** (2016) 74 [arXiv:1602.07688 [hep-lat]].

[125] M. Hoferichter, J. Ruiz de Elvira, B. Kubis and U.-G. Meißner, Phys. Rev. Lett. **115** (2015) no.19, 192301 [arXiv:1507.07552 [nucl-th]].

# In-medium properties of mesons

*Volker* Metag[1,2,*], *Mariana* Nanova[1,**], and *Kai-Thomas* Brinkmann[1,***]

[1] *II. Physikalisches Institut, Universität Gießen, Germany*
[2] *Helmholtz Institut für Strahlen- und Kernphysik, Universität Bonn, Germany*

**Abstract.** In the project **B.4**, the modification of meson properties (mass, width) in a nuclear medium has been studied in photoproduction of mesons off nuclear targets. This work has been motivated by theoretical expectations of in-medium modifications of hadrons based on the conjecture of a partial restoration of chiral symmetry in a strongly interacting medium. It has been shown that these in-medium changes can be discussed in a compact form in terms of an optical potential describing the meson-nucleus interaction. Experimental approaches to determine the real and imaginary part of the meson-nucleus potential have been developed. The experiments have been performed with the Crystal Barrel/TAPS detector at the electron accelerator ELSA (Bonn) and the Crystal Ball/TAPS detector at MAMI (Mainz). Measuring the excitation function and momentum distribution for photo production of $\omega$ and $\eta'$ mesons, the real parts of the $\omega$ and $\eta'$-nucleus potential, given by the in-medium mass shift, have been determined. For the $\eta'$ meson a lowering of the mass at normal nuclear matter density by -(39±7(stat)±15(syst)) MeV is observed, while for the $\omega$ meson a slightly smaller mass shift is found, however, with much larger uncertainties, not excluding a zero mass shift. The imaginary part of the potentials has been extracted from the measurement of the transparency ratio which compares the meson production cross section per nucleon within a nucleus to the production cross section off the free proton. For the $\eta'$ meson the imaginary part of the potential is found to be smaller than the real part. In case of the $\omega$ meson the opposite is observed. This makes the $\eta'$ meson a good candidate for the search for meson-nucleus bound states while no resolved $\omega$ mesic states can be expected. The results are compared with theoretical predictions. An outlook on future experiments is given.

## 1 Introduction

The interaction of mesons with nuclei is an important testing ground for our understanding of Quantum Chromodynamics (QCD) as the theory of the strong interaction in the non-perturbative regime. Research in this field was motivated in particular by theoretical predictions that meson properties might change within nuclei due to a partial restoration of chiral symmetry [1–3]. Mesons are considered to be excitations of the QCD vacuum which has a complicated structure with non-vanishing chiral-, gluon- and higher order quark-condensates. These condensates are predicted to change within a strongly interacting medium and, as a consequence, also the excitation energy spectrum, i.e. the mass spectrum of mesons is expected to be modified. This idea fostered widespread theoretical and experimental activities which have been summarized in recent reviews [4–6].

Pseudoscalar mesons are particularly suited for studying in-medium modifications, as shown in Fig. 1 [7]. Spontaneous chiral symmetry breaking generates a pseudo scalar nonet ($\pi, K, \bar{K}, \eta, \eta'$) of massless Goldstone bosons. The explicit breaking of the $U(1)_A$ symmetry selectively shifts up the $\eta'$ mass, leaving the SU(3) flavor octet of pions, kaons and $\eta$ massless. Explicit chiral symmetry breaking by introducing non-zero quark masses then leads to the experimentally observed meson masses [1, 7, 8]. Since symmetry breaking has such a big effect on meson masses one would expect corresponding effects in case of

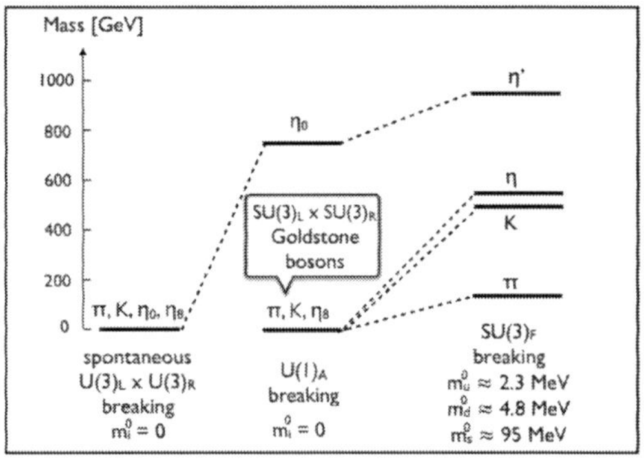


**Figure 1.** Symmetry breaking pattern for pseudoscalar mesons as described in the text, adapted from [7].

*e-mail: volker.metag@exp2.physik.uni-giessen.de
**e-mail: mariana.nanova@exp2.physik.uni-giessen.de
***e-mail: kai-thomas.brinkmann@exp2.physik.uni-giessen.de

a partial restoration of this symmetry in a strongly interacting medium.

If spontaneously broken chiral symmetry were restored at high nuclear densities one would expect masses of vector mesons with spin parity $J^{\pi} = 1^{-}$ to become degenerate in mass with their chiral partners with the same spin but opposite parity, the axial vector mesons with $J^{\pi} = 1^{+}$. Traces of this tendency should already show up at normal nuclear matter density and should thus become observable in photonuclear experiments

Many theory groups, using QCD-sum rules[3, 9] as well as various hadronic models such as the Nambu-Jona-Lasinio (NJL)-model [1, 10–13], the linear sigma model [14], the quark-meson-coupling model [15, 16], and coupled channel resonance models [17, 18] have made quantitative, but partially conflicting predictions for changes in mass and width of mesons in nuclei. As an example for the many calculations, the results obtained for the $\eta'$ meson by Nagahiro et al. and for the $\omega$ meson by Mühlich et al. are shown in Fig. 2 and Fig 3, respectively. Nagahiro et al.[11], applying the NJL model, predict a mass drop by $\approx$ 150 MeV for the $\eta'$ meson at normal nuclear matter density. Within the coupled channel resonance model, Mühlich et al.[18] find a strong broadening but almost no mass shift for the $\omega$ meson at saturation density. The broad range of predictions is summarized in table 1 for the $\eta'$ meson and in table 2 for the $\omega$ meson, respectively. This spread in theoretical predictions calls for an experimental clarification.

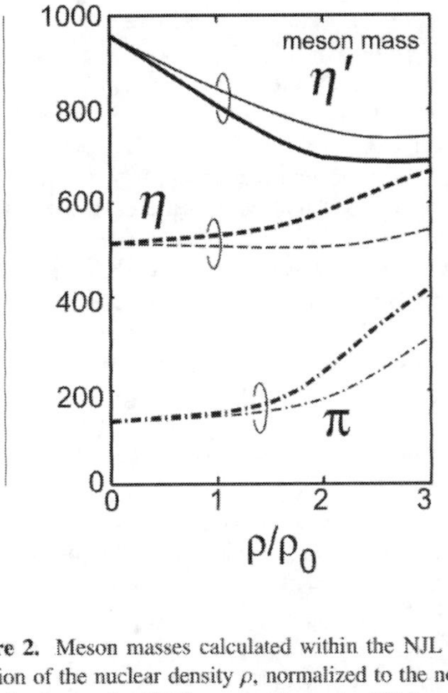


**Figure 2.** Meson masses calculated within the NJL model as function of the nuclear density $\rho$, normalized to the normal nuclear density $\rho_0$, for SU(2) symmetric matter (thick curves) and SU(3) symmetric matter (thin curves) [11].

It should be noted that most of these theoretical predictions have been calculated under idealized assumptions, e.g. for a meson at rest in infinitely extended nuclear matter with constant density, a scenario far from reality. In photoproduction experiments in the 1-3 GeV energy range, the kinematics of the reaction leads to meson recoil momenta on average comparable to their mass; on the way out of the nucleus mesons see the nuclear density profile with a fall-off at the surface. Any density dependent mass shift or broadening is thereby smeared out due to the variation in density. A link between the theoretical predictions and the experimental observables is provided by transport calculations like, e.g. GiBUU [21]. These calculations take nuclear many-body effects into account which evolve dynamically in the course of the reaction. In particular they treat

- initial state effects: the absorption of incoming beam particles
- non equilibrium effects: varying density and temperature
- absorption and regeneration of mesons
- fraction of decays outside of the nucleus

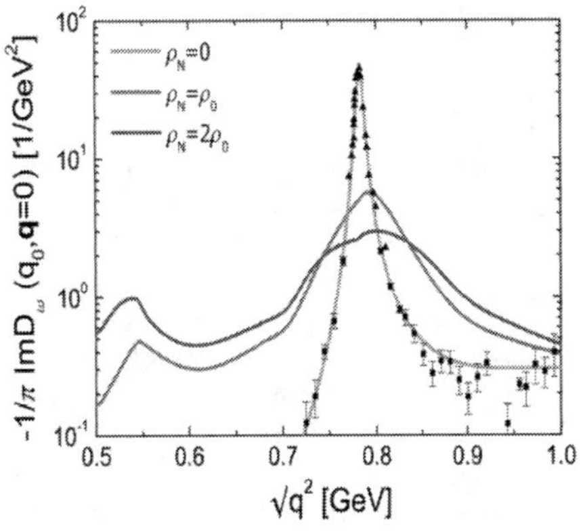


**Figure 3.** The spectral function for an $\omega$ at rest, calculated within the coupled-channel resonance model for normal nuclear matter density $\rho = \rho_0$ (solid curve) and for $\rho = 2\rho_0$ (dashed curve) in comparison to the spectral function of a free $\omega$ meson (dotted) [18].

**Table 1.** predictions for mass modifcations of the $\eta'$ meson at normal nuclear matter density.

| model | ref. | $\Delta m$ ($\rho = \rho_0$) |
|---|---|---|
| NJL | [10] | $\approx$ 0 MeV/c$^2$ |
| NJL | [11] | $\approx$ -150 MeV/c$^2$ |
| lin $\sigma$ | [14] | $\approx$ -80 MeV/c$^2$ |
| QMC | [16] | $\approx$ -40 MeV/c$^2$ |

**Table 2.** predictions for mass modifications of the $\omega$ meson at normal nuclear matter density.

| model | ref. | $\Delta m$ ($\rho = \rho_0$) | $\Gamma(\rho = \rho_0)$ |
|---|---|---|---|
| NJL | [1] | $\approx$ 0 MeV/c$^2$ | |
| QCD SR | [3] | $\approx$ -120 MeV/c$^2$ | |
| L$_{eff}$ | [13] | $\approx$ -(100-150) MeV/c$^2$ | $\approx$ 40 MeV |
| res. coupl. | [17] | $\approx$ -50 MeV/c$^2$ | $\approx$ 40 MeV |
| res. coupl. | [18] | $\approx$ 0 MeV/c$^2$ | $\approx$ 60 MeV |
| $\chi$ unitary | [19] | | 100-200 MeV |
| many-body | [20] | | 150-200 MeV |

- final state interactions: distortion of momenta of decay products.
It is indispensable to investigate with transport calculations how the above mentioned effects change the theoretically predicted initial signals before a meaningful comparison with experimental data can be performed to extract medium modifications of hadrons.

For long-lived mesons such as pions or kaons, meson beams can be used to study the meson-nucleus interaction experimentally. This is not possible for short-lived mesons like $\eta, \omega$, and $\eta'$ mesons where such beams are not available. Here, one has to produce the mesons in a nuclear reaction and to study their interaction with nucleons or nuclei in the final state. The meson-nucleus interaction and the in-medium modifications of mesons can be described by an optical potential

$$U(r) = V(r) + iW(r), \tag{1}$$

comprising a real (V(r)) and an imaginary (W(r)) part [11]; $r$ is the distance between the meson and the centre of the nucleus. In this presentation we discuss the determination of the real and imaginary part of the $\omega$- and $\eta'$- nucleus potential in photo production experiments off nuclei.

## 2 Experimental approaches to study mesons in the medium

The strength of the real part of the meson-nucleus potential is connected to the meson in-medium mass shift $\Delta m(\rho_0)$ at saturation density $\rho_0$ [11]

$$V(r) = \Delta m(\rho_0) \cdot c^2 \cdot \frac{\rho(r)}{\rho_0}. \tag{2}$$

The imaginary part of the potential describes the meson absorption in the medium via inelastic channels and is related to the in-medium width $\Gamma_0$ of the meson at nuclear saturation density by [22]

$$W(r) = -\frac{1}{2}\Gamma_0 \cdot \frac{\rho(r)}{\rho_0}. \tag{3}$$

The experimental approaches to determine the real part of the potential listed below are applied for $\omega$ and $\eta'$ mesons and the results will be discussed later in this paper:
- line shape analysis
- excitation function of the meson
- momentum distribution
- meson-nucleus bound states

The theoretical predictions discussed in the introduction inspired many experiments searching for medium modifications like broadening, structures or mass shift. The experiments have focused on the light vector mesons $\rho$, $\omega$ and $\phi$ with decay lengths comparable to nuclear dimensions. The mass distribution $\mu(\vec{p},\rho)$ depends on the 3-momentum $\vec{p}$ of the vector meson and on the density $\rho$ of the nuclear medium at the decay point. Only mesons decaying inside the nucleus carry information on in-medium properties which are to be studied. An in-medium mass shift of the meson could be observed by comparing the mass calculated from Eq. (4) in the limit of low meson momenta with the vacuum mass of this meson listed in [23].

$$\mu(\vec{p},\rho) = \sqrt{(p_1 + p_2)^2}. \tag{4}$$

The light vector mesons $\rho$, $\omega$ and $\Phi$ are particularly suited for the mass distribution measurements since their lifetimes of 1.3 fm/c, 23 fm/c and 46 fm/c, respectively, are so short that they decay within the nuclear medium with some probability after production in a nuclear reaction. Nevertheless, momentum cuts have to be applied for the longer lived $\omega$ and $\Phi$ mesons to achieve decay lengths comparable to nuclear dimensions. Obviously the line shape analysis is not applicable for $\eta'$ meson in the medium since its decay length is much larger than nuclear dimensions. In case of mesons with long lifetime decaying outside of the nucleus this method is not applicable since the information gained is the vacuum spectral function. The experimental results on the line shape analysis from the B4 project will be presented and discussed in section 5.2.1.
Weil et al. [24] discussed the possibility to extract information on the in-medium meson mass and the real part of the meson-nucleus potential from a measurement of the excitation function and/or momentum distribution of mesons in the photo production off a nucleus. Due to a lowering of the meson mass in the medium the meson production threshold will decrease and the enlarged phase space will lead consequently to increasing the production cross section for a given incident beam energy as compared to a scenario without mass shift. The dropping of the meson mass in the medium also affects the momentum distribution of the produced meson in the final state. When a meson is produced with a lower mass, then its total energy is on average also reduced due to kinematics. In addition, mesons produced within the nuclear medium must regain their free mass upon leaving the nucleus. Thus, in case of an in-medium mass drop, this mass difference has to be compensated at the expense of their kinetic energy. GiBUU transport-model calculations [24] demonstrated that there will be a downward shift in the momentum distribution for near-threshold energies as compared to a scenario without mass shift. A mass shift can thus be indirectly inferred from a measurement of the excitation function as well as from the momentum distribution of the meson. This idea, initially worked out for $\omega$ mesons [24], has independently been pursued on a quantitative level for $\eta'$ mesons by E. Paryev [25]. The experimental results about the determination of the real part of $\omega$ and $\eta'$ mesons from excitation functions and momentum distributions will be presented and discussed in sections 5.2.2 and 5.2.3.
A necessary condition for the experimental observation of meson-nucleus bound states is that $|V| >> |W|$. It is therefore important to study experimentally the relative strengths of the real and imaginary part of the meson-nucleus interaction for the meson of interest. Then by the observation of meson-nucleus bound states information about the real and imaginary part of the meson-nucleus optical potential can be gained. The search for $\omega$- and $\eta'$-

nucleus bound states will be discussed in section 6.
The experimental observable for the imaginary part of the meson-nucleus optical potential is the transparency ratio measurement. As a definition, the transparency ratio compares the meson production cross section off a nucleon within a nucleus with the one off the free nucleon [26]:

$$T_{\mathrm{A}}^{\mathrm{m}} = \frac{\sigma_{\gamma\mathrm{A}\to\mathrm{mX}}}{A \cdot \sigma_{\gamma\mathrm{N}\to\mathrm{mX}}}. \quad (5)$$

Here, the nucleus serves as a target and at the same time as an absorber. If nuclei were completely transparent to the mesons the transparency ratio would be unity, as long as secondary production processes can be ignored. The results on the imaginary part of the $\omega$- and $\eta'$-nucleus optical potential will be discussed in section 5.3.

## 3 Experimental setup

The data on $\omega$ and $\eta'$ photoproduction were taken in a series of experiments in 2003, 2007, 2009 and 2013/14 with the Crystal Barrel/TAPS detector system at the ELSA accelerator in Bonn and in 2007, 2008 with the Crystal Ball and TAPS detector system at MAMI-C in Mainz.

### 3.1 CBELSA/TAPS Detector system

The electron stretcher facility ELSA [27, 28] provides electrons with energy $E_0$ which hit a primary radiation target, a thin copper or diamond crystal, and produce bremsstrahlung [29]. The energy of the bremsstrahlung photons is determined eventwise from the deflection of the scattered electrons in a magnetic field. From the energy of the scattered electron $E_e^-$ the energy of the photon impinging on the nuclear target is given by $E_\gamma = E_0 - E_e^-$. Photons were tagged in the energy range from 0.5 GeV up to 3.1 GeV for an incoming electron energy of 3.2 GeV. The total tagged photon intensity was about $10^7\ s^{-1}$ in this energy range. The part of the electron beam that did not produce any bremsstrahlung photons was deflected by the magnet as well. Since these electrons retained their full energy the curvature of their track is smaller and they passed over the tagger into a beam dump.
The Crystal Barrel and TAPS detector system at ELSA accelerator in Bonn used for data taking in 2003 is shown in Fig. 4 (left). The Crystal Barrel (CB) detector, a photon calorimeter consisting of 1290 CsI(Tl) crystals ($\approx$16 radiation lengths), covered the complete azimuthal angle and the polar angle from 30° to 168° [30]. The scintillating fibre-detector, surrounding the target in the center of the CB, was used to register charged particles [31]. The CB was combined with a forward detector - the TAPS calorimeter - consisting of 528 hexagonal $BaF_2$ crystals ($\approx 12\ X_0$), covering polar angles between 5° and 30° and the complete azimuthal angle [32, 33]. In front of each $BaF_2$ module a 5 mm thick plastic scintillator was mounted for the identification of charged particles.
The CB and TAPS detector system at ELSA used for data taking in 2007, 2009, 2013 and 2014, shown in Fig. 4 (right) is a new updated version of the first setup (Fig. 4 (left)). The main difference is the smaller TAPS detector (MiniTAPS), consisting of 216 $BaF_2$ modules, covering the angular range in the forward direction of 1° to 11°. In the range of 11°-28° the CsI(Tl) modules from CB (Forward Plug FP) were read out by photomultipliers, providing energy and time information while the rest of the CB crystals were read out by photodiodes with energy information only. A more detailed description of the detector setups and the running conditions can be found in [34–38].
Because of the high granularity and the large solid angle coverage of $\approx$96%, both detector system versions were ideally suited for the detection and reconstruction of multi-photon events.
Solid targets of $^{12}$C, $^{40}$Ca, $^{93}$Nb and $^{208}$Pb, used in the experiments, were mounted in the center of the CB.

### 3.2 CB/TAPS@MAMI

In a series of experiments with tagged photon beams at MAMI-C, using the Crystal Ball and TAPS detector systems (Fig. 5), $\omega$ photoproduction on $^{12}$C and $^{93}$Nb targets has been measured in the energy range close to the production threshold 900-1300 MeV [39]. The Crystal Ball detector, comprising 672 NaI(Tl) crystals with 15.7 $X_0$, covered the polar angle range from 20° to 160°. The forward region was subtended by the TAPS calorimeter in a forward wall configuration, consisting of 384 $BaF_2$ crystals. The two innermost rings were replaced by 24 $PbWO_4$ ($\approx$22 $X_0$) crystals to handle higher event rates.
For the charged particle identification 5mm thick plastic scintillators, mounted in front of the $BaF_2$ crystals, were used. The particle identification detector (PID) surrounded the target area in the center of the Crystal Ball and consisted of 24 plastic scintillators. The MWPCs shown in Fig. 5 were not used in the experiments with solid targets. The combination of these detector systems was as suited for the detection of multi-photon final states as the CBELSA/TAPS detector.

## 4 Data analysis

The high segmentation and the almost full solid angle coverage of the detector systems used allowed the reconstruction of $\omega$ and $\eta'$ mesons from multi photon final states by invariant mass analysis. The decay modes used in the analyses were $\omega \to \pi^0\gamma \to 3\gamma$ and $\eta' \to \pi^0\pi^0\eta \to 6\gamma$ with branching ratios of 8.2% and 8.5%, respectively [23].

For the $\omega$ analysis, events with exactly three photons and one charged hit in the detector system were selected. The invariant mass of all photon pairs was calculated and the one combination closest to the $\pi^0$ mass of 135 MeV/$c^2$ was taken to be the $\pi^0$. For the $\eta'$ analysis, events with exactly 6 photons and any number of charged hits and with an energy sum of neutral clusters larger than 600 MeV were selected. The 6 photons were combined in 2 pairs of 2 photons with invariant masses in the range 115 MeV/$c^2$ $\leq m_{\gamma\gamma} \leq$ 155 MeV/$c^2$ (corresponding to a

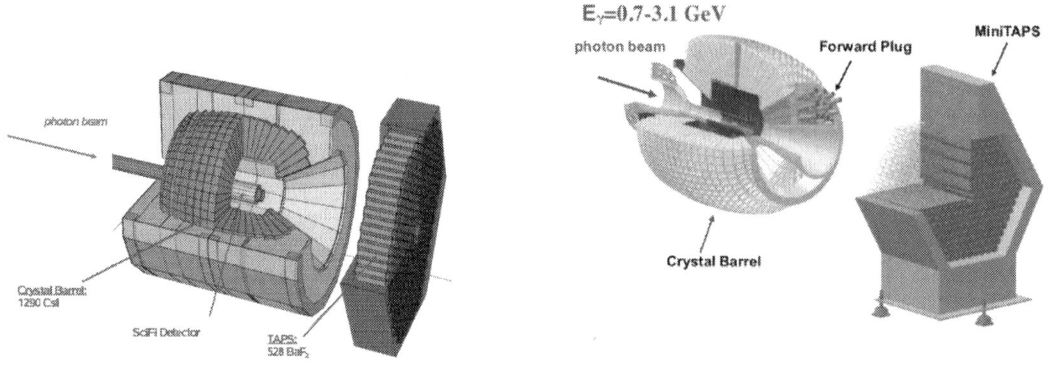


**Figure 4.** Left: Setup used in the experiment in 2003. Right: Improved setup used in the experiments with $^{12}$C (2007, 2009) and $^{93}$Nb (2013/14) targets. See text for more details.

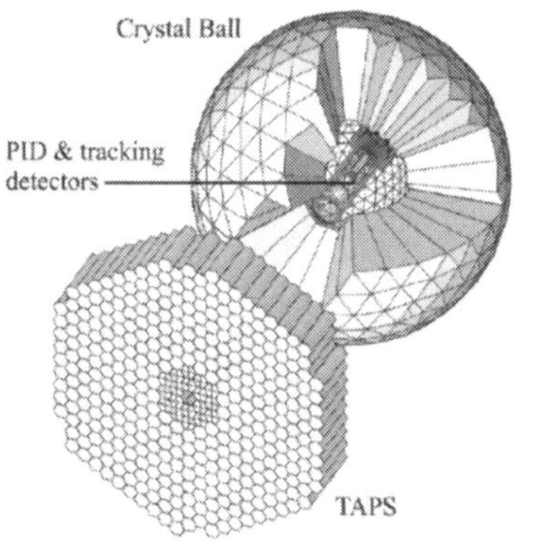


**Figure 5.** Crystal Ball/TAPS setup at MAMI in Mainz. Around the target area in the center of the Crystal Ball the plastic scintillator array (PID) and the multi-wire proportional chamber (MWPC) are shown. The $BaF_2$ calorimeter TAPS in a forward wall configuration, subdivided into six sectors, wth $PbWO_4$ crystals in the two innermost rings, was placed in a distance of 1.47 m behind the target.

$\pm 3\sigma$ cut around $m_{\pi^0}$) and one pair with invariant mass in the range 510 MeV/$c^2$ $\leq m_{\gamma\gamma} \leq$ 590 MeV/$c^2$ (roughly corresponding to a $\pm 2\sigma$ cut around $m_\eta$). The best photon combination was selected based on a $\chi^2$ minimization. To suppress the background from $\eta \to 3\pi^0$ decays and direct $3\pi^0$ production, events with 3 $\gamma$ pairs with an invariant mass within the limits for the pion mass $m_{\pi^0}$ given above were removed from the data set.

Examples for the invariant mass spectra, the obtained mass resolution and signal/background ratio are shown in Fig. 6 and Fig. 7. For the $\omega$ meson the relative mass resolution $\frac{\sigma_m}{m}$ is about 3.3%, while for the $\eta'$ meson a relative mass resolution of 1.2% has been achieved. The better $\eta'$ mass resolution is due to the fact that the nominal mass values [23] of the three mesons in the final state can be used in the calculation of the invariant mass. The signal/background ratios are of the order of 1/2.2 for the $\omega$ and 1/1.2 for the $\eta'$ meson, respectively.

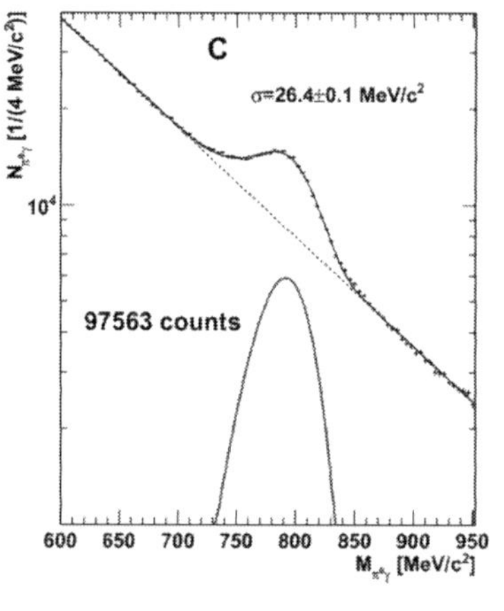


**Figure 6.** $\pi^0\gamma$ invariant mass distribution measured for a carbon target [40].

The conversion of the observed meson yields into cross sections requires - apart from the incident photon flux - the knowledge of the meson reconstruction efficiency which is determined by Monte Carlo simulations. Applying the GEANT3 package [42] with a full implementation of the detector system, the nuclear reaction of interest is simulated, using as input the measured angular distributions of $\omega$ and $\eta'$ mesons produced off protons and neutrons bound in deuterium [43, 44]. In addition, the Fermi motion of nucleons in the target nucleus, as parameterized by [45], has been taken into account. The reconstruction efficiency is determined as a function of the laboratory angle and the momentum of the mesons. This approach ensures that the appropriate acceptance is used even if the angle

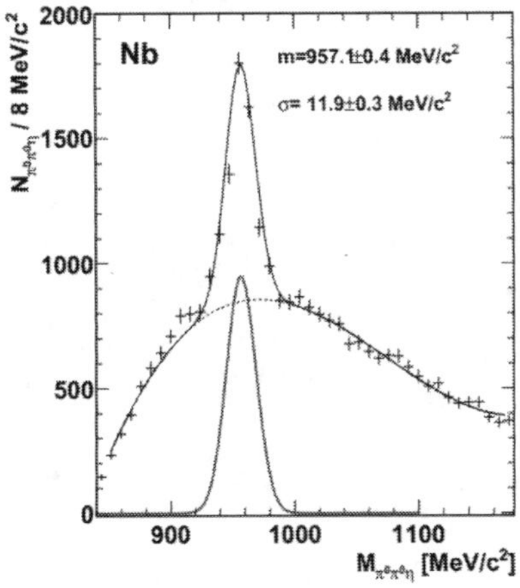


**Figure 7.** $\pi^0\pi^0\eta$ mass distribution measured for a niobium target [41].

and momentum of the meson deviates from the kinematics of the reaction because of elastic scattering in the nucleus. The reconstruction efficiency is determined by taking the ratio of the number of reconstructed and the number of generated events for each angular- and momentum bin. As an example, the resulting reconstruction efficiency $\epsilon_{\gamma\text{Nb}\to\eta'X}(p^{\text{lab}}_{\eta'}, \theta^{\text{lab}}_{\eta'})$ for the $\gamma$Nb $\to \eta'$X reaction is shown in Fig. 8 for the incident photon energy range of 1.2-2.9 GeV. In the Monte Carlo simulations, the same trigger conditions as in the experiment are applied.

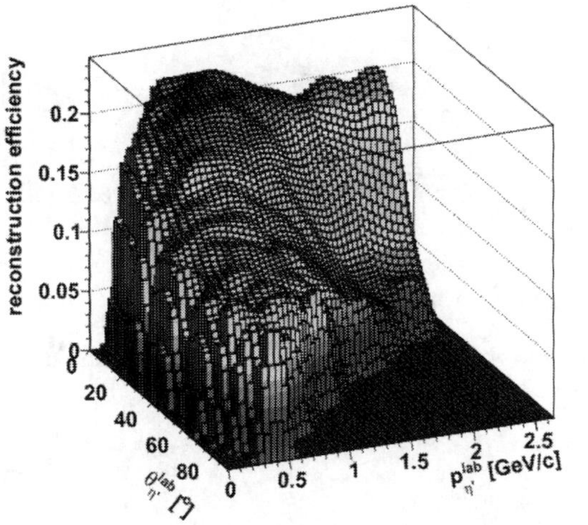


**Figure 8.** Two-dimensional reconstruction efficiency for $\eta'$ photoproduction off Nb as a function of the $\eta'$ momentum and angle in the laboratory system for the incident photon energy range of 1.2-2.9 GeV [41].

## 5 Results and discussion

### 5.1 Meson photoproduction in elementary reactions

Before extracting information on in-medium modifications of mesons from experiments, the photoproduction of mesons off the free proton as well as off the quasifree proton and neutron in deuterium has to be studied to provide a reference for the results obtained from nuclei. The photoproduction of $\omega$ and $\eta'$ mesons has been measured with the CBELSA/TAPS detector by Dietz et al. [43] and Jaegle et al. [44], respectively. Detailed results on the angular distributions and total cross sections have been obtained. As an example, the total cross section for $\omega$ and $\eta'$ production off the proton and neutron in deuterium are shown in Fig. 9. These data are essential as input for calculations of meson production off nuclei. Deviations from these calculations are then interpreted as in-medium effects.

### 5.2 The real part of the meson-nucleus optical potential

#### *5.2.1 Lineshape Analysis*

As discussed in section 2, the most direct way to determine an in-medium mass shift from meson photoproduction off nuclear targets is to calculate the invariant mass from the measured 4-vectors of the meson decay products, using Eq. (4), and to compare this invariant mass with the meson mass listed in [23]. This approach has, however, several, disadvantages:
- The method is sensitive to the nuclear density at the *decay* point. Due to the kinematics of the reaction, mesons obtain a rather broad recoil momentum distribution with an average momentum comparable to their mass. Thus, $\beta \cdot \gamma = \frac{p}{m} \approx 1$ and consequently, the average decay length $\beta\gamma c\tau \approx c\tau$ is several times larger than nuclear dimensions even for short-lived vector mesons, except for the $\rho$. Most of the meson decays thus occur outside of the nuclear medium, yielding a free spectral function. Even when applying cuts to select low momentum mesons to enhance in-medium decays, there will always be two mass peaks for decays inside and outside of the nucleus which can only be separated if the mass shift is larger than the mass resolution.
- an in-medium mass peak will be smeared out depending on the broadening of the meson through inelastic reactions in the nucleus. Broad structures, however, will be more difficult to distinguish from background.
- ambiguities in the subtraction of the background in the invariant mass spectra leads to additional uncertainties in the determination of the meson line shape, as discussed in [36, 47, 49].
- the above problems exist for any decay mode of the meson. If, however, hadrons are among the decay products, as in the $\omega$ and $\eta'$ decays discussed here, the 4-momentum vectors of the decay products and thus also the calculated invariant mass may be distorted by elastic scattering or the mesons may even be absorbed within the nucleus due

 DOI: 10.1051/epjconf/201713403003

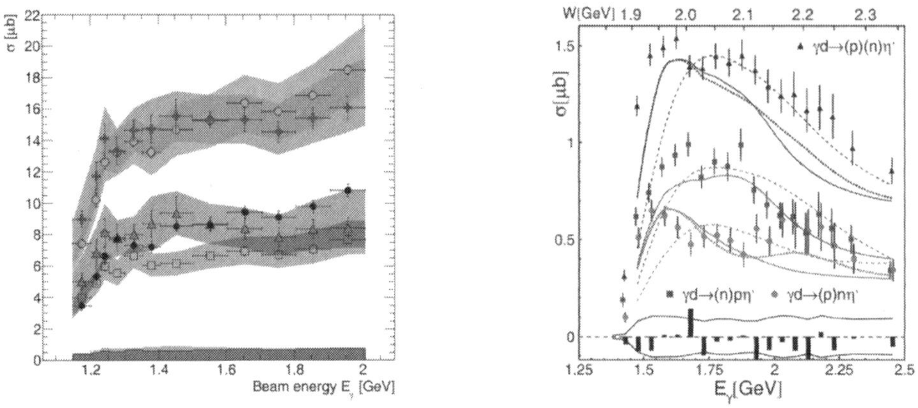


**Figure 9.** Left: Total cross section as a function of the incident photon energy for $\omega$ mesons produced off the bound proton (open squares) and off the bound neutron (red triangles), the sum of the exclusive cross sections (purple crosses), the quasifree inclusive production cross section (green circles) and the cross section off the bound neutron calculated from $\sigma_{incl} - \sigma_p^{bound}$ (black circles). The bands indicate the systematic errors [43]. Right: Total cross section as a function of the incident photon energy for $\eta'$ production off the bound proton (blue squares) and neutron (red circles) and the inclusive cross section (black triangles). The curves are reaction model fits, see [44].

to the strong interaction. Information on decays near the center of the nucleus will thus not reach the detector. Only decays near the nuclear surface will be registered which occur at low nuclear densities where mass shifts - assuming a density dependent mass modification - are also smaller. The sensitivity of the meson lineshape to in-medium modifications is thus reduced.

These caveats of the lineshape analysis have only been recognized in the course of the project. As an example, the results obtained with the Crystal Ball and TAPS detector system (Fig. 5) for $\omega$ photoproduction off $^{12}C$ and $^{93}Nb$ targets will be discussed. High statistics have been accumulated in the energy range close to the production threshold [39], leading to an enhanced yield of low momentum $\omega$ mesons.

Fig. 10 shows the $\omega$ signal after subtraction of the background mostly arising from $2\pi^0$ and $\pi^0\eta$ events where one of the 4 decay photons was not registered in the Crystal Ball/TAPS detector system (for details of the background subtraction techniques, see [36]). The $\omega$ signal in the Nb run is slightly broader than the reference signal obtained with the $LH_2$ target [39]. In Fig. 10 the $\omega$ signal on Nb is compared to GiBUU transport calculations [24, 35] for different in-medium scenarios. Although the statistics of the experiment has been considerably improved compared to previous studies [35, 36, 49] it is still not sufficient to distinguish different scenarios which differ only very little in the predicted line shape. This reflects the limited sensitivity of the lineshape analysis as discussed above. Only the scenario with a strong mass shift seems to be disfavored.

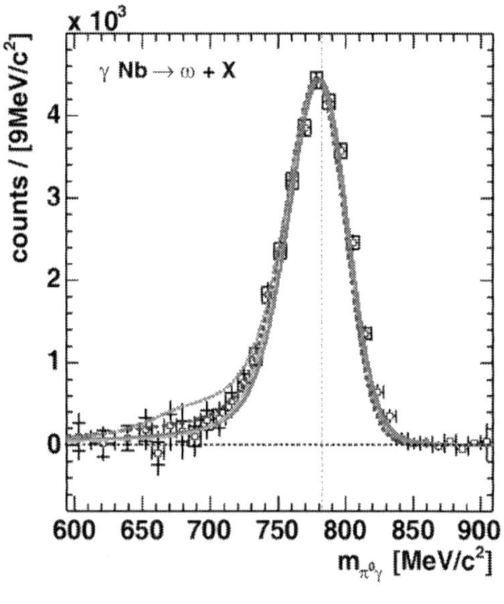


**Figure 10.** Comparison of the observed $\pi^0\gamma$ invariant mass spectrum for the Nb target for inclusive analysis with GiBUU calculations [24] for different in-medium modification scenarios, assuming no in-medium modifications (red solid curve), only collisional broadening (green dashed curve), collisional broadening and mass shift by -14% at normal nuclear matter density (short dashed, blue curve) and mass shift without broadening (dotted, magenta curve) [39].

### 5.2.2 *Excitation function of the $\omega$ and $\eta'$ mesons*

As an alternative to the lineshape analysis, which showed a reduced sensitivity to in-medium modifications, information on the in-medium meson mass and the real part of the meson-nucleus potential can be extracted from the excitation function of the meson [24, 25]. In contrast to the lineshape analysis this method is sensitive to the nuclear density at the *production* point and hence it is applicable for all mesons, irrespective of their lifetime.

The experimental results for the $\omega$ excitation function in comparison to theoretical calculations are shown in Fig. 11. Data were taken on a carbon target with the Crystal Ball and TAPS detector system at MAMI-C. The GiBUU transport model [21] has been applied to calculate the $\omega$ excitation function for 6 different scenarios allowing for mass shifts up to -125 MeV at normal nuclear density as it is shown in Fig. 11. It is again obvious that the data are not consistent with a strong mass shift scenario, V>100 MeV as predicted in [3, 13]. A $\chi^2$-fit of the data with the calculated excitation functions for the different scenarios gives a potential depth of $V_{\omega A}(\rho = \rho_0) = -(42 \pm 17(\text{stat})\pm 20(\text{syst}))$ MeV [50].

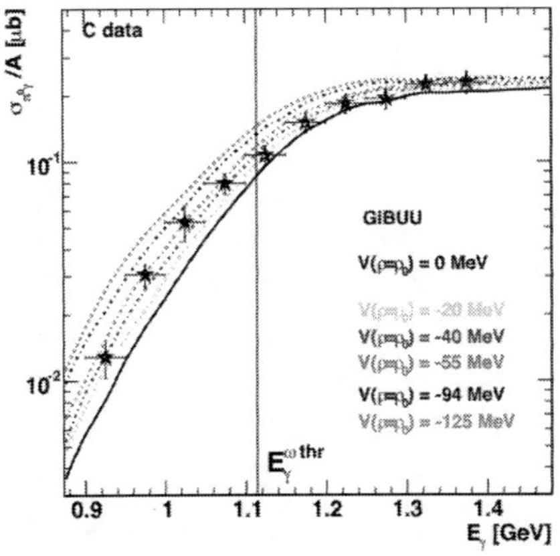


**Figure 11.** Measured excitation function for $\omega$ meson in comparison to GiBUU transport calculations for several in-medium modification scenarios [50].

Experimental data for the $\eta'$ excitation function in comparison to theoretical calculations are shown in Fig. 12 [37, 41]. In order to study the possible dependence of the real part of the $\eta'$-nucleus optical potential on the mass of the nucleus, the data were taken on $^{12}C$ and $^{93}Nb$ targets with the CBELSA/TAPS detector system. In both cases the data disfavor a deep real potential of V>100 MeV, as predicted in [11, 12]. While the $\eta'$ cross section drops dramatically near the production threshold of $E_\gamma$=1447 MeV in case of the proton target [51], there is appreciable yield below this threshold in the reaction on the solid targets. There are several effects which can cause a non-zero cross section below 1447 MeV. This is on the one hand due to the Fermi motion of nucleons in the Nb target which gives rise to a distribution of the energy $\sqrt{s}$ available in the centre-of-mass system for a given incident photon energy. On the other hand, also the mass of the meson might drop in a nuclear medium - as theoretically predicted - which lowers the production threshold and increases the phase space for meson production below the free threshold energy. The calculations have been performed within the first collision model based on the nucleon spectral function and described in detail in [25]. Starting from the measured differential cross sections for $\eta'$ production off the free proton and neutron [44, 51], the cross section for $\eta'$ photoproduction off C and Nb is calculated in an eikonal approximation, taking the effect of the nuclear $\eta'$ mean-field potential into account. The $\eta'$ final-state absorption is determined by the inelastic in-medium $\eta' N$ cross section taken to be $\sigma_{inel}$=11 mb, consistent with the result of transparency ratio measurements [22]. In the calculations for the Nb target $\sigma_{inel}$=13 mb has been assumed, consistent with the new results from [40]. The total nucleon spectral function is taken into account in the parametrization given in [52]. Thereby, the contribution of $\eta'$ production from two-nucleon short-range correlations is taken into account. The calculations are improved in [37, 41] with respect to [25] as the momentum-dependent optical potential from [53], seen by the nucleons emerging from the nucleus in coincidence with the $\eta'$ mesons, is accounted for as well.

The experimental data were compared with six different scenarios assuming depths of the $\eta'$ real potential at normal nuclear matter density of $V$=0, -25, -50, -75, -100 and -150 MeV, respectively (Fig. 12). The reduction in the incident photon flux due to photon shadowing has been taken into account by multiplying the observed $\eta'$ yield by 1.17 [54–56]. The calculated cross sections for Carbon have been further scaled down - within the limits of the systematic uncertainties - by a factor of 0.83 to match the experimental excitation function data at incident photon energies above 2.2 GeV, where the difference between the various scenarios is very small. In the corresponding analysis of the Nb data [41] a rescaling of the theoretical calculations by 0.91 had to be applied. We are not aware of any missing physics in the calculations which might explain this systematic difference between data and calculations. In view of the systematic errors of the cross section data (23%) and the calculations (10-15%) a discrepancy cannot be claimed. The highest sensitivity to the $\eta'$ potential depth is found for incident photon energies near and below the production threshold on the free nucleon. The excitation function data on both nuclei appear to be incompatible with $\eta'$ mass shifts of -100 MeV and more at normal nuclear matter density, as clearly seen in Fig. 12. A $\chi^2$-fit of the data with the calculated excitation functions for the different scenarios over the full range of incident energies gives a potential depth of -(40±6) MeV and -(40±12) MeV for C and Nb, respectively [37, 41].

It has been investigated whether the observed cross section enhancement relative to the $V$=0 MeV case could also be due to $\eta'$ production on dynamically formed compact nucleonic configurations - in particular, on pairs of correlated nucleon clusters - which share energy and momentum. These effects have been studied experimentally [57] and theoretically [58, 59] in very near-threshold $K^+$ production in proton-nucleus reactions. Applying the parametrization of the spectral function given by [52], in Fig. 12 (left: dot-dashed blue curve) has been shown that correlated high momentum nucleons contribute only about 10-15% to the $\eta'$ yield in the incident energy

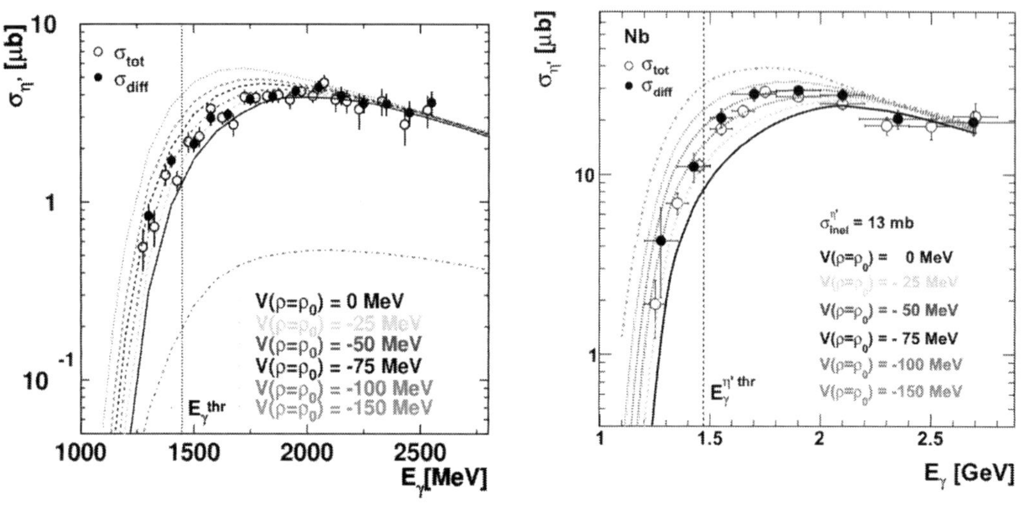


**Figure 12.** Measured excitation function for $\eta'$ meson off $^{12}$C (left) and $^{93}$Nb (right), in comparison to theoretical calculations for different scenarios [37, 41]. The experimental data are extracted by integrating the differential cross sections (full circles) and by direct measurement of the $\eta'$ yield in incident photon energy bins (open circles). The calculations are for $\sigma_{\eta' N}$=11 mb (for C data) and for $\sigma_{\eta' N}$=13 mb (for Nb data), and for potential depths: $V$=0 MeV (black line), -25 MeV (green), -50 MeV (blue), -75 MeV (black dashed), -100 MeV (red) and -150 MeV (magenta) at normal nuclear density, respectively, and using the full nucleon spectral function. The dot-dashed blue curve (in the left spectrum) is calculated for correlated intranuclear nucleons only.

regime above 1250 MeV [37]. The observed cross section enhancement can therefore be attributed mainly to the lowering of the $\eta'$ mass in the nuclear medium.

### *5.2.3 Momentum distribution of the $\omega$ and $\eta'$ mesons*

As discussed in section 2, the momentum distribution of the mesons is also sensitive to the potential depth and has been investigated as well. Fig. 13 shows the momentum distribution of $\omega$ mesons in comparison to GiBUU simulations for different in-medium modification scenarios. The $\omega$ momentum distributions measured for near-threshold photo production off $^{12}$C and $^{93}$Nb do not support calculations predicting mass shifts as large as -16% [3, 13].

The measured momentum differential cross sections for $\eta'$ meson photoproduction off C [37] and Nb [41] are shown in Fig. 14. The average momentum in both cases is $\approx$1.1 GeV/$c$. Bin sizes of 0.1 GeV/$c$ for C data and $\geq$ 0.2 GeV/$c$ for Nb data have been chosen which are large compared to the momentum resolution of 25-50 MeV/$c$ deduced from the experimental energy resolution and from Monte Carlo simulations. The $\eta'$ momentum distributions have been calculated for the incident photon energy range 1.3-2.6 GeV and for different potential depths V = 0, -25, -50, -75, -100 and -150 MeV. The comparison of these calculations with the data again exclude strong $\eta'$ mass shifts. A $\chi^2$-fit of the data with the calculated momentum distributions for the different scenarios over the full range of incident energies gives a potential depth of -(32±11) MeV and -(45±20) MeV off C and Nb, respectively [41].

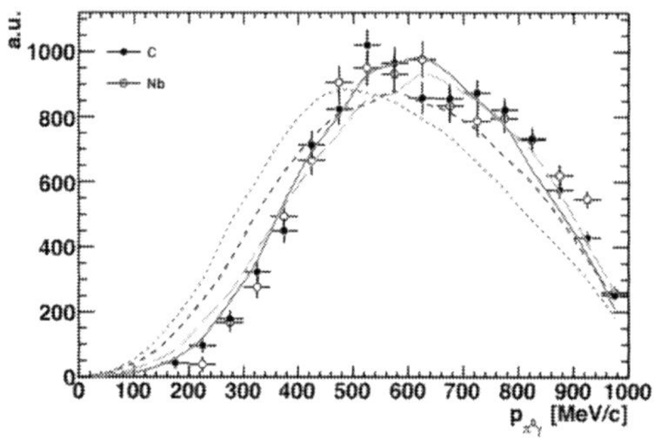


**Figure 13.** Acceptance corrected $\omega$ momentum distribution for incident photon energies of 900 to 1300 MeV and for $^{12}$C and $^{93}$Nb targets, compared to the theoretical predictions for different in-medium modifications scenarios: no modification (solid red line), collisional broadening (dashed green line), collisional broadening plus mass shift (dashed blue line) and mass shift (magenta line). All distributions are normalized to the same area [39].

### *5.2.4 The $\omega$-nucleus real potential at low momenta*

As has been pointed in 5.2.3, the average momentum of $\omega$ and $\eta'$ mesons was around 0.6 - 1 GeV/$c$ (see Fig. 13, Fig. 14). To study the possible momentum dependence of the real part of the meson-nucleus optical potential an experiment has been performed to produce $\omega$ mesons with low momenta (< 500 MeV/$c$). In the reaction $\gamma C \rightarrow \omega p \rightarrow \pi^0 \gamma p$ the proton has been measured in the TAPS

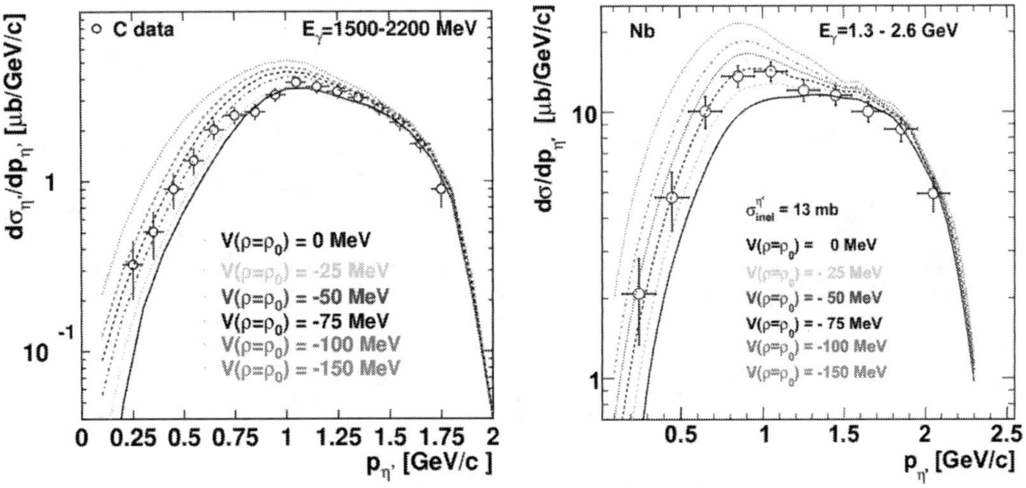


**Figure 14.** Left: Momentum distribution for $\eta'$ photoproduction off C for the incident photon energy range 1500-2200 MeV. The calculations are for $\sigma_{\eta' N}$=11 mb and have been reduced by a factor 0.75 (see text) [37]. Right: Momentum distribution for $\eta'$ photoproduction off Nb for the incident photon energy range 1.3-2.6 GeV. The calculations are for $\sigma^{\eta'}_{inel}$=13 mb and have been multiplied by a factor 0.83 (see text) [41]. In both spectra the theoretical curves are for potential depths V = 0, -25, -50, -75, -100 and -150 MeV at normal nuclear density. The color code is identical to the one in Fig. 12.

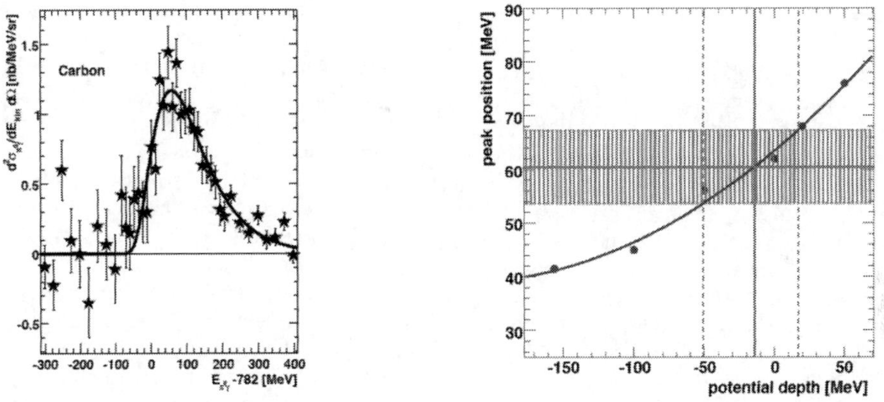


**Figure 15.** Left: Differential cross section for the photo production of $\omega$-mesons off C in coincidence with protons in $\Theta_p$ = 1° − 11° as a function of the total energy of the $\pi^0\gamma$ pairs minus 782 MeV. The data have been fitted with the Novosibirsk function [62]. Right: Correlation between the potential depth and the peak position in the total energy distribution. The (blue) points represent the peak positions in the total energy distribution for the different scenarios [11, 63–65]. The (blue) solid curve is a fit to the points. The red dashed area corresponds to the peak position of (60.5 ± 7) MeV [61].

detector (see Fig. 4 right) covering the polar angle range $\Theta_p$ = 1° − 11°. In this case the forward going proton takes over most of the momentum of the incoming photon beam. Being so low in energy, the $\omega$-mesons are particularly sensitive to the $\omega$ nucleus potential. In case of a strong repulsive (attractive) interaction one would expect the peak in the kinetic energy distribution to be shifted to higher (lower) energies for the C target. The sensitivity of the peak position in the kinetic energy distribution on the potential depth, studied in [61], can be exploited to deduce the depth of the real part of the $\omega$-nucleus potential.

The experimental data for the $\omega$ kinetic energy distribution are shown in Fig. 15 (left). The data have been fitted with the Novosibirsk function [62], giving a peak position of (60.5 ± 7) MeV [61]. The correlation between the potential depth and the peak in the kinetic energy distribution is plotted in Fig. 15 (right). A comparison with the experimentally determined peak position indicates a potential depth of -(15±35) MeV which appears to be too small to allow for the formation and population of $\omega$ mesic states, in particular in view of the large in-medium width of ≈ 90 MeV (see 5.3).

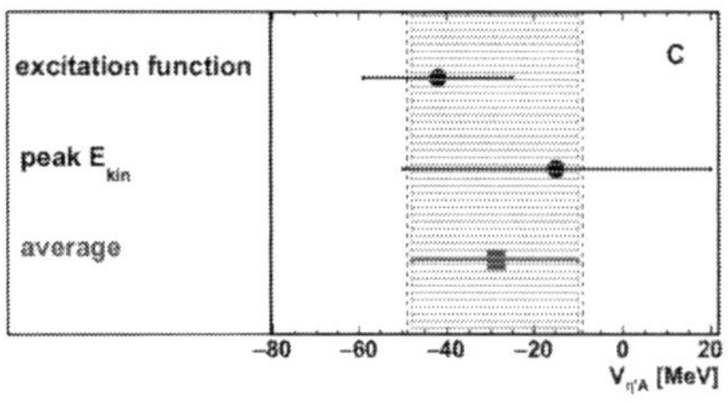


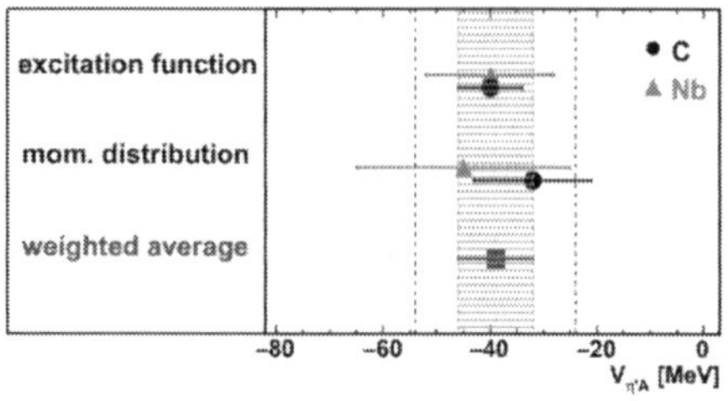


**Figure 16.** Depths of the real part of the $\omega$ (left) and $\eta'$ (right) -nucleus potential determined by analyzing the excitation function and the momentum distributions for C [37, 50] (full black circles) and for Nb [41] (red triangles). The weighted overall average is indicated by a blue square and the shaded area. The vertical hatched lines mark the range of systematic uncertainties

### 5.2.5 Conclusions on the real part of the $\omega$- and $\eta'$-nucleus optical potential

Combining the results from the analysis of the excitation function and the peak position from the kinetic energy distribution an average depth of the real part of the $\omega$-C optical potential of $V(\rho = \rho_0)$ = -(29 ±19(stat)±20(syst)) MeV is obtained Fig. 16 (left).

From the analysis of the excitation functions and the momentum distributions and by proper weighting of the errors a depth of the real part of the $\eta'$-C and -Nb optical potential of $V(\rho = \rho_0)$ = -(39 ±7(stat)±15(syst)) MeV is obtained (see Fig. 16 (right)). The systematic error quoted is mainly due to uncertainties in normalizing the calculations to the data. The sensitivity of the result on this normalization has been studied by varying the normalization factor between 0.7 to 1.0 - well within the systematic errors of the cross section determinations.
This result for $V(\rho = \rho_0)$ is consistent with predictions of the $\eta'$-nucleus potential depth within the Quark-Meson Coupling model (QMC) [16] and with calculations in [60] but does not support larger mass shifts as discussed in [11, 14].

## 5.3 The imaginary part of the meson-nucleus optical potential

Figure 17 presents the differential cross sections per nucleon for $\omega$ and $\eta'$ photoproduction off carbon and niobium as a function of the meson momentum for incident photon energies of 1.2-2.9 GeV [40]. The cross sections include a 15% correction for absorption of the incoming photon beam (photon shadowing) for both nuclear targets [54–56]. The distributions show a maximum at around 800 MeV/$c$ and 1000 MeV/$c$ for the $\omega$ and $\eta'$, respectively, and fall off towards higher momenta. The average momenta, given in Fig. 17, are close to 1000 MeV/$c$. To determine the meson attenuation in nuclei and the inelastic meson-nucleon cross sections as a function of the meson momentum, the transparency ratio [26] is deduced from the data as defined in Eq.5. To suppress nuclear effects not related to meson absorption, the transparency ratio is not taken relative to the free nucleon but rather to a nucleon in a light nucleus like carbon as suggested in [47]. The momentum dependence of the transparency ratio for a meson m is thus obtained by dividing the differential inclusive meson production cross sections (see Fig. 17) for niobium by the one for carbon according to

$$T^{\rm m}_{\rm Nb/C} = \frac{12 \cdot \sigma_{\gamma {\rm Nb}\to {\rm mX}}}{93 \cdot \sigma_{\gamma {\rm C}\to {\rm mX}}}, \qquad (6)$$

where 12 and 93 are the nuclear mass numbers of carbon and niobium, respectively. The resulting transparency ratios as a function of the meson momentum are shown in Fig. 18. Consistent with earlier measurements [37, 48], a slight increase with momentum is observed for the $\omega$ meson while for the $\eta'$ meson the transparency ratio is almost independent of momentum. Differences between the transparency ratios in the present measurements and the earlier ones reflect the systematic uncertainties of the measurements. For the present data the thick error bars (red) in Fig. 18 and subsequent figures represent the statistical errors while the thinner error bars (black) include the systematic errors added in quadrature.

The interpretation of the transparency ratio in terms of meson absorption will only give reliable results if two-step production processes are negligible, where e.g. a pion is produced in an initial step followed by production of the meson of interest in a subsequent pion-induced reaction on another nucleon within the nucleus. As shown in [22], two-step processes are negligible for $\omega$ and $\eta'$ mesons since the spectral distribution of pions falls off towards higher energies and - at the required pion momenta of ≈ 1.3 and 1.5 GeV/$c$ - pion induced meson production cross sections are only ≈ 2.5 mb and 0.1 mb, respectively, compared to the total reaction cross section of about 30-40 mb [23]. Consequently, two-step processes are neglected in the subsequent analysis of the transparency ratio.

Following [22, 25, 48] the in-medium meson width is deduced from the measured transparency ratio within a Glauber model in the high energy eikonal approximation (for details see [40]). Hereby, the in-medium width $\Gamma(\rho)$ is assumed to depend linearly on the nuclear density $\rho$:

$$\Gamma(\rho) = \Gamma_0 \cdot \frac{\rho}{\rho_0}, \qquad (7)$$

where $\Gamma_0$ is the momentum dependent width at normal nuclear matter density. As a result, the in-medium width $\Gamma_0$

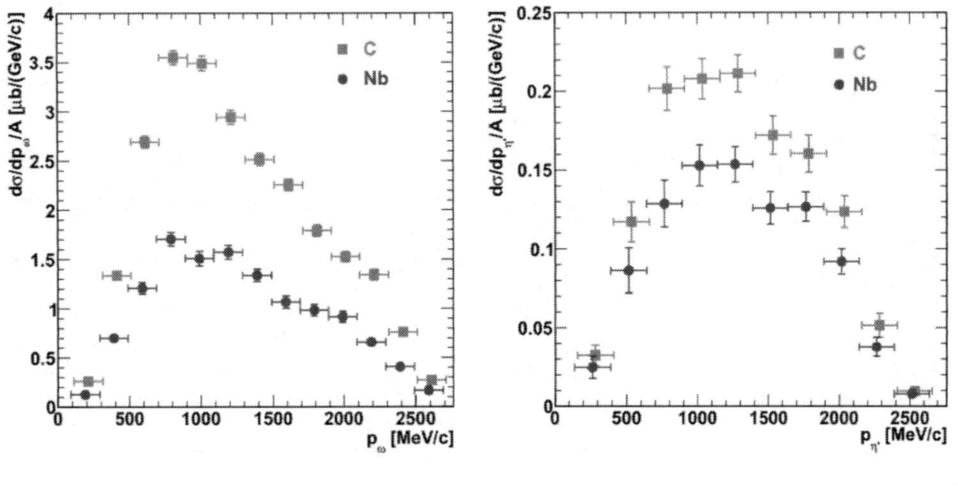


**Figure 17.** Differential cross section per nucleon for (Left) $\omega$ and (Right) $\eta'$ photoproduction off carbon (red squares) and niobium (blue circles) for the incident photon energies of 1.2-2.9 GeV. The data points for carbon are shifted by $+10$ MeV/$c$ and for niobium by $-10$ MeV/$c$ to avoid an overlap of the error bars.

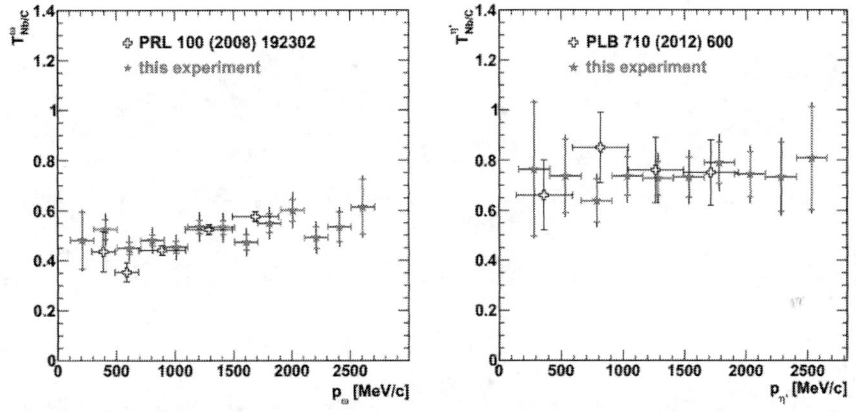


**Figure 18.** The transparency ratio according to Eq. 6 for (Left) $\omega$ and (Right) $\eta'$ photoproduction off carbon and niobium (red stars) in comparison to earlier measurements (open crosses) [22, 48]. The data points of the present work are shifted by $+10$ MeV/$c$ and the previuosly published data by $-10$ MeV/$c$ to avoid an overlap of the error bars. For the present data the thick error bars (red) represent the statistical errors. The thin error bars (black) include the systematic errors added in quadrature.

is shown in Fig. 19 as a function of the $\omega$ and $\eta'$ momentum, respectively. The non-linearity in the correlation between transparency ratio and in-medium width introduces strongly asymmetric error bars. Within errors, the results are consistent with previous measurements [22, 48, 66]. It is immediately apparent that the $\omega$ widths are larger than the $\eta'$ widths by about a factor three. The finer binning in the present data also reveals more clearly a variation of the widths with momentum, indicating a rise with momentum at small momenta and a fall-off towards higher momenta.

The data for the $\omega$ meson are compared to calculations of the in-medium $\omega$ width. Cabrera and Rapp [20] and Ramos *et al.* [19] have studied the width of the $\omega$ meson in cold nuclear matter as a function of the nuclear density and the meson 3-momentum. Both groups independently find that the main contribution to the in-medium $\omega$ width is determined by the $\omega \to \rho\pi$ channel whereby the dressing of the $\pi$ and $\rho$ propagator in the medium is essential. Thus, they argue that the $\pi\rho$ cloud is the main agent for the in-medium broadening of the $\omega$ meson. Both groups obtain in-medium $\omega$ widths of the order of 100-200 MeV, somewhat larger than observed experimentally. They differ in the partitioning into the $\pi$ and $\rho$ modifications and obtain differences in the 3-momentum dependence of the

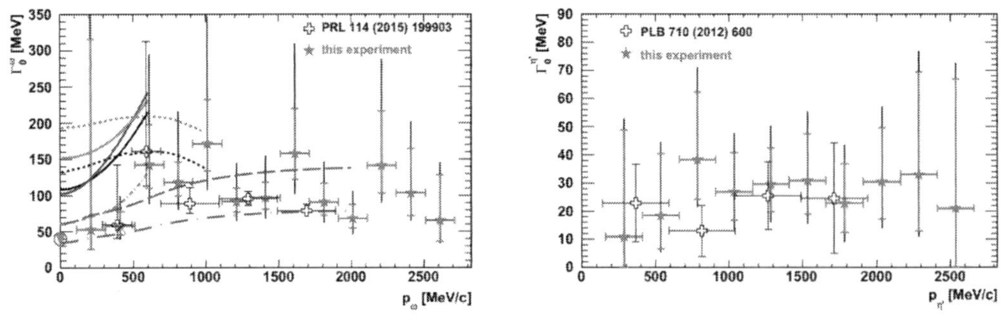


**Figure 19.** In-medium width $\Gamma_0$ of (Left) $\omega$ and (Right) $\eta'$ mesons as a function of the meson momentum (red stars), derived from the data presented in Fig. 18, in comparison to earlier measurements (open crosses) [22, 48, 66]. The data points of the present work are shifted by +10 MeV/$c$ and the previously published data by −10 MeV/$c$ to avoid an overlap of the error bars (symbols as in Fig. 18). The solid curves correspond to calculations by Ramos *et al.* [19], the short dashed curves to calculations by Cabrera and Rapp [20] for different model assumptions (see text). The brown dashed-dotted curve shows the momentum dependence of the $\omega$ in-medium width calculated in a coupled-channel resonance model [18]. The long dashed green curves correspond to different options for the in-medium $\omega$ width used in GiBUU simulations [21]. The blue triangle and the red open circle represent the widths calculated for an $\omega$ meson at rest in the nuclear medium in [13, 17], respectively.

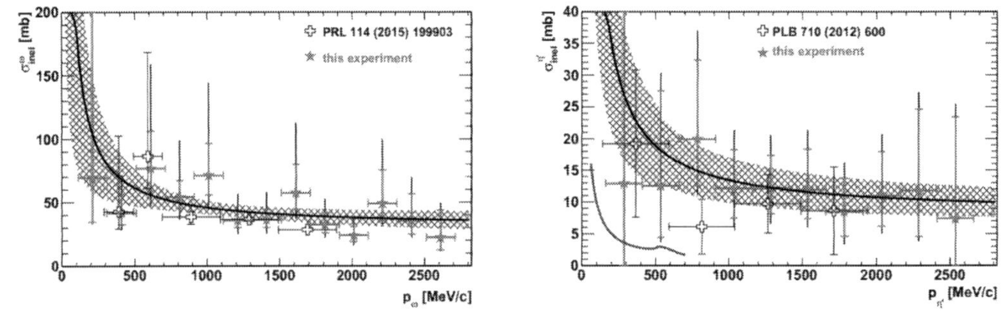


**Figure 20.** Left: $\omega$ inelastic cross section and Right: $\eta'$ cross section as a function of the momentum (red stars) [40] in comparison to earlier measurements (open crosses) [22, 48, 66].

$\omega$ width. While Cabrera and Rapp [20] find a moderate momentum dependence, Ramos *et al.* [19] get an almost linear increase of the width with momentum. Alternatively, Klingl et al. [13], Lutz et al. [17] and Mühlich et al. [18] consider the coupling of the $\omega$ meson to nucleon resonances as the main effect driving the in-medium broadening of the $\omega$ meson. In fact, the data are closer to the momentum dependence of the $\omega$ in-medium width calculated in a coupled-channel resonance model [18] and are only slightly larger at low momenta than the width of 40 MeV calculated for the $\omega$ at rest in the nuclear medium [13, 17]. The statistics of the present experiment is unfortunately not sufficient to clearly discriminate between these different theoretical approaches.

From the in-medium widths $\Gamma_0$ of Fig. 19 inelastic cross sections $\sigma_{\text{inel}}$ can be derived, using the low-density approximation which relates the in-medium meson width $\Gamma(\rho = \rho_0) = \Gamma_0$ at normal nuclear matter density $\rho_0$ to the inelastic meson-nucleon cross section $\sigma_{\text{inel}}$ by

$$\Gamma(\rho = \rho_0) = \Gamma_0 = \hbar c \cdot \rho_0 \cdot \sigma_{\text{inel}} \cdot \beta. \tag{8}$$

Here, $\beta$ is the relative velocity of the meson in the nuclear restframe. The resulting inelastic cross sections are shown in Fig. 20 as a function of the meson momentum. To compare the data to a parametrization frequently used in the literature, the data for both mesons have been fitted with an ansatz

$$\sigma_{\text{inel}}[mb] = a + \frac{b}{p[\text{GeV}/c]}, \tag{9}$$

as proposed by Lykasov et al. [67] and used as parametrisation in GiBUU transport simulations [21]. The present inelastic $\eta'$ cross section data (Fig. 20 (right)) shows a mean value of (13 ± 3) mb, slightly larger but consistent with the earlier result of (10.3 ± 1.4) mb reported in [22]. The experimental data are compared to calculations by Oset and Ramos [68]. They have studied the $\eta'$-nucleon interaction within a chiral unitary approach, including $\pi N$ and $\eta N$ coupled channels, which yields a very weak $\eta' N$ interaction. The $\eta' N$ amplitude is substantially enhanced when vector meson-baryon states are included in the coupled channel scheme via normal and anomalous couplings

DOI: 10.1051/epjconf/201713403003

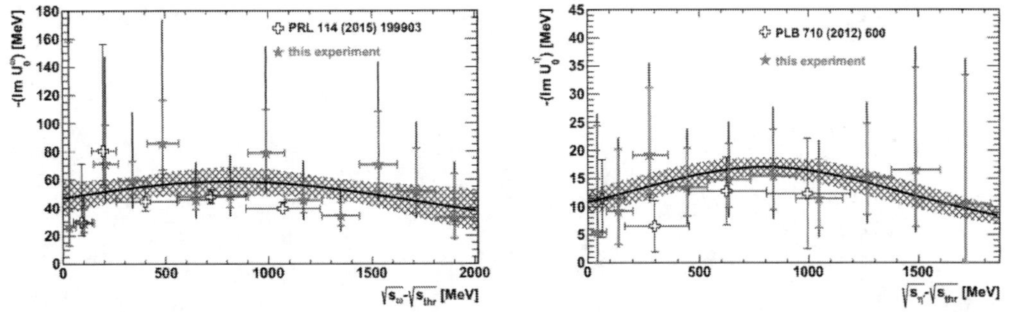


**Figure 21.** Imaginary part of the (Left) $\omega$-nucleus and (Right) $\eta'$-nucleus optical potential as a function of the available energy in the meson-$^{93}$Nb system (red stars) [40] in comparison to earlier measurements (open crosses) [22, 48, 66]. The solid curves are Breit-Wigner fits to the present data. The shaded areas indicate a confidence level of $\pm 1\sigma$ of the fit curve taking statistical and systematic errors into account.

of pseudo-scalar to vector mesons. In this approach inelastic $\eta' N$ cross sections rising from about 3 mb at $p_{\eta'}$ = 600 MeV/$c$ to about 20 mb at $p_{\eta'}$ = 50 MeV/$c$ are predicted. The calculations seem to underestimate the experimentally determined inelastic $\eta'$ cross section. This may not be surprising since multi-particle production, probably dominant because of the large $\eta'$ mass, has not been considered in [68].

As a final step, the momentum dependence of the in-medium $\omega$ and $\eta'$ widths from Fig. 19 can be converted into the dependence of the imaginary part of the $\omega$- and $\eta'$-nucleus potential as a function of the available energy in the meson-$^{93}$Nb system, as shown in Fig. 21. The imaginary part of the potential Im$U$ at normal nuclear matter density is just half of the in-medium width $\Gamma_0$ (see Fig. 19). The finer binning of the present data allows a more reliable extrapolation towards the production threshold by fitting the data. Several fit functions have been applied (polynomial of 1st. and 2nd. order, Gaussian, Breit-Wigner). The range of -Im$U(0)$ values obtained for different fit functions reflects the systematic uncertainties. For the $\omega$ meson the modulus of the imaginary part of the meson nucleus potential near threshold is found to be (48 ± 12(stat)± 9(syst)) MeV comparable to the modulus of the real part of about 30 MeV, determined in [50, 61, 69]. For the $\eta'$ meson the extrapolation towards the production threshold yields an imaginary potential of (13 ± 3(stat)±3(syst)) MeV, corresponding to an imaginary part of the $\eta'$ scattering length Im($a_{\eta' N}$) = (0.16 ± 0.05) fm. This is about a factor two smaller than obtained in the direct determination of the $\eta' N$ scattering length from an analysis of near threshold $\eta'$ production in the $pp \rightarrow pp\eta'$ reaction [70]. The error bars of both completely independent determinations do, however, almost overlap.

## 6 Meson-nucleus bound states

The existence of deeply bound pionic states is clearly established [71–74]. These systems are bound by the attractive Coulomb interaction between a negatively charged pion and the positively charged nucleus. The superposition with the strong interaction, which is repulsive at low pion momenta, leads to a potential pocket near the nuclear surface and consequently to a halo-like $\pi^-$ distribution around the nucleus [75]. Thus pions are only weakly absorbed, giving rise to rather narrow bound states which facilitated their experimental observation.

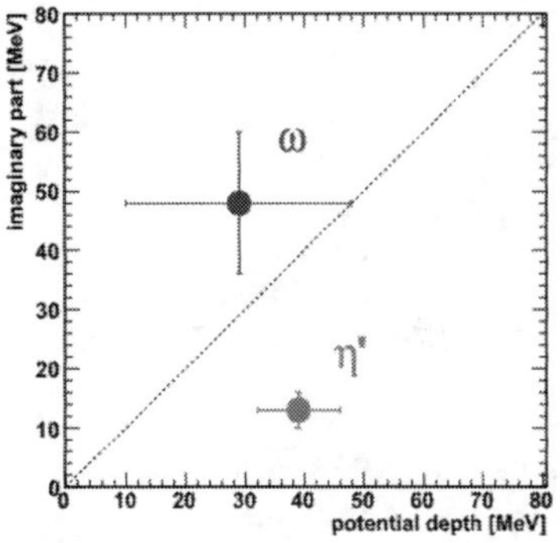


**Figure 22.** The imaginary part versus the real part of the meson-nucleus potential for $\omega$ and $\eta'$ mesons

As outlined in the previous sections, the interaction of neutral mesons with nuclei has been studied to find out whether meson-nucleus states might exist as well, only bound by the strong interaction. Here, the meson-nucleus interaction has to be sufficiently attractive and the meson absorption in nuclei should be relatively weak to allow the formation of relatively narrow states. Considering the results obtained for the $\omega$ and $\eta'$-nucleus potentials it is evident that the $\eta'$ meson is a much better candidate for the formation of meson-nucleus bound states than the $\omega$ meson since we found for the $\eta'$ meson that $|W| << |V|$, while the opposite was observed for the $\omega$ meson, as illustrated in Fig. 22. For the $\eta'$ meson there is a possibility for the existence of relatively narrow bound states while for the $\omega$ meson the imaginary potential is comparable to the

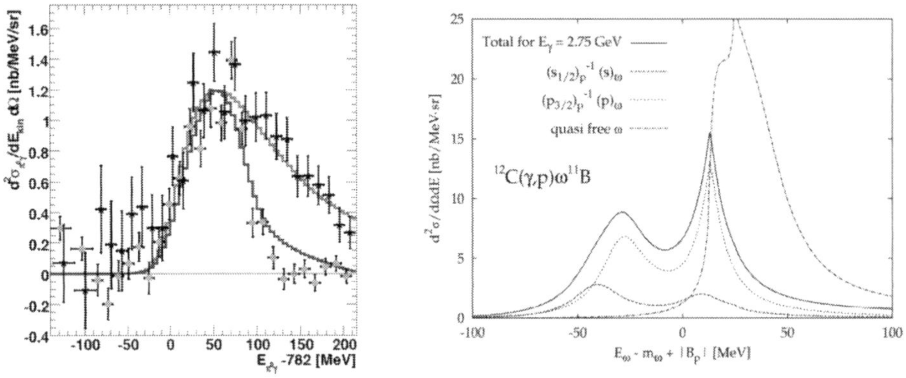


**Figure 23.** Left: Kinetic energy distribution of the $\omega$ meson off Carbon (black starts) compared with the kinetic energy distribution of the $\omega$ meson off the free proton (full blue circles). The $LH_2$ data are normalized to the C data in the peak of the total energy distribution. The experimental distributions are compared to Monte Carlo simulations ($LH_2$ : blue histogram; C: red histogram), taking the Fermi motion of nucleons into account for the C target. All distributions request the detection of a proton in the polar angular range 1° – 11° and are normalized to the fitted peak height for C. The Monte Carlo simulations are folded with the experimental resolution of $\sigma_E \approx 16$ MeV [61]. Right: Missing energy spectra calculated by Marco and Weise [76] for the $^{12}C(\gamma, p)\omega^{11}B$ reaction at $E_\gamma$ = 2.75 GeV. Dotted lines represent the contributions from two particular combinations of bound $\omega$ and proton-hole states.

real one and only broad structures can be expected which makes it difficult to detect them experimentally.

Nevertheless, an attempt to search for $\omega$ mesic states has been performed [61]. Using the $\gamma^{12}C \rightarrow \pi^0\gamma$ +p +X reaction with the proton detected at forward angles in TAPS, the distribution of the total energy $E_\omega$ of $\pi^0\gamma$ pairs with an invariant mass in the $\omega$ mass range has been measured and compared to the reaction $\gamma p \rightarrow \pi^0\gamma$ p on a liquid hydrogen target as shown in Fig. 23 (Left) [61]. Positive values $E_\omega - m_\omega$ give the kinetic energy of the $\omega$ in quasi-free production, while events with negative values may signal the decay of $\omega$ -$^{11}$B bound states via $\pi^0\gamma$ emission. Comparing the energy distributions measured for both targets, some tailing is observed for the carbon target in the fall off towards lower energies which is not observed for the reaction on the free proton, even if one accounts for the smearing due to Fermi motion which is taken into account in the Monte Carlo simulation. The $\pi^0\gamma$ cross section in the energy range from - 100 to 0 MeV is on average (0.3± 0.1) nb $MeV^{-1}sr^{-1}$. Correcting for the effective branching ratio for in-medium $\omega \rightarrow \pi^0\gamma$ decays of $\approx$ 1.5% [61] a population cross section of (22±7) nb $MeV^{-1}sr^{-1}$ is deduced which is of the order of magnitude for the formation of $\omega$ mesic states as calculated by Marco and Weise [76] (see Fig. 23 (Right)). Structures at negative energies which would indicate the population and decay of bound $\omega$ -$^{11}$B states are, however, not observed. The yield at negative energies may arise from the strong in-medium broadening of the $\omega$ meson, discussed in section 5.3.

Since the $\eta'$ meson is the better candidate for the search of mesic states several corresponding experiments have been launched. At the Fragment Separator (FRS) at GSI an experiment to search for $\eta'$ bound states has been performed via missing mass spectroscopy in the $^{12}$C(p,d) reaction in almost recoil free kinematics [77] and is being analysed. An alternative approach is the photoproduction of $\eta'$ mesons in the $^{12}$C($\gamma$,p) reaction, again in almost recoil-free kinematics. In the ongoing experiment at the LEPS2 facility (Spring8) [78] the missing mass spectroscopy is combined with detecting the decay of the $\eta'$ mesic state in a semi-exclusive measurement. An analogous photoproduction measurement has been proposed for the BGO-OD setup at the ELSA accelerator in Bonn [79]. A semi-exclusive measurement has also been considered for the Super-FRS at FAIR [65].

## 7 Conclusion

The photoproduction of $\omega$ and $\eta'$ mesons off nuclei has been studied in the 1-3 GeV energy range to extract information on the in-medium properties of these mesons. The interaction of mesons with nuclei and the in-medium effects have been described with an optical model, comprising a real part, associated with the mass modification in the nuclear medium, and an imaginary part, accounting for the absorption of mesons in nuclei. For the $\omega$- and $\eta'$- nucleus optical potential values of $U_{\omega A}(\rho = \rho_0)$ = -((29 ±19(stat)±20(syst)) + i(48±12(stat)±9(syst))) MeV and $U_{\eta' A}(\rho = \rho_0)$ = -((39 ±7(stat)±15(syst)) + i(13±3(stat)±3(syst))) MeV have been obtained. The latter measurement establishes for the first time an in-medium mass drop of a pseudoscalar meson at normal nuclear matter density. Although not very deep, the real part of the $\eta'$ - nucleus potential is about three time larger than the imaginary part and may thus allow the existence and observation of $\eta'$-nucleus states, only bound by the strong interaction. Corresponding experiments are ongoing.

The results reported here have been obtained within the CBELSA/TAPS collaboration and would not have been achieved without the work of strongly engaged and highly motivated PhD-students and postdocs, in particular F. Dietz, S. Friedrich, M. Kotulla, K. Makonyi, M. Thiel, and D. Trnka. M. Kotulla acted as co-projectleader in July 2008 before leaving our group. The support by the staff members of the ELSA accelerator and collaborating universities is highly appreciated. This project benefitted a lot from the close collaboration with project B5 (U. Mosel, P. Mühlich and J. Weil) regarding the interpretation of model calculations and the link between theoretical predictions and experimental observables, provided by transport calculations. The work reported here would not have been possible without the support from the Deutsche Forschungsgemeinschaft within the SFB/TR16.

## References

[1] V.- Bernard and Ulf-G. Meißner, Nucl. Phys. A **489**, 647 (1988).
[2] G. E. Brown and M. Rho, Phys. Rev. Lett. **66**, 2720 (1991).
[3] T. Hatsuda and S. Lee, Phys. Rev. C **46**, R34 (1992).
[4] R. Rapp, J. Wambach, H. van Hees, in: R. Stock (Ed.), Relativistic Heavy-Ion Physics, in: Landolt Börnstein (Springer), New Series, vol. I/23A, 4-1 (2010)
[5] R. Hayano and T. Hatsuda, Rev. Mod. Phys. **82**, 2949 (2010)
[6] S. Leupold, V. Metag, and U. Mosel, Int. J. Mod. Phys. E **19**, 147 (2010)
[7] S. Klimt et al., Nucl. Phys. A **516**, 429 (1990).
[8] V. Bernard, R. L. Jaffe and U.-G. Meißner, Nucl. Phys. B **308**, 753 (1988).
[9] S. Leupold, W. Peters, and U. Mosel, Nucl. Phys. A **628**, 311 (1998).
[10] V. Bernard and Ulf-G. Meißner, Phys. Rev. D **38**, 1551 (1988).
[11] H. Nagahiro, M. Takizawa, and S. Hirenzaki, Phys. Rev. C **74**, 045203 (2006).
[12] H. Nagahiro and S. Hirenzaki, Phys. Rev. Lett. **94**, 232503 (2005).
[13] F. Klingl et al., Nucl. Phys. A **610**, 297 (1997), Nucl. Phys. A **650**, 299 (1999).
[14] S. Sakai and D. Jido, Phys. Rev. C **88**, 064906 (2013).
[15] K.Saito, K.Tsushima, and A. W. Thomas, Prog. Part. Nucl. Phys. **58**, 1 (2007).
[16] S. D. Bass and A. W. Thomas, Phys. Lett. B **634**, 368 (2006).
[17] M. Lutz et al., Nucl. Phys. A **706**, 431 (2002).
[18] P. Mühlich et al., Nucl. Phys. A **780**, 187 (2006).
[19] A. Ramos, L. Tolos, R. Molina, and E. Oset, Eur. Phys. J. A **49**, 148 (2013).
[20] D. Cabrera and R. Rapp, Phys. Lett. B **729**, 67 (2014).
[21] O. Buss *et al.*, Phys. Rep. **512**, 1 (2012).
[22] M. Nanova *et al.*, CBELSA/TAPS Collaboration, Phys. Lett. B **710**, 600 (2012).
[23] K. A. Olive et al., Particle Data Group, Chin. Phys. C **38**, 090001 (2014).
[24] J. Weil, U. Mosel, and V. Metag, Phys. Lett. B **723**, 120 (2013).
[25] E. Ya. Paryev, J. Phys. G: Nucl. Part. Phys. **40**, 025201 (2013).
[26] D. Cabrera *et al.*, Nucl. Phys. A **733**, 130 (2004).
[27] D. Husmann and W. J. Schwille, Phys. Bl. **44**, 40 (1988).
[28] W. Hillert, Eur. Phys. J. A **28**, 139 (2006).
[29] D. Elsner *et al.*, CBELSA/TAPS Collaboration, Eur. Phys. J. A **33**, 147 (2007).
[30] E. Aker *et al.*, Nucl. Instr. and Meth. A **321**, 69 (1992).
[31] G. Suft *et al.*, Nucl. Instr. and Meth. A **538**, 416 (2005).
[32] R. Novotny *et al.*, IEEE Trans. Nucl. Sci. **38**, 392 (1991).
[33] A. R. Gabler *et al.*, Nucl. Instr. and Meth. A **346**, 168 (1994).
[34] T. Mertens *et al.*, CBELSA/TAPS Collaboration, Eur. Phys. J. A **38**, 195 (2008).
[35] M. Nanova *et al.*, CBELSA/TAPS Collaboration, Eur. Phys. J. A **47**, 16 (2011).
[36] M. Nanova *et al.*, CBELSA/TAPS Collaboration, Phys. Rev. C **82**, 035209 (2010).
[37] M. Nanova *et al.*, CBELSA/TAPS Collaboration, Phys. Lett. B **727**, 417 (2013).
[38] A. Thiel *et al.*, CBELSA/TAPS Collaboration, Phys. Rev. Lett. **109**, 102001 (2012).
[39] M. Thiel et al., Eur. Phys. J. A **49**, 132 (2013).
[40] S. Friedrich *et al.*, CBELSA/TAPS Collaboration, Eur. Phys. J. A **52**, 297 (2016).
[41] M. Nanova *et al.*, CBELSA/TAPS Collaboration, Phys. Rev. C **94**, 025205 (2016).
[42] R. Brun *et al.*, GEANT, CERN/DD/ee/84-1 (1986).
[43] F. Dietz et al., CBELSA/TAPS Collaboration, Eur. Phys. J. A **51**, 6 (2015).
[44] I. Jaegle *et al.*, CBELSA/TAPS Collaboration, Eur. Phys. J. A **47**, 11 (2011).
[45] C. Cioffi degli Atti and S. Simula, Phys. Rev. C **53**, 1689 (1996).
[46] J. G. Messchendorpl *et al.*, Eur. Phys. J. A **11**, 95 (2001).
[47] M. Kaskulov *et al.*, Phys. Rev. C **75**, 064616 (2007).
[48] M. Kotulla *et al.*, CBELSA/TAPS Collaboration, Phys. Rev. Lett. **100**, 192302 (2008).
[49] D. Trnka *et al.*, CBELSA/TAPS Collaboration, Phys. Rev. Lett. **94**, 192203 (2005).
[50] V. Metag *et al.*, A2 Collaboration, Prog. Part. Nucl. Phys. **67**, 530 (2012).
[51] V. Crede *et al.*, CBELSA/TAPS Collaboration, Phys. Rev. C **80**, 055202 (2009).
[52] S. V. Efremov and E. Ya. Paryev, Eur. Phys. J. A **1**, 99 (1998).
[53] Z. Rudy *et al.*, Eur. Phys. J. A **15**, 303 (2002).
[54] T. Falter, S. Leupold, and U. Mosel, Phys. Rev. C **64**, 024608 (2001).
[55] N. Bianchi *et al.*, Phys. Rev. C **54**, 1688 (1996).

[56] V. Muccifora *et al.*, Phys. Rev. C **60**, 064616 (1999).
[57] M. Debowski *et al.*, Acta Phys. Polonica B **27**, 3015 (1996).
[58] A. Sibirtsev, W. Cassing, and U. Mosel, Z. Phys. A **358**, 357 (1997).
[59] E. Ya. Paryev, Eur. Phys. J. A **5**, 307 (1999).
[60] H. Nagahiro, S. Hirenzaki, E. Oset, and A. Ramos, Phys. Lett. B **709**, 87 (2012).
[61] S. Friedrich *et al.*, CBELSA/TAPS Collaboration, Phys. Lett. B **736**, 26 (2014).
[62] B. Aubert *et al.*, Phys. Rev. D **70**, 112006 (2004).
[63] H. Nagahiro *et al.*, Phys. Rev. C **80**, 025205 (2009).
[64] H. Nagahiro *et al.*, Nucl. Phys. A **761**, 92 (2005).
[65] H. Nagahiro *et al.*, Phys. Rev. C **87**, 045201 (2013).
[66] M. Kotulla *et al.*, CBELSA/TAPS Collaboration, Phys. Rev. Lett. **114**, 199903 (2015).
[67] G. I. Lykasov *et al.*, Eur. Phys. J. A **6**, 71 (1999).
[68] E. Oset and A. Ramos, Phys. Lett. B **704**, 334 (2012).
[69] V. Metag, Hyp. Int. **234**, 25 (2015).
[70] E. Czerwinski *et al.*, Phys. Rev. Lett. **113**, 062004 (2014).
[71] H. Gilg *et al.*, Phys. Rev. C **62**, 025201 (2000).
[72] K. Itahashi *et al.*, Phys. Rev. C **62**, 025202 (2000).
[73] H. Geissel *et al.*, Phys. Rev. Lett. **88**, 122301 (2002).
[74] K. Suzuki *et al.*, Phys. Rev. Lett. **92**, 072302 (2004).
[75] P. Kienle and T. Yamazaki, Prog. Part. Nucl. Phys. **52**, 85 (2004).
[76] E. Marco and W. Weise, Phys. Lett. B **502**, 59 (2001).
[77] K. Itahashi *et al.*, Prog. Theo. Phys. **128**, 601 (2012).
[78] N. Muramatsu *et al.*, arXiv:1307.6411.
[79] V. Metag *et al.*, approved proposal ELSA/3-2012-BGO.

EPJ Web of Conferences **134**, 03004 (2017) DOI: 10.1051/epjconf/201713403004
*Subnuclear Structure of Matter: Achievements and Challenges*

# Hadrons in Nuclear Matter

*Ulrich* Mosel[1,*]

[1] *Institut für Theoretische Physik, Justus-Liebig-Universität Giessen, 35392 Giessen, Germany*

**Abstract.** We review the achievements of the project **B.5**, that deals with the calculation of in-medium properties of vector mesons and an analysis of their experimental signals, with a particular emphasis on the $\omega$ photoproduction data from CBELSA/TAPS. Other topics addressed include color transparency, pion electroproduction on nucleons, the Primakoff effect for nuclear targets and studies of hadronization at the EIC.

## 1 Introduction and motivation

The project **B.5** has been funded for the first and second funding period of the SFB/TR 16. This research project aimed at an investigation of selfenergies of vector mesons in medium and an analysis of their experimental signals. On one hand, these selfenergies were calculated using state-of-the-art many-body techniques. On the other hand, in close collaboration with the CBELSA/TAPS collaboration, the observable signals of any in-medium changes of hadronic properties were studied. Since the ELSA experiments use a semi-hadronic channel to look for the in-medium properties of the omega meson close to threshold particular focus was the study of the in-medium production cross section and the effects of final state branching ratios and interactions. Relevant data from JLAB and KEK were also studied.

Our contribution is organized as follows: Section 2 summarizes the work done in the first funding period. The results obtained in the second funding period are presented in Section 3. Due to the retirement of the PI in 2011, this very successfull project was terminated at the end of the second funding period.

## 2 First funding period 2004-2008

The interest in in-medium properties of hadrons has been growing over the last decade because of a possible connection with broken symmetries of QCD and their partial restoration inside nuclear matter. At the start of the present funding period there were some sophisticated theoretical predictions that masses of vector mesons should generally decrease in medium as a function of density due to a partial restoration of chiral symmetry [1, 2]. Specifically, there existed well-worked out predictions that the scalar $2\pi$ strength should decrease in medium and that the vector meson masses should drop [3]. All of these calculations were performed for idealized situations (infinite cold nuclear matter at rest) and little attention was paid to the actual observability of these predicted changes. At the same time experiments (CERES, TAPS) seemed to show the predicted behavior. The CERES results indicated a significant broadening of the $\rho$ meson in medium, whereas the TAPS results on the $2\pi$ strength exhibited the predicted lowering of the $\sigma$ strength inside nuclei.

During the first stage of the SFB/TR we have concentrated our theoretical work on in-medium properties along two different lines. First, we have performed state-of-the-art calculations of vector meson spectral functions in cold nuclear matter. Second, we have followed closely the CBELSA/TAPS experiment and have performed various feasibility studies and analyses of this experiment searching for in-medium changes of the $\omega$ meson in medium.

### 2.1 Vector meson spectral functions

On the first aspect we have initially finished a major calculation on the in-medium properties of the $\pi, \rho$ and $\eta$ mesons [4]. In this work we have generated the in-medium selfenergies of these mesons by nucleon-hole and resonance-hole excitations which in turn are affected by the changed in-medium properties of the mesons. This self-consistency problem has been solved here for the first time. Special care was taken to respect the analyticity of the spectral functions and to take into account effects from short-range correlations both for positive and negative parity states.

Our model has been shown to produce sensible results for pion and $\Delta$ dynamics in nuclear matter, as a test. For the $\rho$ meson we find a strong interplay with the D13(1520), which moves spectral strength of the $\rho$ spectrum to smaller invariant masses and simultaneously leads to a broadening of the baryon resonance. The strong interplay between the $\rho$ meson and the D13(1520)-nucleon hole excitation leads to a dominant lower hump in the rho spectral function also in this relativistic and selfconsistent calculation; it confirms our earlier result obtained in a more simplified approach. Whereas the longitudinal component of the

*e-mail: ulrich.mosel@theo.physik.uni-giessen.de

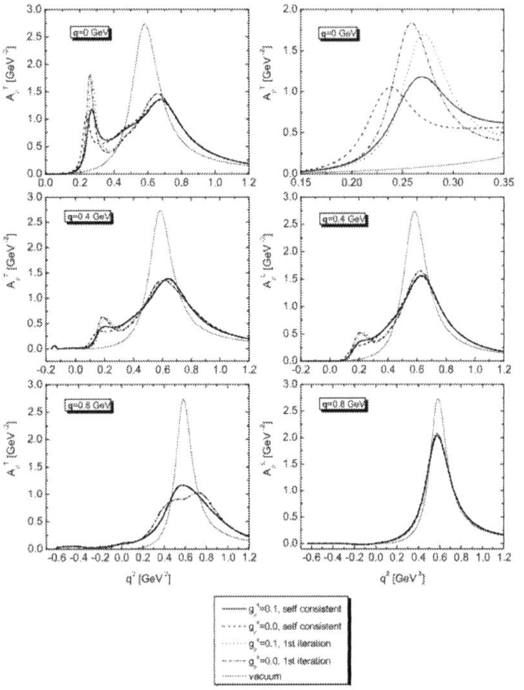


**Figure 1.** Spectral function of the $\rho$ meson in nuclear matter at density $\rho_0$ for various momenta indicated in the figure. The left column shows the transverse spectral function, the right column that of longitudinally polarized $\rho$ mesons. The thin dotted line in each figure is the vacuum spectral function, the other curves give the effect of selfconsistency and short-range correlations (from [4]).

$\rho$ meson only broadens somewhat, the transverse component shows a major distortion which evolves as a function of the $\rho$ momentum (see Fig. 1). At the same time, the D13(1520) resonance broadens considerably. For the $\eta$ meson the optical potential resulting from our model is rather attractive whereas the in-medium modifications of the S11(1535) are found to be quite small.

These studies also allow us to assess the validity range of the often used low-density approximation. We find that this depends very much on the special couplings involved and thus varies from meson to meson. Whereas for the $\eta$ meson the validity ranges up to a density of $\rho_0$, for the $\rho$ meson it already breaks down at about 0.3 $\rho_0$. This may serve as a warning sign for many in-medium calculations that use the low-density approximation without any further proof of its reliability.

Bearing this in mind we have recently also performed a calculation of the selfenergy of the $\omega$ meson in medium. This calculation is based on a unitary coupled channel analysis of all existing $\pi N$ and $\gamma N$ data up to an invariant mass of 2 GeV [5]. The coupled channel character of this calculation is of utmost importance here because it is the only way to include experimental constraints on the $2\pi$ decay channel that was found to be dominant in [2]. This analysis and thus also the selfenergy of the $\omega$ meson extracted from the $\omega N$ scattering amplitude gives a broadening of about 60 MeV at $\rho_0$ and a small upward shift of

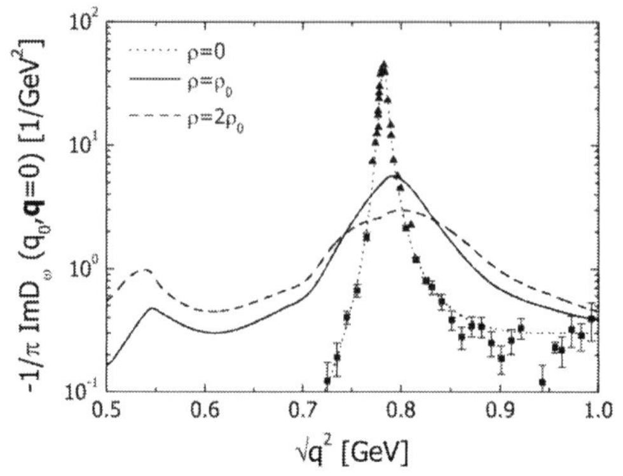


**Figure 2.** Spectral function of the $\omega$ meson in nuclear matter at rest, at densities 0, $\rho_0$ and 2 $\rho_0$ (from [5]).

the peak mass. In addition, due to a nonzero coupling of the $\omega$ to the S11(1535) resonance the $\omega$ spectral function exhibits a small peak at a mass of around 550 MeV. This calculation is in two aspects novel and unique: it gives, for the first time, the $\omega$ selfenergy also for nonzero momenta (which corresponds to the experimental situation) and it takes the experimental constraints better into account than any other calculation of the selfenergy because it is based on a unitary K-matrix analysis of 'real' data. The result of this calculation is shown in Fig. 2.

### 2.2 Analysis of CBELSA/TAPS $\omega$ photoproduction data

Our second line of approach to the problem of in-medium selfenergies has concentrated on an analysis of the recent CBELSA/TAPS data and was followed in intensive collaboration – with nearly daily contacts – with the experimental colleagues of the TAPS collaboration at Giessen. Since the experiment looks for the channel $\gamma + A \rightarrow A^* + \omega \rightarrow A^* + \gamma + \pi^0$ it is mandatory to control the effects of final state interactions on the $\pi^0$ in a quantitative way. The only method available for this is that of coupled channel semi-classical transport calculations which – as we had shown earlier in extensive work – can give a consistent description of many experimental phenomena, both in heavy-ion as well as in nucleon-, pion- and photon-induced reactions. For any reaction on nuclei with hadrons in the final state a state-of-the-art transport calculation is an indispensible part of the theory. We have, therefore, spent significant effort on developing a new code, dubbed 'GiBUU', for the transport calculations. This code is written in object-oriented FORTRAN 95/2003 and incorporates all the experience we have gained with earlier numerical implementations at Giessen over the last 20 years [6]. It should be noted that this code is unique, we do not know about any other competitive effort for the description of inclusive or semi-inclusive incoherent reactions with elementary probes on nuclei.

With this method we have first analysed both results on the experimental determination of the nuclear transparency ratio for $\phi$ mesons [7], measured by a Japanese group at SPRING8 that lead to an unexpectedly large inelastic cross section for $\phi N$ interactions. We have found that indeed cross sections about a factor 3 larger than those theoretically expected are needed to explain the mentioned data, in line with a simple Glauber analysis by the SPRING8 group.

A major effort has gone into an analysis of the $\omega$ photoproduction experiment at CBELSA/TAPS [8, 9]. Our simulations give a full event analysis and thus allow to calculate also background contributions on the same footing as the actual signal. They also allow insight into the effects of rescattering of the pions produced in the decay of the $\omega$ meson and have suggested a method to suppress the rescattered pion background that has actually been adopted by the experimental group. A problem in this context is that the experiment does not determine the spectral function of the $\omega$ meson itself. Instead, we have noted that the result of the experimental analysis is the product of the spectral function with the partial decay probability into the channel under study ($\pi^0\gamma$ here). If the latter depends strongly on the invariant mass itself, as it does for the CBELSA/TAPS experiment, then significant distortions of the spectral function may arise. This is a topic under intensive study by us presently and also planned for the second period of the SFB/TR (see below).

The CBELSA/TAPS collaboration has also measured the nuclear transparency for $\omega$ mesons. This transparency gives directly the imaginary part of the meson's selfenergy in medium; using a low-density approximation one can then extract the inelastic $\omega N$ cross section. We have shown that our calculations reproduce the measured attenuation quite well if the inelastic cross section is increased by about a factor of 2 beyond theoretical expectations. A fit to the data can actually also determine the momentum-dependence of this cross section which at low momenta comes out to be larger than the usually used parametrizations [9].

Finally, we have analyzed the TAPS data on $2\pi^0$ photoproduction on nuclei. The motivation for this experiment was a prediction that – due to chiral symmetry restoration in nuclei – the scalar strength of the $\sigma$ meson should be lowered in nuclei. The TAPS collaboration had indeed initially seen an effect as predicted in the $2\pi^0$ data, whereas a comparison measurement in the $\pi^0\pi^{\pm}$ channel did not seem to show such an effect. Various explanations for these findings have been advanced by the Valencia group and by a group in Lyon in terms of chiral symmetry restoration or $\pi - \pi$ correlations in nuclei, based on a chiral effective field theory model.

None of these calculations, however, did look into the simplest possible explanation of the observed effects in terms of mundane pion rescattering. We have, therefore, performed such calculations using the GiBUU method which is ideally suited for this task. These calculations, which did not use any effects connected with $\pi\pi$ correlations, could reproduce the observed effect for the $\pi^0\pi^0$ channel, but they also predicted a similar effect in the semi-charged channel where it had not been seen experimentally. However, a more recent analysis with higher statistics by Bloch et al. [10] yielded a result for both

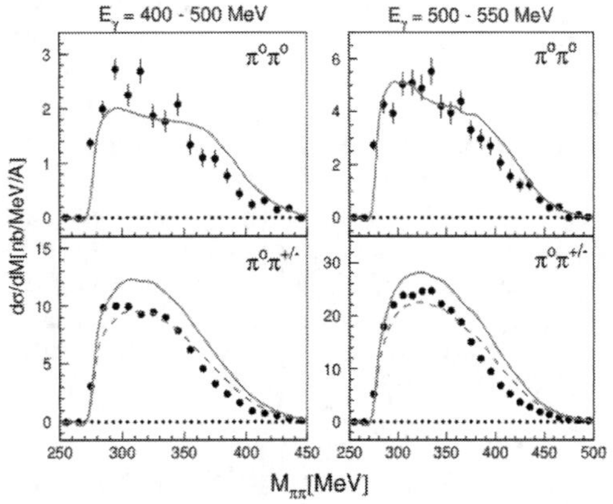


**Figure 3.** Data of the TAPS collaboration (Bloch et al.) for $2\pi^0$ photoproduction on a $^{40}Ca$ target for two different photon energies. The solid curve gives the result of a GiBUU calculation, the dashed curves in the semicharged $2\pi$ channel are normalized to the data (see text). Data from [10].

charge channels that is in perfect agreement with our calculations (see Fig. 3; the dashed lines in this figure are normalized in height to the data, this normalization reflects uncertainties in the elementary cross sections). In particular the yield in the semi-charged channel is strongly influenced by a coupled-channel effect, the charge transfer in $\pi N$ interactions; Glauber based absorption models miss this contribution. This illustrates that a very sophisticated treatment of final state interactions is absolutely mandatory when looking for more 'exotic' effects in nuclei. We conclude that any analysis of the $2\pi^0$ data with respect to a lowering of the scalar strength in nuclei has to take the pion rescattering effects into account. Present day's data are all consistent with simple rescattering.

The capability we have in the field of coupled channel transport calculations for reactions with elementary probes on nuclei, absolutely essential for an understanding of actual observables, is worldwide unique. This outstanding role of the GiBUU effort is reflected in the many publications in international refereed journals and the large number of invited talks on major international conferences and workshops.

All this work has been done in very close collaboration and nearly daily discussions with the local members of the TAPS group, in particular M. Kotulla and V. Metag. It has a direct physics connection to the experimental subproject **B.4**. We also have a close working relationship with the g7 group at JLAB/North Carolina which has resulted already in one joint publication [11] and to the RIKEN/U of Tokyo group working at KEK.

## 3 Second funding period 2008-2012

### 3.1 Development of the project

The central aim of the present project was to explore if any of the theoretically predicted in-medium changes of hadronic properties could actually be observed in ongoing experiments. With this aim in mind we have kept close contacts to the relevant experiments, in particular, to CBELSA/TAPS at ELSA. With the Giessen members of the CBELSA/TAPS group we have had nearly weekly discussions meetings (project **B.4**, V. Metag, M. Nanova). Also to the experiment HADES at GSI, that aims for a determination of in-medium properties of vector mesons in hadronic reactions, we are keeping close contacts. We have also provided guidance to the analysis of the g-7 experiment at JLAB; one of our graduate students, J. Weil, spent time there to help in the evaluation of these data. In both cases the transport theory and code dubbed GiBUU, developed by us over the last twenty year, played an important role as 'work horse'.

At the start of this funding period a contradictory situation existed: on one hand, theory predicted nearly no mass shifts of the vector mesons in medium, but a sizeable collisional broadening. On the other hand, experiments (CBELSA/TAPS and KEK) seemed to show some mass-shift which was significantly bigger than theoretically expected. This discrepancy was discussed on various conferences and meetings. In both cases a possible resolution was reached when it was observed that the background subtraction process, needed to isolate, e.g., the $\omega$ signal, was unreliable. This seems to be the accepted explanation for the mass shifts observed by the KEK experiment (which did not observe any broadening, the latter is hard to accept since dispersion effects link real and imaginary parts of the self energies). Also for the CBELSA/TAPS experiment the background effects came under closer scrutiny. In Ref. [12] M. Kaskulov, the part-time postdoc in the present project, showed that the mass-shift signal depended strongly on the background subtraction method. This was also realized by the CBELSA/TAPS collaboration when a reanalysis of the data and the background determination became available [13].

From that point on we had an intensive collaboration with the Giessen experimental colleagues and have provided both theoretical and practical, computational input to their project. The specific contributions will be listed below. In addition, we broadened our approach by also analyzing other experiments at JLAB that aimed for the same physics, i.e. in-medium properties of hadrons. An important insight gained in these studies was that a reliable analysis of these experiments, that were also set up to discover color transparency, was not possible without a good understanding of the elementary production vertex. In the course of that work we could solve an old outstanding problem, i.e. that of transverse vs. longitudinal strength in hard exclusive reactions. Also these developments will be discussed in some more detail below.

Much work was also spent, as already envisaged in the application, on 'apparatus development', i.e. on further development of the GiBUU event generator. Here we improved various aspects relevant to the present work, among others the description of quasielastic lepton scattering, a better treatment of groundstate stability problems and the implementation of cross sections relevant for electroweak reactions.

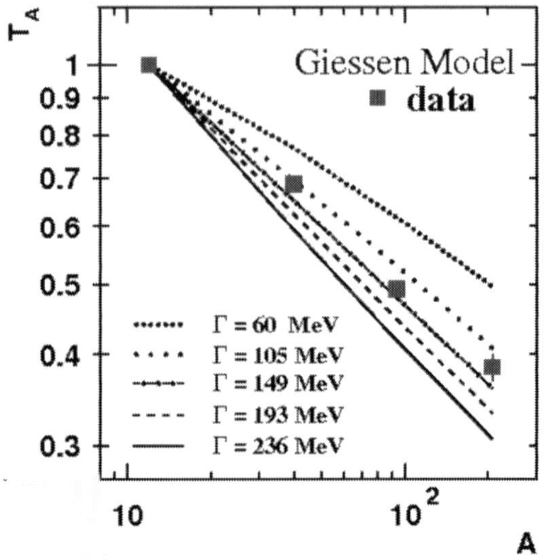


**Figure 4.** The nuclear transparency $T_A$, normalized to $^{12}C$, as a function of mass number $A$ for various assumed widths Γ. Also shown are the datapoints from the CBELSA/TAPS experiment (project **B.4**).

### 3.2 Analysis of transparency ratio

In a new series of experiments the CBELSA/TAPS group had explored the transparency of nuclei for $\omega$ mesons. We have analyzed these data within the framework of GiBUU, a semiclassical coupled channel transport calculation (for details see later subsection). In a joint publication with the experimental group [14] we have shown that the measured transparency ratios require an $\omega$- width in medium of about 150 MeV at a momentum of about 1 GeV (see Fig. 4). Thus, considerable collisional broadening takes place, since the free width of the $\omega$ meson is only 9 MeV. The numbers extracted agreed with the ones obtained in a simplified analysis by the Valencia group.

Interesting about this analysis is that it allows the extraction of an $\omega - N$ inelastic cross section, under the assumption of low density, i.e. prevalence of two-body collisions. These inelastic cross sections are generally unknown, but play a role in any in-medium scenario. The values extracted all lie relatively high ($\approx 60 - 70$ mb) for momenta of about 1 GeV, see detailed discussion in project **B4**.

### 3.3 Lineshape analysis

We had shown already in [15] that nucleon spectral functions are quite insensitive to collisional broadening. The reason is that observed line-shapes always represent an average over many nuclear densities and that the contribution of resonances at the largest density, where they experience most of the collisional broadening, tends to become indistinguishable from background contributions. The same physics holds also for the observability of collisional broadening of meson spectral functions. An ex-

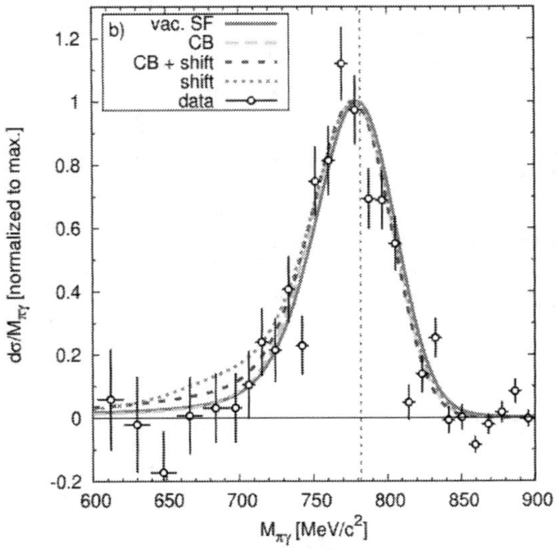

**Figure 5.** $\omega$ signal for the *Nb* target from the CBELSA/TAPS collaboration in comparison to recent GiBUU simulations for the following scenarios: no medium modification (solid), in-medium broadening of $\Gamma_{coll}$ = 140 MeV at nuclear saturation density (long dashed), an additional mass shift by -16% (short dashed), mass shift without broadening (dotted).

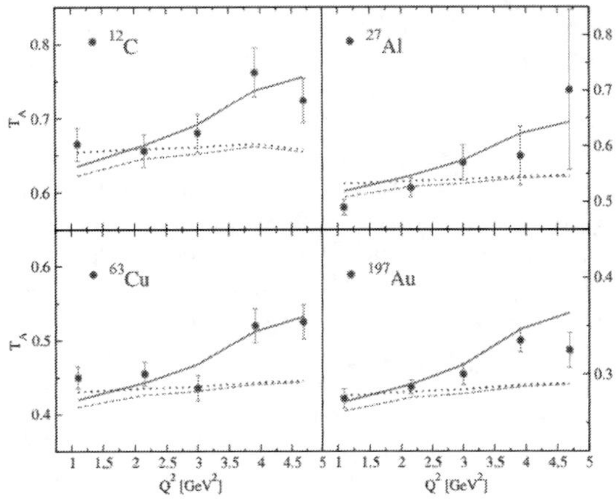

**Figure 6.** Transparency, $T_A$, vs. $Q^2$ for $^{12}$C (left, top panel), $^{29}$Al (right, top), $^{63}$Cu (left, bottom) and $^{197}$Au (right, bottom). Data from JLAB experiment $\pi$CT.

ample of a calculation is given in Fig. 5, together with the data from CBELSA/TAPS (project **B.4**).

The figure shows clearly that the observable effects of assumed in-medium modifications are relatively minor and that any experiment that aims to distinguish between the different scenarios shown in the figure (collisional broadening alone, mass shift alone, both collisional broadening and mass shift) would require systematic and statistical uncertainties well below those of presently feasible experiments. The present data are clearly compatible with some collisional broadening, but any mass-shift can not be inferred.

These results are all compatible with the observations of in-medium hadronic properties in the g-7 experiment at JLAB. Here we have – in close collaboration with that group – analyzed data on photon-induced dilepton production on nuclei which lead to a joint publication [16]. This is in principle an ideal experiment since it involves neither initial state nor final state interactions. The experiment was sensitive to the $\rho$ meson's spectral shape. Because of the $\rho$'s considerable broadening in medium (see our report for the last funding period) this was indeed observable and confirmed previous theoretical predictions.

### 3.4 Color transparency

In the projects discussed before the analysis relied on the transport generator GiBUU. A very interesting application of it, that found widespread interest, is the analysis of pion and $\rho$-meson electroproduction on nuclei with the aim to look for possible effects of color transparency (CT). In [17] we have investigated the onset of color transparency in semi-exclusive pion electroproduction on nuclei. We have found that the data of the experiment $\pi$CT at JLAB could be well reproduced only if hard partonic processes dominate the transverse channel. The needed formation times were found to be compatible to our earlier analyses of HERMES and EMC data [18]. Fig. 6 illustrates the CT effect. The result of a calculation in which the formed hadrons interact without any delay with the full cross section is shown by the dotted curves. These curves are nearly flat and thus do not exhibit the observed rise with $Q^2$. The same is true for the short–dashed curves which include in addition shadowing; the influence of the latter is very small due to the very small coherence lengths in this kinematical regime. Finally, to show the effect of diminished prehadronic cross sections independent of $Q^2$ during their formation time we plot the solid curves which provide a very satisfactory agreement with data.

For the $\rho$ meson production experiment [19], also at JLAB, GiBUU allowed a full event simulation. This was essential since the experiment 'suffered' from various kinematical cuts that had to be reproduced by the calculations for any meaningful comparison. We have compared various different hadronization and CT scenarios and have found that a detailed analysis of elementary cross section, nuclear effects and experimental cuts is needed to reveal the early onset of $\rho$-CT at the presently available JLAB energies [20]. Here the major strength of an analysis based on a full event-generator becomes obvious again: the entanglement of kinematic cuts and Fermi-motion effects in this experiment could not be treated in any other calculation.

### 3.5 Pion electroproduction on nucleons

In the CT analyses of pion production on nuclei we have shown that a proper distinction between the soft hadronic and the hard partonic components of the electroproduction amplitude is of utmost importance. Here the partonic components show up primarily in the transverse part of the reaction amplitude while the hadronic components dominate the longitudinal components. It turned out that the available, widely used model for the calculation of elementary electroproduction of pions on nucleons was seriously deficient: while the longitudinal part of the cross section was described very well, the transverse part was significantly underestimated (by about 1 order of magnitude!). We have, therefore, developed a new model that describes both reaction components over a very wide kinematical range extremely well [17, 21, 22]. In particular, the new model is able to describe the observed onset of transverse strength with increasing momentum transfer $Q^2$, which is in contradition to simple pQCD based expectations for exclusive reaction channels. Essential for this agreement is the consideration of $s$-channel processes in the exclusive electroproduction that involve high-lying nucleon resonances. The publication was listed by the APS on a webpage for 'exceptional research' (http://physics.aps.org/synopsis-for/10.1103/PhysRevC.81.045202).

### 3.6 Hadronization in the EIC

The projects just listed gave us all the necessary experience and tools to also contribute to studies of hadronization, measured in the EMC experiment and with HERMES. Our calculations of hadronic attenuation as a function of energy transfer, rapidity and momentum transfer for these two experiments are still the only ones available that can describe the measurements for all hadron species [18]. Based on this success we had already earlier made predictions for the 12 GeV beam at JLAB. This led to an invitation to contribute to the plans for an electron-ion collider (EIC). On two workshops we presented our results and contributed to the first major write-up of a scientific program for such a machine [23].

### 3.7 Primakoff effect

As a direct application both of the development for in-medium production of pions in the two preceding subsections we have performed a new calculation of the Primakoff effect which relies on a very reliable calculation of the amplitude for photoproduction of $\pi^0$ on nuclear targets [24]. Required is not only a calculation of the Primakoff amplitude itself, but in addition also of the nuclear amplitude which interferes with the former. Essential for a good agreement with recent JLAB data is that our calculations allow for photon shadowing and final state interactions of the outgoing pions. In this paper we also make predictions for planned experiments on $\eta$ and $\eta$' Primakoff production.

### 3.8 Review articles

During the present funding period 2 major review articles have been written.
*Hadrons in strongly interacting matter:*
In this review in International Journal of Modern Physics, published together with S. Leupold and V. Metag in [25], we have summarized the current status of theories and and experiments aiming at an understanding and a determination of the properties of light vector and scalar mesons inside strongly interacting hadronic matter . Starting from a discussion of the relevant symmetries of QCD and their connection with the hadronic description through QCD sum rules we have then discussed hadronic models used to calculate the in-medium self-energies of hadrons and their spectral functions. Finally, we have reviewed in detail all the running experiments searching for in-medium changes of vector and scalar mesons, both with relativistic heavy-ion reactions as well as with elementary reactions on (cold) nuclei. Inconsistencies among experimental results were discussed.
*Transport-theoretical description of nuclear reactions:*
In this review in Physics Reports, published with a number of Giessen coauthors [26], we have first outlined the basics of transport theory and its recent generalization to off- shell transport. We have then presented in some detail the main ingredients of the Giessen Boltzmann-Uehling-Uhlenbeck (GiBUU) implementation of this theory. The central part of this review covered applications of GiBUU to a wide class of reactions, starting from pion-induced reactions over proton and antiproton reactions on nuclei to heavy-ion collisions (up to about 30 AGeV). A major part concerns also the description of photon-, electron- and neutrino-induced reactions (in the energy range from a few 100MeV to a few 100 GeV). For this wide class of reactions GiBUU gives an excellent description with the same physics input and the same code being used. We have argued in this review that GiBUU is an indispensable tool for any investigation of nuclear reactions in which final-state interactions play a role. Studies of pion-nucleus interactions, nuclear fragmentation, heavy-ion reactions, hypernucleus formation, hadronization, color transparency, electron-nucleus collisions and neutrino-nucleus interactions are all possible applications of GiBUU and are described well.

We would like to thank all our collaborators that have contributed to the success of this project, in particular also to the experimental colleagues from project **B.4**. The work reported here would not have been possible without the support from the Deutsche Forschungsgemeinschaft within the SFB/TR16.

## References

[1] T. Hatsuda, S. H. Lee and H. Shiomi, Phys. Rev. C **52**, 3364 (1995) [arXiv:nucl-th/9505005].
[2] F. Klingl and W. Weise, Nucl. Phys. A **606** (1996) 329; F. Klingl, N. Kaiser and W. Weise, Nucl. Phys. A **624** (1997) 527 [arXiv:hep-ph/9704398].
[3] K. Yokokawa, T. Hatsuda, A. Hayashigaki and T. Kunihiro, Phys. Rev. C **66** (2002) 022201 [arXiv:hep-ph/0204163].
[4] M. Post, S. Leupold and U. Mosel, Nucl. Phys. A **741**, 81 (2004) [arXiv:nucl-th/0309085].
[5] P. Muehlich, V. Shklyar, S. Leupold, U. Mosel and M. Post, Nucl. Phys. A **780**, 187 (2006) [arXiv:nucl-th/0607061].

[6] for details see: http://gibuu.physik.uni-giessen.de/GiBUU/

[7] P. Muhlich and U. Mosel, Nucl. Phys. A **765**, 188 (2006) [arXiv:nucl-th/0510078].

[8] D. Trnka *et al.* [CBELSA/TAPS Collaboration], Phys. Rev. Lett. **94**, 192303 (2005) [arXiv:nucl-ex/0504010].

[9] P. Muehlich, "Mesons in Nuclei and Nuclear Reactions", PhD Dissertation, University of Giessen, 2007, http://theorie.physik.uni-giessen.de/documents/dissertation/muehlich.pdf

[10] F. Bloch *et al.*, Eur. Phys. J. A **32** (2007) 219 [arXiv:nucl-ex/0703037].

[11] R. Nasseripour, M. H. Wood, C. Djalali, D. P. Weygand, C. Tur, U. Mosel, P. Muehlich, CLAS Collaboration, Phys. Rev. Lett. **99** (2007) 262302 [arXiv:0707.2324 [nucl-ex]].

[12] M. Kaskulov, E. Hernandez and E. Oset, Eur. Phys. J. A **46** (2010) 223 [arXiv:1003.2363 [nucl-th]].

[13] M. Nanova *et al.* [CBELSA/TAPS Collaboration], Phys. Rev. C **82** (2010) 035209 [arXiv:1005.5694 [nucl-ex]].

[14] M. Kotulla *et al.* [CBELSA/TAPS Collaboration], Phys. Rev. Lett. **100** (2008) 192302 [arXiv:0802.0989 [nucl-ex]].

[15] J. Lehr and U. Mosel, Phys. Rev. C **64** (2001) 042202 [nucl-th/0105054].

[16] M. H. Wood *et al.* [CLAS Collaboration], Phys. Rev. C **78** (2008) 015201 [arXiv:0803.0492 [nucl-ex]].

[17] M. M. Kaskulov, K. Gallmeister and U. Mosel, Phys. Rev. C **79** (2009) 015207 [arXiv:0808.2564 [nucl-th]].

[18] K. Gallmeister and U. Mosel, Nucl. Phys. A **801** (2008) 68 [nucl-th/0701064].

[19] L. El Fassi *et al.* [CLAS Collaboration], Phys. Lett. B **712** (2012) 326 [arXiv:1201.2735 [nucl-ex]].

[20] K. Gallmeister, M. Kaskulov and U. Mosel, Phys. Rev. C **83** (2011) 015201 [arXiv:1007.1141 [hep-ph]].

[21] M. M. Kaskulov and U. Mosel, Phys. Rev. C **80** (2009) 028202 [arXiv:0904.4442 [hep-ph]].

[22] M. M. Kaskulov and U. Mosel, Phys. Rev. C **81** (2010) 045202 [arXiv:1001.1952 [hep-ph]].

[23] D. Boer, M. Diehl, R. Milner, R. Venugopalan, W. Vogelsang, D. Kaplan, H. Montgomery and S. Vigdor *et al. (K. Gallmeister, U. Mosel),*, arXiv:1108.1713 [nucl-th].

[24] M. M. Kaskulov and U. Mosel, Phys. Rev. C **84** (2011) 065206 [arXiv:1103.2097 [nucl-th]].

[25] S. Leupold, V. Metag and U. Mosel, Int. J. Mod. Phys. E **19** (2010) 147 [arXiv:0907.2388 [nucl-th]].

[26] O. Buss, T. Gaitanos, K. Gallmeister, H. van Hees, M. Kaskulov, O. Lalakulich, A. B. Larionov and T. Leitner *et al.*, Phys. Rept. **512** (2012) 1 [arXiv:1106.1344 [hep-ph]].

# Neutron properties from light nuclei

*Evgeny* Epelbaum[1,*], *Hans-Werner* Hammer[2,3,**] and *Ulf-G.* Meißner[4,5,***]

[1] *Institut für Theoretische Physik II, Ruhr-Universität Bochum, D-44780 Bochum, Germany*
[2] *Institut für Kernphysik, Technische Universität Darmstadt, 64289 Darmstadt, Germany*
[3] *ExtreMe Matter Institute EMMI, GSI Helmholtzzentrum für Schwerionenforschung GmbH, 64291 Darmstadt, Germany*
[4] *Universität Bonn, Helmholtz Institut für Strahlen- und Kernphysik, Bethe Center for Theoretical Physics, D-53115 Bonn, Germany*
[5] *Forschungszentrum Jülich, Institut für Kernphysik (IKP-3), Institute for Advanced Simulation (IAS-4) and Jülich Center for Hadron Physics, D-52145 Jülich, Germany*

**Abstract.** We review the achievements of the project **B.6**. Topics addressed include pion photoproduction of three-nucleon systems, electromagnetic and axial currents in chiral nuclear EFT, aspects of EFTs with spin-3/2 fields, nuclear structure and dynamics from the pionless EFT, the development and applications of high-precision two-nucleon forces and pion production in nucleon-nucleon collisions.

## 1 Introduction

The project **B.6** has been funded over the second and third period of the SFB/TR 16. It grew out of the project **B.3** from the first funding period, as explained in the corresponding contribution to that project. In the second funding period, the PIs were Evgeny Epelbaum and Hans-Werner Hammer. After about one year in the third funding period, Hans-Werner Hammer moved to Darmstadt and was replaced as PI by Ulf Meißner, however, the collaboration continued. The project focused on few-nucleon physics in various settings of nuclear EFT (pionless and pionfull) as well as on electroweak currents in nuclei, pion production in nucleon-nucleon collisions and formal aspects related to gauge invariance and the $\Delta(1232)$ resonance.

Our contribution is organized as follows: Section 2 summarizes the results on pion photoproduction off three-nucleon systems. In Sec. 3, we display the results obtained for the electromagnetic currents in chiral nuclear EFT and a first application to deuteron disintegration. Sec. 4 is devoted to some formal developments of including spin-3/2 particles in chiral EFTs. In Sec. 5, the applications of the so-called pionless EFT to nucleon-deuteron scattering, the halo nucleus $^{11}$Be and finite volume methods in the three-particle system are discussed. A novel approach to the two-nucleon system based on a coordinate-space representation and a novel method for error quantification are described in Sec. 6. The rather involved calculation to generate consistent axial currents is briefly summarized in Sec. 7 and aspects of pion production in nucleon-nucleon collisions are displayed in Sec. 8. Finally, an **important disclaimer:** This is the review of the project **B.6** but not a general review of the various fields to which this research contributed. Therefore, only the papers that resulted from this research are quoted, with the exception of a few external papers and of course, in case data were involved, with proper reference to the experimental papers. The links to the other projects are mentioned in passing, but no effort is made to discuss these in any detail.

*e-mail: evgeny.epelbaum@rub.de
**e-mail: hammer@theorie.ikp.physik.tu-darmstadt.de
***e-mail: meissner@hiskp.uni-bonn.de

## 2 Pion photoproduction off three-nucleon systems

Threshold neutral pion photo- and electroproduction off the nucleon is one of the finest reactions to test the chiral QCD dynamics. While elementary proton targets are accessible directly in experiment, pion production off neutrons requires the use of nuclear targets like the deuteron or three-nucleon bound states like $^3$H (triton) or $^3$He. Of particular interest is to test the counterintuitive chiral perturbation theory (CHPT) prediction that the elementary neutron S-wave multipole $E_{0+}^{\pi^0 n}$ is larger in magnitude than the corresponding one of the proton, $E_{0+}^{\pi^0 p}$ [1, 2]. This prediction was already successfully tested in neutral pion photo- [3] and electroproduction off the deuteron [4]. However, given the scarcity and precision of the corresponding data, it is mandatory to study also pion production off three-nucleon bound states, that can be calculated nowadays to high precision based on chiral nuclear effective field theory (EFT), that extends CHPT to nuclear physics, for a recent detailed review, see Ref. [5]. Note that $^3$He appears to be a particularly promising target to extract the information about the neutron amplitude. Its wave function is strongly dominated by the principal "s-state" component which suggests that the spin of $^3$He is largely driven by the one of the neutron.

In Ref. [6], we have presented a calculation of neutral pion production off $^3$H and $^3$He at threshold to leading one-loop order for the production operator in the framework of chiral nuclear effective field theory. We used the chiral wave functions from Epelbaum, Glöckle and Meißner, which are consistent with the pion production operator to calculate the S-wave 3N multipoles $E_{0+}$ and $L_{0+}$. To this order, the production operator gets both one- and two-body contributions. Our calculation shows that the two-body contributions are of the same order of magnitude as the one-body contributions. A similar behavior was observed in the deuteron case [3]. The theoretical uncertainty resulting from the cutoff variation in the employed wave functions appears to be small (of the order of 3%). The dominant theoretical error at this order stems from the threshold pion production amplitude off the proton and the neutron, which is estimated to be about 5%. We explored the sensitivity of neutral pion photoproduction on $^3$He to the elementary neutron multipole $E_{0+}^{\pi^0 n}$ and found a large sensitivity. This makes $^3$He a promising target to test the counterintuitive CHPT prediction for $E_{0+}^{\pi^0 n}$ [1, 2]. The cutoff variation estimate leads to a very small error for the 2N contribution. If the error of this contribution is artificially enlarged by a factor of 10, the extraction of the neutron multipole is still feasible experimentally. We have shown that our prediction for the $^3$He S-wave multipole $E_{0+}$ is roughly consistent with the value deduced from the old Saclay measurement of the threshold cross section [7]. A new measurement using modern technology and better methods to deal with few-body dynamics is urgently called for.

In Ref. [8], we have extended this calculation to fourth order in the chiral expansion, including consistently *all* next-to-next-to-leading order contributions in the standard heavy baryon expansion. This amounts to a complete (i.e. subleading) one-loop calculation in the one-nucleon sector. Apart from the fourth order corrections already included in the single nucleon multipoles $E_{0+}$ and $L_{0+}$, there are various fourth order corrections that arise due to the presence of the other two nucleons in the tri-nucleon system considered here. First, there are the boost corrections that arise from the observation that within a nucleus the threshold for pion production is shifted compared to the free threshold. For light systems as considered here, this essentially induces P-wave contributions. Second, there are further corrections to the two-nucleon production operator, that can be grouped in two categories, namely the so-called static and recoil corrections. The static corrections involve – as the leading 2N corrections do – static propagators but one insertion from the dimension two chiral effective pion-nucleon Lagrangian. The recoil contributions feature corrections to the static propagators with only insertions from the leading order (dimension one) chiral Lagrangian. With the fourth-order single nucleon input from heavy baryon CHPT [1, 2, 9] (the latter result obtained in project **B.3**), we obtain for the threshold multipoles on $^3$He and on $^3$H the values listed in Table 1. We find that the fourth order corrections for the electric dipole amplitude $E_{0+}$ for both tri-nuclear systems come out to be very small,

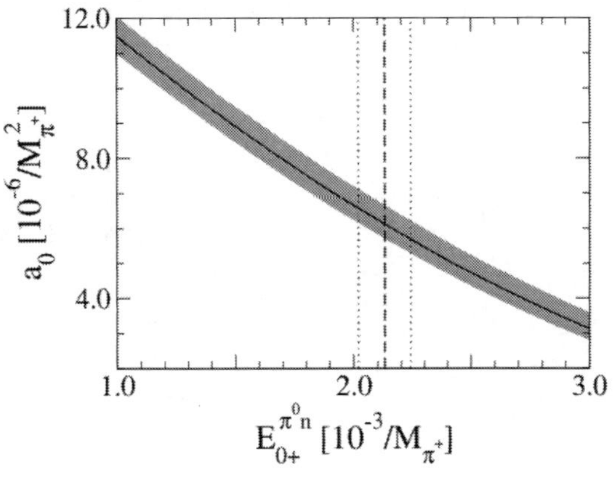


**Figure 1.** Sensitivity of the S-wave cross section $a_0$ for $^3$He in units of $10^{-6}/M_{\pi^+}^2$ to the single-neutron multipole $E_{0+}^{\pi^0 n}$ in units of $10^{-3}/M_{\pi^+}$. The vertical dashed line gives the CHPT prediction for $E_{0+}^{\pi^0 n}$ and the vertical dotted lines indicate the 5% error in the prediction. The shaded band indicates the theory error estimated from the cutoff variation and a 5% error in $E_{0+}^{\pi^0 p}$.

much smaller than in case of the deuteron. In Fig. 1, we illustrate the sensitivity of the S-wave cross section $a_0$ to the single-neutron multipole $E_{0+}^{\pi^0 n}$. The shaded band indicates the theory error estimated from the cutoff variation as described above and a 5% error in $E_{0+}^{\pi^0 p}$. As shown above, the uncertainties related to the nuclear effects are of the order of one percent, i.e. completely negligible. So our estimate of a 10% uncertainty of the 2N contributions in Ref. [6] driven by the analogy to the deuteron case [3] turns out to be much too conservative.

## 3 Electromagnetic currents in chiral nuclear EFT

One of the main advantages of chiral nuclear EFT is the possibility of deriving a set of consistent forces and electroweak currents. This connection was first explored for the electromagnetic currents. A powerful tool to perform such consistent calculations is the so-called method of unitary transformations, that had been used earlier by the Bochum group in the context of the meson-theoretical description of nuclear forces and currents.

In Ref. [10], we calculated the leading two-pion exchange two-nucleon four-current operator in chiral EFT based on the method of unitary transformation which we used to derive nuclear forces in Refs. [11–15], see the pertinent diagrams in Fig. 2. Our work provides an important check of the results presented in Refs. [16, 17] but also differs from these works in several important respects. First, we use a completely different method to compute the current operator. Secondly, we also give results for the exchange charge density which, to the best of our knowledge, have not yet been published before. Finally, we evaluate analytically all loop integrals to obtain a representation in momentum space in terms of the standard loop functions and the three-point functions. The latter are reduced to a form which can be easily treated numerically. We also

**Table 1.** Numerical results for the 3N multipoles. The first error is our estimation of the theoretical uncertainty resulting from the truncation of the chiral expansion while the second one is the statistical error from the Monte Carlo integration, see Ref. [8] for more details. For the total result only the combined error is given.

| $^3$He | 1N ($q^4$) | 2N ($q^3$) | 1N-boost | 2N-static ($q^4$) | 2N-recoil ($q^4$) | total |
|---|---|---|---|---|---|---|
| $E_{0+}$ [$10^{-3}/M_{\pi^+}$] | +1.71(4)(9) | −3.95(3) | −0.23(1) | −0.02(0)(1) | +0.01(2)(1) | −2.48(11) |
| $L_{0+}$ [$10^{-3}/M_{\pi^+}$] | −1.89(4)(9) | −3.09(2) | −0.00(0) | −0.07(1)(1) | +0.07(7)(0) | −4.98(12) |
| $^3$H | 1N ($q^4$) | 2N ($q^3$) | 1N-boost | 2N-static ($q^4$) | 2N-recoil ($q^4$) | total |
| $E_{0+}$ [$10^{-3}/M_{\pi^+}$] | −0.93(3)(5) | −4.01(3) | −0.35(1) | −0.02(1)(1) | +0.01(2)(0) | −5.28(7) |
| $L_{0+}$ [$10^{-3}/M_{\pi^+}$] | −0.99(4)(5) | −3.13(1) | −0.02(0) | −0.07(0)(1) | +0.07(7)(0) | −4.14(10) |

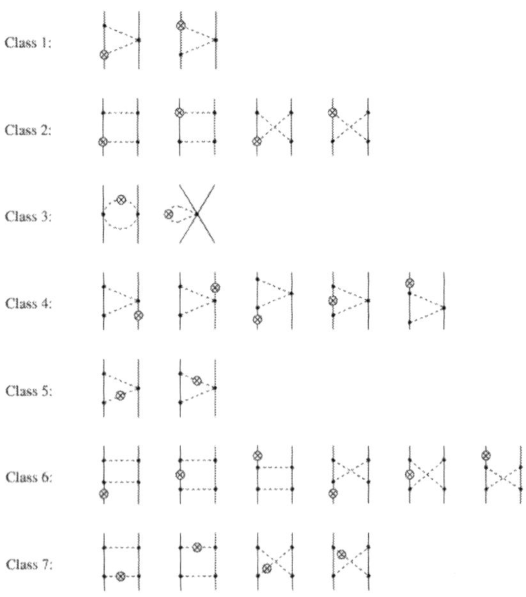


**Figure 2.** Diagrams showing contributions to the leading two-pion exchange currents. Solid and dashed lines refer to nucleons and pions, respectively. Solid dots are the lowest-order vertices from the effective Lagrangian while the crosses represent insertions of the electromagnetic vertices. Diagrams resulting from interchanging the nucleon lines are not shown.

succeeded to analytically carry out the Fourier transformation for all contributions leading to an extremely compact representation of the current and charge densities in coordinate space.

In Ref. [18], we applied the method of unitary transformation to work out the leading loop contributions to the one-pion exchange and short-range two-nucleon electromagnetic current and charge densities. The *renormalized* expressions for the one-pion exchange charge and current operators are given for the first time. We have discussed in detail renormalization of the one-pion exchange contributions which provides a stringent test of our theoretical approach. More precisely, *all* emerging ultraviolet divergences have to be absorbed into redefinition of the low-energy constants $l_i$ and $d_i$ entering the Lagrangians $\mathcal{L}_{\pi\pi}^{(4)}$ and $\mathcal{L}_{\pi N}^{(3)}$, respectively. There is no freedom in this procedure as the corresponding $\beta$-functions of all these LECs in dimensional regularization are fixed and well known. We have demonstrated that it is indeed possible to renormalize the one-loop contributions provided one makes use of the freedom to employ additional unitary transformations. Further, we succeeded to obtain compact, analytical expressions for the current and charge densities both in momentum and coordinate spaces which can be used in future numerical calculations. Finally, we provide a detailed comparison between our results and the ones obtained by Pastore et al. [19] within a different framework, and we spelled out the deficiencies of that calculation.

The long-range two-pion exchange (TPE) contributions to the nuclear current operator which appear at next-to-leading order (NLO) of the chiral expansion were used to describe electromagnetic processes in Ref. [20]. Their role in the photodisintegration of $^2$H and $^3$He was studied and the corresponding predictions were compared with experimental data. The bound and scattering states are calculated using five different parametrizations of the chiral next-to-next-to-leading order nucleon-nucleon potential which allows to estimate the theoretical uncertainty at a given order in the chiral expansion. For some observables the results are very close to the predictions based on the AV18 NN potential. In the most cases, the addition of the long-range TPE currents improved the description of the experimental data. Last but not least, in Ref. [21] we have used the novel contributions to the exchange current operator to analyze the magnetic form factors of the deuteron.

## 4 Formal aspects of spin-3/2 fields

In the paper [22], we have explicitly constructed a nonlinear field transformation that brings the interaction Lagrangian of spin-3/2 fields as given by

$$\mathcal{L} = \bar{\psi}(i\partial\!\!\!/ - m + V)\psi\,,$$

in a form that is invariant under gauge transformation,

$$\psi_\mu \to \psi_\mu + \partial_\mu \epsilon\,.$$

This proves a conjecture made in Ref. [23] where it was shown how a gauge non-invariant Lagrangian with linear couplings in the spin-3/2 fields can be brought into a gauge-invariant form via a suitably chosen field redefinition. We have also given a diagrammatic explanation of our main result, which allows to one split the truly propagating spin-3/2 components from the non-propagating

DOI: 10.1051/epjconf/201713403005

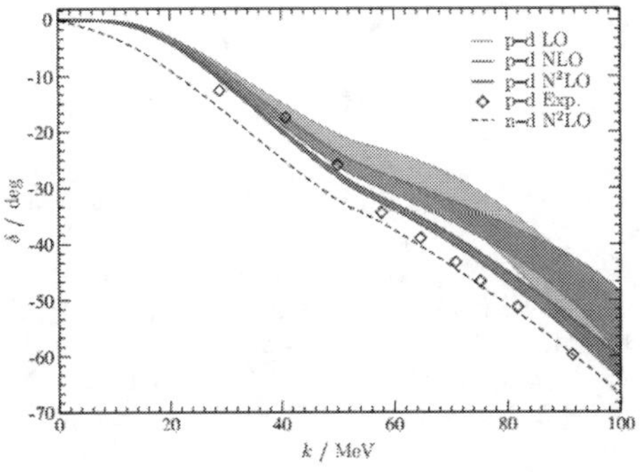


**Figure 3.** *N-d* doublet channel S-wave scattering phase shifts as functions of the center-of-mass momentum $k$ at various orders in the EFT expansion compared to the data. Error bands are generated by cutoff variation from 200 to 600 MeV.

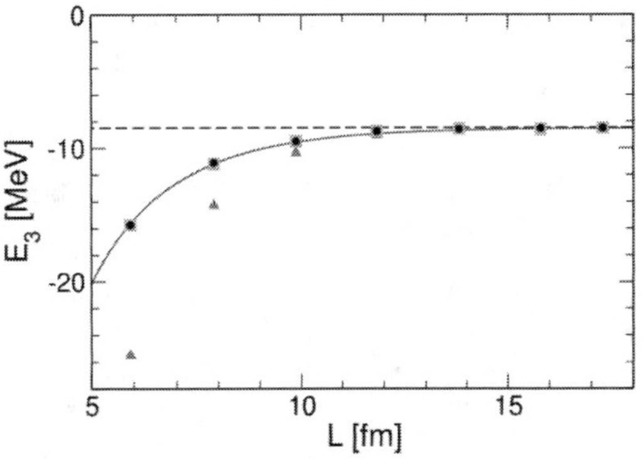


**Figure 4.** Triton energy $E_3$ for different edge lengths $L$ of the cubic volume. Squares and circles are data sets for different cutoffs. The triangles are from a nuclear lattice calculation using chiral EFT at N$^2$LO [31]. The horizontal dashed line indicates the physical triton binding energy, while the solid line is a fit to our results.

spin-1/2 contributions that are subsumed in local operators existing in the most general Lagrangian of deltas coupled to nucleons, pions and external fields.

In Ref. [24], we have further proven to all orders in the small scale expansion that all off-shell parameters which appear in the chiral effective Lagrangian with explicit Δ(1232)-isobar degrees of freedom can be absorbed into redefinitions of certain low-energy constants and are therefore redundant.

## 5 Nuclear structure and dynamics from the pionless EFT

At energies well below the pion mass, one can describe nuclear interactions and dynamics based on a non-relativistic Lagrangian with nucleon contact interactions only, the so-called pionless EFT. In Ref. [25], low-energy proton-deuteron scattering was calculated in this framework. In the quartet channel, the elastic scattering phase shift was worked out up to next-to-next-to-leading order in the power counting. In the doublet channel, a next-to-leading order calculation was performed. We obtained good agreement with the available phase shift analyses down to the scattering threshold. The phase shifts in the region of the non-perturbative Coulomb interactions were calculated by using an optimized integration mesh. Moreover, the Coulomb contribution to the $^3$He-$^3$H binding energy difference was evaluated in first order perturbation theory. In Ref. [26], we have computed $E1$ transitions and electric radii in the Beryllium-11 nucleus using an EFT that exploits the separation of scales in this halo system. We fix the leading-order parameters of the EFT from measured data on the $\frac{1}{2}^+$ and $\frac{1}{2}^-$ levels in Be-11 and the $B(E1)$ strength for the transition between them. We then obtain predictions for the $B(E1)$ strength for Coulomb dissociation of the Be-11 nucleus to the continuum. We also compute the charge radii of the $\frac{1}{2}^+$ and $\frac{1}{2}^-$ states. Agreement with experiment within the expected accuracy of a leading-order computation in this EFT is obtained. We also discuss how next-to-leading-order (NLO) corrections involving both s-wave and p-wave neutron-Be-10 interactions affect our results, and display the NLO predictions for quantities which are free of additional short-distance operators at this order. Information on neutron-Be-10 scattering in the relevant channels is inferred.

Another interesting set of studies, that is closely linked to the finite volume studies of baryon resonances in project **A.6** was published in Refs. [27–29]. The starting point is the observation that three bosons with large scattering length show universal properties that do not depend on the details of the interaction at short distances. In the three-boson system, these properties include a geometric spectrum of shallow three-body states called "Efimov states" and log-periodic dependence of scattering observables on the scattering length. In Ref. [27] the modification of the Efimov states in a finite cubic box was analyzed and the dependence of their energies on the box size was calculated using effective field theory. The renormalization of the effective field theory in the finite volume was also explicitly verified. These studies were extended in Ref. [28]. Previous calculations for positive scattering length that were perturbative in the finite volume energy shift are extended to arbitrarily large shifts and negative scattering lengths. The renormalization of the effective field theory in the finite volume is explicitly verified. Moreover, we investigate the effects of partial wave mixing and study the behavior of shallow trimers near the dimer energy. Further, we provide numerical evidence for universal scaling of the finite volume corrections. Finally, understanding the volume dependence of the triton binding energy is an important step towards lattice simulations of light nuclei. In Ref. [29] the triton binding energy in a finite cubic box with periodic boundary conditions to leading order in the pionless EFT was calculated. Higher order corrections are estimated and the proper renormalization of our results is verified explicitly. We present results for the physical triton, see Fig. 4, as well as for the pion-mass dependence of the triton spectrum near the "critical" pion mass, $M_\pi^c \sim 197\,\mathrm{MeV}$, where chiral effective field theory suggests that the nucleon-nucleon scattering lengths in the singlet- and triplet-channels may diverge simultaneously.

An extension of the Lüscher formula to the three-body system is implicit in these results. Such an explicit formula was supplied only much later in Ref. [30].

## 6 High-precision NN forces: Foundations and applications

In the paper [32], we have presented a new generation of NN potentials derived in chiral EFT up to N$^3$LO. The new chiral forces offer a number of substantial improvements as compared to the widely used N$^3$LO potentials of Refs. [33, 34] introduced a decade ago. First of all, we employ a local regularization scheme for the pion exchange contributions which, differently to the standard nonlocal regularization applied e.g. in Refs. [33, 34], does not distort the low-energy analytic structure of the amplitude and, as a consequence, leads to a better description of phase shifts and experimental data. The employed regulator, by construction, removes the short-range part of the chiral two-pion exchange and thus makes the additional spectral function regularization used in the potential of Ref. [33] obsolete. This is a particularly welcome feature given that the expressions for the three-nucleon force at N$^3$LO and N$^4$LO are only available in the framework of dimensional regularization. Further, in contrast to the earlier studies of Refs. [33, 34], we have taken all pion-nucleon LECs and especially the subleading LECs $c_i$ from pion-nucleon scattering without any fine tuning. The LECs accompanying NN contact interactions were determined by fits to the Nijmegen phase shifts and mixing angles for five different values of the coordinate-space cutoff $R$ chosen in the range of $R = 0.8 \ldots 1.2$ fm and appear to be of natural size in all cases. The new N$^3$LO potentials allow for an excellent description of the Nijmegen np and pp phase shifts at energies below 200 MeV and, for the cutoff choices of $R = 0.9$ fm and $R = 1.0$ fm, even up to $E_{\mathrm{lab}} = 300$ MeV. Furthermore, the deuteron properties are accurately described. Moreover, the deuteron wave functions are free from distortions at distances larger than $r \sim 2 \ldots 3$ fm which appear for the N$^3$LO potentials of Refs. [33, 34] due to the employed form of the regulator. We found that the description of the Nijmegen phase shifts improves substantially when going from LO to NLO, from NLO to N$^2$LO and from N$^2$LO to N$^3$LO as one expects for a convergent expansion. It is worth to emphasize in this connection that the short range part of the NLO and N$^2$LO potentials involves the same set of operators. Our findings therefore provide yet another evidence of the subleading two-pion exchange which was also observed in earlier studies. As an important consistency check of our approach, we have studied the residual cutoff dependence of phase shifts at different orders in the chiral expansion. We found, in particular, that the cutoff dependence is strongly reduced at N$^3$LO compared to N$^2$LO in the whole considered range of energies. To illustrate these features, we consider the chiral expansion of the np total cross section at the energies of $E_{\mathrm{lab}} = 50$, 96, 143 and 200 MeV based on the interactions introduced in the previous sections as shown in Fig. 5

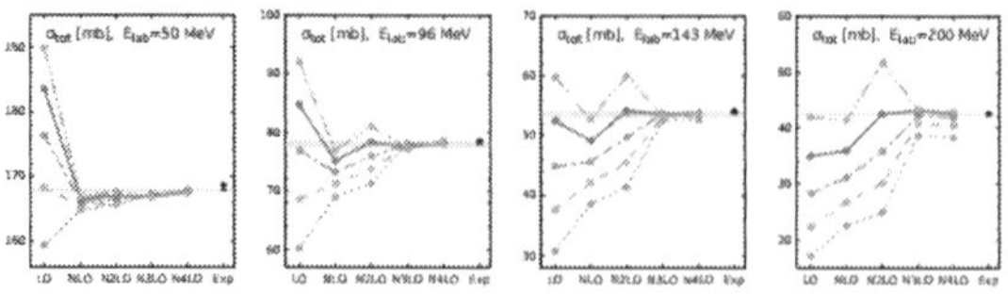

**Figure 5.** Order-by-order convergence of the chiral expansion for the np total cross section at energies of $E_{\mathrm{lab}} = 50$ MeV, $E_{\mathrm{lab}} = 96$ MeV and $E_{\mathrm{lab}} = 143$ MeV and $E_{\mathrm{lab}} = 200$ MeV. Dotted (light brown), dashed (green), dashed-dotted (blue), solid (red) and dashed-double-dotted (pink) lines show the results based on the cutoff $R = 1.2$ fm, $R = 1.1$ fm, $R = 1.0$ fm, $R = 0.9$ fm and $R = 0.8$ fm, respectively. The horizontal band refers to the result of the Nijmegen PWA with the uncertainty estimated by means of deviations from the results based on the Nijmegen I, II and Reid 93 potentials as explained in the text. Also shown are experimental data of Ref. [35].

We have also addressed the issue of the uncertainty quantification of our results due to the truncation of the chiral expansion at a given order. In particular, we have argued that the standard procedure for error estimation based on a cutoff variation is not reliable and employed a simple alternative approach by directly estimating the expected size of higher-order contributions at a given energy. Such a procedure has the advantage of being applicable for any fixed value of the cutoff so that calculations based on different cutoffs can be used to provide additional consistency checks. Furthermore, disentangling the error analysis from the cutoff variation allows one to avoid an unnecessary increase of uncertainty due to softening the interaction. Notice that the versions of the potential corresponding to soft choices of the cutoff $R$ may still be useful for certain kinds of applications including, in particular, many-body calculations. We have applied this approach to the total np cross section at several energies and have verified that the results at different chiral orders and for different values of the cutoff are indeed consistent with each other. We have furthermore used this method to quantify the theoretical uncertainty in the description of the np phase shifts as well as differential cross sections and selected polarization observables in np scattering. In particular, we found the N$^3$LO results for np scattering to be very accurate at energies below $\sim$ 100 MeV with the corresponding error bands being barely visible and still rather accurate at the energy of $E_{\mathrm{lab}} = 200$ MeV. In all considered cases, our results agree with the ones based on the Nijmegen PWA within the estimated theoretical accuracy. This gives us additional confidence in the reliability of the suggested way of quantifying the uncertainty. We have furthermore analyzed the uncertainties associated with making a specific choice of the functional form of the local regulator and employing the additional spectral function regularization of the TPEP and found them to be negligible at the level of the estimated theoretical accuracy at N$^3$LO.

In Ref. [36], we have presented a nucleon-nucleon potential at fifth order in chiral effective field theory. We find a substantial improvement in the description of nucleon-nucleon phase shifts as compared to the fourth-order re-

DOI: 10.1051/epjconf/201713403005

**Table 2.** $\chi^2$/datum for the description of the Nijmegen np (upper two rows) and pp (lower two rows) phase shifts [37] at different orders in the chiral expansion for the cutoff $R = 0.9$ fm. Only those channels are included which have been used in the N$^3$LO/N$^4$LO fits, namely the S-, P- and D-waves and the mixing angles $\epsilon_1$ and $\epsilon_2$.

| $E_{\mathrm{lab}}$ bin | LO | NLO | N$^2$LO | N$^3$LO | N$^4$LO |
|---|---|---|---|---|---|
| 0–100 | 360 | 31 | 4.5 | 0.7 | 0.3 |
| 0–200 | 480 | 63 | 21 | 0.7 | 0.3 |
| 0–100 | 5750 | 102 | 15 | 0.8 | 0.3 |
| 0–200 | 9150 | 560 | 130 | 0.7 | 0.6 |

sults of Ref. [32], see Tab. 2. This provides clear evidence of the corresponding two-pion exchange contributions with all low-energy constants being determined from pion-nucleon scattering. The fifth-order corrections to nucleon-nucleon observables appear to be of a natural size which confirms the good convergence of the chiral expansion for nuclear forces. Furthermore, the obtained results provide strong support for the novel way of quantifying the theoretical uncertainty due to the truncation of the chiral expansion proposed in Ref. [32], as shown for the np phase shifts in Fig. 6. This work opens up new perspectives for precision *ab initio* calculations in few- and many-nucleon systems and is especially relevant for ongoing efforts towards a quantitative understanding the structure of the three-nucleon force in the framework of chiral effective field theory.

In Ref. [39], we have applied these improved nucleon-nucleon potentials up to fifth order in chiral EFT, along with a new analysis of the theoretical truncation errors, to study nucleon-deuteron (*Nd*) scattering and selected low-energy observables in $^3$H, $^4$He, and $^6$Li. Calculations beyond second order differ from experiment well outside the range of quantified uncertainties, providing truly unambiguous evidence for missing three-nucleon forces within the employed framework. The sizes of the required three-nucleon force contributions agree well with expectations based on Weinberg's power counting. We identify the energy range in elastic *Nd* scattering best suited to study three-nucleon force effects and estimate the achievable accuracy of theoretical predictions for various observables.

This new semi-local improved chiral nucleon-nucleon interaction was used for the first time to study several electromagnetic and weak processes at energies below the pion production threshold in Ref. [40]. Cross sections and selected polarization observables for deuteron photo-disintegration, nucleon-deuteron radiative capture, three-body $^3$He photo-disintegration as well as capture rates for decays of the muonic $^2$H and $^3$He atoms were calculated. The Lippmann-Schwinger and Faddeev equations in momentum space are solved to obtain nuclear states. The electromagnetic current operator is taken as a single nucleon current supplemented by many-body contributions induced via the Siegert theorem. For muon capture processes the non-relativistic weak current together with the dominant relativistic corrections is used. Our results compare well with experimental data, demonstrating the same

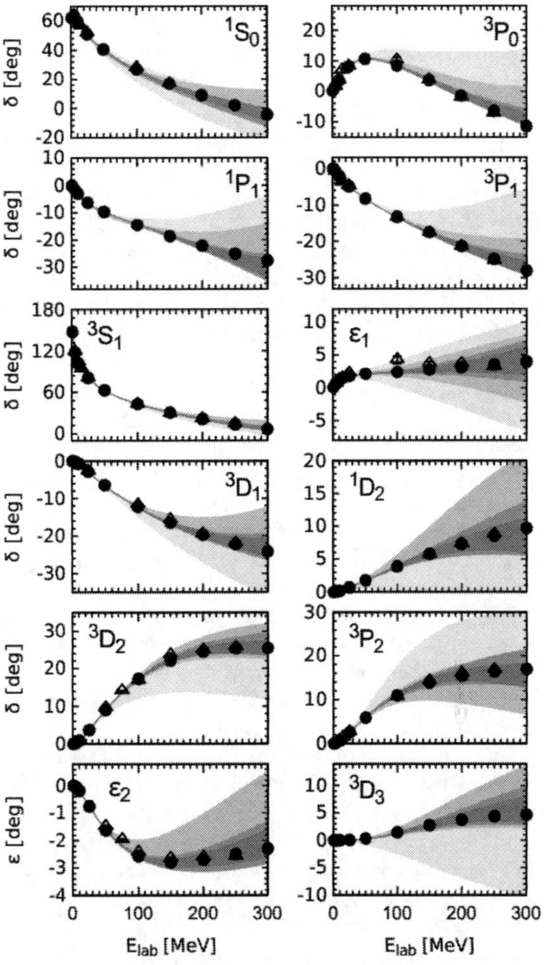


**Figure 6.** Results for the np S-, P- and D- waves and the mixing angles $\epsilon_1$, $\epsilon_2$ up to N$^4$LO based on the cutoff of $R = 0.9$ fm in comparison with the Nijmegen PWA [37] (solid dots) and the GWU single-energy PWA [38] (open triangles). The bands of increasing width show estimated theoretical uncertainty at N$^4$LO (red), N$^3$LO (blue), N$^2$LO (green) and NLO (yellow).

quality as is observed for the semi-phenomenological AV18 potential. Compared to the older version of the chiral potential with a nonlocal regularization, a much smaller cut-off dependence is found for the state-of-art chiral local interaction employed in this paper. Finally, estimates of errors due to the truncation of the chiral expansion are given.

## 7 Axial currents from chiral EFT

Few-nucleon electroweak reactions constitute an exciting frontier for nuclear physics and astrophysics. Being an excellent testing ground for chiral EFT, they are intimately related through the same LO short-range operator to such different processes as three-nucleon interactions, P-wave pion production in NN collisions as well as pion photoproduction and radiative capture reactions in the 2N system.

A precise determination of the strength of this operator is the main task of the ongoing MuSun experiment at PSI [41] which will measure the rate of muon capture on the deuteron with a precision of 1.5%. This will allow one to improve the accuracy and reliability of theoretical predictions for nuclear reactions involving neutrinos such as e.g. the solar proton-proton fusion and the hep processes which figure importantly in astrophysics.

The NN contributions to the axial charge and current operators have been first considered in chiral EFT by Park et al. [42, 43] who have, however, omitted certain contributions and ignored pion-pole terms. Recently, the exchange axial currents have been re-considered by Baroni et al. [44] in the framework of time-ordered perturbation theory. These authors have, however, ignored non-static corrections to the NN operators and the contributions of reducible-like three-nucleon diagrams. In Ref. [45], we have performed the *complete* derivation of the one-, two- and three-nucleon axial charge and current operators up to fourth order relative to the leading one-body contribution using the method of unitary transformation, see section 3 for more details. Notice that the same approach has been used to derive the nuclear forces developed in our group, which are, therefore, by construction consistent with the current operators. Using the freedom to employ additional unitary transformations, we have succeeded to remove all ultraviolet divergences generated by the loop diagrams and to bring the pion-pole contributions into the form which matches exactly the corresponding expressions for the nuclear forces. We expect this feature to be particularly useful for maintaining consistency between regularized expressions for the nuclear forces and current operators. Interestingly, we obtain different results from those of Baroni et al. [44] even for the long-range part of the NN current operator in the static limit. It will be very interesting to apply the resulting expressions to explicitly check consistency between the short-range axial currents and the three-nucleon forces and to study weak few-nucleon processes such as the triton-$\beta$ decay and muon capture reactions on $^2$H and $^3$He.

## 8 Pion production in NN collisions

The reaction $NN \to NN\pi$ offers a challenging testing ground for chiral EFT at intermediate energies, see Ref. [46] for a review, and offers important insights into symmetry-breaking pattern of hadronic reactions. In particular, the charge-symmetry-breaking (CSB) process $pn \to d\pi^0$ probes the small but calculable CSB amplitudes which, to leading order in chiral EFT, are proportional to the strong-interaction and electromagnetic contributions to the nucleon mass difference $\delta m_N^{\rm str}$ and $\delta m_N^{\rm em}$, respectively. The strong-interaction mass shift extracted at LO from the TRIUMF data [47] of the forward-backward asymmetry for the reaction $pn \to d\pi^0$, $\delta m_N^{\rm str,\, LO} = 1.5 \pm 0.8_{\rm exp} \pm 0.5_{\rm th}$ [48], is in a good agreement with the available empirical information on this quantity. However, given the large momenta involved in the pion production reactions, it is important to analyze in detail the convergence of the chiral EFT for such processes. This provides a strong motivation for our N$^2$LO study of the near-threshold cross section for the isospin-conserving channel $pp \to d\pi^+$, which is precisely measured in hadronic atom experiments. The various contributions to the corresponding pion production operator up to N$^2$LO have already been worked out in Refs. [49, 50] where, however, no convolutions with the nuclear wave functions have been performed. Notice that the power counting for pion production reactions differs from the standard one in chiral EFT due to the need to explicitly account for the corresponding momentum scale $p \sim \sqrt{M_\pi m_N} \simeq 340$ MeV. The modified counting is referred to as the momentum counting scheme (MCS). Further, given $m_\Delta - m_N \sim \sqrt{M_\pi m_N}$, one expects it to be advantageous to treat the $\Delta$-isobar as an explicit degree of freedom.

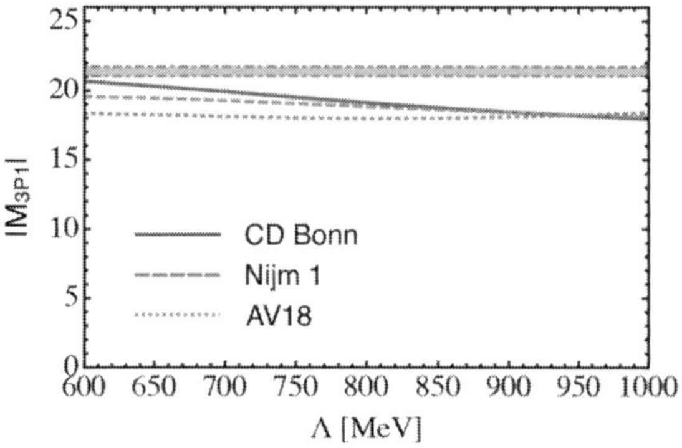


**Figure 7.** The $pp \to d\pi^+$ amplitude $|M_{^3P_1}|$ at threshold as a function of the cutoff $\Lambda$ imposed in the convolution integrals with various phenomenological NN wave functions. The calculated amplitude includes the complete set of LO, NLO and N$^2$LO operators except for the N$^2$LO five-point contact term. The horizontal grey band between the dot-dashed lines shows the experimental value of the amplitude.

In Ref. [51], we have analyzed in detail the convergence of the MCS for the reaction $pp \to d\pi^+$. At threshold, only S-wave pion production has to be considered, and the only contribution to the amplitude emerges from the initial $pp$ $^3$P$_1$ partial wave. We found it to be advantageous to employ phenomenological NN wave functions since the available chiral potentials tend to remove a part of the intermediate-range physics relevant for this reaction.[1] The amplitude $M_{^3P_1}$ is dominated by the longest-range Weinberg-Tomozawa operator at LO. The NLO corrections are known to undergo significant cancellations and yield a small contribution comparable in size with the N$^2$LO correction. We found that the N$^2$LO short-range term needed to reproduce the experimental value of the amplitude of $|M_{^3P_1}|^{\rm exp} = 21.5^{+0.2}_{-0.5}$, extracted from the total cross section in the $nn \to d\pi^-$ channel measured at PSI [52], agrees well with the expected size of N$^2$LO terms in the MCS, see Fig. 7. Also the dependence of the N$^2$LO contribution on the LECs $c_i$ appears to be consistent with the estimated uncertainty at this order. Finally, we have

[1]This effect is less pronounced for the new generation of chiral potentials of Ref. [32, 36] for the two hardest values of the regulator.

studied the role of the Δ-isobar using different strategies to account for its contributions. We found that the main effects of the Δ-isobar for the considered observable are fairly well represented via saturation of the LECs $c_i$. The coupled-channel effects turn out not to exceed the magnitude of N$^2$LO effects.

We thank our collaborators Vadim Baru, Arseniy Filin, Jacek Golak, Johann Haidenbauer, Christoph Hanhart, Hiroyuki Kamada, Stefan Kölling, Sebastian König, Hermann Krebs, Simon Kreuzer, Alexander Kudryatsev, Dean Lee, Mark Lenkewitz, Fred Myhrer, Andreas Nogga, Daniel Phillips, Giullermo Rios, Dagmara Rozpędzik, Roman Skibiński, Henryk Witała and all members of the LENPIC Collaboration. The work reported here would not have been possible without the support from the Deutsche Forschungsgemeinschaft within the SFB/TR16.

## References

[1] V. Bernard, N. Kaiser and U.-G. Meißner, Z. Phys. C **70** (1996) 483 [arXiv:hep-ph/9411287].
[2] V. Bernard, N. Kaiser and U.-G. Meißner, Eur. Phys. J. A **11** (2001) 209 [arXiv:hep-ph/0102066].
[3] S. R. Beane, V. Bernard, T.-S. H. Lee, U.-G. Meißner and U. van Kolck, Nucl. Phys. A **618** (1997) 381 [arXiv:hep-ph/9702226].
[4] H. Krebs, V. Bernard and U.-G. Meißner, Eur. Phys. J. A **22** (2004) 503 [arXiv:nucl-th/0405006].
[5] E. Epelbaum, H.-W. Hammer and U.-G. Meißner, Rev. Mod. Phys. **81** (2009) 1773 [arXiv:0811.1338 [nucl-th]].
[6] M. Lenkewitz, E. Epelbaum, H.-W. Hammer and U.-G. Meißner, Phys. Lett. B **700** (2011) 365 [arXiv:1103.3400 [nucl-th]].
[7] P. Argan *et al.*, Phys. Lett. **206B** (1988) 4 [Erratum-ibid. B **213** (1988) 564].
[8] M. Lenkewitz, E. Epelbaum, H. W. Hammer and U.-G. Meißner, Eur. Phys. J. A **49** (2013) 20 [arXiv:1209.2661 [nucl-th]].
[9] V. Bernard, B. Kubis and U.-G. Meißner, Eur. Phys. J. A **25** (2005) 419 [nucl-th/0506023].
[10] S. Kölling, E. Epelbaum, H. Krebs and U.-G. Meißner, Phys. Rev. C **80**, 045502 (2009) [arXiv:0907.3437 [nucl-th]].
[11] E. Epelbaum, W. Glöckle and U.-G. Meißner, Nucl. Phys. A **637**, 107 (1998) [arXiv:nucl-th/9801064].
[12] E. Epelbaum, W. Glöckle and U.-G. Meißner, Nucl. Phys. A **671**, 295 (2000) [arXiv:nucl-th/9910064].
[13] E. Epelbaum, U.-G. Meißner and W. Glöckle, Nucl. Phys. A **714**, 535 (2003) [arXiv:nucl-th/0207089].
[14] E. Epelbaum, Phys. Lett. B **639**, 456 (2006).
[15] E. Epelbaum, Eur. Phys. J. A **34**, 197 (2007) [arXiv:0710.4250 [nucl-th]].
[16] S. Pastore, R. Schiavilla and J. L. Goity, Phys. Rev. C **78**, 064002 (2008) [arXiv:0810.1941 [nucl-th]].
[17] S. Pastore, L. Girlanda, R. Schiavilla, M. Viviani and R. B. Wiringa, Phys. Rev. C **80**, 034004 (2009) [arXiv:0906.1800 [nucl-th]].
[18] S. Kölling, E. Epelbaum, H. Krebs and U.-G. Meißner, Phys. Rev. C **84** (2011) 054008 [arXiv:1107.0602 [nucl-th]].
[19] S. Pastore, L. Girlanda, R. Schiavilla and M. Viviani, Phys. Rev. C **84** (2011) 024001 [arXiv:1106.4539 [nucl-th]].
[20] D. Rozpędzik, J. Golak, S. Kölling, E. Epelbaum, R. Skibiński, H. Witała and H. Krebs, Phys. Rev. C **83** (2011) 064004 [arXiv:1103.4062 [nucl-th]].
[21] S. Kölling, E. Epelbaum and D. R. Phillips, Phys. Rev. C **86**, 047001 (2012) [arXiv:1209.0837 [nucl-th]].
[22] H. Krebs, E. Epelbaum and U.-G. Meißner, Phys. Rev. C **80** (2009) 028201 [arXiv:0812.0132 [hep-th]].
[23] V. Pascalutsa, Phys. Lett. B **503**, 85 (2001) [arXiv:hep-ph/0008026].
[24] H. Krebs, E. Epelbaum and U.-G. Meißner, Phys. Lett. B **683** (2010) 222 [arXiv:0905.2744 [hep-th]].
[25] S. Koenig and H.-W. Hammer, Phys. Rev. C **83** (2011) 064001 [arXiv:1101.5939 [nucl-th]].
[26] H.-W. Hammer and D. R. Phillips, Nucl. Phys. A **865** (2011) 17 [arXiv:1103.1087 [nucl-th]].
[27] S. Kreuzer and H.-W. Hammer, Phys. Lett. B **673** (2009) 260 [arXiv:0811.0159 [nucl-th]].
[28] S. Kreuzer and H.-W. Hammer, Eur. Phys. J. A **43** (2010) 229 [arXiv:0910.2191 [nucl-th]].
[29] S. Kreuzer and H.-W. Hammer, Phys. Lett. B **694** (2011) 424 [arXiv:1008.4499 [hep-lat]].
[30] U.-G. Meißner, G. Rios and A. Rusetsky, Phys. Rev. Lett. **114** (2015) 091602 [arXiv:1412.4969 [hep-lat]].
[31] E. Epelbaum, H. Krebs, D. Lee and U.-G. Meißner, Eur. Phys. J. A **41**, 125 (2009) [arXiv:0903.1666 [nucl-th]].
[32] E. Epelbaum, H. Krebs and U.-G. Meißner, Eur. Phys. J. A **51** (2015) 53 [arXiv:1412.0142 [nucl-th]].
[33] E. Epelbaum, W. Glöckle and U.-G. Meißner, Nucl. Phys. A **747**, 362 (2005) [nucl-th/0405048].
[34] D. R. Entem and R. Machleidt, Phys. Rev. C **68**, 041001 (2003) [nucl-th/0304018].
[35] W. P. Abfalterer, F. B. Bateman, F. S. Dietrich, R. W. Finlay, R. C. Haight and G. L. Morgan, Phys. Rev. C **63**, 044608 (2001).
[36] E. Epelbaum, H. Krebs and U.-G. Meißner, Phys. Rev. Lett. **115** (2015) 122301 [arXiv:1412.4623 [nucl-th]].
[37] V. G. J. Stoks *et al.*, Phys. Rev. C **48**, 792 (1993).
[38] R. A. Arndt, I. I. Strakovsky and R. L. Workman, Phys. Rev. C **50**, 2731 (1994).
[39] S. Binder *et al.* [LENPIC Collaboration], Phys. Rev. C **93** (2016) 044002 [arXiv:1505.07218 [nucl-th]].
[40] R. Skibinski *et al.*, Phys. Rev. C **93** (2016) 064002 [arXiv:1604.03395 [nucl-th]].
[41] V. A. Andreev *et al.* [MuCap Collaboration], Phys. Rev. Lett. **99**, 032002 (2007) [arXiv:0704.2072 [nucl-ex]].
[42] T. S. Park, D. P. Min and M. Rho, Phys. Rept. **233**, 341 (1993) [hep-ph/9301295].

[43] T. S. Park *et al.*, Phys. Rev. C **67**, 055206 (2003) [nucl-th/0208055].

[44] A. Baroni, L. Girlanda, S. Pastore, R. Schiavilla and M. Viviani, Phys. Rev. C **93**, no. 1, 015501 (2016) Erratum: [Phys. Rev. C **93**, no. 4, 049902 (2016)] [arXiv:1509.07039 [nucl-th]].

[45] H. Krebs, E. Epelbaum and U.-G. Meißner, "Nuclear axial current operators to fourth order in chiral effective field theory," arXiv:1610.03569 [nucl-th].

[46] V. Baru, C. Hanhart and F. Myhrer, Int. J. Mod. Phys. E **23**, no. 4, 1430004 (2014) [arXiv:1310.3505 [nucl-th]].

[47] A. K. Opper *et al.*, Phys. Rev. Lett. **91**, 212302 (2003) [nucl-ex/0306027].

[48] A. Filin, V. Baru, E. Epelbaum, J. Haidenbauer, C. Hanhart, A. E. Kudryavtsev and U.-G. Meißner, Phys. Lett. B **681**, 423 (2009) [arXiv:0907.4671 [nucl-th]].

[49] A. A. Filin, V. Baru, E. Epelbaum, H. Krebs, C. Hanhart, A. E. Kudryavtsev and F. Myhrer, Phys. Rev. C **85**, 054001 (2012) [arXiv:1201.4331 [nucl-th]].

[50] A. A. Filin, V. Baru, E. Epelbaum, H. Krebs, C. Hanhart and F. Myhrer, Phys. Rev. C **88**, no. 6, 064003 (2013) [arXiv:1307.6187 [nucl-th]].

[51] V. Baru, E. Epelbaum, A. A. Filin, C. Hanhart, H. Krebs and F. Myhrer, Eur. Phys. J. A **52**, no. 5, 146 (2016) [arXiv:1602.07333 [nucl-th]].

[52] T. Strauch *et al.*, Phys. Rev. Lett. **104**, 142503 (2010) [arXiv:1003.4153 [nucl-ex]].

EPJ Web of Conferences **134**, 03006 (2017) DOI: 10.1051/epjconf/201713403006
*Subnuclear Structure of Matter: Achievements and Challenges*

# Coupled channels approach to photo-meson production on the nucleon

*Horst* Lenske[1,*]

[1] *Institut für Theoretische Physik, Justus-Liebig-Universität Gießen*

**Abstract.**
The coupled channels Lagrangian approach of the Giessen model (GiM) for meson production on the nucleon is discussed and applied to a selected set of meson production channels on the nucleon, ranging from $\pi N \to \pi N$ and $\gamma N \to \pi N$, eta-production and associated strangeness production to $2\pi N$ channels in the resonance energy region. We present an updated coupled-channel analysis of eta-meson production including all recent photo-production data on the proton. The dip structure observed in the differential cross sections at c.m. energies W=1.68 GeV is explained by destructive interference between the $S_{11}(1535)$ and $S_{11}(1560)$ states, not confirming the postulated sharp state. Kaon production on the nucleon is investigated in $K\Lambda$ and $K\Sigma$ exit channels. The approach to $2\pi N$ production has been significantly improved by using the isobar approximation with $\sigma N$ and $\pi\Delta_{1232}$ intermediate states. Three-body unitarity is maintained up to interference between the isobar subchannels. We obtain $R_{\sigma N}(1440) = 27^{+4}_{-9}$ % and $R_{\pi\Delta}(1440) = 12^{+5}_{-3}$ % for the $\sigma N$ and $\pi\Delta_{1232}$ decay branching ratios of $N^*(1440)$ respectively. The extracted $\pi N$ inelasticities and reaction amplitudes are consistent with the results of other groups.

## 1 Introduction

The discovery of nucleon resonances in the early pion-nucleon scattering experiments provided first indications for a complicated intrinsic structure of the nucleon. With establishing the quark picture of hadrons and developments of the constituent quark models the interest in the study of the nucleon excitation spectra was renewed. Soon after, it was realized that there was an obvious discrepancy between the number of resonances predicted by theory and those identified experimentally. Since then, the problem of *missing resonances* is a major issue of baryon spectroscopy. Final answers about the number of excited states of the nucleon and their properties are still pending. Solutions are searched for both experimentally and theoretically. On the theory side constituent quark (CQM) models, lattice QCD and Dyson-Schwinger approaches have been developed to describe and predict the nucleon resonance spectra. The main problem remains, however, a serious disagreement between the theoretical calculations and the experimentally observed baryon spectra. This concerns both the number and the properties of excited states.

The investigation of properties of nucleon resonances remains one of the primary goals of modern hadron physics. The main information about the hadron spectra comes from the analysis of scattering data. Coupled-channel approaches have proven to be an efficient tool to extract baryon properties from experiment. The Giessen coupled-channel model (GiM) has been developed for a combined analysis of pion- and photon-induced reactions on the nucleon, $(\pi/\gamma) + N$, for extracting properties of nucleon resonances. The applications range from investigations of the elastic and inelastic $\pi N$ and $\pi N^*$ channels [1] to $\omega N$ [2], $\eta N$ [3, 4] production as well as the strangeness channels $K\Lambda$ [5] and $K\Sigma$ [6]. The $2\pi N$ channels were investigated recently in [7]. Here we review central issues of the Giessen approach and present results for selected reactions. In section 2 we introduce the underlying field theoretical model, being based on a phenomenological Lagrangian density for baryons, mesons, and the photon and their interactions. General theoretical aspects of high-spin matter fields are discussed in section 3 with special emphasis on the gauge properties of $s = \frac{3}{2}$ and $s = \frac{5}{2}$. Applications and results of our approach on the production of kaon- and eta-mesons are discussed in section 4 and section 5, respectively. Our latest investigations on double-pion production channels are discussed in section 6. In section 7 we summarize the achievements of the Giessen model. As an illustrative overview we present already here in Fig. 1 our results on the total cross sections in the various hadronic reaction channels.

## 2 The Giessen coupled channels model for baryon spectroscopy

Here we briefly outline the main ingredients of the model. More details can be found in [1, 3–7]. We need to solve the Bethe-Salpeter (BS) equation for the scattering amplitude:

$$M(p', p; w) = V(p', p; w) + \int \frac{d^4q}{(2\pi)^4} V(p', q; w) G_{BS}(q; w) M(q, p; w), \quad (1)$$

*e-mail: horst.lenske@physik.uni-giessen.de

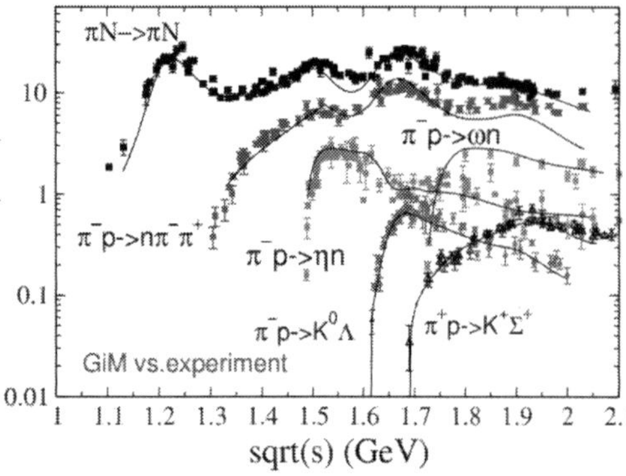


**Figure 1.** Total cross sections for pion-induced reactions. Results of the Giessen model are compared to experimental data.

where $w = \sqrt{s}$ is the available center-of-mass energy. Here, $p$ ($k$) and $p'$ ($k'$) are the incoming and outgoing baryon (meson) four-momenta. After splitting the two-particle BS propagator $G_{BS}$ into its real and imaginary parts, we introduce the $K$-matrix given schematically by $K = V + \int V\text{Re}G_{BS}M$. Then $M$ is given by $M = K + i\int M\text{Im}G_{BS}K$. Since the imaginary part of $G_{BS}$ is given by the on-shell part, the reaction matrix $T$, defined via the scattering matrix $S = 1 + 2iT$, can now be calculated from $K$ after a partial wave decomposition ($PWD$) into total spin $J$, parity $P$, and isospin $I$ via matrix inversion:

$$T(p',p;w) = \left(1 - i\rho(w)K(p',p;w)\right)^{-1} K(p',p;w), \quad (2)$$

where $\rho(w)$ is an appropriately chosen phase space factor. Hence unitarity is fulfilled as long as $K$ is Hermitian. For simplicity we apply the so-called $K$-matrix Born approximation, which means that we neglect the real part of $G_{BS}$ and thus $K$ reduces to $K = V$. The validity of this approximation was tested a long time ago by Pearce and Jennings [8].

The potential $V$ is built up by a sum of $s$-, $u$-, and $t$-channel Feynman diagrams by means of effective Lagrangians which are not shown here but can be found in very detail in [9] and the previously cited references. Certain aspects of the more involved cases of spin-$\frac{3}{2}$ and spin-$\frac{5}{2}$ resonances are discussed below in section 3. In all reaction channels the non-resonant background contributions to the scattering amplitudes are consistently derived from the $u$- and $t$-channel diagrams, thus reducing the number of free parameters greatly. In addition, each vertex is multiplied by a cutoff form factor:

$$F(q^2) = \frac{\Lambda_q^4}{\Lambda_q^4 + (q^2 - m_q^2)^2}, \quad (3)$$

where $m_q$ ($q^2$) denotes the mass (four-momentum squared) of the off-shell particle. To reduce the number of parameters the cutoff value $\Lambda_q$ is chosen to be identical for all final states. We only distinguish between the nucleon cutoff ($\Lambda_N$), the resonance cutoffs ($\Lambda_s$), where a common cutoff is used for all baryons within the same spin-multiplet $s \in \{\frac{1}{2}, \frac{3}{2}, \frac{5}{2}...\}$, and the $t$-channel cutoff ($\Lambda_t$).

A considerable numerical simplification is obtained by the afore mentioned $K-$matrix approximation. Using the partial wave decomposition the integral over $d\Omega_q$ can be calculated analytically. Then Eq. (1) reduces to a linear system of coupled equations for the partial wave scattering amplitudes:

$$\begin{aligned} T^{JP}_{fi}(w) &= K^{JP}_{fi}(w) \\ &+ i\sum_j \int_{\mu^2_{j0}}^{\infty} d\mu_j^2 A_j(\mu_j) T^{JP}_{fj}(\mu_j) K^{JP}_{ji}(\mu_j) \quad (4) \end{aligned}$$

where $K = V$ and $f, i, j$ denote the final, initial, and intermediate $meson - baryon$ channels, respectively. The photo-production reaction channels are treated perturbatively in leading order of the $\gamma N$ and $\gamma N^*$ vertices, respectively. The spectral functions in the meson-meson ($\pi, K, \sigma, \rho$...) and the meson-baryon isobar channels, as e.g the $\Delta$ resonance, are denoted by $A_j(\mu_j^2)$, being integrated over the energy $\mu_j$ available in the isobar subsystem starting at the isobar threshold energy $\mu_{j0}$. For stable particles the spectral functions reduce to trivial delta-functions, projecting the integrand on the particles's mass. Since the two-pion discontinuities are taken into account the three-body unitarity in the form of the optical theorem if fulfilled up to interference terms between different isobar subchannels [7].

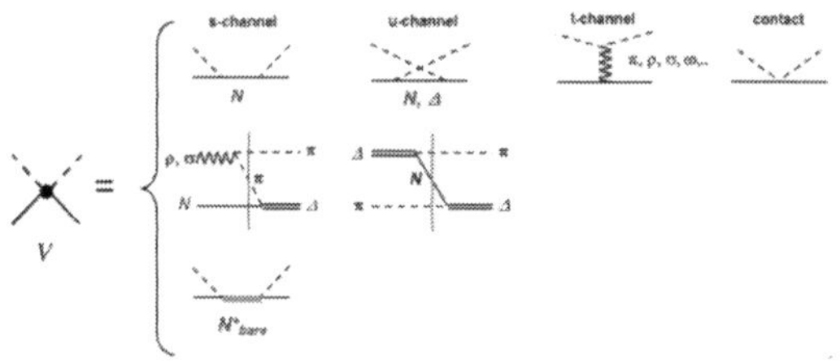


**Figure 2.** The structure of the tree-level interaction potential $V$. $s-$,$u-$, and $t-$ channel interactions defining the non-resonant background contributions are shown in the first line, including contact terms which are chosen such that gauge invariance is assured. The so-called *z-diagrams*, displayed in the second line, are generic for the double-pion channels. $s-$ channel resonance interactions are depicted in the last line. Time is running from left to right.

## 3 Theory of high-spin fermionic fields

### 3.0.1 Gauge properties of spin-3/2 fields

It is well known that the wave equation for the free spin-$\frac{3}{2}$ field [10] being written in a general form depends on one free parameter $A$ (see e.g. [15]). The commonly used Rarita-Schwinger theory [10] corresponds to the special choice $A = -1$. While the so-called Pascalutsa-coupling removes the unwanted degrees of freedom from the Rarita-Schwinger propagator it leaves the problem unsolved in the more general case $A \neq -1$ resulting in the appearance of 'off-shell' components, for example in the $\pi N$ scattering amplitude. Hence, further investigations of the general properties of the interacting spin-$\frac{3}{2}$ fields are of great importance. We have discussed the origin of this problem and shown how it to solved it. Two alternative approaches were indicated:

- the first solution consists in constructing a coupling which includes higher derivatives of the spin-$\frac{3}{2}$ field.
- alternatively, advantage may be taken on the generalization of the original gauge invariant interaction to arbitrary values of $A \neq -1$.

In the latter case the obtained Lagrangian depends on one free parameter which also appears in the free field formalism. However, the physical observables should not depend on this parameter. Hence, the matrix element corresponding to the $\pi N$ scattering at tree level does not contain an *off-shell background*. Rarita and Schwinger suggested a set of constraints which the free spin-$\frac{3}{2}$ field should obey [10]

$$\gamma^\nu \psi_\nu(x) = 0, \quad \partial^\nu \psi_\nu(x) = 0, \tag{5}$$

provided that also the Dirac equation $(\not p - m)\psi_\nu(p) = 0$ is fulfilled. In a consistent theory the set of equations Eq. (5) should follow from the equation of motion obtained from the corresponding Lagrangian. The Lagrangian of the free spin-$\frac{3}{2}$ field can be written in a general form as follows (see, e.g., [15] and references therein)

$$\mathcal{L}_0^{\frac{3}{2}} = \bar\Delta_\mu(x)\, \Lambda^{\mu\nu}\, \Delta_\nu(x), \tag{6}$$

where $\Delta_\nu(x)$ stands for the spin-$\frac{3}{2}$ field and the $\Lambda^{\mu\nu}$-operator is

$$\begin{aligned}\Lambda^{\mu\nu} = \quad & (i\not\partial - m)g^{\mu\nu} + iA(\gamma^\mu\partial^\nu + \gamma^\nu\partial^\mu) \\ & + \frac{i}{2}(3A^2 + 2A + 1)\gamma^\mu\not\partial\gamma^\nu \\ & + m\,(3A^2 + 3A + 1)\gamma^\mu\gamma^\nu,\end{aligned} \tag{7}$$

where $A$ is the afore mentioned arbitrary free parameter, subject only to the restriction $A \neq -\frac{1}{2}$. The propagator of the free spin-$\frac{3}{2}$ field can be obtained as a solution of the equation, e.g. in momentum space,

$$\Lambda_{\mu\rho}(p)\, g^{\rho\sigma}\, G_{\sigma\nu}(p) = g_{\mu\nu}. \tag{8}$$

The propagator $G_{\sigma\nu}(p)$ can be written as an expansion in terms of the spin projection operators $\mathcal{P}^{\frac{3}{2}}_{\mu\nu}(p)$, $\mathcal{P}^{\frac{1}{2}}_{11;\mu\nu}$, $\mathcal{P}^{\frac{1}{2}}_{22;\mu\nu}$, $\mathcal{P}^{\frac{1}{2}}_{12;\mu\nu}(p)$, and $\mathcal{P}^{\frac{1}{2}}_{21;\mu\nu}(p)$ [11]. The first three operators correspond to different irreducible representations of spin-vector whereas the last two account for a mixing between two spin-$\frac{1}{2}$ representations.

Without going too deep into the mathematical details, we constrain the discussion here to the consequences for interaction vertices. A commonly used $\Delta N\pi$-coupling is $\mathcal{L}_{\rm int} \sim \bar\psi_N \theta(z)^{\nu\mu}\Delta_\mu\partial_\nu\pi$ with $\theta^{\mu\nu}(z) = g^{\mu\nu} + z\gamma^\mu\gamma^\nu$. The free parameter $z$ is used to control the *off-shell* contributions to the interaction vertex but does not affect the pole term. In order to remove the dependence on $z$ (or, likewise, $A$) we eliminate the unwanted degrees of freedom by using a gauge-invariant coupling to the spin-$\frac{3}{2}$ field as explained in [11]. The modified $\Delta N\pi$ interaction Lagrangian can be written as follows[1]

$$\begin{aligned}\mathcal{L}_P &= \frac{g_{\Delta N\pi}}{m_\pi m_N}\bar\psi_N(x)\gamma_5\gamma_\mu T^{\mu\nu}_\Delta(x)\partial_\nu\pi(x) + \text{h.c.}, \\ T^{\mu\nu}_\Delta(x) &= \frac{1}{2}\epsilon^{\mu\nu\rho\sigma}\left(\partial_\rho\Delta_\sigma(x) - \partial_\sigma\Delta_\rho(x)\right),\end{aligned} \tag{9}$$

where $\epsilon^{\mu\nu\rho\sigma}$ is the fully antisymmetric Levi-Civita tensor. The tensor $T_{\mu\nu}(x)$ is invariant under the gauge-transformations $\Delta_\nu(x) \to \Delta_\nu(x) + \partial_\nu\xi(x)$ where $\xi(x)$ is an arbitrary spinor field. Hence, $T_{\mu\nu}(x)$ behaves like a *conserved current* with the constraint $\partial_\mu T^{\mu\nu}_\Delta(x) = 0$. The coupling defined in Eq. (9) guarantees that the so-called *off-shell background* does not contribute to the physical observables provided that the free spin-$\frac{3}{2}$ propagator is chosen in the special form corresponding to $A = -1$. This, however, does not hold in the general case for arbitrary values of $A$.

The problem reported above can be solved in different ways. The straightforward one is to use a coupling with higher order derivatives of the spin-$\frac{3}{2}$ field which explicitly involves the $\mathcal{P}^{\frac{3}{2}}_{\mu\nu}(p)$ projection operator:

$$\frac{g_{\Delta N\pi}}{m_\pi m_N^4}\bar\psi_N(x)\left[\Box\mathcal{P}^{\frac{3}{2}}_{\mu\nu}(\partial)\Delta^\nu(x)\right]\partial^\mu\pi(x) + \text{h.c.}. \tag{10}$$

The use of $\mathcal{P}^{\frac{3}{2}}_{\mu\nu}(\partial)$ ensures that only the spin-$\frac{3}{2}$ part of the propagator contributes and the d'Alembert-operator guarantees that no other singularities except the mass pole term $(p^2 - m^2)^{-1}$ appear in the matrix element. Note, that the coupling written in the form of Eq. (10) restores the invariance of the full Lagrangian under the point-like transformations $\Delta_\mu \to \Delta_\mu + z\gamma_\mu\gamma^\nu\Delta_\nu$.

To keep the interaction term in the full Lagrangian as simple as possible we propose here another coupling

$$\begin{aligned}\mathcal{L}_I \;=\; & \frac{g_{\Delta N\pi}}{m_\pi m_N} \\ \times \; & \bar\psi_N(x)\left[\Gamma_{\nu\eta}(A,\partial)\,\Delta^\eta(x)\right]\partial^\nu\pi(x) + \text{h.c.}\end{aligned} \tag{11}$$

with the modified vertex operator $\Gamma_{\nu\eta}(A,\partial)$ depending on the parameter $A$:

$$\Gamma_{\nu\eta}(A,\partial) = \gamma_5\,\gamma^\mu\,\epsilon_{\mu\nu\rho\sigma}\,\theta^\sigma_\eta(A)\,\partial^\rho, \tag{12}$$

$$\theta_{\sigma\eta}(A) = g_{\sigma\eta} - \frac{A+1}{2}\gamma_\sigma\gamma_\eta. \tag{13}$$

In momentum space at $A = -1$ the vertex function Eq. (12) reduces to that suggested by Pascalutsa. The $\theta_{\mu\nu}(A)$-operator has a simple physical meaning: it relates the Rarita-Schwinger theory to the general case of arbitrary A. Hence, the R-S propagator, can be obtained from the general propagator by means of the transformation $G^{\rm RS}_{\mu\nu}(p) = \theta_{\mu\rho}(A)\,G_{\rho\sigma}(p,A)\,\theta_{\sigma\nu}(A)$.

Using the coupling Eq. (11) the final result for the matrix element of $\pi N$ scattering is independent on the none-pole spin-$\frac{1}{2}$ terms in the full propagator

$$\begin{aligned}&\left[\Gamma_{\mu\rho}(A,p_\Delta)\,G^{\rho\sigma}(p_\Delta,A)\,\Gamma^\dagger_{\sigma\nu}(A,p_\Delta)\right] \\ &\qquad = \frac{\not p + m}{p^2 - m^2}\,\frac{p_\Delta^2}{m_N^2}\,\mathcal{P}^{\frac{3}{2}}_{\mu\nu}(p_\Delta)\end{aligned} \tag{14}$$

[1]We omit isospin indices.

and coincides with that obtained for the case $A = -1$. The coupling Eq. (11) can be written in a more compact form which does not contain the Levi-Civita tensor explicitly

$$\mathcal{L}_I = \frac{\mathrm{i}g_{\Delta N\pi}}{4\, m_\pi m_N} \times \bar{\psi}_N(x) \left[ \gamma^{\sigma\rho\nu} \theta_{\sigma\eta}(A) \partial^\rho \, \Delta^\eta(x) \right] \partial^\nu \pi(x) + \text{h.c.}, \quad (15)$$

where $\gamma^{\sigma\rho\nu} = \{\gamma^{\sigma\rho}, \gamma^\nu\}$ and $\gamma^{\sigma\rho} = [\gamma^\sigma, \gamma^\rho]$ and $\theta_{\sigma\eta}(A)$ is defined in Eq. (13).

The full Lagrangian for the interacting $\Delta N\pi$ fields can be written in the form

$$\mathcal{L}^{\frac{3}{2}} = \mathcal{L}_0^{\frac{3}{2}} + \mathcal{L}_I + \mathcal{L}_0^{\pi} + \mathcal{L}_0^{N}, \quad (16)$$

where $\mathcal{L}_0^{\pi} = \pi(\Box + m^2)\pi$ and $\mathcal{L}_0^{N} = \bar{\psi}_N(i\partial\!\!\!/ - m)\psi_N$ stand for the free Lagrangians of pion and nucleon fields, respectively. The free spin-$\frac{3}{2}$ Lagrangian $\mathcal{L}_0^{\frac{3}{2}}$ and $\Delta N\pi$ coupling $\mathcal{L}_I$ are given by expressions Eq. (6) and Eq. (11). The Lagrangian Eq. (16) depends on one arbitrary parameter $A$ which points to the freedom in choosing the 'off-shell' content of the theory. Although $\mathcal{L}_0^{\frac{3}{2}}$ contains one free parameter the physical observables should not depend on it. Without going into the details, we mention that similar conclusion can be made for the electromagnetic coupling and refer to ref. [11] for details.

Summarizing this work, we could show that the gauge-invariant $\Delta N\pi$ coupling, originally suggested by Pascalutsa for spin-$\frac{3}{2}$ fields, removes the *off-shell* degrees of freedom only for a specific choice of the spin-$\frac{3}{2}$ propagator but not in the general case. In the general case the spin-$\frac{3}{2}$ propagator contains a term associated with the $\left[\frac{1}{2} \otimes 1\right]_{\frac{1}{2}}$ irreducible representation. We have shown that the problem can be solved by introducing higher order derivatives to the interaction Lagrangian or by generalizing the original $\Delta N\pi$ coupling suggested by Pascalutsa. In the latter case the full Lagrangian of the interacting $\Delta N\pi$ fields depends on one free parameter which reflects the freedom in choosing an *off-shell* content of the theory.

### 3.0.2 Gauge properties of spin-5/2 fields

The description of pion- and photon-induced reactions in the resonance energy region is faced with the problem of proper treatment of higher spin states. In 1941 Rarita and Schwinger (R-S) suggested a set of equations which a field function of a higher spin should obey [10]. Another formulation has been developed by Fierz and Pauli [16] where an auxiliary field concept is used to derive subsidiary constraints on the field function. Regardless of the procedure used the obtained Lagrangians for free higher-spin fields turn out to be always dependent on arbitrary free parameters. For the spin-$\frac{3}{2}$ fields this issue is widely discussed in the Literature (see e.g. [11, 17, 18] for a modern status of the problem). The case of the spin-$\frac{5}{2}$ fields is even less studied. First attempts were made in [19, 20]. The authors of [20] deduced an equation of motion as a decomposition in terms of corresponding projection operators with additional algebraic constraints on parameters of the decomposition.

The free particle propagator is a central quantity in most of the calculations in quantum field theory. In [20] the authors deduced a spin-$\frac{5}{2}$ propagator written in operator form. In practical calculations, however, one needs an explicit expression of the propagator. An attempt to construct a propagator only from the spin-$\frac{5}{2}$ projection operator has been made in [21, 22]. We demonstrated that such a quantity is not consistent with the equation of motions for the spin-$\frac{5}{2}$ field. In addition, hermiticity can be violated, as was pointed in [14]. Clearly, it is important to derive the propagator and investigate its properties in detail. To the best of our knowledge our study was the first attempt in that direction. Hence, the aim of the work was to deduce an explicit expression for the spin-$\frac{5}{2}$ propagator and study its properties. Guided by the properties of the free spin-$\frac{3}{2}$ Rarita-Schwinger theory one would expect the equation of motion for the spin-$\frac{5}{2}$ field has two arbitrary free parameters which define the non-pole spin-$\frac{3}{2}$ and-$\frac{1}{2}$ contributions to the full propagator. The coupling of the spin-$\frac{5}{2}$ field to the (e.g.) pion-nucleon final state is therefore defined up to two *off-shell* parameters which scale the non-pole contributions to the physical observables. Hence, one can ask whether such an arbitrariness can be removed from the theory.

The possibility to construct consistent higher-spin massless theories has been pointed out already a while ago by Weinberg and Witten [23]. As we demonstrated in [11] the demand for gauge-invariance may not be enough to eliminate the extra degrees of freedom at the interaction vertex. The problem appears when the theory does not have a massless limit. However, a coupling which removes non-pole terms from the spin-$\frac{5}{2}$ propagator can be easily constructed by using higher order derivatives. A corresponding interaction Lagrangian has been deduced in [11] for the case of spin-$\frac{3}{2}$ fields and can be easily extended to higher spins too, as exercised in [14].

The field function of higher spins in a spinor-tensor representation is a solution of the set of equations suggested by Rarita and Schwinger in [10]. In a consistent theory the description of the free field is specified by setting up an appropriate Lagrange function $\mathcal{L}(\psi_{\mu\nu}, \partial_\rho\psi_{\mu\nu})$. The spin-$\frac{5}{2}$ Lagrangian in the presence of the auxiliary spinor field $\xi(x)$ can be written in the form

$$\mathcal{L} = \mathcal{L}^{(1)} + \mathcal{L}^{(2)} + \mathcal{L}^{(\text{aux})}, \quad (17)$$

where the lengthy and mathematically involved expressions for the three pieces are found in Ref. [14]. An important observation is that the Lagrangian in Eq. (17) in general depends on only three independent real parameters $a$, $b$, and $c$.

By variations with respect to $\psi_{\mu\nu}$ and $\xi$ two equations of motion are obtained which in momentum space are given as

$$\left(\Lambda^{(1)}_{\mu\nu;\rho\sigma}(p) + \Lambda^{(2)}_{\mu\nu;\rho\sigma}(p)\right)\psi^{\rho\sigma}(p) + c\, m\, g^{\mu\nu}\xi(p) = 0, \quad (18)$$

$$m\, c\, g^{\rho\sigma}\psi_{\rho\sigma}(p) + B(a,b,c)\,(p\!\!\!/ + 3m)\,\xi(p) = 0, \quad (19)$$

where the explicit forms of the operators $\Lambda^{(1)}_{\mu\nu;\rho\sigma}(p)$, $\Lambda^{(2)}_{\mu\nu;\rho\sigma}(p)$ are found in [14]. Here, it is of interest that the operator $\Lambda^{(1)}_{\mu\nu;\rho\sigma}(p)$ would give an equation of motion $\Lambda^{(1)}_{\mu\nu;\rho\sigma}(p)\psi^{\mu\nu} = 0$ for the spin-$\frac{5}{2}$ fields provided $g^{\mu\nu}\psi_{\mu\nu} = 0$, where the later property is assumed *a priori*. However, the corresponding inverse operator $[\Lambda^{(1)}_{\mu\nu;\rho\sigma}(p)]^{-1}$ has additional non-physical poles in the spin-$\frac{1}{2}$ sector. This indicates that the constraint $g^{\mu\nu}\psi_{\mu\nu} = 0$ should also follow from the equation of motion and cannot be assumed *a priori*. The second operator $\Lambda^{(2)}_{\mu\nu;\rho\sigma}(p)$ acts only in the spin-$\frac{1}{2}$ sector of the spin-tensor representation. This can be checked by a direct decomposition of the operator in terms of projection operators [14]. The same conclusion can be drawn from the observation that $\Lambda^{(2)}_{\mu\nu;\rho\sigma}(p)$ is orthogonal to all $\mathcal{P}^{\frac{5}{2}}_{\rho\sigma;\tau\delta}(p)$, $\mathcal{P}^{\frac{3}{2}}_{ij;\rho\sigma;\tau\delta}(p)$ projection operators, where $i, j = 1, 2$. Hence the parameter $b$ is related only to the spin-$\frac{1}{2}$ degrees of freedom whereas $a$ scales both spin-$\frac{3}{2}$ and -$\frac{1}{2}$ ones.

Of particular interest for spectroscopic research is the coupling of resonances to meson-nucleon channels. In the case of the spin-$\frac{5}{2}$ field in the spinor-tensor representation we deal with a system $(\psi_{\mu\nu}, \xi)$ which contains auxiliary degrees of freedom. The question arises whether the nonphysical degrees of freedom could be eliminated from physical observables. Here we consider a simple case of spin-$\frac{5}{2}$ resonance contribution to $\pi N$ scattering. The corresponding $\pi NN^*_{\frac{5}{2}}$ coupling can be chosen as follows

$$\begin{aligned}\mathcal{L}_I &= \frac{g_{\pi NN^*}}{4m_\pi^2} \\ &\times \left(\bar{\psi}_N(x), 0\right)\Gamma_{\mu\nu;\rho\sigma}\left[\mathfrak{P}\begin{pmatrix}\psi^{\rho\sigma}\\ \xi\end{pmatrix}\right]\partial^\mu\partial^\nu\pi(x) \\ &+ \text{h.c.}, \end{aligned} \tag{20}$$

where the nucleon field is written as $(\bar{\psi}_N(x), 0)$ which implies the absence of auxiliary fields in the final state. The operator

$$\mathfrak{P} = \begin{pmatrix}1 & 0\\ 0 & 0\end{pmatrix}$$

projects out the spin-$\frac{5}{2}$ field and ensures that there is no coupling to $\xi$. Hence, only the spin-$\frac{5}{2}$ component of the propagator $G^{\frac{5}{2}}_{\mu\nu;\rho\sigma}(p)$ contributes to physical observables at any order of perturbation theory. In [14] we could demonstrate that the inclusion of auxiliary degrees of freedom in the vector field does not affect the physical observables. To the best of our knowledge this statement is not generally proven for the $(\psi_{\mu\nu}, \xi)$ system beyond the perturbation expansion. The reason is that the equation of the motion for massive spin-$\frac{5}{2}$ field in the spinor-tensor representation is defined only in the presence of an auxiliary field. This is unlike the case of the vector field where auxiliary degrees of freedom can be removed by proper field transformations. These degrees of freedom contribute due to $\psi_{\mu\nu} - \xi$ mixing. The mixing takes place only between the spin-$\frac{1}{2}$ sector of the spinor-tensor and the auxiliary spinor fields. One may therefore hope that the use of a coupling which suppresses the spin-$\frac{1}{2}$ contributions would also prevent the appearance of the auxiliary degrees of freedom in the physical observables in the non-perturbative regime.

The solution to the problem is following closely the results for spin-3/2 fields, presented in [11]. According to our previous findings the interaction vertex fulfills the condition $\gamma \cdot \Gamma = \Gamma \cdot \gamma = 0$. With this constraint one finds that only the $\mathcal{P}^{\frac{5}{2}}_{\mu\nu;\rho\sigma}(q)$ projector fulfills the desired property [14], ensuring that only the spin-$\frac{5}{2}$ part of the propagator contributes. The formalism also guarantees that no other singularities except the mass pole term $(p^2 - m^2)^{-1}$ appear in the amplitude. Thus, the physical observables no longer depend on the arbitrary parameters $a$ and $b$ of the free Lagrangian. Finally, the $\pi N$ scattering amplitude reads

$$\begin{aligned}\mathcal{M} &= \left(\frac{g_{\pi NN^*}}{m_\pi^2}\right)^2 \bar{u}_N(p') \\ &\left[\left(\frac{q^2}{m_R^2}\right)^4 \mathcal{P}^{\frac{5}{2}}_{\mu\nu;\lambda\tau}(q)\right] u_N(p)\, k'^\mu k'^\nu k^\lambda k^\tau, \end{aligned} \tag{21}$$

Summarizing this part of the project, we have investigated the general properties of the free spin-$\frac{5}{2}$ fields in the spinor-tensor representation. We could show that the Lagrangian in general depends on three arbitrary parameters; two of them are associated with the lower spin-$\frac{3}{2}$ and -$\frac{1}{2}$ sector of the theory whereas the third one is linked to the auxiliary field $\xi$. We have deduced a free propagator of the system in form of a 2 x 2 matrix in the $(\psi_{\mu\nu}, \xi)$ space. The diagonal elements stand for the propagation of the spin-$\frac{5}{2}$ and $\xi$ fields whereas the non-diagonal ones correspond to $\psi_{\mu\nu} - \xi$ mixing. The mixing takes place between the spin-$\frac{1}{2}$ sector of the spinor-tensor representation and an auxiliary spinor field. An important result was that the auxiliary degrees of freedom do not contribute to the physical observables calculated within the perturbation theory provided there is no coupling to $\xi$. As an application to hadron spectroscopy, the $\pi NN^*_{\frac{5}{2}}$ interaction vertex was discussed. Gauge invariance was obtained by constructing a coupling with higher order derivatives. In the latter case the amplitude of the $\pi N$ scattering does not depend on the arbitrary parameters of the free Lagrangian. The suggested coupling is generalized to the Rarita-Schwinger fields of any half-integer spin.

## 4 Strangeness production on the nucleon

Strangeness production on the nucleon by excitation of resonances which decay into kaon-hyperon channels is an important spectroscopic tool giving access to the SU(3) flavour structure of baryons. Moreover, such *exotic* channels like the kaon-hyperon final states are expected to play a cental role in identifying hitherto undetected excited states of the nucleon, thus addressing the still open question of the notorious problem of *missing resonances*. In [5] we have performed a study of the pion- and photon-induced $K\Lambda$ reactions within our unitary coupled-channel effective Lagrangian approach. A major goal of those investigations was to address the at that time still open question on the major contributions to the associated strangeness production channels. Since $K\Lambda$ photoproduction data [24, 25] gave an indication for *missing* resonance

 DOI: 10.1051/epjconf/201713403006

contributions, a combined analysis of the $(\pi, \gamma)N \to K\Lambda$ reactions was expected to identify clearly these states. Assuming small couplings to $\pi N$, these *hidden* states should not exhibit themselves in the pion-induced reactions and, consequently, in the $\pi N \to K\Lambda$ reaction. The aim of our calculations was to explore to what extent the data available at that time can be explained by known reaction mechanisms without introducing new resonances. Our results for total cross sections are displayed in Fig.3 and further results on differential cross sections, polarization observables and angular distributions are found in [5]. As discussed in [5] the SAPHIR [24] and the CLAS [25] data sets, in fact, are leading to two slightly different sets of interaction parameters, reflecting and emphasizing the differences among the two measurements. Below, that point is discussed again.

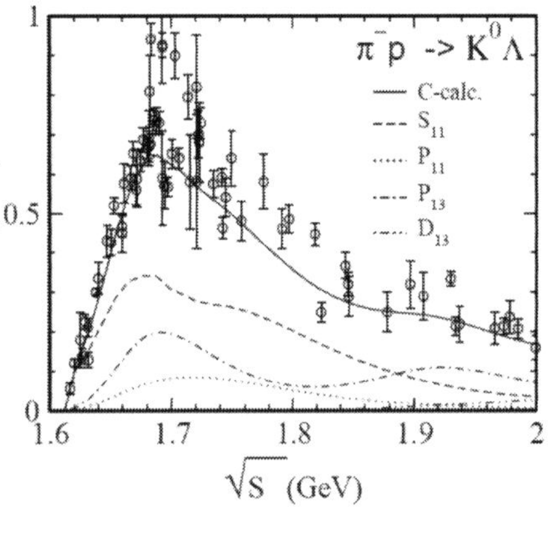


**Figure 3.** $\pi^- p \to K^0\Lambda$ total partial wave cross sections, predicted by parameter set *C* of Ref. [5], obtained from a fit to the CLAS data [25]. The experimental cross section data are taken from [26–28].

More recent CLAS-data on $K\Sigma$ production inspired us to a revised updated large scale coupled-channels analysis of associated strangeness production on the nucleon. Based on the coupled-channel effective Lagrangian formalism underlying the Giessen model (GiM) a combined analysis of $(\pi, \gamma)N \to K\Sigma$ hadro- and photo-production reactions were performed. The analysis covered a center-of-mass energy range up to 2 GeV. The recent photoproduction data obtained by the CLAS, CBELSA, LEPS, and GRAAL groups were included into our calculations. The central aim was to extract the resonance couplings to the $K\Sigma$ state. In [6] the Giessen model was used to reanalyze newly released data from various experimental groups for $K\Sigma$ production on the nucleon. Both s-channel resonances and t,u-channel background contributions are found to be important for an accurate description of angular distributions and polarization observables, assuring a high quality description of the data. The extracted properties of isospin $I = 3/2$ resonances were discussed in detail. We found that the $I = 1/2$ resonances are largely determined by the non-strangeness channels.

Our calculations included 11 isospin $I = 1/2$ resonances and 9 isospin $I = 3/2$ resonances, respectively. In this work we continued the investigations of the $I = 1/2$ and $3/2$ sectors with the parameters fitted to newly published $K\Sigma$ photoproduction data together with the previous $\pi N \to K\Sigma$ measurements in the energy region $\sqrt{s} \leq 2.0$ GeV. The included $K\Sigma$ photoproduction data are those of the $\gamma p \to K^+\Sigma^0$ published by the LEPS [29–31], CLAS [32, 33] and GRAAL [43] group, and those of $\gamma p \to K^0\Sigma^+$ released by the CLAS [44] and CBELSA [45] collaboration, respectively. The SAPHIR data have been left out here because of the known inconsistencies of the $K^+\Sigma^0$ data [24] with the corresponding CLAS and GRAAL data (for the details, see Ref. [33]). Also, the $K^0\Sigma^+$ SAPHIR data [24] have much bigger error bars than those of the CBELSA and CLAS group. Here, the data before 2002 are also no longer used.

Strangeness production on the nucleon plays a key role for our understanding of baryon structure and elementary reactions with hadrons. In addition, such production reactions are also an appropriate tool to identify excited states $N^*$ of the nucleon which decay preferentially into hyperon-kaon channels, thus adding to solve the problem of missing resonances. In [6] we have analysed the latest CLAS-, CBELSA-data sets and re-analysed the earlier SAPHIR-data on photoproduction of kaons on the nucleon. The Giessen model was used, describing meson production on the nucleon by a coupled channels K-matrix approach including meson production by photoproduction and pion-induced reactions as initial channels. The Giessen model is obeying the elementary symmetries of hadron physics and conserving unitarity. Meson production proceeds through s-channel resonances and t- and u-channels re-scattering processes. Results for total cross sections are shown in Fig.4. Up to a total center-of-mass energy of about $\sqrt{s} = 2$ GeV the data are well described.

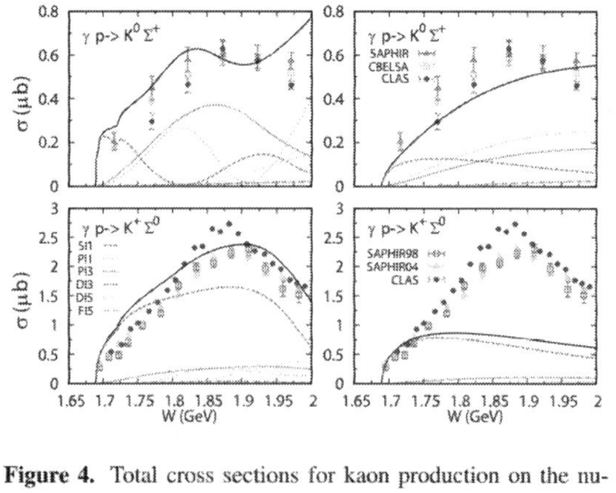


**Figure 4.** Total cross sections for kaon production on the nucleon. Results of the Giessen model [6] are compared to CLAS, CBELSA, and SAPHIR data.

The analysis included all charge channels, $K^\pm\Sigma^\mp$ and $K^0\Sigma^0$. We achieved a quite satisfactory description of the $\gamma p \to K^+\Sigma^0$ data ($\chi^2 = 1.8$) and the $\gamma p \to K^0\Sigma^+$ data ($\chi^2 = 2.0$). However, the pion-induced strangeness production reactions are described slightly less accurate as indicated by the corresponding $\chi^2$ values of $\chi^2 = 4.1, 3.2$ and

DOI: 10.1051/epjconf/201713403006

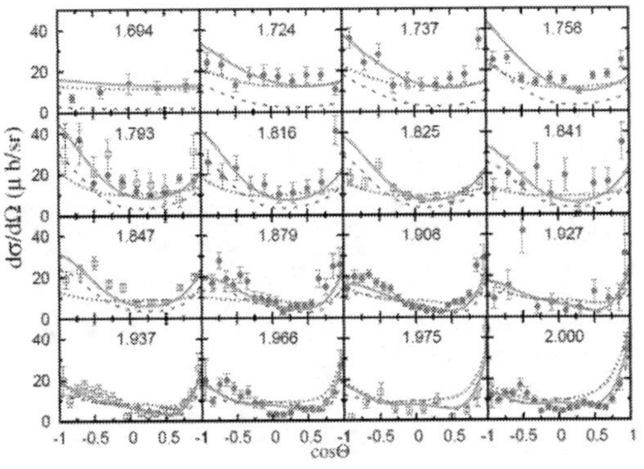


**Figure 5.** The differential cross section of $\pi^- p \to K^0\Sigma^0$ reaction. The solid (green), dashed (blue) and dotted (magenta) lines are the full model calculation, the model calculation with the $S_{11}(1650)$ and $F_{15}(1680)$ turned off, respectively. The numerical labels denote the center of mass energies in units of GeV. CBELSA and CLAS data are shown for comparison.

2.8 for the $\pi^+ p \to K^+\Sigma^+$, $\pi^- p \to K^0\Sigma^0$ and $\pi^- p \to K^+\Sigma^-$ reactions, respectively. The parameters have been varied in our fit simultaneously to the $I = 1/2$ and $3/2$ sectors. Because of the smallness of the $N^*K\Sigma$ couplings, all previously obtained Breit-Wigner masses, branching ratios and couplings corresponding to non-strangeness production [2] are hardly affected by the additional $K\Sigma$ photoproduction data, so we could concentrate on the properties of the $I = 3/2$ resonances. Although the new data are available with reduced total errors the refitted model parameters were changed only very little. A typical result is displayed in Fig. 5, illustrating the quality of the description on the example of $\pi^- p \to K^0\Sigma^0$ reaction. The complete set of results, including partial wave cross sections, angular distributions of cross sections and polarization observables for the full set of $K\Sigma$ exit channels are found in [6].

## 5 $\eta$-meson production

Understanding the dynamics of eta-meson production and, *vice versa*, the decay of nucleon resonances into the nucleon-eta exit channel is of ongoing interest in hadron spectroscopy. The $\eta$-meson photoproduction on the proton has been measured with high precision by the Crystal Ball collaboration at MAMI [34]. These high-resolution data provide a new step forward in understanding the reaction dynamics and in the search for a signal from the 'weak' resonance states. The main result reported in [34] is a very clean signal for a dip structure around $W = 1.68$ GeV, seemingly confirming older data [35–38]. This raised the question on the origin of that structure, eventually indicating the appearance of a new narrow, possibly exotic, resonance state.

The aim of the study was to extend our previous coupled-channels analysis of the $\gamma p \to \eta p$ reaction by including the data from the new high-precision measurements [34]. The main question is whether the $\eta p$ reaction dynamics can be understood in terms of the established resonance states or whether a new state has to be introduced, thus confirming previous conjectures. A major issue for the analysis is unitarity and a consistent treatment of self-energy effects as visible in the total decay width of resonances. Since the latter are driven by hadronic interactions the analysis of photo-production data requires the knowledge of the hadronic transition amplitudes as well. Hence, a coupled-channels description as in the Giessen model is an indispensable necessity.

As discussed in very detail in [4] various relevant meson-baryon coupling constants were newly determined at the occasion of this work in large scale coupled-channels calculations. This gave rise to improved constraints on the interaction parameters and the derived resonance parameters, i.e. masses and widths. As an representative example we mention here the mass and width of the $D_{13}(1520)$ resonance. Our results confirm the values obtained by Arndt *et al.* [40]: mass $M = 1516 \pm 10$ MeV and width $\Gamma = 106 \pm 4$ MeV, respectively. It is interesting to note that the mass of this resonance deduced from pion photoproduction tends to be 10 MeV lower that the values derived from the pion-induced reactions [41]. The second $D_{13}(1900)$ state has a very large decay width. We associate this state with the $D_{13}(2080)$ two-star state, proposed by PDG. The GiM results for the $\eta$-photo production channel are shown in Fig. 6 together with the experimental data.

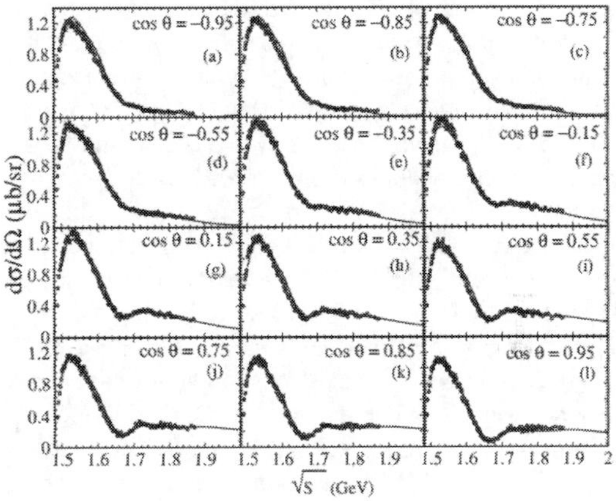


**Figure 6.** Differential $\eta p$ cross section compared to MAMI data [34].

We obtain a very satisfactory agreement with the experimental data in the whole kinematical region. The first peak is related to the $S_{11}(1535)$ resonance contribution. Similar to the $\pi^- p \to \eta n$ reaction the $S_{11}(1650)$ and $S_{11}(1650)$ states interfere destructively producing a dip around W=1.68 GeV. The coherent sum of all partial waves leads to the more pronounced effect from the dip at forward angles. We also corroborate our previous findings [3] where a small effect from the $\omega N$ threshold was found.

We also do not find any strong indication for contributions from a hypothetic narrow $P_{11}$ state with a width of 15-20 MeV around W=1.68 GeV. It is natural to assume that the contribution from this state would induce a strong modification of the beam asymmetry for energies close to

DOI: 10.1051/epjconf/201713403006

the mass of this state. This is because the beam asymmetry is less sensitive to the absolute magnitude of the various partial wave contributions but strongly affected by the relative phases between different partial waves. Thus even a small admixture of a contribution from a narrow state might result into a strong modification of the beam asymmetry in the energy region of W=1.68 GeV.

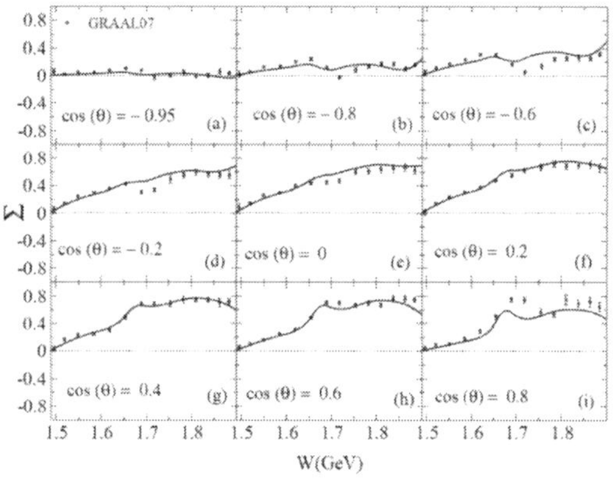


**Figure 7.** (Differential $\eta p$ cross section compared to GRAAL data [38].

In Fig. 7 we show results for the photon-beam asymmetry $\Sigma$ in comparison with the GRAAL data. One can see that even close to the $\eta N$ threshold where our calculations exhibit a dominant $S_{11}$ production mechanism the beam asymmetry is non-vanishing for angles $\cos(\theta) \geq -0.2$. This shows that this observable is very sensitive to very small contributions from higher partial waves. At W=1.68 GeV and forward angles the GRAAL measurements show a rapid change of the asymmetry behavior. We explain this effect by a destructive interference between the $S_{11}(1535)$ and $S_{11}(1650)$ resonances which induces the dip at W=1.68 GeV in the $S_{11}$ partial wave. The strong drop in the $S_{11}$ partial wave modifies the interference between $S_{11}$ and other partial waves and changes the asymmetry behavior. Note that the interference between $S_{11}(1535)$ and $S_{11}(1650)$ and the interference between different partial waves are of different nature. The overlapping of the $S_{11}(1535)$ and $S_{11}(1650)$ resonances does not simply mean a coherent sum of two independent contributions, but also includes rescattering (coupled-channel effects). Such interplay is hard to simulate by the simple sum of two Breit-Wigner forms since it does not take into account rescattering due to the coupled-channel treatment.

## 6 Double-pion production on the nucleon

In certain energy regions the $\pi N \to 2\pi N$ reaction accounts for up to 50% of the $\pi N$ inelasticity as seen in Fig.1. Therefore, this production channel had been included from the very beginning into the GiM approach, albeit in a schematic manner. An improved and considerably extended description of double-pion production within our coupled channels scheme was started recently and first results are found in [7]. The inclusion of multi-meson con-

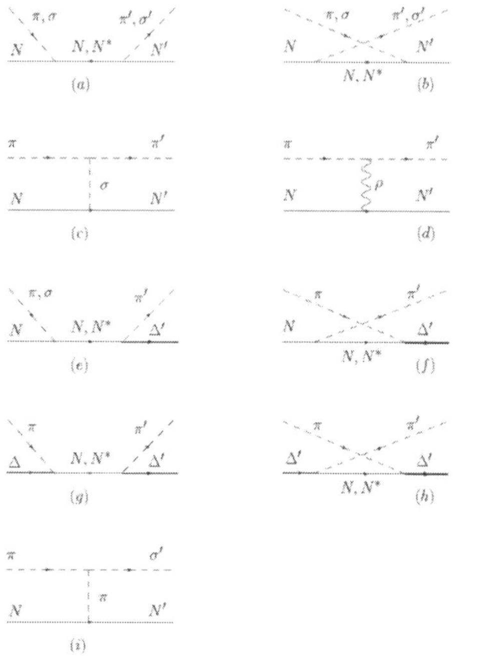

**Figure 8.** Diagrammatic structure of the tree-level interactions contributing to double-pion production on the nucleon.

figurations into a coupled channels approach is a highly non-trivial exercise in three-body dynamics and beyond. In view of the complexities physically meaningful approximations are necessary, retaining the essential dynamical aspects but making numerical calculations feasible. For that goal the *ansatz* used in [7] relies on an isobar description of intermediate two-pion configurations and their decay into the final double pion states on the mass shell. In Fig.8 the tree-level interactions for two-pion production are displayed diagrammatically. The derived processes contributing to the T-matrix of double-pion production on the nucleon in that energy region are depicted in Fig.9.

This approach allows for the direct analysis of the $2\pi N$ experimental data. Since the corresponding Dalitz plots are found to be strongly non-uniform it is natural to assume that the main effect to the reaction comes from the resonance decays into isobar subchannels [42]. The most important contributions are expected to be from the intermediate $\sigma N$, $\pi\Delta_{1232}$, and $\rho N$ states. The analysis of the $\pi N \to 2\pi N$ reaction would therefore provide very important information about the resonance decay modes into different isobar final states. Presently, lattice simulations, e.g. [47, 48], and functional approaches as in [49] are applied to approach the baryon spectrum from the QCD-side. However, despite considerable progress a number of open problems are persisting like the incorporation of the dispersive self-energies from the coupling of the QCD configurations to the meson-nucleon decay channels. Similar to the constituent quark models [50, 51] the QCD-inspired

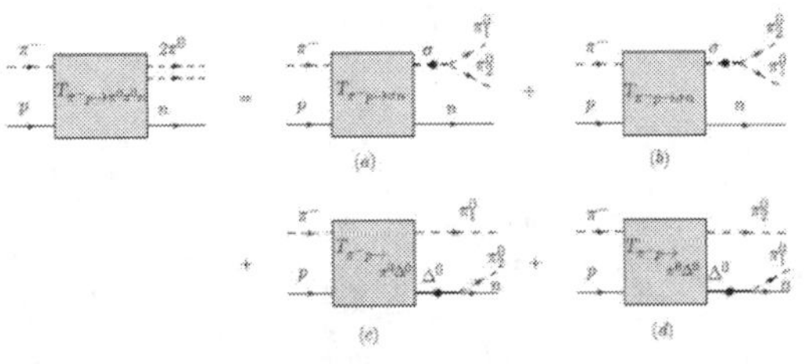

**Figure 9.** The processes contributing to double-pion production T-matrix are depicted diagrammatically: (a) and (b) production through the $\sigma$-isobar, (c) and (d) production through the $\Delta^0$-isobar. Symmetrization is indicated.

calculations predict a much richer baryon spectrum [48] than observed experimentally so far.

Clearly, an unambiguous identification of the excitation spectrum of baryons would provide an important link between theory and experiment. On the experimental side most of the non-strange baryonic states have been identified from the analysis of elastic $\pi N$ data [40, 52, 53]. However, as pointed out by Isgur [50], the signal of excited states with a small $\pi N$ coupling is likely to be suppressed in elastic $\pi N$ scattering. As a solution to this problem a series of photoproduction experiments has been done in order to accumulate enough data for detailed studies of the nucleon excitation spectrum. However, the results from the photoproduction reactions are still controversial. While recent investigations of the photoproduction reactions presented by the BoGa group [54] reported indications for some new resonances not all of these states are found in other calculations [41]. This raises a question about an independent confirmation of the existence of such states by other reactions. Because of the smallness of the electromagnetic couplings the largest contribution to the resonance self-energy comes from the hadronic decays. If the $N^* \to \pi N$ transition is small one can expect sizable resonance contribution into remaining hadronic decay channels. As a result the effect from the resonance with a small $\pi N$ coupling could still be significant in the inelastic pion-nucleon scattering: here the smallness of resonance coupling to the initial $\pi N$ states could be compensated by the potentially large decay branching ratio to other different inelastic final states. Such a scenario is realized e.g. in the case of the well known $N^*(1535)$ state. While the effect from this resonance to the elastic $\pi N$ scattering is only moderate at the level of total cross section its contribution to the $\pi N \to \eta N$ channel turns out to be dominant [4]. Since the $\pi N \to 2\pi N$ reaction could account for up to 50 % of the total $\pi N$ inelasticity this channel becomes very important not only for the investigation of the properties of already known resonances but also for the search for the signals of possibly unresolved states.

Another important issue in studies of the $2\pi N$ channel is related to the possibility to investigate cascade transitions like $N^{*\prime} \to \pi N^* \to \pi\pi N$, where a massive state $N^{*\prime}$ decays via intermediate excited $N^*$ or $\Delta^*$. It is interesting to check whether such decay modes are responsible for the large decay width of higher lying mass states. So far only the $\pi N^*(1440)$ isobar channel has been considered in [42] in a partial wave analysis (PWA) of the $\pi N \to 2\pi N$ experimental data [42].

There are several complications in the coupled-channel analysis of $2 \to 3$ transitions. The first one is the difficulty to perform the partial-wave decomposition of the three-particle state. The second complication is related to the issue of three-body unitarity. For a full dynamical treatment of the $2 \to 3$ reaction the Faddeev equations have to be solved. This makes the whole problem quite difficult for practical implementations. Here we address both issues and present a coupled-channel approach for solving the $\pi N \to 2\pi N$ scattering problem in the isobar approximation. In this formulation the $(\pi/\pi\pi)N \to (\pi/\pi\pi)N$ coupled-channel equations are reduced to the two-body scattering equations for isobar production. Such a description accounts by construction for the full spectroscopic strength of intermediate channels and, in addition provides a considerable numerical simplification. Three-body unitarity leads to a relation between the imaginary part of the elastic scattering amplitude and the sum of the total elastic and inelastic cross sections by the well known optical theorem. Since in the isobar approximation the pions in the $\pi\pi N$ channel are produced from the isobar subchannels all contributions to the total $\pi N \to \pi\pi N$ cross section are driven by the isobar production. The optical theorem can be fulfilled if all discontinuities in isobar subchannels are taken into account. In the present work the three-body unitarity is maintained up to interference term between the isobar subchannels.

As a first application of our model we apply the developed approach for the study of the $\pi^- p \to \pi^0\pi^0 n$ data in the first resonance energy region assuming the dominant $S_{11}$ and $P_{11}$ partial wave contributions in the $\sigma N$ and $\pi\Delta$ reaction subchannels. The main purpose of this paper is to introduce the model and demonstrate the feasibility of treating two-pion dynamics in the framework of a large-scale coupled channels approach. For this aim, we restrict the calculations to the $\pi^0\pi^0 n$ channel, taking advantage of the fact that only isoscalar two-pion and $\pi\Delta$ isobar channels are contributing to the process. We emphasize that this restriction is not a matter of principle but is only for the sake of a feasibility study. In particular, this means that at this stage we do not consider the $\rho N$ state but postpone its inclusion into the numerical scheme to a later stage. Naturally, the results presented in the following are most meaningful for the energy region of the $N^*(1440)$ Roper resonance.

The first resonance energy region is of particular interest because of the sizable effect from $N^*(1440)$. The dynamics of the Roper resonance turns out to be rich because of the two-pole structure reported in earlier studies [55, 56], (see [40, 57, 58] for the recent status of the problem. ) The origin of the Roper resonance is also controversial. For example the calculations in the Jülich model explain this state as a dynamically generated pole due to the strong attraction in the $\sigma N$ subchannel. At the same time the Crystal Ball collaboration finds no evidence of strong $t$-channel sigma-meson production in their $\pi^0\pi^0$ data [59].

 DOI: 10.1051/epjconf/201713403006

From the further analysis of the $\pi^0\pi^0$ production the effect of the sigma meson was found to be small [60]. On other hand the $pp \to pp\pi^0\pi^0$ scattering experiment by CELSIUS-WASA collaboration [61] finds the $\sigma N$ decay mode of the Roper resonance to be dominant.

In view of these problems we have performed an analysis of the $\pi^0\pi^0$ data assuming dominant contributions from the $S_{11}$ and $P_{11}$ amplitudes in the isospin $I = \frac{1}{2}$ channel. Since the effect from $N^*(1520)$ is expected to be important above 1.46 GeV we have limited the present calculations up to $\sqrt{s}$ =1.46 GeV energy region.

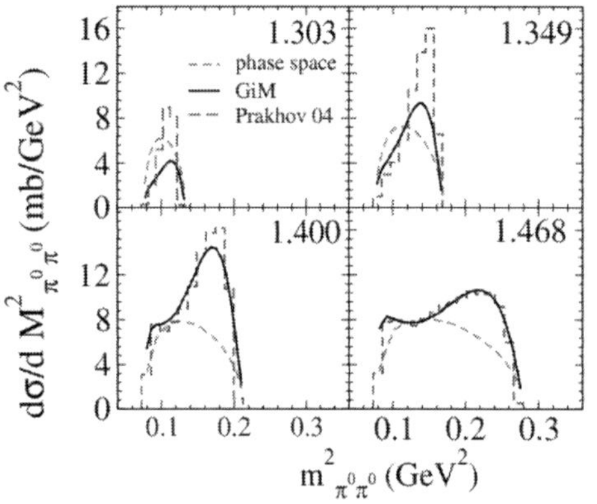


**Figure 10.** Reaction $\pi^- p \to \pi^0\pi^0 n$: differential cross section and bare phase space distribution (short dashed) as a function of $m^2_{\pi\pi}$ at fixed c.m. energies are compared to the Crystal Ball data (dashed) [60].

The difference between $\sigma N$ and $\pi\Delta$ production mechanism is seen in the invariant mass distributions, Fig. 10. Close to threshold the Crystal Ball data demonstrate a shift to the higher invariant masses for all energies up to 1.5 GeV whereas the three-body phase space tends to have a maximum at lower $m^2_{\pi^0\pi^0}$. In the present calculations the main contributions to the $\pi^- p \to \pi^0\pi^0 n$ reaction close to threshold are driven by $t$-channel pion exchange. This mechanism produces the invariant distributions which are shifted to the higher $\pi^0\pi^0$ invariant masses. However, the present calculations do not completely follow the experimental data at 1.303 and 1.349 GeV.

In the region of the Roper resonance our calculations are able to describe the mass distributions rather satisfactorily. Also in this region the production strength is shifted to higher invariant masses $m^2_{\pi^0\pi^0}$. At the same time a peak at small $m^2_{\pi^0\pi^0}$ becomes also visible. In the present calculations the fit tends to decrease the magnitude of the $\pi\Delta_{1232}$ production and compensate it by enhancing the strength into $\sigma N$. The obtained decay branching ratio of $N^*(1440)$ for the $\sigma N$ channel is about twice as large as for the $\pi\Delta_{1232}$.

Both the small peak at small and the broad structure at large invariant masses are well reproduced indicating an important interplay between the $\sigma N$ and $\pi\Delta_{1232}$ production mechanism. It is interesting that the isoscalar correlations in the $\pi\pi$ rescattering are also found to be necessary in order to reproduce the asymmetric shape of the mass

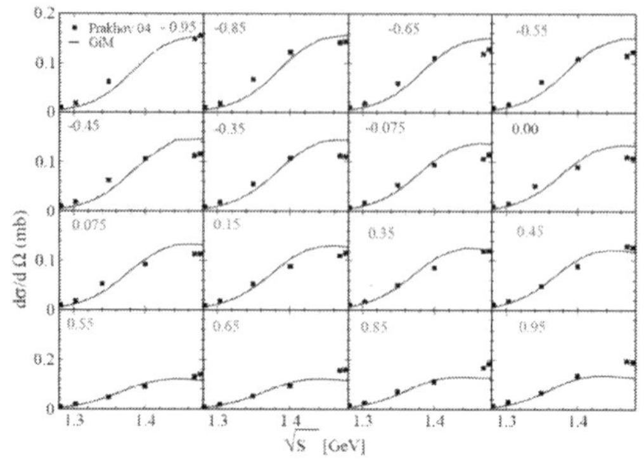


**Figure 11.** $\pi^0\pi^0$ differential cross sections for the reaction $\pi^- p \to \pi^0\pi^0 n$ at fixed $t_{\pi^0\pi^0} = \cos\theta_{\pi^0\pi^0}$, shown in the upper left corner of the the panels. Energy distributions for $-0.95 \le t_{\pi^0\pi^0} \le +0.95$ are shown. The experimental data are from [60].

distributions. Though the $\pi\Delta_{1232}$ production gives rise to a two-peak structure only the first one at small $m^2_{\pi^0\pi^0}$ is visible at energies 1.4-1.468 GeV. Within the present calculation the second peak at high $m^2_{\pi^0\pi^0}$ is not seen because of the large $\sigma N$ contributions. In the present study $\pi^0\pi^0 n$ production is calculated as a coherent sum of isobar contributions. Though the interference effect are important they are found to be very small at the level of the total cross sections.

We briefly discuss the reaction data base used in the calculations. To simplify the analysis the $S_{11}$ and $P_{11}$ $\pi N$ partial waves are directly constrained by the single energy solutions (SES) derived by GWU(SAID) [40]. The experimental data on the $\pi^- p \to \pi^0\pi^0 n$ reaction are taken from [60]. These measurements provide high statistics data on the angular distributions $d\sigma/d\Omega_{\pi\pi}$, where $\Omega_{\pi\pi}$ is the scattering angle of the $\pi\pi$ pair (or the final nucleon in c.m.). This data are accompanied by the corresponding statistical and systematical errors. No such information is available for the mass distributions in [60]. These observables are provided in a form of weighted events without systematic and statistical uncertainties. In the data analysis we impose the constraint that the integrated distributions must reproduce the total cross section of the $\pi^- p \to 2\pi^0 n$ reaction. We also have assigned about 10% error bars to each mass bin to perform the $\chi^2$ minimization. Starting from 1.46 GeV the excitation of $N^*(1520)$ starts to be important. Already at this energy a small contribution from the spin $J = \frac{3}{2}$ partial wave could modify the angular and mass distributions. Because of this reason we do not try to fit the data above 1.46 GeV.

The calculated $\pi^0\pi^0$ differential cross sections are shown in Fig. 11 and compared to the Crystal Ball data as a function of the c.m. energy. The measurements demonstrate a rapid rise of the cross sections at the energies 1.3-1.46 GeV. We identify this behavior as an indication for the strong contribution coming from the Roper resonance. Indeed, the resulting $\pi N$ inelasticities from the GWU(SAID) [40] analysis indicate that the $P_{11}$ partial wave dominates

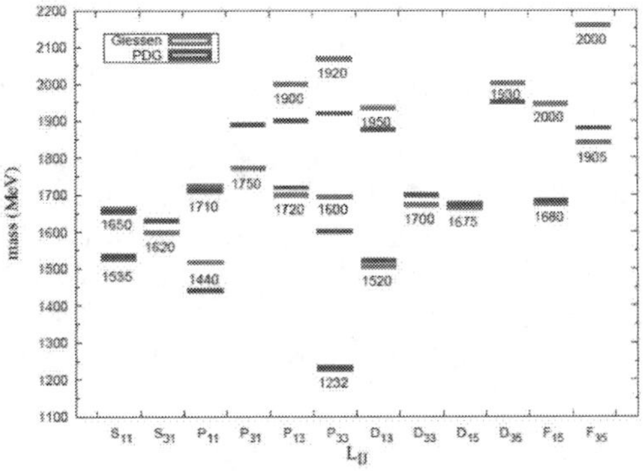


**Figure 12.** (Color online) Comparison of the GiM $N^*$ resonance level scheme in various partial waves to the recent PDG compilation [41].

the inelastic transitions at these energies. The inelasticity from the $S_{31}$ channel is about three times less than that from $P_{11}$. At the same time the $\Delta(1620)$ is strongly coupled to the $2\pi N$ final state through the $\pi\Delta_{1232}$ decay [41]. Since the contribution from the $\sigma N$ subchannel is found in the present work to be about twice as large than that of $\pi\Delta_{1232}$ possible effects from the $\Delta(1620)$ resonance can safely be neglected in first approximation. We also allow the $N^*(1535)$ resonance decays to the $\pi\Delta_{1232}$ and $\sigma N$ isobar final states which are however found to be negligible. At energies close to 1.5 GeV the obtained cross section slightly overestimates the experimental data at backward and underestimates them at forward scattering angles. This is a region where the $N^*(1520)$ starts to play a dominant role. We conclude that the contribution from the $D_{13}$ partial wave should be included for the successful description of the data at 1.5 GeV.

The invariant $\pi^0\pi^0$ mass distributions play a crucial role in the separation of the isobar contributions. The $\pi^- p \to \pi^0\pi^0 n$ reaction close to threshold is dominated by the $\sigma N$ production due to the $t$–channel pion exchange. The nucleon Born term contribution to the $\pi\Delta_{1232}$ channel is found to be less significant. For the decay branching ratios of $N^*(1440)$ we obtain $R^{N(1440)}_{\sigma N} = 27^{+4}_{-9}\%$ and $R^{N(1440)}_{\pi\Delta_{1232}} = 12^{+5}_{-3}\%$.

The comparison of our results with the parameters extracted by the BoGa group $R^{N(1440)}_{\sigma N} = 17^{+7}_{-7}\%$ and $R^{N(1440)}_{\pi\Delta_{1232}} = 21^{+8}_{-8}\%$ demonstrates that despite on the visible difference in the central values these quantities could still coincide within their error bars. The extended analysis of the $\pi\pi N$ which includes higher partial waves would help to reduce the uncertainties of the extracted resonance properties.

## 7 Summary

Excited states of the nucleon as observed in $\pi N$ and $\gamma N$ reactions have been described in the GiM coupled channels approach. The theoretical background was discussed. The GiM approach is based on a phenomenological field theory described the baryons, mesons, and their interactions by a Lagrangian density conserving by construction the fundamental symmetries of QCD, including also chiral symmetry. Resonance and background contributions are generated consistently out of the tree-level interactions obtained from the Lagrangian. The scattering amplitudes are determined by linear system of coupled equations which is solved numerically in K-matrix approximation and partial wave representation. The gauge-invariant description of high-spin resonances was discussed in detail. Applications to selected reaction channels have been presented, ranging from single pion, eta, and kaon production to double-pion production. The spectroscopic results are summarized in Fig.12 where the GiM spectrum of resonances is compared to the recent PDG resonance compilation.

The work reported here would not have been possible without the support from the Deutsche Forschungsgemeinschaft within the SFB/TR16 and the major contributions by Vitaliy Shklyar, and Xu Cao, now at IMP Lanzhou and CAS Beijing.

## References

[1] V. Shklyar, G. Penner, and U. Mosel, Eur. Phys. J. **A21**, 445 (2004), nucl-th/0403064.

[2] V. Shklyar, H. Lenske, U. Mosel and G. Penner, Phys. Rev. C **71**, 055206 (2005) Erratum: [Phys. Rev. C **72**, 019903 (2005)] doi:10.1103/PhysRevC.72.019903, 10.1103/PhysRevC.71.055206 [nucl-th/0412029].

[3] V. Shklyar, H. Lenske, and U. Mosel, Phys.Lett. **B650**, 172 (2007), nucl-th/0611036.

[4] V. Shklyar, H. Lenske and U. Mosel, "Eta-meson production in the resonance energy region," Phys. Rev. C **87**, 015201 (2013).

[5] V. Shklyar, H. Lenske and U. Mosel, Phys. Rev. C **72**, 015210 (2005) doi:10.1103/PhysRevC.72.015210 [nucl-th/0505010].

[6] X. Cao, V. Shklyar, and H. Lenske, Phys.Rev. **C88**, 055204 (2013), 1303.2604.

[7] V. Shklyar, H. Lenske and U. Mosel, Phys. Rev. C **93**, no. 4, 045206 (2016)

[8] B.C. Pearce and B.K. Jennings, Nucl. Phys. **A528** (1991) 655.

[9] T. Feuster and U. Mosel, Phys. Rev. **C58**4571998; T. Feuster and U. Mosel, *ibid.* **59**, 460 (1999).

[10] W. Rarita and J. S. Schwinger, Phys. Rev. **60**, 61 (1941).

[11] V. Shklyar and H. Lenske, Phys. Rev. **C80**, 058201 (2009).

[12] V. Pascalutsa and R. Timmermans, Phys. Rev. **C60**, 042201 (1999), nucl-th/9905065.

[13] V. Pascalutsa, Phys. Lett. **B503**, 85 (2001), hep-ph/0008026.

[14] V. Shklyar, H. Lenske and U. Mosel, Phys. Rev. C **82**, 015203 (2010).

[15] M. Benmerrouche, R. M. Davidson, and N. C. Mukhopadhyay, Phys. Rev. **C39**, 2339 (1989).

[16] M. Fierz and W. Pauli, Proc. Roy. Soc. Lond. **A173**, 211 (1939).

[17] V. Pascalutsa, Phys. Rev. **D58**, 096002 (1998), hep-ph/9802288.

[18] H. Krebs, E. Epelbaum and U.-G. Meißner, Phys. Rev. C **80**, 028201 (2009) doi:10.1103/PhysRevC.80.028201 [arXiv:0812.0132 [hep-th]].

[19] J. S. Schwinger, PARTICLES, SOURCES AND FIELDS, Reading 1973, 459p.

[20] F. A. Berends, J. W. van Holten, P. van Nieuwenhuizen, and B. de Wit, Nucl. Phys. **B154**, 261 (1979), Note, that there is an obvious missprint in definition of the spin $\mathcal{P}^{\frac{3}{2}}_{22\rho\sigma;\tau\delta}(p)$ projector in (A.1): the last term $(-1/15)\theta_{\mu\nu}\theta_{\rho\sigma}$ should read $(-2/15)\theta_{\mu\nu}\theta_{\rho\sigma}$.

[21] J. C. David, C. Fayard, G. H. Lamot, and B. Saghai, Phys. Rev. **C53**, 2613 (1996).

[22] M. Zetenyi and G. Wolf, Heavy Ion Phys. **17**, 27 (2003), nucl-th/0202047.

[23] S. Weinberg and E. Witten, Phys. Lett. **B96**, 59 (1980).

[24] K. H. Glander, et al. (SAPHIR Collaboration), Eur. Phys. J. **A19**, 251 (2004).

[25] The CLAS collaboration, J. W. C. McNabb et al., Phys. Rev. C69, 042201 (2004)

[26] R. D. Baker *et al.*, Nucl. Phys. **B141**, 29 (1978).

[27] D. H. Saxon *et al.*, Nucl. Phys. **B162**, 522 (1980).

[28] T. M. Knasel *et al.*, Phys. Rev. **D11**, 1 (1975).

[29] R. G. T. Zegers et al. (LEPS Collaboration), Phys. Rev. Lett. **91**, 092001 (2003).

[30] M. Sumihama et al. (LEPS Collaboration), Phys. Rev. C **73**, 035214 (2006).

[31] H. Kohri et al. (LEPS Collaboration), Phys. Rev. Lett. **97**, 082003 (2006).

[32] R. Bradford et al. (CLAS Collaboration), Phys. Rev. C **73**, 035202 (2006).

[33] B. Dey et al. (CLAS Collaboration), Phys. Rev. C **82**, 025202 (2010).

[34] Crystal Ball at MAMI, E. F. McNicoll *et al.*, Phys. Rev. **C82**, 035208 (2010).

[35] CLAS Collaboration, M. Dugger *et al.*, Phys.Rev.Lett. **89**, 222002 (2002).

[36] CBELSA/TAPS Collaboration, V. Crede *et al.*, Phys.Rev. **C80**, 055202 (2009).

[37] CB-ELSA Collaboration, O. Bartholomy *et al.*, Eur.Phys.J. **A33**, 133 (2007).

[38] The GRAAL collaboration, O. Bartalini *et al.*, Eur.Phys.J. **A33**, 169 (2007).

[39] R. M. Davidson and R. Workman, Phys. Rev. **C63**, 025210 (2001).

[40] R. Arndt, W. Briscoe, I. Strakovsky, and R. Workman, Phys.Rev. **C74**, 045205 (2006), nucl-th/0605082.

[41] Particle Data Group, J. Beringer *et al.*, Phys.Rev. **D86**, 010001 (2012).

[42] D. M. Manley, R. A. Arndt, Y. Goradia, and V. L. Teplitz, Phys. Rev. **D30**, 904 (1984).

[43] A. Lleres et al. (GRAAL-ESRF Collaboration), Eur. Phys. J. **A31**, 79 (2007).

[44] B. Carnahan, Ph.D. thesis, The Catholic University of America, 2003.

[45] R. Castelijns et al. (CBELSA/TAPS Collaboration), Eur. Phys. J. **A35**, 39 (2008).

[46] CB-ELSA Collaboration, V. Crede *et al.*, Phys.Rev.Lett. **94**, 012004 (2005), hep-ex/0311045. ev. **D84**, 074508 (2011).

[47] S. Dürr *et al.*, Science **322**, 1224 (2008).

[48] R. G. Edwards, J. J. Dudek, D. G. Richards, and S. J. Wallace, Phys.R

[49] H. Sanchis-Alepuz, G. Eichmann, S. Villalba-Chavez, and R. Alkofer, Phys.Rev. **D84**, 096003 (2011).

[50] R. Koniuk and N. Isgur, Phys. Rev. D **21**, 1868 (1980).

[51] S. Capstick and W. Roberts, Phys.Rev. **D58**, 074011 (1998).

[52] R. Cutkosky, C. Forsyth, R. Hendrick, and R. Kelly, Phys.Rev. **D20**, 2839 (1979).

[53] G. Hoehler, PiN Newslett. **9**, 1 (1993).

[54] A. Anisovich *et al.*, Eur.Phys.J. **A48**, 15 (2012).

[55] R. A. Arndt, J. M. Ford, and L. D. Roper, Phys.Rev. **D32**, 1085 (1985).

[56] R. Cutkosky and S. Wang, Phys.Rev. **D42**, 235 (1990).

[57] M. Döring, C. Hanhart, F. Huang, S. Krewald, and U.-G. Meißner, Nucl.Phys. **A829**, 170 (2009).

[58] N. Suzuki *et al.*, Phys.Rev.Lett. **104**, 042302 (2010).

[59] Crystal Ball Collaboration, K. Craig *et al.*, Phys.Rev.Lett. **91**, 102301 (2003).

[60] Crystal Ball Collaboration, S. Prakhov *et al.*, Phys.Rev. **C69**, 045202 (2004).

[61] T. Skorodko *et al.*, Eur.Phys.J. **A35**, 317 (2008).

# Recoil polarization measurements

*Kai-Thomas* Brinkmann[1,*]

[1] *II. Physikalisches Institut, Justus-Liebig-Universität Gießen*

**Abstract.** Polarization observables in photon-induced meson production off nucleons have long been recognized to hold the promise of a detailed understanding of the excited states in the excitation spectrum of the nucleon. Photon beam and proton target polarization are routinely used at the ELSA facility in the Crystal Barrel/TAPS experiment and have yielded a wealth of data on contributing partial waves and nucleon resonances. A detector study on how to complement these ongoing studies by recoil polarization measurements that offer an orthogonal approach with otherwise unmeasurable observables in the field of non-strange meson photoproduction has been performed.
Building on experience with silicon detectors operated in the photon beamline environment, first possible layouts of Si detector telescopes for recoil protons were developed. Various geometries, e.g. Archimedean spiral design of annular sensors, sector shapes and rectangular sensors were studied and have been used during test measurements. A prototype for the recoil polarimeter was built and subjected to performance tests in proton-proton scattering at the COSY-accelerator in Jülich.

## 1 Introduction

Polarization observables in photon-induced meson production off nucleons hold the promise of a detailed understanding of the excited states in the nucleon spectrum. Photon beam and proton target polarization are routinely used at the ELSA facility [1] in the Crystal Barrel/TAPS experiment [2] and have yielded a wealth of data on contributing partial waves and nucleon resonances. These studies could be complemented by recoil polarization studies that offer an orthogonal approach with otherwise unmeasurable observables in the field of non-strange meson photoproduction. These can provide a crucial test of as well as further constraints on the partial wave analyses of the reaction channels with the largest cross sections.

In this project, experience with silicon detectors operated in the photon beamline environment was gathered and first possible layouts of Si recoil detector telescopes for recoil protons have been developed. Various geometries, e.g. Archimedean spiral design of annular sensors, sector shapes and rectangular sensors were studied and used during test measurements. Possible use of the hardware in experiments such as those involving strangeness degrees of freedom can be envisaged, since the very high spatial resolution of the silicon sensors greatly improves the identification of secondary decay vertices.

The goal of this project was the development of instrumentation for the improvement of the data base on polarization observables that require measurement of the proton polarization in the final state, e.g. in $\gamma p \to p\pi^0$ and $\gamma p \to p\eta$. These data may have significant impact on the PWA analyses since the recoil polarization degree of freedom in non-strange photoproduction has not yet been explored in detail.

*e-mail: Kai-Thomas.Brinkmann@exp2.physik.uni-giessen.de

Available data on pion and eta production to date cover total and differential cross sections over a wide range in energy with high statistical accuracy and small systematic errors as well as single polarization data with linearly polarized photons. Experiments using polarized targets and beams to measure double-polarization observables have, among others, been performed by CBELSA/TAPS [2].

As an example for the possible impact of polarization observables, Fig. 1 shows predictions for the proton recoil polarization of $\gamma p \to p\pi^0$ for MAID [3], SAID [4] and the Bonn-Gatchina partial wave analysis solutions for photon energies from 300 MeV to 3.2 GeV in 100 MeV steps. Large differences are visible at high energies beyond the reach of MAID, while existing data [for a recent summary, also refer to the SAID homepage at: http://gwdac.phys.gwu.edu/] at the lower energies drive the agreement of all three approaches. This effect becomes even more pronounced when polarization transfer observables that connect the spin orientation of the incoming photon with that of the outgoing proton are considered. In short, sensitivity to the solutions is large in particular for the double-polarization observables where almost no data are available. Obviously, a measurement of these observables can have a large impact on the PWA solutions and in turn on the physics in various partial waves and, thus, a large impact on our understanding of baryon resonances.

Because of the complete lack of experimental data for the $\gamma p \to p\eta$ recoil polarization (Fig. 2), huge differences between the different approaches become visible already

 DOI: 10.1051/epjconf/201713403007

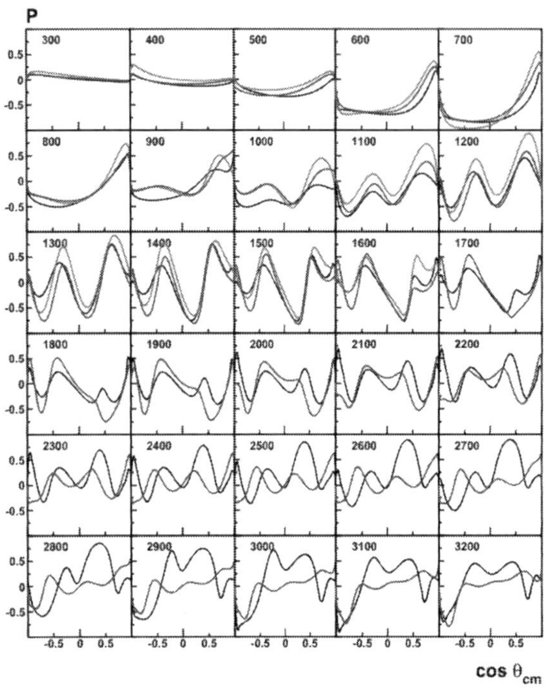


**Figure 1.** Angular distribution of recoil polarization P in $\gamma p \rightarrow p\eta$ as predicted by different models. MAID (red line), SAID (blue line) and Bonn-Gatchina PWA (black line). Differences increase with increasing beam energy, in particular where experimental data become increasingly sparse.

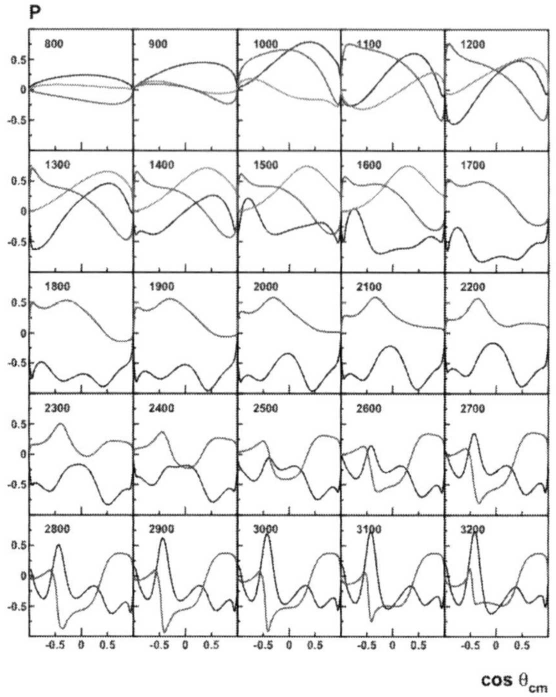


**Figure 2.** Angular distribution of recoil polarization P in $\gamma p \rightarrow p\eta$ as predicted by different models. MAID (red line), SAID (blue line) and Bonn-Gatchina PWA (black line).

at small excess energies above threshold. The discrepancy is not only in the overall magnitude of the predicted polarization of the recoil proton but often completely different shapes with large absolute values and different signs occur.

The present project resulted in the realization of a detector prototype, which was used to demonstrate the feasibility of proton detection in the ELSA environment. The very high position resolution of the sensors used may prove advantageous in a variety of cases from recoil studies to hyperon identification. The detector setup that was developed can be deployed at CBElsa as well as the BGO-OD set-up [5].

## 2 Detector Hardware Studies

A first set of studies was aimed at sensor development and validation for measurements of recoil protons under the special conditions of a tagged-photon beam facility. To this end, several detector layouts were studied in a variety of experimental conditions. In a first campaign of experiments the overall applicability of Si technology close the photon beam was established. Then, a first sensor, whose shape and dimensions are well suited for the geometry at hand, was brought into operation but found ill-suited for the harsh environment. Using large-area double-sided silicon strip sensors, which proved to be sufficiently radiation-hard for operation in or very close to the photon beam, an optimized layout was developed and a prototype setup built. This was subjected to extensive testing and its suitability was verified in simulations. Parallel to the detector studies, readout and data acquisition for the new addition to the experimental capabilities at ELSA was developed.

### 2.1 Photon Beam Compliance

Small silicon test sensors with an area of $2 \times 2\,cm^2$ and a thickness of $300\,\mu m$ were placed in the photon beam behind the Crystal Barrel detector and their long-term properties were studied. Estimates of the photon flux and the radiation load on the detectors had been made prior to these measurements. According to these, the aging of the detectors was expected to be of no concern on typical experiment timescales, but the charged-particle background due to conversion is very difficult to estimate theoretically. The development of the detector reverse current during a long irradiation period is depicted in Fig. 3. It is dominated by the short-term changes induced by variations of the ambient temperature. Two regions with a clear correlation of photon flux (rapidly increased integral, green line) and reverse current are marked with red circles. The diagram also shows that the increase in reverse current in direct response to the irradiation is at least partly reversable through annealing.

DOI: 10.1051/epjconf/201713403007

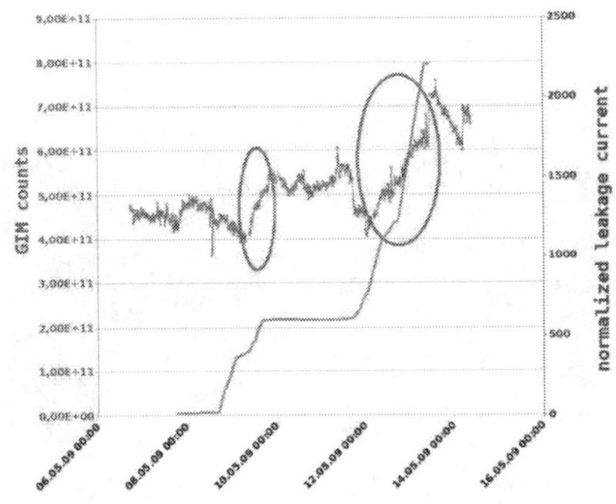


**Figure 3.** Development of the reverse current of a typical test detector during an irradiation period. The short-term fluctuations dominate. These are mainly due to the temperature changes. Two regions with rapid changes in reverse current (indicated by the red circles) show the correlation with the photon flux as also shown in the figure (green line, measured with the forward flux monitor GIM.)

## 2.2 Test of Annular Sensors

Having thus proven the basic applicability of Si sensors in the experiments, several detector technologies were selected for further investigation. Annular detectors with $r/\phi$ granularity are best suited to match the geometrical constraints of an experiment that aims at detection of particles close to 0°. Hence, an Archimedean spiral design available from a vendor (Micron Semiconductors, GB) and used in similar setups at COSY [6] was chosen as one of the possible viable options.

The sensors feature an active diameter of almost 70 mm and a small central hole of 8 mm diameter. At a nominal thickness of 300 $\mu$m there are 256 Archimedean spirals with opposite senses of rotation on each side. The sensors are DC-coupled and thus have rather high noise floors. Fig. 4 shows a photograph of one of the sensors, while Fig. **??** shows a typical charge spectrum. Minimum ionizing particles cannot be discriminated from the large noise component at low amplitudes but particles are clearly visible above charges of about 3 to 5 times that of average minimum ionizing. Therefore, these sensors will allow measurements of protons with few hundreds of MeV of kinetic energy, albeit with non-unity efficiencies due to the pedestal cuts needed. Readout has to be done using charge-sensitive preamplifiers, which require significant electrical power and are difficult to mount in close vicinity to the detectors because of the space needed and their contribution to the material budget.

However, based on the large noise floor in a laboratory environment, it is to be expected that the sensors will exhibit rapid changes of the properties and, thus, unrecoverable time-dependent radiation damage effects during operation close to the photon beam at ELSA. After qualitative confirmation of these problems, this option was not followed up further.

**Figure 4.** Photograph of one of the annular sensors.

## 2.3 Setup with Square Double-Sided Sensors

Rectangular and square double-sided AC-coupled sensors of a thickness of 300 $\mu m$ and a pitch (strip-to-strip distance) of 65 $\mu m$ offer an alternative to the DC-coupled sensors. Their layout is well advanced and measures are available to produce them with properties that make them very well suited for the task at hand. A set of these sensors was submitted to extensive electrical testing. Fig. 5 shows the capacitance measurement for a small sensor of $2 \times 2cm^2$ Full depletion and, hence, optimum charge collection is reached at a voltage of 60 V, and depletion can be achieved even after an irradiation with a total proton dose equivalent to 10 Mrad. Annealing of the sensor for increased times recovers some of the original performance with respect to small capacitance, but is not needed for proper operation. The electronic noise level of these sensors is well below 1,000 electrons compared to about 25,000 electrons in a signal for a minimum ionizing particle.

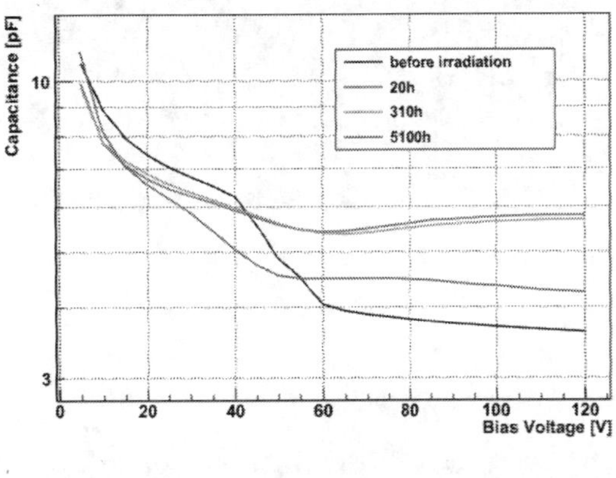


**Figure 5.** Bias curve for a small test sensor from the sensor run at CiS Erfurt. The sensor has been subjected to a radiation dose of 10 Mrad after the first measurement (black curve) and then measured at different times after irradiation.

Because of the superior performance of these detectors, a setup using the square sensors was developed. It allows deployment in forward direction inside a tube small enough to avoid interference with the outer wire chamber tracking in the BGO-OD set-up at the cost of covering only about 60% of the area of an equivalent circular area. The conceptual layout of the holding structure and the sensors in a realistic scenario including a carbon rescatterer is shown in Fig. 6.

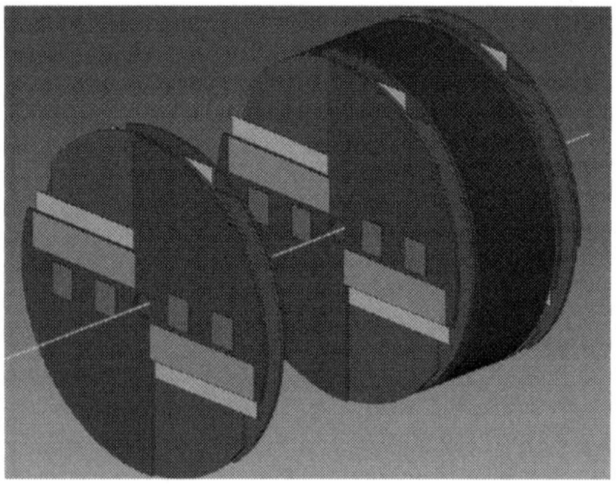

**Figure 6.** Layout of a mounting and support structure for double-sided square silicon sensors. All electrical services and the readout of the strips will be placed on the boards. The beam direction is indicated by the yellow line. PCBs that contain all electrical connections and serve as holding structures are shown in green, readout ASICs in dark red, and sensors are indicated by the interleaved grey areas. The black cylinder is a carbon rescatter.

This setup has been realized as shown in Fig. 7 for one sensor layer using two sensors. It is fully equipped with double-sided readout of all strips.

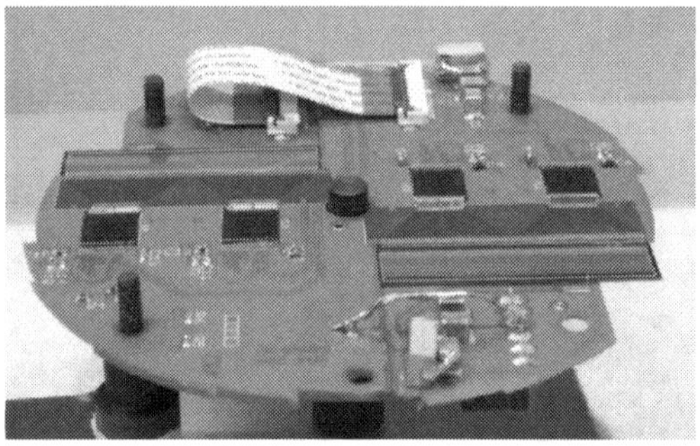

**Figure 7.** Photograph of one layer of the prototype recoil detector.

### 2.4 Experimental Validation of the Setup

The double-sensor design shown in the previous chapter was validated in a proton-proton scattering experiment at the COSY accelerator in Jülich [7]. The detector was placed in a symmetric position with respect to the impinging proton beam of 800 MeV/c as shown in Fig. 8. A two mm thick plastic sheet serves as hydrogen-rich target and the mounting plate is arranged so that the two outgoing protons from elastic scattering events will each hit one of the sensors. Two plastic scintillator detectors read out with photomultipliers serve as trigger detectors for the coincidence requirement between both protons.

**Figure 8.** Layout of the COSY proton-proton scattering experiment. The protons from the accelerator hit a plastic target, behind which the sensor mount plate is placed so that the beam exits through the central hole. Two plastic scintillators (dark rectangles) serve to detect the two protons from elastic scattering events and generate a fast readout trigger.

The quality of the position measurement can be inferred from Fig. 9, where the angles of both outgoing protons detected in coincidence and derived from the position in each of the silicon detectors are shown in a 2-dimensional representation. The upper frame shows data from the experiment and the lower frame is from a Monte-Carlo simulation of the setup. The level of agreement is high, which demonstrates the quality of the detector setup chosen. Alongside the detector setup, all components for the readout and integration in experiments at ELSA have been developed [8] [9].

## 3 Conclusions

A detector system for the measurement of recoil polarization of protons by double-scattering was developed for use at experiments currently under way at the ELSA facility in Bonn. The sensors and their spatial arrangement were optimized for the experimental requirements of the existing large installations and the photon beam environment. All components needed for deployment including data acquisition and high-level hit reconstruction were made available. The viability of high-precision position measurements was shown in a scattering experiment at COSY.

The work reported here would not have been possible without the support from the Deutsche Forschungsgemeinschaft within the SFB/TR16.

## References

[1] W. Hillert, Eur. Phys. J. A **28**, 139 (2006).
[2] J. Hartmann et al., Phys. Lett. B **748**, 212 (2015).
[3] D. Drechsel, S.S. Kamalov, L. Tiator, Nucl. Phys. **A645**, 145 (1999), [http://www.kph.uni-mainz.de/MAID/].

DOI: 10.1051/epjconf/201713403007

[4] SAID webpage, [http://gwdac.phys.gwu.edu/].
[5] B. Bantes et al., Int. J. Mod. Phys. Conf. Ser. 26, 1460093 (2014).
[6] COSY proposal 179, "Commissioning of the COSY-TOF Straw Tube Tracker and the silicon-microstrip Quirl", P.Wintz, available online at http://www2.fz-juelich.de/ikp/publications/PAC34/Propo_179.pdf as of June 2012.
[7] R. Maier, Nucl. Inst. Meth. A, **390**, 1 (1997).
[8] M. Becker, PhD thesis, Bonn (2014).
[9] K. Koop, PhD thesis, Bonn (2014).

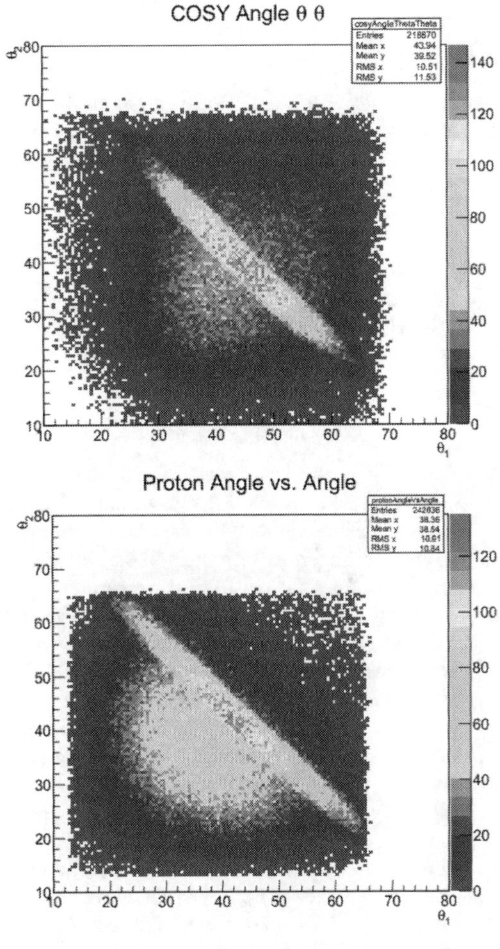


**Figure 9.** Data from the COSY proton-proton scattering experiment (upper frame) compared to a Monte-Carlo simulation (bottom). The angular correlation of the two outgoing protons shows a clear structure along the elastic line. Inelastic events yield smaller relative energy and thus, in general, smaller correlation angle.

# Dispersion theoretical analysis of the nucleon form factors

*Hans-Werner* Hammer[1,2,⋆] and *Ulf-G.* Meißner[3,4,⋆⋆]

[1] *Institut für Kernphysik, Technische Universität Darmstadt, 64289 Darmstadt, Germany*
[2] *ExtreMe Matter Institute EMMI, GSI Helmholtzzentrum für Schwerionenforschung GmbH, 64291 Darmstadt, Germany*
[3] *Universität Bonn, Helmholtz Institut für Strahlen- und Kernphysik, Bethe Center for Theoretical Physics, D-53115 Bonn, Germany*
[4] *Forschungszentrum Jülich, Institute für Kernphysik (IKP-3), Institute for Advanced Simulation (IAS-4) and Jülich Center for Hadron Physics, D-52145 Jülich, Germany*

**Abstract.** We review the achievements of the project **C.2**, that deals with a dispersion theoretical analysis of the nucleon form factors. We have analyzed the world data for the space-like and time-like form factors to extract nucleon radii and vector meson coupling constants. Our proton charge radius $r_E^p \simeq 0.85$ fm has gained prominence in the so-called proton radius puzzle, see Ref. [1]. We also have analyzed the strong energy-dependence of the $e^+e^- \to \bar{p}p$ amplitude at the nucleon-antinucleon threshold, two-photon exchange effects and isospin violation in the isovector nucleon form factors.

## 1 Introduction and motivation

The project **C.2** has been funded just for the first funding period of the SFB/TR 16. It grew out of an "Einzelverfahren" of the DFG and was then included in the SFB/TR 16. Despite many interesting results, foreseeing e.g. the much debated proton radius puzzle, it was terminated after one funding period. The PIs were Hans-Werner Hammer and Ulf Meißner, who have collaborated on the dispersive analysis of the nucleon form factors since 1996.

The project **C.2** dealt with the dispersion-theoretical analysis of the nucleon form factors. These encode important information about perturbative and non-perturbative aspects of QCD manifest in the nucleon structure. They are not only interesting by themselves, but a precise knowledge is mandatory for many other experiments such as the measurement of the nucleon's strange form factors and fundamental tests of Quantum Electrodynamics. Dispersion relations provide a powerful tool to analyze the form factors in a model-independent way and include constraints from multi-meson continua and chiral perturbation theory. In this project, we have made considerable progress towards our goal of understanding the nucleon structure as parameterized by the form factors. We have reanalyzed the contribution of the $2\pi$ continuum and included the contributions from the $K\bar{K}$ and $\rho\pi$ continua to constrain the spectral functions. For the first time, we have given an estimate for the theoretical errors in the extracted form factors. The form factor data in the space- and time-like regions are well described within our error bands. The strong rise of the time-like form factors could be explained by well-known properties of the nucleon-antinucleon ($N\bar{N}$) interaction. Finally, we have quantified the effects of isospin violation in the nucleon form factors, which is an important ingredient in the extraction of strange nucleon form factors from the measurements of parity violating asymmetries in electron-proton scattering.

Our contribution is organized as follows: Section 2 summarizes the results on the isovector spectral functions of the nucleon formfactors. Based on this, a full-fledged dispersive analysis including meson continua was performed, as summarized in Sec. 3. The strong energy dependence of the $e^+e^- \to p\bar{p}$ amplitude near threshold is analyzed in Sec. 4, followed by a model-independent calculation of the so important two-photon corrections, see Sec. 5. Finally, isospin violation in the vector form factors of the nucleon is discussed in Sec. 6. For convenience, we collect some basic definitions in the Appendix.

Note the **important disclaimer:** This is the review of the project **C.2** but not a general review of the various fields to which this research contributed. Therefore, only the papers that resulted from this research are quoted, with the exception of a few external papers and of course, in case data were involved, with proper reference to the experimental papers. The links to the other projects are mentioned in passing, but no effort is made to discuss these in any detail.

## 2 Novel evaluation of the two-pion contribution to the nucleon isovector form factors

The spectral functions of the nucleon form factors are constrained by input from meson continua. The most important constraint on the isovector spectral function comes

⋆e-mail: hammer@theorie.ikp.physik.tu-darmstadt.de
⋆⋆e-mail: meissner@hiskp.uni-bonn.de

DOI: 10.1051/epjconf/201713404001

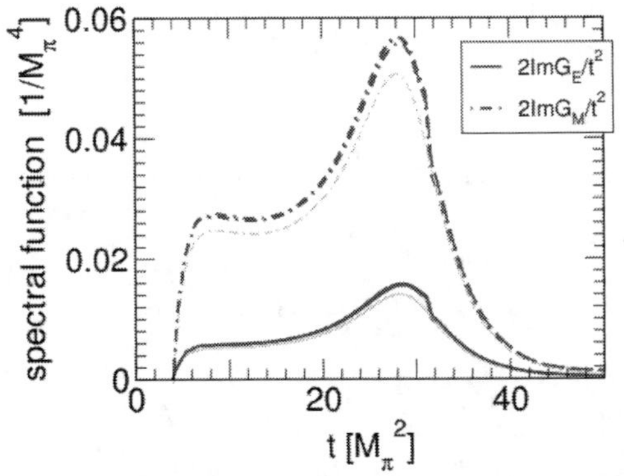


**Figure 1.** The $2\pi$ spectral function using the new high statistics data for the pion form factor [3–5]. The spectral functions weighted by $1/t^2$ are shown for $G_E$ (solid line) and $G_M$ (dash-dotted line). The previous results by Höhler et al. [6] (without $\rho$-$\omega$ mixing) are shown for comparison by the gray/green lines.

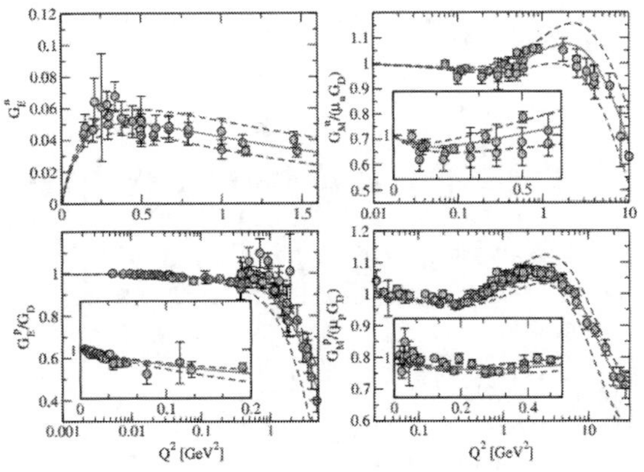

**Figure 2.** The nucleon electromagnetic form factors for space-like momentum transfer in the SC approach. The solid line gives our best fit while the dashed lines indicate the $1\sigma$ error band.

from the $2\pi$ continuum. Approximating this continuum by a simple $\rho$ pole leads to a severe underestimation of the isovector radii of the nucleon. This contribution was evaluated more than 40 years ago be Höhler and Pietarinen. We have performed a new calculation of this contribution drawing upon the new high statistics measurements of the pion form factor by the CMD-2, KLOE, and SND collaborations, see Fig. 1. The general structure of the spectral functions remains unchanged, but the magnitude increases by about 10%. Using the updated spectral functions, we have calculated the contribution of the $2\pi$ continuum to the nucleon isovector form factors and radii. Moreover, we have given a convenient parametrization of the result for use in future form factor analyses [2]. For the isoscalar spectral functions, we rely on the determinations of the $K\bar{K}$ and $\rho\pi$ continua in the literature.

## 3 Dispersion analysis of the nucleon form factors including meson continua

Dispersion theory simultaneously describes all four nucleon form factors over the whole range of momentum transfers in both the space-like and time-like regions. It allows for the inclusion of constraints from other physical processes, unitarity, and chiral perturbation theory and therefore is an ideal tool to analyze the form factor data. Using the $2\pi$ continuum discussed in the previous section as well the $K\bar{K}$ [9, 10] and $\rho\pi$ continua [11] as input, we have analyzed the world data for the electromagnetic form factors of the nucleon in the space- and time-like regions [7, 8]. The central objects of the dispersive analysis of the nucleon form factors are the *spectral functions*, i.e. the imaginary parts of the form factors given by

$$F(t) = \frac{1}{\pi} \int_{t_0}^{\infty} \frac{\operatorname{Im} F(t')}{t' - t - i\epsilon} dt' ,$$

where $t_0$ is the threshold of the lowest cut of $F(t)$ and the $i\epsilon$ defines the integral for values of $t$ on the cut. In all our fits the spectral function includes the $2\pi$, $K\bar{K}$, and $\rho\pi$ continua from unitarity and the $\omega$ pole. Note that we also include a pole at the $\phi$ mass to account for explicit $\phi$ strength not included in the $K\bar{K}$ and $\rho\pi$ continua. In addition to that there are a number of effective poles at higher momentum transfers in the isoscalar ($s_1, s_2, ...$) and isovector channels ($v_1, v_2, ...$). Note that for the dispersive analysis, the isospin basis is most natural rather than the particle basis. The spectral function has the general structure

$$\begin{aligned} \operatorname{Im} F_i^s(t) &= \operatorname{Im} F_i^{(s,K\bar{K})}(t) + \operatorname{Im} F_i^{(s,\rho\pi)}(t) \\ &\quad + \sum_{V=\omega,\phi,s_1,...} \pi a_i^V \delta(M_V^2 - t), \quad i = 1,2, \\ \operatorname{Im} F_i^v(t) &= \operatorname{Im} F_i^{(v,2\pi)}(t) \\ &\quad + \sum_{V=v_1,...} \pi a_i^V \delta(M_V^2 - t), \quad i = 1,2. \end{aligned}$$

The vector meson residua $a_i^V$ (for the $\omega$, $\phi$, $s_i$ and $v_i$) are fit parameters. Constraints are derived from the normalizations of the four form factors at zero momentum transfer as well as from the perturbative QCD behavior of the form factors. We have used two different methods to include the asymptotic constraints from perturbative QCD: (a) applying superconvergence relations only (SC) and (b) adding an explicit pQCD term to the spectral function. The results for the form factors were similar in both approaches. In general, we have 14 – 16 fit parameters, for details see Ref. [8]. Moreover, we have provided a $1\sigma$ error band for our analysis and generally found good agreement with the world data within the errors. This method to generate theoretical errors was later used in other projects, like **B.3** and **B.6**. As an example, we show our best fit in the SC approach in Figs. 2 and 3. In particular, no evidence for a bump-dip-structure in $G_E^n$ was found [21]. We have also extracted the nucleon radii, see Tab. 1, and the $\omega NN$ coupling constants. For the radii, we generally found good agreement with other determinations with the exception of the electric charge radius of the proton which comes out smaller. While the Lamb shift in electronic hydrogen

**Table 1.** Nucleon radii extracted from our fits in the SC (2nd column) and explicit pQCD approaches (3rd column). The first number gives the value for our best fit, while the numbers in parentheses indicate the range from the $1\sigma$ band. For comparison, we give the results of Ref. [12] (4th column) and other recent determinations from low-momentum transfer data [13–18] (5th column).

| | SC approach | explicit pQCD app. | Ref. [12] | recent determ. |
|---|---|---|---|---|
| $r_E^p$ [fm] | 0.844 (0.840 ... 0.852) | 0.830 (0.822 ... 0.835) | 0.848 | 0.886(15) [13–15] |
| $r_M^p$ [fm] | 0.854 (0.849 ... 0.859) | 0.850 (0.843 ... 0.852) | 0.857 | 0.855(35) [14, 16] |
| $(r_E^n)^2$ [fm$^2$] | −0.117 (−0.11 ... − 0.128) | −0.119 (−0.108 ... − 0.13) | −0.12 | −0.115(4) [17] |
| $r_M^n$ [fm] | 0.862 (0.854 ... 0.871) | 0.863 (0.859 ... 0.871) | 0.879 | 0.873(11) [18] |

gives $r_E^p \simeq 0.88$ fm, we find a "small" radius in the range from 0.84 ... 0.85 fm. We are not able to obtain good fits when we force the proton charge radius to the large value. This has preceded the now famous "proton radius puzzle" by many years and was unfortunately not appreciated by the reviewers of this project. The $\omega NN$ vector coupling constant is determined relatively well by the fits, but for the tensor coupling constant even the sign can not be determined. Some challenges remain in the time-like region.

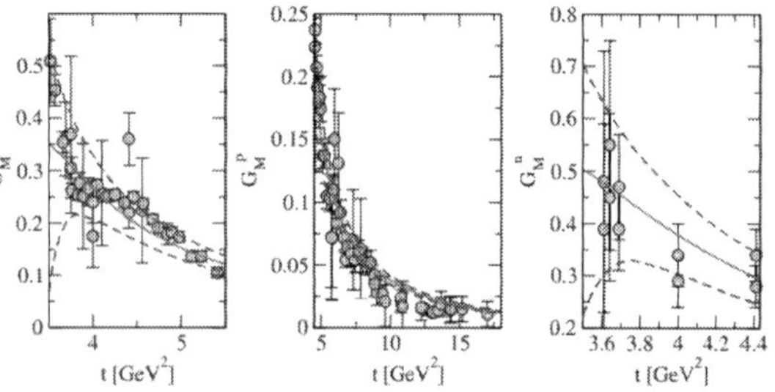


**Figure 3.** Left panel: The nucleon electromagnetic form factors for time-like momentum transfer in the SC approach (right panel: $G_M^n$, left panels: $G_M^p$). The proton data participate in the fit while the neutron data are a genuine prediction. The solid line gives our best fit while the dashed lines indicate the $1\sigma$ error band.

## 4 On the strong energy dependence of the $e^+e^- \to p\bar{p}$ amplitude near threshold

It is often claimed that the strong energy dependence of the nucleon time-like form factors, cf. Fig. 3, requires the presence of a new resonance in the $N\bar{N}$-threshold region. In order to resolve this puzzle, we have studied the energy dependence of the $e^+e^- \to p\bar{p}$ cross section close to the two-nucleon threshold, recently reported by the BaBar collaboration. Our analysis also included the $\bar{p}p \to e^+e^-$ data collected by PS170 collaboration and the $e^+e^- \to N\bar{N}$ data from the FENICE collaboration. We have shown that the near-threshold enhancement in the $e^+e^- \to p\bar{p}$ cross section can be explained by the final-state interaction between proton and antiproton in the $^3S_1$ partial wave [22], utilizing the Jülich nucleon-antinucleon model [23, 24], see Fig. 4. As a consequence, the strong dependence of the proton electromagnetic form factors on the momentum transfer close to the two-nucleon threshold is then likewise driven by this well-known final-state interaction effect. This result is in line with our previous studies [25, 26] of the near-threshold enhancement of the $p\bar{p}$ invariant mass spectrum seen in the $J/\Psi \to \gamma p\bar{p}$ decay by the BES collaboration and in the $B^+ \to p\bar{p}K^+$ decay by the BaBar collaboration.

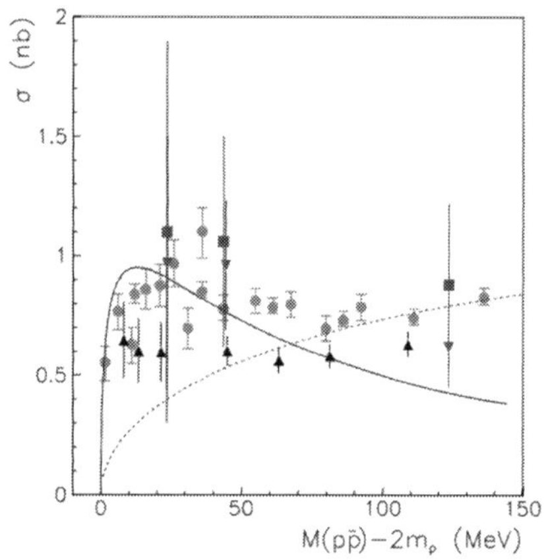


**Figure 4.** Cross section of the $e^+e^- \to p\bar{p}$ and $e^+e^- \to n\bar{n}$ reactions as a function of the excess energy. The data are from the FENICE (inverse triangles and squares) [27, 28] and BaBar (circles) collaborations [29]. Triangles represent results obtained by applying detailed balance to the $\bar{p}p \to e^+e^-$ cross section measured by the PS170 collaboration [30]. The dashed line indicates the energy dependence of the two-body phase space. The solid line is the scattering amplitude squared predicted by the Jülich $N\bar{N}$ model A(OBE) for the $^3S_1$ partial wave, multiplied by appropriate phase-space factors.

## 5 A model-independent estimate of two-photon effects in unpolarized elastic electron-proton scattering

In order to investigate the importance of two-photon effects we have directly analyzed the experimental cross sections for the proton. When the discrepancy between the Rosenbluth and the polarization transfer measurement results was first observed, it was noted that the values of $G_E^p$ extracted using the Rosenbluth technique are not consistent with each other. It was often assumed that the difference between the results of the two techniques can be explained by systematic uncertainties in the Rosenbluth

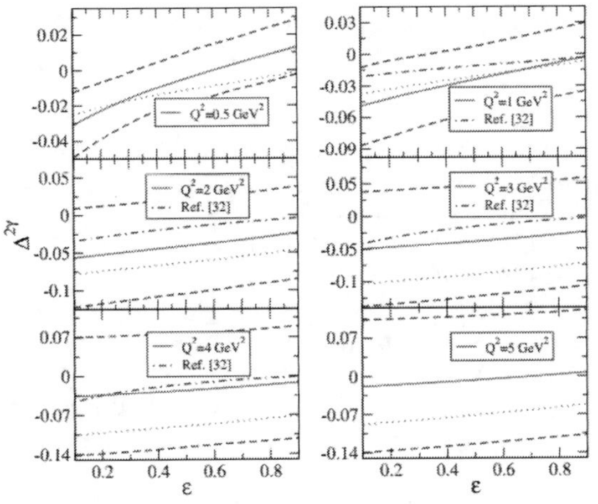


**Figure 5.** The extracted two-photon correction $\Delta^{2\gamma}$ for $Q^2$ =0.5, 1.0, 2.0, 3.0, 4.0,and 5.0 GeV$^2$. Our correction $\Delta^{2\gamma}$ corresponds to the sum of the extracted hard two-photon correction $\delta^{2\gamma}$ and the Coulomb correction $\delta^C$. The solid lines show our results for $\Delta^{2\gamma}$ within the SC approach, while the dashed lines gives the theoretical $1\sigma$ uncertainty bands. The dotted lines give the central value for $\Delta^{2\gamma}$ within the pQCD approach. For comparison, the direct full calculation by Blunden *et al.* [32] is shown by the dash-dotted lines.

extractions. A global reanalysis of the Rosenbluth data showed that only the data from individual experiments are consistent but the discrepancy persisted. Applying the standard radiative corrections consistently, we have performed a global analysis of the Rosenbluth cross section data for the proton and form factor data for the neutron. Comparing the result for the proton (electron-proton scattering) to the polarization data for the proton, we have obtained an estimate of the two-photon corrections defined by

$$\left(\frac{d\sigma}{d\Omega}\right)_{\rm Ros} = \left(\frac{d\sigma}{d\Omega}\right)(1+\delta^{2\gamma}),$$

in Ref. [31], with $(d\sigma/d\Omega)_{\rm Ros}$ the standard Rosenbluth cross section in terms of the electric and magnetic nucleon form factors. The full two-photon correction $\Delta^{2\gamma}$ is obtained by adding the Coulomb correction $\delta^C$. The results from our work are in good agreement with the direct calculations of the two-photon exchange corrections by Blunden et al. [32] but the error bars leave room for improvement, see Fig. 5. Our results also show that the discrepancy between the Rosenbluth and the polarisation transfer measurement results becomes significant only above $Q^2 \approx 4\,\mathrm{GeV}^2$.

## 6 Isospin violation in the vector form factors of the nucleon

A quantitative understanding of isospin violation is an increasingly important ingredient for the extraction of the nucleon's strange vector form factors from experimental data. We have calculated the isospin violating electric and magnetic form factors in chiral perturbation theory to leading and next-to-leading order, respectively [33]. We have extracted the low-energy constants from resonance saturation using the $\omega NN$ coupling constants from our dispersion analysis [8]. The uncertainties in the isospin violating form factors are largely dominated by limitations in the current knowledge of some vector meson couplings. The resulting bounds on isospin violation are sufficiently precise to be of value to on-going experimental studies of the strange form factors. As an example, we show the isospin breaking form factor $G_E^{u,d}(t)$ in Fig. 6.

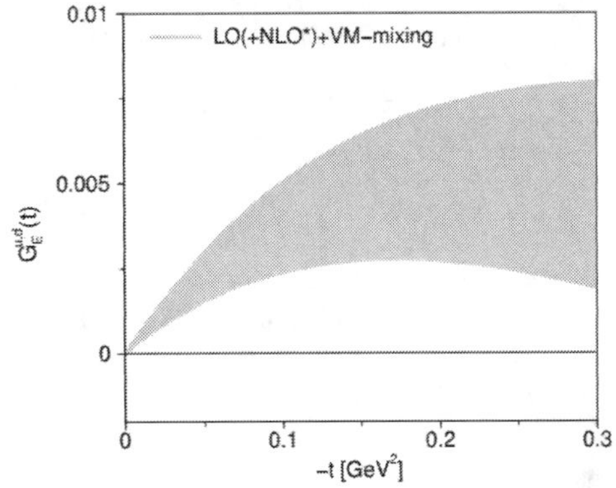


**Figure 6.** The isospin breaking form factor $G_E^{u,d}(t)$. The band combines uncertainties in the vector meson mixing amplitude as well as from higher-order chiral corrections.

We would like to thank our collaborators Maxim Belushkin, Johan Haidenbauer, Siegfried Krewald, Anthony Thomas, Randy Lewis and the late Alexander Sibirtsev. The work reported here would not have been possible without the support from the Deutsche Forschungsgemeinschaft within the SFB/TR16.

## A Basic definitions

The electromagnetic (em) structure of the nucleon is determined by the matrix element of the vector current operator $j_\mu^{\rm em}$ between nucleon states. Using Lorentz and gauge invariance, this matrix element can be expressed in terms of two form factors,

$$\langle p'|j_\mu^{\rm em}|p\rangle = \bar u(p')\left[F_1(t)\gamma_\mu + i\frac{F_2(t)}{2M}\sigma_{\mu\nu}q^\nu\right]u(p), \quad (1)$$

where $M$ is the nucleon mass and $t = (p'-p)^2$ the four-momentum transfer squared. For data in the space-like region, it is often convenient to use the variable $Q^2 = -t > 0$. The functions $F_1(t)$ and $F_2(t)$ are the Dirac and Pauli form factors, respectively. They are normalized at $t = 0$ as

$$F_1^p(0) = 1\,,\ F_1^n(0) = 0\,,\ F_2^p(0) = \kappa_p\,,\ F_2^n(0) = \kappa_n\,, \quad (2)$$

with $\kappa_p = 1.793$ and $\kappa_n = -1.913$ the anomalous magnetic moment of the proton and the neutron, respectively, in units of nuclear magnetons.

It is convenient to work in the isospin basis and to decompose the form factors into isoscalar and isovector parts,

$$F_i^s = \frac{1}{2}(F_i^p + F_i^n), \quad F_i^v = \frac{1}{2}(F_i^p - F_i^n),$$

where $i = 1, 2$. The experimental data are usually given for the Sachs form factors

$$\begin{aligned} G_E(t) &= F_1(t) - \tau F_2(t), \\ G_M(t) &= F_1(t) + F_2(t), \end{aligned} \tag{3}$$

where $\tau = -t/(4M^2)$. In the Breit frame, $G_E$ and $G_M$ may be interpreted as the Fourier transforms of the charge and magnetization distributions, respectively.

The nucleon radii $\sqrt{\langle r^2 \rangle}$ can be defined from the low-$t$ expansion of the form factors,

$$F(t) = F(0)\left[1 + t\langle r^2 \rangle/6 + \ldots\right],$$

where $F(t)$ is a generic form factor. In the case of the electric and Dirac form factors of the neutron, $G_E^n$ and $F_1^n$, the expansion starts with the term linear in $t$ and the normalization factor $F(0)$ is dropped.

## References

[1] R. Pohl *et al.*, Nature **466** (2010) 213.

[2] M. A. Belushkin, H.-W. Hammer and U.-G. Meißner, Phys. Lett. B **633** (2006) 507 [arXiv:hep-ph/0510382].

[3] R. R. Akhmetshin *et al.* [CMD-2 Collaboration], arXiv:hep-ex/9904027, Phys. Lett. B **527** (2002) 161 [arXiv:hep-ex/0112031], Phys. Lett. B **578** (2004) 285 [arXiv:hep-ex/0308008].

[4] A. Aloisio *et al.* [KLOE Collaboration], Phys. Lett. B **606** (2005) 12 [arXiv:hep-ex/0407048].

[5] M. N. Achasov *et al.*, J. Exp. Theor. Phys. **101** (2005) 1053 [Zh. Eksp. Teor. Fiz. **101** (2005) 1201] [hep-ex/0506076].

[6] G. Höhler, "Pion-Nucleon Scattering", Landolt-Börnstein Vol. I/9b, ed. H. Schopper, Springer, Berlin, 1983.

[7] H.-W. Hammer, Eur. Phys. J. A **28**, s01 (2006) 49 [arXiv:hep-ph/0602121].

[8] M. A. Belushkin, H.-W. Hammer and U.-G. Meißner, Phys. Rev. C **75** (2007) 035202 [arXiv:hep-ph/0608337].

[9] H.-W. Hammer and M.J. Ramsey-Musolf, Phys. Rev. C **60**, 045205 (1999) [Erratum-ibid. C **62**, 049903 (2000)] [arXiv:hep-ph/9812261].

[10] H.-W. Hammer and M.J. Ramsey-Musolf, Phys. Rev. C **60**, 045204 (1999) [Erratum-ibid. C **62**, 049902 (2000)] [arXiv:hep-ph/9903367].

[11] U.-G. Meißner, V. Mull, J. Speth and J. W. van Orden, Phys. Lett. B **408**, 381 (1997) [arXiv:hep-ph/9701296].

[12] H.-W. Hammer and U.-G. Meißner, Eur. Phys. J. A **20**, 469 (2004) [arXiv:hep-ph/0312081].

[13] R. Rosenfelder, Phys. Lett. B **479**, 381 (2000) [arXiv:nucl-th/9912031].

[14] I. Sick, Phys. Lett. B **576**, 62 (2003) [arXiv:nucl-ex/0310008].

[15] K. Melnikov and T. van Ritbergen, Phys. Rev. Lett. **84**, 1673 (2000) [arXiv:hep-ph/9911277].

[16] I. Sick, private communication.

[17] S. Kopecky, M. Krenn, P. Riehs, S. Steiner, J. A. Harvey, N. W. Hill, and M. Pernicka, Phys. Rev. C **56**, 2229 (1997).

[18] G. Kubon et al., Phys. Lett. B **524**, 26 (2002) [arXiv:nucl-ex/0107016].

[19] P.G. Blunden and I. Sick, Phys. Rev. C **72**, 057601 (2005) [arXiv:nucl-th/0508037].

[20] J. Arrington, private communication.

[21] U.-G. Meißner, AIP Conf. Proc. **904** (2007) 142 [nucl-th/0701094].

[22] J. Haidenbauer, H.-W. Hammer, U.-G. Meißner and A. Sibirtsev, Phys. Lett. B **643** (2006) 29 [arXiv:hep-ph/0606064].

[23] T. Hippchen, J. Haidenbauer, K. Holinde, V. Mull, Phys. Rev. C **44**, 1323 (1991); V. Mull, J. Haidenbauer, T. Hippchen, K. Holinde, Phys. Rev. C **44**, 1337 (1991).

[24] V. Mull and K. Holinde, Phys. Rev. C **51** (1995) 2360 [nucl-th/9411014].

[25] A. Sibirtsev, J. Haidenbauer, S. Krewald, U.-G. Meißner and A. W. Thomas, Phys. Rev. D **71** (2005) 054010 [arXiv:hep-ph/0411386].

[26] J. Haidenbauer, U.-G. Meißner and A. Sibirtsev, Phys. Rev. D **74** (2006) 017501 [arXiv:hep-ph/0605127].

[27] A. Antonelli et al., Phys. Lett. B **334**, 431 (1994).

[28] A. Antonelli et al., Nucl. Phys. B **517**, 3 (1998).

[29] B. Aubert et al., Phys. Rev. D **73** (2006) 012005 [hep-ex/0512023].

[30] G. Bardin et al., Nucl. Phys B **411** (1994) 3.

[31] M. A. Belushkin, H.-W. Hammer and U.-G. Meißner, Phys. Lett. B **658** (2008) 138 [arXiv:0705.3385 [hep-ph]].

[32] P. G. Blunden, W. Melnitchouk and J. A. Tjon, Phys. Rev. C **72** (2005) 034612 [arXiv:nucl-th/0506039].

[33] B. Kubis and R. Lewis, Phys. Rev. C **74** (2006) 015204 [arXiv:nucl-th/0605006].

EPJ Web of Conferences **134**, 04002 (2017) DOI: 10.1051/epjconf/201713404002
*Subnuclear Structure of Matter: Achievements and Challenges*

# Spin structure of the nucleon (theory)

*Ulf-G. Meißner*[1,2,*] and *Akaki Rusetsky*[1,**]

[1] *Universität Bonn, Helmholtz Institut für Strahlen- und Kernphysik, Bethe Center for Theoretical Physics, D-53115 Bonn, Germany*
[2] *Forschungszentrum Jülich, Institute für Kernphysik (IKP-3), Institute for Advanced Simulation (IAS-4) and Jülich Center for Hadron Physics, D-52145 Jülich, Germany*

**Abstract.** We review the achievements of the project **C.3**, that dealt with the formulation and application of effective field theories including the Δ and Roper resonances as well as vector mesons. A particular focus was the calculation of finite volume effects and chiral extrapolations for lattice QCD.

## 1 Introduction and motivation

The project **C.3** has been funded just for the first funding period of the SFB/TR 16. The PIs were Ulf Meißner and Akaki Rusetsky. The project was focused on the theory of the nucleon spin structure and expected to support the electroproduction experiments of the proposed project **C.1**. As this project was not funded, the focus **C.3** shifted to the study of chiral extrapolations and finite volume effects, with a special emphasis on the inclusion of the Δ resonance and performing ground-breaking work of effective field theories (EFTs) with spin-1 fields and on the inclusion of the Roper resonance. The work on finite volume effects was continued in the second and third funding period in project **A.6** "Baryon resonances in a finite volume" (PIs: Ulf Meißner and Akaki Rusetsky) whereas the work on unstable states in EFT was taken up again in the third funding period in the project **C.9** "Chiral dynamics of/with unstable states" (PIs: Evgeny Epelbaum, Hermann Krebs and Ulf Meißner). Therefore, despite the re-orientation of the original project **C.3**, it has contributed significantly to the success of the SFB/TR 16.

Our contribution is organized as follows: Section 2 summarizes the results on including the Δ-resonance in baryon chiral perturbation theory and making contact to lattice QCD through chiral extrapolations and finite volume methods. Sec. 3 is devoted to the studies of virtual Compton scattering off the nucleon, that were originally at the heart of this project and have later been taken up again in project **C.9**. The groundbreaking work on chiral extrapolations for the the low-lying baryons and the nucleon axial-vector coupling is discussed in Sec. 4. Sec. 5 deals with the developments of EFTs for the inclusion of vector mesons and the Roper resonance, that have been the trigger of many works by other groups in later years.

Note the **important disclaimer:** This is the review of the project **C.3** but not a general review of the various fields to which this research contributed. Therefore, only the papers that resulted from this research are quoted, with the exception of a few external papers and of course, in case data were involved, with proper reference to the experimental papers. The links to the other projects are mentioned in passing, but no effort is made to discuss these in any detail.

## 2 Effective field theory studies for spin-3/2 particles

To discuss the spin structure of the nucleon in baryon chiral perturbation theory (CHPT), the inclusion of the Δ(1232) resonance as an effective degree of freedom is mandatory, see e.g. the detailed discussion in the review [1]. In Ref. [2], we have given an improved formulation of the covariant small scale expansion (SSE) , i.e. an effective field theory (EFT) [3] with explicit spin-3/2 (delta) degrees of freedom, counting the nucleon-delta mass splitting $m_\Delta - m_N$ as an additional small parameter, thus $\epsilon \in \{p/\Lambda, M_\pi/\Lambda, (m_\Delta - m_N)/\Lambda\}$ (with $\Lambda \approx 1\,\text{GeV}$ the scale of chiral symmetry breaking), and requiring consistency with the chiral limit of QCD (decoupling constraints). Within this approach, we have calculated the nucleon and the delta mass to fourth order $O(\epsilon^4)$ (complete one-loop calculation). We have analyzed lattice data from the MILC [4] and QCDSF [5] collaborations and demonstrated that the available lattice data combined with our knowledge of the physical values for the nucleon and delta masses lead to consistent chiral extrapolation functions for both observables up to fairly large pion masses, see Fig. 1. This holds in particular for very recent data on the delta mass from the QCDSF collaboration. The resulting pion-nucleon sigma term is $\sigma_{\pi N} = 48.9$ MeV. This first quantitative analysis of the quark-mass dependence of the structure of the Δ(1232) in full QCD within chiral effective field theory suggests that the nucleon-delta mass-splitting in the chiral limit, $(m_\Delta - m_N)_0 = 0.33$ GeV, is slightly larger

*e-mail: meissner@hiskp.uni-bonn.de
**e-mail: hammer@theorie.ikp.physik.tu-darmstadt.de

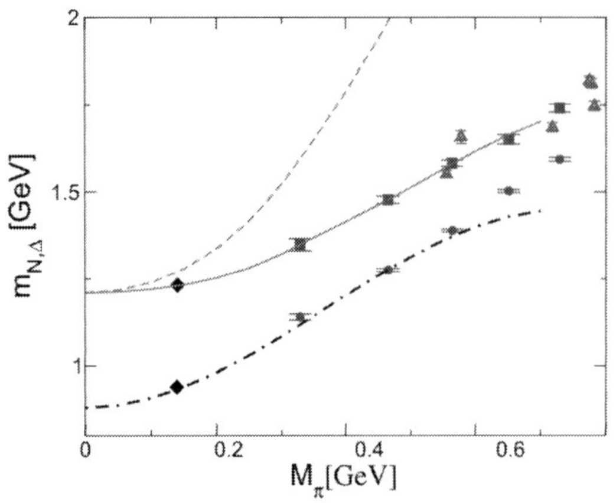


**Figure 1.** The nucleon mass (dot-dashed line) and the (real part of the) delta mass (solid line) as a function of the pion mass. The filled diamonds denote their physical values at the physical pion mass. The dashed line is the chiral extrapolation for the Δ based on SU(4). The filled squares and circles are the MILC data. The filled triangles are the recent data from QCDSF.

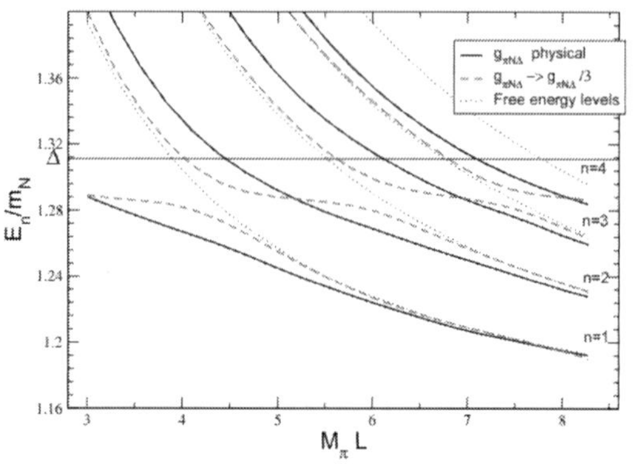


**Figure 2.** The dependence of the energy levels in a finite box on the box size $L$ for different values of the coupling constant $g_{\pi N\Delta}$. The avoided level crossing, which is clearly seen at small values of $g_{\pi N\Delta}$ (dashed lines), is washed out for the physical value of this coupling constant (solid lines). For comparison, we also display the free energy levels (dotted lines). It is seen that the energy levels in the presence of the interaction interpolate between different free energy levels. As expected, an abrupt change emerges in the vicinity of the resonance energy (the resonance position corresponds to the solid horizontal line).

than at the physical point, $(m_\Delta - m_N)_{\rm phys} = 0.29$ GeV. Further analysis of simultaneous fits to nucleon and delta lattice data are needed for a precision determination of the properties of the first excited state of the nucleon. Furthermore, the large deviation from SU(4) symmetry found in this work needs to be understood better and theoretical error bars have to be supplied, based on lattice results at lower quark masses. Clearly, these results should be considered indicative since the lattice data have been obtained at too high quark masses for a truely quantitative comparison with a small theoretical uncertainty.

As pointed out by Lüscher [6], the extraction of resonance properties from finite volume simulations is tricky. In particular, in a finite volume one only has discrete energy levels, and resonances can not be connected to any particular energy level measured on the lattice. Rather, in the vicinity of a resonance, the phenomenon of an avoided level crossing appears [7]. This allows one to extract the resonance mass and width (for well isolated resonances). This framework had before only be considered for the most simple cases like scalar field theories. To connect to the physical baryon spectrum, we have studied in Ref. [8] the extraction of Delta-resonance parameters from lattice data for small quark masses, corresponding to the case of an unstable Delta. To this end, we calculated the spectrum of the correlator of two Delta-fields in a finite Euclidian box up-to-and-including $\mathcal{O}(\epsilon^3)$ in the small scale expansion using infrared regularization. On the basis of our numerical study, we argue that the extraction of the parameters of the Delta-resonance (in particular, of the mass and the pion-nucleon-delta coupling constant) from the measured volume dependence of the lowest energy levels should be feasible. For illustration, the dependence of the energy levels on the pion-nucleon-delta coupling is displayed in Fig. 2. This work has laid the foundation for the new project **A.6** (see that project for a more detailed discussion of these issues).

## 3 Virtual Compton scattering off the nucleon

In earlier work with Bernard and Hemmert [9], we had calculated the spin structure functions of the nucleon, in particular their lower moments, as measured in double virtual Compton scattering off nucleons based on covariant nucleon CHPT (using infrared regularization) to fourth order in the chiral expansion, supplemented by tree level delta and vector meson contributions. At that order, no undetermined parameters appear and the predictions are scrutinized by the data from Jefferson Laboratory for the proton and neutron spin structure function $g_1$ down to photon virtualities as low as $Q^2 \simeq 0.05\,{\rm GeV}^2$ (experiments with even lower values of $Q^2$ having been performed and are presently being analyzed). We have extended these calculations based on our covariant version of the SSE. This work was done in collaboration with M. Dorati from Pavia and T. Hemmert from Munich, and was later published by the Munich group in Ref. [10]. In that paper, the first moments of the parity-even Generalized Parton Distributions (GPDs) in a nucleon, corresponding to six (generalized) vector form factors, were calculated. The analysis was performed at leading-one-loop order in baryon CHPT, predicting both the momentum and the quark-mass dependence for the three (generalized) isovector and (generalized) isoscalar form factors, which are currently under investigation in lattice QCD analyses of baryon structure. We also studied the limit of vanishing four-momentum transfer where the GPD-moments reduce to the well known moments of Parton Distribution

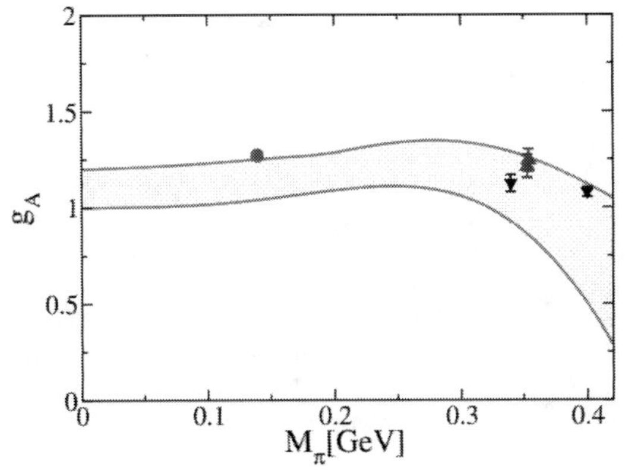


**Figure 3.** Pion mass dependence of the axial-vector coupling – lattice results from LHPC/MILC (triangles) [12] and QCDSF (inverted triangles) [5] in comparison to the two–loop CHPT result (filled area). The circle is the physical point.

Functions (PDFs). For the isovector moment $\langle x \rangle_{u-d}$ this baryon CHPT calculation predicted a new mechanism for chiral curvature, connecting the high values for this moment typically found in lattice QCD studies for large quark masses with the smaller value known from phenomenology. Likewise, we analyzed the quark-mass dependence of the isoscalar moments in the forward limit and extracted the contribution of quarks to the total spin of the nucleon.

## 4 Chiral extrapolations

Related to the work on the delta and the nucleon mass, we have also considered the chiral expansion of various other observables. In Ref. [11] we have analyzed the nucleon axial-vector coupling $g_A$ to two loops in chiral perturbation theory making use of renormalisation group methods. This was motivated by the fact that while the one-loop representation of $g_A$ shows a strong pion mass dependence, the lattice results of various collaborations (MILC/LHPC, QCDSF, RBC, ...) are essentially independent of the pion mass for a large range of (unphysical) values. We show that chiral extrapolations based on this representation require lattice data with pion masses below 300 MeV for a small theoretical uncertainty, see Fig. 3 (the band is obtained by varying the low-energy constants at dimension four and five within the bounds given by naturalness and assuming that $g_A(M_\pi)$ stays flat as indicated by all lattice data). This clearly is a benchmark calculation of interest for upcomimg lattice data in the true chiral regime.

In Ref. [13], we have calculated the baryon octet masses to fourth order in CHPT employing dimensional and cut-off regularization. We analyze the pion and kaon mass dependences of the baryon masses based on the MILC data from 2001 [14] and 2004 [15]. We show that chiral perturbation theory gives stable chiral extrapolation functions for pion (kaon) masses below 450 (500) MeV. The pion mass dependence of the nucleon mass is shown in Fig. 4. The pion-nucleon sigma term in SU(3) is also investigated, we find $\sigma_{\pi N}(0) = 50.7 - 53.7$ MeV. We note that some of the MILC 2004 data for the nucleon can not be described when one insists on the physical point (see Fig. 4) and that the pion mass dependence of the cascade mass of the MILC data is more pronounced than given by our extrapolation formula (having fixed the parameters on the nucleon).

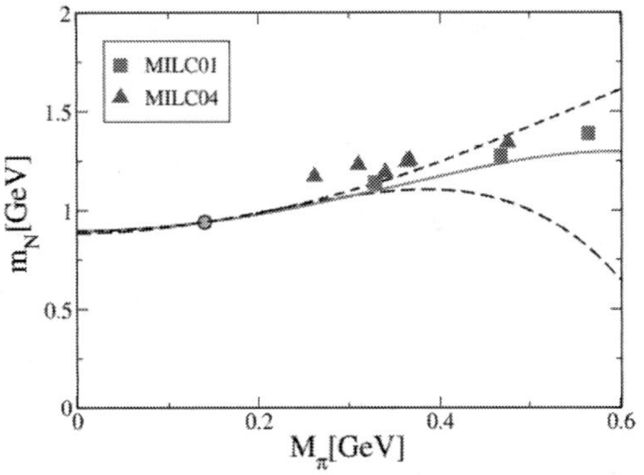


**Figure 4.** Pion mass dependence of the nucleon mass in SU(3). The red solid line gives the best fit as discussed in the text and the black dashed lines give the theoretical uncertainty under the constraint that the nucleon mass takes its physical value for the physical quark masses. Green squares: MILC 2001 data [14], purple triangles: MILC 2004 data [15]. The circle stands for the physical values.

## 5 Effective field theories for excited states

The experience obtained in dealing with the $N\Delta$-splitting in the SSE allowed us to consider also the pertinent EFTs for other excited hadrons. In Ref. [16], we have extended the method of infrared regularization to spin-1 fields. This is an important step in the formulation of an EFT for such instable particles with a consistent power counting. As applications of this scheme, we have worked the chiral extrapolation of the rho meson mass, which takes the form

$$M_\rho = M_\rho^0 + c_1\, M_\pi^2 + c_2\, M_\pi^3 + c_3\, M_\pi^4\, \ln\left(\frac{M_\pi^2}{M_\rho^2}\right) + O(M_\pi^4)\,, \quad (1)$$

with $M_\rho^0$ the rho-meson mass in the chiral limit and the $c_i$ are combinations of low-energy constants of the effective chiral Lagrangian of Goldstone bosons and vector mesons. We have performed various fits of the existing lattice data from the CP-PACS collaboration, see Fig. 5. The chiral extrapolation function Eq. (1) has been used by many lattice groups after it was proposed in our work. We have also analyzed the pion-rho sigma term.

In the paper [18], we have presented an extension of the infrared regularization scheme that allows for an inclusion of explicit (vector and axial-vector) meson resonances in the single-nucleon sector of CHPT. For the processes we have considered here, the meson resonances do not appear as external particles, and the corresponding power counting rules for the internal resonance lines is set up such that the resonance four-momentum is considered

DOI: 10.1051/epjconf/201713404002

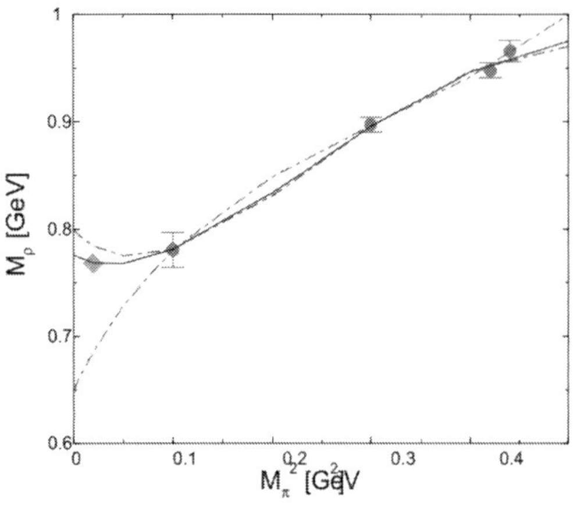


**Figure 5.** The rho meson mass $M_\rho$ as a function of the light quark mass, $M_\pi^2 \sim (m_u + m_d)$. The solid (dot-dashed) line(s) refers to fit 1 (2,3) as described in the text. The lattice data are from CP-PACS [17]. The diamond denotes the physical rho mass.

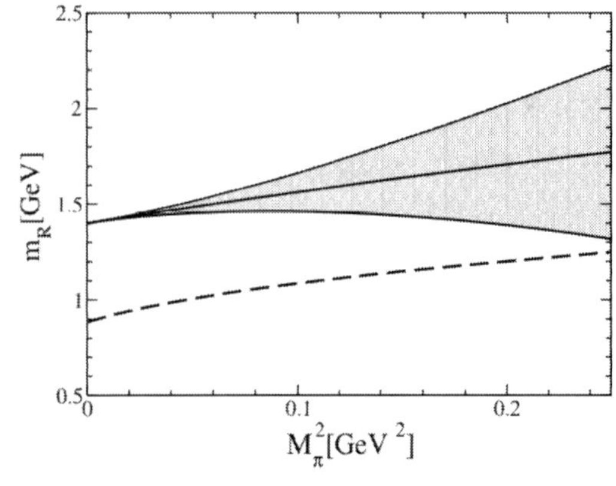


**Figure 6.** Quark mass dependence of the Roper mass for different parameter sets of the dimension two and four LECs within natural bounds (hatched area bounded by the solid curves). The lower dashed curve represents the quark mass dependence of the nucleon for standard values of the corresponding LECs: $c_1 = -0.9, c_2 = 3.2, c_3 = -3.45, e_1 = -1.0$ (in appropriate units of 1/GeV). The theoretical uncertainty for the nucleon mass is not shown in this plot (it is similar to the one for the Roper).

to be small compared to its mass. The infrared regularization scheme extracts the part of the one-loop graphs to which this power counting scheme applies (for any value of the dimension parameter $d$ used in dimensional regularization), while the remaining parts of the loop graphs will in general violate the power counting requirements, but can be absorbed in a renormalization of the local terms of the effective Lagrangian.

In Ref. [19], we have analyzed the quark mass dependence of the Roper mass (the real part of the self-energy) to one-loop order in covariant baryon chiral perturbation theory. The pertinent effective Lagrangian is spelled out in the Appendix. The loop integrals are evaluated using an extension of infrared regularization for two well separated large mass scales (as it is roughly fulfilled in nature since $m_N^2/m_R^2 \simeq 1/2.4$). This method preserves chiral symmetry and establishes a chiral counting scheme. The derived chiral expansion of the Roper mass may prove useful for chiral extrapolations of lattice data. For couplings of natural size the quark mass dependence of the Roper mass is similar to the one of the nucleon, see Fig. 6. In particular, the derived quark mass dependence does not show a strong variation for small masses as indicated by some lattice results. We are presently extending these studies to a three–scale problem to also analyse the decay $S_{11}(1535) \to N\eta$ and the relative quark mass dependence of the Roper and the $S_{11}$ mass.

We would like to thank our collaborators Véronique Bernard, Bugra Borasoy, Peter Bruns, Matthias Frink, Thomas Hemmert, Randy Lewis and Ilka Scheller. The work reported here would not have been possible without the support from the Deutsche Forschungsgemeinschaft within the SFB/TR16.

## A Effective Lagrangian of the nucleon-Roper-pion system

Since the Roper is the first even-parity excited state of the nucleon, the construction of the chiral SU(2) effective Lagrangian follows standard procedures, see e.g. [20]. The effective Lagrangian relevant for our calculation is

$$\mathcal{L} = \mathcal{L}_0 + \mathcal{L}_R + \mathcal{L}_{NR} ,$$

with the free part

$$\mathcal{L}_0 = i\bar{N}\gamma_\mu D^\mu N - M_N \bar{N}N + i\bar{R}\gamma_\mu D^\mu R - M_R \bar{R}R ,$$

where $N, R$ are nucleon and Roper fields, respectively, and $M_N, M_R$ the corresponding baryon masses in the chiral limit. $D_\mu$ is the chiral covariant derivative, for our purpose we can set $D_\mu = \partial_\mu$, see e.g. [20] for definitions. The pion-Roper coupling is given to leading chiral order by

$$\mathcal{L}_R^{(1)} = \frac{1}{2} g_R \bar{R}\gamma_\mu\gamma_5 u^\mu R ,$$

with an unknown coupling $g_R$ and the superscript denotes the chiral order. We have use $g_R = 1.0$ as explained in Ref. [19]. The pion fields are collected in $u_\mu = -\partial_\mu \boldsymbol{\pi}/f_\pi + O(\boldsymbol{\pi}^3)$, where $f_\pi$ is the pion decay constant in the chiral limit. At next–to–leading order, the relevant terms in $\mathcal{L}_R$ are (we work in the isospin limit $m_u = m_d$ and neglect electromagnetism)

$$\mathcal{L}_R^{(2)} = c_1^* \langle \chi_+ \rangle \bar{R}R - \frac{c_2^*}{8M_R^2} \bar{R}\left(\langle u_\mu u_\nu \rangle \{D^\mu, D^\nu\} + \text{h.c.}\right) R + \frac{c_3^*}{2} \langle u_\mu u^\mu \rangle \bar{R}R ,$$

where $\chi_+$ is proportional to the pion mass and induces explicit chiral symmetry breaking. Further, $\langle\ \rangle$ denotes the trace in flavor space. For a complete one loop calculation we also need the fourth order effective Lagrangian, more precisely, the term

$$\mathcal{L}_R^{(4)} = -\frac{e_1^*}{16} \langle \chi_+ \rangle^2 \bar{R}R .$$

DOI: 10.1051/epjconf/201713404002

The interaction piece between nucleons and the Roper reads

$$\mathcal{L}_{NR}^{(1)} = \frac{1}{2} g_{NR} \bar{R} \gamma_\mu \gamma_5 u^\mu N + \text{h.c.}\,.$$

The coupling $g_{NR}$ can be determined from the strong decays of the resonance $R$, its value is $g_{NR} = 0.3 \ldots 0.4$. In principle a term of the form

$$i\lambda_1 \bar{R} \gamma_\mu D^\mu N - \lambda_2 \bar{R} N + \text{h.c.}$$

is possible, but applying the equations of motion removes the first term (and its hermitian conjugate) such that we are left with the terms $\bar{R}N$ and $\bar{N}R$. These terms induce mixing between the nucleon and Roper fields, but diagonalization of the $N$-$R$ mass matrix does not lead to new operator structures and its effect can be completely absorbed into the couplings already present in the Lagrangian. We can thus safely work with the Lagrangian in Eq. (2). A complete one-loop calculation involves tree graphs with insertion of chiral dimension two and four and one-loop graphs with at most one insertion from $\mathcal{L}_R^{(2)}$.

## References

[1] V. Bernard, Prog. Part. Nucl. Phys. **60** (2008) 82 [arXiv:0706.0312 [hep-ph]].

[2] V. Bernard, T. R. Hemmert and U.-G. Meißner, Phys. Lett. B **622** (2005) 141 [arXiv:hep-lat/0503022].

[3] T. R. Hemmert, B. R. Holstein and J. Kambor, J. Phys. G **24** (1998) 1831 [hep-ph/9712496].

[4] C. W. Bernard *et al.*, Phys. Rev. D **64**, 054506 (2001) [arXiv:hep-lat/0104002].

[5] G. Schierholz, private communication; QCDSF collaboration, in preparation.

[6] M. Lüscher, Commun. Math. Phys. **105** (1986) 153.

[7] U.-J. Wiese, Nucl. Phys. Proc. Suppl. **9** (1989) 609.

[8] V. Bernard, U.-G. Meißner and A. Rusetsky, Nucl. Phys. B **788** (2008) 1 [hep-lat/0702012 [HEP-LAT]].

[9] V. Bernard, T. R. Hemmert and U.-G. Meißner, Phys. Rev. D **67** (2003) 076008 [hep-ph/0212033].

[10] M. Dorati, T. A. Gail and T. R. Hemmert, Nucl. Phys. A **798** (2008) 96 [nucl-th/0703073].

[11] V. Bernard and U.-G. Meißner, Phys. Lett. B **639** (2006) 278 [arXiv:hep-lat/0605010].

[12] R. G. Edwards *et al.* [LHPC Collaboration], Phys. Rev. Lett. **96** (2006) 052001 [arXiv:hep-lat/0510062].

[13] M. Frink, U.-G. Meißner and I. Scheller, Eur. Phys. J. A **24** (2005) 395 [arXiv:hep-lat/0501024].

[14] C. W. Bernard *et al.*, Phys. Rev. D **64** (2001) 054506 [arXiv:hep-lat/0104002].

[15] C. Aubin *et al.*, Phys. Rev. D **70** (2004) 094505 [arXiv:hep-lat/0402030].

[16] P. C. Bruns and U.-G. Meißner, Eur. Phys. J. C **40** (2005) 97 [arXiv:hep-ph/0411223].

[17] S. Aoki *et al.* [CP-PACS Collaboration], Phys. Rev. D **60** (1999) 114508 [arXiv:hep-lat/9902018].

[18] P. C. Bruns and U.-G. Meißner, Eur. Phys. J. C **58** (2008) 407 [arXiv:0808.3174 [hep-ph]].

[19] B. Borasoy, P. C. Bruns, U.-G. Meißner and R. Lewis, Phys. Lett. B **641** (2006) 294 [arXiv:hep-lat/0608001].

[20] N. Fettes, U.-G. Meißner, M. Mojžiš and S. Steininger, Annals Phys. **283** (2000) 273 [Erratum-ibid. **288** (2001) 249] [arXiv:hep-ph/0001308].

# Study of QCD-dynamics in $\eta$ and $\eta'$ production and decays

*Reinhard* Beck[1,⋆] and *Ulrich* Wiedner[2,⋆⋆]

[1] *Helmholtz-Institut für Strahlen- und Kernphysik, Rheinische Friedrich-Wilhelms-Universität Bonn, Nußallee 14-16, 53115 Bonn*
[2] *Institut für Experimentalphysik I Ruhr-Universität Bochum Universitätsstraße 150, 44780 Bochum*

**Abstract.** One aim of this project lies in studying QCD dynamics using photoproduction of $\eta$ and $\eta'$ mesons with the CBELSA/TAPS experiment at the accelerator facility ELSA. Hadronic decay modes of $\eta$ and $\eta'$ are analyzed that allow the study of symmetry breaking effects in QCD. Furthermore, excited $\eta$-states produced with a photon beam are studied within the scope of this project. Of special interest is hereby the nature of the $\eta(1405)$. The high data samples for $\eta$- and $\eta'$-photoproduction off the proton allow in addition the determination of polarization observables.

## 1 Introduction

The scientific goal of this project is the study of QCD dynamics in the photoproduction of pseudoscalar and vector mesons and their subsequent decay pattern. Especially in focus are the ground-state $\eta$ and $\eta'$ mesons and excited $\eta$ states. Various decays of the $\eta$ and $\eta'$ are forbidden by conservation laws. For example, the hadronic decay modes $\eta \to 3\pi$ only occur due to the isospin-violating quark-mass difference $m_u - m_d$ or due to electromagnetic effects, and the decays $\eta, \eta' \to \pi\pi\gamma$ are sensitive to the QCD box anomaly. A systematic study of such decays offers a way to investigate symmetries and symmetry breaking effects in QCD (see project B.3). The second goal within the project is related to the question of the production cross-section of excited $\eta$-states with a photon beam. In particular, the nature of the $\eta(1405)$, discovered in antiproton-proton annihilation, has been discussed extensively and controversially in the literature. It is considered as a good candidate for the pseudoscalar glueball, even though its mass is significantly lower than lattice calculations predict. Recent studies of $\eta(1405)$ decays did lead to unexpected results, thus adding to the ambiguity in the interpretation of its nature. More light on the nature of the $\eta(1405)$ can be shed by studying the processes that lead to its production. While the $\eta(1405)$ is clearly produced in gluon-rich processes like radiative $J/\Psi$ decays or antiproton-proton annihilation, the production cross section with a photon beam is unclear. Data from the CBELSA/TAPS experiment give only a very weak signal for the production of the $\eta(1405)$ – perhaps an indication for a gluonic nature of the particle, because glue is not expected to be produced copiously by photon beams. However, the statistical evidence does not allow any decisive conclusion. In addition, we are carrying out analyses of the production of $\eta(1405)$ in radiative $J/\psi$ and $\psi'$ decays measured with the BES III detector in Beijing and in antiproton-proton annihilation at different momenta. The latter data were taken with the Crystal Barrel detector at the LEAR accelerator at CERN.

⋆e-mail: beck@hiskp.uni-bonn.de
⋆⋆e-mail: wiedner@ep1.rub.de

## 2 Project Results

### 2.1 Determination of the Dalitz plot parameter $\alpha$ and precise measurement of the $\eta$ mass

The Dalitz plot parameter $\alpha$ for the $\eta \to 3\pi^0$ decay was precisely measured, see the PhD-thesis of M. Unverzagt [1] and [2, 3]. The experiment was performed with the Crystal Ball and TAPS large acceptance photon detectors at the tagged photon beam facility of the MAMI-B electron accelerator in Mainz. High statistics of $1.8 \cdot 10^6$ $\eta \to 3\pi^0$ events were obtained, giving the result $\alpha = -0.032 \pm 0.002_{\text{stat.}} \pm 0.002_{\text{syst.}}$ which is presented in Fig. 1(a). The most recent values for the slope parameter are: $\alpha = -0.044 \pm 0.004$ [5], $\alpha = -0.025 \pm 0.004$[6], $\alpha = -0.0302 \pm 0.0011$[7].

In addition, a new precise determination of the $\eta$ meson mass was performed, see the PhD-thesis of A. Nikolaev [8] and [9]. It is based on a measurement of the threshold for the $\gamma p \to p\eta$ reaction using the tagger focal-plane microscope detector at the MAMI-B facility in Mainz. In Mainz the real photon beam was produced by bremsstrahlung of the 883 MeV electrons from MAMI-B [10, 11] on a 100 $\mu$m thick diamond radiator. The absolute electron energy, $E_0$, of the incident beam was precisely determined [12, 13] in the third race-track microtron of MAMI-B with a total uncertainty of about $\sigma_0 = 140$ keV. The photon energies were determined using the Glasgow photon tagging spectrometer (tagger) [14, 15], which provided a tagged photon flux of roughly $10^5$ s$^{-1}$ MeV$^{-1}$ at a beam current of about 35 nA.

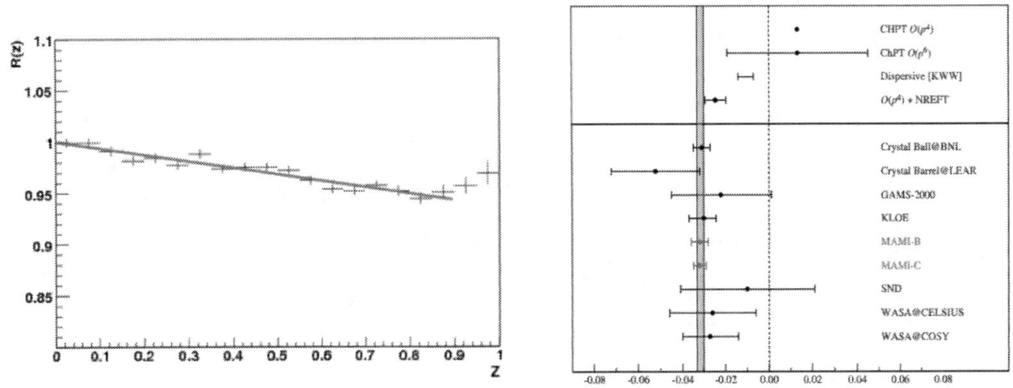


(a) (b)

**Figure 1.** (a) Ratio R(z) of the simulated and the measured z distributions. The Dalitz plot parameter, $\alpha$, was obtained directly from the fit of the function $c(1 + 2\alpha z)$ (red line). (b) World data and prediction for the slope parameter $\alpha$ (PDG 2010 value shaded gray) [4] and the new precise $\alpha$ values from MAMI-B [2] and MAMI-C [3].

For the first time the tagger focal-plane microscope detector [16] was used to improve the tagged photon energy resolution. The microscope detector was placed in front of the main focal-plane spectrometer (see Fig. 2(a)), so that it covered the region around the $\eta$ production threshold ($E_{\text{thr}} \approx 707$ MeV) from $E_\gamma = 674$ MeV to $E_\gamma = 730$ MeV at an electron beam energy $E_0 = 883$ MeV. Made of 96 scintillator strips overlapping to one third, it provided 191 tagging channels with a higher energy resolution of about 0.29 MeV per channel compared to approximately 1.8 MeV available from the main focal-plane detector.

The 4.76 cm long liquid hydrogen target was located at the center of the Crystal Ball photon spectrometer. The Crystal Ball [17, 18], covering polar angles between $\theta = 20°$ and $\theta = 160°$, consisted of 672 NaI(Tl) crystals (see Fig. 2(b)) and had two openings for the beam in forward and backward directions. Each NaI(Tl) crystal had the form of a truncated 41 cm long pyramid and was equipped with an individual photomultiplier. In order to distinguish between neutral and charged particles detected by the Crystal Ball, the system was equipped with a particle identification detector (PID) [19]. The PID was a cylindrical detector, consisting of 24 plastic scintillator strips of 2 mm thickness, arranged parallel to the photon beam axis.

The forward wall detector, TAPS [20], had 510 $BaF_2$ hexagonally shaped crystals, each equipped with a 5 mm thick plastic scintillator for identifying charged particles. A single $BaF_2$ crystal was 25 cm long and had an inner diameter of 5.9 cm. The TAPS detector, intended for detecting particles in the forward direction ($\theta = 4° - 20°$), was located at a distance of 173 cm from the Crystal Ball center, making it possible to use a time-of-flight analysis for the particle identification.

The experimental trigger comprised two levels. In the first level, the total energy deposited in the Crystal Ball was checked. If the sum of all photomultiplier analog signals exceeded a threshold corresponding to about 390 MeV, the event was accepted. The second-level trigger included a condition on the Crystal Ball sector multiplicity. The 672 crystals of the spectrometer were grouped into 45 sectors of up to 16 crystals each. If at least one of the 16 signals exceeded a threshold of 20 to 40 MeV, depending on the relative energy calibration of the photomultiplier signals, the sector contributed to the multiplicity. All events with multiplicity $M \geq 3$ and every third event with $M \geq 2$ were recorded for further analysis. The latter condition was used for the detection of the $\eta \to 2\gamma$ decay, while the first enhanced events from the six-photon decay.

The tagger microscope has a higher energy resolution than the standard tagging spectrometer and, hence, allowed an improvement in the accuracy compared to the previous $\eta$ mass measurement at MAMI-B (see Fig. 3(a)). The result $m_\eta = (547.851 \pm 0.031_{\text{stat.}} \pm 0.062_{\text{syst.}})$ MeV agrees very well with the precise values of the NA48, KLOE and CLEO collaborations and deviates by about $5\sigma$ from the smaller, but also very precise value obtained by the GEM collaboration at COSY. A comparison of all current measurements is given in Fig. 3(b).

There were significant improvements in the $\eta'$ event selection for the CBELSA/TAPS experiment, which is described in project A.1, especially concerning the $\eta' \to \pi^0\pi^0\eta$ channel, see the Diploma thesis of P. Nuhn [21]. This was achieved by implementing a kinematic fit and improvements in the background description. For every event, the kinematic fit returns a confidence level for the event hypothesis (in this case $\pi^0\pi^0\eta$). Rejecting events with a confidence level lower than 0.0001 eliminates a lot of background (cf. 5(a) red line). Without this cut the $\eta'$-peak would not be visible. The still considerable background stems from $3\pi^0$; it can be further reduced by using the kinematic fit with a $3\pi^0$-hypothesis and applying an anti-cut (cf. 5(a) blue line). The remaining background (cf. 5(b)) is attributed to further misidentified $3\pi^0$ final states and to direct $\pi^0\pi^0\eta$ production. For further analysis, it can be subtracted by methods such as multivariate

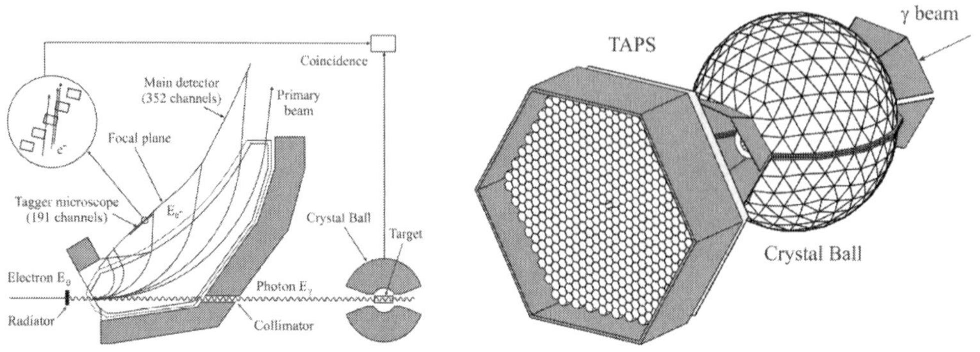


(a) (b)

**Figure 2.** (a) Plan view of the tagging bremsstrahlung facility [14, 15] and the Crystal Ball detector (not to scale) at Mainz. The tagger microscope detector [16], giving improved resolution, was installed in the focal plane in front of the main detector at the position indicated. The inset shows the geometry of the microscope scintillators. (b) The Crystal Ball photon spectrometer and TAPS. The individual photomultipliers attached to the crystals are not shown; however, those supporting steel structures that were included in the Monte Carlo simulations are shown. Taken from [9].

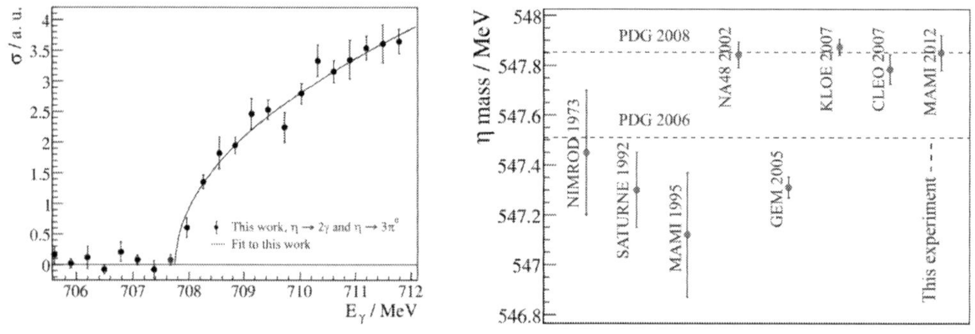


(a) (b)

**Figure 3.** (a) Final results $\gamma p \to p\eta$ in the threshold region. (b) Overview of previous $\eta$ mass measurements in comparison to the world average reported by the Particle Data Group in 2006 and 2008 (dashed lines) and the result of the new analysis [9].

side band subtraction or by determining the background contribution in each bin using a suitable fit function.

## 2.2 The double polarization observable E in $\eta'$ photoproduction

Apart from the $\eta' \to \pi^0\pi^0\eta$ decay mode, the CBELSA/TAPS detector setup can be used to study the $\eta' \to \gamma\gamma$ decay mode with a branching ratio of only 2.2% as well. Both decay modes were analyzed in the Master thesis of F. N. Afzal [22, 23] with the aim to study baryon resonances by determining the double polarization observable E. Unlike the unpolarized cross section, polarization observables are sensitive to interference terms of multipoles and hence are sensitive to weakly contributing resonances, see project A.1. The $\eta'$ meson has an isospin of $I = 0$, thus $N^*$ resonances with isospin $I = 1/2$ become exclusively accessible; Δ resonances can not contribute. Furthermore it is possible to investigate the high mass region ($m_{N^*} > 1890$ MeV) of the nucleon excitation spectrum due to the high mass of the $\eta'$ meson (958 MeV). In this region many predicted but so far not observed $N^*$ states should be located. For the measurement of the double polarization observable E, a circularly polarized photon beam in combination with a longitudinally polarized butanol target was used. Fig. 5 shows the results for the double polarization observable E as a function of the photon beam energy. These results together with the already measured unpolarized cross section $\sigma_0$ data [24, 25] can be used to determine the helicity dependent cross sections $\sigma_{1/2}$ and $\sigma_{3/2}$ [23]. The positive sign of the double polarization observable E over the entire energy range of 1447 MeV - 2320 MeV indicate that the $A_{1/2}$ helicity amplitude is favored by the contributing resonances. For a complete disentanglement of all contributing resonances in this mass region more statistics and the measurement of more observables is needed however. In the future it is planned to take more data with a circularly polarized photon beam and a longitudinally polarized target with an electron beam of 3.2 GeV from ELSA.

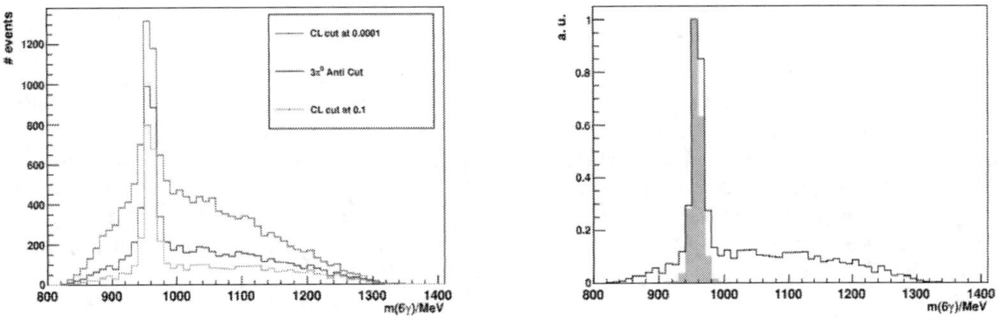


(a) (b)

**Figure 4.** (a) Cuts to improve $\eta'$ event selection for $6\gamma$ final states (b) Comparison between Monte Carlo simulations (gray) and measured $\eta'$ events.

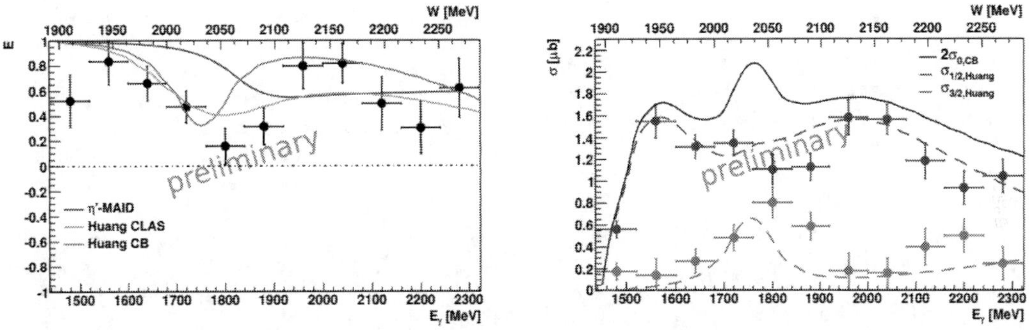


(a) (b)

**Figure 5.** (a)The double polarization observable E in $\eta'$ photoproduction was determined in Master thesis of F.N. Afzal and is shown here as a function of the beam energy $E_\gamma$ and the center of mass energy W. The data (black points) are compared to partial wave analysis predictions [27, 28].(b) The helicity dependent cross section $\sigma_{1/2}$ and $\sigma_{3/2}$ together with the fit to the unpolarized cross section $\sigma_0 = \frac{1}{2}(\sigma_{1/2} + \sigma_{3/2})$ data by [27]. Taken from [23].

## 2.3 Measurement of the beam asymmetry in $\eta$- and $\eta'$-photoproduction

From July to October 2013 the so far largest data sample for $\eta$- and $\eta'$-photoproduction off the proton with the CBELSA/TAPS experiment at ELSA was taken. An electron beam of 3.2 GeV from ELSA was incident on a thin diamond radiator. Linearly polarized photons were produced through coherent bremsstrahlung. The diamond radiator was oriented relative to the beam so that the coherent edge was located at 1750 MeV or at 1850 MeV, giving a maximum beam polarization degree of $p_\gamma = 40\%$ or 35% at 1750 MeV or 1850 MeV, respectively. As target a liquid hydrogen target of 10 cm thickness was chosen.
This beamtime can be used for several purposes, e.g. the determination of the beam asymmetry $\Sigma$ in $\eta$ and $\eta'$-photoproduction which is subject of an ongoing analysis. Within the energy range of 1100 MeV – 1800 MeV, $8.3 \cdot 10^6$ $\pi^0$ and $8.2 \cdot 10^5$ $\eta$ events in the decay modes $\pi^0 \to \gamma\gamma$ and $\eta \to \gamma\gamma$ were selected [36] using kinematic cuts on the data. These events were used to determine the beam asymmetry $\Sigma$ by fitting the $\phi$-asymmetries $\Delta N(\phi)$ for each energy and angular bin:

$$\Delta N(\phi) = \frac{N_{-45^\circ} - N_{+45^\circ}}{N_{-45^\circ} + N_{+45^\circ}} = p_\gamma \Sigma \sin(2\phi), \qquad (1)$$

where $N_{-45^\circ}$ and $N_{+45^\circ}$ are the yields obtained with the polarization plane being at $\pm 45^\circ$ relative to the reaction plane, respectively and $\phi$ being the azimuthal angle of the selected meson in the final state. The preliminary results of the beam asymmetry $\Sigma$ in $\eta$-photoproduction are shown in Fig. 6 for selected energy bins as a function of $\cos\theta$, where $\theta$ is the polar angle of the $\eta$ meson in the center of mass system. The data are in good agreement with existing data from CB-ELSA [29] and GRAAL [30] (see Fig. 6(a) and 6(b)). Furthermore the different partial wave analyses (solutionsBnGa2014-01, BnGa2014-02 [35], the SAID-GE09 [34] and the JüBo2015 fit B [32]) agree well with the data since the existing data have been included in all solutions. Above 1500 MeV no data exist until now. Here our new data offer new input that will constrain the different partial wave analysis solutions which differ immensely at 1670 MeV (see Fig. 6(d)).
The beam asymmetry in $\eta'$-photoproduction has been measured so far only near the $\eta'$ photoproduction threshold

for photon energies of 1461 MeV and 1480 MeV reported by the GRAAL collaboration [37]. These results show a zero crossing from positive to negative from forward to backward angles for both energy bins. This is surprising since near the $\eta$ threshold, no such zero crossing was observed. This could indicate that the dynamics of $\eta$ and $\eta'$ photoproduction processes are different near threshold. With our collected data we will be able to extend the measured energy range for the beam asymmetry to 1800 MeV and investigate the dynamics of $\eta$ and $\eta'$ mesons further.
Another aim of this beamtime is the analysis of rare $\eta$ and $\eta'$ decays which will be done in future analyses. In addition it is possible to look for excited $\eta$ states such as the $\eta(1405)$ using this and future beamtimes.

### 2.4 $\eta$-mesons and the quest for glueballs

The self-interaction of gluons that is central to QCD leads to a flux tube of gluons exchanged between the quarks that are bound by them. In consequence, quarks cannot be separated infinitely and are subject to confinement. It is a fact that individual free quarks have never been observed experimentally to date. The self-interaction in phenomenological models and in lattice gauge theory calculations clearly points to the existence of bound states of pure glue, known as glueballs.
Glueballs pose a very interesting facet of nature. The gluons are the mediator particles of the strong interaction and themselves massless. However, they do carry color, i.e., the charges of the strong interaction. It is generally believed that there is an attractive interaction between gluons leading for example in lattice calculations based on the QCD Lagrangian to a flux tube of gluons between the quark and the antiquark in a meson. The mutual gluon attraction should also allow for the formation of meson-like bound states of gluons, even if no quarks are present. Lattice calculations predict a whole spectrum of bound gluon states, the glueballs.
While the Higgs mechanism might be responsible for the masses of the elementary particles, the mass-creation mechanism for hadrons is quite different. Only a few percent of the mass of the proton is due to the Higgs mechanism. The rest is created, in a to date not very well determined way, by the strong interaction. In particular, glueballs themselves would be massless without the strong interaction and their predicted masses arise solely from the strong interaction. Glueballs thus offer a unique way to study the mass creation of strongly interacting particles.
How can the unknown structure of glueballs and their properties be addressed and studied? First of all, it would be important to identify and study their behavior as detailed as possible. What criteria do we have to identify glueballs? The experience from previous searches for the scalar glueball tells us that their decay pattern might not be decisive. The scalar glueball with the quantum numbers $J^{PC} = 0^{++}$ is predicted by lattice calculation around at a mass $\approx$ 1.6 GeV/c$^2$. In the mass region, however, many mesons are present with the same quantum numbers. Mixing between overlapping resonances might occur and destroy a distinctive decay pattern, so the interpretation as glueball is complicated by the mixing of the glueball with the broad light-quark states. But we should have other criteria.
A glueball does not fit into the normal meson nonet with identical quantum numbers, it should be surplus. Glueballs should be produced in so-called gluon-rich reactions like the antinucleon-nucleon annihilation process. The annihilations of the antiquarks with quarks leads to a gluon-rich environment favoring the formation of gluonic degrees of freedom. The central region of a high energetic hadron-hadron scattering process should favor the production of glueballs since it is neither populated by target nor by beam quarks. The radiative decay of quarkonium states and here especially the radiative decay of $J/\psi$ or $\psi'$ mesons should be a prime searching ground for glueballs. Glueballs should not be produced in two photon collisions since gluons do not carry electric charge, which also would suppress their production in a photon beam. The best candidate for the scalar glueball is nowadays the $f_0(1500)$, because it fulfills all of the above criteria.
A consequence of the above mentioned model for glueballs, with the structure of a closed loop of gluonic flux, is the existence of a mass-degenerate glueball state with opposite parity, i.e. a pseudoscalar glueball $J^{PC} = 0^{-+}$. Indeed, the experimental data provide evidence for such a glueball. The particle usually noted as $\eta(1440)$ may indeed be two particles, which are called $\eta(1405)$ and $\eta(1475)$. Especially the $\eta(1405)$ has a strong affinity to glue [38].
We should now see, if the $\eta(1405)$ fulfills our criteria for being a glueball in a similar way as the $f_0(1500)$. Provided that the $\eta(1295)$ does exist, only one of the two states, $\eta(1405)$ and $\eta(1475)$, could be the first radial excitation of the $\eta'$. The strong coupling to kaons make the $\eta(1475)$ the prime candidate for being the strangonium member of the nonet. This would leave the $\eta(1405)$ surplus. The key question to answer is the existence of the $\eta(1295)$.
In the past years we have studied the reaction

$$\gamma p \to p(\pi^0\pi^0\eta) \to p(6\gamma) \tag{2}$$

with the Crystal Barrel detector. All appropriate available beam times were analyzed. The resulting invariant mass spectrum is shown in Fig. 7(a):

A hint for a structure at 1405 MeV/c$^2$ is visible. Also a structure around $\approx$ 1300 MeV/c$^2$ is present. Selecting mainly $a_0(980)\pi^0$ events by a mass cut of the $\eta\pi$ system between 900 – 1100 MeV/c$^2$ makes the structure at 1405 MeV/c$^2$ more evident (Fig. 7(b)). Recent results from PhD theses of Tobias Triffterer and Cathrina Sowa [42, 43] support the assumption, that the $\eta(1405)$ is produced in $\gamma p$. Using the data acquired by the CBELSA/TAPS experiment with a liquid hydrogen target in 2013, a relative cross section for the production of $\eta(1405)$ candidates can be calculated to 0.16 $\pm$ 0.05 by comparing the data to the $\eta'$ (see Fig. 11)). To determine if the signal around 1300 MeV/c$^2$ is the $\eta(1295)$ or the well-known $f_1(1285)$ requires to resolve its spin in a partial wave analysis. However, the existing statistics is not sufficient enough to perform such a partial wave analysis.

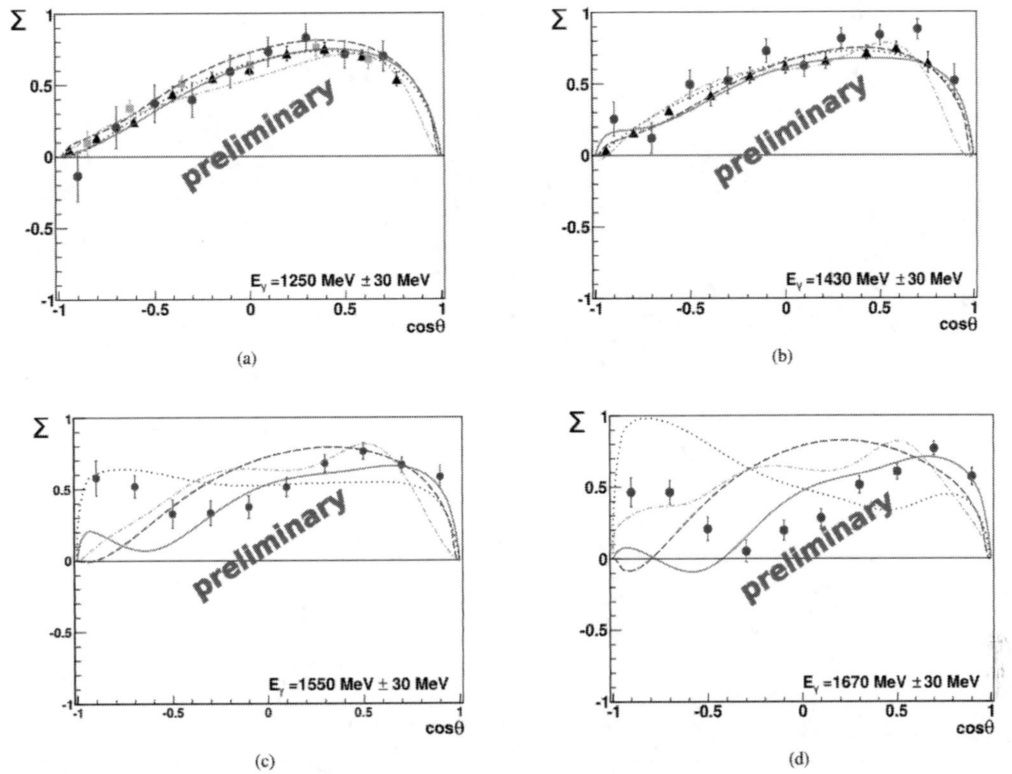


**Figure 6.** (a)-(d) The beam asymmetry Σ in $\eta$-photoproduction is shown for four different energy bins as a function of $\cos\theta_{CMS}$ (blue points) [26]. The data is compared to CB-ELSA data [29] (green squares) and GRAAL data [30] (black triangles) and to the following partial wave analysis solutions: BnGa2014-01 (black dotted line), BnGa2014-02 [35] (red solid line), SAID-GE09 [34] (blue dashed line) and JüBo2015 fit B [32] (green dash-dotted line). (a), (b) and (d) taken from [36].

The recent analysis of the $\eta\pi^+\pi^-$ final state collected by the CLAS collaboration [44] comes to the conclusion, that the mass, width, and an amplitude analysis of the Dalitz distribution are consistent with the axial-vector $f_1(1285)$ identity, rather than the pseudoscalar $\eta(1295)$. The production mechanism is more consistent with s-channel decay of a high-mass $N^*$ state and not with t-channel meson exchange.
For the future, after the successful trigger upgrade of the Crystal Barrel calorimeter (see project D.3) , it is planned to take a dedicated beam time, which will increase the statistics up to two orders of magnitude. This will be sufficient to determine if the $\eta(1405)$ is produced at a cross section comparable to ordinary mesons or with a reduced production cross section, as expected for a glueball. Also, a partial wave analysis will tell us about the existence of the $\eta(1295)$ in the photoproduction process on protons.
While a non-observation, or an observation with suppressed production cross section, in the photoproduction process might serve as indication for the gluonic nature of the $\eta(1405)$, gluon-rich reactions should contain a significant number of $\eta(1405)$ if it has a strong gluonic component. We have studied therefore in the past years in parallel the production of $\eta(1405)$ and $\eta(1295)$ in antiproton-proton annihilations at different energies and radiative charmonia decays.

The BESIII experiment in Beijing has taken in recent years several hundred million $J/\psi$ and $\psi'$ data on resonance. We have looked at gluon-rich radiative decay channels:

$$J/\psi \to \gamma(\pi\pi\eta) \quad \text{and} \quad \psi' \to \gamma(\pi^+\pi^-\eta) \qquad (3)$$

The invariant mass spectra (Fig. 8(a) and 8(b)) show a signal around 1400 MeV/c$^2$, which contains in case of the $J/\psi$ decays much more events than the signal at $\approx$ 1300 MeV/c$^2$. This is in sharp contrast to our observations with the photoproduction data. Why the $\psi'$ data look different in the relative ratios between 1405 and 1300 is currently not understood, but certainly worth investigating further. A partial wave analysis of the BESIII data to determine the spin composition of the particles at the respective masses is underway.

Adding additional questions on the nature of the $\eta(1405)$ does another paper published recently by the BES collaboration. In this paper, for the first time the observation of the decay mode of the $\eta(1405)$ into $f_0(980)\pi$ is reported, at a rate completely unexpected by theory [39]. Previous data show that in antiproton-proton annihilations

DOI: 10.1051/epjconf/201713404003

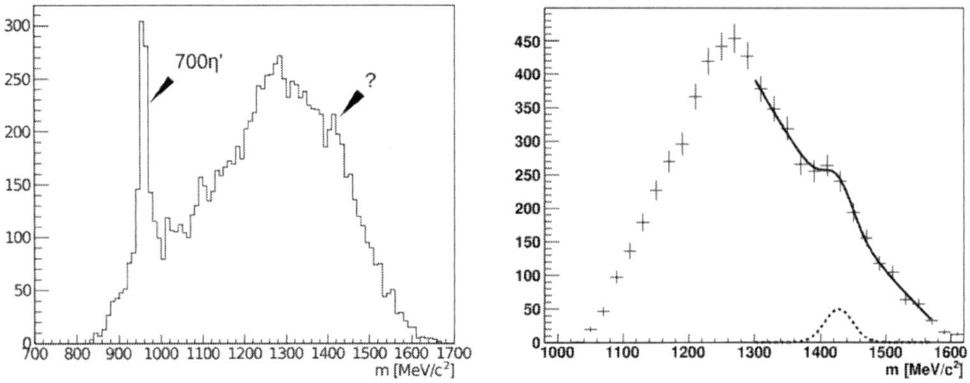


(a) (b)

**Figure 7.** (a) The $\pi^0\pi^0\eta$ invariant mass spectrum in photon-nucleon reactions for $E_{CM} - m_p$ > 1450 MeV with Crystal Barrel at ELSA. (b) The $\pi^0\pi^0\eta$ invariant mass spectrum in photon-nucleon reactions with Crystal Barrel at ELSA requiring that the $\pi^0\eta$ mass is in the $a_0(980)$ (900 MeV< $m(\pi^0\eta)$ < 1100 MeV) mass region.

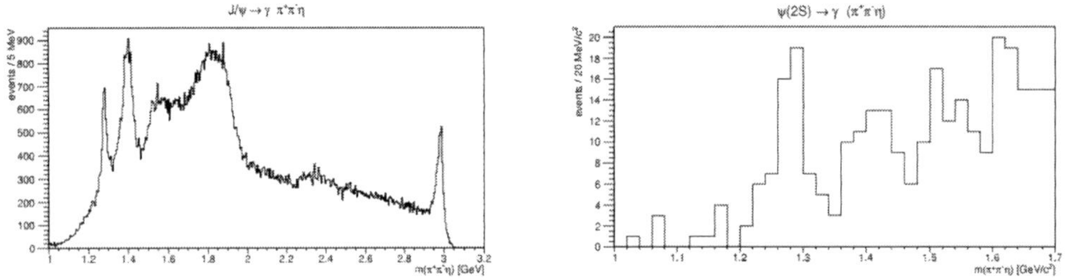

(a) (b)

**Figure 8.** (a) The $\pi^+\pi^-\eta$ invariant mass in radiative $J/\psi$ decays measured with the BESIII experiment. (b) The $\pi^+\pi^-\eta$ invariant mass in radiative $\psi'$ decays measured with the BESIII experiment.

at rest there is clear evidence that both states, $\eta(1405)$ and $\eta(1475)$, are produced [40, 41]. The $\eta(1405)$ has decay channels both into up and down quark mesons and into strange quark mesons, as expected from a glueball, while the $\eta(1475)$ decays mainly into open strange mesons. These were data at rest and we have now started to look for excited $\eta$ states in Crystal Barrel data at LEAR from annihilations at 900 and 1800 MeV/c antiproton momentum to study any dependence on the participating initial waves.
The statistics is again rather low, as could be seen in the invariant mass spectra. Nevertheless indications for the presence of signals round 1300 and 1405 MeV/c$^2$ are there (Fig. 9 and Fig. 10).

To summarize: The comparison between the photoproduction process, radiative charmonia decays and antiproton-proton annihilations shows that the signals at $\approx$ 1300 MeV/c$^2$ and $\approx$ 1400 MeV/c$^2$ are produced with very different ratios. The $\approx$ 1400 MeV/c$^2$ signal in the photoproduction process still needs confirmation in the upcoming years by new data. Partial wave analyses will show if the $\eta(1295)$ is present in the $\approx$ 1300 MeV/c$^2$ signal or if it is due to the $f_1(1285)$. Such a partial wave analysis, crucial for the interpretation of the data, is currently only possible in the radiative charmonia decays but becomes feasible with higher statistics in the photonuclear reaction with Crystal Barrel experiment.

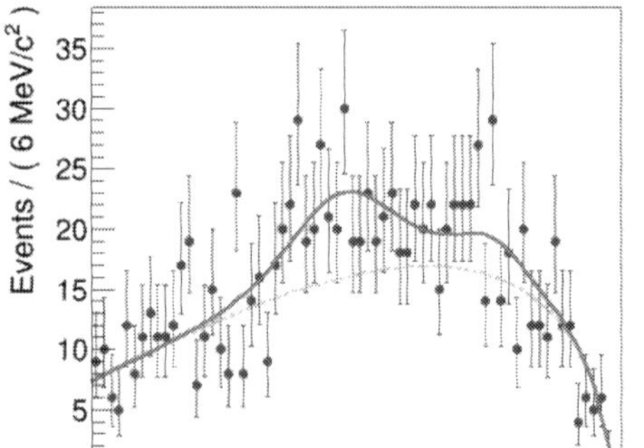


**Figure 11.** The $\pi^0\pi^0\eta$ invariant mass measured with the CBELSA/TAPS experiment using a liquid hydrogen target in 2013 (blue points) [42]. Additionally, the red curve represents a fit with the hypothesis of $f1(1285)$ and $\eta(1405)$ and the orange curve the background contribution.

DOI: 10.1051/epjconf/201713404003

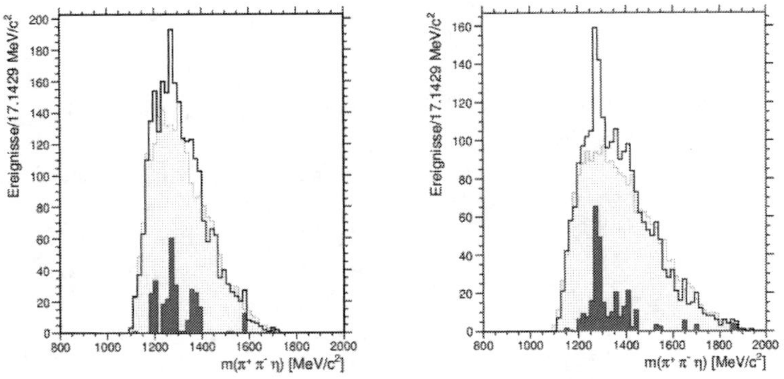


**Figure 9.** The $\pi^+\pi^-\eta$ invariant mass in antiproton-proton annihilations measured with Crystal Barrel at LEAR. The left side shows data from antiproton momenta of 900 MeV/c and the right side of antiproton momenta of 1642 MeV/c. The non-resonant background calculated by Monte Carlo is shown in yellow and the difference between data and Monte Carlo is shown in blue. Cuts on the invariant $\pi\eta$ mass requiring the events to be within an $a_0(980)$ mass window are applied.

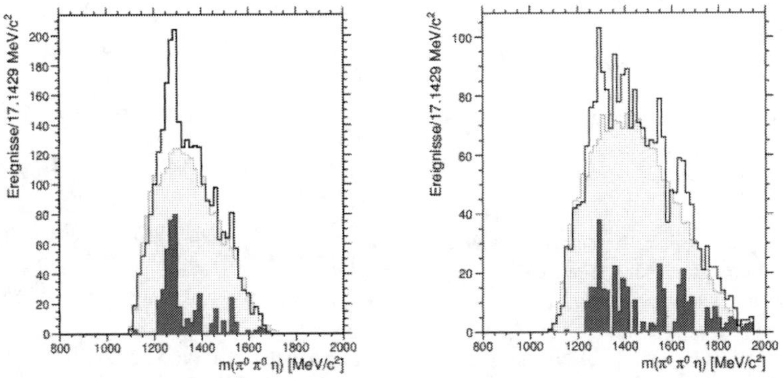


**Figure 10.** The $\pi^0\pi^0\eta$ invariant mass in antiproton-proton annihilations measured with Crystal Barrel at LEAR. The left side shows data from antiproton momenta of 900 MeV/c and the right side of antiproton momenta of 1642 MeV/c. The non-resonant background calculated by Monte Carlo is shown in yellow and the difference between data and Monte Carlo is shown in blue. Cuts on the invariant $\pi\eta$ mass requiring the events to be within an $a_0(980)$ mass window are applied.

## 3 Outlook

After the successful trigger upgrade of the Crystal Barrel calorimeter (see project D.3), we plan to have a dedicated beam time with the ELSA tagged photon beam at the highest energies to produce a large $\eta'$ event sample, to study the $\eta'$ decays and the $\eta(1405)$ excitation. The cross section for the reaction $\gamma p \rightarrow p\eta'$ lies in the vicinity of 1 $\mu$b. The Crystal Barrel is a calorimeter with a wide range in energy and delivers in combination with the TAPS forward wall a solid angle coverage of up to $\Omega = \varepsilon_\Omega \cdot 4\pi$ sr with $\varepsilon_\Omega = 0.98$. The photon beam is produced via the bremsstrahlung process on a thin iron radiator of typical 50 $\mu$m thickness. For the reaction $\gamma p \rightarrow p\eta'$ we intend to measure in a photon energy interval from 1.4 GeV to 3.0 GeV expecting an average cross section of $\sigma_T = 1\,\mu$b. By increasing the beam current of the accelerator ELSA and by disabling tagger channels outside the energy range of interest, the integral flux of photons can be increased by a factor of 3 so that the photon flux in the energy range ($\Delta\omega$) between 1400 MeV and 3000 MeV yields $\dot{N}_\gamma(\Delta\omega) = 6 \cdot 10^6\ \gamma$/s.

As target, we intend to use liquid hydrogen in a cylindrical capton container of a length of $l$ = 10.2 cm. The target density can be calculated using

$$N_T = \rho_{\mathrm{H}} l N_A / A\,, \tag{4}$$

where $N_T$ represents the target density, $\rho_{\mathrm{H}} = 0.0708$ g/cm$^3$ as the specific density of liquid hydrogen, $N_A = 6.022 \cdot 10^{23}$ mol$^{-1}$ as the Avogadro number and the atomic mass $A = 1.00794$ g/mol. This yields $N_T = 4.315 \cdot 10^{23}$ 1/cm$^2$. The total rate of produced $\eta'$ for this set up can be calculated according to

$$\dot{N}_{\eta'} = \dot{N}_\gamma \cdot N_T \cdot \sigma_{\eta'}. \tag{5}$$

This yields for $\dot{N}_{\eta'} = 2.589$ s$^{-1}$ over the given 1.6 GeV range between 1400 MeV and 3000 MeV for the total cross section. This results in a total rate of $\eta'$ per hour

$$\dot{N}_{\eta'} = 9\,320\,\frac{\text{events}}{\text{h}}\,, \tag{6}$$

According to the branching ratios for $\eta'$ decays published by the PDG we expect for the six most dominant modes:

**Table 1.** The branching ratios and number of $\eta'$ per hour or 1000 h are listed for the most dominant $\eta'$ decay modes.

| Mode | Fraction | $\frac{N_{\eta'}}{\text{h}}$ | $\frac{N_{\eta'}}{10^3\text{h}}$ |
|---|---|---|---|
| $\pi^+\pi^-\eta$ | $43.2 \pm 0.7$ % | 4 026 | 4 026 240 |
| $\rho^0\gamma$ | $29.3 \pm 0.5$ % | 2 730 | 2 730 760 |
| $\pi^0\pi^0\eta$ | $21.7 \pm 0.8$ % | 2 022 | 2 022 440 |
| $\omega\gamma$ | $2.75 \pm 0.22$ % | 256 | 256 300 |
| $\gamma\gamma$ | $2.22 \pm 0.08$ % | 206 | 206 904 |
| **Total** | 99.17 % | 9 240 | 9 242 644 |

The first column of Tab. 1 shows the decay branch, the second column gives the branching ratio, the third column represents the rate of the $\eta'$ per hour and the fourth column represents the number of $\eta'$ that can be produced in 1 000 hours of beam time. The sum of "rare" $\eta'$ decay modes starting with the $\pi^0\pi^0\eta$ contribute 26.7% to the total. Already with the current acceptance of the Crystal Barrel setup, which is ~60 % for the $\gamma\gamma$ and ~17 % for the $\pi^0\pi^0\eta$ decay mode, we expect 467 events/h.

The work reported here would not have been possible without the enduring enthusiasm of many bachelor, master, and PhD students and the continuous technical support from staff members of the participating Universities. The support of the ELSA staff members and the polarized target group is gratefully recognized. We would like to thank F. Afzal for contributions to this report. We appreciate the support of the Schweizerischer Nationalfonds, the US National Science Foundation, and the Russian Foundation for Basic Research, in particular the generous support from the Deutsche Forschungsgemeinschaft (SFB/TR16).

## References

[1] M. Unverzagt, PhD thesis HISKP, University Bonn (2008).
[2] M. Unverzagt et al., Eur. Phys. J. A **39**, 169 (2009).
[3] S. Prakhov et al., Phys. Rev. C **79**, 035204 (2009).
[4] S. P. Schneider, B. Kubis and C. Ditsche, JHEP **1102**, 028 (2011).
[5] K. Kampf, M. Knecht, J. Novotny and M. Zdrahal, Phys. Rev. D **84**, 114015 (2011).
[6] P. Guo, I. V. Danilkin, C. Fernández-Ramírez, V. Mathieu and A. P. Szczepaniak, arXiv:1608.01447 [hep-ph].
[7] G. Colangelo, S. Lanz, H. Leutwyler and E. Passemar, arXiv:1610.03494 [hep-ph].
[8] A. Nikolaev, PhD thesis HISKP, University Bonn (2011).
[9] A. Nikolaev et al., Eur. Phys. J. A **50**, 58 (2014).
[10] H. Herminghaus et al., Nucl. Inst. Meth. A **138**, 1 (1976).
[11] Th. Walcher, Prog. Part. Nucl. Phys. **24**, 189 (1990).
[12] A. Jankowiak, *Compilation of: Measurement of the electron beam energy at the Mainz Microtron*, MAMI Internal Report 03/2006, Institut für Kernphysik, Mainz (2006).
[13] A. Jankowiak, Eur. Phys. J. A **28**, 149 (2006).
[14] I. Anthony et al., Nucl. Inst. Meth. A **301**, 230 (1991).
[15] S. J. Hall et al., Nucl. Inst. Meth. A **368**, 698 (1996).
[16] A. Reiter et al., Eur. Phys. J. A **30**, 461 (2006).
[17] M. Oreglia et al., Phys. Rev. D **25**, 2259 (1982).
[18] A. Starostin et al., Phys. Rev. C **64**, 055205 (2001).
[19] D. Watts, *Calorimetry in Particle Physics: Proceedings of the 11th International Conference, Perugia, Italy, 2004*, edited by C. Cecchi, P. Lubrano, M. Pepe (World Scientific, Singapore), 560 (2005).
[20] R. Novotny, IEEE Trans. on Nucl. Sc. **38**, 392 (1991).
[21] P. Nuhn, Diploma thesis HISKP, University Bonn (2012).
[22] F. N. Afzal, Master thesis HISKP, University of Bonn (2012).
[23] F. N. Afzal, EPJ Web of Conferences **73**, 04005 (2014).
[24] V. Crede et al., Phys.Rev. **C80**, 055202 (2009).
[25] M. Dugger et al., Phys.Rev.Lett. **96**, 062001 (2006).
[26] F. N. Afzal, PhD thesis HISKP, University of Bonn (2017).
[27] F. Huang et al., arXiv:1208.2279v1.
[28] L. Tiator, Int.J.Mod.Phys. A **22**, 297 (2007).
[29] D. Elsner et al., Eur. Phys. J. A **33**, 147 (2007).
[30] O. Bartalini et al., Eur. Phys. J. A **33**, 169 (2007).
[31] A. V. Anisovich et al., arXiv:1503.05774 [nucl-ex].
[32] D. Rönchen et al., Eur. Phys. J. A **51**, 70 (2015).
[33] R. L. Workman, M. W. Paris, W. J. Briscoe, I. I. Strakovsky, Phys. Rev. C **86**, 015202 (2012).
[34] SAID: http://gwdac.phys.gwu.edu/
[35] E. Gutz et al., Eur. Phys. J. A **50**, 74 (2014).
[36] F. N. Afzal, JPS Conf. Proc. **10**, 032006 (2016).
[37] P. Levi Sandri et al., Eur.Phys.J. **A51**, no.7,77 (2015).
[38] M. Bargiotti et al., Phys. Rev. D **65**, 012001 (2002).
[39] M. Ablikim et al., Phys. Rev. Lett. **108**, 182001 (2012).
[40] C. Amsler et al., Phys. Rev. D **66**, 058101 (2002).
[41] C. Amsler et al., Phys. Lett. B **358**, 389 (1995).
[42] T. Triffterer, PhD thesis, University of Bochum (2016).
[43] C. Sowa, PhD thesis, University of Bochum (2016).
[44] R. Dickson, R. A. Schumacher et al., Phys. Rev. C **93**, 065202 (2016).

# Properties of flavour-singlet pseudoscalar mesons from lattice QCD

*Carsten* Urbach[1,*]

[1] *HISKP (Theorie), Universität Bonn, Nussallee 14-16, 53115 Bonn, Germany*

**Abstract.** We report on the status of the determination of properties of flavour-singlet pseudoscalar mesons using Wilson twisted mass lattice QCD at maximal twist. As part of project **C7**, a large number of phenomenologically relevant quantities could be extracted from first principle, from $\eta$ and $\eta'$ masses to decay widths of pseudoscalar mesons to two photons.

## 1 Introduction

In the early days of the theory of strong interactions, the peculiar mass hierarchy found experimentally for the nine light pseudoscalar mesons led to the formulation of the "$U(1)$ problem" [1]. If also the axial $U(1)$ symmetry was spontaneously broken in the SU(3) chiral limit of QCD, one would expect the ninth pseudoscalar meson to have a mass similar or a bit lighter than the kaons and the $\eta$ meson. The solution to this puzzle comes from the anomalous breaking of the corresponding axial current [2, 3]

$$\partial^\mu A^0_\mu = \partial^\mu(\bar{q}\gamma_\mu\gamma_5 q) = \frac{\alpha_s}{4\pi}G\tilde{G}$$

at the quantum level triggered by a non-trivial topological structure of QCD [4–6], which is known as the $U_A(1)$-*anomaly*. Here, $G$ denotes the SU(3) field strength tensor and $\tilde{G}$ its dual.

The connection between the $U_A(1)$-anomaly and the larger-than-expected mass of the $\eta'$ meson is given by the famous Witten-Veneziano formula [7, 8], which is derived assuming the anomaly to be in place and in the large $N_c$ limit of QCD:

$$\frac{f_0^2}{4N_f}(M_\eta^2 + M_{\eta'}^2 - 2M_K^2) = \chi_\infty, . \tag{1}$$

Here, $\chi_\infty$ is the topological susceptibility in the pure Yang-Mills (YM) gauge theory, $N_f$ the number of flavours, $f_0$ the flavour singlet decay constant and $M_\eta$, $M_{\eta'}$ and $M_K$ are the $\eta$, the $\eta'$ and the kaon masses, respectively. It relates non-trivial topological fluctuations of the gauge field in YM to the corresponding meson masses and singlet decay constant $f_0$ in QCD with $N_f$ active quark flavours. Using lattice gauge theory techniques we have the unique possibility to verify this relation.

In lattice QCD, the computation of $\eta$ and $\eta'$ meson related quantities represents a challenge because of so-called fermionic disconnected diagrams. These are usually very noisy and, therefore, the results available from lattice QCD were limited.

In the DFG funded project within the collaborative research centre CRC16 "Subnuclear Structre of Matter" we could increase the available lattice results in the flavour-singlet pseudoscalar sector significantly. This was possible on the one hand due to the lattice discretisation we are using: the Wilson twisted mass formulation at maximal twist is particularly well suited to study $\eta$ and $\eta'$ mesons because of a powerful noise reduction method we developed in Ref. [9]. On the other hand we have applied and developed other methods which allow us to extract the quantities of interest with less uncertainty. Finally, we could perform this study employing a wide range of pion mass values, several strange quark mass values, several volumes and three values of the lattice spacing. Therefore, continuum and chiral extrapolations could be performed in a controlled way.

The results presented in the remaining sections represent an summary, update and extension of what was published in Refs. [10–15].

## 2 Lattice Action

The lattice QCD action for $N_f = 2 + 1 + 1$ Wilson twisted mass fermions reads

$$S = S_g + \bar{\chi}_\ell D_\ell \chi_\ell + \bar{\chi}_h D_h \chi_h . \tag{2}$$

For the gauge action $S_g$ we use the Iwasaki gauge action [16]. The twisted mass Dirac operator for the light – i.e. up/down quark – doublet reads [17]

$$D_\ell = D_W + m_0 + i\mu_\ell\gamma_5\tau^3 \tag{3}$$

and for the strange/charm doublet [18]

$$D_h = D_W + m_0 + i\mu_\sigma\gamma_5\tau^1 + \mu_\delta\tau^3 , \tag{4}$$

where $D_W$ is the Wilson Dirac operator and the $\tau^i$ are the Pauli matrices acting in flavour space. The value of $m_0$ was

*e-mail: urbach@hiskp.uni-bonn.de

**Table 1.** compilation of values for the Sommer parameter $r_0/a$, the lattice spacing $a$ and $Z \equiv Z_P/Z_S$ determined with methods **M1,M2** the three values of the lattice spacing. See Ref. [28] for details.

| $\beta$ | $Z$ (**M1**) | $Z$ (**M2**) | $a$ [fm] | $r_0/a$ |
|---|---|---|---|---|
| 1.90 | 0.699(13) | 0.651(06) | 0.0885(36) | 5.31(8) |
| 1.95 | 0.697(07) | 0.666(04) | 0.0815(30) | 5.77(6) |
| 2.10 | 0.740(05) | 0.727(03) | 0.0619(18) | 7.60(8) |

tuned to its critical value $m_{\rm crit}$ as discussed in Refs. [19, 20] in order to realise automatic $O(a)$ improvement at maximal twist [21]. The bare twisted masses $\mu_\sigma$ and $\mu_\delta$ are related to the bare strange and charm quark masses via the relation

$$m_{c,s} = \mu_\sigma \pm Z\mu_\delta\,, \quad Z \equiv \frac{Z_P}{Z_S}\,, \tag{5}$$

where $Z$ denotes the ratio of non-singlet, pseudoscalar and scalar renormalisation constants $Z_P$ and $Z_S$. Quark fields in the twisted basis are denoted by $\chi_{\ell,h}$ and in the physical basis by $\psi_{\ell,h}$. They are related via the axial rotations

$$\begin{aligned} \psi_\ell &= e^{i\pi\gamma_5\tau^3/4}\chi_\ell\,, \quad \bar\psi_\ell = \bar\chi_\ell\, e^{i\pi\gamma_5\tau^3/4}\,, \\ \psi_h &= e^{i\pi\gamma_5\tau^1/4}\chi_h\,, \quad \bar\psi_h = \bar\chi_h\, e^{i\pi\gamma_5\tau^1/4}\,. \end{aligned} \tag{6}$$

With automatic $O(a)$ improvement being the biggest advantage of twisted mass lattice QCD (tmQCD) at maximal twist, the downside is that flavour (and parity) symmetry is broken at finite values of the lattice spacing. This was theoretically and numerically shown to affect mainly the mass value of the neutral pion [22–24]. For this reason we use masses of charged pions throughout this paper denoted as $M_{\rm PS}$. However, in the case of $N_f = 2+1+1$ dynamical quarks, the flavour symmetry breaking implies the complication of mixing between strange and charm quarks. Note that the kaons are still degenerate, though.

We use gauge configurations as produced by the European Twisted Mass Collaboration (ETMC) with action described above [20, 25, 26]. The details of the configurations are described in Ref. [20] and we follow the notation adopted there: we use ensembles denoted with $A$, $B$ and $D$ with values of the lattice spacing $a_A = 0.0885(36)$fm, $a_B = 0.0815(30)$fm and $a_D = 0.0619(18)$fm, corresponding to $\beta_A = 1.90$, $\beta_B = 1.95$ and $\beta_D = 2.10$, respectively [26]. The physical volumes are – with only a few exceptions – larger than 3 fm.

Throughout this proceeding we will use the Sommer parameter $r_0$ [27] to investigate the scaling of our results. The actual numbers are taken from Ref. [28] and compiled in Table 1. In physical units we use $r_0 = 0.474(14)$ fm taken from the same reference.

For fixing the light and strange quark masses to their physical values we will use the experimental values of $M_{\pi^0} = 135$ MeV and $M_{K^0} = 498$ MeV. We use the masses of the neutral mesons to reduce uncertainties from the fact that we do not include electromagnetic effects in our simulation. See also ref. [29] and references therein for a discussion.

For every $\beta$-value the bare values of $a\mu_\sigma$ and $a\mu_\delta$ were kept fixed, and only the value of the light twisted mass parameter $a\mu_\ell$ is varied. The kaon masses measured on these ensembles are close to the physical value with a deviation of up to 10% [20] in particular for the $A$ and $D$ ensembles. The D-meson mass values have a large uncertainty, but they are also close to physical [20].

For the $A$ lattice spacing we have two additional ensembles with re-tuned values of strange and charm-quark masses, denoted by $A80.24s$ and $A100.24s$, which reproduce the physical kaon mass value more accurately than the original $A80.24$ and $A100.24$ ensembles. We will use these ensembles to estimate the strange quark mass dependence of the $\eta$ mass and other quantities.

## 3 Flavour-Singlet Pseudoscalar Mesons from Lattice QCD

In order to compute masses of pseudoscalar flavour singlet mesons we have to include light, strange and charm contributions to build the appropriate correlation functions. In the light sector, one appropriate operator is given by [30]

$$\frac{1}{\sqrt{2}}(-\bar\chi_u\chi_u + \bar\chi_d\chi_d) \equiv O_\ell\,, \tag{7}$$

in the twisted basis. With twisted mass fermions we have to work with doublets of quarks, hence, in the strange and charm sector the corresponding operators read

$$\begin{pmatrix}\bar\chi_c \\ \bar\chi_s\end{pmatrix}^T \frac{-\tau^1 \pm iZ\gamma_5\tau^3}{2}\begin{pmatrix}\chi_c \\ \chi_s\end{pmatrix} \equiv O_{c,s}\,. \tag{8}$$

Note that the sum of pseudoscalar and scalar contributions appears with the ratio of renormalisation factors $Z$, which needs to be taken into account properly. The renormalisation factors for Eq. (8) are again the non-singlet $Z_P$ and $Z_S$. It is another advantage of the Wilson twisted mass formulation at maximal twist that only non-singlet renormalisation factors are required to obtain renormalised matrix elements. Defining $\vec{O} = (O_l, O_c, O_s)^T$, the correlation function matrix is given by

$$C(t) = \sum_{\mathbf{x}} \left\langle \vec{O}(x) \otimes \vec{O}(0) \right\rangle\,. \tag{9}$$

Scalar and pseudoscalar contributions can be disentangled by changing the operator basis via an appropriate orthogonal matrix $\mathcal{R}$

$$C_{\mathcal{R}}(t) = \left\langle \mathcal{R}\vec{O}(t) \otimes \mathcal{R}\vec{O}(0)\right\rangle = \mathcal{R}C(t)\mathcal{R}^T\,, \tag{10}$$

where

$$C_{\mathcal{R}} = \begin{pmatrix} \eta_{O_\ell O_\ell} & \eta_{O_\ell S_h} & \eta_{O_\ell \mathcal{P}_h} \\ \eta_{S_h O_\ell} & \eta_{S_h S_h} & \eta_{S_h \mathcal{P}_h} \\ \eta_{\mathcal{P}_h O_\ell} & \eta_{\mathcal{P}_h S_h} & \eta_{\mathcal{P}_h \mathcal{P}_h} \end{pmatrix}\,, \tag{11}$$

is again a symmetric, real and positive definite correlation matrix with

$$\begin{aligned} \mathcal{P}_h &\equiv (\bar\chi_c i\gamma_5\chi_c - \bar\chi_s i\gamma_5\chi_s)/2\,, \\ \mathcal{S}_h &\equiv Z^{-1}(\bar\chi_s\chi_c + \bar\chi_c\chi_s)/2\,, \end{aligned} \tag{12}$$

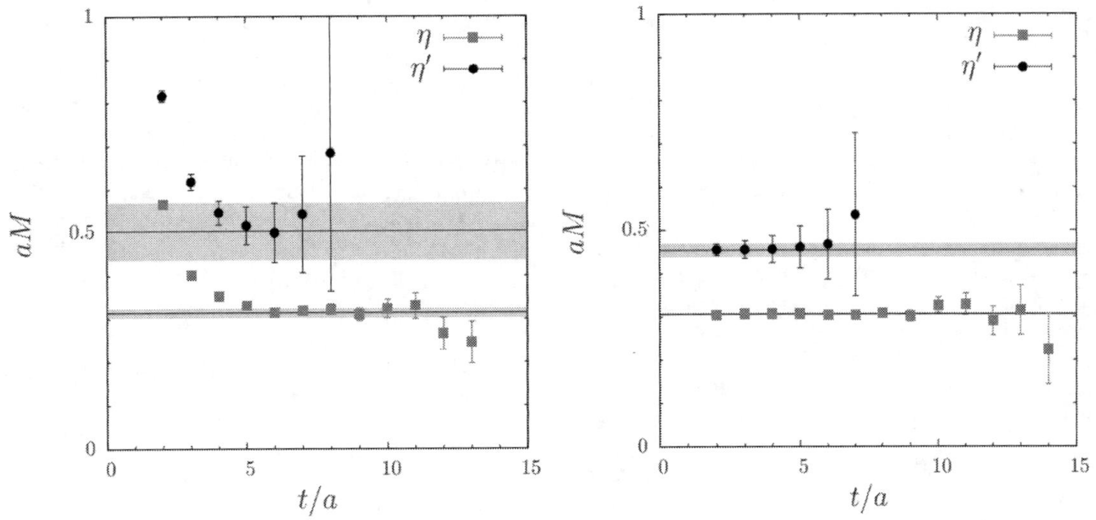


**Figure 1.** Left: Effective masses in lattice units determined from solving the GEVP for a $6 \times 6$ matrix with $t_0/a = 1$ for ensemble A100. Right: the same, but after removal of excited states in the connected contributions.

and $\eta_{XY}$ denoting the corresponding correlation function. The rotation matrix $\mathcal{R}$ is given by

$$\mathcal{R} = \begin{pmatrix} 1 & 0 & 0 \\ 0 & -\frac{1}{\sqrt{2}} & -\frac{1}{\sqrt{2}} \\ 0 & +\frac{1}{\sqrt{2}} & -\frac{1}{\sqrt{2}} \end{pmatrix}. \tag{13}$$

Now we can drop the factor $Z^{-1}$, which appears only as a constant scaling factor for $\mathcal{S}_h$. We denote the corresponding correlation matrix by $\tilde{C}_{\mathcal{R}}$. Solving the generalised eigenvalue problem (GEVP) [31–33]

$$\tilde{C}_{\mathcal{R}}(t)\,\eta^{(n)}(t,t_0) = \lambda^{(n)}(t,t_0)\,\tilde{C}_{\mathcal{R}}(t_0)\,\eta^{(n)}(t,t_0), \tag{14}$$

and taking into account the periodic boundary conditions for a meson, we can determine the effective masses $M_{\rm eff}^{(n)}$ by solving

$$\frac{\lambda^{(n)}(t,t_0)}{\lambda^{(n)}(t+1,t_0)} = \frac{e^{-M_{\rm eff}^{(n)}t} + e^{-M_{\rm eff}^{(n)}(T-t)}}{e^{-M_{\rm eff}^{(n)}(t+1)} + e^{-M_{\rm eff}^{(n)}(T-(t+1))}} \tag{15}$$

for $M_{\rm eff}^{(n)}$, where $n$ counts the eigenvalues. The state with the lowest mass should correspond to the $\eta$ and the second state to the $\eta'$ meson.

### 3.1 Excited State Subtraction

The result of solving the GEVP for a $6 \times 6$ matrix for ensemble A100 including fuzzed operators is shown as effective masses $M_{\rm eff}$ in the left panel of Figure 1. It is visible that the ground state, the $\eta$ meson, can be extracted with good precision, while for the $\eta'$ meson it is unclear that a plateau is reached before the signal is lost in noise.

However, there is a possibility to obtain a significant improvement for the extraction of $\eta'$ mass (and further observables) using a powerful method to separate ground and excited states, which has first been proposed in Ref. [34] and that has already been successfully employed for the case of the $\eta_2$ for two dynamical quark flavours in Ref. [30], and we have first used it for the $N_f = 2+1+1$ flavour case in Ref. [11].

The method is based on the assumption that the quark disconnected diagrams give a sizeable contribution only to the $\eta$ and $\eta'$ state, but are negligible for any heavier state with the same quantum numbers. Considering the fluctuations of the topological charge which are expected to give a dominant contribution to the mass of the $\eta'$, this assumption would be valid if these fluctuation mainly couple to the $\eta$ and $\eta'$ states. Still, the validity of this assumption needs to be carefully checked from our data and may introduce systematic uncertainties.

Since the quark connected contributions exhibit a constant signal-to-noise ratio, it is in principle possible to determine the respective ground states at sufficiently large $t/a$ with high statistical accuracy and without any significant contamination from higher states. After fitting the respective ground states of the connected correlators, we can use it to subtract the excited state contributions such that the full connected correlators are replaced by correlators that contain only the ground state. Note that for sufficiently large $t/a$ this reproduces the original ground state by construction.

Now, if the aforementioned assumption holds, i.e. the disconnected diagrams are relevant only to the two lowest states $\eta$, $\eta'$ one should obtain a plateau in the effective mass at very low values of $t/a$ after solving the GEVP. The result of the procedure is shown in the right panel of Figure 1. Indeed, one observes a plateau for both states starting basically at the lowest possible value of $t/a$. Furthermore, a comparison with the effective masses from the standard $6 \times 6$ matrix in the left panel of Figure 1 reveals that the plateau values agree very well within their respective errors. Most importantly, the data in the right panel allows for a much better accuracy in the determination of both masses as the point errors are much smaller at such low values of $t/a$. Therefore, we will use this method for all the results presented in this proceeding contribution.

However, we remark that in the twisted mass formulation with the non-degenerate doublet this procedure is in practice restricted to the connected correlation functions corresponding to physical light and strange quarks. This is due to the violation of flavour symmetry in the heavy sector of the twisted mass formulation, implying that the four connected contributions in the heavy sector will all yield the same ground state. This ground state corresponds to an artificial particle, i.e. a connected-only, neutral pion-like particle made out of strange quarks. Therefore, we will restrict ourselves in the following discussion to the analysis of a $2\times 2$ correlation function matrix corresponding to (local) physical operators made of light and strange quarks.

We remark that this procedure modifies correlators only at small $t/a$ and per construction the limit of large Euclidean times is not modified. The method only serves to reduce statistical uncertainties by extending the plateaus to smaller values of $t/a$.

### 3.2 Pollution by topological effects

It is a well known problem, in particular for small lattice spacing values, that in lattice QCD simulations the topological charge is properly sampled [35]. This can lead in extreme cases to a freeze of the topological charge. We found in our ensembles that the topological charge has zero expectation value, however, the second moment of the charge distribution might not be sampled fully correctly. While being a finite volume effect, this can lead to time independent contributions to the flavour-singlet correlation function [36, 37]. As a consequence, the corresponding correlation functions do not decay to zero at large time separations.

Such constants present in correlation functions can be subtracted by fitting the appropriate functional form to the data [37]. However, we found that shifting the correlation functions – invented to cancel so-called thermal pollution in multi-hadron correlation functions [38] – subtracts the constant without the need of any additional input. Moreover, shifting decorrelates neighbouring time slices significantly, which gives an additional advantage.

We define the shifted correlation matrix as

$$\tilde{C}^s_{\mathcal{R}}(t) = \tilde{C}_{\mathcal{R}}(t) - \tilde{C}_{\mathcal{R}}(t+\delta t)\,. \qquad (16)$$

While all time independent terms are cancelled in $\tilde{C}^s_{\mathcal{R}}(t)$, every matrix element possesses now the following functional form

$$\propto \sum_n \left[e^{-E_n(t+\delta t/2)} - e^{-E_n(T-(t+\delta t/2))}\right],$$

with $T$ the time extent of the lattice. The shift $\delta t$ can be adjusted for optimal results.

We did not observe obvious cases in our data where the correlation functions themselves do not decay to zero. Only on a few ensembles with small volume and lattice spacings (e.g. D45) such a residual behaviour was observed for the resulting principal correlator of the $\eta'$-state. However, in these cases the effect is of the size of a few standard deviations at most. Still, the shifting procedure turns out to systematically improve the signal for all and in particular for $\eta'$ related quantities significantly. Final results of this procedure are currently in preparation [39], and we will only show $\eta$ and $\eta'$ masses extracted with this procedure.

### 3.3 Decay Constants and Mixing

In general, decay constants are defined for any pseudoscalar meson P from axial vector matrix elements

$$\langle 0|A^a_\mu|\mathrm{P}(p)\rangle = i f^a_{\mathrm{P}} p_\mu\,, \qquad (17)$$

which leads to

$$\langle 0|\partial^\mu A^a_\mu|\mathrm{P}(0)\rangle = f^a_{\mathrm{P}} M^2_{\mathrm{P}}\,, \qquad (18)$$

for projection to zero momentum. Assuming that $\eta$ and $\eta'$ are not flavour eigenstates, each of them exhibits a coupling to the singlet and octet axial vector current $A^0_\mu$ and $A^8_\mu$, respectively. Therefore, one ends up with four independent decay constants for the $\eta$,$\eta'$-system, which are commonly parametrised in terms of two decay constants $f_0$, $f_8$ and two mixing angles $\theta_0$, $\theta_8$

$$\begin{pmatrix} f^8_\eta & f^0_\eta \\ f^8_{\eta'} & f^0_{\eta'} \end{pmatrix} = \begin{pmatrix} f_8\cos\theta_8 & -f_0\sin\theta_0 \\ f_8\sin\theta_8 & f_0\cos\theta_0 \end{pmatrix} \equiv \Xi(\theta_8,\theta_0)\,\mathrm{diag}(f_8, f_0)\,. \qquad (19)$$

The singlet decay constant $f_0$ needs renormalisation, determined by the anomalous dimension of the axial singlet current [40]. The dependence on the scale is $O(1/N_C)$ and can, therefore, be dropped in the following discussion. For a detailed discussion see Refs. [41, 42].

We work in the quark flavour basis, with the axial vector currents $A^0_\mu$ and $A^8_\mu$ replaced by the combinations

$$A^l_\mu = \frac{2}{\sqrt{3}}A^0_\mu + \sqrt{\frac{2}{3}}A^8_\mu = \frac{1}{\sqrt{2}}\left(\bar{u}\gamma_\mu\gamma_5 u + \bar{d}\gamma_\mu\gamma_5 d\right), \qquad (20)$$

$$A^s_\mu = \sqrt{\frac{2}{3}}A^0_\mu - \frac{2}{\sqrt{3}}A^8_\mu = \bar{s}\gamma_\mu\gamma_5 s\,, \qquad (21)$$

in which the light quarks and the strange quark contributions are disentangled. In exact analogy to the singlet-octet basis this basis again allows for a parametrisation in terms of two decay constants and two mixing angles

$$\begin{pmatrix} f^l_\eta & f^s_\eta \\ f^l_{\eta'} & f^s_{\eta'} \end{pmatrix} = \Xi(\phi_l,\phi_s)\,\mathrm{diag}(f_l, f_s)\,, \qquad (22)$$

where the mixing matrix $\Xi$ has the same form as the one defined in Eq. (19). In this basis it is also expected that we have [41–45]

$$\left|\frac{\phi_l - \phi_s}{\phi_l + \phi_s}\right| \ll 1\,, \qquad (23)$$

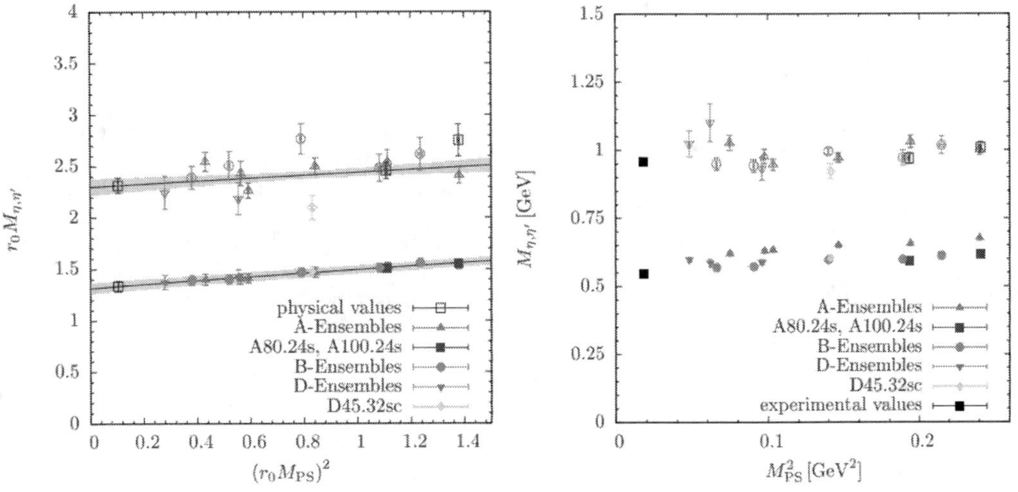


**Figure 2.** LEFT: Our results for $r_0\overline{M}_\eta$ (filled symbols) (corrected for the mismatch in $r_0M_K$) and $r_0M_{\eta'}$ (open symbols). The fitted curves are linear functions in $(r_0M_{PS})^2$ as discussed in the text. RIGHT: $M_\eta$ (filled) and $M_{\eta'}$ (open) in physical units (*not* corrected for the mismatch in $M_K$) determined from the shifted correlation matrix as a function of $M_{PS}^2$.

motivating a simplified mixing scheme in the quark flavour basis with only one angle $\phi$

$$\begin{pmatrix} f_\eta^l & f_\eta^s \\ f_{\eta'}^l & f_{\eta'}^s \end{pmatrix} = \Xi(\phi)\,\mathrm{diag}(f_l, f_s) + O(\Lambda_1)\,, \tag{24}$$

where $\Xi(\phi) \equiv \Xi(\phi, \phi)$ and $\Lambda_1$ parametrises residual OZI violating terms. The mixing angle $\phi$ is related to the double ratio of amplitudes

$$\tan^2(\phi) = -\frac{f_{\eta'}^l f_\eta^s}{f_\eta^l f_{\eta'}^s}\,, \tag{25}$$

and hence independent of the ratio of renormalisation constants $Z$. Axial vector current matrix elements turn out to be difficult to measure in actual simulations due to noise. This is why it is most convenient to consider pseudoscalar currents in the quark flavour basis in analogy to Eqs. (20), (21).

$$P^l = \frac{1}{\sqrt{2}}\left(\bar{u}\gamma_5 u + \bar{d}\gamma_5 d\right)\,, \tag{26}$$

$$P^s = \bar{s}i\gamma_5 s\,, \tag{27}$$

such that the matrix elements for pseudoscalar mesons P are given by

$$h_P^i = 2m_i \langle 0| P^i |P\rangle\,, \tag{28}$$

which are free from renormalisation. Making use of $\chi$PT and dropping sub-leading terms leads to the following expression [46]

$$\begin{pmatrix} h_\eta^l & h_\eta^s \\ h_{\eta'}^l & h_{\eta'}^s \end{pmatrix} = \Xi(\phi)\,\mathrm{diag}\left(M_\pi^2 f_l,\ \left(2M_K^2 - M_\pi^2\right) f_s\right)\,. \tag{29}$$

This expression allows access to the decay constants $f_s$ and $f_l$ from pseudoscalar matrix elements under the assumption that $\chi$PT can be applied. In terms of pseudoscalar matrix elements the mixing angle $\phi$ is obtained as

$$\tan^2(\phi) = -\frac{h_{\eta'}^l h_\eta^s}{h_\eta^l h_{\eta'}^s}\,, \tag{30}$$

where actually quark masses and renormalisation constants drop out in the ratio. Expanding again to two angles, $\phi_l$ and $\phi_s$ are written as

$$\tan(\phi_l) = \frac{h_{\eta'}^l}{h_\eta^l}\,, \qquad \tan(\phi_s) = -\frac{h_\eta^s}{h_{\eta'}^s}\,. \tag{31}$$

Finally, we remark that in order to compute the matrix elements $h_P^i$ in Wilson twisted mass lattice QCD the ratio of renormalisation constants $Z = Z_P/Z_S$ is required.

## 4 Results

### 4.1 Masses

We determine the masses $\tilde{C}_R(t)$ using the excited state removal method described in the previous section. It turns out that in $aM_\eta$ we see a strange quark mass dependence that we can resolve well within our statistical errors. As the physical values of the strange quark masses vary by about 10% between the different lattice spacing values, we have to correct for this. This is described in detail in Ref. [10]. We use the ensembles A80 and A80s (A100 and A100s), which differ in the bare strange quark mass value, to estimate $D_\eta = dM_\eta^2/dM_K^2$. Next we correct all $aM_\eta$ values to correspond to a line of $M_K[M_{PS}^2]$ values which reproduces the physical kaon mass at the physical pion mass point $M_{PS} = M_{\pi^0}^{\mathrm{phys}}$. The such corrected values we denote with $\overline{M}_\eta$. Note that for this procedure we ignore any dependence of $D_\eta$ on the quark masses and the lattice spacing. For the $\eta'$ mass we do not resolve quark mass or

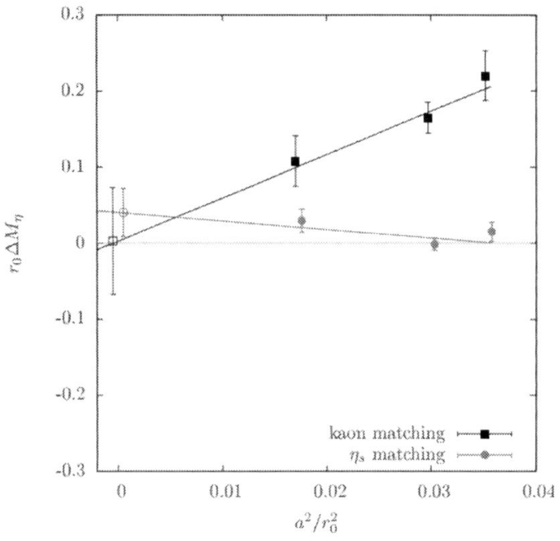


**Figure 3.** $r_0\Delta M_\eta$ as a function of $(a/r_0)^2$ for two matching procedures of OS valence strange quark to the unitary one [14].

lattice spacing dependence within our errors, so we do not attempt to correct for those.

The results for $\overline{M}_\eta$ and $M_{\eta'}$ are summarised in Figure 2, where we show $r_0\overline{M}_\eta$ as filled and $M_{\eta'}$ as open symbols, respectively, both as functions of $(r_0 M_{\mathrm{PS}})^2$. For both mesons the data fall on a single line within errors such that in both cases we model the data for $(r_0 M)^2$ inspired by chiral perturbation theory as constant plus a linear term in $(r_0 M_{\mathrm{PS}})^2$. The corresponding best fit to the data and its error band is shown as lines with shaded bands. The error band for $M_\eta$ is mainly due to the error of $D_\eta$.

After extrapolating to the physical point and converting to physical units we obtain

$$M_\eta(M_\pi) = 556(8)_{\mathrm{stat}}(5)_{\mathrm{sys}}\,\mathrm{MeV}\,, \qquad (32)$$

where the systematic error has been estimated from fitting to the data at each value of the lattice spacing separately. In addition, for the SU(2) chiral limit we find $r_0\mathring{M}_\eta^{\mathrm{SU(2)}} = 1.32(5)_{\mathrm{stat}}(1)_{\mathrm{sys}}$, which yields

$$\mathring{M}_\eta^{\mathrm{SU(2)}} = 548(9)_{\mathrm{stat}}(6)_{\mathrm{sys}}\,\mathrm{MeV}\,. \qquad (33)$$

We may extrapolate further quantities in order to check the validity of our correction procedure for mis-tuned values of the strange quark mass. First, we consider the GMO relation

$$3M_\eta^2 = 4M_{\mathrm{K}}^2 - M_\pi^2\,,$$

and build ratios left hand side over right hand side. This ratio is evaluated directly on our data. Next we perform an extrapolation in $(r_0 M_{\mathrm{PS}})^2$. However, it turns out that taking the uncorrected values of $M_\eta$ in the ratio the extrapolation misses the experimental value

$$(3M_\eta^2/(4M_{\mathrm{K}}^2 - M_\pi^2))^{\mathrm{exp}} = 0.926$$

considering only the statistical error by more than $2\sigma$, i.e. we obtain

$$\left(3M_\eta^2/(4M_{\mathrm{K}}^2 - M_\pi^2)\right)_{M_\pi} = 0.965(15)_{\mathrm{stat}}(24)_{\mathrm{sys}}\,.$$

This may be seen as a hint that the significantly increased statistical precision of the improved analysis strategy allows to resolve a residual strange quark mass dependence which is not cancelled in the ratio. This was not possible with the statistical accuracy we could obtain in Ref. [10]. Note that compared to the direct extrapolation the statistical precision is even further enhanced for the case of dimensionless ratios because the physical value of $r_0$ is only required for fixing the physical point, but not for the conversion to physical units. Considering only the $B$ ensembles for which the value of the strange quark mass is close to physical yields

$$\left(3M_\eta^2/(4M_{\mathrm{K}}^2 - M_\pi^2)\right)_{M_\pi}^{B} = 0.928(26)_{\mathrm{stat}}\,,$$

indicating that the systematic error is mainly caused by such a residual effect. Using the corrected values $\bar{M}_\eta$ and the corresponding values of the kaon mass to calculate the GMO ratio the extrapolation gives

$$\left(3M_\eta^2/(4M_{\mathrm{K}}^2 - M_\pi^2)\right)_{M_\pi} = 0.960(19)_{\mathrm{stat}}(22)_{\mathrm{sys}}\,,$$

which is slightly closer to the experimental value. For the physical value of the $\eta$ mass we obtain

$$M_\eta = 558(6)_{\mathrm{stat}}(7)_{\mathrm{sys}}\,\mathrm{MeV}\,, \qquad (34)$$

in agreement with the result from the direct extrapolation and the experimental value.

A similar picture arises from the light quark mass extrapolation of the ratio $M_\eta/M_K$. Using the uncorrected values of $M_\eta$ which yields

$$\left(M_\eta/M_{\mathrm{K}}\right)_{M_\pi} = 1.117(8)_{\mathrm{stat}}(23)_{\mathrm{sys}}\,,$$

missing the experimental value

$$\left(M_\eta/M_{\mathrm{K}}\right)^{\mathrm{exp}} = 1.100$$

again by roughly $2\sigma$ if taking only the statistical error into account. Like for the case of the GMO ratio taking only the $B$ ensembles into account gives nice agreement with the experimental value, i.e.

$$\left(M_\eta/M_{\mathrm{K}}\right)_{M_\pi}^{B} = 1.093(14)_{\mathrm{stat}}\,,$$

whereas $A$ and $D$ ensembles give significantly larger values. Therefore, we have also repeated the extrapolation using the corrected values $\bar{M}_\eta$ and the corresponding kaon masses. Within the statistical errors one obtains excellent agreement with experiment, i.e.

$$\left(M_\eta/M_{\mathrm{K}}\right)_{M_\pi} = 1.109(11)_{\mathrm{stat}}(09)_{\mathrm{sys}}\,,$$

which results in

$$M_\eta = 0.552(6)_{\mathrm{stat}}(4)_{\mathrm{sys}}\,\mathrm{MeV}\,, \qquad (35)$$

compatible with the results from direct and GMO ratio extrapolations.

In order to obtain our final result for the physical mass of the $\eta$ meson we take the weighted average from the

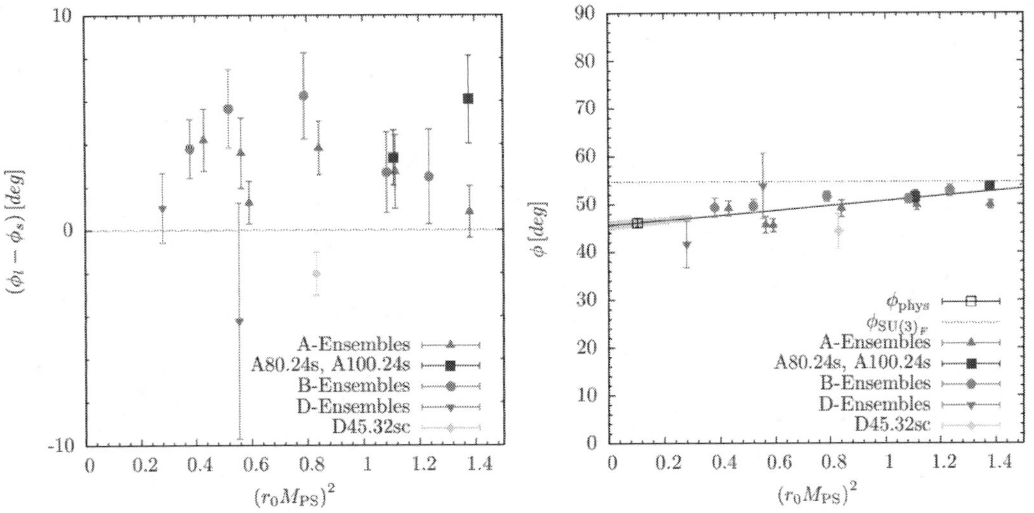


**Figure 4.** The difference $\phi_l - \phi_s$ (LEFT) and $\phi$ (RIGHT) in degrees as functions of $(r_0 M_{PS})^2$. For $\phi$ we also show our best fit to the data and the extrapolation to the physical pion mass value. The dotted horizontal line represents the value 54.2° one expects for $\phi$ in the SU(3) flavour symmetric case.

three previously discussed methods used for the extrapolation in the light quark mass. Accounting for any correlations, this yields

$$M_\eta = 555(6)_{\text{stat}}(5)_{\text{sys}} . \quad (36)$$

which is in excellent agreement with experiment and exhibits substantially smaller errors compared to Ref. [10].

For the $\eta'$ meson we obtain

$$M_{\eta'} = 964(41)_{\text{stat}}(34)_{\text{sys}} , \quad (37)$$

from a linear extrapolation in $(r_0 M_{PS})^2$ of all the (uncorrected) data. The corresponding fit is shown in the left panel of Figure 2. Again, the systematic error has been determined from fits to the data at single values of the lattice spacing. Within the larger errors this result is again in very good agreement with experiment, confirming that QCD indeed accounts for the significantly larger mass of the $\eta'$ that is observed experimentally.

In the right panel of Figure 2 we show $M_{\eta,\eta'}$ again as a function of $M_{PS}^2$, but now $M_\eta$ and $M_{\eta'}$ have been determined using the shifting procedure explained in section 3.2. The data is not corrected for the mismatch in the strange quark mass, and, therefore, the plot in the right panel is not directly comparable to the one in the left pane. But comparing it qualitatively to the left panel in the same figure shows that the shifting procedure greatly reduces the uncertainty in $M_{\eta'}$, possibly to a level where strange quark and lattice spacing effects must be taken into account. A corresponding analysis is in progress [39].

#### *4.1.1 Mixed Action*

It is an interesting question whether or not quantities with a significant contribution of fermionic disconnected diagrams can be reliably determined in a so-called mixed action approach. In such an approach discretisation of valence and sea quark are different. Prime examples of quantities with large disconnected contribution are $\eta$ and $\eta'$ meson masses.

We have investigated this question using the so-called Osterwalder-Seiler (OS) discretisation [47] for the valence strange quark combined with unitary up/down quarks on top of the Wilson twisted mass gauge configurations. The additional advantage of this mixed action approach is the avoided flavour mixing in the strange/charm sector. The conclusion of this investigation [14] is that indeed $\eta$ and $\eta'$ masses can be determined reliably in this mixed action approach. But we observed that lattice artefacts strongly depend on the chosen matching quantity. As an example we show in Figure 3 the difference between the OS and the unitary $\eta$ meson mass

$$\Delta M_\eta = M_\eta^{\text{OS}} - M_\eta^{\text{unitary}}$$

as a function of $(a/r_0)^2$. Results for two matching quantities are shown: one where the valence strange quark mass was fixed by matching the masses of the OS and the unitary $\eta_s$ meson. For the second the OS kaon was matched to the unitary for fixing the strange quark mass. As visible in Figure 3, kaon matching has significantly larger lattice artefacts, while they appear to be negligible for $\eta_s$ matching. Performing a continuum extrapolation of both matchings, however, gives identical results within error bars.

### 4.2 Mixing Angles

The mixing angles $\phi_l$ and $\phi_s$ can be extracted using Eq. (31) with a mixing model based on the pseudoscalar matrix elements $h_P^q$. In the left panel of Figure 4 we show $\phi_l - \phi_s$ in degrees as a function of $(r_0 M_{PS})^2$. One observes that this quantity is consistent with zero within errors. Also an extrapolation to the physical point yields

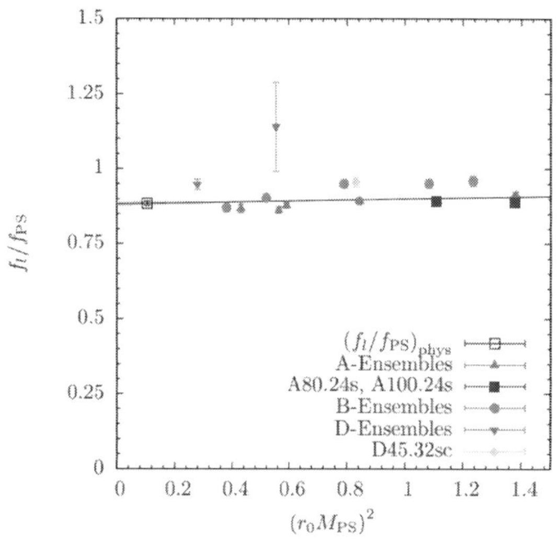


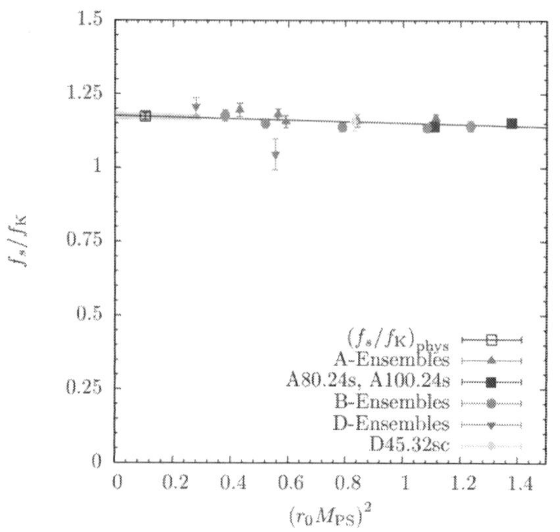


**Figure 5.** $f_l/f_{PS}$ (Left) and $f_l/f_K$ (Right) as functions of $(r_0M_{PS})^2$. The lines represent fits to our data with error band as explained in the text.

$3(1)_{\text{stat}}(3)^{\circ}_{\text{sys}}$, where the systematic error is estimated from the maximal difference compared to extrapolating the data sets for the three different lattice spacings separately.

In the right panel of Figure 4 we show the average angle $\phi$ (Eq. (30)) in degrees as a function of $(r_0M_{PS})^2$, with smaller statistical errors than $\phi_l$ and $\phi_s$ separately, because of correlation in the matrix elements. Our precision is not sufficient to resolve any residual lattice spacing or strange quark mass dependence. Hence, we extrapolate linearly in $(r_0M_{PS})^2$ and obtain

$$\phi = 46(1)_{\text{stat}}(3)^{\circ}_{\text{sys}}\,, \tag{38}$$

where the first error is statistical and the second systematic from fitting the three values of the lattice spacing separately.

Besides the mixing angle $\phi$, we consider the angles $\phi_l$, $\phi_s$ which are relevant to cross-check the assumptions entering our mixing scheme. Again, we have performed linear fits in $(r_oM_{PS})^2$ and obtain at the physical value of the pion mass

$$\phi_l = 48(1)_{\text{stat}}(4)^{\circ}_{\text{sys}}\,, \qquad \phi_s = 44(1)_{\text{stat}}(3)^{\circ}_{\text{sys}}\,, \tag{39}$$

where the systematic uncertainties have been determined in the same way as for the angle $\phi$ itself. The results are compatible within errors. Notably, for $\phi_s$ there is very good agreement for the results within each of the two set ($A80.24$, $A80.24s$) and ($A100.24$, $A100.24s$), indicating that the influence of the strange quark is smaller for this quantity and in general more of the data points lie within the error band of the linear fit.

### 4.3 Decay Constants

The application of the excited state removal method discussed in the previous section allows to extract the decay constants $f_l$ and $f_s$ to a rather high statistical precision by means of the pseudoscalar matrix elements $h^i_P$ and using Eq. (29). Of course, one needs to keep in mind that this is based on the assumption that the underlying chiral perturbation theory analysis is valid. Currently, we cannot estimate a corresponding systematic uncertainty.

In Figure 5 we show $f_l/f_{PS}$ and $f_s/f_K$ as functions of $(r_0M_{PS})^2$ in the left and right panel, respectively. We have chosen to plot these ratios because it appears that most of the quark mass and lattice spacing dependence cancels. A linear extrapolation of $f_l/f_{PS}$ in $(r_0M_{PS})^2$ to the physical point results in

$$f_l/f_\pi = 0.884(7)_{\text{stat}}(63)_{\text{sys}}\,. \tag{40}$$

However, from the plot it appears that there is still a rather sizeable dependence on the lattice spacing present while the strange quark mass dependence seems to cancel in the ratio as all $A$-ensembles fall on one single curve. In fact, the systematic error estimated from fitting the data at each value of the lattice spacing separately is one order of magnitude larger than the statistical error and there is clear trend towards larger values of $f_l/f_{PS}$ extrapolated to the physical pion mass for decreasing values of the lattice spacing. Therefore, we additionally quote the result of a linear fit restricted to the data at the finest lattice spacing, which yields

$$(f_l/f_\pi)^D = 0.947(23)_{\text{stat}}\,. \tag{41}$$

For the ratio $f_s/f_K$ most of the strange quark mass dependence is cancelled and the data seem almost perfectly linear in the light quark mass, exhibiting only a moderate slope. Moreover, for this case there are no discernible scaling artefacts within errors and the data are well described by a linear fit which gives

$$f_s/f_K = 1.176(12)_{\text{stat}}(28)_{\text{sys}} \tag{42}$$

at the physical value of the pion mass. Clearly, the systematic error is significantly smaller than the one obtained

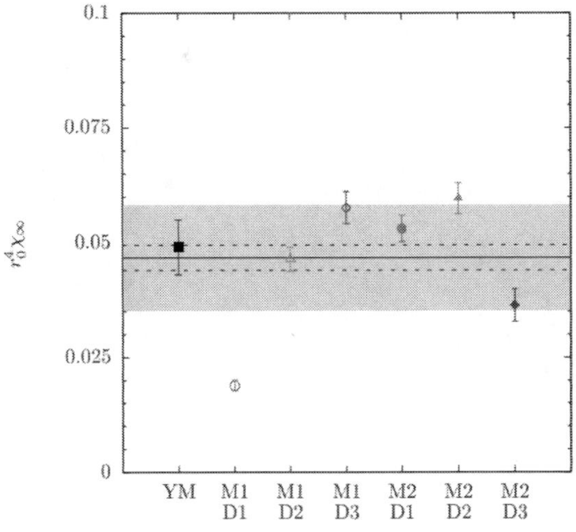


**Figure 6.** A comparison of the YM topological susceptibility $\chi_\infty$ to the right hand side of the Witten-Veneziano formula. The latter is determined using two different renormalisation procedures (**M1, M2**) and three different ways to estimate $f_0$ (**D1, D2, D3**).

for the physical value of $f_l/f_\pi$, confirming the smallness of any residual lattice artefacts or strange quark mass dependence for $f_s/f_K$.

Finally we have to remark that finite volume effects might play an important role for the decay constants, but they hopefully cancel in the ratios we used.

### 4.4 Witten-Veneziano Formula

Now we have all ingredients at hand to investigate the Witten-Veneziano formula in more detail. As we have shown in Ref. [15], one can relate the decay constants $f_\ell$ and $f_s$ to the singlet decay constant $f_0$ and the kaon and pion decay constants $f_K$ and $f_\pi$ using $\chi$PT again

$$f_0^2 = -7/6f_\pi^2 + 2/3f_K^2 + 3/2f_\ell^2\,, \quad (43)$$
$$f_0^2 = +1/3f_\pi^2 - 4/3f_K^2 + f_\ell^2 + f_s^2\,, \quad (44)$$
$$f_0^2 = +3f_s^2 - \frac{8}{3}(2f_K^2 - f_\pi^2)\,. \quad (45)$$

respectively. We denote these three formulae for $f_0$ with **D1, D2, D3**. $f_0$ is not uniquely determined because of higher order terms that have been dropped, see Ref. [15] for details. These three equations together with the methods **M1** and **M2** to determine the renormalisation constants [28], lead to six determinations of the left hand side of Eq. (1), which are compared to the YM susceptibility in Figure 6. We have determined the susceptibility $\chi_\infty$ in a dedicated simulation of the YM theory with three values of the lattice spacing applying the so-called spectral projector method [48].

This comparison is only meaningful up to higher order contributions in $1/N_c$. Therefore, one cannot expect exact agreement. However, qualitatively the observed agreement is very good.

### 4.5 Decay Widths $\Gamma_{P\to\gamma\gamma}$

The decay constants $f_l$ and $f_s$ are important low energy constants. However, they can also be used to estimate phenomenologically interesting quantities, most prominently the decay widths of $\eta, \eta' \to \gamma\gamma$. To the same order in the effective theory we worked until here, one can relate the decays widths with the mixing parameters in the quark flavour basis as follows [50, 51] (see Refs. [41, 42] for how to include these quantities into the effective field theory framework)

$$\Gamma_{\eta\to\gamma\gamma} = \frac{\alpha^2}{288\pi^3}M_\eta^3\left[\frac{5}{f_l}\cos\phi - \frac{\sqrt{2}}{f_s}\sin\phi\right]^2,$$
$$\Gamma_{\eta'\to\gamma\gamma} = \frac{\alpha^2}{288\pi^3}M_{\eta'}^3\left[\frac{5}{f_l}\sin\phi + \frac{\sqrt{2}}{f_s}\cos\phi\right]^2, \quad (46)$$

where again OZI suppressed terms have been dropped and our normalisation is that $f_\pi = 130.7$ MeV. Using our lattice data for $M_{\eta,\eta'}$, $\phi$, $f_l$ and $f_s$, we have computed the decay widths and show them in units of the Sommer parameter as a function of $(r_0 M_{\rm PS})^2$ in Figure 7. We also include the PDG value for convenience [49]. For $\Gamma_{\eta'\to\gamma\gamma}$ we observe a flat dependence of the light quark mass and agreement between our data and the PDG value within our errors. However, the data show a tendency for lower values with decreasing values of the lattice spacing.

The situation is not so clear for $\Gamma_{\eta\to\gamma\gamma}$, where the lattice data is of the right magnitude compared to the PDG value. But there are clearly strange quark mass and light quark mass effects visible that we cannot control at the moment. In addition there might be lattice artefacts.

Two remarks are in order: first of all the mass dependence of the widths has not been computed in effective field theory. Hence, we have no rigorous means to extrapolate our data to the physical point. Eqs. (46) are strictly speaking only valid in the chiral limit. Second, the PDG value for $\Gamma_{\eta\to\gamma\gamma}$ does not include Primakoff experiments, which give a significantly smaller value[1].

Related quantities are the pseudoscalar transition form factors $F_{\eta\gamma\gamma^*}$ and $F_{\eta'\gamma\gamma^*}$ for large momentum transfer $q^2$, which can be expressed as [52]

$$\lim_{q^2\to\infty} q^2 F_{\eta\gamma\gamma^*}(q^2) = \frac{10}{3\sqrt{2}}f_l\cos\phi - \frac{2}{3}f_s\sin\phi$$
$$= 100(5)_{\rm stat}(25)_{\rm sys}\,\text{MeV}\,,$$
$$\lim_{q^2\to\infty} q^2 F_{\eta'\gamma\gamma^*}(q^2) = \frac{10}{3\sqrt{2}}f_l\sin\phi + \frac{2}{3}f_s\cos\phi$$
$$= 277(3)_{\rm stat}(20)_{\rm sys}\,\text{MeV}\,. \quad (47)$$

These results have to be understood to be very preliminary. They are obtained by using the values of $f_l/f_\pi$ and $f_s/f_K$ in Eqs. (41) and (42), respectively, together with the physical values of the pion and kaon decay constants $f_\pi = 130.42$ MeV and $f_K = 156.2$ MeV[53]. The value for $\phi$ has been taken from Eq. (38). The systematic uncertainty is again calculated from the maximal difference to results from the separate lattice spacing values.

[1] We thank P. Masjuan for drawing our attention to this fact.

DOI: 10.1051/epjconf/201713404004

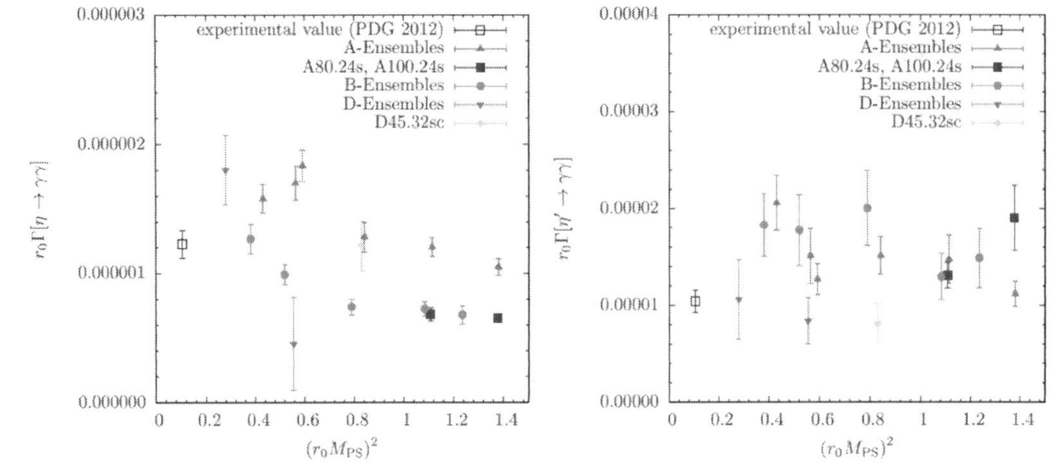


**Figure 7.** We show the decays widths $\Gamma_{\eta\to\gamma\gamma}$ (Left) and $\Gamma_{\eta'\to\gamma\gamma}$ (Right) in units of $r_0$ as functions of $(r_0 M_{PS})^2$. In addition we show the corresponding estimate from the PDG [49].

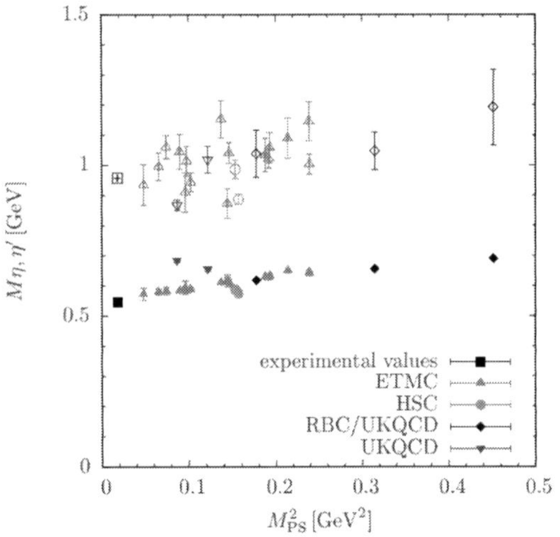


**Figure 8.** Results for $M_\eta$ and $M_{\eta'}$ in physical units as a function of $M^2_{PS}$ compared to results from other lattice QCD computations (HSC: Ref. [54], RBC/UKQCD: Ref. [55] and UKQCD Ref. [56]). Open symbols represent the $\eta'$ and filled symbols the $\eta$ mass, respectively.

## 5 Summary

In this proceeding we have summarised the results obtained for properties of flavour-singlet pseudoscalar mesons from lattice QCD as part of the collaborative research centre CRC16.

In Figure 8 we show a compilation of our results for $\eta$ and $\eta'$ masses together with results available in the literature for $N_f = 2 + 1$ flavour lattice QCD. For $M_\eta$ we show the values corrected for the mismatch in $M_K$. We remark that in Ref. [57] $\eta$ and $\eta'$ meson masses have been computed using $N_f = 2 + 1$ flavours of overlap quarks at one value of the lattice spacing and large values of the pion mass, however, in this reference not enough details are given to be included in our comparison Figure 8. The results in Ref. [55] have been obtained using $N_f = 2 + 1$ flavours of domain wall fermions and again for a single value of the lattice spacing $a \approx 0.1$ fm but for three values of the pion mass in a range from $\sim$ 400 MeV to $\sim$ 700 MeV. The corresponding data points in Figure 2 are labelled "RBC/UK QCD". Another single data point is added from Ref. [54] by the Hadron Spectrum Collaboration (HSC) for which Wilson fermions have been employed. Again, it was not possible to include more recent results by the HSC [58] due to the lack of explicit numerical values in this reference for the relevant masses. Finally, in Ref. [56] data from staggered fermions are presented for two different values of the lattice spacing with each of them also at a different value of the pion mass. In Figure 8 the corresponding data points are labelled "UKQCD". The figure suggests an overall agreement between all collaborations.

In Table 2 we compare our results for various quantities to experimental results from the PDG [53] and to phenomenological results from the so-called FKS-scheme taken from Ref. [46] (see also Ref. [52]). We observe mostly quite good agreement with the exception of $f_\ell/f_\pi$. This exception is likely a systematic uncertainty coming from the fact that in this proceeding we did not yet include lattice artefacts in the analysis of $f_\ell/f_\pi$. A corresponding analysis is ongoing [39]. Note that in general the spread in the phenomenological determinations is substantial.

As discussed in Ref. [52], this determination of mixing parameters can be used to better estimate the hadronic light-by-light contribution to the anomalous magnetic moment of the muon. Of course, eventually a computation of the corresponding transition form factors is desired as recently performed for the neutral pion in Refs. [59, 60].

Finally, we have investigated the Witten-Veneziano formula by computing all for the formula relevant quantities non-perturbatively. The results indicate clearly that the axial anomaly in QCD is able to explain the large mass of the $\eta'$ meson (see also Ref. [61]).

DOI: 10.1051/epjconf/201713404004

**Table 2.** Summary of results. The literature values are taken from Ref. [53] for $M_\eta$ and $M_{\eta'}$ and from Ref. [46] from the so-called FKS scheme for the others.

| | this work | literature |
|---|---|---|
| $M_\eta$ [MeV] | $555(6)_{\text{stat}}(5)_{\text{sys}}$ | 547.86(2) |
| $M_{\eta'}$ [MeV] | $964(41)_{\text{stat}}(34)_{\text{sys}}$ | 957.78(6) |
| $f_\ell/f_\pi$ | $0.884(7)_{\text{stat}}(63)_{\text{sys}}$ | 1.07(2) |
| $f_s/f_K$ | $1.176(12)_{\text{stat}}(28)_{\text{sys}}$ | 1.11(5) |
| $\phi$ | $46(1)_{\text{stat}}(3)^{\circ}_{\text{sys}}$ | $39(1)^{\circ}$ |

The author wishes to thank K. Ottnad and C. Michael for their essential contributions to this project. Thanks also to Falk Zimmermann for his work on the mixed action approach. The work reported here would not have been possible without the support from the Deutsche Forschungsgemeinschaft within the SFB/TR16. We thank the members of ETMC for the most enjoyable collaboration. The computer time for this project was made available to us by the John von Neumann-Institute for Computing (NIC) on the JUDGE, Jugene and Juqueen systems in Jülich. Support by the Bonn-Cologne Graduate School (BCGS) of Physics and Astronomie is acknowledged. The open source software packages tmLQCD [62–64], Lemon [65] and R [66] have been used.

## References

[1] S. Weinberg, Phys.Rev. **D11**, 3583 (1975)
[2] S.L. Adler, Phys. Rev. **177**, 2426 (1969)
[3] J.S. Bell, R. Jackiw, Nuovo Cim. **A60**, 47 (1969)
[4] J.B. Kogut, L. Susskind, Phys. Rev. **D11**, 3594 (1975)
[5] A.A. Belavin, A.M. Polyakov, A.S. Schwartz, Yu.S. Tyupkin, Phys. Lett. **B59**, 85 (1975)
[6] G. 't Hooft, Phys. Rev. Lett. **37**, 8 (1976)
[7] E. Witten, Nucl.Phys. **B156**, 269 (1979)
[8] G. Veneziano, Nucl.Phys. **B159**, 213 (1979)
[9] C. Michael, C. Urbach (ETM), PoS **LAT2007**, 122 (2007), 0709.4564
[10] K. Ottnad, C. Michael, S. Reker, C. Urbach, C. Michael, S. Reker, C. Urbach (ETM), JHEP **11**, 048 (2012), 1206.6719
[11] C. Michael, K. Ottnad, C. Urbach (ETM), Phys. Rev. Lett. **111**, 181602 (2013), 1310.1207
[12] C. Michael, K. Ottnad, C. Urbach (ETM), PoS **LATTICE2013**, 253 (2014), 1311.5490
[13] K. Ottnad, Ph.D. thesis, Bonn U. (2013), hss.ulb.uni-bonn.de/2014/3506/3506.htm
[14] K. Ottnad, C. Urbach, F. Zimmermann (OTM), Nucl. Phys. **B896**, 470 (2015), 1501.02645
[15] K. Cichy, E. Garcia-Ramos, K. Jansen, K. Ottnad, C. Urbach (ETM), JHEP **09**, 020 (2015), 1504.07954
[16] Y. Iwasaki, Nucl. Phys. **B258**, 141 (1985)
[17] R. Frezzotti, P.A. Grassi, S. Sint, P. Weisz (ALPHA), JHEP **08**, 058 (2001), hep-lat/0101001
[18] R. Frezzotti, G.C. Rossi, Nucl. Phys. Proc. Suppl. **128**, 193 (2004), hep-lat/0311008
[19] T. Chiarappa et al., Eur. Phys. J. **C50**, 373 (2007), hep-lat/0606011
[20] R. Baron et al. (ETM), JHEP **06**, 111 (2010), 1004.5284
[21] R. Frezzotti, G.C. Rossi, JHEP **08**, 007 (2004), hep-lat/0306014
[22] R. Baron et al. (ETM), JHEP **1008**, 097 (2010), 0911.5061
[23] C. Urbach (ETM), PoS **LAT2007**, 022 (2007), 0710.1517
[24] P. Dimopoulos, R. Frezzotti, C. Michael, G. Rossi, C. Urbach, Phys.Rev. **D81**, 034509 (2010), 0908.0451
[25] R. Baron et al. (ETM), Comput.Phys.Commun. **182**, 299 (2011), 1005.2042
[26] R. Baron et al., PoS **LATTICE2010**, 123 (2010), 1101.0518
[27] R. Sommer, Nucl.Phys. **B411**, 839 (1994), hep-lat/9310022
[28] N. Carrasco et al. (ETM), Nucl. Phys. **B887**, 19 (2014), 1403.4504
[29] G. Colangelo, S. Durr, A. Juttner, L. Lellouch, H. Leutwyler et al., Eur.Phys.J. **C71**, 1695 (2011), 1011.4408
[30] K. Jansen, C. Michael, C. Urbach (ETM), Eur.Phys.J. **C58**, 261 (2008), 0804.3871
[31] C. Michael, I. Teasdale, Nucl.Phys. **B215**, 433 (1983)
[32] M. Lüscher, U. Wolff, Nucl.Phys. **B339**, 222 (1990)
[33] B. Blossier, M. Della Morte, G. von Hippel, T. Mendes, R. Sommer, JHEP **0904**, 094 (2009), 0902.1265
[34] H. Neff, N. Eicker, T. Lippert, J.W. Negele, K. Schilling, Phys.Rev. **D64**, 114509 (2001), hep-lat/0106016
[35] S. Schaefer, R. Sommer, F. Virotta (ALPHA), Nucl. Phys. **B845**, 93 (2011), 1009.5228
[36] S. Aoki, H. Fukaya, S. Hashimoto, T. Onogi, Phys. Rev. **D76**, 054508 (2007), 0707.0396
[37] G.S. Bali, S. Collins, S. D?rr, I. Kanamori, Phys. Rev. **D91**, 014503 (2015), 1406.5449
[38] X. Feng, K. Jansen, D.B. Renner, Phys.Lett. **B684**, 268 (2010), 0909.3255
[39] K. Ottnad, C. Urbach (2016), in preparation
[40] J. Kodaira, Nucl.Phys. **B165**, 129 (1980)
[41] R. Kaiser, H. Leutwyler (1998), hep-ph/9806336
[42] R. Kaiser, H. Leutwyler, Eur.Phys.J. **C17**, 623 (2000), hep-ph/0007101
[43] J. Schechter, A. Subbaraman, H. Weigel, Phys.Rev. **D48**, 339 (1993), hep-ph/9211239
[44] T. Feldmann, P. Kroll, B. Stech, Phys.Lett. **B449**, 339 (1999), hep-ph/9812269
[45] T. Feldmann, P. Kroll, B. Stech, Phys.Rev. **D58**, 114006 (1998), hep-ph/9802409

[46] T. Feldmann, Int.J.Mod.Phys. **A15**, 159 (2000), hep-ph/9907491
[47] R. Frezzotti, G.C. Rossi, JHEP **10**, 070 (2004), hep-lat/0407002
[48] L. Giusti, M. Luscher, JHEP **03**, 013 (2009), 0812.3638
[49] J. Beringer et al. (Particle Data Group), Phys.Rev. **D86**, 010001 (2012)
[50] T. Feldmann, P. Kroll, Eur.Phys.J. **C5**, 327 (1998), hep-ph/9711231
[51] T. Feldmann, P. Kroll, Phys.Scripta **T99**, 13 (2002), hep-ph/0201044
[52] R. Escribano, P. Masjuan, P. Sanchez-Puertas (2013), 1307.2061
[53] K.A. Olive et al. (Particle Data Group), Chin. Phys. **C38**, 090001 (2014)
[54] J.J. Dudek, R.G. Edwards, B. Joo, M.J. Peardon, D.G. Richards et al., Phys.Rev. **D83**, 111502 (2011), 1102.4299
[55] N. Christ, C. Dawson, T. Izubuchi, C. Jung, Q. Liu et al., Phys.Rev.Lett. **105**, 241601 (2010), 1002.2999
[56] E.B. Gregory, A.C. Irving, C.M. Richards, C. McNeile (UKQCD), Phys.Rev. **D86**, 014504 (2012), 1112.4384
[57] T. Kaneko et al. (TWQCD/JLQCD), PoS **LAT2009**, 107 (2009), 0910.4648
[58] J.J. Dudek, R.G. Edwards, P. Guo, C.E. Thomas (2013), 1309.2608
[59] X. Feng, S. Aoki, H. Fukaya, S. Hashimoto, T. Kaneko et al., Phys.Rev.Lett. **109**, 182001 (2012), 1206.1375
[60] A. Gérardin, H.B. Meyer, A. Nyffeler (2016), 1607.08174
[61] M. Cè, M.G. Vera, L. Giusti, S. Schaefer (2016), 1607.05939
[62] K. Jansen, C. Urbach, Comput.Phys.Commun. **180**, 2717 (2009), 0905.3331
[63] A. Abdel-Rehim, F. Burger, A. Deuzeman, K. Jansen, B. Kostrzewa, L. Scorzato, C. Urbach, PoS **LATTICE2013**, 414 (2014), 1311.5495
[64] A. Deuzeman, K. Jansen, B. Kostrzewa, C. Urbach, PoS **LATTICE2013**, 416 (2014), 1311.4521
[65] A. Deuzeman, S. Reker, C. Urbach, Comput.Phys.Commun. **183**, 1321 (2012), 1106.4177
[66] R Development Core Team, *R: A language and environment for statistical computing*, R Foundation for Statistical Computing, Vienna, Austria (2005), ISBN 3-900051-07-0, http://www.R-project.org

EPJ Web of Conferences **134**, 04005 (2017) DOI: 10.1051/epjconf/201713404005
*Subnuclear Structure of Matter: Achievements and Challenges*

# Chiral dynamics of/with unstable particles

*Evgeny* Epelbaum[1,*], *Hermann* Krebs[1,**] and *Ulf-G.* Meißner[2,3,***]

[1] *Institut für Theoretische Physik II, Ruhr-Universität Bochum, D-44780 Bochum, Germany*
[2] *Universität Bonn, Helmholtz Institut für Strahlen- und Kernphysik, Bethe Center for Theoretical Physics, D-53115 Bonn, Germany*
[3] *Forschungszentrum Jülich, Institute für Kernphysik (IKP-3), Institute for Advanced Simulation (IAS-4) and Jülich Center for Hadron Physics, D-52145 Jülich, Germany*

**Abstract.** We review the achievements of the project **C.9**. Topics addressed include the magnetic moment of the $\rho$-meson, elastic pion-nucleon scattering in various settings of chiral perturbation theory, the reaction $\pi N \to \pi\pi N$ as well as new directions in chiral perturbation theory.

## 1 Introduction

The project **C.9** has been funded over the third period of the SFB/TR 16. It grew out of the project **C.3**, that was funded in the first period. Within this project, we have carried out detailed investigations of pion-nucleon scattering using various formulations of baryon chiral perturbation theory (ChPT) and addressed the role of the $\Delta(1232)$ isobar. Furthermore, we have explored the possibility of extending the applicability of ChPT beyond the low-energy region, verified the consistency of the complex-mass renormalization scheme at the two-loop level and performed a number of applications in the two-nucleon sector. This project has been quite successful which is partly due to the intelligent use of the lump sum funds.

Our contribution is organized as follows. In section 2, we present the calculation of the pion-mass dependence of the magnetic moment of the $\rho$-meson. Sections 3 and 4 are devoted to elastic $\pi N$ scattering and the reaction $\pi N \to \pi\pi N$, respectively. Next, in section 5, we discuss new directions in ChPT focusing, in particular, on the applicability of the complex-mass renormalization scheme beyond the one-loop level and a possibility to extend the ChPT framework beyond the low-energy region. Finally, section 6 describes the applications in the nucleon-nucleon (NN) sector. Here, we discuss the electromagnetic form factors of the deuteron, a non-perturbative inclusion of the subleading contact interaction in the $^1S_0$ channel within a renormalizable framework and new fixed points of the renormalization-group equation.

## 2 The magnetic moment of the $\rho$-meson

In Ref. [1], we have analyzed the pion mass dependence of the magnetic moment of the $\rho$-meson at the leading one-loop order. To this aim, we have employed the most general chiral effective Lagrangian for $\rho$- and $\omega$-mesons, pions and external sources

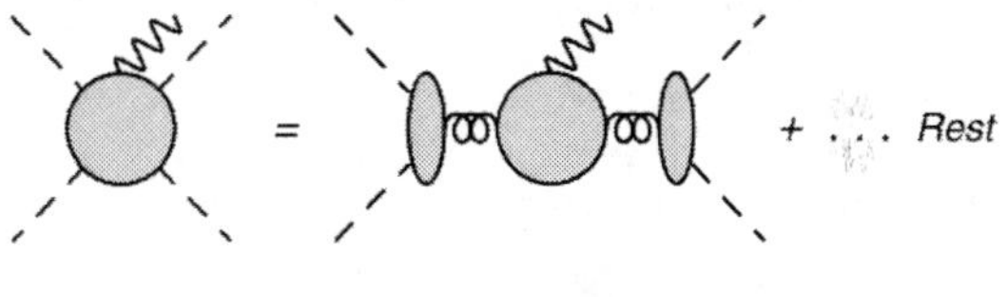


**Figure 1.** The reaction $\pi\pi \to \gamma\pi\pi$. Dashed, wavy and wiggled lines refer to pions, photons and $\rho$-mesons, respectively. "Rest" denotes the non-resonant contributions.

$$\mathcal{L} = \mathcal{L}_\pi + \mathcal{L}_{\rho\pi} + \mathcal{L}_\omega + \mathcal{L}_{\omega\rho\pi}\,, \qquad (1)$$

where the individual terms are taken in the parametrization of the model III of Ref. [2]. For a general review of effective chiral Lagrangians with vector mesons, see Ref. [3]. We made use of the universality of the vector-meson couplings, which was shown in Ref. [4] to follow from the consistency of an effective field theory (EFT) with respect to perturbative renormalization. Furthermore, the coupling constant $g_{\rho\pi\pi}$ obeys the KSFR relation $M_\rho^2 = 2g_{\rho\pi\pi}^2 F_\pi^2$, with $F_\pi$ denoting the pion decay constant.

The magnetic moment of the $\rho$-meson can be extracted from the resonant part of the five-point function in the reaction $\pi\pi \to \gamma\pi\pi$ as visualized in Fig. 1. Assuming that the interaction terms with a higher number of derivatives and/or more fields are suppressed by powers of some large hadronic scale, we have performed a one-loop calculation in terms of the expansion parameter $\xi = g_{\rho\pi\pi}^2/(4\pi)^2 \simeq 0.2$. At tree level, the magnetic moment is found to be real, and the gyromagnetic ratio is $g_\rho = 2$. One-loop diagrams yield corrections of the order of 10% and lead to a small imaginary part of $g_\rho$. The resulting pion mass dependence of $g_\rho$ is shown in Fig. 2. At the physical point, we found $\mathrm{Re}(g_\rho) \simeq 2.2$ in good agreement with the most recent lattice QCD calculation of Ref. [5], $(g_\rho)_{\mathrm{unquenched}} = 2.4(2)$.

*e-mail: evgeny.epelbaum@rub.de
**e-mail: hermann.krebs@rub.de
***e-mail: meissner@hiskp.uni-bonn.de

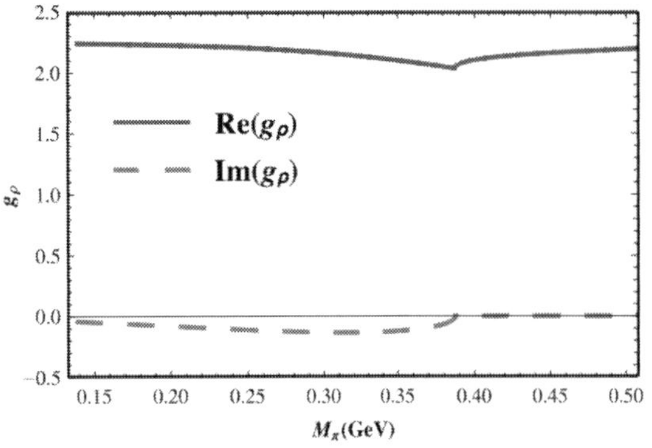


**Figure 2.** The gyromagnetic ratio $g_\rho$ of the $\rho$-meson as a function of the pion mass.

Earlier lattice QCD studies have reported the values of $(g_\rho)_{\rm quenched} \sim 2.3$ [6], $(g_\rho)_{\rm unquenched} = 1.6(1)$ [7].

## 3 Elastic pion-nucleon scattering

Elastic pion-nucleon ($\pi N$) scattering certainly belongs to the the most extensively studied processes in ChPT [8–16]. The $\pi N$ scattering amplitude enters as a subprocess into the interactions between two and three nucleons at large distances. Thus, its detailed understanding is a necessary pre-requisite for deriving high-precision nuclear forces in chiral EFT [14, 17, 18].

In our paper [19], we have analyzed $\pi N$ scattering up to fourth order $Q^4$ within different formulations of ChPT. Specifically, we employed the heavy-baryon (HB) formulations utilizing the standard counting of the nucleon mass as $m_N \sim \Lambda_b$ (HB-$\pi$N) with $\Lambda_b$ denoting the breakdown scale of the chiral expansion, and the counting scheme with $m_N \sim \Lambda_b^2/Q$ (HB-NN) employed in few-nucleon studies. We have also performed calculations within a covariant (Cov) formulation of ChPT based on an extended on-mass-shell (EOMS) renormalization scheme [20–22]. Our analysis differs in several aspects from the already mentioned earlier studied of this topic. First, the determination of the low-energy constants (LECs) was carried out using directly the available $\pi N$ scattering data instead of having to rely on partial wave analyses (PWA), see also Ref. [23] for a similar approach. Secondly, we have benefited from the developments in the project **B.6** by employing a novel approach to estimate the theoretical uncertainty from the truncation of the chiral expansion formulated in Ref. [17] in order to stabilize the fits with respect to the energy range. Finally, we have discussed in detail the renormalization and finite shifts of the various LECs within the covariant formulations and studied the role of the $\Delta$(1232) isobar by treating it as an explicit degree of freedom at lowest order in the chiral expansion.

The main results of our study can be summarized as follows. In the $\Delta$-less case, all three considered approaches are found to lead to a comparable description of low-energy $\pi N$ data, and the extracted values of the LECs compare well with those reported in the literature at the cor-

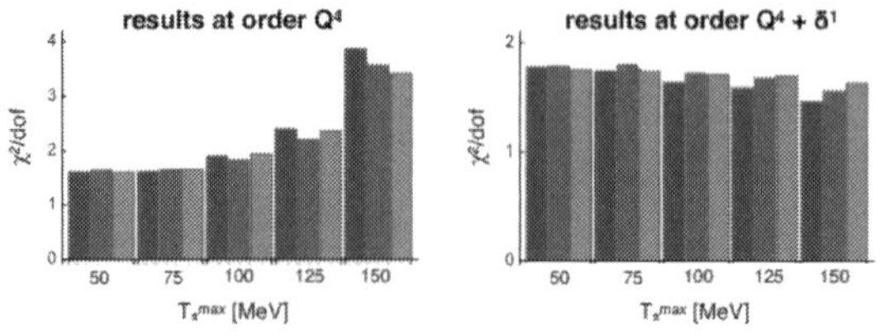


**Figure 3.** Reduced $\chi^2$ for fits up to various values of the maximal energy $T_\pi^{\rm max}$ in the $\Delta$-less (left panel) and $\Delta$-full (right panel) formulations of ChPT. The blue/red/green bars denote the results for the HB-NN/HB-$\pi$N/Cov counting. The theoretical uncertainty is estimated using the approach of Ref. [17] and assuming $\Lambda_b \sim 600$ MeV.

responding chiral orders. While the inclusion of the theoretical uncertainties in the definition of the $\chi^2$ helps to stabilize the fits by decreasing the relative weight of high-energy data, we found clear indications of the lower break-down scale of the chiral expansion than the assumed one of $\Lambda_b \sim 600$ MeV [17]. This conclusion is supported by the increasing behavior of the reduced $\chi^2$ with energy as visualized in the left panel of Fig. 3. It is also reflected in the extracted values of the LECs, which show sizable sensitivity on the energy range used in the fits, see the upper panel of Fig. 4, and to the chiral order. Furthermore, some of the LECs take rather large numerical values. For example, at order $Q^4$ in the covariant approach, we found the values of

$$c_1 \simeq -0.82, \quad c_2 \simeq -3.56, \quad c_3 \simeq -4.59, \quad c_4 \simeq 3.44$$

for the order-$Q^2$ LECs in units of GeV$^{-1}$ and

$$d_{1+2} \simeq 5.43, \quad d_3 \simeq -4.58, \quad d_4 \simeq -0.40, \quad d_{14-15} \simeq -9.94$$

for the relevant order-$Q^3$ LECs in units of GeV$^{-2}$.

Clearly, one expects the convergence of the chiral expansion for $\pi N$ scattering to be slow due to the appearance of the $\Delta$(1232) resonance. The situation can be improved by treating the $\Delta$-isobar as an explicit degree of freedom [24–26]. To quantify the effects of the $\Delta$-isobar, we have repeated our analysis including its lowest-order contributions. The only new parameters, namely the

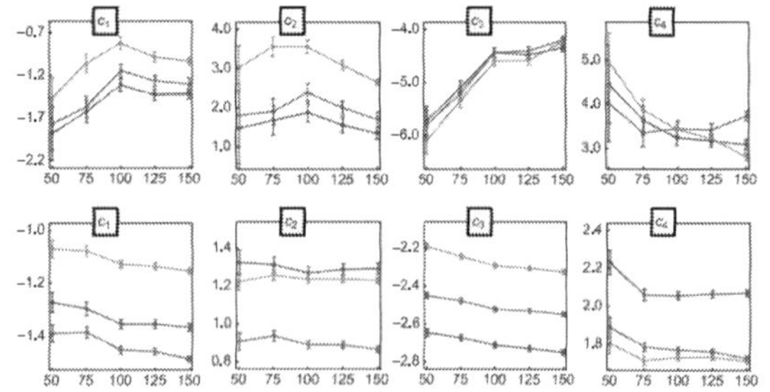


**Figure 4.** The dependence of the order-$Q^2$ $\pi N$ LECs $c_i$, given in units of GeV$^{-1}$, on the maximal incoming pion energy $T_\pi$ used in the fits at order $Q^4$ in the $\Delta$-less approach (upper panel) and at order $Q^4 + \delta^1$ in the $\Delta$-full approach (lower panel). The blue/red/green points and lines denote the results of the HB-NN/HB-$\pi$N/Cov formulations.

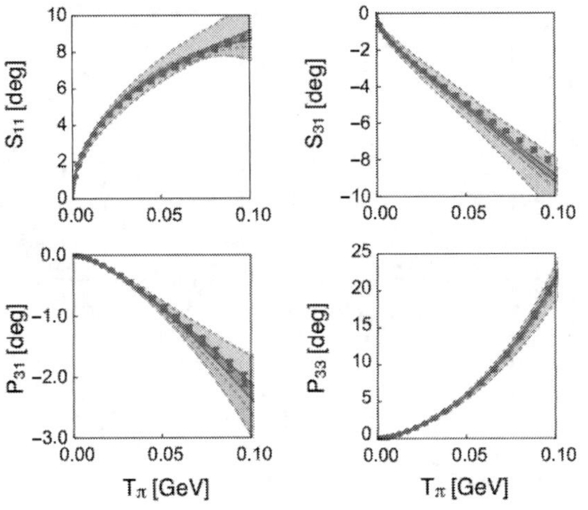


**Figure 5.** Predictions for selected S- and P-wave phase shifts up to $T_\pi = 100$ MeV in the covariant Δ-full formulation of chiral EFT. Orange dotted, pink dashed and red solid bands refer to $Q^2 + \delta^1$, $Q^3 + \delta^1$ and $Q^4 + \delta^1$ results, respectively. The truncation errors are estimated using the approach of Ref. [17] and assuming $\Lambda_b \sim 600$ MeV. Blue points show the results of the RS equation analysis of Ref. [27].

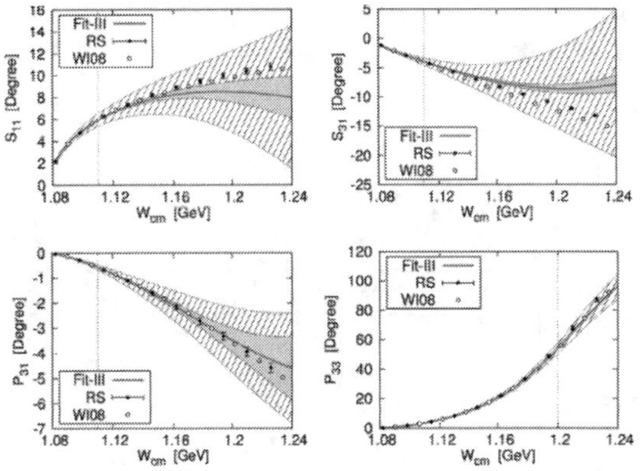


**Figure 6.** Phase shifts obtained from the Δ-full covariant ChPT at leading one-loop order. Dots with error bars stand for the RS phase shifts [27] while circles without error bars represent the phase shifts from the George Washington partial wave analysis [32]. The solid (red) line represents the result of Fit III of Ref. [30]. The red narrow error bands correspond to the uncertainties propagated from the errors of LECs while the wide dashed error bands show truncation uncertainties at order $Q^3$ estimated using the approach of Ref. [17].

mass of the Δ and the $\pi N \Delta$ coupling constant, were fixed to its Breit-Wigner and the large-$N_c$ values, respectively: $m_\Delta = 1.232$ GeV, $g_{\pi N \Delta} = 3/(2\sqrt{2}) g_A = 1.35$. While the quality of the reproduction of the $\pi N$ scattering data well below the Δ-region is not significantly impacted by the explicit inclusion of the Δ-isobar, we indeed found an improved convergence of the resulting EFT. The corresponding breakdown scale is consistent with or even slightly larger than the estimated $\Lambda_b \sim 600$ MeV, see the right panel of Fig. 3. Not surprisingly, the Δ-full formulation allows for a more reliable determination of the LECs, which turn out to be rather stable with respect to choosing the fitting range, see Fig. 4. We also found all LECs in the Δ-full approach to be of a natural size. In particular, at order $Q^4 + \delta^1$ in the covariant approach, we obtained the values

$$c_1 \simeq -1.15, \quad c_2 \simeq -1.57, \quad c_3 \simeq -2.54, \quad c_4 \simeq 2.61$$

for the order-$Q^2$ LECs and

$$d_{1+2} \simeq 1.27, \quad d_3 \simeq -1.83, \quad d_4 \simeq 0.37, \quad d_{14-15} \simeq -2.22$$

for the relevant order-$Q^3$ $\pi N$ LECs, to be compared with the values in the Δ-less approach listed above.

Last but not least, the resulting phase shifts are found to be in a very good agreement with the ones determined within the project **B.3** in the framework of Roy-Steiner (RS) equations, where detailed error estimates of all input quantities, the solution procedure and truncations was performed [27–29]. As a representative example, we show in in Fig. 5 our predictions for selected S- and P-waves along with the estimated truncation errors in comparison with the results from Ref. [27].

While the above results look rather promising and indicate the importance of the explicit treatment of the Δ-isobar, its inclusion beyond the tree level in the covariant framework is a rather challenging task due to a large complexity of the calculations emerging from the Δ propagators. In Ref. [30], we have performed a *complete* analysis at the leading one-loop order (i.e. $Q^3$) in the covariant chiral EFT with explicit Δ degrees of freedom. In this work, all possible nucleon- and Δ-propagator contributions have been calculated up to the leading one-loop order. To achieve this goal, we applied the EOMS (complex-mass [31]) scheme to diagrams involving pion and nucleon lines (delta lines) and determined the corresponding renormalized LECs by performing fits to the phase shifts obtained in the RS equation analysis of Ref. [27] up to the center of mass energy of 1.11 GeV (1.2 GeV for the $P_{33}$ partial wave). In Fig. 6, we show our results for the selected partial waves along with the theoretical uncertainties.

We have also looked at the $\pi N$ sigma term but were unable to obtain an accurate prediction for this quantity. Our results are, however, consistent with the value found in the RS analysis, $\sigma_{\pi N} = 59.1 \pm 3.5$ MeV within the large errors of our determination. We have also extended these studies to the SU(3) sector by calculating the masses and the corresponding sigma terms of the baryon octet and studying the strangeness content of the nucleon.

## 4 The reaction $\pi N \to \pi\pi N$

In addition to the elastic channel of $\pi N$ scattering, we have studied in Ref. [33] single pion production off nucleons from threshold up to the delta resonance region, extending earlier ChPT studies from Refs. [34–37]. With three pions in the initial and final states, the theoretical study of this process provides complimentary information to elastic scattering and offers access to certain LECs which are not probed in the reaction $\pi N \to \pi N$. In particular, the

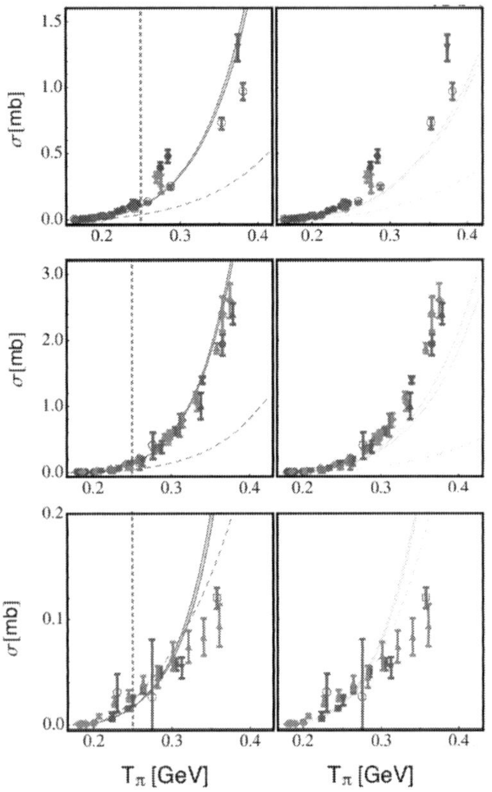


**Figure 7.** Predictions for the total cross section up to $T_\pi \simeq 400$ MeV in the covariant Δ-full (left panel) and Δ-less (right panel) versions of chiral EFT. Dashed and solid lines refer to LO and NLO results, respectively. The maximal energy used in the fit at NLO is marked by the vertical dotted line. The bands at NLO reflect the uncertainty from using different partial wave analyses in the determination of the LECs $c_i$ as explained in the text. Experimental data are from Refs. [45–47].

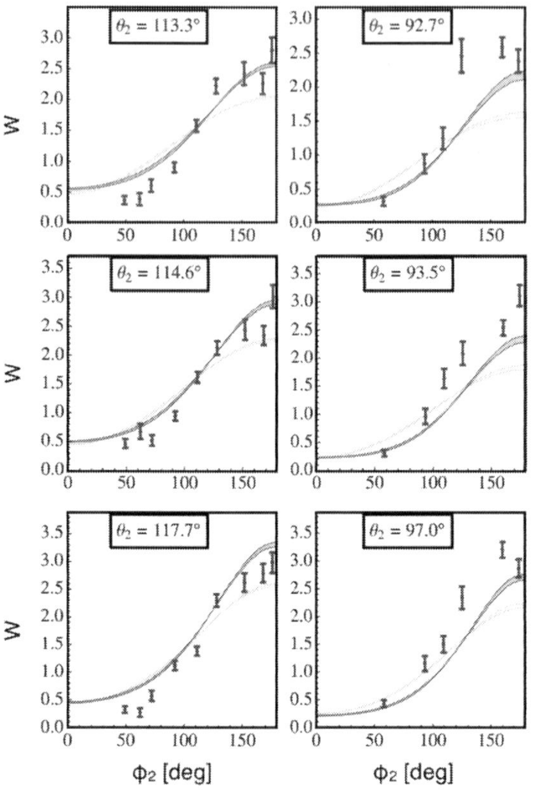


**Figure 8.** NLO predictions for the angular correlation functions in the $\pi^- p \to \pi^+\pi^- n$ channel at fixed values of $\theta_2$ and $\theta_1 = 91°$ (upper panel), $\theta_1 = 76°$ (middle panel) and $\theta_1 = 71°$ (lower panel) for $\sqrt{s} = 1301$ MeV in the Δ-full (red lines) and Δ-less (cyan lines) covariant chiral EFT approaches. The bands at NLO reflect the uncertainty from using different partial wave analyses in the determination of the LECs $c_i$ as explained in the text. Experimental data are from Ref. [48].

LEC $d_{16}$ is of great interest as it governs the quark mass dependence of the axial charge of the nucleon. If fact, the lack of knowledge of the precise value of $d_{16}$ is one of the main sources of theoretical uncertainty in chiral extrapolations of nuclear observables [38–44]. Notice further that a good theoretical understanding of this process is expected to provide useful constraints for multi-channel PWA of $\gamma$- and $\pi$-induced reactions off the nucleon, which is the main task of the project **A.2**.

In our exploratory study, we have restricted ourselves to calculating the leading (LO), i.e. $Q$, and subleading-order (NLO), i.e. $Q^2$ contributions to the scattering amplitude which emerge from tree-level diagrams. We have used both the HB and Cov formulations of baryon chiral EFT with and without explicit Δ degrees of freedom. Using the large-$N_c$ value for the $\pi N\Delta$ coupling constant, the results at LO come as parameter free predictions. At NLO, several LECs have to be determined. For the LECs $c_i$ from $\mathcal{L}^{(2)}_{\pi N}$, we have adopted the values extracted from two different PWA of elastic $\pi N$ scattering. Thus, our Δ-less results at NLO are parameter free, too. In the HB Δ-full approach, one is left with three unknown LECs, namely $b_4$, $b_5$ from $\mathcal{L}^{(2)}_{\pi N\Delta}$ and $g_1$ from $\mathcal{L}^{(1)}_{\pi\Delta}$. In the covariant framework, one, in addition, encounters the contributions from the LECs $b_3$ and $b_6$. These LECs have been determined from the available total cross section data in the reactions $\pi^- p \to \pi^0\pi^0 n$, $\pi^- p \to \pi^+\pi^- n$, $\pi^+ p \to \pi^+\pi^+ n$, $\pi^+ p \to \pi^+\pi^0 p$ and $\pi^- p \to \pi^0\pi^- p$, see Fig. 7 for representative examples. The extracted values of $g_1$ are found to be consistent with its large-$N_c$ value of $g_1 = 2.27$. We also found strong anticorrelations between the LECs $b_4$ and $b_5$ as well as $b_3$ and $b_6$, which prevented a reliable determination of the linear combinations.

With all LECs being determined as described above, we have looked at various differential observables. As a representative example, we show in Fig. 8 our predictions for the angular correlation function $W$, see Ref. [33] for a precise definition. Here $\theta_{1,2}$ denote the angles between the momentum of the incoming and the two outgoing pions, while $\phi_2$ is the azimuthal angle of one of the outgoing pi-

 DOI: 10.1051/epjconf/201713404005

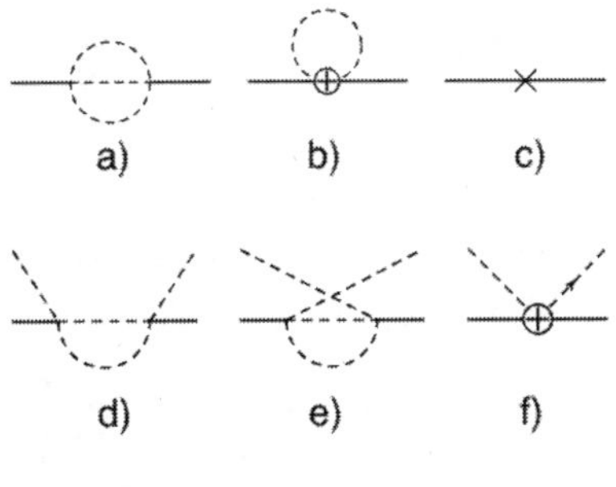


**Figure 9.** Diagrams a)-c): The two-point function at the two-loop order. Solid and dashed lines refer to the heavy and light particles $\Psi$ and $\pi$, respectively. The vertices in graphs b) and c) denote one-loop- and two-loop-order counterterms. Diagrams d)-f): one-loop graphs contributing to $\Psi\pi$ scattering.

ons[1], see Ref. [33] for more details. Generally, the explicit treatment of the delta isobar is found to result in a better description of the data, most notably of the $\pi^- p \to \pi^0\pi^0 n$ and $\pi^- p \to \pi^+\pi^- n$ total cross sections at higher energies and most of the differential cross sections. Still, certain single- and double-differential cross sections could not be properly described at this order in the chiral expansion. This calls for extending these calculations to next-higher order. Work along this line is in progress.

# 5 New directions in chiral perturbation theory

## 5.1 Complex-mass renormalization scheme at the two-loop level

ChPT can be straightforwardly extended to include meson and baryon resonances as dynamical degrees of freedom for *soft* processes, i.e. for reactions with external pion momenta and three-momenta of the nucleons of the order of $M_\pi$, see e.g. the already discussed applications of $\Delta$-full formulations of ChPT. On the other hand, formulating a consistent EFT at higher energy to study e.g. the properties of (axial-) vector mesons is a challenging task, see [1, 49, 50] and references therein for the first steps along this line using the complex-mass renormalization scheme [31]. It is, however, not a priori clear that this scheme is applicable beyond the one-loop level, i.e. that the renormalized expressions fulfill the power counting.

To clarify this issue, we have studied in Ref. [51] the two-point function of a heavy scalar $\Psi$-meson, which can decay into three light scalar fields $\pi$, at the two-loop level, see diagrams (a-c) in Fig. 9. The restriction to a simple model has allowed us to avoid complications due to the spin and chiral structure of the EFT which are not relevant for the discussion of the applicability of the complex-mass scheme. By explicitly calculating the diagrams a)-c) in Fig. 9, with the one-loop counterterm being determined from $\pi\Psi$ scattering graphs d)-f), and by employing the complex-mass scheme, we were able to explicitly

[1]The coordinate frame is defined by choosing the $z$-axis in the direction of the incoming pion and letting one of the outgoing pions to lie in the $xz$-plane.

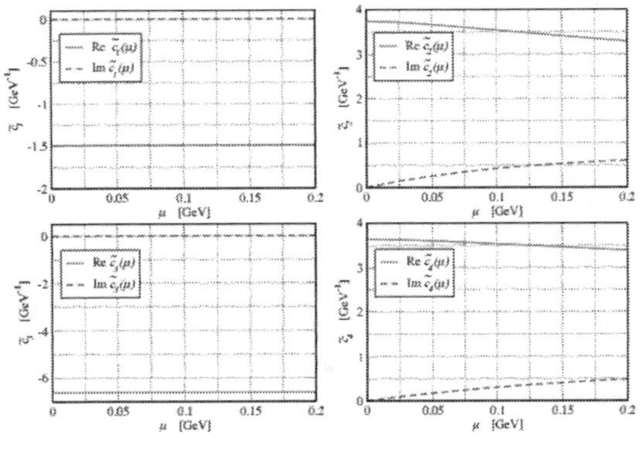

**Figure 10.** The renormalized LECs $c_i$ in the generalized EOMS scheme as functions of the sliding scale $\mu$.

show the absence of power-counting-breaking terms in the final renormalized expressions for the self-energy. Our work opens the way for a systematic analysis of e.g. the $\omega$-meson in chiral EFT.

## 5.2 Baryon chiral perturbation theory beyond the low-energy region

In Ref. [52], we have formulated a new approach to baryon ChPT which is applicable beyond the low-energy region at small scattering angles and applied it to the reaction $\pi N \to \pi N$. Specifically, instead of considering the kinematical region of $t = O(Q^2)$ and $\nu \equiv (s-u)/(4m_N) = O(Q)$ with $s$, $t$ and $u$ being the Mandelstam variables as it is usually the case in ChPT, we have demonstrated that the amplitude can be systematically calculated in the region of $t = O(Q^2)$ and $\nu^2 - \mu^2 = O(Q)$ even for $\nu = O(1)$. In this kinematical region, despite the higher energies, the quark structure of hadrons is still not resolved.

To achieve this goal, we have re-arranged the effective chiral Lagrangian according to a new power counting and exploited the freedom in the choice of renormalization conditions for loop diagrams. The re-arranged effective Lagrangian at any given order in the new counting contains a finite number of terms but involves some vertices of higher orders in the standard counting. Provided all the relevant resonances are treated as explicit degrees of freedom, the re-arranged effective Lagrangian generates a convergent Taylor series of the analytic part of the amplitude. Notice that the LECs of the re-arranged Lagrangian differ from the original ones by (known) $\mu$-dependent shifts.

To ensure that loop diagrams obey the new power counting, we have employed a generalization of the EOMS scheme by shifting the renormalization point into the physical region above the threshold. More precisely, the loop integrals are subtracted at the nucleon momentum $p^2 = m_N^2 + 2m_N\mu$. Since this procedure results in subtracting also the imaginary parts of the integrals, the renormalized LECs become complex in our scheme for $\mu > 0$ as visualized in Fig. 10.

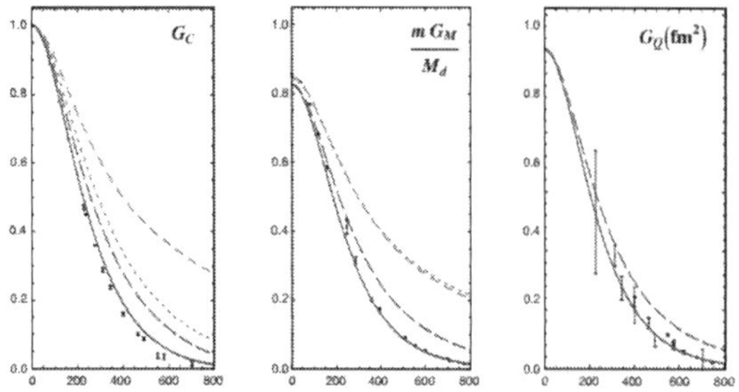


**Figure 11.** Electromagnetic form factors of the deuteron at LO in the renormalizable chiral EFT approach of Ref. [54] as a function of the momentum transfer $q$ in comparison with experimental data from Refs. [60, 61]. Solid and long-dashed violet lines (short- and medium-dashed pink lines) show the predictions in the chiral (pionless) EFT with and without using phenomenological form factors of the nucleon, respectively.

As an application, we have used the resulting framework to analyze $\pi N$ scattering at low energy.[2] With the relevant LECs being fitted to the $\pi N$ PWA, all phase shifts in the threshold region were shown to be well described within the new formulation, with the results being similar to those discussed in section 4. Our new approach is not limited to elastic $\pi N$ scattering and is also applicable to pion photo- and electro-production reactions, Compton scattering and processes involving several pions.

# 6 Applications in the two-nucleon sector

## 6.1 Deuteron electromagnetic form factors in a renormalizable formulation of chiral EFT

In Ref. [53], we have analyzed the electromagnetic form factors $G_C$, $G_M$ and $G_Q$ of the deuteron at LO within a renormalizable formulation of chiral nuclear EFT proposed in Ref. [54], see Refs. [55–58] and references therein for related earlier studies. This approach is based on a manifestly Lorentz invariant form of the effective Lagrangian without performing a non-relativistic expansion for the nucleon propagators. The NN amplitude can then be calculated by solving the corresponding three-dimensional integral equation which satisfies relativistic elastic unitarity [59]. The resulting scattering amplitude at LO in chiral EFT is renormalizable, which allows one to obtain regulator-independent results for observables.

At LO, the form factors of the deuteron only receive contributions from single-nucleon currents, and the results do not involve any free parameters. In Fig. 11, we show by the violet long-dashed lines the resulting LO parameter-free cutoff-independent predictions for $G_C$, $G_M$ and $G_Q$. Notice that the deviations from the experimental at momentum transfers of $q \sim 200$ MeV and higher can be largely traced back to the well-known slow convergence of the chiral expansion for the nucleon form factors [62, 63]. Indeed, replacing the LO single-nucleon current by the dispersion-theoretical form factor parametrization of Ref. [64] as shown by the solid lines in Fig. 11, a very good description of the data is observed even at rather high values of $q$. The resulting predictions for the deuteron magnetic and quadrupole moments, $\mu^{\rm LO} = 0.826(e/(2m))$ and $Q^{\rm LO} = 0.271$ fm are in a good agreement with the experimental values of $\mu^{\rm exp} = 0.85741(e/(2m))$ and $Q^{\rm exp} = 0.2859$ fm.

The renormalizable approach also provides a clear evidence of the importance of the one-pion exchange NN potential by comparing the LO results in $\pi$-less and $\pi$-full versions of the EFT as visualized in Fig. 11. Notice that the quadrupole form factor of the deuteron vanishes at LO in the $\pi$-less approach.

[2] To go beyond the threshold region, one would have to explicitly take into account the relevant contributions of the nucleon resonances. This task is postponed for future work.

## 6.2 $^1S_0$ NN scattering in the modified Weinberg approach

While the LO calculations in the novel renormalizable approach to NN scattering of Ref. [54] look promising, the inclusion of higher-order corrections requires much more extensive calculations than in the standard, non-relativistic framework. In Ref. [65], we made a first step along this line by analyzing the subleading contact interaction in the $^1S_0$ partial wave. Given that the singular tensor part of the one-pion exchange potential does not contribute in spin-singlet channels, it is possible to include the NLO contact interaction both perturbatively and without relying on perturbation theory. In the latter case, the scattering amplitude shows a residual dependence on the subtraction point $\mu$. We have argued that the unnaturally large value of the scattering length in the $^1S_0$ channel requires the subtraction scale $\mu$ corresponding to the LO contact interaction to be chosen of the order of the hard scale in the problem. We have also addressed perturbativeness of the subleading contact interaction within our scheme and found it to be advantageous to treat it nonperturbatively at energies above $E_{\rm lab} \sim 50$ MeV. Finally, we have demonstrated that the predictions for the low-energy theorems for the coefficients in the effective range expansion are strongly improved when going from LO to NLO.

## 6.3 New fixed points of the renormalization group for nonrelativistic two-body systems

The renormalization group (RG) is known to be a powerful tool to study the scale dependence of systems in various areas of physics. For a system of two nonrelativistic particles interacting via short-range forces, the nontrivial fixed point describing a system in the "unitary limit" and analyzed in Ref. [66] has attracted much interest in the context of NN scattering in the $^1S_0$ and $^3S_1$ channels. While the existence of other nontrivial fixed points of the RG equation for two-body scattering has been suggested in the literature, no explicit results were known. In Ref. [67], we have presented a general method for constructing an infinite number of possible fixed points by utilizing a separable ansatz for the potentials that arise in EFT description of such systems. Our results indicate a much richer structure than previously recognized in the RG flows of simple short-range potentials.

We would like to thank our collaborators Véronique Bernard, Michael Birse, Dalibor Djukanovic, Ashot Gasparyan, Jambul Gegelia, Matthias Schindler, Dmitrij Siemens, and De-Liang Yao. The work reported here would not have been possible without the support from the Deutsche Forschungsgemeinschaft within the SFB/TR16.

## References

[1] D. Djukanovic, E. Epelbaum, J. Gegelia and U.-G. Meißner, Phys. Lett. B **730**, 115 (2014) [arXiv:1309.3991 [hep-ph]].

[2] G. Ecker, J. Gasser, H. Leutwyler, A. Pich and E. de Rafael, Phys. Lett. B **223**, 425 (1989).

[3] U.-G. Meisßner, Phys. Rept. **161**, 213 (1988).

[4] D. Djukanovic, M. R. Schindler, J. Gegelia, G. Japaridze and S. Scherer, Phys. Rev. Lett. **93**, 122002 (2004) [hep-ph/0407239].

[5] E. V. Luschevskaya, O. A. Kochetkov, O. V. Teryaev and O. E. Solovjeva, JETP Lett. **101**, no. 10, 674 (2015).

[6] J. N. Hedditch et al., Phys. Rev. D **75**, 094504 (2007) [hep-lat/0703014 [HEP-LAT]].

[7] M. Gurtler *et al.* [QCDSF Collaboration], PoS LATTICE **2008**, 051 (2008).

[8] N. Fettes, U.-G. Meißner and S. Steininger, Nucl. Phys. A **640**, 199 (1998) [hep-ph/9803266].

[9] P. Buettiker and U.-G. Meißner, Nucl. Phys. A **668**, 97 (2000) [hep-ph/9908247].

[10] N. Fettes and U.-G. Meißner, Nucl. Phys. A **676**, 311 (2000) [hep-ph/0002162].

[11] T. Becher and H. Leutwyler, JHEP **0106**, 017 (2001) [hep-ph/0103263].

[12] M. Hoferichter, B. Kubis and U.-G. Meißner, Nucl. Phys. A **833**, 18 (2010) [arXiv:0909.4390 [hep-ph]].

[13] J. M. Alarcon, J. Martin Camalich and J. A. Oller, Phys. Rev. D **85**, 051503 (2012) [arXiv:1110.3797 [hep-ph]].

[14] H. Krebs, A. Gasparyan and E. Epelbaum, Phys. Rev. C **85**, 054006 (2012) [arXiv:1203.0067 [nucl-th]].

[15] J. M. Alarcon, J. Martin Camalich and J. A. Oller, Annals Phys. **336**, 413 (2013) [arXiv:1210.4450 [hep-ph]].

[16] Y. H. Chen, D. L. Yao and H. Q. Zheng, Phys. Rev. D **87**, 054019 (2013) [arXiv:1212.1893 [hep-ph]].

[17] E. Epelbaum, H. Krebs and U.-G. Meißner, Eur. Phys. J. A **51**, no. 5, 53 (2015) [arXiv:1412.0142 [nucl-th]].

[18] E. Epelbaum, H. Krebs and U.-G. Meißner, Phys. Rev. Lett. **115**, no. 12, 122301 (2015) [arXiv:1412.4623 [nucl-th]].

[19] D. Siemens, V. Bernard, E. Epelbaum, A. Gasparyan, H. Krebs and U.-G. Meißner, Phys. Rev. C **94**, no.1, 014620 (2016) [arXiv:1602.02640 [nucl-th]].

[20] J. Gegelia and G. Japaridze, Phys. Rev. D **60**, 114038 (1999) [hep-ph/9908377].

[21] J. Gegelia, G. Japaridze and X. Q. Wang, J. Phys. G **29**, 2303 (2003) [hep-ph/9910260].

[22] T. Fuchs, J. Gegelia, G. Japaridze and S. Scherer, Phys. Rev. D **68**, 056005 (2003) [hep-ph/0302117].

[23] K. A. Wendt, B. D. Carlsson and A. Ekström, arXiv:1410.0646 [nucl-th].

[24] E. E. Jenkins and A. V. Manohar, Phys. Lett. B **259**, 353 (1991).

[25] T. R. Hemmert, B. R. Holstein and J. Kambor, J. Phys. G **24**, 1831 (1998) [hep-ph/9712496].

[26] V. Pascalutsa and D. R. Phillips, Phys. Rev. C **67**, 055202 (2003) [nucl-th/0212024].

[27] M. Hoferichter, J. Ruiz de Elvira, B. Kubis and U.-G. Meißner, Phys. Rev. Lett. **115**, 092301 (2015) [arXiv:1506.04142 [hep-ph]].

[28] M. Hoferichter, J. Ruiz de Elvira, B. Kubis and U.-G. Meißner, Phys. Rev. Lett. **115**, no. 19, 192301 (2015) [arXiv:1507.07552 [nucl-th]].

[29] M. Hoferichter, J. Ruiz de Elvira, B. Kubis and U.-G. Meißner, Phys. Rept. **625**, 1 (2016) [arXiv:1510.06039 [hep-ph]].

[30] D. L. Yao, D. Siemens, V. Bernard, E. Epelbaum, A. M. Gasparyan, J. Gegelia, H. Krebs and U.-G. Meißner, JHEP **1605**, 038 (2016) [arXiv:1603.03638 [hep-ph]].

[31] A. Denner, S. Dittmaier, M. Roth and D. Wackeroth, Nucl. Phys. B **560**, 33 (1999) [hep-ph/9904472].

[32] R. A. Arndt, W. J. Briscoe, I. I. Strakovsky and R. L. Workman, Phys. Rev. C **74**, 045205 (2006) [nucl-th/0605082].

[33] D. Siemens, V. Bernard, E. Epelbaum, H. Krebs and U.-G. Meißner, Phys. Rev. C **89**, no. 6, 065211 (2014) [arXiv:1403.2510 [nucl-th]].

[34] V. Bernard, N. Kaiser and U.-G. Meißner, Phys. Lett. B **332**, 415 (1994) Erratum: [Phys. Lett. B **338**, 520 (1994)] [hep-ph/9404236].

[35] V. Bernard, N. Kaiser and U.-G. Meißner, Nucl. Phys. B **457**, 147 (1995) [hep-ph/9507418].

[36] V. Bernard, N. Kaiser and U.-G. Meißner, Nucl. Phys. A **619**, 261 (1997) [hep-ph/9703218].

[37] N. Fettes, V. Bernard and U.-G. Meißner, Nucl. Phys. A **669**, 269 (2000) [hep-ph/9907276].

[38] E. Epelbaum, U.-G. Meißner and W. Glöckle, Nucl. Phys. A **714**, 535 (2003) [nucl-th/0207089].

[39] E. Epelbaum, U.-G. Meißner and W. Glöckle, nucl-th/0208040.

[40] S. R. Beane and M. J. Savage, Nucl. Phys. A **713**, 148 (2003) [hep-ph/0206113].

[41] S. R. Beane and M. J. Savage, Nucl. Phys. A **717**, 91 (2003) [nucl-th/0208021].

[42] J. C. Berengut, E. Epelbaum, V. V. Flambaum, C. Hanhart, U.-G. Meißner, J. Nebreda and J. R. Pelaez, Phys. Rev. D **87**, no. 8, 085018 (2013) [arXiv:1301.1738 [nucl-th]].

[43] E. Epelbaum, H. Krebs, T. A. Lähde, D. Lee and U.-G. Meißner, Phys. Rev. Lett. **110**, no. 11, 112502 (2013) [arXiv:1212.4181 [nucl-th]].
[44] E. Epelbaum, H. Krebs, T. A. Lähde, D. Lee and U.-G. Meißner, Eur. Phys. J. A **49**, 82 (2013) [arXiv:1303.4856 [nucl-th]].
[45] V. V. Vereshagin et al., Nucl. Phys. A **592**, 413 (1995) [hep-ph/9504361].
[46] M. Kermani *et al.* [CHAOS Collaboration], Phys. Rev. C **58**, 3419 (1998).
[47] J. B. Lange *et al.*, Phys. Rev. Lett. **80**, 1597 (1998).
[48] R. Muller *et al.*, Phys. Rev. C **48**, 981 (1993).
[49] D. Djukanovic, J. Gegelia and S. Scherer, Phys. Lett. B **690**, 123 (2010) [arXiv:0903.0736 [hep-ph]].
[50] T. Bauer, J. Gegelia and S. Scherer, Phys. Lett. B **715**, 234 (2012) [arXiv:1208.2598 [hep-ph]].
[51] D. Djukanovic, E. Epelbaum, J. Gegelia, H. Krebs and U.-G. Meißner, Eur. Phys. J. A **51**, 101 (2015) [arXiv:1507.06771 [hep-ph]].
[52] E. Epelbaum, J. Gegelia, U.-G. Meißner and D. L. Yao, Eur. Phys. J. C **75**, no. 10, 499 (2015) [arXiv:1510.02388 [hep-ph]].
[53] E. Epelbaum, A. M. Gasparyan, J. Gegelia and M. R. Schindler, Eur. Phys. J. A **50**, 51 (2014) [arXiv:1311.7164 [nucl-th]].
[54] E. Epelbaum and J. Gegelia, Phys. Lett. B **716**, 338 (2012) [arXiv:1207.2420 [nucl-th]].
[55] M. Walzl and U.-G. Meißner, Phys. Lett. B **513**, 37 (2001) [nucl-th/0103020].
[56] D. R. Phillips, J. Phys. G **34**, 365 (2007) [nucl-th/0608036].
[57] M. P. Valderrama, A. Nogga, E. Ruiz Arriola and D. R. Phillips, Eur. Phys. J. A **36**, 315 (2008) [arXiv:0711.4785 [nucl-th]].
[58] S. Kölling, E. Epelbaum and D. R. Phillips, Phys. Rev. C **86**, 047001 (2012) [arXiv:1209.0837 [nucl-th]].
[59] V. G. Kadyshevsky, Nucl. Phys. B **6**, 125 (1968).
[60] D. Abbott *et al.* [JLAB t20 Collaboration], Eur. Phys. J. A **7**, 421 (2000) [nucl-ex/0002003].
[61] D. M. Nikolenko *et al.*, Phys. Rev. Lett. **90**, 072501 (2003).
[62] B. Kubis and U.-G. Meißner, Nucl. Phys. A **679**, 698 (2001) [hep-ph/0007056].
[63] T. Bauer, J. C. Bernauer and S. Scherer, Phys. Rev. C **86**, 065206 (2012) [arXiv:1209.3872 [nucl-th]].
[64] M. A. Belushkin, H.-W. Hammer and U.-G. Meißner, Phys. Rev. C **75**, 035202 (2007) [hep-ph/0608337].
[65] E. Epelbaum, A. M. Gasparyan, J. Gegelia and H. Krebs, Eur. Phys. J. A **51**, no. 6, 71 (2015) [arXiv:1501.01191 [nucl-th]].
[66] M. C. Birse, J. A. McGovern and K. G. Richardson, Phys. Lett. B **464**, 169 (1999) [hep-ph/9807302].
[67] M. C. Birse, E. Epelbaum and J. Gegelia, Eur. Phys. J. A **52**, no. 2, 26 (2016) [arXiv:1510.02209 [hep-ph]].

# Polarized Solid State Target

*Hartmut* Dutz[1,*], *Stefan* Goertz[1,**], and *Werner* Meyer[2,***]

[1] *Physikalisches Institut, Universität Bonn, Nussallee 12, 53115 Bonn*
[2] *Institut für Experimentalphysik - AG I, Ruhr-Universität Bochum, Universitätsstraße 150, 44780 Bochum*

**Abstract.** The polarized solid state target is an indispensable experimental tool to study single and double polarization observables at low intensity particle beams like tagged photons. It was one of the major components of the Crystal-Barrel experiment at ELSA. Besides the operation of the 'CB frozen spin target' within the experimental program of the Crystal-Barrel collaboration both collaborative groups of the D1 project, the polarized target group of the Ruhr Universität Bochum and the Bonn polarized target group, have made significant developments in the field of polarized targets within the CRC16. The Bonn polarized target group has focused its work on the development of technically challenging polarized solid target systems towards the so called '$4\pi$ continuous mode polarized target' to operate them in combination with $4\pi$-particle detection systems. In parallel, the Bochum group has developed various highly polarized deuterated target materials and high precision NMR-systems, in the meantime used for polarization experiments at CERN, JLAB and MAMI, too.

## 1 Introduction

Within the recent decades polarized solid targets as well as polarized particle beams evolved into indispensable tools in order to explore the spin structure of the proton and the neutron. Single and double polarization measurements were performed with a variety of different particles beams, among them not only experiments with polarized electrons and real photons. Protons, antiprotons and even polarized muons have been used so far at various laboratories [1]. (Double) polarization experiments at ELSA [2, 3] with tagged photon beam fluxes in the order of some megahertz have to rely on polarized solid targets with high target densities in order to provide sufficient luminosities of the particle reactions of interest. The developments of the polarized targets for particle physics experiments at ESLA within the CRC16 were twofold. On the one side we focussed on the enhancement of the figure of merit (FOM), which is defined as the product of the number $n_t$ of target particles per cm$^3$, the squared target polarization $P_t^2$ and the squared ratio $f^2$ of the number of polarizable nucleons and the total number of nucleons in the target. Improvements in $P_t$ and $f$ are directly related to the search and preparation of suitable target materials and need a detailed understanding of the dynamic nuclear polarization (DNP), as given by the spin temperature theory (for an overview we refer to [1] and for deeper studies see [4]).
On the other hand we worked on a significant enhancement of the luminosity $L$ and angular acceptance $\Delta\Omega$ of the polarized target beyond the standard 'frozen spin technique' towards the so called 'large angular continuous mode target' or '$4\pi$ continuous mode polarized target'. The goal is to enhance the reaction counting rate $N$, which is defined as $N = L \cdot d\sigma/d\Omega \cdot \Delta\Omega$, and thus to improve considerably upcoming polarization experiments. The new scheme led to challenging and outstanding developments in cryogenics and superconducting magnet technology in the field of polarized solid state targets.
Besides the developments of new polarized target components and materials we operated the existing 'CB frozen spin target' (former 'GDH frozen spin target') as a major component of the Crystal Barrel experiment at ELSA for (double) polarization experiments.
Over the period of the CRC the goals and methods of the Bochum polarized target group focused on intense research on polarized target materials and their polarization behavior and the design and construction of more accurate as well as sensitive polarization measurement equipment. The broad spectrum of research covered the following topics:

- **construction of a 'top-loading' $^4$He refrigerator for target material research**
- **pulsed and CW NMR for polarization measurements [5, 6]**
- **polyethylene and other oxygen-free hydrocarbons (alkanes) as polarized target materials**
- **radiation resistance of target materials with spinless spectator nuclei for high intense particle beam experiments**

*e-mail: dutz@physik.uni-bonn.de
**e-mail: goertz@physik.uni-bonn.de
***e-mail: meyer@ep1.rub.de

- **conventional X-band and high field elctron paramagnetic resonance (EPR) measurements for systematic target material studies**
  Apart from the performance of the refrigerator and from the strength of the magnetic field, the achievable maximum polarization in a certain target material strongly depends on the efficiency of the DNP process. This efficiency is intimately related to the properties of the paramagnetic centers, from which the polarization is transferred to the nuclei. In particular the longitudinal Zeeman and the dipolar relaxation times as well as the strength of the electronic non-Zeeman interaction determine the minimum spin temperature and thus the maximum polarization of the nuclear spins. Therefore, to optimize the performance of a given target material, it is essential to characterize the paramagnetic system generated by the various doping methods as completely as possible. Meanwhile, besides a conventional X-band (10 GHz) EPR spectrometer, which has been used since several years and which has produced a lot of very useful results (e.g. radical characterization and structure studies [7] or observation of spin temperature in $^6$LiD [8]), also a V-band (70 GHz) spectrometer is in use in the Bochum polarized target laboratory [9]. This self-built device is implemented in the $^4$He polarization apparatus and enables EPR measurements in the DNP regime at 2.5 T and down to 1 K. Especially for deuterated target materials very important results concerning the correlation between the properties of the paramagnetic species (i.e. EPR linewidth) and the maximum achievable deuteron polarization have been obtained at DNP relevant magnetic fields (see Fig. 1).

- **Magnetic field dependence of the polarization in nitroxyl and trityl doped deuterated alcohol materials**
  With respect to the projected '4$\pi$-continuous mode' polarized solid state target, it is of great importance to obtain an extensive knowledge about the polarization behavior within the moderate magnetic field region between 1.8 T and 2.5 T. Since with internal polarizing field values beyond 2.5 T will probably not be accessible in the near future experiments with such targets will have to rely on a thorough optimization of the polarization conditions in order to achieve polarization values as high as required. Such an optimization means to check material parameters like radical concentration, EPR line-width and electron as well as nuclear relaxation times in dependence of the magnetic field with the goal to pin down an optimal set of parameters for each material. On the other hand the results of such measurements will help to specify the requirements for the construction of an internal polarizing magnet in terms of magnetic field values and magnetic field homogeneity.
  A new top-loading $^4$He-refrigerator capable to be operated up to 7.0 T has been successfully set into operation. Measurements of the electron relaxation time at 2.5 T have been performed in the very homogeneous magnetic field ($\Delta B/B < 10^{-5}$). For this a new pulsed nuclear magnetic resonance (NMR)-system has been built, which is able to measure relaxation times in the order of milliseconds [10]. This is the first time that such measurements have been performed with samples used for particle physics experiments.
  Furthermore a new continuous wave NMR-system was built, with which the degree of polarization of the polarized target materials can be detected with improved precision in the frequency range between 10 MHz and 300 MHz (corresponding magnetic field strengths between 1.8 T and 7.0 T for protons and deuterons, respectively). This development was necessary, as for the existing continuous wave NMR-system (Liverpool box) components are obsolete and other parts not available anymore on the electronic market (see Sec. 7.
  A microwave system with a center frequency of 56 GHz has been set up and is ready to work. Polarization values for protons at the corresponding field strength of 2.0 T and 1 K are expected to be in the order of some percent. To optimize the preparation of the samples for the particle physics experiments at temperatures <80 mK the $^3$He/$^4$He dilution refrigerator in the Bochum laboratory has been modified and successfully operated for target material studies at low temperatures.

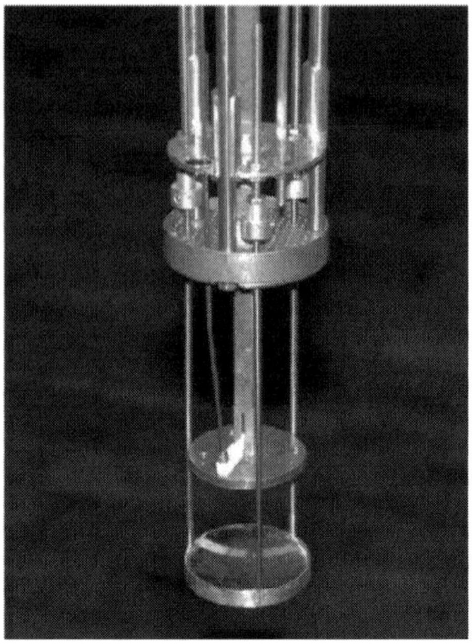

Figure 1: The Fabry-Perot like multimodal resonator of the high field EPR spectrometer (diameter of mirrors: 31 mm). Because at 2.5 T the wavelength of the microwaves equals the dimensions of the solid target samples, a common Single Mode Cavity cannot be used.

Since the Bochum group stressed the target material and NMR topics in polarized target research, the Bonn polarized target group put their emphasis more in the development of cryogenic components, small superconducting magnets as well as the operation of the 'CB frozen spin target':

- **design and construction of a new 'transverse holding-coil' and an 'internal superconducting polarizing magnet'**
- **construction of a new horizontal $^3$He/$^4$He dilution refrigerator for the '4$\pi$-continuous mode' polarized solid state target**
- **set-up of a new irradiation facility for target materials at the 20MeV Linac of the Bonn accelerator facility and target material studies of irradiated polymeric materials.**

In the following sections 2 - 7 we will describe this work and recent results in more detail.

## 2 'CB frozen spin target' at Crystal Barrel

### 2.1 Introduction

The experimental program at the Crystal Barrel detector was focused on the measurement of single and double polarization observables off polarized nucleons using the existing GDH frozen spin target [11, 12], which with the start of the CRC16 has been labeled as 'CB frozen spin target'. Here we will briefly report on the setup and the performance of the 'CB frozen spin target' at the Crystal Barrel detector for polarization experiments at ELSA in the reporting period of CRC16.

### 2.2 Setup and performance

Upon completion of the GDH experiment at Mainz the target was moved back to Bonn and has been prepared for the experimental program at ELSA [13]. The main pumping unit was dismounted for revision and has been reinstalled to have a better access to the roots blowers to reduce the down time in case of a failure. In parallel the control unit was upgraded to guaranty a reliable operation of the system at ELSA. On top of that, the installation included a reconstruction of the mechanical support structure of the target, the railway system to move the polarizing magnet and the Crystal Barrel detector and the main vacuum system.

For the first run a butanol target material with low electron spin density ($1.4 \times 10^{19}$spins/ml) was prepared to guaranty high relaxation times at moderate maximum proton polarizations. The relaxation time of the proton polarization during the first measurement at ELSA exceeded 500 h in a longitudinal 'holding field' of 0.64 T. The nucleon polarization had to be refreshed every three days to minimize the beam time loss by the time consuming repolarizing process and movement of the detector components. The chosen frozen-spin cycle of about 72 h and a mean value of the target polarization at the beginning of the frozen spin mode of about 60% amounts to an average polarization of 57% during the data taking. After more than 8 years of successfull operation in the GDH experiment, the target has proven its known reliability. It is an ideal experimental tool to study spin dependent reactions in combination with a movable 4$\pi$-detection system.

To increase the reliability as well as the operational availability of the existing GDH frozen spin target, we replaced in the second funding period various hardware components of the control and vacuum-system of the refrigerator. Namely, the PLC- and HMI-systems were completely renewed and reprogrammed. Due to super-leaks in the front part of the target inserts, both units have been redesigned and reconstructed. We replaced the existing splitted NMR-coaxial cables in the inserts by only one. This led to a reduced heat load of the mixing chamber by longitudinal heat conduction and enabled us to improve the accuracy of the polarization measurement via the NMR-technique.

From 2008 to 2013 we run the frozen spin target for more than 15900 hours under cold operational conditions and could provide 7500 hours of longitudinally polarized protons up to a maximum polarization of ± 83% and deuterons up to ± 70% for single and double polarization experiments with the Crystal Barrel detector.

### 2.3 Transverse 'holding coil'

One of the major topics of the CRC TR16 was the measurement of the target asymmetry T and the double polarization observables F and H off the proton and neutron. These observables require a polarization direction perpendicular to the scattering plane of the pion nucleon system. To provide such a transverse polarization in the frozen spin target the existing 'longitudinal holding coil' had to be replaced by a new 'transverse holding magnet'. To ensure with the new set-up as long relaxation times as achieved with the 'longitudinal holding magnet', a field strength of at least $B_H = 0.5$ T is needed. And, as a crucial factor for a new magnet design, the maximum load of the current leads inside the dilution refrigerator is limited to $I_H = 33$ A. The symmetry of the refrigerator, the available space in the target area as well as the cylindric form of the bobbin defines the 'transverse holding coil' as a race-track type superconducting magnet. In this design the magnet consists of two individual dipoles bend to a race-track each placed inside the bobbin. Additionally to the electrical parameters the magnet has to fulfill the requirements of a minimized mass distribution to ensure a good detection probability for the outgoing particles and the outer diameter of the magnet which is limited by the geometry of the refrigerator.

Taking these considerations and a superconducting wire with a diameter of 0.2 mm into account, the simulation of the magnet led to a solution for each coil consisting out of 510 windings within 7 layers. The combination of both coils result in a maximum field of the 'holding coil' of $B_H = 0.5$ T at a nominal current of $I_H = 30$ A. In contrast to comparable magnet designs, both coils were wound in a wet winding process. Thereby a cylindric shaped epoxy form was placed in between two race-track like guiding tools and used as a core for the winding of the wire. After hardening of the adhesive the core has been removed and both dipoles were placed and glued inside the bobbin. The fact that the bobbin of the new 'transverse holding coil' is identical in construction with the one of the 'longitudinal holding coil' allows us to change both magnets in short time (3 days). This gives us the opportunity to change the general settings and orientation of the polarization for a measurement of a new polarization observable. On the other hand the special design of the two dipoles inside of the bobbin leaves additional space for an optional solenoid wound on the outer shell of the bobbin. For the future one has with this double coil magnet the possibility to run a transverse and a longitudinal magnetic field to maintain the nucleon polarization in any possible angle to the scattering plane in the outgoing pion-nucleon frame by simply varying the currents in both magnets during the data taking.

In 2010 the new 'transverse holding coil' was used in the measurement of the target asymmetry T and the double polarization observable H at the Crystal Barrel detector. In more than 4300 h of cold operation of the frozen spin target it has proven its reliability. At the nominal holding field of $B_H = 0.5$ Tesla, mean relaxation times in the order of 370 h have been achieved. The maximum proton polarization was of the order of ±83%. In Fig. 2 the polarization history for three typical double polarization runs in 2009, 2010 and 2011 for longitudinal, transverse, proton and deuteron polarization are shown respectively.

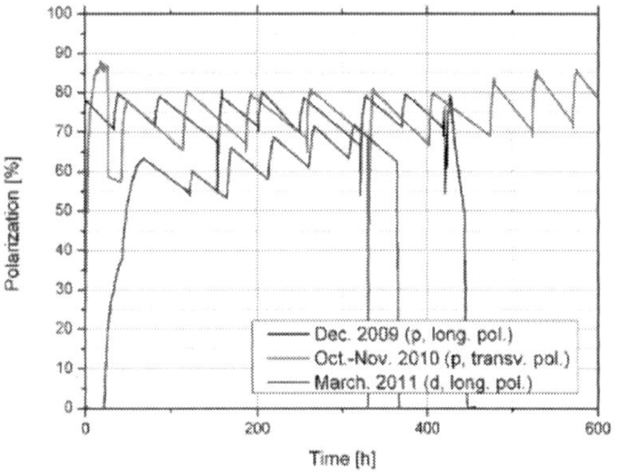


Figure 2: Polarization history plot for three typical double polarization runs in 2009, 2010 and 2011 for longitudinal and transverse, proton and deuteron polarization, respectively.

## 3 A new horizontal $^3$He/$^4$He dilution refrigerator for a $4\pi$-continuous-mode polarized target

### 3.1 Introduction

Besides the advantage of the good angular acceptance, the frozen spin target has in comparison with a 'continuous mode target' the disadvantage of a reduced beam time efficiency. It is caused by the polarization refresh cycles and the movement of the polarizing magnet as well as the detector components. In best cases a beam time efficiency of about 80% is attainable.

The second restriction from the 'figure of merit' point of view is the decay of the polarization during the data taking in all target materials. The overall polarization of an experiment using a frozen spin target is given by the relaxation times and limited to approximately $0.8 \cdot P_{max}$. A reduction of the target temperature to increase the relaxation times is suppressed by the Kapitza-resistance in the target material at a given beam intensity. Commonly used intensities up to $10^7$ particles/sec are setting the effective target material temperature independently of the lowest refrigerator temperatures (e.g. 20 mK) to about 50 mK to 60 mK. Beam intensities above $10^8$ particles/sec are unattainable in the 'frozen spin mode' operation.

To circumvent the disadvantages of the frozen spin technique we have combined the advantages with those of the 'continuous mode' operating target to a so called '$4\pi$ continuous mode' polarized target [14]. It is simply a 'frozen spin target' equipped with an internal superconducting polarizing magnet for permanent DNP. This new polarized target scheme, which can be described as a continuous mode operating target with a large angular acceptance, leads to an improvement in the figure of merit by a factor of 2 compared to the existing 'CB frozen spin target'. The special features of the target are

- large angular acceptance ~ $4\pi$ (homogeneous mass distribution)
- high average polarization during the data taking (continuous DNP at high polarizing field)
- high luminosities $L$ up to $10^{33}\text{cm}^{-2}\text{sec}^{-1}$
- good beam time efficiency (no maintenance work for the moving of components nor repolarization),
- no moving system for the polarization process required (no detector size restriction, fixed target-detector setup)

Once the system will have been installed in the Crystal Barrel detector it will provide highly polarized nucleons with good access to the target region for scattering experiments with real and virtual photons.

Nevertheless, at this point it has to be mentioned that the '$4\pi$ continuous mode' polarized target in the first version only can provide longitudinal polarized nucleons for the scattering experiment. The symmetry and geometry of the horizontal refrigerator anticipate a solenoid as polarizing magnet enclosing the target material and the mixing chamber. In case transverse polarization is needed the '$4\pi$ continuous mode' polarized target has to be operated again in the classic 'frozen spin mode' whereby the transverse field orientation to maintain the nucleon polarization has to be provided by an internal 'transverse holding magnet'. The target material will be polarized via DNP in the internal longitudinal polarizing magnet and then rotated and maintained by the internal holding magnet. It seems to be possible to combine both coils to a hybrid coil, but this will be subject of future developments.

The main tasks towards the new target scheme in the CRC16 was the construction of a new horizontal $^3$He/$^4$He-dilution refrigerator and the development of a suitable

small superconducting high field magnet as internal polarizing magnet implemented into the refrigerator.
Since the geometry of the new refrigerator is comparable with the existing one of the 'CB frozen spin target' the shape and the overall dimensions of the new internal polarizing magnet are the same as the actual internal holding coils. It has to fulfill the requirements of the high homogeneity of the external polarization magnets ($\Delta B/B \leq 10^{-4}$) and the low mass distribution of the internal holding coils to ensure a good detection probability for the outgoing particles. A more detailed description of the development and the constrains of the internal polarizing magnet is given in section 5.

### 3.2 Design of the new horizontal $^3He/^4He$ dilution refrigerator

The design, construction and setup of vertical as well as horizontal dilution refrigerators has a long and experienced tradition in the Bonn polarized target group and the precision workshops of the institute. Most of the components of the new refrigerator has been machined and built by staff members of our institute. The outer shape of the new refrigerator follows the design of the existing horizontal 'CB frozen spin target' dilution refrigerator. Various design aspects of our new cryostat have partly influenced the construction of the new horizontal dilution refrigerator of the 'Mainz Polarized Target' built by JINR Dubna. Differently from the existing type we focused on high cooling power (200 mW) in the temperature range at 250 mK to 300 mK to dynamically polarize an approximately 7 cm$^3$ large target sample (DNP-mode). Nevertheless, to run the system optionally as a frozen spin target if transverse polarization is needed, we set the design base temperature to about 30 mK to guaranty long relaxation times. As distinguished from the 'CB frozen spin target' dilution refrigerator we have added an additional $^4$He-evaporation refrigerator which encloses completely the main dilution refrigerator. This unit plays in first order the role as individual cooling sink for the internal superconducting magnet and its currend leads, in second order it is the thermal shielding for the complete system. The $^4$He-cryostat will be feed with liquid $^4$He from the inner dilution refrigerator but nevertheless it can be operated independently from the inner cooling stages.
The beam line is the central part of the cryostat and has to fulfill the function of the internal isolation vacuum. The target material will be inserted into the refrigerator along the beam line by an insert that will have the same length as the one of the 'CB frozen spin target'. It will be vacuum tight against the mixing chamber by means of a cold indium seal in the still region. The precooling of the incoming $^3$He/$^4$He mixture, liquefying at the thermal sink (1.2 K sink) will be comparable to the GDH-type. In contrast to the old refrigerator the first precooling stage is designed as a turbine like heat exchanger to improve the cooling power for the precooling and concurrently to reduce the $^4$He-consumption. The evaporator (1 K pot) is clasped around the still and has a volume of about 0.5 l of liquid helium. The simulation of the precooling stages of the refrigerator is described in section 4. It is the first computational fluid dynamics (CFD) simulation of a dilution refrigerator which takes all in- and backstreaming gases and fluids into account. The final heat exchanger and the still are in first order conditioned by the outer shape of the cryostat. From the outside, the system will look like a typical 'Roubeau - cryostat', the cylindric mixing chamber which is surrounded by the internal polarizing magnet is followed by the final heat exchanger. The low temperature heat exchanger was designed and delivered by the low temperature working group of the JINR Dubna. It is a 10 staged large surface sintered copper heat exchanger with an expected base temperature of about 30 mK. To the backward region the diameter opens conical to the pumping tee. The smooth transition from the larger diameter of the 70 K region to the low temperature part of the cryostat allows one to use high temperature superconducing (HTSC)-current leads for the internal polarizing magnet.
An additional heat sink cooled by a separated $^4$He-circuit provide the required cooling power to cool the normal conducting current leads at the transition point to the HTSC down to 50 K. Taking the parameter of a standard HTSC wire into account, a maximum current of 120 A can be accepted for an internal magnet. A cut-view of the new dilution refrigerator is shown in Fig. 3.

### 3.3 Status

We started with the design and first technical drawings of the refrigerator in 2006 and expected a construction time of 10 man years. After a successful start of the construction the history of this project is marked by various setbacks due to technical difficulties and staff changes in the precision workshops of the institute. The most crucial setback was caused by material faults in the thin wall stainless steel tubings of the already assembled precooling stages in 2013. This led to a series of vacuum leaks in the $^4$He to $^3$He heat exchangers although all components have been leak checked during the production process and later on during the assembling of the heat exchangers. We could reconstruct the leakages which were evoked by cavitation shrink holes (due to carbon cavities) in the tubings and the soft solder flux, as well as thermal stress during the TIG-welding. Approximately 3 to 4 month after the production process the shrink holes opened and cause the leakage. Unpleasantly nearly all leaks were unattainable for repair work, like additional soft or hard soldering. All this happened shortly before the first cool down of the cryostat. Consequently we decided to rebuild the complete precooling stage including a new central beam tube. All tubings have been replaced by new ones and each one was tested before the production process to build the new heat exchangers. As long as necessary, we have renewed soft solder connections and changed from TIG welding to laser welding. Since most of the components of the refrigerator as well as the custom made tooling for the heat exchangers, which had to be rebuild, were already built and tested we were able to complete the renewing and assembling of the

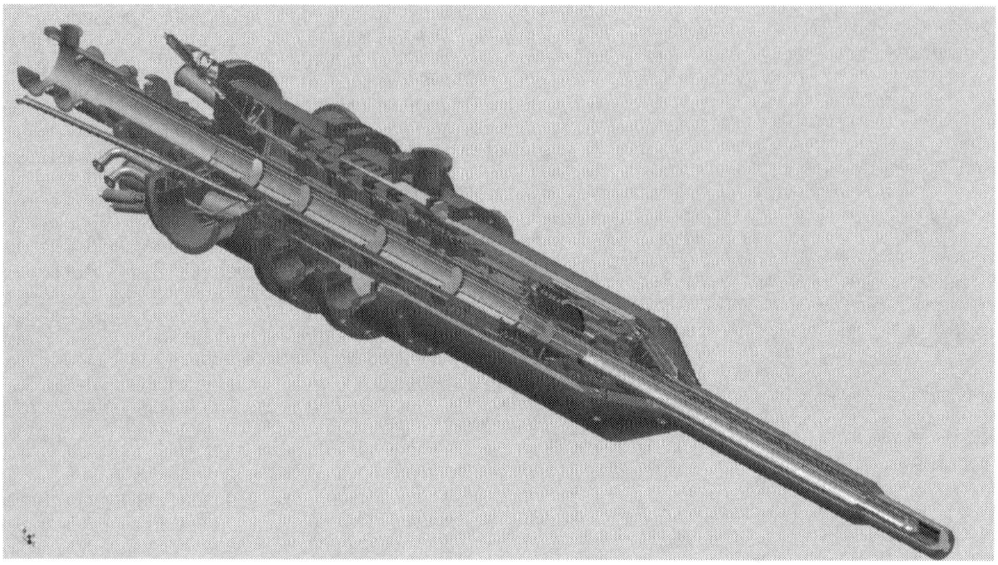

Figure 3: Cut view (technical drawing) of the new horizontal $^3$He/$^4$He dilution refrigerator

precooling stages in fall 2015. Subsequently we started with the overall assembling of the precooling units, the dilution unit (still and low temperature heat exchanger), cooling shields and the outer vacuum envelope to complete the refrigerator. The first cool down of the completed system is now foreseen for September this year. We are still optimistic to get an innovative and reliable working dilution refrigerator for a new class of (double) polarization experiments within the scheme of the so called '4$\pi$ continuous mode' polarized target.
To cool down and cold test the refrigerator we have installed a complete test facility in our traget laboratory to avoid conflicts and overlaps with experimental program of the Crystal-Barrel experimental at ELSA. The test set-up is identical with the one at the CB area and it allows us to operate the new dilution refrigerator under typical run conditions at the beam area. As soon as we have reached the designed parameters and a stable operation at reasonable parameter setting, we will dynamically polarize the target material by means of the internal superconducting polarizing magnet. Once we have successfully demonstrated the feasibility of the new target scheme, we will move the target to the Crystal-Barrel experimental area for upcoming double polarization experiments.

## 4 CFD Simulations of the new dilution refrigerator

### 4.1 Introduction

Parallel to the design and construction of the new dilution refrigerator we started a program to study in more detail the fluid dynamics of a dilution refrigerator more general. Here the precooling stages of the refrigerator were of special interest. Since only rough calculations of the various precooling stages and types of heat exchangers can be found in the literature, we decided to perform a computational fluid dynamics (CFD) simulation of the single units and the combined precooling stage. The goal of the CFD-simulations is to estimate the characteristic flow parameter for every in and out going stream and to simulate the heat exchange between them for a better insight to the operational characteristics of this dilution refrigerator. Simulations for the different parts of the precooling stages were performed and will presented here in more detail [15].

### 4.2 The precooling stages

The precooling unit consists of two heat sinks and three heat exchangers which are shown in Fig. 4. One additional heat exchanger is located in the second heat sink. It is called evaporator and is working at a temperature of 1.2 K. HE1 and HE2 have to precool the incoming coolant (highly concentrated $^3$He gas ($\approx$ 90%) and 10% $^4$He impurity at room temperature) by the usage of $^4$He from the separator and the $^3$He back stream coming from the still down to the separator temperature of 3.8 K. Fig. 6 shows the necessary $^4$He flow rate to cool the incoming $^3$He gas down to this temperature. HE3 is supported by the liquid $^4$He of the evaporator and liquefy the incoming coolant ($^3$He gas). The heat exchanger in the evaporator should prevent the incoming stream from being cooled down to 1.2 K before it enters the dilution unit. Fig. 5 shows a flow scheme for the precooling unit.
The initial calculations for the heat stages of the precooling were done similar to [18–22] based on energy conservation for each stream. The main challenge for the calculation of heat exchangers used in horizontal dilution refrigerators is the determination of the heat exchange between

 DOI: 10.1051/epjconf/201713405001

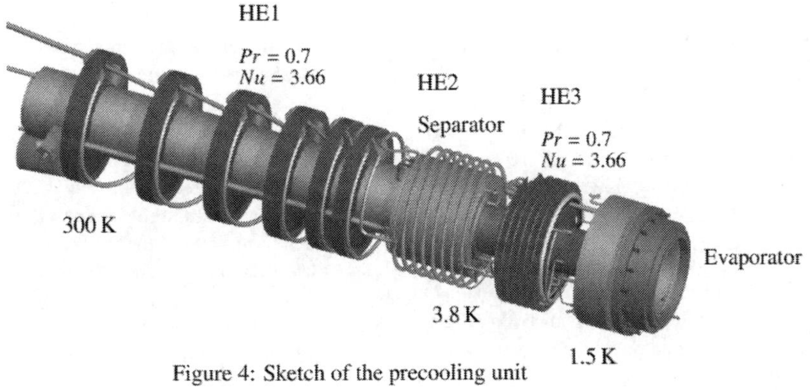


Figure 4: Sketch of the precooling unit

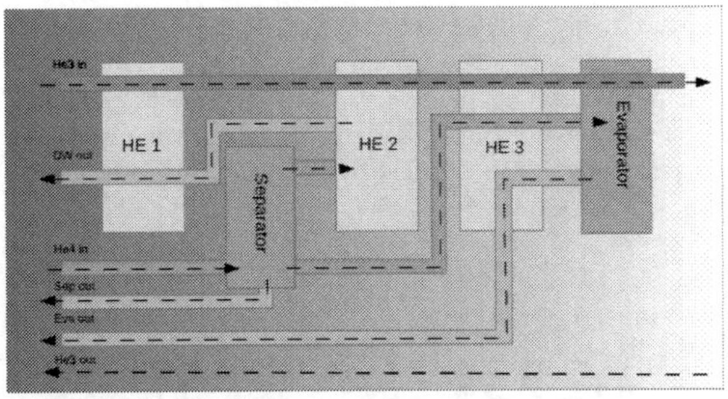


Figure 5: Scheme of the precooling unit and the different streams

the different streams through the solid

$$\dot{Q}_{\text{solid}}(\Delta T_{\text{m}}) = \alpha \cdot A \Delta T_{\text{m}} \text{ with } \alpha \propto Nu(Re, Pr) \cdot \frac{\lambda}{L}. \quad (1)$$

$Nu$ is the Nusselt number, which depends on the Reynolds ($Re$) and the Prandtl ($Pr$) number [20]. These three values are used to characterize the stream behavior. Equation 1 shows that the heat exchange depends on the characteristic flow parameter, the geometry (the surface $A$ and lenght $L$ of the heat exchanger as well as the thermal conductivity $\lambda$ of the material) and the temperature difference $\Delta T_m$. Unfortunately it can be a great challenge to determine this numbers. For the initial calculations the Prandlt and the Reynolds numbers for the backstreaming $^3$He were estimated to $Pr = 0.7$ and $Nu = 3.66$. For the other streams the fluid flows through tubes and the calculations from [18] were used, for the calculation of $Re$ and $Pr$ see [19].

### 4.3 CFD-Simulations

CFD simulations were performed to estimate the influence of the backstreaming $^3$He-$^4$He on the cooling power of the precooling stages. Almost all CFD simulations are based on [16, 17]

$$\frac{\partial}{\partial t}(\rho\phi) + \underbrace{\nabla \cdot (\rho\vec{u}\phi)}_{F_\phi} = D_\phi + Q_\phi. \quad (2)$$

whereupon $\rho$ is the density, $\phi$ a fluid dynamic parameter (e.g. fluid velocity $\vec{u}$, pressure $p$ or temperature $T$...), $F_\phi$ and $D_\phi$ are the convective and the diffusive flow and $Q_\phi$ represents all other distributions given by $\phi$. For a compressible fluid this leads to the

- continuity equation

$$\frac{\partial}{\partial t}(\rho) + \nabla \cdot (\rho\vec{u}) = 0, \quad (3)$$

- momentum equation

$$\frac{\partial}{\partial t}(\rho\vec{u}) + \nabla \cdot (\rho\vec{u}\vec{u}) = \nabla \cdot \boldsymbol{\tau} - \nabla p + \rho\vec{g}, \quad (4)$$

  where $\vec{g}$ represents body forces acting on the fluid contained in the control volume and $\boldsymbol{\tau}$ the stress tensor which depends on viscous and pressure effects of the fluid.

- energy equation

  Defining the specific enthalpy $h = u + p/\rho$ and introducing Fourier's law for heat transfer by conduction $q = -k\nabla T$, with the thermal diffusivity $k = (\mu c_P)/Pr$, the dynamic viscosity $\mu$, the specific heat at constant pressure $c_P$ and the Prandtl ($Pr$) number one can write the energy equation in the enthalpy form

$$\frac{\partial}{\partial t}(\rho h) + \nabla \cdot (\rho\vec{u}h) = -\nabla \cdot q'' + \frac{\partial p}{\partial t} + \nabla \cdot (\boldsymbol{\tau} \cdot \vec{u}). \quad (5)$$

Depending on the specific case additional conditions like the equation of state or different specific material equations lead to a solvable problem. This simulations were performed with OpenFOAM [23]. The meshes for the calculation were created with the including mesh generating code called snappyHexMesh. There are two possible ways of simulation:

1. **Single mesh simulations:** a simulation of a single stream to get additional information about the characteristic flow parameters. This simulations are based on the solver *rhoPorousSimpleFoam* [23].
   The advantage of this method is that higher mesh resolutions are possible and it is easier to determine the characteristic flow parameters like the Prandtl,

the NUSSELT and the REYNOLDS number. These values characterize the thermal coupling between the fluid and the solid in a system and if the flow is laminar or turbulent. The simulation can be performed for all streams until the flow parameters are known. In addition the characteristic flow parameters can be used to perform the calculations from section 4.2 again.

2. **Multi mesh simulations:** a simulation with a different submesh for the calculation of the heat exchange between the different fluid streams and solids. This simulations are based on the solver *chtMultiRegionSimpleFoam* [23].
By the usage of a multi mesh it is possible to set the boundary conditions for each stream in one simulation and run the solvers for each submesh representing a solid or a fluid. In this case the data necessary for the connecting surfaces of the submeshes can be replaced in each iteration and couple the solvers for each region. With this method the heat exchange between the different streams of the heat exchanger can be calculated in one simulation.

For both solvers we added the low temperature data from [25–27] to the thermo physical library implemented in these solvers. The viscosity of the helium gas was modulated with the SUTHERLAND equation [20]. The heat capacity $c_p$ was assumed to be constant for the gaseous streams.

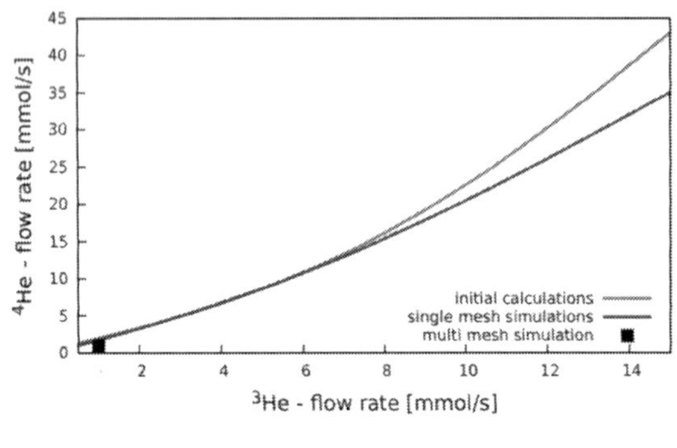


Figure 6: $^4$He flow rate necessary to cool the incoming coolant to the temperature of the separator

### 4.4 Single Stream Simulations

In the beginning single stream simulations of the $^3$He backstream were performed due to the lack of knowledge about the flow parameter of the backstream. Figure 7 shows the pressure and temperature gradient and the velocity profile as an example of this simulations. Here the backstream around heat exchanger 1, 2 and the separator is shown. The beamline and the outer wall are expected to be adiabatic. The massflow is set to 1 mmol s$^{-1}$and the temperature is set to 3 K at the inlet. The heat transfer through the solids is given by the FOURIER equation. For this simulation a constant mean temperature given by the calculations of section 4.2 was estimated for each cooling stage. These initial boundary conditions for the temperature are given in Fig. 7a.
The simulations were performed until a residual of $\leq 10^{-7}$ for all parameters was reached. Due to the heat exchange between the different streams these single mesh simulations have to be repeated until convergence for the complete system is reached. It is also possible to extract *Re*, *Pr* and *Nu* from these data. Afterwards the necessary $^4$He flow-rate to precool the circulating coolant can be recalculated, this is displayed in Fig. 6 by the blue line.

### 4.5 Simulation of a Heat Exchanger

Multi region simulations were performed to calculate the heat transfer between the different streams and the important flow parameters of a heat exchanger in one simulation. In Fig. 8 and 9 the pressure distribution for the incoming and outgoing $^3$He (coolant) is shown for HE2 as an example of this simulation. The temperature distribution for the different streams is given in Fig. 10 and 11. For the simulation shown in Fig. 8 to 11 a $^3$He circulation rate of 1 mmol s$^{-1}$ was estimated. The inlet temperatures for the $^3$He backstream and the $^4$He are set to 3.8 K, the inlet temperature of the incoming $^3$He is estimated to 5 K. The simulations for all heat exchangers of the precooling stages with the flow rates for the different operation modes are ongoing. With this data a prediction for the necessary $^4$He flow rate to precool the circulating $^3He$ under the influence of the $^3$He backstream at a given circulating rate can be done. Fig. 6 shows the prediction from the multi region simulations for the flow parameter of successfully completed simulations at this time. Now it is possible to access the velocity profile, the temperature and the pressure gradient for all parts of a heat exchanger within one simulation. This information can also be used to optimize the performance of the different cooling stages by changing the geometry and by optimizing the flow impedances.

### 4.6 Conclusion

The polarized target group in Bonn developed a new horizontal continuous mode target for the CB experiment at ELSA. To increase the knowledge about the flow characteristics of a horizontal dilution refrigerator CFD simulations were set up. It is possible to use multi region simulations to calculate the performance of the different heat exchangers of a dilution refrigerator (see Sec. 4.5). This simulations are ongoing with all flow parameters, which can be expected for the new continuous mode dilution refrigerator for the CB-ELSA experiment. With this additional information we are able to optimize the concept of our dilution refrigerator which is under construction. After finishing the construction performance tests will be started and the experimental data will be compared with the simulations.

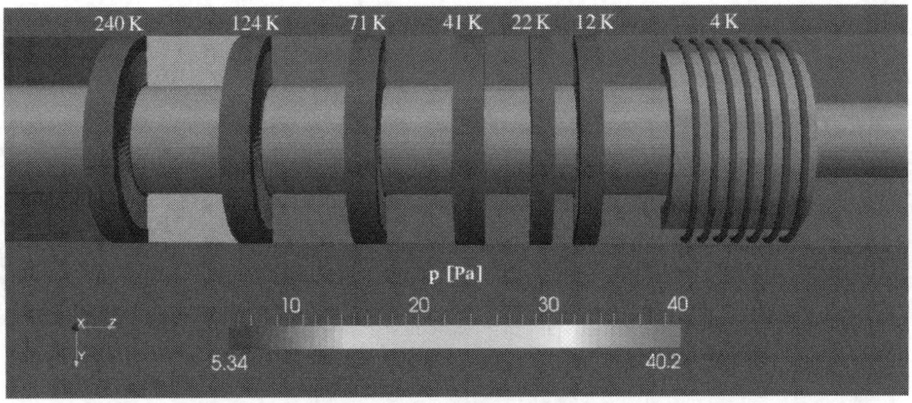


(a) Pressure distribution of the backstreaming $^3$He

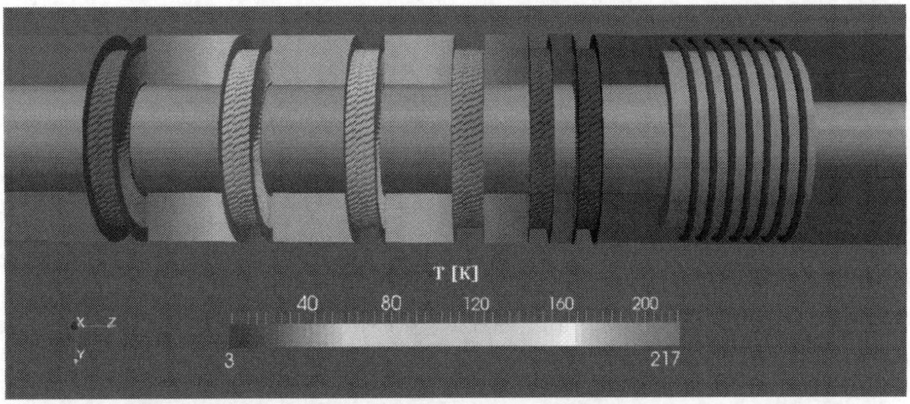


(b) Temperature distribution for the backstreaming $^3$He

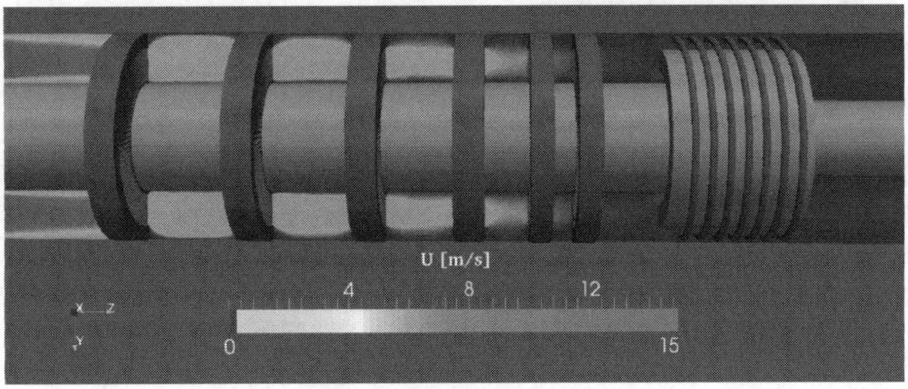


(c) Velocity profile of the backstreaming $^3$He

Figure 7: Simulation of the $^3$He backstream for HE1 and HE2 at a mass flow rate of 1 mmol s$^{-1}$

## 5 Development of a thin, internal superconducting polarization magnet for the Polarized Target

### 5.1 Introduction

For improving the figure of merit of double-polarization experiments at CB-ELSA in Bonn the development of a novel *Continuous Mode-Target* has been proposed in 2004 [14]. One key element for this concept is a thin, high field polarization magnet, which is placed within the refrigerating structure. With such a magnet DNP could be used during the measurement of the experiment. A very first version of a 'thin high field internal magnet' (total thickness 1.3 mm [28]) has been wound for the existing 'CB frozen spin target' as holding coil and was in operation in nearly all double polarization measurements at ELSA.

Summarizing the first experience and tests with the inter-

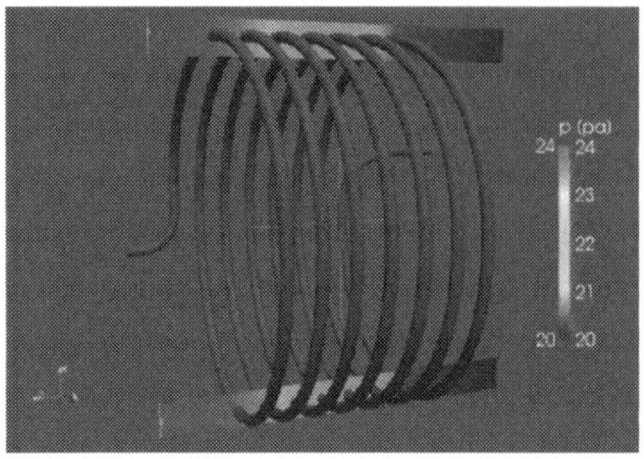


Figure 8: Pressure of the $^3$He backstream in HE2

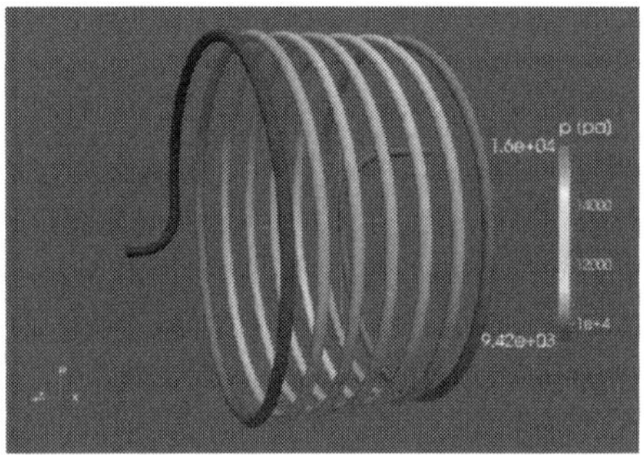


Figure 9: Pressure for the incoming $^3$He stream (coolant) in HE2

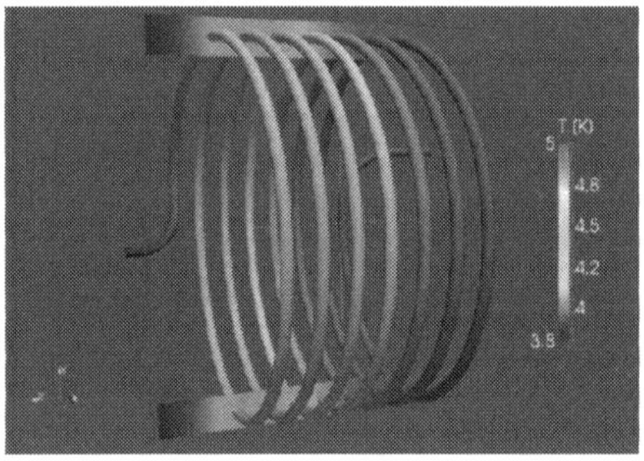


Figure 10: Temperature of the incoming and the outgoing $^3$He stream in HE2

nal superconducting high field magnet, magnetic fields of 2 T are achievable in a thin coil with only 4 to 6 layers of superconducting wire at currents up to 100 A. But with regard to the homogeneity of the magnet, the test measurements led to $\Delta B/B \approx 10^{-3}$, a more dedicated magnet simulation was required. The problems one has to deal with and which had to be solved within the simulation to improve the homogeneity are twofold. The specified minimized dimensions of the solenoid yield a ration between the magnetic volume and the target volume only of 0.05 which presuppose a calculation with respect to the accuracy of the positioning of the field correction elements as well as the positioning of each wire loop on the bobbin.

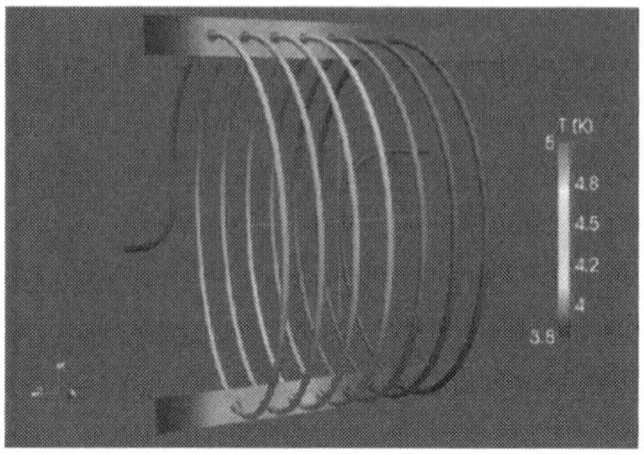


Figure 11: Temperature of the outgoing $^3$He and $^4$He stream in HE2

Finally one has to take the (limited) mechanical accuracy of the winding machine for the winding process of the superconducting wire into account.
The investigation of this problem in the framework of a diploma thesis has led to an 'inverse notched coil' corrected solenoid for the design of the magnet [29]. The simulated solenoid consist of 6 layers of superconducting wire (Ø254 $\mu$m) each with 590 windings. The corrector coils wound on the inner diameter of the solenoid ('inverse notched coil') (Fig. 12) consists of 16 windings within 2 layers. At a nominal current of 90 A a maximum field of 2.5 T should be achievable. The calculated homogeneity is of the order of $\Delta B/B \leq 5 \times 10^{-5}$.
The practical realization of the first design failed due to the limited mechanical accuracy of the winding process which was required to fulfill the high sensitivity of a stable solution of the homogeneity calculation. Therefore a new design concept with respect to the winding accuracy on the bases of the existing coil design was needed [30, 31].

### 5.2 Boundary condition and design concept

A technical drawing of the basic coil carrier can be seen in Fig. 12. The overall thickness of 2 mm should not be exceeded otherwise the detecting threshold will be raised by absorption of particles in the structure. For the DNP process a magnetic field of 2.5 T is needed. To fulfill both requirements a superconducting wire[1] is used. The main field is created by a 6-layer solenoid with 586 windings per layer.
For polarizing deuterons the inhomogeneity (Eq. 8) of the field within the target area must be smaller than $\Delta B/B_0 = 10^{-4}$. Because the field of a perfect solenoid is not sufficient ($\Delta B/B_0 > 10^3$), two correction coils are added. The homogeneity is very sensitive to any misalignment of these coils. In contrast to the *Notched Solenoid* [32], where the correctors are wound on the outer shell of the main solenoid, they are placed in the inner shell of the solenoid. Two notches in the bobbin give maximum preciseness for positioning the coils (Fig. 12). Hence this concept is called the **Inverse Notched Solenoid**. Each correction coil has 2 layers with 9 windings each.

[1] multifilament NbTi, Ø=0.254 mm including insulation

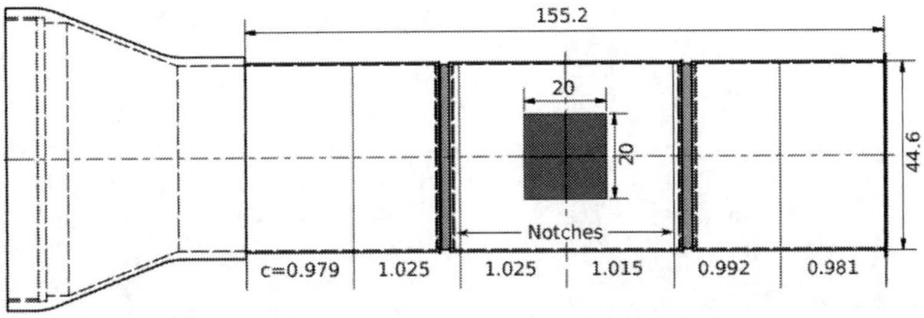


Figure 12: Sketch of the bobbin of an internal polarization magnet. The target area is located in the magnetic center. Within this area the magnetic field must fulfill the homogeneity criteria. The grey filled areas indicate the notches. The six values on the bottom side are the factors $c$ used for the density calculation (see Sec. 5.4.2). Length-units are in mm.

The whole magnet has an inductance of 160 mH and a stored energy of 650 J at 2.5 T. The wires are fixed to the carrier structure by a cryogenic epoxy glue in a wet wiring process.

### 5.3 Calculation

#### 5.3.1 Biot-Savart law

For calculating the magnetic field of a solenoid and its correction coils the Biot-Savart-Law is used (eq. 6).

$$\vec{B}(\vec{x_0}) = \frac{\mu_0}{4\pi} I \int \frac{(\vec{\gamma}(t) - \vec{x_0}) \times \frac{\dot{\vec{\gamma}}(t)}{|\dot{\vec{\gamma}}(t)|}}{|(\vec{\gamma}(t) - \vec{x_0})|^3} \mathrm{d}t \tag{6}$$

Figure 13 shows all necessary vectors for a certain path $\vec{\gamma}(t)$. For this design the path is given by a loop parametrization (Eq. 7) with $r$ as the radius of each loop and $d_{eff}$ (effective diameter, see Sec. 5.3.2) as the distance between two wires.

$$\vec{\gamma} = (r\cos(t), r\sin(t), n \cdot d_{eff}) \tag{7}$$

The product of the counting variable $n$ and the effective diameter $d_{eff}$ gives the position of each loop. Due to the superposition principle the field of each loop may be added up to an overall field at a position $x_0$. To find the best position of the correction coils an algorithm loops over all possible positions and calculates the maximum field inhomogeneity (Eq. 8) within the target area. The central field $B_0 = B(\vec{0})$ is used here as the reference value.

$$\frac{\Delta B(\vec{x_0})}{B_0} = \frac{B(\vec{x_0}) - B_0}{B_0} \tag{8}$$

At the best position of the correction coils the field inhomogeneity can be calculated to $\Delta B/B_0 < 0.4 \times 10^{-4}$. Figure 14 shows the field map calculated for this case. The slight asymmetric structure comes from the fact that the magnetic center does not coincides with the geometric center. This is caused by the orthocyclic winding scheme (see fig.15) in combination of a constant loop number for each layer.

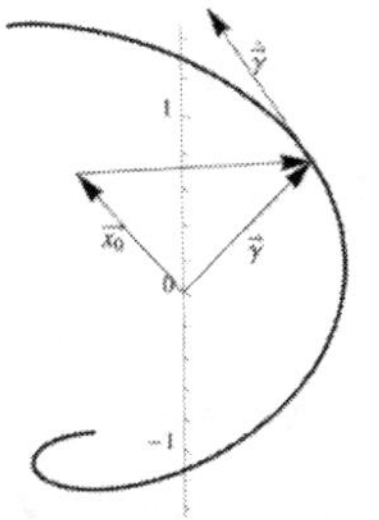

Figure 13: Electric current on a hectically path. The Biot-Savart law is used for calculating the magnetic field $B$ at the position $\vec{x_0}$.

#### 5.3.2 Effective diameter

The *effective diameter* $d_{eff}$ is defined as the distance between one wire and its neighbor. Because there is some space filled up with glue it does not coincide with the mechanical wire diameter $d$. As can be seen in Fig. 15 [33], in orthocyclic wiring the wire forms an equilateral triangle with $d_{eff}$ as edge length. The effective diameter mainly depends on the glue's viscosity, the tensile stress and (lateral) pressing forces on the wire. The latter is caused by the hand tool which applies the glue onto the surface and the efficiency of the wire guiding mechanism.

### 5.4 First prototype

#### 5.4.1 Estimation of the field homogeneity

A first prototype of an internal polarization magnet was produced and tested in liquid helium at a temperature of 4.2 K and a cnetral field of 1.73 T at a nominal current of 65 A [30]. A hall probe could be moved along the central axis of the magnet for measuring the magnetic field $B$. The data are plotted in Fig. 16 with $B$ against the position $z$ on the magnet's central axis. As it can be seen there

DOI: 10.1051/epjconf/201713405001

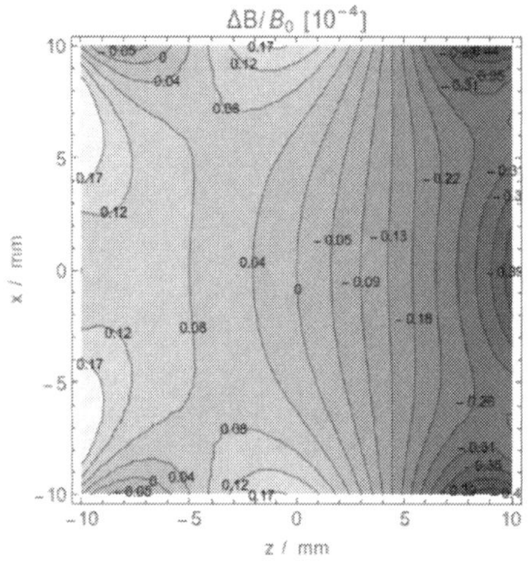


Figure 14: Field map of the target area of the 'Inverse Notched Solenoid'.

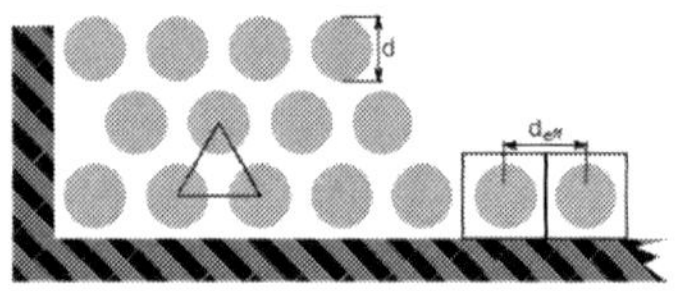


Figure 15: Schematic drawing of an orthocyclic wiring.

is a significant discrepancy between the measured and the calculated field. An asymmetric structure with two local maxima can be seen near the position of the correction coils. From this data it is clear, that the criterion of homogeneity has not been reached.
An explanation for this behavior can be found by a precise analysis of the wire position on the solenoid.

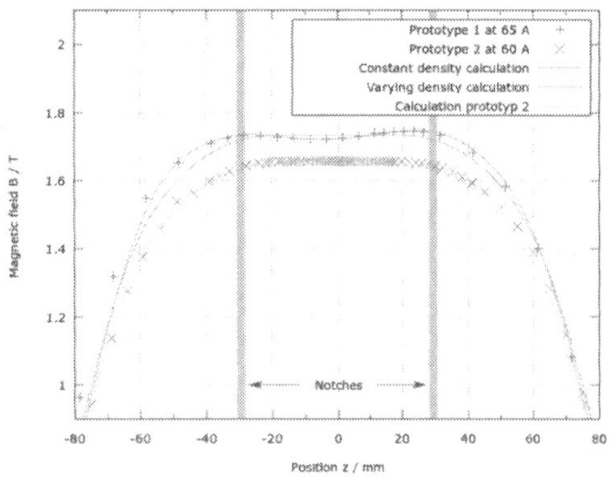


Figure 16: Comparison of the magnetic field measurement of the first and second prototype as well as the calculated (simulated) field for constant and distributed current density of prototype 1. The grey filled rectangles illustrate the position and size of the two correction coils.

### 5.4.2 Density calculation

When calculating the magnetic field a homogeneous current density is assumed over the whole solenoid which presuppose a constant effective diameter. In practice the effective diameter can vary along the solenoid during the wiring process due to a non constant application of the glue to the wire. This lead to a local variation of the wire position compared to the calculated position and finally to a wire density variation which is equivalent to a local variation of the current density.
A simple model considering this effect is introduced in Fig. 12. The solenoid is split up into 6 parts, where the effective distance is modulated by a factor $c$. A factor $c < 1$ means the effective diameter is reduced, the current density in this region is raised. As it can be seen in Fig. 16 this simple model can reproduce the measured asymmetric structure by varying the effective diameter only by about 2 %. The mispositioning of the wires and the non constant effective diameters could be verified by an optical microscopic scan of the magnet surface. The position of each wire has been determined afterwards with high precision by an image analyzing system.
As conclusion from these measurements one have to state, that the local variation of the wire has to be reduced otherwise the homogeneity will be diminished.

## 5.5 Second protoype

The variation of the effective diameter could be reduced to a value better than 1 % by improving the wiring technique. Most important for a high quality winding process towards an orthocyclic wiring (Fig. 15 is a high precision wire guiding. In our case we used a custom made tool to guide the wire with a constant lateral force toward its neighboring loop. The feed speed of the guide was adjusted by a feedback control unit to the coiler speed, the viscosity of the used adhesive and the actual wired layer. All improvements of the new wiring technique lead to a second prototype of the "Inverse Noched Coil".

### 5.5.1 Estimation of the field homogeneity

### 5.5.2 Field measurement

The second prototype has been tested in liquid helium at 4.2 K with a current of 60 A. The magnetic field distribution within the target area on the z-axis has been measured using a standard hall probe at a maximum field of $B_{max} = 1.65$ T. The data are plotted in Fig. 16. In contrast to the first prototype there is, as expected, a flat plateau within the target area. But due to the known limited resolution of the hall probe the homogeneity of the solenoid could be estimated not better than $\Delta B/B_0 \leq 10^{-3}$.
For a further test the NMR-signals of a polyethylen-sample at 4.2 K and a center field of $B_0 = 1.599$ T have been recorded. The cylindric sample with a diameter of 5 mm and a length of 22 mm could be moved along the the z-axis. Figure 17 shows the normalized signals at different position. A frequency shift of each peak caused by

the field inhomogeneity was compensated by adjusting the current resp. magnetic field to align each peak to the center frequency. As it can be seen in Fig. 17 this alignment could only be archived within a field band of about 0.35 mT around the center frequency. The field deviation $\Delta B$ has been calculated from this current adjustment. The maximum field deviation close to the margin of the target area can be found at z=10.28mm and is of the order of 0.20 mT. With regard to the uncertainty of the alignment the measurement lead to an estimated value for the homogeneity of about $\Delta B/B_0 = 3.5 \times 10^{-4}$.

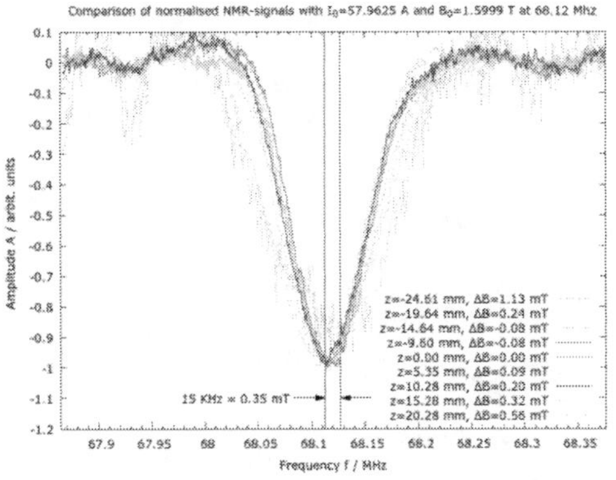


Figure 17: Normalized NMR-Signals of a polyethylene sample in thermal equilibrium in different positions within the magnet. All solid lines correspond to a position within the target area.

### 5.6 Conclusion

A thin, internal superconducting polarization magnet for the Polarized Target is actually in development. A magnetic field of 2.5 T with a field inhomogeneity better than $10^{-4}$ within the target area is necessary for the DNP process with deuterons. The basic concept - the Inverse Notched Solenoid - works with a solenoid for the main field and two correction coils which are added in the inner radius. This gives an accurate positioning of these correction coils.

Tests with the first prototype showed that the field inhomogeneity will definitely be influenced by a local variation of the wire position in the solenoid. To reduce the variation and to improve the accuracy of the wire positioning, a modified high precision wiring technique has been established. A second prototype has been build and passed first tests: The field measurement along the z-axis shows the expected flat plateau within the target area. NMR-Signals of polyethylene at 4.2 K could be seen with this magnet and indicate a sufficient homogeneity.

For the next step the magnet will be installed in our $^4$He-refrigerator. The maximum possible current will be determined at 1 K. Additionally the magnet will be tested for dynamically polarizing a butanol-target.

Since the experimental program of the Crystal-Barrel experiment at ELSA has foreseen the measurement of double polarization observables on transversely polarized target nucleons, it is planed to design a combined coil providing a longitudinal high magnetic field for the dynamic polarization process and a weak transverse magnetic field to maintain the polarization in the so called 'holding' or 'frozen spin mode' of the polarized target. First preliminary design studies indicate the feasibility of this concept.

## 6 Polarization and relaxation characteristics of irradiated polymeric materials at 1 K and at 2.5 T

### 6.1 Introduction

Polymers have been discussed as candidates for solid state target material for at least the last two decades. Unlike the commonly used alcohols e.g. butanol or inorganics e.g. ammonia, polymer targets have a major advantage in the fact that they can be formed to practically any desired geometry and can thus provide a filling factor close to 1. In addition, the materials are solid at room temperature and handling is unproblematic. The defining point of a good solid state target material is a high polarization of the proton or deuteron components. The polarized target group at the Paul Scherrer Institute (PSI) showed dynamic nuclear polarization of protons in thin polyethylene (PE) foils and small tubes, chemically doped via diffusion of the highly stable free nitroxyl radical 2,2,6,6-tetramethyl-piperidine-1-oxyl (TEMPO) to a spin density of $2 \times 10^{19}$ spins/cm$^3$, to values of 70% in a 2.5 T magnetic field at a temperature below 0.3 K [34, 35]. Ideally one would like to achieve polarizations in polymer materials comparable to the classic materials of over 90%. It has also been proposed that the crystallinity of polymers may in fact be a major factor in obtainable polarization values, as polarizations seem to be higher in amorphous materials than in crystalline materials at least with regards to chemically doped samples [36]. Kumada et al. [37] showed that for PE doped in the same manner as the PSI materials this was indeed the case. Their EPR studies showed that the chemical dopant TEMPO only diffused into the amorphous regions of the polymer. Whether or not the crystallinity affects the polarization if the radicals can indeed permeate the crystalline region, as is the case if the materials are irradiated, is still an open question [38].

### 6.2 Polymer sample materials

For the ongoing irradiation studies the focus was placed on two types of long chained polymer molecules: Polyethylene (PE) and polypropylene (PP). The density of both is comparable to that of butanol with a slightly higher dilution factor $f$ (see Sec. 1). PE is a long chain of carbon atoms, each with two associated hydrogen atom substituents ($CH_2$ monomer - see Fig. 18). For PP the situation is slightly different: The backbone of the molecule consists of carbon atoms, however next to every polyethylene monomer there is a carbon atom associated with a single hydrogen atom and a $CH_3$ group ($CH_2$-CH-$CH_3$ monomer - see Fig. 19). The arrangement of the chains can be amorphous, the chains are random and unoriented, crystalline, the chains are folded and packed, or more commonly semi-crystalline in which crystalline structures are dispersed within the amorphous regions. Three different types of polyethylene were used.The samples consisted of 1) highly amorphous, low density PE (LDPE) with lots of branching, 2) a low branching, high density PE (HDPE) with a high amount of linearly arranged chains (crystalline) and 3) linear low density PE (LLDPE), a high branching material interspersed with small crystalline regions. Polypropylene is differentiated by its stereo regularity, the relative orientation of the $CH_2$ and CH-$CH_3$ groups. Three isotactic materials of varying chain length were used for which all the substituents are located on the same side of the macromolecular backbone.

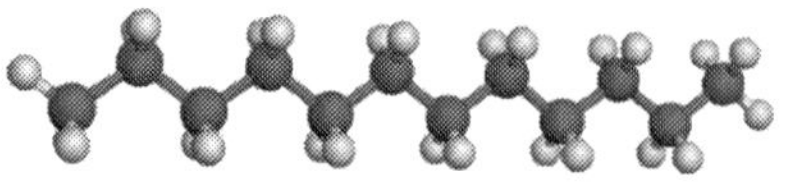

Figure 18: Polyethylene $(CH_2)_n$

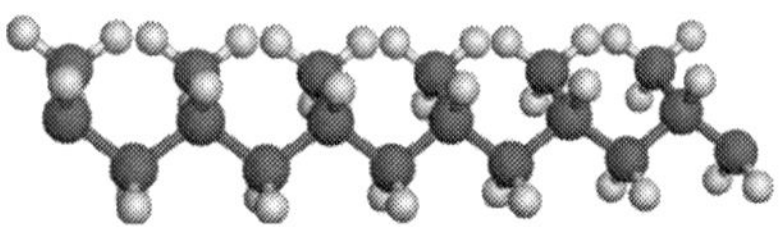

Figure 19: Polypropylene $(CH_2\text{-}CH\text{-}CH_3)_m$

## 6.3 Classification of samples after irradiation

To achieve high polarization of nuclei, DNP is used to transfer spin polarization from electrons to nuclei. In order for this to be possible electron spins must be introduced into sample materials e.g. by the creation of persisting structural defects by irradiation. In theory hydrogen extraction and chain scissions can occur, however the former is by far more likely in polymer materials. After the irradiation the first step in the analysis of the materials is EPR spectroscopy. The comparison of the spectra of the irradiated material with spectra of a known spin density taken under the same experimental conditions gives access to the spin density of the irradiated material. This essentially tells us how many defects per unit weight were created in the irradiation process. Typical X-band spectra measured at a temperature of 77 K for PE and PP are given in figures 20 and 21. The EPR spectrum also gives insight into the type of paramagnetic centers that have been created. Understanding the underlying radical structure can help to optimize the temperature dependent irradiation process with regards to polarization. An alkyl-type radical (see Fig. 22) is created in irradiated PE and gives the sextet of the EPR spectrum: The electron is localized on a p-orbital of the single carbon atom and has a hyperfine interaction with the single $\alpha$- and four surrounding $\beta$-protons [39],

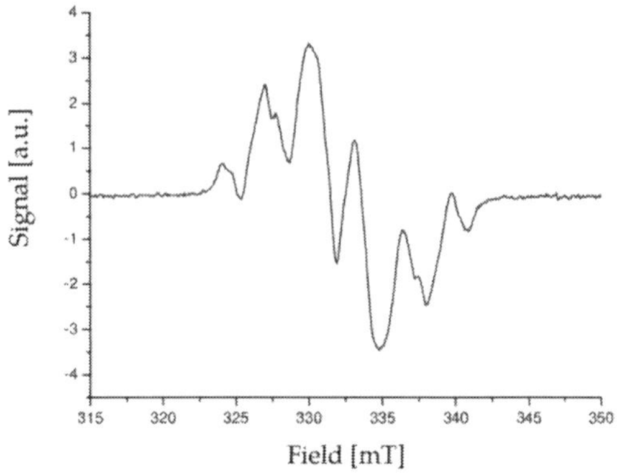


Figure 20: EPR of irradiated polyethylene.

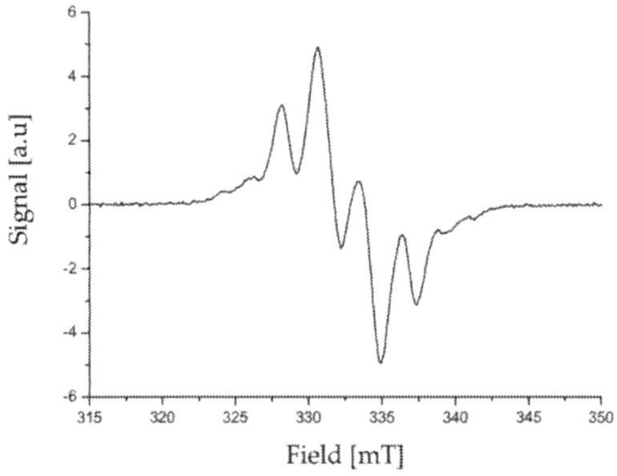


Figure 21: EPR of irradiated polypropylene.

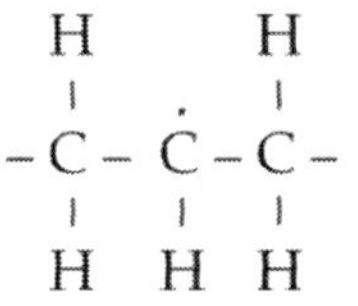

Figure 22: Primary radicals created by irradiation of PE.

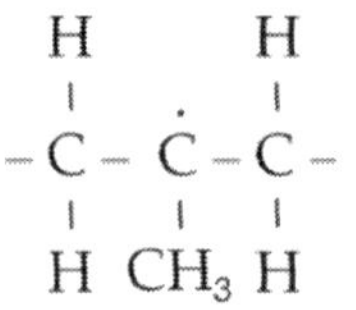

Figure 23: Primary radicals created by irradiation of PP.

[40]. On the other hand many structures have been proposed for the radicals resulting from the irradiation of PP, however a general consensus does not seem evident from the literature and major differences have been reported between spectra resulting from different types of irradiation e.g. $e^-$, $\gamma$ and UV-sources. The main obstacle in determining the constituents of the spectra is the fact that irradiation

of the material is thought to lead to a superposition of radical structures that are primarily obtained by the removal of hydrogen atoms. In polypropylene there are 3 sites from which a hydrogen atom can be removed, each of which would be expected to give an individual and distinct EPR spectrum. A hydrogen atom can be removed from the carbon backbone opposite of the methyl group or from the methylene bridge, as well as from the side-chain methyl itself. The most likely explanation is the dominance of the alkyl-type radical shown in Fig. 23 with additional structure resulting from the methyl group [41–44].

### 6.4 Preliminary irradiations at the LINAC1

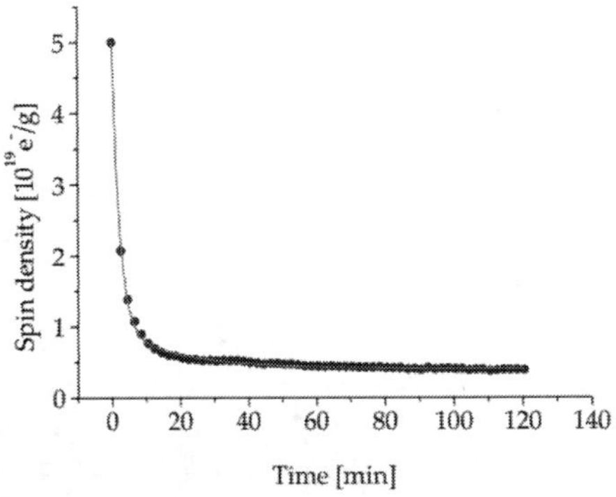


Figure 24: Decay of spin density in irradiated PP caused by heating of sample in nitrogen.

The electron irradiations in this study have so far been conducted in two consecutive data taking periods. Each of which will be described separately. The LINAC1 of the ELSA facility was used for the initial irradiation. A batch of PP pellets was irradiated in an argon cooled cryostat [45] and subsequently stored in liquid nitrogen. The initial spin density of the material was very close to 5 x $10^{19}e^-/g$ and in the order of magnitude in which one would typically expect DNP to be possible. As illustrated in Fig. 24 a subsequent heating of the samples in nitrogen at room temperature leads to a significant drop in the spin density. The initial rapid decay of the alkyl-type radical is followed by an almost stable region that lasts for some hours in which the alkyl-type radical is slowly converted into a allyl-type radical [46]. The conversion of the primary radical is evident from the slow increase of a sub-structure in the EPR spectrum, however the contribution of the latter to the spectrum on a whole is minimal and extremely difficult to extract due to the underlying noise. In this regard it is to be expected that any change in polarization characteristics of the material by heating are related to the primary radical. Two of the characteristics of interest for target materials are the nucleon relaxation time and the maximum obtainable proton polarization values. The change of these characteristics are shown in figures 25 and 26. Samples were analyzed at our standard $^4$He-refrigerator experimental conditions at a

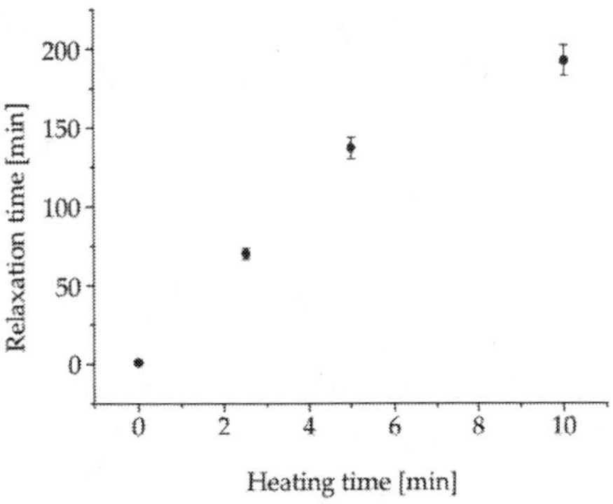


Figure 25: Relaxation times of irradiated PP after heat treatment.

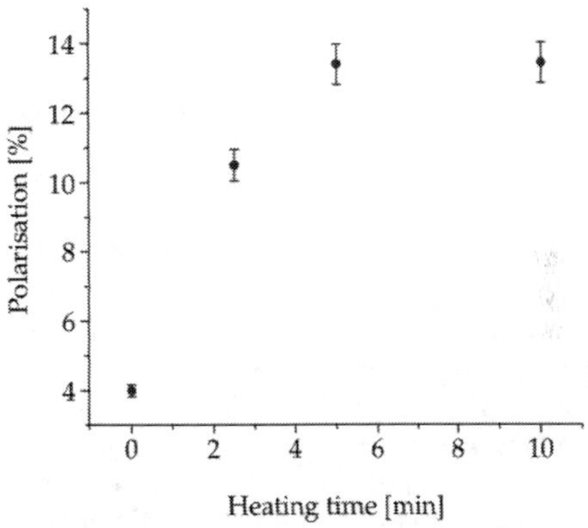


Figure 26: Polarization of irradiated PP after heat treatment

temperature of 1 K and a magnetic field of 2.5 T. Samples were heated for 2.5, 5 and 10 min. In these short periods of heating the change of the relaxation times is enormous. The non heated sample has a relaxation time in the region of a minute in comparison to approximately 70 min for the sample heated for only 2.5 min. Additional heating increases the relaxation time further. The non heated sample gave maximum polarization values of $(4 \pm 0.2)\%$, with additional heating this was increased to $(13.4 \pm 0.6)\%$.
Unfortunately shortly after our introductory efforts the LINAC1 entered a prolonged period of maintenance and it was not possible to conduct any further irradiations of materials at this site: An alternative was needed.

### 6.5 Present irradiations at the LINAC2

In the past all irradiations of target materials in Bonn have taken place at the LINAC1 of the ELSA facility, however at present it is not possible to use this accelerator. Luckily, the second linear accelerator LINAC2 is well maintained and primarily provides electron beams for the Crystal Barrel and BGO groups. Modifications e.g. to the current beamline were needed, however it is now possible to extract the electron beam into the irradiation area. The end energy of LINAC2 is slightly higher than that of LINAC1 with 24 MeV instead of 20 MeV, yet the beam intensity is 20x lower, meaning the irradiation times are significantly longer than in the previous irradiation of the polymeric materials (10 min), yet still not even close to the irradiation times of the non-organic materials such as $NH_3$ or LiD (>10 h). A typical irradiation time needed to obtain a nominal spin density of $10^{19}e^-/g$ is conveniently in the region of a few hours.

As described in the previous section, a heat treatment of irradiated samples enhances the maximum polarization values and increases the relaxation times. This positive effect should also be obtained by irradiating the materials above a material dependent specific temperature. In this way the "reorganization" of the electronic structure is obtained directly during the irradiation. In 20 min the change of the heated sample is more or less completed by the transition of the primary radical into a quasi-stable state. In the case of an irradiation the same (or a similar) radical can be created over a period of a few hours in a much more controlled way. To find the optimal temperature of irradiation a wide range cryostat was used [47],[48]. The cryostat uses a dual regulating system, implementing a slow cycled liquid nitrogen cooled heat exchanger coupled to a closed helium gas system with a fast regulating 1000 W heater. In this way it is possible to irradiate samples at temperatures from 90 K to room temperature with a stability of ±1 K. The lower beam current of the LINAC2, in comparison to the LINAC1, works in favour of the temperature stability.

At present the work is very preliminary as we have only recently gained access to the facility and a broad outline of the programme will be given. Three different polyethylene samples, differing in density and linearity, and 3 different polypropylene samples, with varying chain length, have been irradiated simultaneously in the above mentioned cryostat. This ensures comparability of the samples: Any differences within a sample batch are thus due to differences in the materials. Irradiations were conducted at temperatures of 140 K, 180 K and 210 K. The injected charge of approximately 3 mC was sufficient to produce radical densities in the order of magnitude necessary for DNP.

As described in the previous section the first analysis of the materials was done with an EPR spectrometer and is summarised in Fig. 27. It was found that the spin densities of the different types of polyethylene are indistinguishable at the irradiated temperatures and thus the average was taken as a polyethylene batch. The same applies to the polypropylene samples. It is evident that the radical production mechanism in polypropylene is much more effective and the yield of radicals was around 40% higher. Over the analyzed temperature range there is a drop in the radical yield to higher temperatures. This is likely mostly due to the temperature dependent recombination process of the primary alkyl radical, as well as the conversion into a secondary radical to a far smaller proportion. As with the heated samples the latter can be seen the in EPR spectrum of the materials, however plays a subordinate role. The alkyl radical is still the dominant radical in all spectra. The analysis of the polarization values of the irradiated samples at a temperature of 1 K and a magnetic field of 2.5 T is currently in progress.

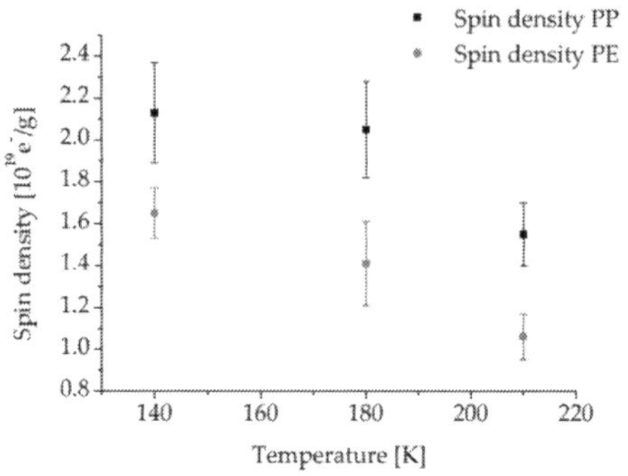


Figure 27: Spin density of PP and PE after irradiation at various temperatures.

### 6.6 Conclusion

As has been proposed in many papers, polymers can be used in particle physics experiments in which high polarization values are obtained using DNP. To what extent the polarization values are adequately high for practical use in such experiments remains to be seen, but previous results from chemically doped samples in the past and current results from the ongoing electron beam irradiations are very promising. The heating of the sample after irradiation has an effect on the relaxation and polarization characteristics of the samples. Significantly higher values of polarization are obtained after heating the sample, whilst relaxation rates become longer. The yield of radicals created as paramagnetic centers in PP by irradiation are in the region of 40% higher than in PE at the investigated irradiation temperatures. What influence the irradiation temperature has on the polarization characteristics is presently being studied: The change of the electronic structure initiated by heating should be obtained directly by irradiation at higher temperatures. Polarization measurements at very low temperatures are planned for the near future.

## 7 Development of a new Q-meter module

### 7.1 Introduction

In the research field of polarized target physics the Q-meter is a well established technique to determine the polarization of solid target materials [49]. One of the most common Q-meter modules is represented by the so called Liverpool-NMR [53], which was originally developed for the EMC experiment in the 1970s. However, since the module was built nearly 40 years ago, many components are no longer available. Thus the repair of existing modules and reproduction of new modules is impossible. Starting from that point, a new Q-meter module was developed and investigated for the use in polarized target experiments

[50].
For the operation and evaluation of measurements in particle physics experiments with polarized solid targets a precise knowledge of the sample polarization $P_t$ is essential. This can already be seen by the so-called figure of merit

$$FOM \sim f^2 P_t^2 \sim \frac{1}{t_{meas}} \quad (9)$$

of a polarized solid target, which characterizes the quality of the used target material with a dilution factor $f^2$ and is anti-proportional to the required measurement time $t_{meas}$ for a given precision of these measurements.
In general, for the detection of target polarization the method of nuclear magnetic resonance is used. By means of a resonant circuit whose inductance is located in proximity to the sample, an alternating field is generated, which is oriented orthogonally to the external magnetic field $B_0$. If the frequency of the alternating field matches the Larmor frequency $\omega_L$ of the investigated spin system with the gyromagnetic factor $\gamma$,

$$\omega = \omega_L = \gamma B_0 \quad (10)$$

transitions between the Zeeman levels are induced. However, this interaction only occurs for a narrow frequency range in the region of the Larmor frequency of the nuclei. Therefore, this effect is also called resonance absorption. Macroscopically, this results in a change of the sample magnetization, which is caused by the different occupied Zeeman levels of the sample.
Based on this change in magnetization, the NMR spectrum of the investigated sample can be determined. The area $A$ under the NMR line is proportional to the target polarization

$$P_t = k \cdot A \quad (11)$$

with the so called enhancement factor[3] $k$.

### 7.2 NMR methods

In principle, there are two different methods for the detection of the NMR spectra of a sample. For the so-called pulsed NMR method, which is the current standard in many scientific fields, the sample magnetization is deflected from its original position parallel to the external field $B_0$ by a short and powerful RF pulse. The subsequent relaxation of the magnetization back to equilibrium can be described by means of the Bloch equations [51, 52], leading to a voltage induction into the NMR coil. This time-resolved voltage signal is commonly known as free induction decay signal FID and can be converted by a Fourier transformation into the desired NMR spectrum. However, for the exact determination of polarization by this method there arise different problems, which are attributable to a non-uniform signal excitation and a signal loss at the beginning of the FID detection.
Therefore, the well established continuous wave NMR technique is still the method of choice for the accurate detection of the undistorted NMR spectra and so for the measurement of the target polarization. Usually, this method is preformed at constant external field[4] $B_0$, while the frequency of the applied radio-frequency alternating field is swept over the resonance. Depending on the number and direction of the excited transitions between the Zeeman levels of the spin system, the power consumption or output of the resonant circuit changes. Since also the quality of the resonant circuit is affected, this set-up is also called Q-meter. The change of the quality of the resonant circuit can be detected as a variation of the radio-frequency power reflected from the oscillating circuit.
However, by this means there are two difficulties. On the one hand, radio-frequency signals are difficult to handle, on the other hand, the signal to detect consists of two signal parts and is complex. Only the area below the imaginary part of the signal, the so-called absorptive signal component, is proportional to the target polarization $P_t$. To determine the amplitude of the radio-frequency signal one uses a diode detector followed by a low-pass filter which converts the RF signal to a DC signal proportional to the RF amplitude. For the detection of the absorptive signal component, a phasesensitive detector is used, which is in phase with the exiting signal.

### 7.3 Developement of a new cw NMR module

Therefore, the requirement for a newly developed cw NMR module for polarization detection of solid targets was given, which should serve as a substitute for the Liverpool module. However, possible improvements should not be disregarded. To enable maximum compatibility with existing systems, the new development [54] is based in many ways on the construction of the Liverpool module. Specifically, these are the following:

- exciting RF signal via external generator
- two demodulation channels
  - diode channel for receiving the magnitude spectrum and to tune the NMR resonant circuit
  - phase-sensitive detector for the detection of the absorption spectrum
- layout of the connector for power supply and signal transmission
- geometric dimensions of the module

The main differences between the modules can be attributed to various improvements and the necessary replacement of no longer available components:

- outsourcing of the resonant circuit components – ensures greater flexibility in experimental set-up of the measuring circuit (e.g. $\lambda/2$-cable)

[2]The ratio of polarizable nuclei to the total number of nuclei in the target material

[3]This factor is usually determined by a calibration in thermal equilibrium, because the corresponding polarization $P_{t,TE}$ can be calculated for a given magnetic field and a given temperature, so that a simple assignment $k = P_{t,TE}/A_{TE}$ can be made. This method is also called TE-calibration

[4]For the operation of a polarized target, this results in the advantage that the necessary dynamic nuclear polarization frequency $\omega_{DNP}$ is kept constant

- the capacitive elements of the resonant circuit have been replaced by a varicap diode – this allows electronic tuning of the resonant circuit and an access to a measurement based on synchronous resonance tuning (see [55])
- except the resonant circuit components, all components are on one circuit board with a system-wide impedance of 50 ohms – this allows a simple construction and repair of the modules, also possible signal reflections should be minimized
- replacement of the radio-frequency amplifier by common models from Mini Circuits – good signal-to-noise ratio and price-performance ratio
- exchange of rectifier diodes - granted a higher linearity of this channel (see below)

To shield the components against internal and external noise, the boards were placed in specially designed housings of solid brass, which were adjusted to each other.

### 7.4 Test measurements

For the warranty of functionality of the newly developed cw NMR module various test measurements were carried out. The most important ones are represented here briefly.

#### *7.4.1 Linearity*

A basic requirement for the exact detection of the target polarization over the area units $A$ under the NMR-line is only possible, if the amplification of the signals in the relevant performance range[5] is constant. Therefore, measurements at different frequencies[6] were performed to test the linearity of the two channels in this region. The NMR resonant circuit was replaced by a variable attenuator, so that different input power levels (-62dBm to -31dBm) could be simulated.

Considering the resulting voltage profiles of the matched BRM channel in Fig. 28, a high degree of linearity can be seen in the work area for the different frequencies. The varying slopes between these traces at the given frequencies are due to different gains. This fact can be attributed to the frequency dependencies of the used components and is not important for the detection of the polarization at a fixed frequency.

Checking the linearity of the diode-channel based on BAT85-diodes, also used in the Liverpool module, there was a high deviation in a linear amplification at 212.6MHz. This effect could ultimately be attributed only to the diodes used. For this reason, various diodes were tested with respect to their linearity in the work region. It was found that the diode type SMS7630 which, according to data, is especially designed for mixers, shows the most promising behavior. Also measurements in the other relevant frequencies demonstrated the advantages of this type of diode, so that the linearity of this channel could be increased significantly.

[5]The input power of the NMR module during a polarization measurement is, according to experience, at about -41dBm.

[6]Larmor frequency of deuterons and protons at 2.5 T and 5 T.

#### *7.4.2 Detection of real spectra*

In addition to linearity, the possibility for the exact detection of the TE polarization plays an important role. For this reason, the TE signals of protons and deuterons of different samples were measured. As can be seen in Fig. 29, even smallest signals become clearly apparent from the predominantly white noise of the circuit, so that an exact determination of area units in the thermal equilibrium is possible.

A particular challenge for the polarization detection is represented by the TE signals of quadrupole broadened deuterons, since these spectra are split over a wide frequency range and the corresponding signal amplitudes are small. Nevertheless, as illustrated by fig. 30, also these spectra can be well detected with the new module, so that a polarization determination based on the TE method is possible.

By the characteristic waveform, which is for example described in [56], a further access to the polarization by the so-called asymmetry method is possible, directly from the line shape. Plotting the determined polarization against the respective area units, as shown in Fig. 30, leads to a linear correlation, where the slope is the known enhancement factor $k$. Comparing the two enhancement factors ($k_{TE} = 12.22 \pm 0.57$ and $k_{asy} = 12.39 \pm 0.07$) identified for this sample, shows no major contradictions and can thus be used for a confirmation of the correct operation of the module.

### 7.5 Conclusion

The various tests have shown that the newly developed cw NMR-module is ready for experimental use and even smallest signals can be well detected. First measurements were performed with a trityl-doped D-butanol target in Mainz and have shown no discrepancies. The high compatibility with the Liverpool module also ensures an easy deployment in existing test set-ups, therefore an additional alternative for the polarization detection is available.

Finally with the new NMR module we have build an experimental tool to measure the nucleon polarization with high presission for single or double polrization experiments. It will be the NMR 'standard' for upcoming polarized target set-ups and polarization experiments.

## 8 Conclusions and Outlook

Here we reported on latest developments of the collaborative work of the polarized target groups at the Ruhr-Universität Bochum and the Universität Bonn within the CRC16 subproject D1 'Polarized solid state target'. Over the last 12 years both groups always pursued the objective to increase the 'figure of merit' and the luminosity of the polarized target for polarization experiments at various particle beams and experiments. While the Bochum group concentrated on dedicated target material research to maximize the nucleon polarization especially in deuterated samples, the Bonn group focused its research program on the development of the so called '4$\pi$ continuous

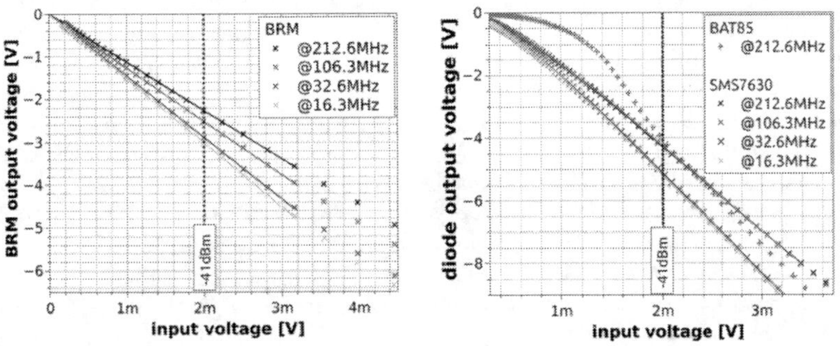


Figure 28: Measuring results for both channels for the observation of the linearity at selected frequencies. As can be seen, a constant gain of the diode channel based on BAT85 diodes was not given. However, by the exchange by SMS7630 diodes, the effect could be eliminated, so that both channels show a high degree of linearity in the measuring range at about 2mV input voltage (~ -41dBm input power).

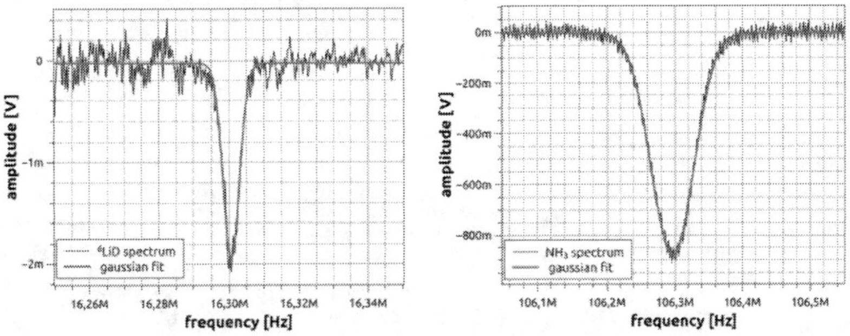


Figure 29: NMR spectra of deuterons ($^6$LiD @77K: $P_t$ = 0.0007%) and protons (ammonia @1K: $P_t$ = 0.255%) in thermal equilibrium at 2.5 T. As can be seen, even smallest signals can be well detected, enabling a sufficiently accurate measurement for a TE-calibration.

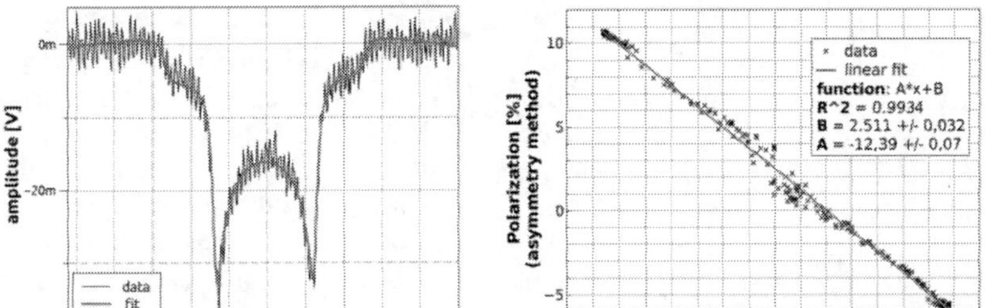


Figure 30: The TE spectra of quadrupole broadened D-butanol and the calculated polarization versus the determined area units of signals. As can be seen, there is an almost linear relationship, thereby providing a high linearity of this channel is illustrated.

mode polarized target'. Nowadays the research activities led to highest proton and deuteron polarization values up to $P_p \sim 90\%$ and $P_d \sim 85\%$ in scattering experiments and a frozen spin target facility with large angular acceptance (close to $4\pi$) which provides all possible polarization directions by means of the internal superconducting holding coils. This technique has been developed by the Bonn polarized target group in the early 1990s, long before the CRC16, and is nowadays the standard frozen spin spin technique at the accelerator facilities in Bonn, Mainz and JLAB. The '$4\pi$ continuous mode polarized target' extend the range of the frozen spin technique to higher beam intensities (by a factor of 1000) and data taking efficiency and will set the new standard for the application of the polarized solid state target in scattering experiments using $4\pi$-detection systems.

The work reported here has been supported by the Deutsche Forschungsgemeinschaft within the SFB/TR16.

## References

[1] S. Goertz, W. Meyer and G. Reicherz, *Polarized H, D and $^3$He targets for particle physics experiments*, Prog. Part. Nucl. Phys. **49**, 403-489 (2002)

[2] W. Hillert, *The Bonn Electron Stretcher Accelerator ELSA: Past and future*, EPJ **A28**, 139 (2006)

[3] W. Hillert, *in this proceedings*, (2016)

[4] M. Goldman, *Spin Temperature and Nuclear Magnetic Resonance in Solids*, Oxford University Press, (1970)

[5] G. Reicherz et al., *Pulsed NMR for the determination of the nuclear polarization*, NIM **A526**, 96 (2004)

[6] Ch. Hess, *Ein gepulstes NMR-System zur Polarizationsmessung an Festkörpertargets*, Diploma thesis, RUB (2005)

[7] St. Goertz et al., *Highest polarizations in deuterated compounds*, NIM **A526**, 43 (2004)

[8] St. Goertz, *Spintemperatur und magnetische Resonanz verdünnter elektronischer Systeme - ein Weg zur Optimierung polarizierter Festkörper-Targetmaterialien*, Habilitation thesis, RUB (2002)

[9] J. Heckmann, *Electron spin resonance and its implication on the maximum nuclear polarization of deuterated solid target materials*, Phys. Rev. **B74** , 134418 (2006)

[10] J. Heckmann et al., *Recent Progress in the Dynamic Nuclear Polarization of Solid Deuterated Butanol Targets*, Appl. Magn. Reson. **34**, 461 (2008)

[11] Ch. Bradtke et al., *A new frozen-spin target for $4\pi$ particle detection*, NIM **A436** 430 (1999)

[12] H. Dutz et al., *First measurement of the Gerasimov-Drell-Hearn sum rule for $^1$H from 0.7 to 1.8 GeV at ELSA*, Phys. Rev. Lett. **91**, 192001 (2003)

[13] Ch. Rohlof and H. Dutz, *Effective densities and polarizations of the targets for the GDH-experiments at MAMI and ELSA*, NIM **A526**, 126 (2004)

[14] H. Dutz, *Summary of the 9$^{th}$ International Workshop on polarized solid state targets and techniques*, Trieste, Italy 2004, World Scientific **221** (2005)

[15] S. Runkel, *CFD-Simulations of a '$4\pi$-continuous-mode' dilution refrigerator for the CB-ELSA experiment*, XVI$^{th}$ International Workshop in Polarized Sources, Targets, and Polarimetry, **PoS(PSTP2015)018**, Bochum (2015)

[16] J. H. Ferziger, M. Peric, *Computational Methods for Fluid Dynamics*, Springer, Berlin (2002)

[17] H.K. Veersteg, W. Malalasekera, *An introduction to Computational Fluid Dynamics*, Longman Scientific & Technical, Harlow (1995)

[18] C.Bradtke, *A New Frozen-Spin Target for the Measurement of Helicity Asymmetry of the Total Photoabsorption Cross-Section*, PHD-Thesis, Universität Bonn (2000)

[19] W. Polifke and J.Kopitz, *Wärmeübertragung: Grundlagen, analytischer und numerische Methoden*, Pearson Studium, 2nd edition, **chapter 8**, Munich, Germany (2009)

[20] H. Oertel, *Prandtl - Führer durch die Ströhmungslehre*, Springer, Wiesbaden (2012)

[21] G. Frosatti, *Experimental Techniques: Methods for Cooling Below* 300 mK, Journal of Low Temperature Physics **Vol. 87**, 3 (1992)

[22] Frank Pobell, *Mater and Methods at Low Temperatures*, Springer 3rd. edition, Berlin (2007)

[23] OpenFOAM Foundation, *OpenFOAM User Guide*, Version **2.3.0** (2014)

[24] J. P. Van Doormal and G. D. Raithby, *Enhancement of the SIMPLE Method for Predicting compressible Flows*, Num. Heat Trans **7**, 147-163 (xxxx)

[25] NIST, *Cryogenics Technologies Group*, Website (2010), http://cryogenics.nist.gov/MPropsMAY/material-properties.htm

[26] NIST , *Chemical Web book*, Website (2010), http://webbook.nist.gov/chemistry/fluid/

[27] W. E. Keller, *$^3$He and $^4$He*, 2 ed., Plenum Press (1969).

[28] Ch. Rohlof, *Entwicklung polarisierter Targets zur Messung der Gerasimov-Drell-Hearn-Summenregel an ELSA*, PhD thesis, Universität Bonn (2003)

[29] B. Feher, *Entwicklung eines supraleitenden, kryostatinternen Polarizationsmagneten für polarizierte Targets*, Diploma thesis, Universität Bonn (2010)

[30] M. Bornstein, *Bau und Test eines kryostatinternen supraleitenden Polarizationsmagneten*, Diploma-Thesis, Universität Bonn (2013).

[31] M. Bornstein, *Development of a thin, internal superconducting polarisation magnet for the Polarised Target*, XVI$^{th}$ International Workshop in Polarized Sources, Targets, and Polarimetry, **PoS(PSTP2015)006**, Bochum (2015)

[32] Y. Iwasa, *Case Studies in Superconducting Magnets*, Second Edition, Springer US, New York (2009).

[33] E. Kallenbach, et al. *Elektromagnete -Grundlagen, Berechnung, Entwurf*, 4. Edition, Vieweg + Teubner Verlag, Wiesbaden (2012).

[34] B. van den Brandt et al. *Dynamic nuclear polarization in thin polymer foils and tubes*, NIM **A356**, 36 (1995)

[35] B. van den Brandt et al., *Dynamic nuclear polarization in thin polyethylene foils cooled via a superfluid $^4$He film*, NIM **A381**, 219 (1996)

[36] E. I. Bunyatova, *Free radicals and polarized targets*, NIM **A526**, 22 (2004)

[37] Kumada et al., *Dynamic nuclear polarization of high- and low-crystallinity polyethylenes*, NIM **A606**, 669 (2009)

[38] S. Reeve, *Polarisation and relaxation characteristics of irradiated polymeric materials at 1 K and at 2.5 T*, XVI$^{th}$ International Workshop in Polarized Sources, Targets, and Polarimetry, **PoS(PSTP2015)017**, Bochum (2015)

[39] D. Libby and M.G. Ormerod, *Electron spin resonance spectrum of stretched polyethylene*, Journal of Physics and Chemistry of Solids **18**, 316 (1961)

[40] E.J. Lawton et al., *Paramagnetic-Resonance Studies of Irradiated High-Density Polyethylene. I. Radical Species and the Effect of Environment on Their Behavior*, The Journal of Chemical Physics **33**, 395 (1960)

[41] H. Fischer and K.-H. Hellwege, *Elektronenspinresonanz-Untersuchungen an bestrahltem Polypropylen*, Journal of Polymer Science **56**, 33 (1962)

[42] L. J. Forrestal and W. G. Hodgson, *Electron spin resonance studies of irradiated polypropylene*, Journal of Polymer Science **Part A 2.3**, 1275 (1964)

[43] M. Iwasaki et al., *Electron spin resonance of trapped free radicals and the photoinduced radical conversion in $\gamma$-irradiated isotactic polypropylene*, Journal of Polymer Science **Part B: Polymer Letters 5.5**, 423 (1967)

[44] S. Nara et al., *Configurations of the free radicals in irradiated polypropylene and the relation of their decay reactions to the molecular motion*, Journal of Polymer Science **Part A-2: Polymer Physics 8.9**, 929 (1970)

[45] S. Runkel, *Herstellung und Untersuchung von Ammoniakkristallen als polarisiertes Target für das COMPASS-Experiment*, Univerität Bonn (2011)

[46] T. Fujimura et al., *Conversion of alkyl radicals to allyl radicals in irradiated single crystal mats of polyethylene*, Polymer **19**, 1031 (1978)

[47] A. Meier, *Entwicklung eines Kryo-Thermostaten hoher Kühlleistung zur Targetpräparation für das COMPASS-Experiment*, Ruhr-Universität Bochum (1997)

[48] S. Reeve, *A Wide Temperature Range Cryostat for Polarized Target Material Preparation*, Univerität Bonn (2011)

[49] G. Reicherz et a., *The Bonn polarized target NMR-system*, NIM **A356**, 74 (1995)

[50] J. Herick, *Development of a new Q-meter module*, XVI$^{th}$ International Workshop in Polarized Sources, Targets, and Polarimetry, **PoS(PSTP2015)011**, Bochum (2015)

[51] F. Bloch, *Nuclear Induction*, Phys. Rev., **70**, 460-474 (1946).

[52] R.K. Harris, *Nuclear Magnetic Resonance Spectroscopy*, Longman Scientific & Technical, Harlow, 1986.

[53] G.R. Court et al., *A high precision Q-meter for the measurement of proton polarization in polarized targets*, NIM **A324**, 433-440 (1993).

[54] J. Herick, *Ein neues cw-NMR-Modul und Trityl-dotierte D-Butanol-Targets für Streuexperimente*, PhD thesis, Ruhr-Universität Bochum (2015).

[55] P. Hautle, *Detection of small NMR signals*, Proceedings of the Workshop on NMR in Polarized Targets, University of Virginia (1998).

[56] Ch.M. Dulya et al., *A line-shape analysis for spin-1 NMR signals*, NIM **A398**, 109-125 (1997).

# Beam and spin dynamics in the fast ramping storage ring ELSA

## Concepts and measures to increase beam energy, current and polarization

*Wolfgang* Hillert[1,⋆], *Andreas* Balling[2], *Oliver* Boldt[2], *Andreas* Dieckmann[1], *Maren* Eberhardt[2], *Frank* Frommberger[1], *Dominik* Heiliger[3], *Nikolas* Heurich[1], *Rebecca* Koop[1], *Fabian* Klarner[2], *Oliver* Preisner[3], *Dennis* Proft[1], *Thorsten* Pusch[4], *André* Roth[2], *Dennis* Sauerland[1], *Manuel* Schedler[5], *Jan Felix* Schmidt[1], *Michael* Switka[1], *Jens-Peter* Thiry[1], *Jürgen* Wittschen[2], and *Sven* Zander[6]

[1] *University of Bonn, Physics Institute, ELSA, Nussallee 12, D-53115 Bonn, Germany*
[2] *Varian Medical Systems Particle Therapy GmbH, Mottmannstr. 2, D-53842 Troisdorf, Germany*
[3] *TÜV Rheinland Industrie Service GmbH, Am Grauen Stein, D-51105 Köln, Germany*
[4] *Fraunhofer-Institut für Naturwissenschaftlich-Technische Trendanalysen, Appelsgarten 2, D-53879 Euskirchen, Germany*
[5] *Karlsruhe Institute of Technology, LAS, Kaiserstraße 12, D-76131 Karlsruhe, Germany*
[6] *TÜV Nord EnSys GmbH & Co. KG, Große Bahnstraße 1, D-22525 Hamburg, Germany*

**Abstract.** The electron accelerator facility ELSA has been operated for almost 30 years serving nuclear physics experiments investigating the sub-nuclear structure of matter. Within the 12 years funding period of the collaborative research center SFB/TR 16, linearly and circularly polarized photon beams with energies up to more than 3 GeV were successfully delivered to photoproduction experiments. In order to fulfill the increasing demands on beam polarization and intensity, a comprehensive research and upgrade program has been carried out. Beam and spin dynamics have been studied theoretically and experimentally, and sophisticated new devices have been developed and installed. The improvements led to a significant increase of the available beam polarization and intensity. A further increase of beam energy seems feasible with the implementation of superconducting cavities.

## 1 Introduction

Accelerator research, development and operation has been performed at the University of Bonn for more than 60 years. In the beginning a small 500 MeV strong focussing synchrotron provided the first electron beam in 1958. A larger fast cycling 2.5 GeV alternating gradient synchrotron was constructed and is in operation since 1967. It was extended in 1987 by the electron stretcher and 3.5 GeV storage ring ELSA, leading to increased energy and improved beam characteristics. The complete accelerator facility (see Fig. 1) is based on the storage ring and an injector chain consisting of two linear accelerators and the fast cycling booster synchrotron [1].

Within the collaborative research center SFB/TR16, which has been established in 2004, ELSA is serving two photoproduction experiments (CBELSA/TAPS [2] and BGO-OD [3]) with a quasi-continuous electron beam. The external electron beam is generated via slow resonance extraction from the storage ring over a periodically recurring time frame of a few seconds and converted to a photon beam via the process of bremsstrahlung. The energy of the photons is determined by dedicated tagger spectrometers that measure the energy of those electrons which radiated a photon [2, 4].

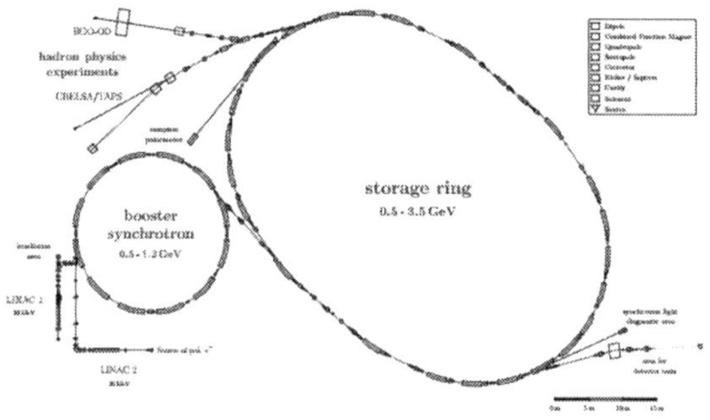


**Figure 1.** The accelerator facility ELSA at Bonn University.

The experimental program pursued in the SFB/TR 16 is based on the availability of linearly and circularly polarized photons. Whereas linearly polarized photons are produced by coherent bremsstrahlung off a diamond crystal [5], circularly polarized photons are generated by polarization transfer from longitudinally polarized electrons in the bremsstrahlung process.

Therefore both experimental setups demand a stable and reliable delivery of an electron beam with sufficient intensity, polarization and energy. Over the past 12 years, intense accelerator research and development has been car-

⋆e-mail: hillert@physik.uni-bonn.de

DOI: 10.1051/epjconf/201713405002

ried out to increase the degree of beam polarization and intensity as well as to enhance the stability of the external electron beam. In addition, several aspects of a maximum beam energy increase to 5 GeV have been investigated. In this paper we report on the related upgrade program of the ELSA accelerator and present the results obtained.

## 2 Polarized electrons

Polarized electrons are generated in a dedicated source [6] which has been developed and set up primarily for the measurement of the Gerasimov-Drell-Hearn sum rule [7–9]. In order to enhance the uptime of the source, the existing load-lock has been extended, implementing a newly developed photocathode surface cleaning procedure using atomic hydrogen [10–13]. By optimizing the optical settings of the transfer line to the linear accelerator, the available beam current has been increased significantly [11, 14].

Beam depolarization due to crossing of depolarizing resonances on the fast energy ramp in the storage ring has been successfully suppressed by different measures [15]. A newly developed orbit correction system has been implemented, allowing for a fast setting of harmonic corrections [16–20]. In combination with a novel harmonic spin-orbit correction scheme, integer depolarizing resonances have been effectively compensated [21, 22]. Intrinsic depolarizing resonances are routinely suppressed by increasing the crossing speed utilizing pulsed tune jump quadrupole magnets [23].

Depolarizing effects have been investigated theoretically using numerical simulation techniques [24–27]. A dedicated simulation tool *polematrix* has been developed which allows detailed studies of beam depolarization when crossing depolarizing resonances [27].

The measurement of the polarization degree (polarimetry) is typically performed using a Møller polarimeter in the external beam line to the CBELSA/TAPS experiment. In order to directly measure the polarization of the stored electron beam, a Compton polarimeter has been set up at the storage ring [28–30]. First measurements show the expected build-up of polarization caused by the Sokolov-Ternov mechanism.

### 2.1 Source of polarized electrons

Polarized electrons are produced by irradiating a strained-layer superlattice photocathode with circularly polarized laser light from a flashlamp pumped Ti:sapphire laser. In order to enable the electrons to escape into vacuum, the photocathode has to be activated first by cleaning the surface from remaining contaminations and thereafter depositing a monomolecular layer of cesium and oxygen. Due to the adsorption of contaminating molecules the initial quantum efficiency $QE_0$ after an activation procedure decreases with time $t$ according to

$$QE(t) = QE_0 \cdot \exp\left(-\frac{t}{\tau}\right).$$

The time constant $\tau$ is called the lifetime of the photocathode and is strongly dependent on the quality of the vacuum. It is therefore indispensable to achieve a total pressure of $10^{-11}$ mbar and partial pressures of the most contaminating gases like water vapor, oxygen and carbon dioxide in the range of $10^{-13}$ mbar in the operating chamber of the source. This extreme ultrahigh vacuum (XHV) has to be preserved by an effective isolation from the moderate vacuum of the subsequent linear accelerator. Beam loss in the transfer line will cause additional gas desorption and has to be prevented by careful beam steering and focussing, based on numerical simulations of the beam dynamics. Insertion of new photocathodes as well as heat cleaning and activation of used photocathodes require additional vacuum chambers and a transfer to the operating chamber without breaking the XHV conditions. Such a load-lock system has been installed at the operating chamber becoming an integral part of the source.

#### 2.1.1 Beam transport

For the first years of operation, the beam current in standard operation was set to 100 mA. After changing from a Be-InGaAs/Be-AlGaAs to a GaAs/GaAsP photocathode and increasing the usable cathode diameter from 8 mm to 8.9 mm, a considerably higher emission current of up to 200 mA has been produced reliably but not safely transported to the linac. To benefit from this increased intensity, numerical simulations of the beam transport have been carried out to derive optimal settings of the focussing elements of the transfer line for minimizing beam loss. The experimental optimization has been performed by additionally detecting the transverse beam intensity profiles using dedicated wire scanners located at different positions in the transfer line.

*Transfer line*

The transfer line connects the operation chamber with the linear accelerator (see Fig. 2).

Along the entire transfer line, 41 magnets are installed. There are two symmetry planes: one at the second wire scanner and one at the Mott polarimeter. The most important elements of the folded beam line are an electrostatic deflector and two $\alpha$-magnets for deflecting as well as solenoids and quadrupoles for the focussing of the beam.

To ensure a conservation of polarization in the following circular accelerators, the longitudinal spin is rotated into the transverse plane with respect to the momentum by the electrostatic deflector. To avoid a further influence on the spin orientation, mainly double solenoids are used to focus the beam behind the deflector. The polarization is measured using a Mott polarimeter.

*Beam dynamics and simulation*

The transversal beam dynamics of a low-energy electron beam with a homogeneous, elliptical charge distribution[1]

[1]The so called Kapchinsky-Vladimirsky distribution is assumed, whose projection is a sharp-edged, homogeneous and elliptical charge distribution [31].

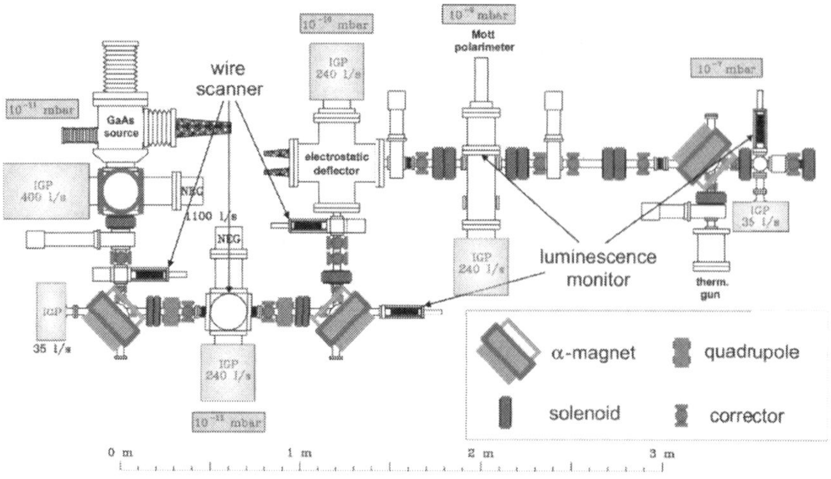


**Figure 2.** The transfer line between the operating chamber and the linear accelerator [10].

in presence of external electromagnetic fields assuming a "laminar" flow are described by the so called paraxial differential equations [31]:

$$\frac{d^2x}{ds^2}+[k_x(s)+S(s)+T(s)]\cdot x-\frac{\varepsilon^2}{x^3}-\frac{2K}{x+z}=0$$

$$\frac{d^2z}{ds^2}+[k_z(s)+S(s)+T(s)]\cdot z-\frac{\varepsilon^2}{z^3}-\frac{2K}{x+z}=0\,.$$

The increase of the beam envelope due to the emittance $\varepsilon$ and the space charge are expressed in the third and fourth term. $K$ is called the generalized perveance and depends on the beam current and energy. In the first three terms the restoring forces of quadrupoles ($k_{x,z}(s)$), solenoids ($S(s)$) and the electrostatic deflector ($T(s)$) are included.

Due to the low beam energy and the high intensity, the beam transport is strongly space charge dominated. Hence, all focal lengths of the magnets have to be adjusted according to the actual beam current. To examine the feasibility of a beam transport of 200 mA using the existing magnets in the transfer line, the focussing strengths of the magnets were optimized by iteratively feeding adjusted parameter sets into an in-house developed software package used for the numerical integration of the differential equations. The optimization criteria were a beam focus in the symmetry planes and a minimal beam envelope along the whole transfer line in order to approach a lossless beam transport to the linear accelerator and not to degrade the vacuum.

The final results of the simulations for currents of 100 mA and 200 mA are shown in Fig. 3. The solid lines indicate the beam envelope for the optimal settings of the focussing strengths of the magnets. The initial point of the simulation is the first beam waist close to the anode, whose position varies for different currents and diameters of the emitting surface and is caused by the focussing geometry of the electrodes. The origin of the coordinate system was set to the position of this beam waist for 100 mA. The initial parameters, such as the beam envelope at the waist

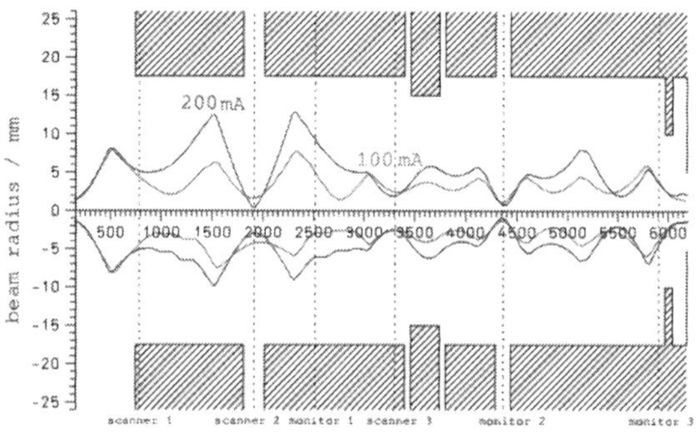


**Figure 3.** Final results of the numerical simulations for a beam current of 100 mA (diameter of the photocathode $\O_{pc}$ = 8.0 mm) and 200 mA ($\O_{pc}$ = 8.9 mm). The optimization criterion of a minimum beam envelope at the symmetry points is fulfilled for both currents [10].

and the position of the waist as well as the emittance, were numerically simulated for each emission current using the software *EGUN*[32].

Even in case of a current of 200 mA the radius is always smaller than one half of the aperture except near the $\alpha$-magnets, indicating that a quasi lossless transport appears to be feasible with 200 mA.

*Achieved performance*

To experimentally optimize the optics of the transfer line, measurements of the beam profiles were performed under variation of the focussing strength of the magnets based on the simulation results for a current of 200 mA. The adjustments for a current of 100 mA were used as initial values.

The two beam profiles shown in Fig. 4 were recorded with the first wire scanner in the transfer line at a beam current of 200 mA. The measurements are in reasonable

 DOI: 10.1051/epjconf/201713405002

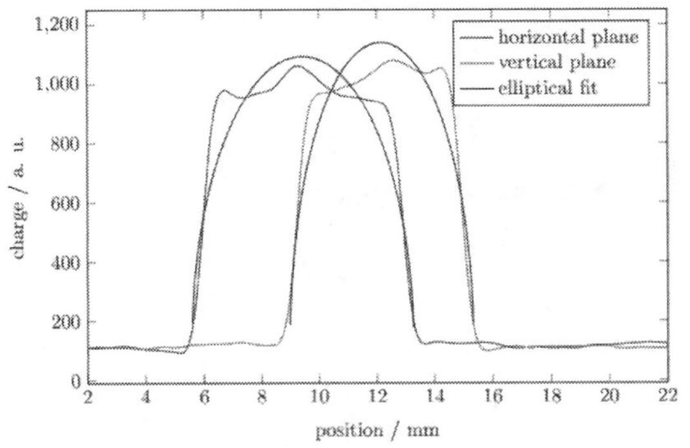


**Figure 4.** Measurement of the beam profile with the first wire scanner and a beam current of 200 mA [33].

agreement with a fitted elliptical profile expected for a homogeneous cylinder symmetric charge distribution. So the assumptions made for the paraxial differential equations seem justified. In order to ensure an elliptical charge distribution also at the end of the transfer line, beam profile measurements were performed with the third wire scanner. Figure 5 shows this measurement at a beam current of 200 mA proving the expected beam profile. The slight charge redistribution visible in Fig. 4 and 5 can be explained by an inhomogeneous fringe field of a permanent magnet belonging to an ion getter pump.

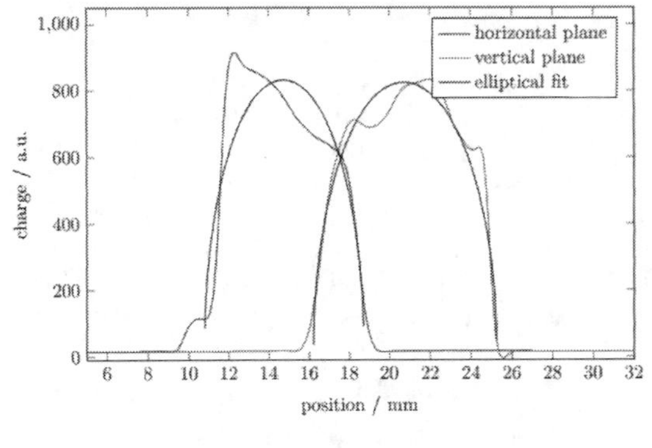


**Figure 5.** Measurement of the beam profile with the third wire scanner and a beam current of 200 mA [33].

The beam profile measurements verify the assumption of a nearly cylinder symmetric, homogeneous and sharp-edged beam profile over the whole transfer line. Finally, an optimized optics setting of the transfer line with an overall beam loss of less than 1 % for beam currents of almost 200 mA has been determined experimentally.

### *2.1.2 New load-lock system*

*Overview*

The new load-lock system consists of three XHV chambers (see Fig. 6). The loading chamber allows the insertion

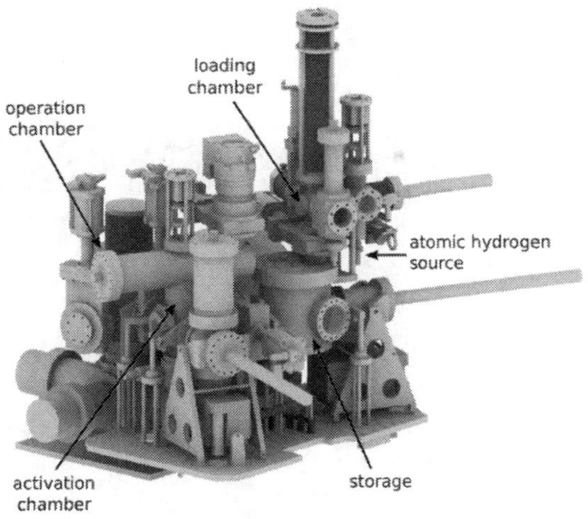


**Figure 6.** Overview of the new load-lock system with its three XHV chambers [13].

of a new photocathode and cleaning it with atomic hydrogen for a better removal of surface oxidations. The photocathode can be transported to the storage chamber via an elevator, where several photocathodes can be stored under extreme high vacuum conditions. In the activation chamber the photocathode is prepared for operation by evaporating cesium and oxygen on its surface. All chambers were baked out for two weeks to reduce the desorption of molecules from the walls and achieve the XHV.

*Storage chamber*

In the storage chamber XHV conditions (total pressure $5 \times 10^{-12}$ mbar) are achieved by four NEG modules (330 L/s each) and an ion getter pump (75 L/s). Photocathodes can be stored inside the chamber on a rotary plate (see Fig. 7), which is adjustable in height by a bellow. The force transmission between the vacuum and non-vacuum side is carried without contact by magnets.

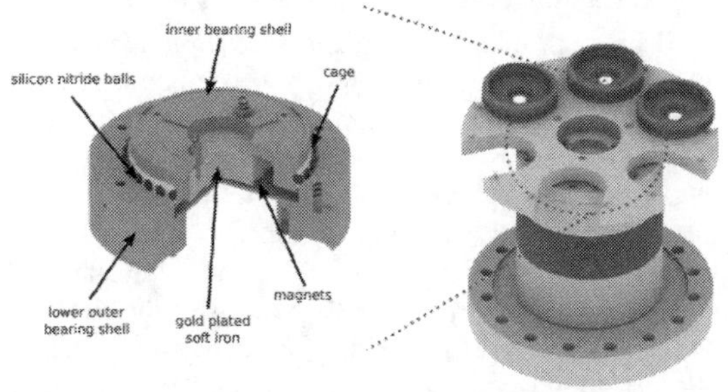


**Figure 7.** The rotary plate in the storage chamber and a cut through the ball bearing [33].

After one year of pumping resulting in a high degree of surface purification and possible cold shut of the materials, the ball bearing of the rotary plate moved tightly. As a precaution the rotary plate was clamped in a fixed

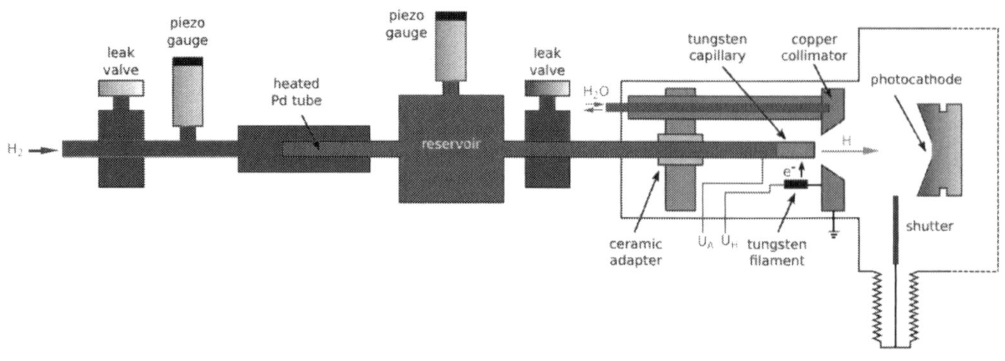


**Figure 8.** Inlet of hydrogen into the loading chamber for H-cleaning [13].

position to ensure that at least two storage areas are accessible. Further investigations have to be performed in order to understand the exact mechanism and to guarantee a long time low friction movement of the rotary plate. Nevertheless up to three photocathodes can be stored and used for operation within the new load-lock system even if the rotary plate is fixed.

*Activation chamber*

In this chamber a photocathode can be heat cleaned and activated with cesium and oxygen. An ion getter pump (75 L/s), two NEG-pumps (2000 L/s each) and a turbo molecular pump (300 L/s) provide an ultra high vacuum during these processes. By alternately heating a silver tube[2] and cesium dispensers, a sub-mono molecular layer of CsO-dipoles is evaporated on the surface of the photocathode. The quantum efficiency and the emitted current during the activation process is measured by irradiating the photocathode with laser light from a helium-neon laser and applying an accelerating voltage between the photocathode and the ground potential. A countervoltage within the distance of acceleration prevents a backstreaming of ions evaporated from the cesium dispensers.

For heat cleaning two cylindrical radiation shields with a heated filament encase the photocathode and heat it by thermal radiation up to 550 °C. Most of the oxidations desorb from the photocathode's surface at this temperature, but especially strongly ligated molecules like galliumoxides do not leave the photocathode's surface below its damage threshold of about 600 °C. Thus, the quantum efficiency does not reach its initial value after the cleaning and decreases over several activations until the photocathode can no longer be used for operation, because not all impurities are removed. Additionally, heat cleaning produces a rough surface, which reduces the optimal orientation of the CsO-dipoles, the quantum efficiency as well as the polarization degree [34, 35]. Due to these disadvantages a source for atomic hydrogen was installed inside the loading chamber.

[2]Heated silver is permeable for oxygen only.

*Loading chamber*

The loading chamber fulfills two main functions: it allows the insertion of new photocathodes and the cleaning of the photocathode's surface by thermal heating with additional supply of atomic hydrogen.

Atomic hydrogen converts strongly ligated molecules (like $Ga_2O_3$, $Ga_2O$ and $As_2O_x$) on the photocathode's surface, that are resistant to heat cleaning, into volatile ones [34, 36, 37]:

$$\begin{aligned} As_2O_x + 2xH &\rightarrow As_2 + xH_2O\uparrow \\ Ga_2O + 2H &\rightarrow 2Ga + H_2O\uparrow \\ Ga_2O_3 + 4H &\rightarrow Ga_2O + 2H_2O\uparrow \ . \end{aligned}$$

The arrows in the equations above indicate the oxidations (ligated as $H_2O$), which are removed by additionally heating the photocathode to around 550 °C. The gallium and arsenic reaction products remain on the surface. The atomic hydrogen beam does not contain any hydrogen ions, because of thermal cracking. This results in an atomically clean and flat surface with only a few defects near to the surface [35, 38].

*Hydrogen filtering, inlet and cracking*

A maximum degree of purity of hydrogen is desired for the cleaning procedure. Thus, besides using highly purified molecular hydrogen (99.9999 %) from a gas bottle, an additional filter is utilized. It consists of a palladium tube (5 cm in length, inner and outer diameter of 5 mm and 6 mm). Heated palladium above temperatures of 200 °C is permeable for hydrogen. The tube is heated using a wire with high oxidation resistance, the temperature is measured with a thermocouple. Both sides of the tube can be evacuated before and after hydrogen cleaning, in order to prevent the formation of a hydride phase while cooling the tube below 150 °C in presence of hydrogen [39]. The inner side of the tube is clean and evacuated anyway, because it was baked out together with the loading chamber.

The filtered hydrogen accumulates on the baked-out vacuum side of the tube in a reservoir, where the pressure is measured with a piezo gauge. Afterwards the hydrogen is piped through a capillary, which is heated by electron bombardment up to 1900 °C [40] controlling the flow rate with a calibrated leak valve. In this capillary the thermal energy cracks the molecular hydrogen into atomic hydrogen with a high dissociation rate (see Fig. 8). The photocathode is heated to 550 °C while atomic hydrogen cleaning is applied to assure a rapid evaporation of the volatile reaction products, mainly $H_2O$. A turbo molecular pump and a NEG-pump keep the pressure during the hydrogen inlet below $3 \times 10^{-6}$ mbar.

Several measurements with different photocathodes demonstrated the success of this cleaning procedure (see [33]).

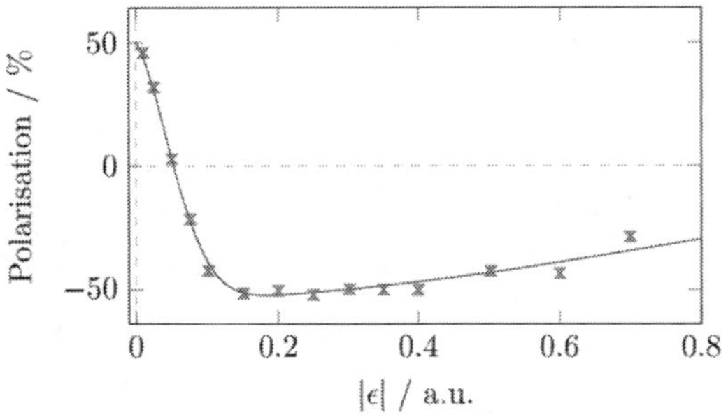


**Figure 9.** Measurement of the Froissart-Stora curve for the integer resonance $\gamma a = 5$ in the ELSA storage ring.

## 2.2 Acceleration of polarized electrons

Polarized electrons are accumulated in the ELSA storage ring via the 1.2 GeV booster synchrotron, then accelerated utilizing a fast energy ramp with up to 6 GeV/s and delivered to the experiments via slow extraction. This concept requires to preserve the polarization during the whole acceleration process.

During acceleration, the crossing of several depolarizing resonances may cause severe beam depolarization. Even in case of fast ramping speeds of up to 6 GeV/s, first order effects like integer and intrinsic resonances have to be compensated by dedicated measures [24]. Depolarizing resonances occur whenever the electron spin precession around the vertical fields of the bending magnets is in phase with any component of the accelerator's horizontal magnetic field distribution. According to the Thomas-BMT equation [41], the spins precess $\gamma a$ times per turn in a flat circular electron accelerator (neglecting longitudinal magnetic fields), where $a = (g_s - 2)/2$ is the gyromagnetic anomaly. $\gamma a$ is also called the spin tune and increases linearly with beam energy due to the Lorentz factor $\gamma$. Several depolarizing resonances are crossed during the acceleration in the ELSA storage ring. Their correction is the major challenge to reach high degrees of polarization. Without any compensation all polarization might be lost at 2.35 GeV. Disabling all dedicated countermeasures merely up to 30 % can be reached.

There are two main types of depolarizing resonances. Integer spin tune resonances occur about every 440 MeV, when the spin tune $\gamma a$ becomes an integer number. Accordingly, these resonances are driven by those horizontal magnetic fields that act on the beam in the same way each turn. In other words, the resonance $\gamma a = n$ is driven by the field component changing with the $n$th revolution harmonic around the ring. These are e.g. magnet misalignments as well as quadrupole fields caused by a vertical closed orbit displacement. Thus, the first measure against integer resonances is a preferably good vertical closed orbit correction during the fast energy ramp (see Section 2.2.1). Additionally, the remaining depolarizing field component has to be compensated.

In proton accelerators, usually integer resonances are deliberately excited to induce a full spin flip during resonance crossing and preserve the absolute value of polarization. This is not possible for electron beams, because of synchrotron radiation. The radiative energy loss requires permanent accelerating rf fields leading to higher synchrotron frequencies compared to proton beams. Therefore, the synchrotron sidebands of depolarizing resonances have a larger distance from the central resonance and are crossed separately during the energy ramp. This results in additional depolarization and a complete spin flip is not possible according to the Froissart-Stora formula [42]. The measurement shown in Fig. 9 illustrates this effect. Here, the polarization after crossing the integer resonance $\gamma a = 5$ in the ELSA storage ring is shown depending on the resonance strength $\epsilon$. It was measured with the Møller polarimeter at the CBELSA/TAPS experiment. The resonance strength was varied using the vertical orbit correction magnets. Because the absolute value of polarization decreases for high resonance strengths, integer resonances are compensated at ELSA by minimizing the resonance strength with a properly adjusted field distribution. An improved scheme for this correction method is discussed in Section 2.2.2.

The second type of resonances are intrinsic resonances, which occur if the spin tune is a linear combination of the betatron tunes. Hence, these resonances are driven by betatron motion. Certainly, the resonance strength can be reduced by aiming for small betatron oscillation amplitudes. Yet the more adequate method is to force a fast change of the vertical tune at the resonance specific time, such that the resonance crossing speed is increased. At ELSA, this method is successfully applied by using Panofsky type tune jump quadrupoles (for details see [23]).

During the SFB/TR16 the polarization in the external beam line of the CBELSA/TAPS experiment was increased to $(71 \pm 1)$ % at 2.35 GeV and $(62 \pm 2)$ % at 2.92 GeV.

### 2.2.1 Closed orbit correction update

Reducing remaining horizontal magnetic fields plays a major role in polarization conservation. Vertical orbit displacements in quadrupole magnets will lead to horizontal magnetic field components, and hence the central goal is to keep especially the vertical orbit as flat as possible. Therefore, the beam should be steered through the magnetic centers of all quadrupole magnets.

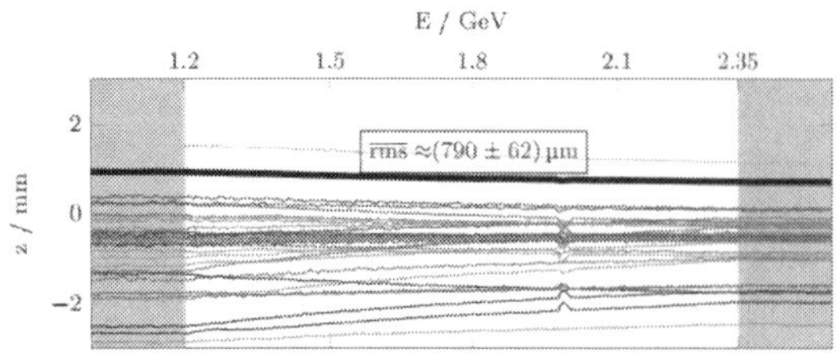


**Figure 10.** Development of the uncorrected vertical beam displacements at each of the 32 BPMs (bold black: $\overline{rms}$). The energy is increased from 1.2 GeV to 2.35 GeV with 6 GeV/s.

*Beam position monitors*

Each of ELSA's 32 quadrupole magnets is equipped with an in-house developed four-button monitor chamber (see Fig. 24 in Section 3.2). During booster mode operation the measurement of beam positions can be switched to a special readout mode: the internal readout rate is increased to 1 kHz and position data is stored locally in the readout electronics of each beam position monitor (BPM). Up to 4095 data points can be stored, thus the development of beam displacements during the first four seconds (covering the injection, ramping and a large period of the extraction phase) of the ELSA cycle is available. Figure 10 shows the development of uncorrected beam positions at every quadrupole magnet during the fast energy ramp of 6 GeV/s.

*Corrector magnets*

For horizontal correction the trim windings of the 24 dipole magnets are used. For correction in the vertical plane dedicated corrector dipole magnets had to be installed.

For the harmonic correction scheme (see Section 2.2.2) additional harmonic kick distributions varying sinusoidally $a\gamma$ times along one revolution are applied every 440 MeV during acceleration. Based on a precisely corrected vertical closed orbit (average rms value below 60 µm during the energy ramp) vertical steerer magnet kicks are added. The vertical corrector system must be capable of changing corrector kicks between two consecutive integer spin tune resonances. At the maximum acceleration speed of 6 GeV/s this leads to a minimal time between two resonances of:

$$\Delta\, t_{\text{int}} = \frac{440\,\text{MeV}}{6\,\text{GeV/s}} \approx 73\,\text{ms}\,.$$

In order to preserve the ability to perform orbit and harmonic corrections at energies up to 3.2 GeV a new design for the vertical corrector magnets became necessary. In 2003, a new C-shape geometry for the magnet yokes was developed [17]. In the design phase two important parameters needed to be optimized: An integral corrector field strength $\left(\int B\,\text{d}s\right)$ greater than 9.6 mT m is needed. Additionally, the field rising time must not extend 70 ms to be capable to deal with the worst case situation where the harmonic correction demands a complete field reversal between two integer resonances. Further consideration was given to field homogeneity and the fact that in several of ELSA's half cells the overall dimensions for installation of the magnets are quite narrow. The yokes are composed out of more than 300 thin metal sheets isolated against each other to prevent for eddy currents when fast field changes are requested. All 30 of these new magnets were installed during the last years.

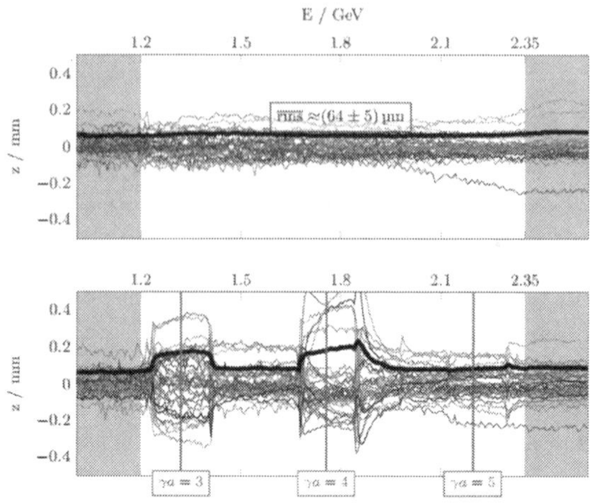


**Figure 11.** Development of the corrected vertical beam displacements at each of the 32 BPMs (bold black: $\overline{rms}$). Top part: interpolated correction, lower part: additional harmonic corrections applied.

Based on the measurements of the uncorrected beam positions (see Fig. 10) closed orbit corrections at the beginning and end of the energy ramp are calculated. According to these, linearly interpolated corrections are obtained and applied during the ramp. In the top part of Fig. 11 the development of the resulting corrected beam positions is shown. In the lower part additional kicks for harmonic correction of three integer resonances are applied.

*Power supplies*

Based on the requirements and a comprehensive analysis of commercially available power supplies it was decided to design and assemble new power supplies for all corrector magnets in-house. A key decision in the design phase was to use a pulsed full H-bridge consisting of 4 NMOS-transistors. Together with a dedicated PI controller unit and a CAN bus driven communication device all these are placed in one 19 inch crate [43].

All required power supplies (24 powering the trim coils in the 24 bending dipoles for horizontal correction and 30 for vertical orbit correction) were installed in 2012 and have been in reliable operation since then.

*Closed orbit correction bandwidth*

In order to verify that the new orbit correction setup fulfills the desired requirements dedicated bandwidth studies

DOI: 10.1051/epjconf/201713405002

were performed [20]. To determine the closed orbit correction bandwidth, which is the limiting bandwidth for resonance crossing, we measured the cutoff frequency of the whole system. This has been carried out by applying a sinusoidal excitation of a single steerer magnet and measuring the amplitude of the closed orbit distortion at that excitation frequency. The bandwidth of the system can then be defined as the frequency $f_{\mathrm{cutoff}}$ at which the measured amplitude is reduced to $\frac{1}{\sqrt{2}}$. The correction bandwidth was determined to $(170 \pm 5)$ Hz.

High bandwidth closed orbit control is essential for acceleration of polarized electrons in a circular accelerator. The bandwidth measurements confirm that the closed orbit correction system fulfills the requirements and ensures the correction of all arising resonances even at the maximum acceleration speed.

### 2.2.2 Improved correction scheme for integer resonances

In the ELSA storage ring, integer resonances are compensated by destructive superposition of the depolarizing field component with a harmonic field distribution applied as kick angles with the orbit correction magnets (VC) around the ring:

$$\alpha^{\mathrm{VC}}(\theta) = S \cdot \sin(\gamma a \cdot \theta) + C \cdot \cos(\gamma a \cdot \theta) \ . \qquad (1)$$

Here $\theta = \int_0^s 1/R(s')\mathrm{d}s' \in [0, 2\pi]$ is the spin phase advance changing only within the 24 bending magnets (finite curvature $1/R$), where the spins precess around the vertical field. In between there are 24 segments $i$ of constant equidistant spin phase advance $\theta_i = i \cdot 2\pi/24$ with $i \in \{0, 1, \ldots, 23\}$.

This correction procedure is called harmonic correction. The two parameters of the harmonic correction field $(S, C)$ have to be adjusted empirically for each resonance by iteratively optimizing the degree of polarization at the beginning of a beam time, because the resonance driving misalignments are not known precisely and may change over time. Figure 12 shows an example for such an optimization from March 2014. Here up to 25 % polarization was gained from the harmonic correction of $\gamma a = 3, 4, 5$.

*Previous correction scheme*

Naturally, the harmonic correction applied by the vertical correction magnets changes the vertical closed orbit. Thus, the horizontal fields acting on the beam in the quadrupoles also change. That is why the actual change of the resonance driving field distribution $\alpha^{\mathrm{H}}(\theta)$ by the harmonic correction not only consists of the set correction kicks $\alpha_m^{\mathrm{VC}}$ according to Eq. 1, but also implies quadrupole contributions $\alpha_j^{\mathrm{Q}}$ caused by the $\alpha_m^{\mathrm{VC}}$:

$$\alpha_i^{\mathrm{H}} := \alpha^{\mathrm{H}}(\theta_i) = \sum_{m \in \mathrm{VC}} \delta_{m,i}^{\mathrm{VC}} \, \alpha_m^{\mathrm{VC}} + \sum_{j \in \mathrm{Q}} \delta_{j,i}^{\mathrm{Q}} \, \alpha_j^{\mathrm{Q}} \ . \qquad (2)$$

Here the abbreviations $\delta_{m,i}^{\mathrm{VC}}$ and $\delta_{j,i}^{\mathrm{Q}} \in \{0, 1\}$ are used, which equal 1 if the corrector $m$ or quadrupole $j$ is located within

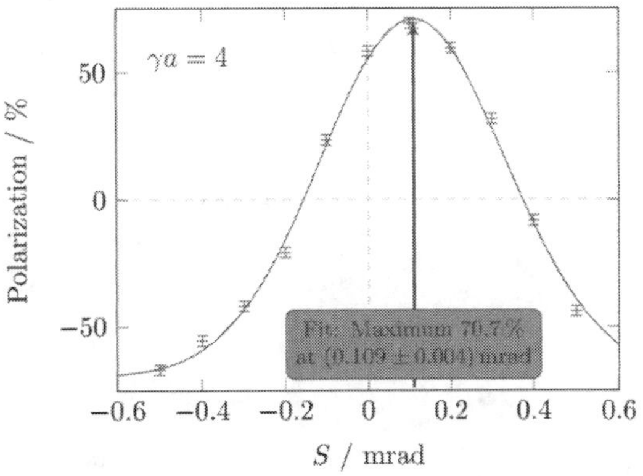


**Figure 12.** Møller polarimeter measurement from polarization optimization of integer resonance $\gamma a = 4$ in March 2014. It took about 10 min to obtain one polarization value with a statistical error of 2 %. A Gaussian fit is used to determine the maximum [22].

the accelerator segment corresponding to the spin phase advance $\theta_i$.

Figure 13 demonstrates the impact of the previous correction scheme on the harmonic field distribution. For different set parameter pairs $(S, C)$ the change of the closed orbit $\Delta z$ was measured [22]. In each case the total kick distribution $\alpha^{\mathrm{H}}(\theta)$ was calculated and the actual measured parameter pair $(S, C)$ was determined from the corresponding frequency component via a Fast Fourier Transformation. Finally, each measured parameter is plotted against its corresponding set value. The plot shows no significant correlation between the set harmonic correction and the actual resulting field distribution. Although the resulting field changes with $(S, C)$, it is not the desired one. Nevertheless, the empirical polarization optimization is possible with this correction scheme (e.g. Fig. 12), but two problems may occur: First, a part of the parameter space may not be accessible at all, which could inhibit the optimization. Second, the consecutive optimization of both parameters may not result in the global maximum of polarization.

*Improved correction scheme*

An improved correction scheme was implemented [21] to achieve an accurate behavior of the harmonic correction. It does not apply the harmonic field distribution (Eq. 1) directly to the correction magnets, but considers the orbit response $\Delta z_j$ at the quadrupole $j$ to each correction magnet $m$ to determine the correction magnet kicks $\alpha_m^{\mathrm{VC}}$ needed for a proper sinusoidal harmonic distribution $\alpha^{\mathrm{H}}(\theta)$. Therefore, the quadrupole kicks are expressed via the orbit response matrix $O_{jm} = \partial z_j / \partial \alpha_m^{\mathrm{VC}}$, which is also used for the orbit correction, as $\alpha_j^{\mathrm{Q}} = l_j k_j \cdot \vec{O}_j \cdot \vec{\alpha}^{\mathrm{VC}}$. Now the so called harmonic correction matrix $\mathcal{H}$ can be defined to write the total kick caused by each correction magnet $m$ in each seg-

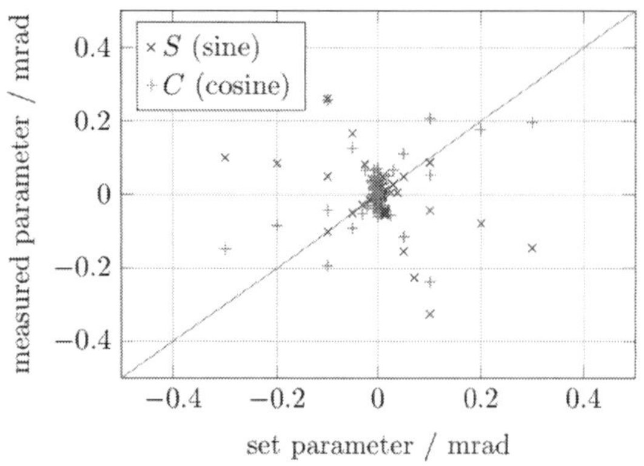


**Figure 13.** Parameters of the previous harmonic correction scheme for $\gamma a = 3, 4, 5$. They deviate from the set parameters due to the quadrupole contributions caused by the orbit response to the correction fields [22].

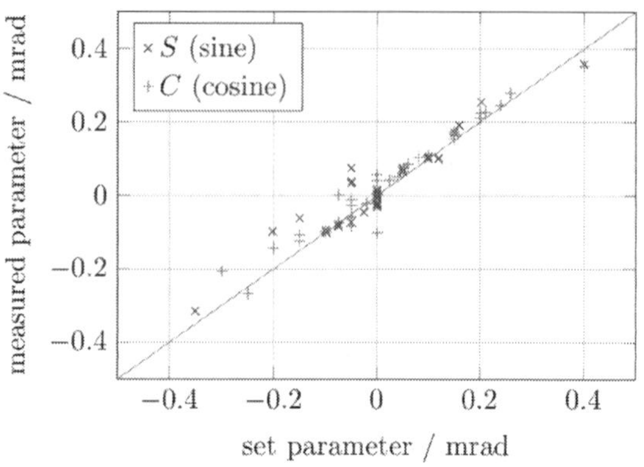


**Figure 14.** Parameters of the improved harmonic correction scheme for $\gamma a = 3, 4, 5$. They are consistent with the set parameters, because the orbit response is included in the calculation of the correction fields [22].

ment $i$ as a matrix equation

$$\mathcal{H}_{im} := \delta^{\mathrm{VC}}_{m,i} + \sum_{j \in \mathrm{Q}} \delta^{\mathrm{Q}}_{j,i}\, l_j\, k_j\, O_{jm}$$

with quadrupole lengths $l_j$ and strengths $k_j$. Using this definition, Eq. 2 can be expressed as

$$\vec{\alpha}^{\mathrm{HC}} = \mathcal{H}\, \vec{\alpha}^{\mathrm{VC}} \quad \Rightarrow \quad \vec{\alpha}^{\mathrm{VC}} = \mathcal{H}^{-1}\, \vec{\alpha}^{\mathrm{HC}} .$$

$\mathcal{H}$ only depends on the orbit response and the accelerator lattice and optics. Finally, the pseudoinverse $\mathcal{H}^{-1}$ is determined numerically using a matrix inversion via single value decomposition.

Figure 14 shows the analysis of the actual parameters for the improved correction scheme [22] analog to Fig. 13. In contrast to the previous scheme, the harmonic field distribution is in good agreement with the set parameters. The improved correction scheme allows for accurate adjustment of harmonic field distributions in the accelerator, which is equivalent to direct control of individual resonances. For instance, it paves the way to a longitudinally polarized beam, which can be achieved if the beam is extracted exactly on an integer resonance and the precession is guided by the adjustable harmonic field distribution [44].

### 2.3 Numerical simulation of the spin dynamics

The final polarization at the experiment after the fast energy ramp in the ELSA storage ring depends on a large number of parameters including beam optics, magnet misalignments, individual setup of the energy ramp and all resonance correction schemes. Therefore, numerical simulations of spin dynamics are important for understanding and enhancement of operation with polarized beam. On the relevant timescale up to seconds, a common approach is tracking many individual 3D spin vectors through the fields of the accelerator and calculating polarization by averaging. Of course, this also implies particle tracking, because both synchrotron and betatron motion affect spin dynamics. Thereby, synchrotron radiation has to be taken into account, but many spin tracking codes do not cover synchrotron radiation, as they are commonly used for proton beams.

The new spin tracking code *polematrix* [27] was developed in-house at ELSA aiming to fulfill all requirements to model the energy ramp in the ELSA storage ring including a realistic synchrotron radiation model. *polematrix* solves the Thomas-BMT equation as

$$\vec{S}(s_0 + l) \approx \mathbf{R}_{3\times 3}(s_0, l)\, \vec{S}(s_0) ,$$

where the spin vector $\vec{S}$ is tracked from the position $s_0$ through a magnet with length $l$ by a three dimensional rotation matrix $\mathbf{R}_{3\times 3}$. The matrix contains the spin precession in the magnet and is constructed from the corresponding energy normalized magnetic field $\vec{\mathcal{B}} := e\vec{B}/(\beta\gamma mc)$. Its direction determines the precession axis and the rotation angle $\theta$ is calculated from the absolute value according to the Thomas-BMT equation:

$$\theta = \vec{\xi} \cdot \left|\vec{\mathcal{B}}(x, z)\right| \cdot l$$

with transverse $\xi_x = \xi_z = \gamma a$ corresponding to the spin tune and longitudinal $\xi_s = a$. Here, $a$ is the gyromagnetic anomaly and $\gamma$ the Lorentz factor. The linear algebra calculations are performed using the *C++* library *Armadillo*[45]. Multiple spins can be tracked in parallel with an adjustable number of threads. Afterwards, the polarization vector is calculated as the average over all spin vectors for each output step. The program options are set in an xml configuration file.

Synchrotron motion and energy ramps are exclusively contained in $\gamma$ as a function of time for each particle. Transverse beam dynamics are included in the magnetic field $\vec{\mathcal{B}}(x, z)$, which depends on the horizontal ($x$) and vertical ($z$) particle trajectory. Several models for both longitudinal and transverse dynamics can be selected in the configuration file. One option is using the particle tracking from *elegant* as basis for the spin tracking. *elegant* is an established 6D particle tracking code, which includes synchrotron radiation and is actively developed at the Advanced Photon Source [46]. The particle trajectories are recorded during *elegant* execution and then imported by the spin tracking code.

The C++ particle accelerator lattice library *libpalattice* is used as interface to *elegant*. It is also developed at ELSA and was published as open source software [47]. *libpalattice* can be used to import any output of *elegant* or *MAD-X*[48] into C++ programs. The tracking is executed automatically if needed. Basic parameters, like energy or number of turns, can be set directly in C++. Furthermore, *libpalattice* provides a convenient data structure to access and edit lattices with all element properties, which can also be automatically imported from and exported to *elegant* or *MAD-X*. So, a lattice file conversion tool is already included. Additionally, lattices can be exported to LaTeX to sketch an accelerator with the *tikz-palattice* package [49].

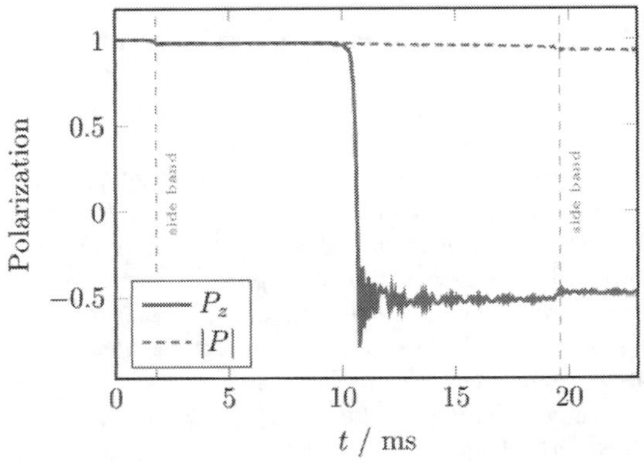


**Figure 15.** *polematrix* simulation of crossing an integer resonance with 4 GeV/s with 100 particles, spins starting in vertical direction.

Figure 15 shows a *polematrix* example of crossing the integer resonance $\gamma a = 4$ with a linear energy ramp of 4 GeV/s in the ELSA storage ring. The vertical polarization $P_z$ is decreased by the resonance. Additionally, the individual synchrotron oscillations cause synchrotron side bands of the resonance. They reduce both $P_z$ and $|P|$, because the crossing speed of these sidebands is different for each particle resulting in incoherent spin motion. After the resonance, when the polarization vector is tilted out of the vertical direction ($P_z < |P|$), synchrotron radiation leads to slow depolarization [27]. Combining many of these simulations allows to perform systematic studies of the resonance compensation measures. Figure 16 shows an exemplary 3D analysis of the harmonic correction discussed in Section 2.2.2. It visualizes the impact of the parameter correlation in the old correction scheme.

*polematrix* is a versatile tool for further optimizations to provide high degrees of polarization over the full energy range of ELSA. It will also be published as open source software soon.

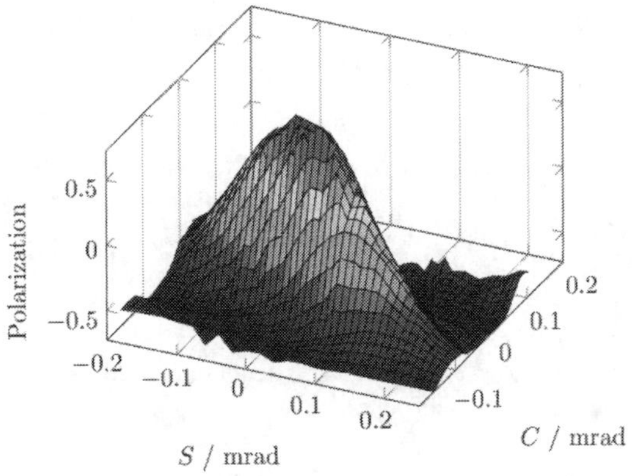


**Figure 16.** Simulation of the polarization after crossing $\gamma a = 6$ depending on the setup of the previous harmonic correction scheme as discussed in Section 2.2.2.

## 2.4 Compton polarimetry

In addition to the Mott- and Møller polarimeters adjacent to the source and at the experimental stations, the electron polarization degree in the storage ring is observable via Compton scattering. As previous Compton polarimeter designs provided unsatisfactory measurement results due to insufficient laser power, pointing stability and detector noise, a revised polarimeter beamline was designed [28] and implemented.

### *2.4.1 Polarization measurement technique*

A Yb:YAG disk laser provides two circularly polarized green laser beams which are brought to head-on collision with the transversely polarized electron beam (see Fig. 17). The intensity profile of the backscattered photons is detected with a silicon microstrip detector where the position of the profile's vertical centroid is a function of photon and electron polarization $P_\gamma$ and $P_e$.

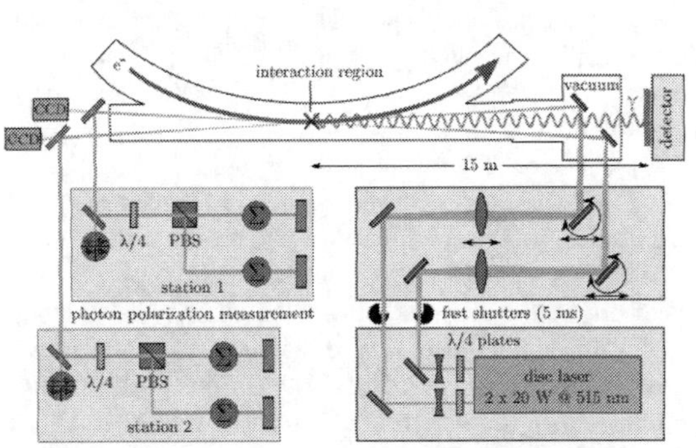


**Figure 17.** Layout of the Compton polarimeter. The microstrip detector measures the vertical profile of the backscattered photons. A laser beam polarization measurement station includes photo diodes for the measurement of the photon polarization [30].

When reversing $P_\gamma$ by the rotation of $\lambda/4$ retarder plates, the profile centroid is displaced by

$$\Delta z = \Delta z_{\text{max}} \cdot P_\gamma P_e \ .$$

The electron polarization degree $P_e$ is therefore obtained by measuring $P_\gamma$ and $\Delta z$. $P_\gamma$ is evaluated behind the interaction region where the photon polarization is re-linearized and characterized by a polarizing beam splitter (PBS) and position sensitive diode detectors (PSD). The analyzing power $\Delta z_{\text{max}}$ is dependent on beam energy and its value is obtained by calibration utilizing the time-dependent self-polarization of an initially unpolarized, stored electron beam [50]:

$$P_e = P_{\text{max}} \cdot \left(1 - \exp\left(-\frac{t}{\tau(E)}\right)\right) , \qquad (3)$$

where the relaxation time $\tau$ is only dependent on the beam energy $E$ and $P_{\text{max}}$ [51]. The maximum achievable polarization amounts to $P_{\text{max}} \approx 92.4\,\%$ if depolarizing effects are not considered, reducing the value otherwise.

Numerical simulations of the Compton-Process with the particle tracking toolkit *Geant4*[52] were carried out, taking into account the Gaussian nature of the laser beam, the dimensions of the silicon detector as well as ELSA's beam parameters. If the electron beam is extracted, ELSA's beam parameters differ from the case of stored electrons. The simulations showed, however, that the calibration factor $\Delta z_{\text{max}}$ will not be altered during extraction.

The intensity profile of backscattered photons on the detector is overlapped by beam gas radiation with similar photon energy. This background is evaluated and subtracted in a separate measurement with blocked laser beams. Signal as well as background are proportional to the electron current.

### 2.4.2 Detector system

A silicon microstrip detector in combination with a lead conversion target of two radiation lengths was chosen for measuring the backscattered photons [30]. The silicon sensor is AC-coupled and has a thickness of 300 µm. It covers a surface of approximately 4 cm × 4 cm with 768 horizontally orientated strips of 50 µm pitch. With an expected width of the intensity profile of backscattered photons of about 1 cm and $\Delta z_{\text{max}} = 70\,\mu\text{m}$, a polarization measurement error of less than 1 % is expected.

The detector readout electronics is FPGA based and allows the processing of the expected photon rate of about 1 MHz. For this purpose, dedicated ASIC chips were developed in close cooperation with the SiLab/ATLAS group of the University of Bonn. The charge threshold for which a hit is counted was set to 3500 electrons (with a minimum ionizing particle producing about 23 000 electrons in 300 µm silicon). This results in a noise rate of less than 1 count/min.

### 2.4.3 First polarization measurements

The performance of the detector system and optical beamline was tested during several measurements with a self-polarizing stored electron beam. Fig. 18 shows the development of $\Delta z$ as a function of time. Initially, a current of 75 mA at 2.35 GeV was injected into the storage ring. The available laser power varied between 5 W to 12 W and a circular polarization degree of about 98.6 % was estimated. The data fit according to Eq. 3 determines the relaxation time $\tau = (28.9 \pm 6.4)$ min in good agreement with the expected value from theory, neglecting depolarizing effects. Due to the decreasing stored current the measurement error (which is purely statistical) is yet too large to allow conclusions on the strength of depolarizing effects which influence the value of $P_{\text{max}}$. Higher laser power for a better signal to background ratio would therefore be necessary. The displacement offset results from a reproducible angular shift of the laser beam when turning the quarter-wave plates, which was confirmed by the installed laser beam diagnostics.

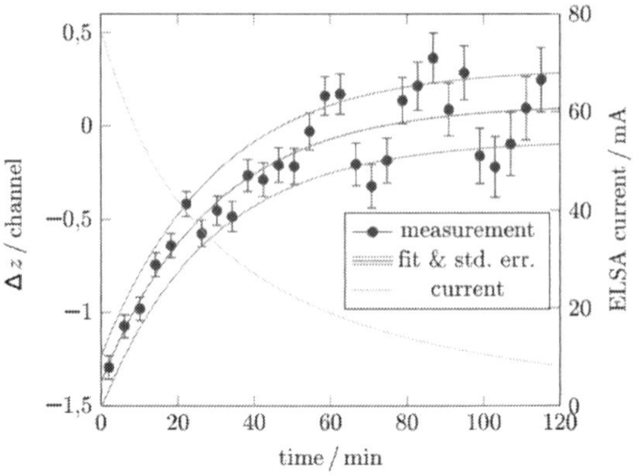


**Figure 18.** Vertical shift $\Delta z$ of the backscattered photons due to self-polarization of the stored electron beam. The Sokolov-Ternov effect is utilized to calibrate the analyzing power of the Compton polarimeter.

## 3 High beam intensities

One major part of the ELSA upgrade program consisted in an increase of the external beam current by one order of magnitude to a maximum value of about 10 nA. When keeping the macroscopic duty cycle of the post accelerating mode at the desired values of about 80 %, the envisaged intensity upgrade can only be achieved by increasing the current in the storage ring to a maximum value of about 200 mA. To supply the required beam current, the existing injector LINAC 1 has been overhauled, which included the redesign and construction of several major components [53–56].

Different measures have been undertaken to suppress the excitation of beam instabilities in the storage ring. The overall impedance was reduced significantly by replacing the majority of the beam pipes by newly developed ones with smooth geometry. A simple higher order modes (HOM) damping system and a new temperature stabilization have been developed and installed at the accelerating PETRA cavities resulting in a significant reduction of HOM driven beam instabilities [57]. To achieve effective damping of the remaining excitation of multi-bunch instabilities, a powerful bunch-by-bunch feedback system has been developed and successfully set up at the storage ring [58–63]. In addition, a precise stabilization of the rf amplitude and phase, as required by the new bunch-by-bunch feedback system, has been achieved by installation of a new low level rf system (LLRF) which is based on digital signal processing [64].

Using the diagnostic capabilities of the feedback system, a fast measurement and stabilization procedure of the transverse tunes has been developed and set up. Monitoring transverse and longitudinal beam dynamics with temporal resolution of picoseconds were made possible with the installation of a streak camera system at the storage ring [65–69].

### 3.1 High current and single bunch injector

The new high current injector is based on the existing normal conducting S-band traveling wave accelerating structure of LINAC 1. It has been equipped with a new pulsed triode thermionic gun, a prebuncher cavity operating at a subharmonic frequency of 500 MHz and a 4-cell S-band traveling wave buncher (TWB) which all were taken from the S-band test facility (SBTF) at DESY. Major parts of the rf power system of the linac and the transfer line to the booster synchrotron were completely rebuilt, and a new energy compression system has been designed and constructed. The final commissioning of the new injector will start right after finishing the installation of the new components. In Fig. 21 an overview of the complete setup is presented.

#### *3.1.1 Gun and bunching system*

The 90 kV electron source uses the original components of the thermionic gun of the SBTF at DESY. According to the considerably lower emission current required for the operation at ELSA, the cathode conus has been redesigned and rebuilt. For this purpose, the emission characteristics of the gun have been optimized by simulating the beam propagation with the software *EGUN*[32] to enable a single pulse operation with 1.5 A beam current as well as long pulse operation with 0.8 A [53]. Further bunching is achieved by the prebuncher and the TWB. To prevent phase shifts, all rf cavities and structures are powered by one 500 MHz 2 W continuous wave rf generator, whose frequency is sextupled before feeding the klystrons that power the S-band structures. Measurements of the gun's and prebuncher's performance in the single bunch mode have been carried out using the wall current monitors installed at different positions in the transfer line to the linac (WCM1 - WCM3) [55] and are presented in Fig. 19. The measured pulse lengths ($2\sigma$) amount to $(1.02 \pm 0.04)$ ns behind the electron gun, $(1.08 \pm 0.04)$ ns before the prebuncher and $(0.41 \pm 0.04)$ ns at the linac entrance, demonstrating that the requirements for a successful single bunch accumulation can be fulfilled.

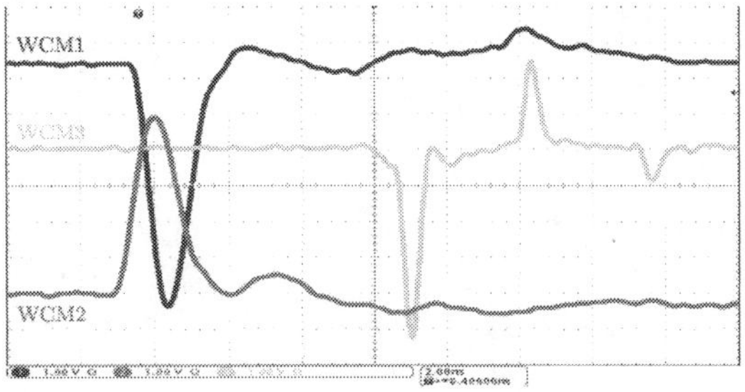


**Figure 19.** Longitudinal intensity profiles measured using the wall current monitors (WCM) installed in the transfer line at the linac. The absolute signal positions depend on the cable lengths, the signal polarities depend only on the way the connector is linked to the WCMs [55].

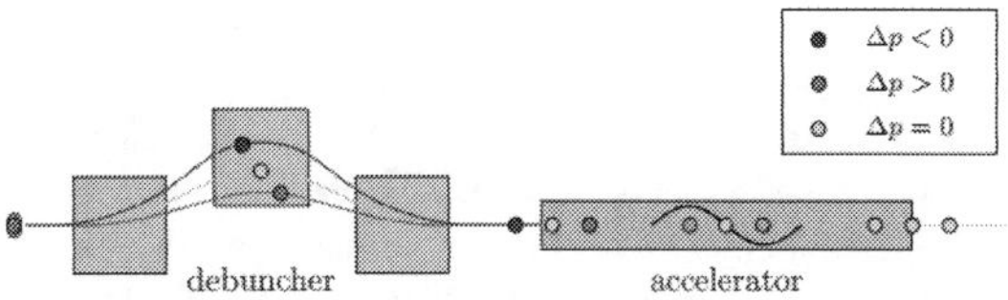


**Figure 20.** Scheme of the energy compression system [70].

#### *3.1.2 Energy compression system*

The redesigned transfer line to the booster synchrotron has been constructed and commissioned using already existing magnets (dipole bends and focussing quadrupoles). The total number and the positions of the individual quadrupole magnets have been changed according to detailed simulations of the optics, which were carried out using the software package *MAD-X*. In order to enhance the injection efficiency into the booster synchrotron which is limited by its energy acceptance of 0.5 %, the energy spread of the electron beam of LINAC 1 has to be reduced by one order of magnitude. Despite using the multi-stage bunching process of the injector, the energy spread has to be further reduced after acceleration in the main linac. This is achieved by adding an energy compression system.

A dispersive debuncher section is used to rotate the phase space distribution of the electron beam in order to sort the electrons by their momentum, using three dipole magnets. Different momenta lead to different curvatures and thus to different path lengths. The resulting longitudinal particle distribution is sorted by the particle's momentum. The adjacent accelerating structure is used to apply a longitudinally alternating electric field to the debunched beam in order to decelerate or re-accelerate electrons depending on their longitudinal position in the bunch. By choosing an appropriate rf phase for this acceleration, the energy spread can be reduced by about one order of magnitude. A scheme of the energy compression is given in Fig. 20.

The accelerating structure of the energy compression system is a normal conducting, 28 cell 3 GHz rf travel-

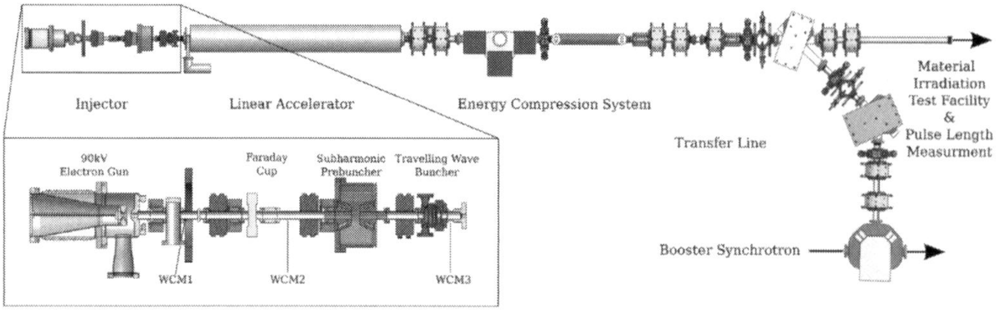


**Figure 21.** The new linear accelerator facility at ELSA [55].

ing wave structure, manufactured by *Research Instruments GmbH*.

The output power of the main klystron is split to simultaneously drive traveling wave buncher, main linac and energy compression structure. The rf amplitude and phase of both, traveling wave buncher and energy compression system, can be changed relatively to the main linac using a hybrid based high power in-vacuum waveguide system to achieve a maximum bunching and energy compression effect.

### 3.1.3 Hybrid based waveguide system

To drive the traveling wave buncher, the main linac and the energy compression accelerating structure, a pulsed 3 GHz klystron is used. The rf power is distributed to the three accelerating structures with individual and adjustable rf amplitudes and phases using a high power WR284 waveguide system based on 3 dB hybrids and phase shifters. Figure 22 shows the schematics of the new waveguide system.

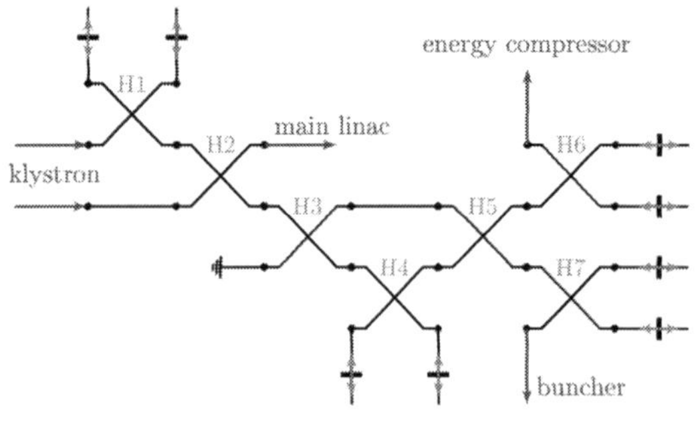


**Figure 22.** Schematics of the new waveguide system including 3 dB couplers and variable waveguide shorts [56].

The two output arms of the main klystron are combined using the hybrid coupler H2. To couple the main part of the rf power to the main linac and a small part to the bunching and energy compression structures, the hybrid H1 is used as a phase shifter by shorting its two output ports, using position variable in-vacuum rf shorts shown in Fig. 23. By varying the position of the shorts synchronously only the output phase of the hybrid with respect to the input one is affected. The waveguide shorts are based on a design of TU Dortmund [71] and have been modified to fulfill ultra high vacuum requirements.

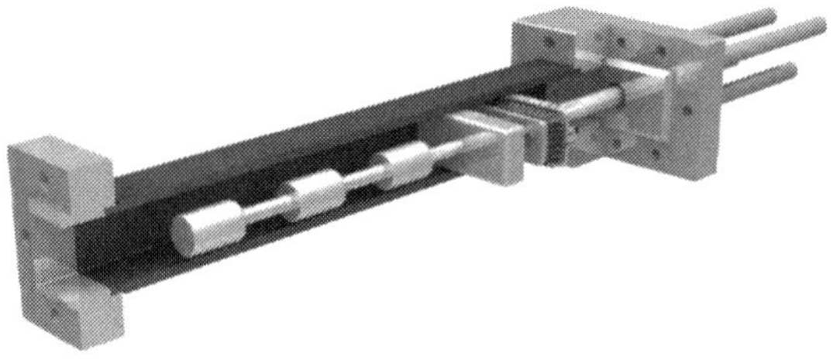

**Figure 23.** Variable in-vacuum high power waveguide short [56].

In the same way, a hybrid can be used as a splitter by driving a single input port, which is used in H3. To avoid damages from undesired reflected powers, the second input port of H3 has to be terminated using a water cooled rf load.

These schemes are used in hybrids H4 and H5 as well, splitting the remaining rf power for traveling wave buncher and energy compression structure. The adjacent hybrids H6 and H7 are connected to the energy compression structure and the traveling wave buncher and are used as phase shifters for both paths individually.

The complete system has been designed to handle powers of up to 30 MW and allows full amplitude and phase control of the three output paths by varying the position of the waveguide shorts.

## 3.2 Low impedance vacuum chambers

In case of sudden changes in the geometry of vacuum chambers, strong wakefields may be excited. Depending on the type of the geometry mismatch, the resulting broadband or narrowband impedances will cause single or multi-bunch instabilities. In order to significantly reduce the ELSA vacuum chambers' total impedance, new elliptical chambers for major parts of the straight sections in the storage ring have been developed and constructed. The new chambers in the quadrupole magnets (see Fig. 24) have been equipped with capacitive pick-up beam position

 DOI: 10.1051/epjconf/201713405002

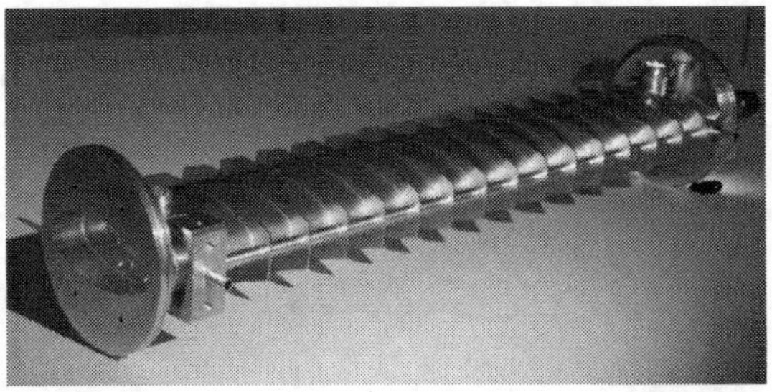

**Figure 24.** Newly developed water-cooled quadrupole chamber with capacitive pick-up beam position monitor (left end) and ion clearing electrode (right end).

monitors (BPM) and clearing electrodes with a bias voltage of up to 3 kV to efficiently remove ions captured by the electron beam.

Almost all pumping crosses were replaced by newly constructed ones, where special effort has been made to maintain a smooth beam pipe geometry (see Fig. 25).

**Figure 25.** Newly developed water-cooled pumping cross with smooth beam pipe geometry.

In addition, the interconnections of the curved dipole and straight quadrupole chambers have been modified to reduce their impedance. Spiral springs have been inserted to ensure electrical contact. The existing ceramic interfaces at both ends of the bended dipole chambers were bridged by dedicated clips made from copper beryllium (see Fig. 26).

With these modifications, the total impedance of the vacuum chambers has been reduced significantly. Nevertheless, it is assumed that still a considerably high impedance is present which is mainly caused by an inadequate construction of vacuum components like kicker tanks and septum magnets.

### 3.3 Higher order modes

Higher order modes (HOM) of the accelerating cavities are one of the dominant sources of multi-bunch instabilities in storage rings. If the eigenfrequency of a specific

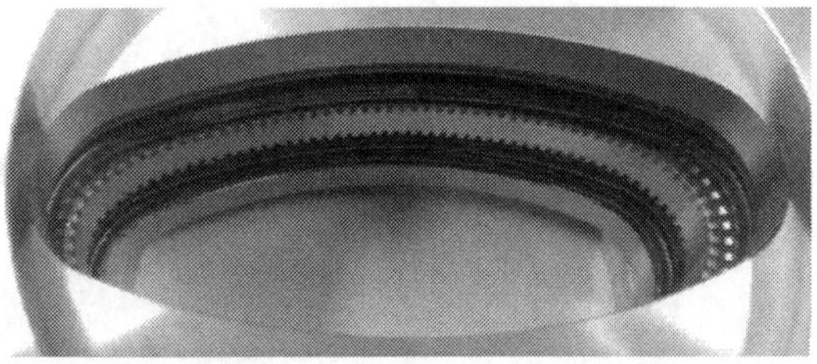

**Figure 26.** Mounted rf clip made from copper beryllium to bridge the ceramic break at the ends of the bended dipole vacuum chamber [70].

HOM overlaps with one of the beam's synchrotron sidebands, this HOM will be excited and may lead to strong beam instabilities which deteriorate the beam quality and can cause severe beam loss when increasing the beam current. Beside an effective damping of all HOMs excited by the circulating beam, a shift of the most harmful HOMs to frequencies apart from the synchrotron sidebands proves to be an effective way of HOM suppression. Both measures have been successfully deployed at ELSA.

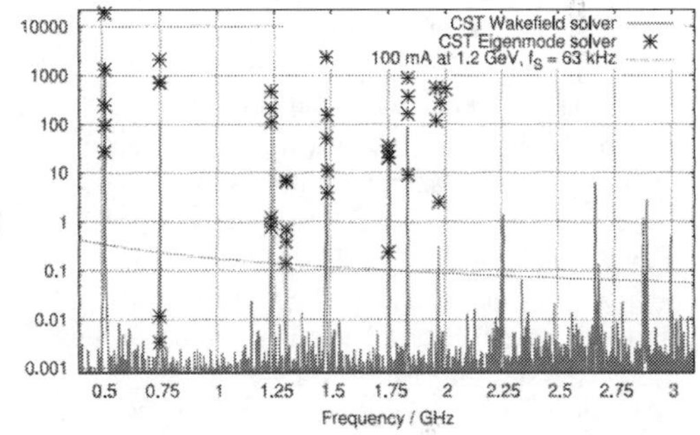


**Figure 27.** Wake impedance and natural damping calculations of a five-cell PETRA cavity using *CST Microwave Studio®* [63].

#### *3.3.1 HOM spectrum*

The most important measure for the perturbing effect of a specific HOM on the beam is its longitudinal impedance. To estimate the HOM impedances of a PETRA cavity, its geometry was modeled within *CST Microwave Studio®*[72]. Using the *CST® wakefield* and *eigenmode solver*, the impedances of all PETRA HOMs up to 3 GHz have been calculated. The resulting spectrum including the natural damping threshold is shown in Fig. 27 which has been calculated for a circulating beam current of 100 mA. Besides the fundamental mode at 500 MHz, which is used for particle acceleration, the $TM_{021}$ mode at 1.46 GHz has an estimated impedance well above 1 MΩ. In addition, many other HOMs show up with impedances exceeding 100 kΩ. A measurement of the impedance of all relevant HOMs has been performed applying the bead-pull technique [73] at a structurally identical spare PETRA cavity. For the $TM_{021}$ mode an impedance of

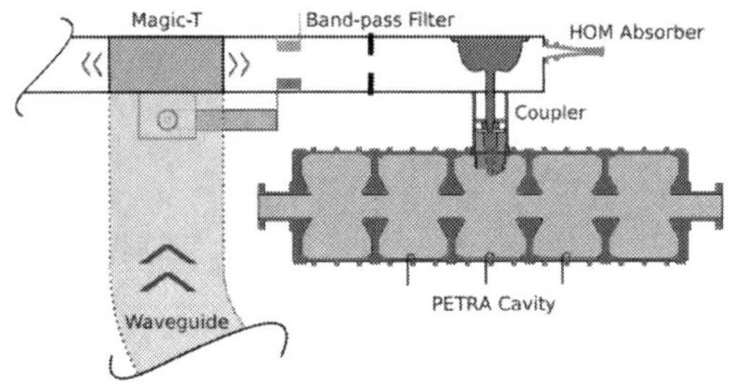


**Figure 28.** Schematic excerpt of the waveguide system with HOM absorbers attached to the PETRA cavity.

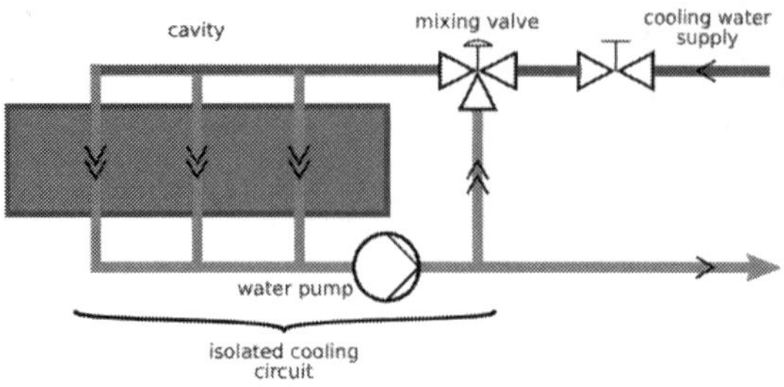


**Figure 29.** Schematic of the temperature stabilization of the PETRA cavities.

$(1.221 \pm 0.191)\,\mathrm{M\Omega}$ was measured, confirming the results of the *CST*® simulations. This indicates that in particular this mode can cause multi-bunch instabilities.

### *3.3.2 HOM damping in the waveguide system*

If a HOM is excited by the electron beam, it can partially propagate into the attached waveguide system, which directs the rf power from the klystron to the cavities. To reduce the impedance of all HOMs, broadband HOM absorbers in combination with 500 MHz band-pass filters were installed at the waveguide system (see Fig. 28). Whereas the 500 MHz rf power from the klystron is transmitted through the band-pass filter, HOMs which propagate inside the waveguide system get reflected back and are attenuated by the HOM absorbers. Due to the absorber's cut-off frequency of about 900 MHz, the 500 MHz rf power is not affected and completely transmitted into the cavity. Thus the absorber provides additional damping for all HOMs which are subject to coupling to the waveguide system via the fundamental input coupler.

### *3.3.3 Temperature stabilization and HOM shifting*

For beam currents up to 100 mA mainly longitudinal multi-bunch instabilities are excited at ELSA. As expected, especially the $\mathrm{TM}_{021}$ mode at 1.46 GHz is causing longitudinal beam instabilities by driving the lower synchrotron sideband of the 21st revolution harmonic. With the harmonic number being 274, this corresponds to an excitation of the multi-bunch mode 252 which showed up as the only and dominating mode in the recorded spectrum of multi-bunch modes [63]. Particular attention was therefore drawn to shift the frequency of this most harmful mode. This can be achieved by rising the operating temperature of the cavities. Since the resonance frequencies of all eigenmodes are defined by the cavity's geometry, a thermal expansion shifts all HOMs to lower frequencies. Whereas the fundamental mode is fully restored by the tuner loop control, a residual frequency shift of the $\mathrm{TM}_{021}$ mode remains.

An increase of the cavities' operating temperature can easily be achieved by reducing the flow rate of the cooling water. To maintain full temperature control, an isolated cooling circuit was integrated for each cavity (see Fig. 29). The body temperature of each cavity is controlled by a mixing valve adjusting the admixture of cool water from the main cooling system into the local circuit. The temperature is stabilized using PID controllers.

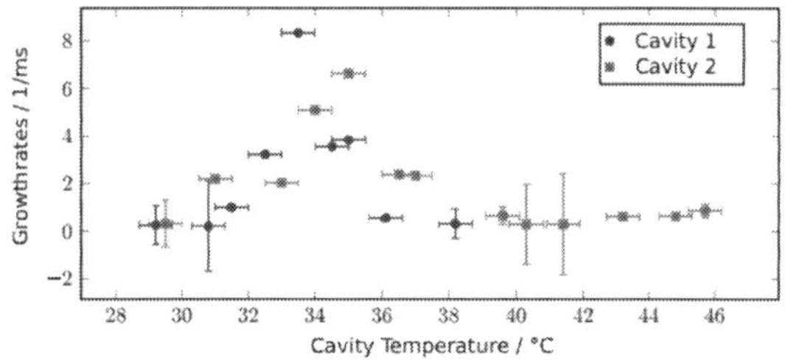


**Figure 30.** Growth rate of the multi-bunch mode 252, depending on the temperature of the PETRA-Cavities at a beam current of about 35 mA [70].

Figure 30 shows the measured growth rate of the multi-bunch mode 252, which is excited by the $\mathrm{TM}_{021}$ mode of the PETRA cavities. The measurement confirms the strong dependence on the cavity temperature. Rising the cavities' temperature to > 38 °C leads to a remarkable decrease of growth rate, which allows a stable storage of significantly higher beam currents.

## 3.4 Bunch-by-bunch feedback system

To damp the multi-bunch oscillations excited in the longitudinal and transverse planes, a state-of-the-art bunch-by-bunch feedback system has been successfully set up at the ELSA storage ring. The system consists of four major parts: A beam position monitor (BPM) picks up a signal corresponding to the bunch's position. Then, the frontend of the feedback system mixes the BPM signal with the third harmonic of the ELSA 500 MHz master rf to isolate a signal only including information about the beam oscillations. By amplitude or phase demodulation the requested properties can be obtained: Using phase demodulation, the longitudinal position of every single bunch can be extracted. Amplitude demodulation allows for measuring the transverse position. A fast ADC converts the signals into digital format. Using a field programmable gate array (FPGA), a digital filter with a maximum of

 DOI: 10.1051/epjconf/201713405002

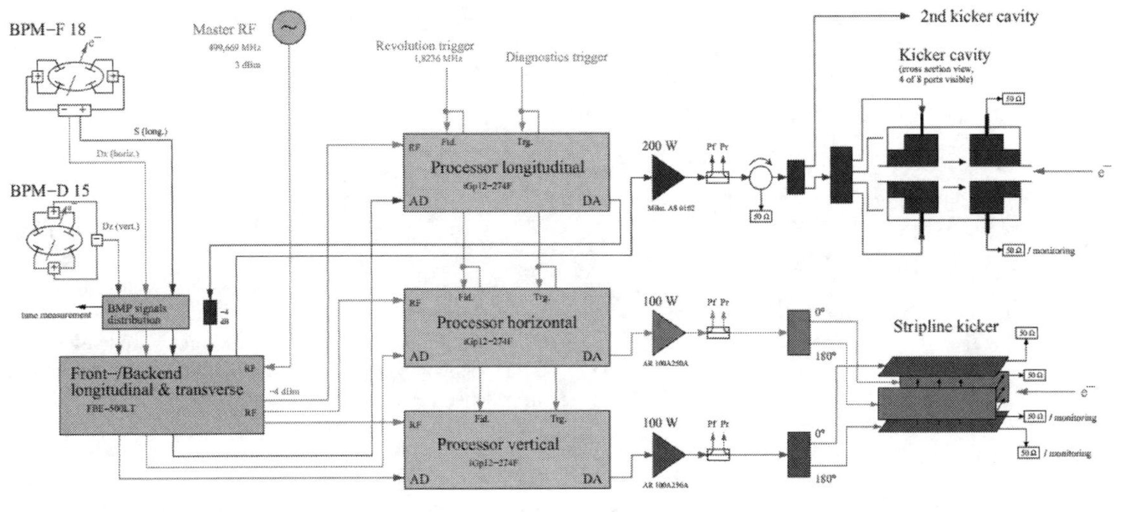


**Figure 31.** Schematic overview of the longitudinal and both transversal bunch-by-bunch FB systems with all main components installed at the ELSA storage ring [58].

32 taps isolates the oscillation frequency of every single bunch, allowing to generate a correction signal which is sent to a fast DAC. The resulting correction signal requires broadband processing, reaching from DC to a maximum of 250 MHz, which is half of the accelerating rf. In the transverse planes, two broadband rf amplifiers in use ensure a sufficient rf power for a proper damping of every single bunch by the connected stripline kicker. In the longitudinal plane, the attached kicker cavity operates at 1 GHz to 1.25 GHz, which requires a mixing of the correction signal with the second rf harmonic before amplification. An overview of the feedback system including all major components is given in Fig. 31. The digital signal processing units (DSP) have been purchased as commercially available modules of *Dimtel Inc.* [74]. Both stripline kicker and kicker cavity are in-house developments adjusted to the ELSA rf frequency and bunch length.

### 3.4.1 Longitudinal kicker cavities

Due to the frequency range of all multi-bunch oscillations, the center frequency $f_c$ of a kicker cavity for the longitudinal feedback system should be configured to

$$f_c = \left(N \pm \frac{1}{4}\right) f_{rf}\,, \quad N \in \mathbb{N}\,.$$

Considering the quite large bunch length of up to 6 cm at an energy of 3.2 GeV in the storage ring, we have chosen $N = 2$ which corresponds to a kicker cavity center frequency of 1.125 GHz and an operating range of 1.0 GHz to 1.25 GHz. The kicker cavity is designed as a waveguide loaded pillbox cavity with a cylindrically symmetrical beam pipe. The geometry of the cavity is based mainly upon several developments at other accelerator facilities [75–77]. The large bandwidth of 250 MHz required to damp all multi-bunch modes is obtained by loading the fundamental mode of the cavity with four single-ridged waveguides on each cavity side which leads to a significantly small cavity's quality factor.

Furthermore, eight broadband waveguide-to-coaxial transitions and rf feedthroughs provide the external connection via type N connectors. Four of them are used for the input signals driving the cavity, the other four are terminated by water-cooled 50 Ω loads or are usable for monitoring. Figure 32 shows a picture of the final installation of both in-house fabricated kicker cavities.

**Figure 32.** Longitudinal kicker cavities installed at the ELSA storage ring [62].

The geometry of the cavity was determined from numerical simulations carried out with *CST Microwave Studio*® (*MWS*®). For details of the geometry and the design parameters we refer to [60]. The cavities' scalar reflection coefficient $S_{11}$ simulated by using the *MWS*® *Transient Solver* is shown in Fig. 33 and is characterized by a large bandwidth of $\Delta f = 254$ MHz and an almost zero reflection (critical rf coupling) at the center frequency of 1.102 GHz.

DOI: 10.1051/epjconf/201713405002

**Table 1.** Selected kicker cavities' design parameters, simulation and measurement results.

| Parameter | Specified | Simulated | Measured cavity 1 | Measured cavity 2 |
|---|---|---|---|---|
| center frequency $f_c$ / MHz | 1125.0 | 1101.0 | 1125 ± 1 | 1126 ± 1 |
| bandwidth $\Delta f$ / MHz | 250.0 | 254.0 | 297 ± 32 | 339 ± 18 |
| shunt impedance $R_S$ / Ω | | 414.0 | 338 ± 15 | 316 ± 12 |
| quality factor $Q$ | 4.5 | 4.3 | 3.8 ± 0.1 | 3.3 ± 0.1 |

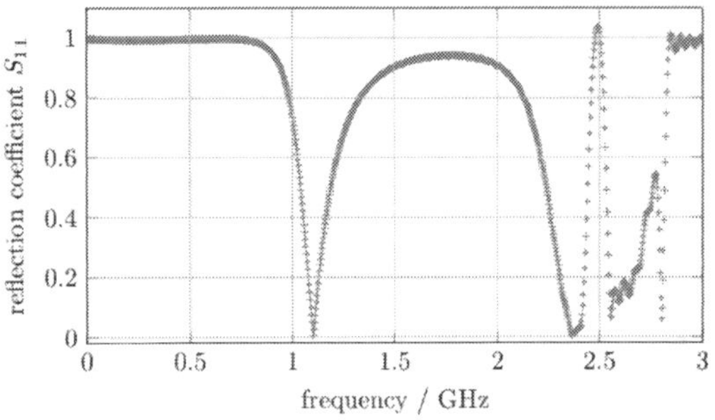


**Figure 33.** Scalar reflection coefficient of the kicker cavity as a function of frequency until 3 GHz simulated by *MWS®Transient Solver*. The broadband fundamental mode at 1.1 GHz is clearly visible. First HOMs do not occur below 2.3 GHz and will be damped due to their propagation through the beam pipe [62].

The cavities' geometry was adapted to this frequency, which is lower than the specified value of 1.125 GHz. This is due to compensation for a frequency shift between the simulation results and the measurement of the resonance curve of a prototype cavity fabricated during the design phase. Thereby, both later fabricated and installed kicker cavities have a center frequency of nearly 1.125 GHz which is confirmed by the measured resonance curve of both cavities presented in Fig. 34. The corre-

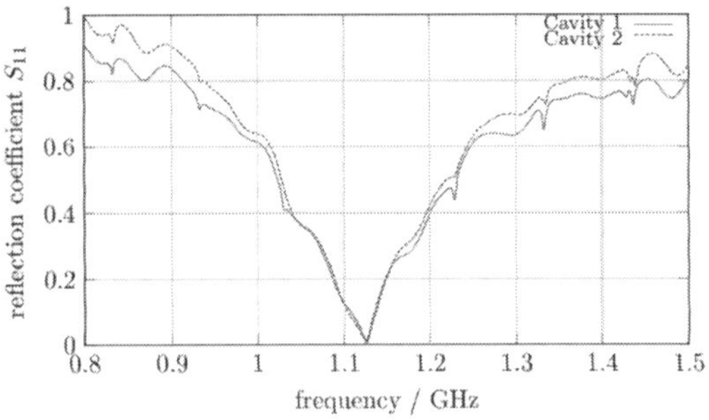


**Figure 34.** Measured resonance curves of both installed kicker cavities.

sponding fundamental mode of the cavity excited at the center frequency is the $TM_{010}$ monopole mode which is characterized by a pure longitudinal electric field being maximum on the beam axis and decreasing with higher distance to the beam axis. The magnetic field of the mode is axially symmetric, transversal to the electric field and zero at the beam axis.

The major design parameters as well as simulation and measurement results are summarized in Tab. 1. The shunt-impedance of the kicker cavities was determined from the longitudinal distribution of the electric field along the beam axis [60] which was measured by using a bead pull perturbation method [73]. Taking into account one 200 W amplifier, the measured attenuation of the rf cables and the used rf splitters, the maximum achievable accelerating voltage is (198 ± 22) V for cavity 1 and (191 ± 21) V for cavity 2 for the setup shown in Fig. 31.

### *3.4.2 Transverse stripline kicker*

Corresponding to the bandwidth and frequency range of transverse multi-bunch dipole oscillations a broadband stripline kicker is used as correction element for the transverse bunch-by-bunch feedback system. The operation frequency ranges from DC to $f_{rf}/2$ and defines the basic layout of the kicker. The kicker is based on a coaxial design of a four quadrant waveguide, matched to a line impedance of 50 Ω. Figure 35 shows a cut through the kicker structure. Eight specially designed rf feedthroughs

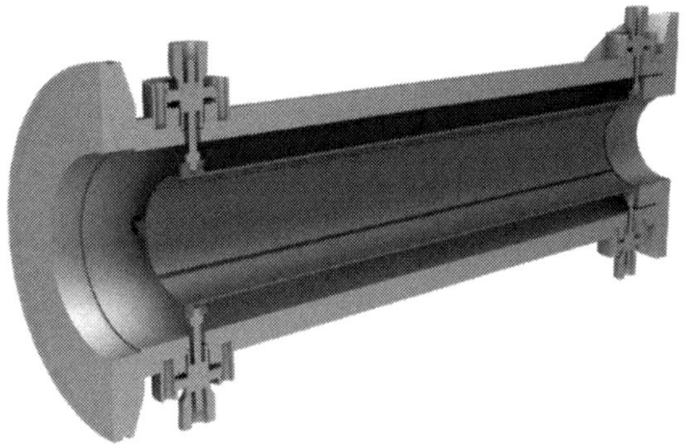

**Figure 35.** Cut through the stripline kicker used at ELSA [70].

connect the inner copper striplines to the outer shell and are used to inject a signal to the waveguide structure. The signal propagates across the striplines, is extracted on the opposite side and terminated in a matched rf load. The impedance matching of the kicker structure allows for affecting only a single bunch with a calculated and amplified correction signal. A 180° splitter is used to split the amplified correction signal into two equal parts with opposite polarity. Each signal drives one of both opposing

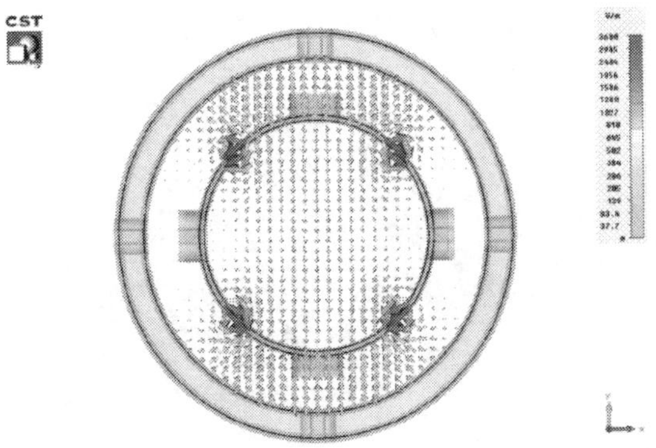

**Figure 36.** Electric dipole field distribution of the stripline kicker in vertical operation calculated with *CST Microwave Studio®* [59].

striplines, which yields to an electric dipole field in the center of the kicker structure corresponding to Fig. 36.

The shunt impedance $R_S$ of a stripline kicker with line impedance $Z_L$ depends on excitation frequency $f$, length $L$ and radius $r$ of the striplines which cover an angle of $\theta$ as [78]

$$R_S(f) = \frac{8Z_Lc^2}{\pi^2r^2f^2}\sin^2\left(\frac{\theta}{2}\right)\sin^2\left(\frac{2\pi f}{c}\cdot L\right).$$

Thus, the length of the striplines defines the bandwidth of the kicker structure. At ELSA, a minimum bandwidth of 250 MHz is required for damping transverse dipole bunch oscillations. Setting the length of the striplines to 30 cm ensures a proper kicking strength over the full bandwidth, as well as a vanishing shunt impedance on the rf harmonics. With this ansatz, the excitation and consequently the heating and beamloading of the kicker structure is minimized. The shunt impedance measurement results for the ELSA stripline kicker are shown in Fig. 37.

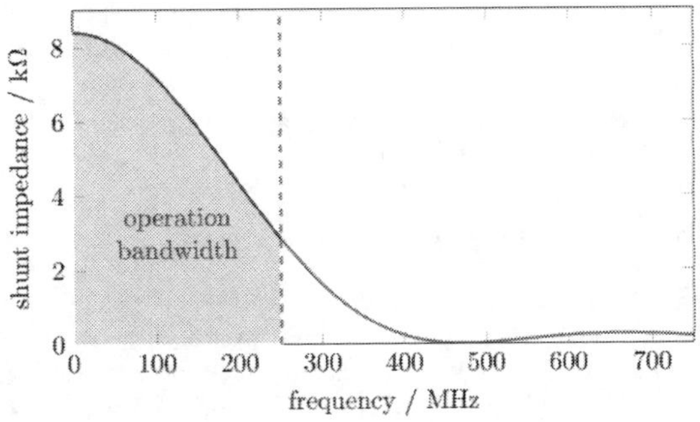


**Figure 37.** Shunt impedance of the ELSA stripline kicker as function of the excitation frequency [70].

### 3.4.3 Feedback performance

*Grow-damp measurement and Fourier spectrum*

To verify the damping capability of the bunch-by-bunch feedback system, so-called grow-damp measurements are of high importance. For that purpose, the system is deactivated for a few milliseconds using a filter with zero gain and an internal software trigger of the DSP unit, resulting in a growth of the bunches' oscillation amplitude in case of non-sufficient synchrotron radiation damping. After the reactivation of the standard filter, the impact of the feedback system on the oscillating bunches can be observed immediately.

Exemplary, such a measurement is shown in Fig. 38 which was conducted after the successful commissioning of the longitudinal feedback system at a beam energy of 2.35 GeV and for a stored beam current of about 20 mA in the storage ring. The digital input data of the longitudinal feedback system for bunch number 162 is presented, as its oscillation amplitude was largest among all filled buckets. The data was recorded turn-by-turn by the ADC of the DSP module resulting in a sampling frequency of 1.8236 MHz. Above noise[3] an exponential growth of the oscillation amplitude can be observed which confirms again that longitudinal coherent oscillations are already present at the storage ring for relatively moderate beam currents. After activation of the feedback system at 5 ms, the reduction of the amplitude within 300 µs and the damping performance of the system are clearly visible.

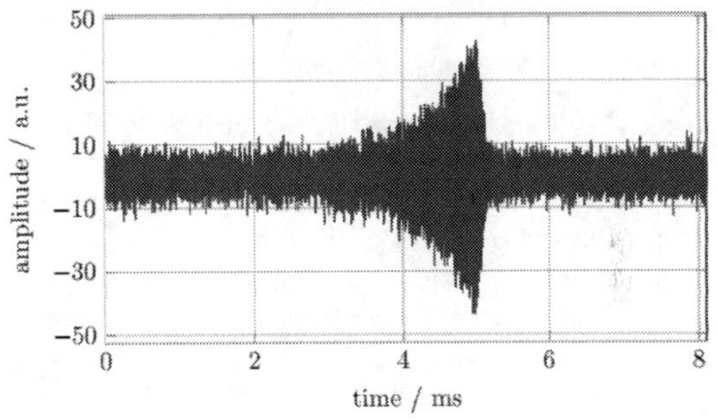


**Figure 38.** Grow-damp measurement for bunch number 162 at a beam energy of 2.35 GeV after the commissioning of the longitudinal feedback system at the storage ring: At 0 ms the feedback system was deactivated, at 5 ms the system was activated again. The fast damping performance of the system is observable [62].

Furthermore, with the Fourier spectrum of such a grow-damp measurement the resistive damping of the feedback system can be verified. First, the Fourier spectrum for each filled bucket is obtained by computing the discrete Fourier transform for each of the bunches' digitized amplitudes. By averaging over all single Fourier spectra, the Fourier spectrum for the complete grow-damp measurements is calculated. In case of the presented example, the Fourier spectrum is shown in Fig. 39.

[3]The noise level of the recorded ADC data amounts to ±10 ADC counts which is also measurable without electron beam.

DOI: 10.1051/epjconf/201713405002

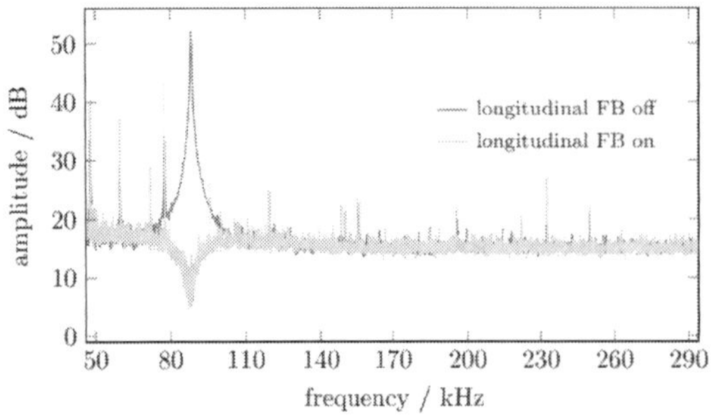


**Figure 39.** Fourier spectrum of the grow-damp measurement presented in Fig. 38: Peak and notch at nearly the same frequency of about 88 kHz indicate a perfect resistive damping of the longitudinal feedback system. The additional narrow peaks are caused by noise signals and are also visible in spectra recorded without beam [62].

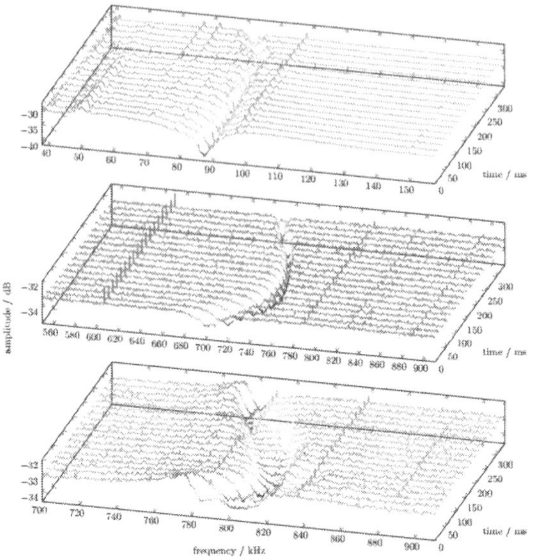


**Figure 40.** Feedback operation on the fast energy ramp. A beam current of about 35 mA has been accelerated from 1.2 GeV to 3.2 GeV with a ramp speed of 6 GeV/s. Fourier spectra for all three planes have been recorded successively every 20 ms. From top to bottom: longitudinal plane, horizontal plane, vertical plane. [70]

Due to the successful damping of the oscillations of all bunches the peak disappears completely and converts to a notch whose depth can be used as a measure for the quality of the damping performance. The notch is a consequence of the additional suppression of noise signals by the feedback system in the range of the synchrotron frequency, which is caused by the strong beam response on resonance. In addition, no shift between the frequencies of peak and notch can be observed, which indicates a purely resistive damping and overall a perfectly working system.

*Operation on the fast energy ramp*

In view of the phase characteristics of the used FIR filters in the DSP units, it is essential that the longitudinal and transverse tunes are kept constant during the energy ramp. Otherwise, a shift of the synchrotron or betatron frequencies would result in a change of the filter's phase and the corresponding phase of the generated correction signal. In this case, the primarily realized resistive damping performance of the feedback system exists no longer and in the worst case, if the phase shift exceeds 90°, an excitation of coherent electron oscillations can occur. To ensure matched feedback settings and stable operation independent of beam energy and operation mode of the storage ring, a new LLRF system has been installed which is capable of stabilizing and controlling amplitude and phase of the acceleration voltage (see Section 3.5).

A successful operation of the feedback system can be monitored best by measuring the Fourier spectra of each filled bucket on the fast energy ramp. As already mentioned the presence and depth of a notch in the spectra at the synchrotron and betatron frequencies indicates the quality of the damping performance. From the changing positions of the notches, the accuracy of the stabilization of the longitudinal and transverse tunes can be verified.

Figure 40 shows the time series of Fourier spectra for the longitudinal and transverse planes successively recorded every 20 ms on the fast energy ramp. In this measurement, a beam current of about 35 mA has been accelerated from 1.2 GeV to 3.2 GeV with a ramping speed of 6 GeV/s. The characteristic suppression (notch) at the synchrotron and betatron frequencies is clearly visible. In the longitudinal plane, the position of the notch and therefore the damping is almost independent of energy, which demonstrates a successful stabilization of the longitudinal tune. Only at the end of the energy ramp the synchrotron frequency is slightly reduced which is caused by a lack of acceleration voltage due to missing rf power required for compensating the beam loading in the accelerating cavities. The situation is different for the transverse planes, where fast varying magnetic fields caused by eddy currents in the vacuum chamber lead to additional magnetic multipoles which affect the beam focussing and therefore the transverse tunes. To allow an effective compensation of these effects, a dedicated tune measurement and correction procedure has been developed based on the experience with the feedback system and using its diagnostic capabilities. It is presented in Section 3.6.

### 3.5 RF amplitude and phase stabilization

To ensure matched feedback settings and stable operation independent of beam energy and operation mode of the storage ring, a new LLRF system has been installed. It is capable of stabilizing and controlling amplitude and phase of the acceleration voltage.

*Low level rf system*

The new *LLRF9/500* system, developed and manufactured by *Dimtel, Inc.* [74] has been successfully commissioned

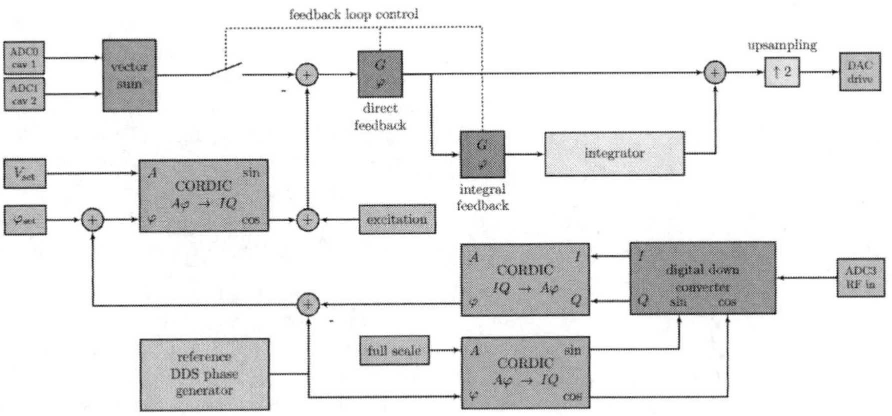


**Figure 41.** Overview of the signal processing of the LLRF system [70].

at ELSA. Nine wideband rf inputs are used to monitor rf cavity and waveguide signals. Two cavity probe signals are used in a digital feedback loop to stabilize the vector sum of the accelerating cavity voltages. The LLRF system is based on a fully digital approach using an FPGA. All of the nine analog rf inputs are digitized after down-conversion to 1/12 of the rf frequency (~ 42 MHz) and then processed by the FPGA. The cavity field stabilization is performed by two feedback loops: the low-latency proportional and the integral loop. The detailed layout and performance of the FPGA feedback processing is described in [57, 64], an overview of the signal processing is given in Fig. 41.

A digital proportional and integral feedback loop on the cavity station voltage and phase allows for compensation of the voltage fluctuations due to beam-loading effects. The fully digital approach provides monitoring capability of amplitude and phase for each input signal with 10 Hz update frequency. The FPGA continuously compares all input signals with the adjustable interlock thresholds. With the intermediate frequency of 42 MHz the maximum latency for detecting an interlock event is limited to 96 ns. Every interlock event is latched and time stamped. Fast data acquisition with pre-trigger capture provides information for post-mortem analysis. The FPGA generates a trip signal which, together with an external interlock input, controls an rf switch disabling the klystron drive output. The new LLRF system implements adjustment of the two tuner motors per accelerating cavity with a PID controller, operating at 10 Hz. The tuner loop acts to keep the cavity probe and forward signals in phase for minimum reflected power. In addition to stabilize the amplitude and phase, the LLRF system allows to ensure a constant synchrotron frequency during the fast energy ramp via hardware triggerable ramp profiles for amplitude and phase setpoints.

*Beam dynamics on the energy ramp*

On the fast energy ramp, the synchrotron frequency $f_s$ of ultra-relativistic electrons oscillating longitudinally with respect to the synchronous phase $\varphi_s$ is mainly given by the accelerating voltage $U$, the mean energy loss per turn $eU_{\rm rev}$, and the beam energy $E$ [79]:

$$f_s = f_{\rm rev}\sqrt{\frac{e\alpha_c h}{2\pi}\frac{U}{E}\sqrt{1-\frac{U_{\rm rev}^2}{U^2}}}\,,$$

where $\alpha_c$ represents the momentum compaction factor, $f_{\rm rev}$ the revolution frequency and $h$ the harmonic number. The revolution voltage is determined by

$$U_{\rm rev} = \frac{e\beta^3}{3\varepsilon_0\,(m_0c^2)^4}\frac{E^4}{R}$$

which defines the acceleration voltage required to compensate for the energy loss of the electrons due to synchrotron radiation in the circular accelerator. A constant synchrotron frequency can be achieved by adjusting

$$\sqrt{U^2-U_{\rm rev}^2} = \frac{f_s^2}{f_{\rm rev}^2}\frac{2\pi E}{e\alpha_c h}$$

which leads to an energy dependent accelerating voltage which has to be set to

$$U(E) = \sqrt{U_{\rm rev}^2 + E^2\,\frac{f_s^4}{f_{\rm rev}^4}\frac{4\pi^2}{e^2\alpha_c^2h^2}}\,.$$

The synchronous phase is then changing by

$$\Delta\varphi_s(E) = \sin^{-1}\left(\frac{U_{\rm rev}}{U(E)}\right)$$

which has to be compensated by shifting the rf phase setpoint value accordingly on the fast energy ramp.

An example for a fast energy ramp profile of the rf station is shown in Fig. 42. Here, the synchrotron frequency of 92 kHz is stabilized on the energy ramp from 1.2 GeV to 3.2 GeV applying a ramp profile of 512 steps within about 333 ms and changing the accelerating voltage from 1.1 MV to 3.1 MV. Applying a phase ramp with a total amplitude of −15.2° allows for compensation of the change of synchronous phase and ensures a matched timing of the bunch-by-bunch feedback loop.

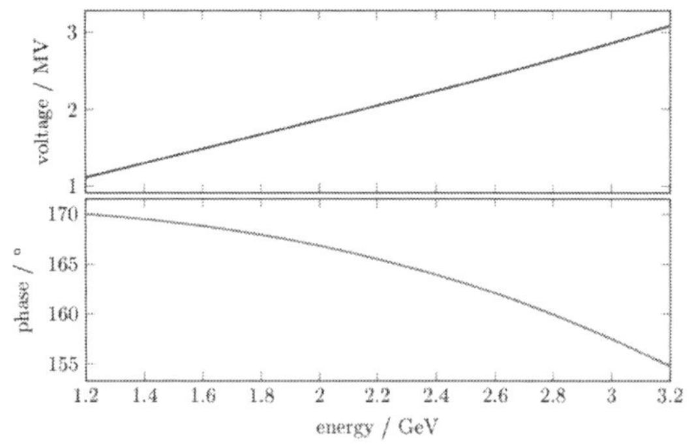


**Figure 42.** Required settings of the accelerating voltage and phase on the fast energy ramp from 1.2 GeV to 3.2 GeV to achieve a constant beam phase and synchrotron frequency of 92 kHz [70].

### 3.6 Fast tune measurement and correction

Fast ramping of magnets like dipoles and quadrupoles requires precise timing and correction techniques for magnet strengths. This has been achieved by measuring horizontal and vertical betatron tunes with a fast tune monitor and applying feed forward tune corrections to the fast energy ramp. The tune is measured using the bunch-by-bunch feedback's data acquisition capability and beam excitation functionality. Applying an offline FFT and data evaluation allows extracting tunes and calculating tune shifts as a correction signal for the main magnets' current ramp profiles. This is used as well for precise tune correction required by the bunch-by-bunch feedback system and to prevent polarized electron beams from depolarization on the fast energy ramp.

#### *3.6.1 Requirements on the tune measurement*

The tunes $Q_{x,z}$ of a circular accelerator cannot be chosen arbitrarily. Due to optical resonances the beam's oscillation amplitude may be driven to grow exponentially. This will cause beam loss or reduce the beam's quality for user operation. These resonances occur if the following condition is fulfilled:

$$j \cdot Q_x + k \cdot Q_z = N \quad \text{with } j, k, N \in \mathbb{N} .$$

In the ELSA storage ring the horizontal tune is placed close to the third integer optical resonance $Q_x = 4\frac{2}{3}$ and is slowly shifted towards it, allowing slow beam extraction. From this it is evident that the betatron tunes should be determined with accuracy better than one percent.

*Timing mismatch of magnet ramps*

Applying a fast energy ramp of 6 GeV/s requires a precise setting of the storage ring's magnets' current ramp profiles and timings. For this purpose, a hardware ramp-start trigger is applied to all of the magnets' power supplies which starts the programmed ramp. Every timing mismatch of the dipole and quadrupole ramps will cause transverse tune shifts, which can cause partial or complete beam loss when crossing an optical resonance. Such timing errors of the storage ring's main magnets' power supplies have been investigated to be in the order of 5 milliseconds. To successfully eliminate those timing errors and to stabilize the corresponding tunes, a tune measurement with temporal resolution of about 1 ms is required.

*Tune shifts induced by eddy currents in vacuum chambers*

A fast changing magnetic field, which is used to generate the fast energy ramp of the storage ring, induces voltages in the conducting vacuum chambers installed inside the dipole and quadrupole magnets. These induced voltages lead to eddy currents which generate additional magnetic field contributions and act on the beam's motion. Amongst other effects, this will lead to transverse tune shifts, generated by additional quadrupole strengths. To correct for those tune shifts a tune measurement and correction scheme with a temporal resolution well below the ramping time of up to 330 ms is needed.

#### *3.6.2 Setup of the measurement*

The fast tune measurement requires a fast readout of bunch position data as well as data evaluation and tune extraction. This fast readout has been realized using the bunch-by-bunch feedback system described in Section 3.4. To extract the beam's oscillation frequency from the raw data of a single bunch, a Fast Fourier Transform (FFT) is applied while exciting the corresponding bunch using a frequency sweep over the expected betatron frequency range. The excitation is generated by the bunch-by-bunch feedback system and applied by the rf kickers.

First, a drive signal is applied to one single bunch. It is set up to sweep over a fixed frequency range centered at the corresponding tune's frequency. The time of the frequency sweep has to be lower than the resulting FFT window to ensure a complete frequency scan over the chosen frequency span. After reading out the raw ADC data from the feedback system's single bunch acquisition buffer, the positions of the bunch collected every turn are split into subsets including a fixed number of turns each. The number of turns corresponding to the length of those subsets is chosen to match the desired temporal step size of the tune measurement settings.

For every subset an FFT is performed to extract the corresponding amplitude spectrum of the beam's motion. Searching for the maximum amplitude in every spectrum allows the extraction of the single bunch's tune in every subset.

#### *3.6.3 First results*

In order to correct for arbitrary tune and timing shifts on the fast energy ramp, the ELSA control system provides an interface for a feed forward method for tune correction. First, a delay $\Delta t^0_{F,D}$ can be added to the start of the magnetic field ramp of the focussing ($F$) and defocussing ($D$) quadrupoles to correct for timing errors. The remaining tune shifts along the energy ramp can be reduced by applying corrected quadrupole strengths which are based on the tune measurement. Using the accelerator model of the control system, the tune shifts are mapped into correction currents for every time step. The resulting current ramps are send to the power supplies. This procedure allows for iterating over several measurements to enhance the accuracy of the resulting tune correction.

The measurements at the ELSA storage ring have been conducted using a temporal resolution of $N = 2000$ revolutions, corresponding to $\Delta t = 1.096\,\mathrm{ms}$. The resulting frequency resolution is given by

$$\Delta f = \frac{f_{\mathrm{rev}}}{N} \approx 0.91\,\mathrm{kHz}\,,$$

which corresponds to $\Delta Q = 5 \times 10^{-4}$. Frequency and tune resolution match the required accuracies.

*Optimizing timing shifts on the fast energy ramp*

To reduce the transverse tune shifts on the fast energy ramp, the timing delays of the focussing and defocussing quadrupole power supplies have been changed and the resulting tune shifts were measured. The minimum vertical tune shift is achieved by setting the delay of the defocussing quadrupoles to $\Delta t_D^0 = 1.5\,\mathrm{ms}$ and $\Delta t_F^0 = -0.25\,\mathrm{ms}$ for the horizontal tune by shifting the focussing quadrupoles' ramp. Figure 43 shows the behavior of the vertical and horizontal tunes during the first 160 ms of the fast energy ramp with and without optimized delay. The negative shift has been achieved by shifting dipole and defocussing quadrupole timing setpoints to positive values.

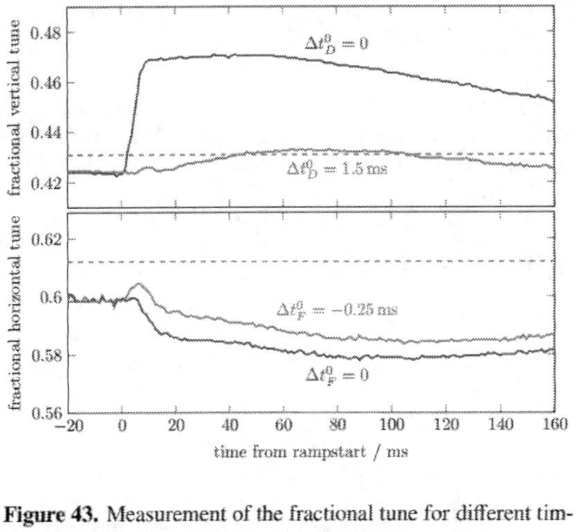


**Figure 43.** Measurement of the fractional tune for different timing shift corrections of the quadrupoles [70].

*Tune Correction on the fast energy ramp*

To achieve a successful and reliable operation of the bunch-by-bunch feedback system and to facilitate the correction of intrinsic depolarizing resonances applying fast tune jumps, both horizontal and vertical tunes have to be corrected to match the setpoints. Figure 44 shows a measurement of the vertical tune spectrum during the fast energy ramp, including four typical tune jumps. The sawtooth like tune jumps with a fall time of a few milliseconds are clearly visible, demonstrating the performance of the fast tune measurement. Since the transverse tune is corrected only during the fast energy ramp, it does not match its setpoint before and after the fast energy ramp. This leads to a minor tune deviation at ramp-start and ramp-stop. In future operation, a new ramp profile generator will allow to correct the tunes at the energy-flattop as well. This would prevent the observed tune deviation which potentially leads to beam loss.

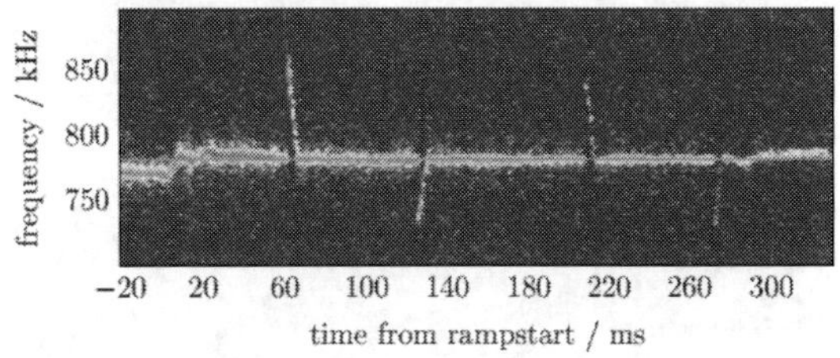


**Figure 44.** Corrected vertical tune on fast energy ramp from 1.2 GeV to 3.2 GeV when operating with polarized electrons and applying four tune jumps. The green line shows the setpoint of the vertical tune [70].

## 3.7 Visualization of beam dynamics

The emission of visible synchrotron radiation allows the imaging of electron beam dynamics through camera systems. At ELSA a streak camera[4] functions as fast time resolving beam diagnostic apparatus, capable of visualizing dynamics in all three dimensions from milliseconds down to the picoseconds time regime. A designated synchrotron radiation diagnostics beamline was installed [66, 85] and commissioned (Fig. 45) in order to provide up to single-shot 3D images of the revolving electron bunches in the storage ring. The system is capable of bunch length and damping time measurements as well as monitoring coherent and incoherent beam dynamics.

### *3.7.1 Experimental setup*

Streak cameras are specialized on measuring longitudinal photon distributions with the consequence of suppressing spatial information of one transverse plane. The ELSA beamline includes an optics set-up which preserves both transverse dimensions by splitting the photon beam (see Fig. 46) and transversally rotating one branch with a Dove prism. An adjustable path length provides a controllable photon time-of-flight delay for the second branch. Interchangeable magnification optics, a photon time-of-flight delay line and motorized mirrors allow proper signal positioning on the streak camera's sensitive measuring area.

[4] *Hamamatsu* C10910

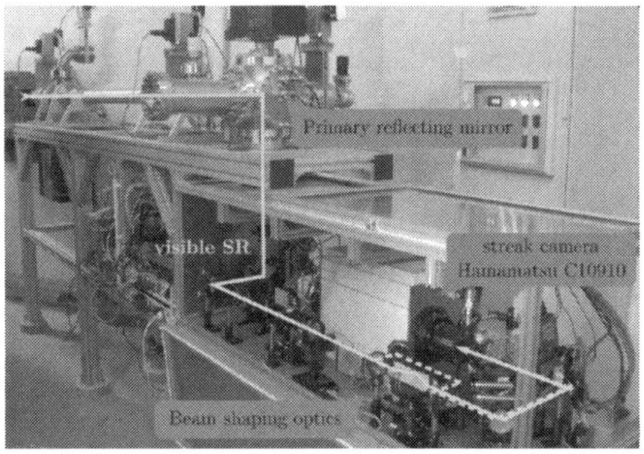


**Figure 45.** Synchrotron radiation diagnostics beamline with streak camera housing outside of the ELSA shielding tunnel [69].

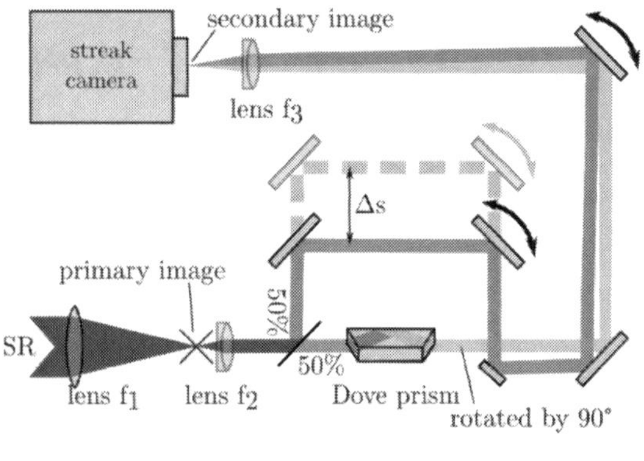


**Figure 46.** A horizontal and vertical beam image is coupled onto the streak camera to provide full 3D imaging capability [69].

### 3.7.2 3D bunch imaging capability

As the intensity of synchrotron radiation scales with the electron current, the observation of instabilities in the high-current regime yields sufficient single-shot images at shortest time scales. Figure 47 shows multiple single-shot electron bunch length measurements in top and side view, where the vertical axis denotes the longitudinal photon distribution. Due to the *synchroscan* measuring technique every fourth bunch is displayed in one row (e.g. bunches no. 1 and no. 5 in the top row and no. 3 and no. 7 in the bottom row). In addition, the dual imaging creates another pair of rows with vertical displacement $\Delta t_{\mathrm{ph}}$ corresponding to the time of flight difference. On the horizontal axis the bunches are 8 ns apart, whereas the horizontal and vertical set differ by an arbitrary distance set by the optics. At 80 mA of stored beam current and 1.2 GeV beam energy the storage ring chromaticity was increased by a factor > 10, creating an instability affecting only a selection of electron bunches (e.g. no. 1). The result is a longitudinal deformation as indicated by the measured bunch length. As the sextupole component was further increased, more bunches were successively affected in the same manner, eventually resulting in complete beam loss. Standard ELSA operation indicates no such instability behavior. However, future machine settings, which may cause abnormal intra-bunch dynamics whose traces result in intensity distortions as e.g. in head-tail instabilities, can now be identified.

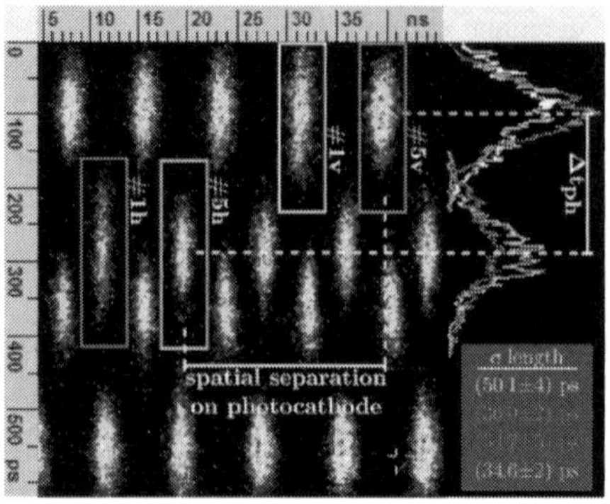


**Figure 47.** Single shot image of multiple electron bunches in top (h) and side view (v). A beam instability causes bunch lengthening for particular bunches before complete beam loss occurs. The actual electron beam dimension is larger horizontally than vertically despite the appearance in the image which is caused by an aperture restriction in the optical beamline [69].

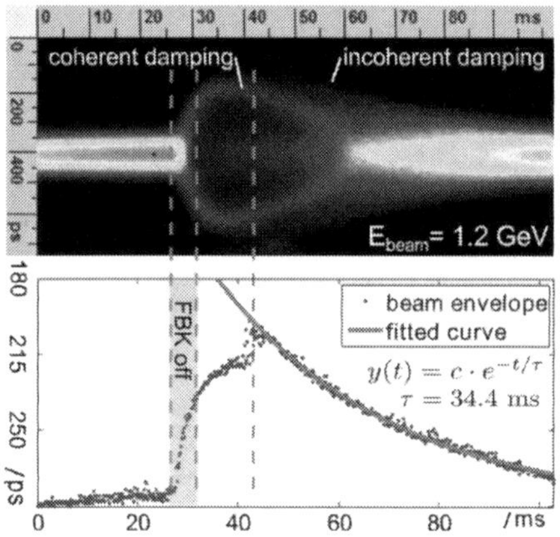


**Figure 48.** Grow-damp measurement of coherent and incoherent synchrotron oscillations. The streak camera is the only device capable of measuring incoherent oscillation amplitudes [67].

### 3.7.3 Multi-bunch beam dynamics

Beam dynamics up to time scales of several milliseconds can be observed in the same manner as described in Section 3.7.2. However, individual bunches cannot be resolved transversally but overlap within a single horizontal signal line containing only longitudinal information. Figure 48 illustrates a grow-damp measurement, where the bunch-by-bunch feedback system (see Section 3.4) is turned off for 5 ms at a beam current of 145 mA. Immediately multi-bunch instabilities start to grow resulting in coherent longitudinal particle oscillations which are damped by the feedback system after it is turned on again. As the decohered particle population is not affected by the feedback correction, natural damping occurs due to the emission of synchrotron radiation. The exponential decay of the envelope can be evaluated from the bitmap image.

Cases could be identified where the feedback correction is not powerful enough to regenerate a stable beam. For example, Fig. 49 shows reoccurring grow-damp behavior as function of the accelerating cavities' temperature. As the re-excitation period peaks near 36.8 °C, it vanishes for decreasing or increasing temperatures [68]. However, the amplitude and envelope shape describing the decoherence trace differs for higher or lower temperatures.

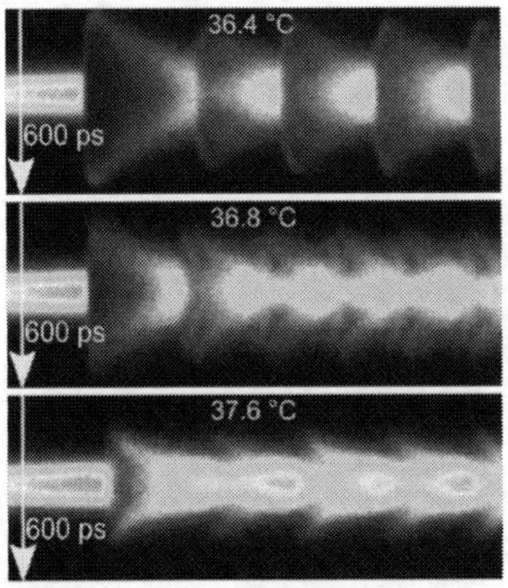


**Figure 49.** Multi-bunch instabilities reoccurring for certain accelerating cavity temperatures [68].

# 4 Slow beam extraction

A quasi continuous (but still bunched) external electron beam is produced using the technique of resonant beam extraction. To achieve a constant external beam current, an intensity feedback is applied generating appropriate current ramps for the extraction quadrupoles. The required small shift of the storage ring's horizontal tune causes a slowly changing beam pointing which has to be precisely monitored and compensated. The emittance of the external beam depends on the settings of the storage ring's optics and shows minor variations during the extraction period.

## 4.1 Resonant beam extraction technique

The slow beam extraction is based on an accelerator operation close to a third integer stop band. Dedicated sextupole magnets located in dispersion free straight sections of the storage ring are powered in order to excite a third-order horizontal betatron resonance. As a consequence, the electrons' motions in the horizontal phase space become unstable except for those located in a small central part which is bound by separatrices (the stable triangle, see Fig. 50). The area of the central part depends on the strength of the

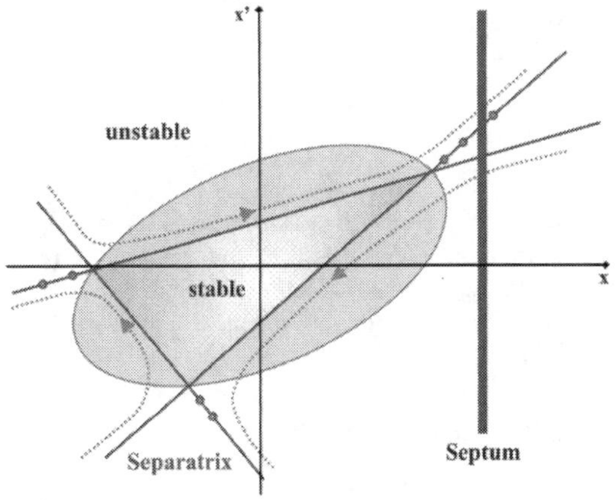


**Figure 50.** Particles' motion in the horizontal phase space when operating the accelerator close to a third-order resonance. The area of the stable triangular central part depends on the distance from the resonance and the resonance strength. Unstable particles will spiral outwards and will be extracted when passing the septum of the extraction magnet.

sextupole fields and the distance of the tune from the third-order resonance. By shrinking the area of the stable triangle, parts of the electron beam become unstable, resulting in an increasing horizontal betatron amplitude. These unstable electrons move along the separatrix lines in phase space and are finally extracted after crossing the septa of two extraction magnets [1].

## 4.2 Position and intensity stabilization

Due to the required variation of the stable phase area, the positions of the separatrix lines are shifted during the extraction period. This in general causes a small shift of the angle of the extracted beam which increases during the extraction period. A compensation method commonly applied at proton accelerators aligning all extraction separatrices (known as Hardt condition [80]) cannot be realized at ELSA due to missing flexibility of the storage ring's optics. Instead, the beam deflection in the first septum magnet is changed appropriately, thus compensating the shift of angular displacement of the extracted beam at the position of the septum blade to first order. The required additional septum control is derived adding a scalable fraction of the extraction quadrupole current setting to the current setpoint of the septum magnet's power supply. The scaling factor is determined empirically by minimizing the shift of the external beam at the tagging systems of the nuclear physics experiments. This procedure requires precise beam position monitoring at the photon tagger which is performed by a newly developed non-invasive monitor system based on two rf diagnostic cavities. This system is described in detail in [81].

A constant external beam current over the extraction period is achieved by applying a feedback loop to the extraction quadrupoles. Either the overall tagging rate or the voltage signal of an rf diagnostic cavity serving as beam intensity monitor (see [81, 82]) can be selected as input for a PI controller of the quadrupole power supplies.

### 4.3 Emittance measurements

The emittance measurements are based on recording the transverse intensity distribution of the electron beam with the help of luminescence screens and synchrotron radiation monitors. For this purpose, dedicated systems have been set up at the storage ring and the external beam lines (see [65]).

#### *4.3.1 Measurement principle*

Once knowing the beam's relative momentum deviation $\Delta p/p$, the beta functions $\beta_{x,z}$ and the dispersion $D_x$ at the position of the monitor, the horizontal and vertical emittance can be extracted from

$$\sigma_x(s) = \sqrt{\epsilon_x \cdot \beta_x(s) + \left(D_x(s) \cdot \frac{\Delta p}{p}\right)^2} \quad \text{and} \qquad (4)$$
$$\sigma_z(s) = \sqrt{\epsilon_z \cdot \beta_z(s)}\,,$$

where the rms widths $\sigma_x$ and $\sigma_z$ are determined from fitting a Gaussian function to the measured transverse intensity profiles. Whereas $\Delta p/p$, which is energy dependent, can be taken from the accelerator's optics model, $\beta_x$, $\beta_z$ and $D_x$ depend on the extraction optics and therefore have to be measured.

In a first step the dispersion function is determined by varying the rf frequency $f_{\mathrm{rf}}$ and measuring the corresponding shifts $\Delta x$ of the entire beam:

$$D(s)\left(-\frac{1}{\alpha_c}\frac{\Delta f_{\mathrm{rf}}}{f_{\mathrm{rf}}}\right) = \Delta x(s)\ .$$

This measurement is repeated for different quadrupole strengths, and a dispersion parameter $D_0$ and its derivative $D_0'$ at a position in front of the first quadrupole of the external beamline are calculated from this data.

In a second step, the emittances $\epsilon_x$ and $\epsilon_z$ are determined applying the quadrupole scan technique. Here, intensity profiles are recorded for different settings of a focussing quadrupole in the external beamline and an optics dependent function is fitted to the set of extracted rms beam widths, where contributions caused by the dispersion are subtracted according to Eq. 4. This procedure allows the extraction of both, the beta function and the emittance (for further details see e.g. [83, 84]) and has been repeated for different optics settings of the storage ring.

#### *4.3.2 Results*

To determine the impact of the storage ring optics on the emittance of the extracted beam, two sets of measurements have been carried out: First, when increasing the strength of the extraction sextupoles and thereby increasing the excitation of the third integer resonance, the step size at the septum blade becomes larger leading to an increase of the external beam's horizontal emittance (see

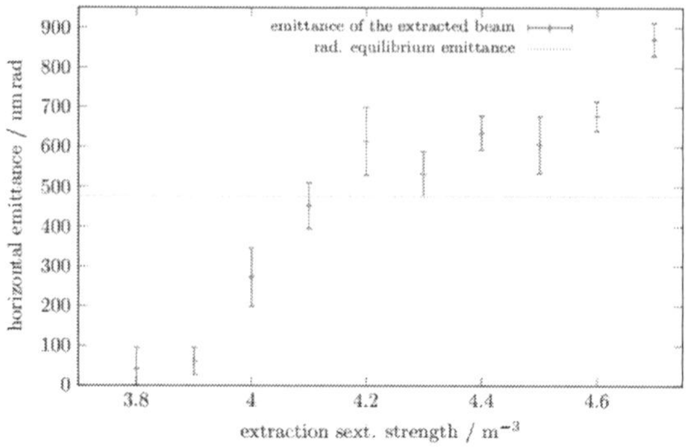


**Figure 51.** Measured horizontal beam emittance as function of the extraction sextupole strength for a beam energy of 2.35 GeV. The green line indicates the radiative equilibrium emittance in the storage ring.

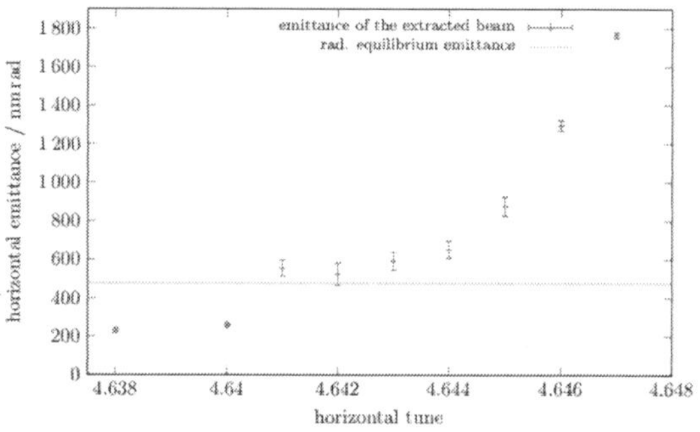


**Figure 52.** Measured horizontal beam emittance as function of the horizontal tune for a beam energy of 2.35 GeV. The green line indicates the radiative equilibrium emittance in the storage ring.

Fig. 51). Second, when increasing the strength of the extraction quadrupoles and thereby shifting the horizontal betatron tune closer to the third integer stop band, again the step size at the septum blade becomes larger leading to an increase of the external beam's horizontal emittance (see Fig. 52).

In addition, time resolved measurements for fixed settings of the storage ring's optics have been performed. Here, the temporal change of the external beam's emittances has been determined and compared to the emittances of the stored beam which have been measured simultaneously (see [84]). Whereas the horizontal emittance is mainly determined by the (fixed) step size at the septum blade and therefore remains constant throughout the extraction period (see Fig. 53), the vertical emittance depends on the coupling of the horizontal and vertical betatron oscillations. The extracted electrons originate from the unstable region in phase space and therefore have larger betatron amplitudes than the stored electrons occupying the central stable triangle. The vertical emittance of the extracted beam is therefore slightly larger than the vertical emittance of the stored beam (see Fig. 54).

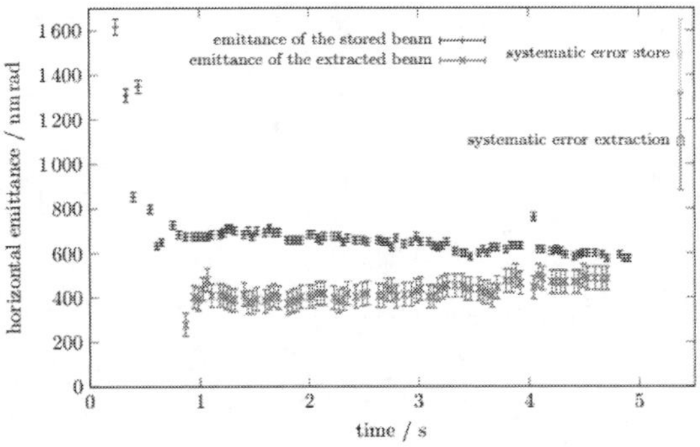


**Figure 53.** Measured temporal change of the horizontal beam emittance over the extraction period. The extraction of a 2.35 GeV beam started at 1 s [85].

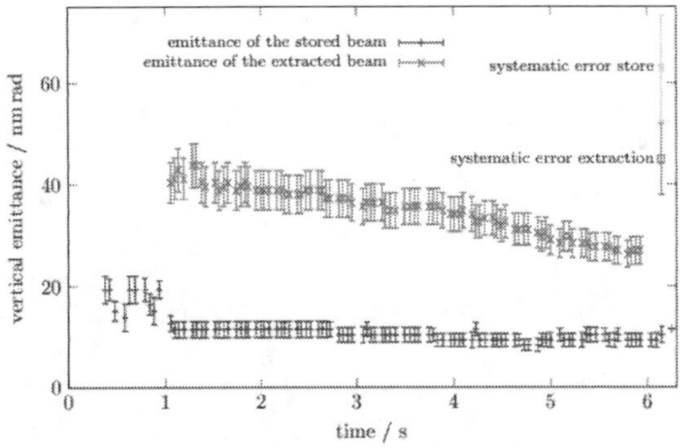


**Figure 54.** Measured temporal change of the vertical beam emittance over the extraction period. The extraction of a 2.35 GeV beam started at 1 s [85].

## 5 Energy increase

Regarding the production of linearly polarized photon beams, an increase of the maximum achievable electron beam energy is highly desirable. Whereas circularly polarized photons, generated by polarization transfer from longitudinally polarized electrons in the bremsstrahlung process, get maximum polarization in case of a full energy transfer from electrons to photons [86], the achievable linear photon polarization via coherent bremsstrahlung drops significantly at high energy transfers (see e.g. [5]) and vanishes for a full energy transfer. In order to optimize the overall figure of merit (in which the photon polarization enters quadratically), the kinematics of the coherent bremsstrahlung are typically chosen such that the coherent peak is set at about half of the electron beam energy. This limits the useful energy range of linearly polarized photons to the same value and makes an increase of the electron beam energy highly attractive.

The maximum beam energy at the ELSA storage ring is mainly limited by the achievable magnetic bending fields and acceleration voltage. In principle, an energy upgrade may be attained by reducing the gap of the dipole magnets and the installation of superconducting rf cavities. Design concepts and the impact on beam dynamics and cavity control at the fast ramping ELSA storage ring have been investigated for a proposed energy increase to 5 GeV and are presented in the following.

### 5.1 Magnets and beam dynamics

#### *5.1.1 Installation of new magnets*

To generate the required bending field of 1.53 T, the existing dipole magnets have to be replaced by new ones. The gap has to be reduced from 50 mm to 38 mm, the number of windings of the coils must be increased from 14 to 20. In addition, the existing dipole power supply has to be modified with respect to the higher total resistance of the magnet coils, the requirements on the current will not change.

The operating point in the tune diagram is fixed to the settings used up to now in order to achieve a dispersion suppression in the straight sections caused by the missing magnet concept of the ELSA lattice. The existing quadrupole and sextupole magnets can be operated up to 5 GeV without modifications. Only minor modifications of the quadrupole power supplies (increase of the maximum current from 910 A to 970 A because of the energy dependence of the quadrupole strength) are required.

No modification of the kicker magnets of the injection system is needed. The extraction magnets (2 septum magnets) are limited in field strength to a maximum beam energy of 3.5 GeV and have to be replaced by stronger ones.

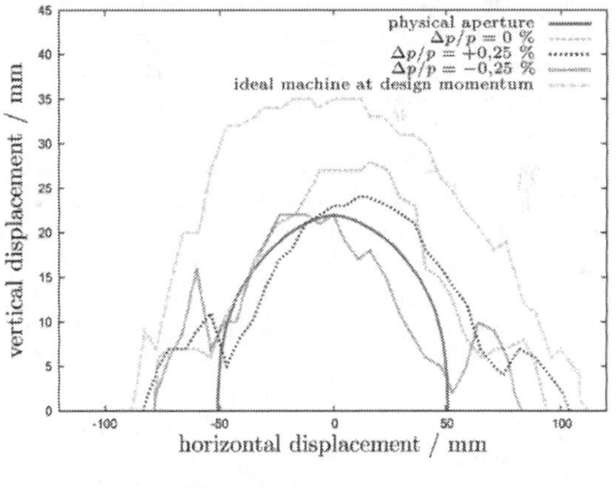


**Figure 55.** Dynamic aperture of the ELSA storage ring with alignment and magnetic field errors influenced by eight sextupole magnets for chromaticity correction: For each line in the plot several error distributions have been simulated in order to obtain sufficient statistics. The minimum value of those apertures was selected and is shown here.

#### *5.1.2 Static aperture and beam lifetime*

The accelerator lattice was investigated up to 5 GeV beam energy. It turned out that the aperture limitations (except

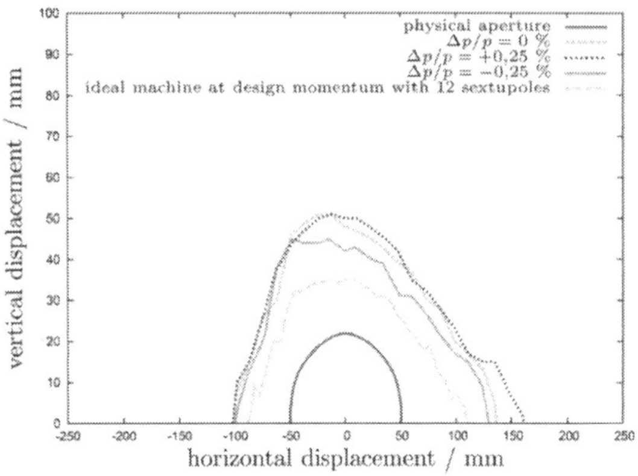


**Figure 56.** Dynamic aperture of the ELSA storage ring with alignment and magnetic field errors using twelve sextupole magnets for chromaticity correction. In this case larger magnetic field errors would be acceptable than the ones used in Fig. 55.

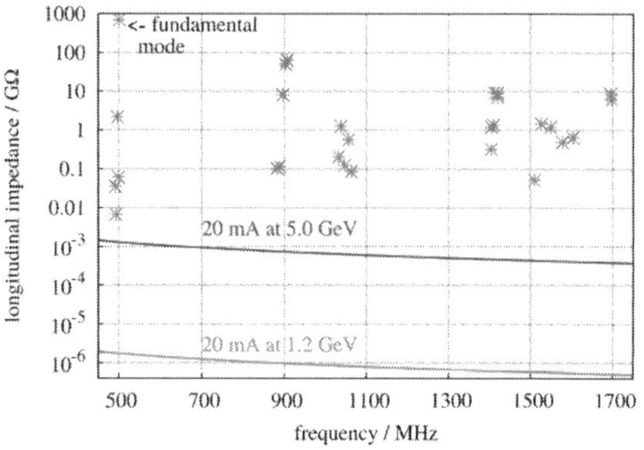


**Figure 57.** Longitudinal shunt impedances of monopole HOMs and typical threshold impedances for longitudinal multi-bunch instabilities at a beam current of 20 mA in the storage ring [87].

for the extraction septa which have to be replaced for reasons mentioned above) of the existing vacuum chambers allow for a beam lifetime of about 50 hours which is several orders of magnitude larger than the requirements for accelerator operation with beam extraction to external experiments.

### 5.1.3 Dynamic aperture

Detailed investigations of the influence of alignment and field errors of the dipole and quadrupole magnets as well as higher multipole components were performed applying particle tracking simulations with the help of dedicated software packages. From these calculations we derived the dynamic aperture which indicates the maximum beam sizes circulating stably without significant beam loss. Even in case of relative momentum deviations of 0.25 % – i.e. twice the beam energy spread at 5 GeV – and compensation of chromatic errors, the existing number of sextupoles seems sufficient in order to obtain a dynamic aperture comparable with the physical one (see Fig. 55).

The situation changes in case of larger multipole contributions which may arise from partial saturation of the magnets at maximum energy. In the worst case this would require an installation of additional four sextupole magnets for chromaticity correction (see Fig. 56). This can be decided only if new dipole magnets are available, so that they can be analyzed regarding their magnetic field errors.

## 5.2 Operation of superconducting cavities

The accelerator operation for hadron physics experiments requires beam life times of at least one minute at flat top energy in order to avoid significant beam losses. This corresponds to an overvoltage factor of 2.69 and a peak accelerating voltage of 13.67 MV which has to be generated. An appropriate 500 MHz accelerating structure is given by the geometry of the superconducting (sc) five-cell cavities constructed for the JAERI FEL linac by the former company *Accel Instruments* in the early '90s [88]. These sc acceleration modules are characterized by an operating gradient of up to 5 MV/m with a cavity quality factor above $10^9$. Using two of these five-cell cavities, the necessary overall cavity voltage of 13.67 MV could be provided.

### 5.2.1 Cavity coupling and detuning

Assuming a shunt impedance of 500 GΩ and a quality factor of $2 \times 10^9$, the beam current dependent optimal coupling factor of the cavities was calculated [87]. It turned out that, using this setup and the existing rf plant (250 kW klystron, waveguides, etc.), an internal beam with up to 50 mA beam current can be accelerated up to 5 GeV and stored for at least some seconds with negligible beam loss (which is sufficient in case of slow beam extraction). The coupling would be adjusted to an optimal coupling factor of 2720 for 50 mA at 5 GeV and does not need to be varied for different beam currents and beam energies.

The cavity detuning defines the difference between the rf frequency and the resonance frequency of a cavity and has to be adjusted during the beam injection and on the energy ramp by fast frequency tuning of the cavities. The maximum cavity detuning was calculated to be less than 16 kHz, the maximum detuning speed is less than about 32 kHz/s [87] which should be achievable using state-of-the-art tuning devices for sc resonators. A typical scenario of an accelerator cycle is illustrated in Fig. 58.

### 5.2.2 HOM and beam instabilities

The investigation of the required higher order mode (HOM) damping was based on the geometry of the JAERI structure. All relevant monopole and dipole HOMs were simulated numerically with the software package *CST Microwave Studio®* neglecting the HOM damping of the main power coupler [87]. Figures 57 and 59 show the results for the longitudinal and transversal impedances of

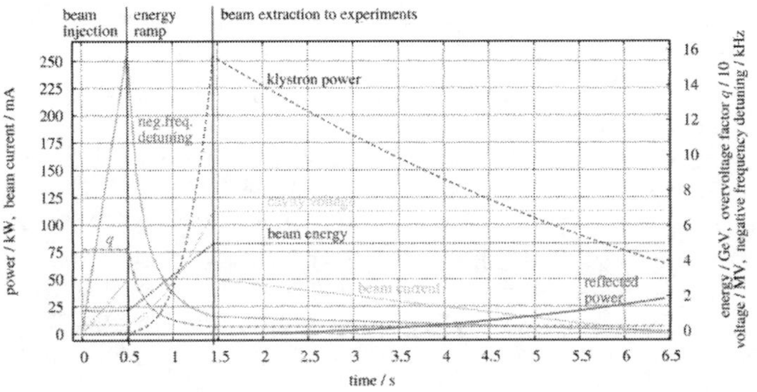


**Figure 58.** Accelerator cycle for a beam current of 50 mA up to 5 GeV using two superconducting cavities with a shunt impedance of $5 \times 10^{11}\,\Omega$ and a quality factor of $Q_0 = 2 \times 10^9$. Time dependence of cavity voltage and overvoltage factor, total klystron power as well as reflected power and negative frequency detuning per cavity are shown [87].

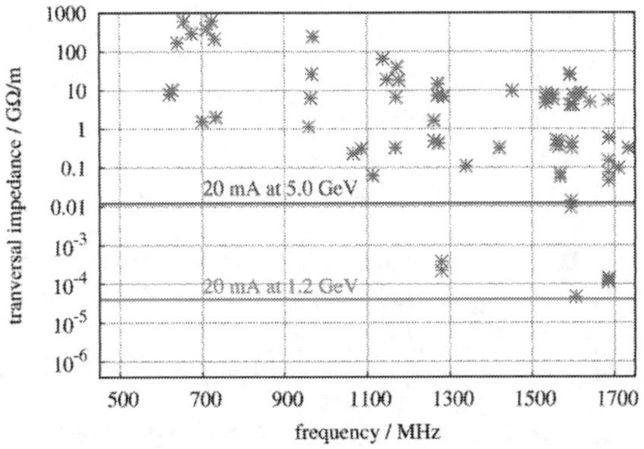


**Figure 59.** Transversal shunt impedances of dipole HOMs and typical threshold impedances for transverse multi-bunch instabilities at a beam current of 20 mA in the storage ring [87].

the HOMs and typical threshold impedances. It turned out that the growth rate of all higher order monopole modes and a large number of dipole modes are well above the stability threshold given by radiative damping and may excite coherent beam instabilities. Therefore sophisticated HOM couplers have to be developed, even in case of partial damping of HOMs by the fundamental mode coupler. The layout of these HOM couplers is considered to be the major problem of a possible future energy upgrade program.

## 6 Summary

Within the past 12 years, the electron stretcher accelerator facility ELSA has been successfully upgraded with respect to beam polarization, intensity and stability. The upgrade program comprised developments and various dedicated measures for an increase of beam current and energy, damping of coherent beam instabilities, compensation of depolarizing resonances and enhancement of the quality of the extracted beam. The envisaged intensity increase of approximately a factor of 10 has been successfully achieved, and polarization losses on the fast energy ramp have been reduced to less than 10 % up to energies of about 3 GeV. The quality and stability of the extracted beam is sufficient for a reliable generation of linearly and circularly polarized photon beams for photoproduction experiments. ELSA is therefore well prepared for a successful continuation of the proposed hadron physics experimental program.

The work reported here would not have been possible without the support from the Deutsche Forschungsgemeinschaft within the SFB/TR16.

## References

[1] W. Hillert, Eur. Phys. J. A **28** s1, 139 (2006)
[2] J. Hartmann et al., Phys. Lett. B **748**, 212 (2015)
[3] B. Bantes et al., Int. J. Mod. Phys. Conf. Ser. **26**, 1460093 (2014)
[4] A. Bella et al., EPJ Web of Conf. **81**, 06002 (2014)
[5] D. Elsner et al., Eur. Phys. J. A **39**, 373 (2009)
[6] W. Hillert et al., AIP Conf. Proc. **570**, 961 (2001)
[7] H. Dutz et al., Phys. Rev. Lett. **91**, 192001 (2003)
[8] H. Dutz et al., Phys. Rev. Lett. **93**, 032003 (2004)
[9] H. Dutz et al., Phys. Rev. Lett. **94**, 162001 (2005)
[10] D. Heiliger et al., PSTP'09 Conf. Proc., 232 (2010)
[11] D. Heiliger et al., J. Phys. Conf. Ser. **298**, 012003 (2011)
[12] D. Heiliger et al., IPAC'12 Conf. Proc., 1521 (2012)
[13] D. Heiliger et al., IPAC'13 Conf. Proc., 294 (2013)
[14] D. Heiliger et al., IPAC'11 Conf. Proc., 1123 (2011)
[15] W. Hillert et al., IPAC'11 Conf. Proc., 2133 (2011)
[16] A. Balling et al., PAC'09 Conf. Proc., 3714 (2009)
[17] A. Balling et al., AIP Conf. Proc. **1149**, 809 (2009)
[18] J.-P. Thiry et al., IPAC'12 Conf. Proc., 789 (2012)
[19] J.-P. Thiry et al., IPAC'13 Conf. Proc., 479 (2013)
[20] J.-P. Thiry et al., IPAC'14 Conf. Proc., 1111 (2014)
[21] O. Boldt et al., PAC'11 Conf. Proc., 1507 (2011)

[22] J.F. Schmidt et al., IPAC'14 Conf. Proc., 1108 (2014)
[23] M. Hoffmann et al., AIP Conf. Proc. **570**, 756 (2001)
[24] O. Boldt et al., IPAC'11 Conf. Proc., 2130 (2011)
[25] J.F. Schmidt et al., IPAC'13 Conf. Proc., 891 (2013)
[26] J.F. Schmidt et al., IPAC'15 Conf. Proc., 808 (2015)
[27] J.F. Schmidt et al., IPAC'16 Conf. Proc., 695 (2016)
[28] W. Hillert et al., AIP Conf. Proc. **1149**, 1160 (2009)
[29] R. Koop et al., IPAC'12 Conf. Porc., 783 (2012)
[30] R. Koop et al., IPAC'16 Conf. Porc., 95 (2016)
[31] A. Septier, *Focusing of Charged Particles* (Acad. Press, New York, 1967)
[32] W.B. Hermannsfeldt, SLAC-PUB **331**, UC28 (1988)
[33] D. Heiliger, Diss., Univ. Bonn (2014)
[34] M. Yamada et al., Appl. Surf. Sci. **70/71**, 531 (1992)
[35] V.E. Andreev et al., AIP Conf. Proc. **570**, 901 (2000)
[36] M. Baylac et al., Phys. Rev. ST Accel. Beams **8**, 123501 (2005)
[37] M. Yamada and Y. Ide., Surf. Sci. **339**, L914 (1995)
[38] P. Tomkiewicz et al., Appl. Surf. Sci. **252**, 7647 (2006)
[39] O.M. Katz and E.A. Gulbransen., Rev. Sci. Instrum. **31**, 615 (1960)
[40] U. Bischler and E. Bertel, J. Vac. Sci. Technol. A **11**, 458 (1993)
[41] V. Bargmann et al., Phys. Rev. Lett. **2**, 435 (1959)
[42] M. Froissart and R. Stora, Nucl. Instr. and Meth. **7**, 297 (1960)
[43] A Dieckmann et al., IPAC'12 Conf. Proc., 3623 (2012)
[44] O. Boldt et al., IPAC'12 Conf. Proc., 1326 (2012)
[45] C. Sanderson and R. Curtin, J. Open Sour. Soft. **1**, 26 (2016)
[46] M. Borland, Advanced Photon Source LS-287, (2000)
[47] J. F. Schmidt, github.com/janfschmidt/palattice
[48] MAD–X, Methodical Accelerator Design, https://mad.web.cern.ch/mad/
[49] J. F. Schmidt, www.ctan.org/pkg/tikz-palattice
[50] A. A. Sokolov and I. M. Ternov, Soviet Physics Doklady **8**, 1203 (1964)
[51] Ya. S. Derbenev and A. M. Kondratenko, Sov. Phys. JETP **37**, no. 6, (1973)
[52] S. Agostinelli et al., Nucl. Instr. Meth. A **506**, 250 (2003)
[53] F. Klarner et al., LINAC'08 Conf. Proc., 392 (2009)
[54] F. Klarner et al., LINAC'10 Conf. Proc., 325 (2010)
[55] S. Mey et al., IPAC'11 Conf. Proc., 1072 (2011)
[56] M. Schedler et al., LINAC'14 Conf. Proc., 847 (2014)
[57] D. Sauerland et al., IPAC'13 Conf. Proc., 2717 (2013)
[58] A. Roth et al., IPAC'11 Conf. Proc., 250 (2011)
[59] M. Schedler et al., IPAC'11 Conf. Proc., 481 (2011)
[60] N. Heurich et al., IPAC'11 Conf. Proc., 484 (2011)
[61] M. Schedler et al., IPAC'12 Conf. Proc., 2840 (2012)
[62] A. Roth, Diss., Univ. Bonn (2013)
[63] M. Schedler et al., IPAC'13 Conf. Proc., 2714 (2013)
[64] M. Schedler et al., IPAC'14 Conf. Proc., 2418 (2014)
[65] S. Zander et al., PAC'11 Conf. Proc., 408 (2011)
[66] S. Zander et al., IPAC'12 Conf. Proc., 795 (2012)
[67] M. Switka et al., IPAC'14 Conf. Proc., 3471 (2014)
[68] M. Switka et al., IBIC'14 Conf. Proc., 144 (2014)
[69] M. Switka et al., IPAC'15 Conf. Proc., 144 (2015)
[70] M. Schedler, Diss., Univ. Bonn (2015)
[71] H. Alberto, Dipl., TU Dortmund (1998)
[72] CST, Computer Simulation Technology, http://www.cst.com/
[73] L.C. Maier Jr. and J.C. Slater, J. Appl. Phys. **23**, 68 (1952)
[74] Dimtel Inc., San Jose, USA, www.dimtel.com
[75] Y. Kim et al., IEEE Trans. Nucl. Sci. **47**, no. 2, 452 (2000)
[76] W.Z. Wu et al., Nucl. Instr. Meth. Phys. Res. A **632**, 32 (2011)
[77] R. Boni et al., EPAC'96 Conf. Proc, 1881 (1996)
[78] D.A. Goldberg and G.R. Lambertson, AIP Conf. Proc. **249**, 537 (1992)
[79] H. Wiedemann, *Particle accelerator physics* 2nd ed. (Springer, Berlin, 2003)
[80] W. Hardt, CERN PS-DL-LEAR-Note, 81-6 (1981)
[81] T. Pusch et al., Phys. Rev. ST Accel. Beams **15**, 112801 (2012)
[82] T. Pusch et al., DIPAC'11 Conf. Proc., 263 (2011)
[83] S. Zander et al., IPAC'11 Conf. Proc., 1168 (2011)
[84] S. Zander et al., IPAC'12 Conf. Proc., 792 (2012)
[85] S. Zander, Diss., Univ. Bonn (2013)
[86] H. Olsen and L.C. Maximon, Phys. Rev. **114**, 887 (1959)
[87] A. Roth et al., IPAC'11 Conf. Proc., 253 (2011)
[88] M. Pekeler et al., APAC'07 Conf Proc., 791 (2007)

EPJ Web of Conferences **134**, 05003 (2017) DOI: 10.1051/epjconf/201713405003
*Subnuclear Structure of Matter: Achievements and Challenges*

# Timing and tracking for the Crystal Barrel detector

*Reinhard* Beck[1,*], *Kai* Brinkmann[2,**], and *Rainer* Novotny[2,***]

[1] *Helmholtz-Institut für Strahlen- und Kernphysik, Rheinische Friedrich-Wilhelms-Universität Bonn, Nußallee 14-16, 53115 Bonn*
[2] *II. Physikalisches Institut, Justus-Liebig Universität Gießen, Heinrich-Buff-Ring 16, D-35293, Gießen*

**Abstract.** The aim of the project D.3 is the upgrade of several detector components used in the CBELSA/TAPS experiment at ELSA. The readout of the Crystal Barrel Calorimeter will be extended by a timing branch in order to gain trigger capability for the detector, which will allow to measure completely neutral final states in photoproduction reactions (see projects A.1 and C.5). Additionally, the readout of the inner crystals of the TAPS detector, which covers the forward opening of the Crystal Barrel Calorimeter, will be modified to be capable of high event rates due to the intensity upgrade of ELSA. Furthermore, a full-scale prototype Time Projection Chamber (TPC) has been built to be used as a new central tracker for the CBELSA/TAPS experiment at ELSA and the FOPI experiment at GSI.

## 1 Introduction

During the last beam times, the CBELSA/TAPS experiment has proven to be a reliable and stable detector system to measure polarization observables in different reactions with high precision (see project A.1). The previous measurements were mostly focused on the photoproduction of single or multiple neutral mesons off the proton in reactions like $\gamma p \to p\pi^0$, $p\eta$ and $p\pi^0\pi^0$. The measured data sets have been included in the partial wave analysis and are used to further constrain the calculations (see project A.2). However, most data sets used so far contain the proton in the initial state. Completely neutral reactions like $\gamma n \to n\pi^0$ pose a higher challenge for the detector system and only sparse data is available right now. Since the CBELSA/TAPS experiment is able to measure these reactions only with a very low efficiency, one of the goals of the project D.3 was the upgrade of the detector system to provide higher trigger capabilities (see section 2). Therefore the readout of the Crystal Barrel Calorimeter was extended by a timing branch.

Additionally, a higher photon rate in the experiment would be desirable, to enlarge the data set taken during a beam time. For the higher rates, an upgrade of the inner crystals of the MiniTAPS detector is necessary. These crystals are close to the photon beam line and are therefore hit by a high rate of particles. The upgrade of the crystals is described in section 4. To achieve a higher data rate in the crystals of the Crystal Barrel detector a new digitization stage of the energy branch is needed. Therefore a new Sampling-ADC (SADC) readout was developed, which is further shown in section 3.

Another goal of the project D.3 is the development of a new central tracker, which could be inserted into the Crystal Barrel detector. A Time Projection Chamber (TPC) based on Gas Electron Multipliers (GEM) for charge amplification was built and the first prototype could be used as a tracker for the FOPI experiment at GSI. The detailed description of the new TPC and the gain calibration can be found in section 5.

## 2 Trigger Upgrade of the Crystal Barrel Calorimeter

### 2.1 Situation before the SFB

The data acquisition of the CBELSA/TAPS experiment is built up in gated mode. This means that the experiment has a global trigger signal, which needs to be distributed to each module involved in digitizing data. The latency of this signal, i.e. the delay from the physical event to the leading edge of the trigger signal needs to fulfill the requirements of the existing read-out modules of the whole experiment and thereby is limited. The MiniTAPS modules allow for a maximum delay of 750 ns. To achieve the allowed latency for trigger signals, the propagation delay on cables and the processing time of the central trigger module has to be taken into account. This results in a maximum latency of 500 ns. Any signal which is supposed to be used in the trigger condition has to arrive at the central trigger module within this time.

In the former setup, the following detectors could be included in the trigger condition:

- The **MiniTAPS** Calorimeter, consisting of $BaF_2$ crystals, which have a fast light output.

* e-mail: beck@hiskp.uni-bonn.de
** e-mail: Kai-Thomas.Brinkmann@exp2.physik.uni-giessen.de
*** e-mail: r.novotny@exp2.physik.uni-giessen.de

- The **Forward Conus**, which is a small part of the Crystal Barrel Calorimeter in the forward region, upgraded with a fast readout.
- The **Inner Detector**, which consists of scintillating plastic fibers.

The fiber detector offers a high angular coverage, but is only sensitive to charged particles. MiniTAPS and the Forward Conus can additionally detect photons with a high efficiency and even neutrons with a limited efficiency. However, those only cover an angular range of 1° to 27.5° relative to the beam axis.

The remaining part of the Crystal Barrel Calorimeter was only included in the second level of the trigger. Therefore, an insufficient number of hits in the Crystal Barrel Calorimeter can abort the digitization of an event to save dead time. However, due to its long latency of 10 μs, the hit information cannot be used to start the acquisition process.

This results in a limited trigger efficiency for reactions with purely neutral final states, like $\gamma n \rightarrow n\pi^0$. The result of a simulation of the current efficiency is shown in Fig. 1.

### 2.2 Upgrade within the SFB

The goal of this subproject was to modify the existing readout of the Crystal Barrel Calorimeter, or to build a new one, which generates hit signals fast enough to be included in the first level of the experiment trigger. The most obvious approach - extending the readout of the Forward Conus to the remaining calorimeter - is not applicable: the Forward Conus uses photomultipliers to convert the scintillation light into an electrical signal. As these are not operational in strong magnetic fields, such a readout would prevent the magnet upgrade which is planned in extension of the TPC upgrade.

### 2.3 Challenges of extracting fast timing information from CsI:Tl

As seen in section 2.1, the hit information of the calorimeter has to be present at the central trigger module within 500 ns after the physical event in the target. This is conflicting with the scintillation properties of the CsI:Tl crystals of the Crystal Barrel Calorimeter. The light emission has a decay time constant of $\tau$ = 1 μs. This means that only a fraction of the light has been emitted, when the hit information has to be present latest. Furthermore, the full 500 ns cannot be used to accumulate the signal. Within that time span, also signal processing and transmission has to take place. The signal processing includes the evaluation of the number of clusters in the whole calorimeter from the hit pattern of each single crystal (cluster finding). Assuming a time span of 150 ns for the signal accumulation results in only ~ 10% of the full signal.

Moreover, the crystals are large compared to a typical photodetector. This means that the scintillation light in average has to be reflected many times inside the crystal before it can be absorbed in the photodetector. This allows photons to be absorbed in passive material, before it can reach the photodetector and contribute to the output signal. The light collection of the finally used readout was measured to be 3000 Photons/MeV [10]. This is an increase compared to the previous readout of 2000 Photons/MeV [11], but small compared to the total amount of light (54000 Photons/MeV [12]).

So, although CsI:Tl is known as a very bright scintillator, the readout has to be able to process comparatively small scintillation signals. Furthermore, it has to be able to process the fully accumulated scintillation signal, as this is used for the energy determination.

The time span actually used for the accumulation of the scintillation signal results from the design of the readout and has to be chosen carefully. It should be large enough to yield a sufficiently high signal for a given energy deposit. On the other hand, the hit information has to fit into the existing trigger scheme. A too short time span will result in poor signal strength, which will increase the detection threshold for the hit information. A too large time span might result in a loss of hit information of other detectors in the experiment.

### 2.4 Implementation

Three approaches have been evaluated during the SFB:

1. Keep the existing readout and extend it by a timing branch, which uses signal filters that generate fast timing signals.
2. Use the above mentioned filters, but also replace the existing readout with one based on avalanche photo diodes.
3. Extend the front-end with silicon photomultipliers, to get fast signals.

Options 1 and 3 resulted in either too high dark count rates or too high detection thresholds and were therefore discarded. Option 2 was developed into a final solution and elaborately tested on prototypes at accelerator facilities. The solution and the tests will be described in the following sections.

#### *2.4.1 The new Front-end*

The new front-end consists of two avalanche photodiodes (APDs), a charge sensitive preamplifier, a high voltage supply circuit, and a line driver.

*Avalanche Photodiode*

The front-end utilizes two APDs of the type S11048(X3) from Hamamatsu, operated at a gain of 50. Each APD has an active area of $6.8 \times 14$ mm$^2$ and has been characterized for temperature and voltage dependence [14]. The average coefficients were found to be $\alpha_T = -2.28$ %/K and $\alpha_V = 2.94$ %/V. The temperature coefficient varies in the range $\alpha_T = -2.16 ... -2.50$ %/K and the voltage coefficient in the range $\alpha_V = 2.80 ... 3.35$ %/V. The main part of the distributions have standard deviations of

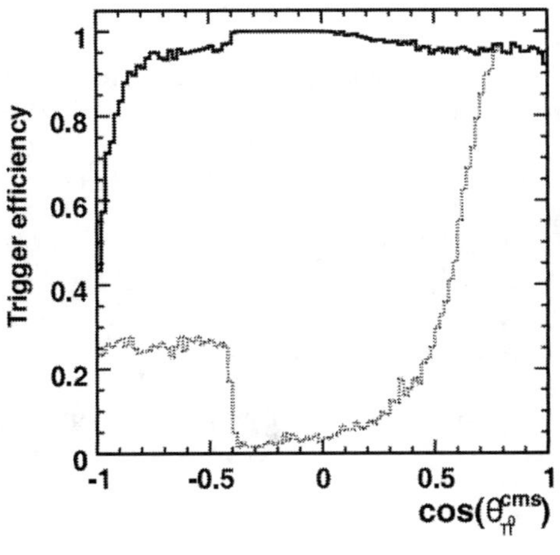


(a) Photonenergy $E_\gamma$ = 700 MeV.

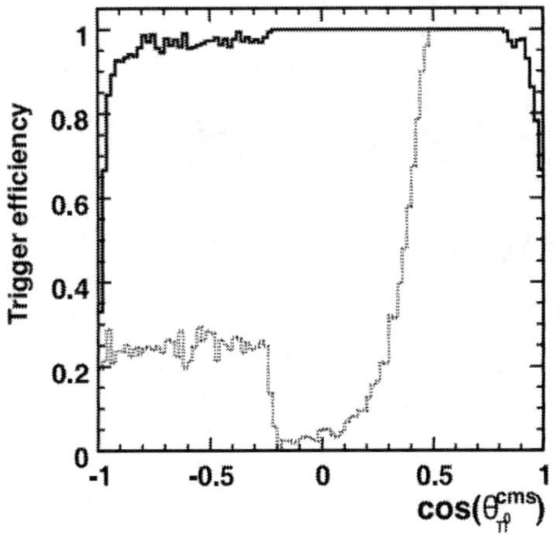


(b) Photonenergy $E_\gamma$ = 2 GeV.

**Figure 1.** Trigger efficiency for the reaction $\gamma N \rightarrow N\pi^0$, for different photon energies. Dashed line: Efficiency for neutrons. Solid line: Efficiency for protons. [9]

0.08 %/K and 0.08 %/V respectively [14]. The dark current was specified by the manufacturer to be in average $I_D$ = 14.8 nA, with a total span of 5.6 nA $\leq I_D \leq$ 33.7 nA at 25 °C. The front-end utilizes two instead of one APD to improve the energy resolution below 10 MeV energy deposit, to lower the threshold in the timing channel, and to enhance failure safety by redundancy.

*Preamplifier*

Each of the two APDs has its own preamplifier, which is a charge sensitive design. The amplification is $G$ = 147 V/nC, which results in a dynamic range of up to $E_{max}$ = 3.7 GeV for the crystals with the highest light yield.

The decay time constant was chosen to be $\tau$ = 54 µs, which is short enough to not cause extensive pile-up on the output signal and long enough to not degrade the signal strength due to ballistic deficit.
The preamplifier was originally developed for the P̄ANDA experiment at the University of Basel and adapted for the usage in the Crystal Barrel Calorimeter. The printed circuit board (PCB) is manufactured in flexrigid technology. The amplifier circuit and the APDs are soldered on rigid parts, which are connected by a flexible part, which is a necessary feature during the assembly process. One of the 1300 produced new preamplifiers is shown in Fig. 2.

**Figure 2.** The new APD preamplifier.

*High Voltage Supply Circuit*

To enable the internal gain mechanism, the APD needs an external bias voltage in the order of $V_B$ = 400 V. Due to variations in the manufacturing process, each APD has a different bias voltage at which the desired gain of 50 is reached. One way to not need a unique high voltage channel for each APD is to group them. However, it was unknown how the desired bias voltage changes due to aging and radiation damage, which could result in the desired bias voltage of a group drifting apart by time. Therefore, another approach was used.
Each front-end module is supplied with a fixed high voltage of 500 V. This voltage is reduced to the desired value in the front-end, independently for each APD. The voltage is set using a digital potentiometer, which can be remotely controlled ([15]) in the experiment setup and can be tuned in the range of 328 V $\leq V_B \leq$ 458 V in steps of $\Delta V$ = 127 mV.
To address the problem of the large temperature coefficient of the APD's gain, one implemented measure amongst others is a gain stabilization circuit, which is part of the HV supply circuit. This circuit utilizes an NTC to measure the temperature and to change the bias voltage in such a way that the gain change due to temperature change is counterbalanced.
A 16 Bit ADC allows to monitor the output voltages and the temperature. The HV supply card is shown in Fig. 3.

*Line Driver*

The signal is differentially transmitted to the back-end. The line driver circuit converts the preamplifier signal to a differential signal. Additionally, the gain of the driver circuit is adjustable in 16 steps to address the variation of the light yield of the crystals. The span between the brightest

**Figure 3.** The new High Voltage Supply Card.

and darkest crystal was known to be around a factor of 2.5. By setting the gain of the line driver to the proper value, the dynamic range on the transmission lines is equalized to $E_{max}$ = 2.5 GeV

For each two preamplifiers in the font-end of one crystal, only one signal is transmitted. Therefore, the preamplifier signals need to be multiplexed. The multiplexer circuit outputs the average of both APDs by default. The averaging reduces the electronic noise of the front-end. Additionally, the multiplexer can be configured to output only the signal of one of the APDs, which increases the reliability, in the case that one of the preamplifiers is broken and outputs a bad signal.

Also, this feature turned out to be very valuable to check the quality of the optical contact between APD and crystal in the assembly process.

### 2.4.2 The new Back-end

The back-end consists of a timing branch, which provides the fast trigger information as well as timing data, and an energy branch, which is used to measure the energy deposit in the crystals. In the first step of the upgrade, the existing energy readout will be retained. In the second step, the existing QDCs will be replaced by SADCs, which are currently in development. The timing branch has to be fully built up.

*The Existing Energy Branch*

The current energy branch uses signal filters which output pulses with a peaking time of ~ 3 μs for the CsI:Tl preamplifier signals. The pulse area is digitized with LeCroy 1885F dual range QDCs.

*The New Timing Branch*

The main parts of the timing branch are a signal filter and a discriminator. The signal filter is used to achieve signals with a rise time, which is short enough to allow for integration into the first level trigger. The discriminator converts the analog timing signal into a digital hit information. The custom designed modules also include TDCs and a part of the cluster finder electronics.

*Timing Filter Modules*

The timing filters are $CR(RC)^2$ filters. The low pass is implemented in a dual feedback topology. In most prototypes, the time constant was chosen such that the peaking time of the step response is 80 ns. In the final design, this time was increased to 100 ns to further increase the signal to noise ratio.

The so achieved timing signal is differentially transmitted to the discriminator module. An additional output allows for a further extension of the module by an analog summation circuit. This is planned to be used as a fast energy sum of the whole calorimeter.

To connect the energy readout, the module has two buffered outputs which forward the preamplifier signal. As there are two outputs available, the upcoming SADC can be operated in parallel to the existing QDC in a commissioning phase.

One PCB covers 23 channels which reflects the detector geometry of 30 subsections of 23 channels in each half (Fig. 4). Two PCBs are housed in a double width NIM cassette.

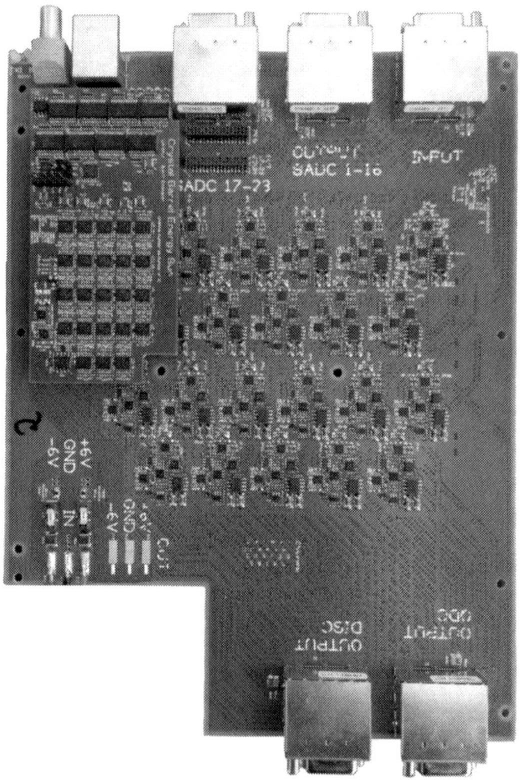

**Figure 4.** Timing filter module with distribution to the QDC/SADC boards.

*Discriminator*

The discriminator is a VME-Module. It contains 92 channels corresponding to 4 subsections of the calorimeter. Each channel has 2 comparators which allow for a time walk correction in the module. A Spartan 6 LX150 FPGA processes the comparator signals. A time information is extracted and as well as the first part of the clusterfinding is processed here (Fig. 5).

MRJ21 connectors are used to achieve the required high input connector density. The 184 TDC channels are implemented utilizing the SERDES2 primitives on the Spartan 6 FPGA. Thereby a sampling rate of 800 MSPS is achieved, corresponding to a binning of 1.25 ns [16]. The resolution achieved is appropriate for the expected time resolution of the fully upgraded detector (see Sec. 2.5.3).
To use this module also on detectors with a much better time resolution, an existing FPGA TDC based on the carry chain approach [17] was adapted to the discriminator design and evaluated. A Time resolution of $\sigma_t \approx 30\ \mathrm{ps_{RMS}}$ was measured for sufficient sharp signals [18] Pulses with a width of 1.2 ns (FWHM) were shown to be reliably detectable, although the threshold has to be set to lower values than the amplitude of the pulse [18].

**Figure 5.** Discriminator board with first part of the cluster finder.

### 2.4.3 Light Pulser System

The former experiment setup used a light pulser system to perform a calibration of the high range versus the low range of the QCDs in the energy readout. This system had issues like low rate capability of a few Hz, a rate dependence of the flash intensity, and a maximum intensity which did not allow to drive the QDC up to the highest channels.
The new light pulser system [14] utilizes a LED instead of a discharge bulb. The color of the LED matches the scintillation light wavelength. 6 LEDs with a power of 5 W each are used for the whole calorimeter. The custom developed driver circuit was tested with flash rates up to $f = 20$ kHz [14], and a maximum energy equivalent of $E_{\mathrm{Eq,max}} = 6$ GeV. The flash intensity can be set using a DAC in the driver circuit. A preliminary analysis found the rate dependence of the intensity to be less than 2% [14], over the range from 2 kHz to 5 kHz. A more precise determination is ongoing.

*APD Gain Monitoring*

Additionally to the QDC calibration, the light pulser system will be used for another purpose which is actually the main purpose of the upgrade: it will be used to measure the gain of the APDs during data taking in a photoproduction measurment.
The gain $G$ will be calculated using the quotient of the intensities with applied bias voltage $A_{\mathrm{HV\,ON}}$ and with bias voltage set to zero $A_{\mathrm{HV\,OFF}}$.

$$G = \frac{A_{\mathrm{HV\,ON}}}{A_{\mathrm{HV\,OFF}}} \tag{1}$$

A necessary key feature of the APD is that the temperature dependence of the signal strength is negligible compared to the temperature dependence of the light yield of the scintillation process in the calorimeter crystals.
As the temperature coefficient at zero bias condition is negligible, this operating mode is used as a reference point. Thus, changes in the light pulser system, like a changed transmission strength of the light fibers, can be cancelled out by repeating the reference measurement. It is planned to run the light pulser system during data taking in bias on mode and perform reference measurements only if necessary. To keep the light pulser as stable as possible, the system has been built up in a temperature stabilized environment. Thereby the brightness temperature coefficient of the LED of 0.25 %/K is addressed [14].

Another property of the APD makes the reference measurement more challenging: the parasitic capacitance of the APD is increased to roughly $C_P = 7$ nF [14] at zero bias condition. In addition to the fact that the signal is smaller as the gain of the APD equals one, this introduces two more problems. First, the noise level is increased, as the capacitance of the APD is a directly influencing factor. Second, the signal strength is decreased even more: the charge sensitive preamplifier decouples the signal from the high potential of the APD's cathode using a capacitor. The signal charge is effectively divided between the input decoupling capacitance of the preamplifier and the parasitic detector capacitance $C_P$.
In the produced design, the decoupling capacitance is chosen to be 47 nF to minimize the effect. The capacitance under operating conditions is reduced to 17 nF due to the DC bias effect, which results in limiting the signal loss to 30%. The high intensity capability of the new LED light pulser system allows to achieve sufficient signal to noise ratios despite the increased noise level.

### 2.4.4 Ultra Fast Cluster Finder

The former cluster finder was implemented on ASICs. The achieved latency is $\Delta t = 0.8\ \mu\mathrm{s} + n \cdot 0.8\ \mu\mathrm{s}$, where $n$ is the number of clusters. In the former operation, the result of

the encoder was evaluated 10 μs after the first level of the trigger had fired. To include the Crystal Barrel Calorimeter into the first level of the trigger, this process has to be improved to give a result within the available latency of 500 ns, sharing that time span with the signal generation in the signal filters, the signal processing in the discriminators, as well as the signal distribution.
To fulfill this constraint, an entirely different encoding algorithm was chosen. Instead of latching the hit pattern and successively identifying the clusters in it, the new algorithm processes the whole hit pattern in parallel. Key idea of this approach is to use a certain hit pattern to identify one cluster, instead of determining connected entries in the hit pattern, like the former algorithm did.

The demanded hit pattern is shown in Fig 6(a). Each box represents a crystal. To be identified as a cluster, the crystal drawn in green has to have a hit, while the crystal drawn in red must not have one. Other crystals are not evaluated for this position. To cover the whole calorimeter, the pattern has to be checked on all possible positions on the detector matrix.
As the pattern only evaluates the local neighborhood of a crystal, it is not necessary to have the full hit information stored in one unique module. The contrary is the case: the algorithm allows by definition to split the encoding in subsectors and to distribute their processing over multiple modules. Only the hit information on the borders has to be exchanged between the processing units.
In the second stage, the cluster information, which is encoded in over 1000 yes/no decisions, has to be processed into a number which represents the number of present clusters. This is done by an adder in hierarchic structure. In each step, the number of interim results is halved, until after 11 steps one single number is present.
Another advantage is that no latching of the hit pattern is required as it was in the former algorithm. Implemented in a pipeline architecture, the algorithm can operate free running and provide a new number of clusters each clock cycle.

The algorithm is being implemented in the discriminator modules, as far as possible regarding the available hit data. The preprocessed information will be transmitted to a central module where the number of clusters is finally determined. According to the current development status, a total latency of less than 100 ns seems achievable.

Simulations have been performed to analyze the potential of the algorithm. The number of clusters found is shown in Fig. 7. The comparison to the former algorithm shows that in few cases a larger number of hits is found, while a lower number is in no case found. So, compared to the former algorithm, the sensitivity is at least the same, while the selectivity is slightly worse.
The supernumerary clusters are believed to originate from complex shaped clusters.

Other studies have shown that the new algorithm is also able to detect overlapping clusters to some degree. Therefore the number of detected clusters was compared to the number of particle energy deposits found in the full analysis. [16]

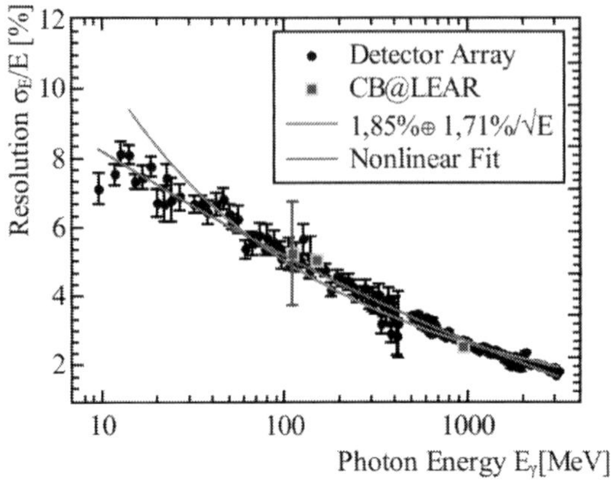


**Figure 8.** Measured energy resolution of the new and the former readout. The curves represents fits to the data [20].

## 2.5 Prototype Evaluation

In total, four generations of prototypes of CsI:Tl crystals with APD readout have been built up within the SFB. In this section, major results will be shown.

### 2.5.1 Test Setup and Test Sites

To be able to perform conclusive measurements on a photon beam, a 3×3 array of detector crystals from the Crystal Barrel Calorimeter has been built up. The array resembles a sub unit of the whole calorimeter. As only one size type of the crystals was available, the array is not perfectly congruent to any sub unit. Simulations showed that around 90% of the electromagnetic shower deposits its energy in the active material in the relevant range of incident photon energy.

To determine the energy and time resolution, measurements at the accelerator facilities MAMI and ELSA have been performed. The array was positioned in the photon beams of the A2 and the CBELSA/TAPS experiments. The tagging systems have been used to obtain the energy information of the incoming photons, as well as their time. Measurements have been performed at incident electron energies of $E_e$ = 180, 450, 800, 1500, 2400, and 3200 MeV. This allowed to investigate the detector performance for photon energies in the range 9 MeV $\leq E_\gamma \leq$ 3.1 GeV.

### 2.5.2 Energy Resolution

The determined energy resolution is shown in Fig. 8 as black circles [20]. Blue squares represent the energy resolution that was found with the former readout, when the calorimeter was installed at LEAR.
The red curve represents a fit to the data:

$$\frac{\sigma_E}{E} = \frac{(0,68 \pm 0,19\%)}{\sqrt{E/\text{GeV}}} \oplus \frac{(2,56 \pm 0,11\%)}{\sqrt[4]{E/\text{GeV}}}. \quad (2)$$

DOI: 10.1051/epjconf/201713405003

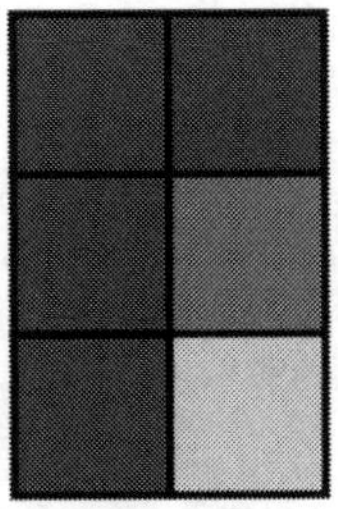

(a) Demanded hit pattern. Dark green: Hit demanded, dark red: hit must not be present, gray: not evaluated.

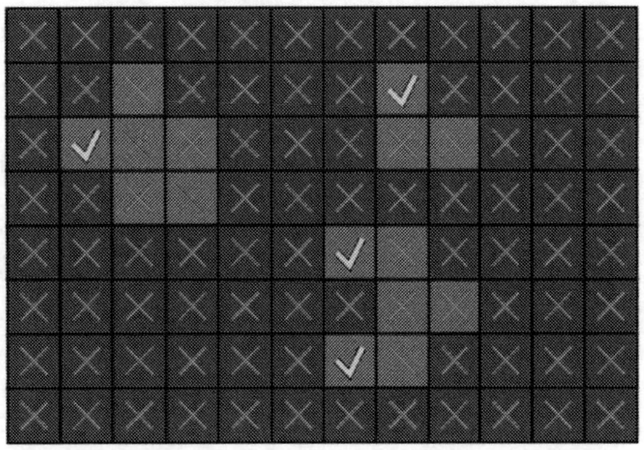

(b) Example hit pattern. Fulfilling of pattern shown in Fig. 6(a) denoted by green tick, not fulfilling by a red cross. The cluster on the lower right is mistakenly detected as two.

**Figure 6.** Illustration of the new cluster finder algorithm.

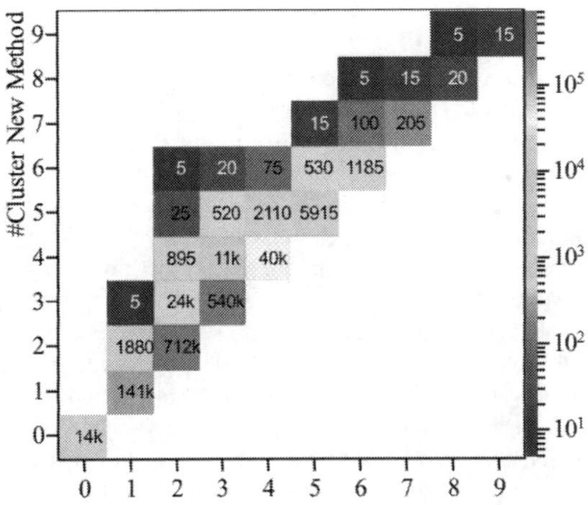


(a) $p\pi^0$

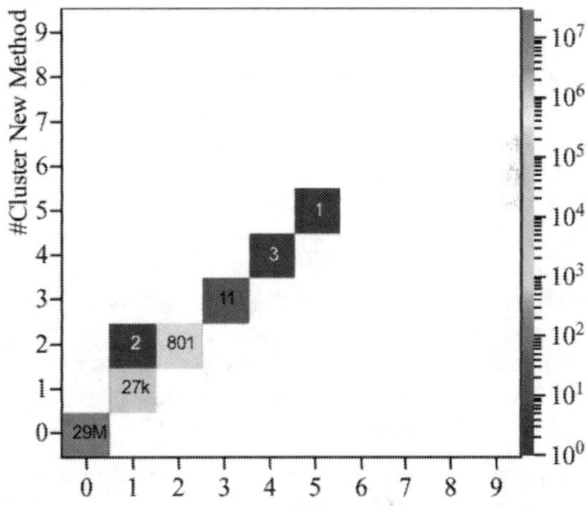


(b) Electromagnetic Background Reactions

**Figure 7.** Number of identified clusters using the former and the new algorithm. Data from Monte Carlo Analysis. [8]

The phenomenological $E^{-\frac{1}{4}}$ term was added to get a more precise description of the data. This term is also used for other CsI:Tl calorimeters to parameterize the energy resolution, and was the only term in an earlier publication about the Crystal Barrel Calorimeter [11]. A further constant term and one proportional to $1/E$ were compatible to zero in the fit. If the energy resolution needs to be parameterized for practical reasons, like applying energy dependent cuts, it seems more reasonable to use the physically unmotivated parametrization $\frac{\sigma_E}{E} = 0.117 - \frac{0.01657}{\log(E/\mathrm{MeV})} + \frac{0.0005}{\log^2(E/\mathrm{MeV})}$, which is shown as a green line in Fig. 8.

### *2.5.3 Time Resolution*

The time resolution was extracted from the same data, which was used to determine the energy resolution. There are few differences one has to be aware of. While the energy of all crystals in one event is summed up for the determination of the energy resolution, crystals are treated independently to extract the time resolution.

The time resolution can be determined for energies much below the low energy end of the tagged photon range, as the electromagnetic shower distributes the energy of the incident photon among multiple crystals.

For each event, the energy entry of each crystal is saved as well as the time information if available. The hit signal of the tagger hodoscope was used as time reference.

The trigger condition during the data acquisition at the photon beams was:

```
(hit in any crystal) AND (hit in the tagger).
```

Fig. 9(a) shows an example of the time information of one crystal in dependence of the energy deposit in that crystal.

The vertical width of the distribution results from the accuracy of the time information, while the energy dependence

 DOI: 10.1051/epjconf/201713405003

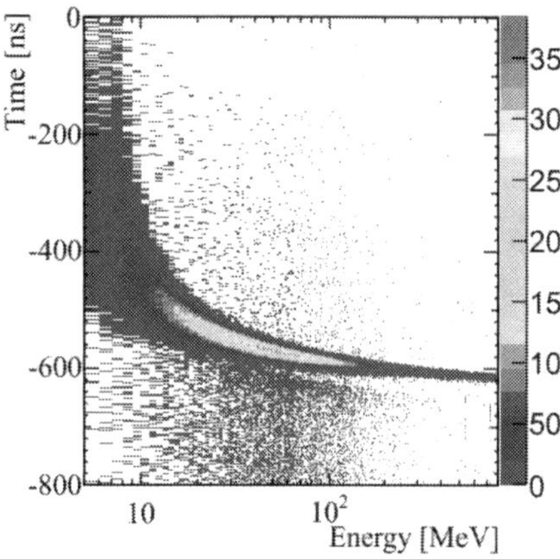


(a) Time information in dependence of energy deposit.

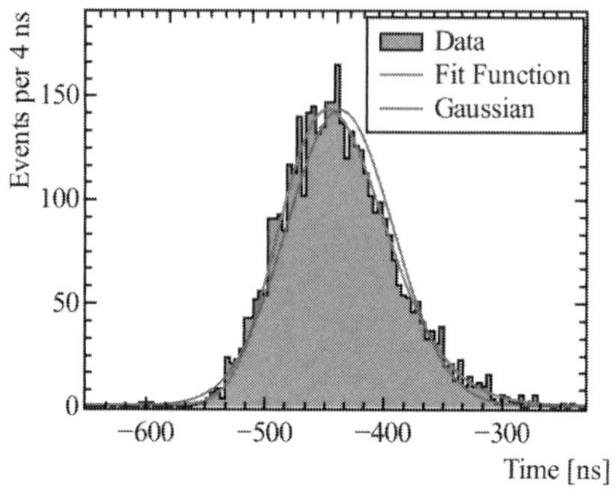


(b) Time information for a fixed energy with fits.

**Figure 9.** Diagrams on the time resolution [20].

of the mean time at a fixed energy results from the walk. As the vertical width of the distribution is energy dependent, the time resolution was determined in dependence of the energy. To achieve the standard deviation from the data, the Novosibirsk fit function was fitted to the data. This function implements a parameter $\tau$ characterizing the asymmetry of the distribution. For $\tau \rightarrow 0$, this function approaches the Gaussian distribution. The function was chosen regarding the slight asymmetry of the time information distribution. Fig. 9(b) shows the data of one energy bin. The used fit is shown in red. The fit of a Gaussian distribution is shown in green for comparison.

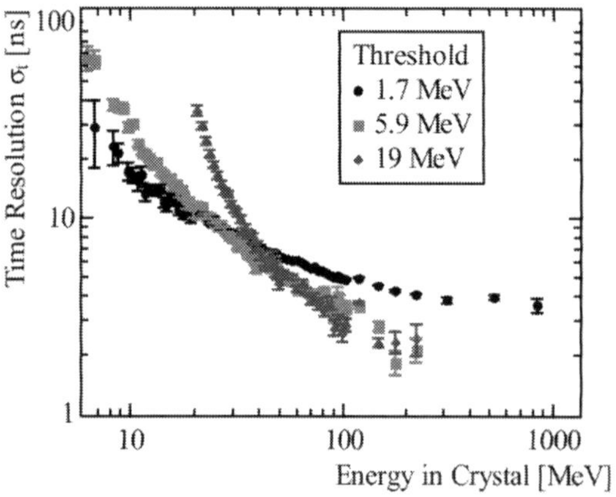


**Figure 10.** Time resolution at different thresholds in dependence of the energy deposit [20]

Fig. 10 shows the time resolution $\sigma_t$ of a typical crystal. The resolution improves with the energy deposit by more than an order of magnitude and reaches values around $\sigma_t = 2\ \mathrm{ns_{RMS}}$.

The threshold of the discriminator has not only influence on the minimal detectable energy, but also on the time resolution. The resolution improves significantly at low energies if the threshold is lowered.

The custom build discriminators utilize two thresholds to compensate for walk of the timing signal. The simplest compensation approach is to forward the information of the lower threshold to the output, but only if the higher threshold was crossed within a given time limit after the lower one was crossed. By this, only pulses are forwarded which have more than a minimum steepness between the thresholds, which corresponds to rejecting pulses below a certain energy. This energy is defined by the thresholds and the allowed delay. In Fig. 9(a) it is visible, that the major time shift of the walk is present at low energies. Also the worst time resolution is found in the low energy

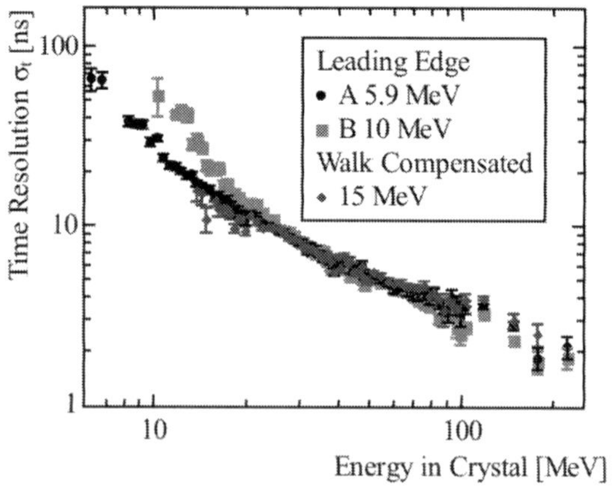


**Figure 11.** Time resolution of two thresholds and the resolution of the compensated signal (blue), which is generated from two thresholds (black, red) [20].

region. The timing signals will be used in the trigger using a coincidence. Therefore, the whole variation of the information has to be taken into consideration when the coincidence window is set.

The compensation algorithm offers two advantages. First, low energy hits which have the worst timing resolution, are discarded (see Fig. 11). Second, the time resolution just above the detection threshold is better for the compensated information than for a leading edge discriminator having the same threshold. This can be seen in Fig 12 in the range 15 MeV $< E <$ 30 MeV. The compensated information has the same time resolution as the lower threshold, which was used to generate it. Only at high energies a slight degradation is visible, which results from the clocked operation of the compensation electronics.

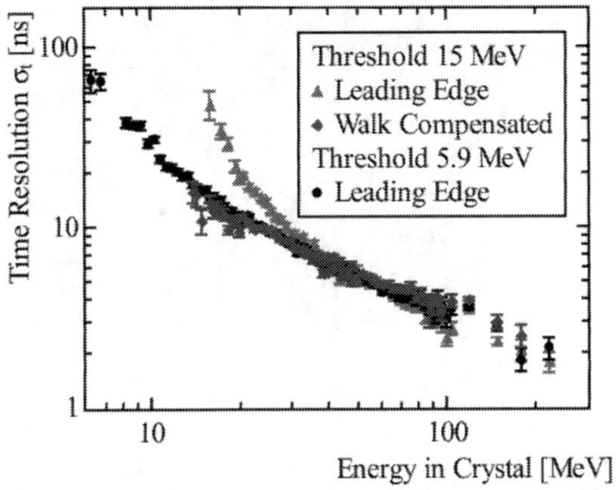


**Figure 12.** Time resolution of two thresholds and the resolution of the compensated signal [20].

The upcoming SADCs will allow to extract the time information also from the energy channel. A measurement has been performed using a digital oscilloscope to estimate the achievable resolution and threshold. For the measurement, a light pulser was used to generate signals in the front-end. The signal of the pulser was chosen such that the leading edge of the pulse in the timing channel resembles that from scintillation events. Fig. 13 shows

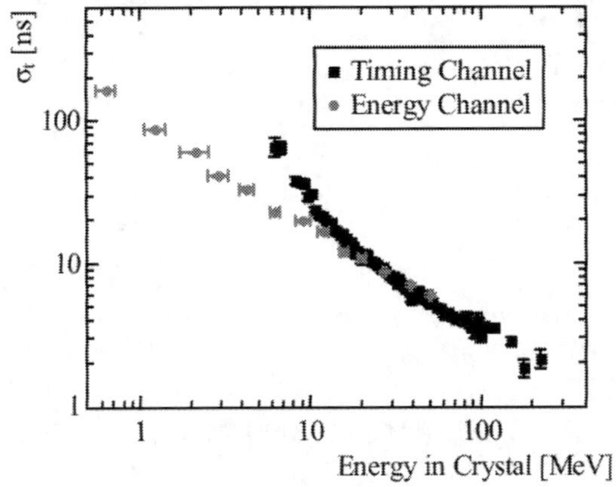


**Figure 13.** Time resolution of two thresholds and the resolution of the compensated signal [20].

the so found time resolution, compared to the resolution found in the timing channel. The most distinct difference is the much lower detection threshold in the energy channel. The paid price for this improvement is an increased latency of the information. The time was extracted at 50% of the full amplitude which is 1.5 µs after the energy deposit. Therefore, the SADCs would not be able to replace the trigger upgrade or the timing channel, still they might provide valuable information for the data analysis.

### 2.5.4 Online Gain Monitoring

The latest prototypes of the sub-detector unit included a light pulser system, which was suited to measure the gain of the APDs during a measurement at a tagged photon beam (see s Sec. 2.4.2). The gain measured during a warm up phase is shown in Fig. 14. Investigations about the

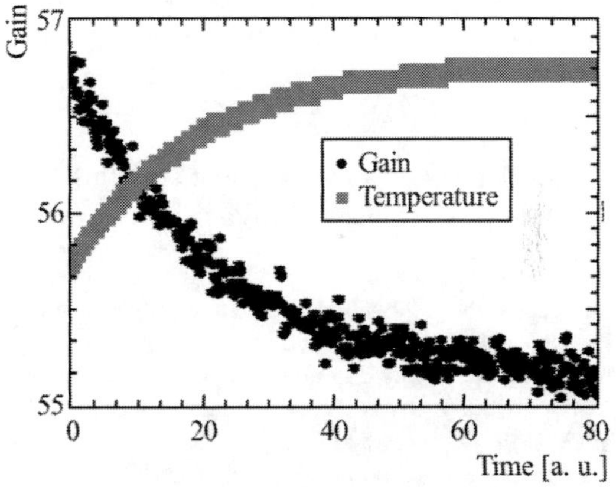


**Figure 14.** Gain variation during a warm up phase [14].

optimum number of flashes to calculate the gain and the required frequency are ongoing [14].

### 2.5.5 Online Gain adjustment

Although the calorimeter will be operated at a stabilized temperature, a way to improve the gain stabilization to work in a wider temperature range was investigated [19]. Fig. 15 shows the APD gain in dependence of the temperature. The compensation circuit, which is part of the bias supply, stabilizes the gain around the desired operating temperature of 27 °C (shown as black circles). Adjusting the high voltage setting in the frontend allows to extend the flat region over a much wider temperature range. The method was tested with a single crystal in a temperature stabilized environment. The relative gain was determined using $^{22}Na$ as reference.

### 2.5.6 Status at the end of the SFB

The calorimeter has been fully upgraded with the new readout. The new backend is being produced. First measurements on a photon beam with a polarized target are planned for 2017.

DOI: 10.1051/epjconf/201713405003

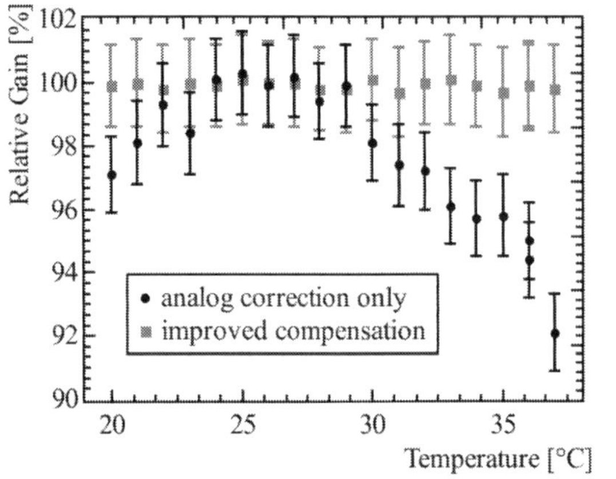


**Figure 15.** Temperature dependence of the gain. The analog correction shows a flat region at the desired operating temperature (black circles). The improved compensation extends the flat region (red squares) [19].

## 3 FPGA-based Sampling-ADC for the Crystal Barrel Calorimeter

The event-rate of the data acquisition system is currently limited by the digitization stage of the energy branch (LeCroy QDC1885F) to ≈ 2 kHz. To overcome this limitation, the implementation of a new Sampling-ADC (SADC) readout is the second important step in the upgrade of the detector system [21].

The prototype design of a 64-channel SADC for the PANDA electromagnetic calorimeter ([23]) was modified, adapting it to the needs of the Crystal Barrel Experiment. The new SADC offers 64 channels in one NIM module with up to 14 bit@80 MHz, accompanied by a custom modular low-noise power supply. Two prototypes are currently under investigation (see Fig 16).

A modular analog input stage allows signal shaping with up to $(CR)^5$-RC in combination with a digitally adjustable baseline and variable pole-zero compensation [22].

The on-board data processing and reduction is realized with two Kintex7 FPGAs. Using an FPGA provides a multitude of possibilities for online feature extraction, such as the determination of the energy deposited in the crystal, determination of the timing, and detection and correction of pile-up events.

The readout system will be realized with gigabit ethernet links and COTS switches for the backend, allowing a sufficiently fast readout with feature-extracted data, or alternatively a slower mode in which full samples of a variable length can be transmitted (oscilloscope-mode).

Preliminary measurements have been conducted with a simulated pulse to compare the energy resolution of the SADC to the current LeCroy digitizer ([24]). Different approaches have been compared (see Fig. 17), showing that the SADC performs well. Seemingly, a dual range setup will not be necessary. Two methods are exploited in [24] in which the digitizer is deliberately overdriven,

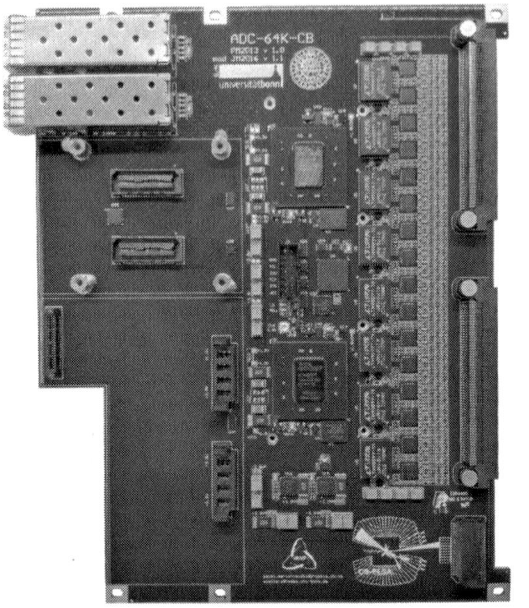


**Figure 16.** Second Prototype of the CB-SADC board.

allowing a tradeoff between better overall resolution and worse resolution starting with the clipping of the pulse.

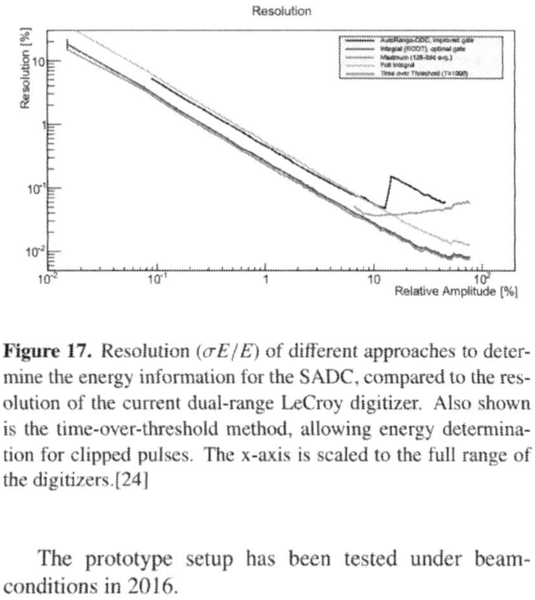


**Figure 17.** Resolution ($\sigma E/E$) of different approaches to determine the energy information for the SADC, compared to the resolution of the current dual-range LeCroy digitizer. Also shown is the time-over-threshold method, allowing energy determination for clipped pulses. The x-axis is scaled to the full range of the digitizers.[24]

The prototype setup has been tested under beam-conditions in 2016.

## 4 MiniTAPS intensity upgrade

The forward opening of the Crystal Barrel Calorimeter is covered by the TAPS detector system (MiniTAPS), consisting of 216 hexagonal $BaF_2$-crystals of 25 cm length, individually read out via photomultiplier tubes with bialkali photocathodes and fused silica windows to be sensitive to the UV-components of the $BaF_2$ luminescence. The

scintillation light consists of two major components based on different mechanisms. The fast component with a decay time of $\tau$ ~0.8 ns is based on a core-valence transition with emission wavelength at 195 nm and 210 nm, respectively. Due to this mechanism, the luminescence yield is independent of temperature. The slow component, based on self-trapped excitons, has a decay time of $\tau$ ~620 ns, peaking at an emission wavelength of 320 nm, representing nearly 90% of the overall scintillation yield. To collect the total scintillation response, a complete integration time of ~2 μs is required due to the slow component and is consequently limiting the rate capability of each individual detector component due to pile-up. Since the Mini-TAPS device is covering the most forward direction, it has to withstand highest rates. The reduced photon statistics due to the lower light output of the fast component has only a relatively small impact on the achievable resolution as shown in Fig. 18(a), which shows the achieved experimental energy resolution of the reconstructed shower. The concept of the upgrade is based on a hardware suppression of the slow scintillation component. Instead of using a photomultiplier equipped with a solar-blind photocathode, appropriate narrow band UV filters have been placed in between the crystal and photomultiplier. In a detailed study [7], various commercially available filters (see Fig. 18(b)) have been tested with respect to the transmitted absolute light yield and the impact on the resulting low energy threshold and achievable energy resolutions even for low energy $\gamma$-rays. The obtained results are very promising. However, the overall performance can be significantly improved by customized semiconductor-based filters with optimized transmission and dimensions adapted to the detector geometry, which requires development in collaboration with the manufacturer.
Another rate limiting factor is the currently used QDC based TAPS readout electronics which introduces significant deadtime due to its self gating design. To alleviate this situation a new readout based on sADCs is planned.

## 5 New Inner Tracker for the Crystal Barrel Detector

A full scale Time Projection Chamber (TPC) prototype was built for experiments in hadron collisions within the existing FOPI spectrometer at GSI and photoproduction experiments within the Crystal Barrel Calorimeter at ELSA. The prototype is composed of the following parts:

(a) An active gas volume of 48 dm$^3$ with a total drift length of 72.78 cm, an inner field cage with a diameter of 10.40 cm and an outer field cage with a diameter of 30.80 cm. In the cylindrical detection volume, traversing particles ionize the gaseous detection medium producing positive and negative charge carriers in an electrical drift field. This field, which is oriented parallel to the beam axis, is established by a cathode endcap downstream of the target position and the anode layer upstream of it. A field-defining system is required to keep the gradient of the field constant throughout the active volume. All electrode structures, made from metallic material patterned on a carrier material, are optimized in terms of achievable field homogeneity and a low material budget.

(b) An amplification stage that serves to multiply the ionization electrons after they have traversed the drift volume. In case of a GEM-TPC this is achieved by a triple GEM stack of circular shape which is composed of a plane-parallel set of GEM foils oriented perpendicular to the beam axis.

(c) A patterned electrode structure on the anode layer, the pad plane, where the signals corresponding to the collected charges are detected.

(d) The electronic readout based on the AFTER-chip of the T2K experiment for the pad plane (10000 channels) and a system of various detector-near sensors monitoring all relevant operation parameters like temperature, pressure and flow of gas and cooling media, high and low voltages.

(e) Gaseous and fluid media lines, low and high voltage supplies and their respective cables, pipes and terminals (panels) needed to operate the TPC, and in addition support, alignment structures and mounting points.

Fig. 19(a) shows a photograph of the full scale prototype that has been built. Furthermore a detailed technical design report (TDR) has been published [2].

### 5.1 TPC Beam Telescope and first Test Measurements

Systematic resolution studies of a GEM-TPC require an excellent external beam reference. For this purpose the TPC tracking beam telescope has been developed [3]. It consists of two planar GEM detectors with a strip readout of 400 μm pitch and 2 silicon strip detectors of 50 μm pitch which allow for a very precise beam definition of the order of ~25 μm. Triggering is done with 2 pairs of plastic scintillators of which one pair can be moved with a step motor to perform beam scans to align the whole detector setup. In the middle of the table, a mounting structure is holding a small prototype TPC which was developed to perform reconstruction studies. It has an outer diameter of 20 cm and a drift length of 7.96 cm. Gas amplification is performed with a triple-GEM-stack and its 1500 pads are read out with 4x6 T2K-AFTER boards. The pad plane consists of two different hexagonal pad types with outer radii of 1.25 mm and 1.5 mm which allow resolution studies depending on pad size and pad response. The testbench has been used at the electron beam at ELSA and is currently located at the beam dump of the COMPASS experiment at CERN to optimize the resolution by varying shaping settings and gas parameters in the halo of the high energetic muon beam of 160 GeV with negligible multiple scattering effects. First in-beam tests show a resolution of the track definition of about 25 – 30 μm and a spatial resolution of the TPC of about 200 μm with Ar/$CO_2$ (70/30) as a drift gas [3] [5]. To monitor and control all critical TPC parameters like high

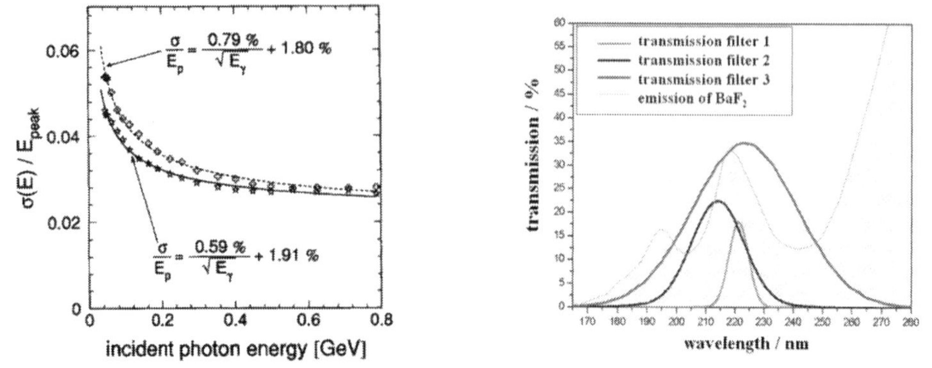


(a)                (b)

**Figure 18.** (a) Energy resolution of a $BaF_2$ sub-matrix of the TAPS detector measured with energy marked photons shown for the total light output (stars) and the fast component (crosses) only (taken from [6]). (b) Optical transmission of three selected optical filters in comparison to the two fast emission components of $BaF_2$. The emission spectrum has been normalized to the transmission of the used filters.

(a)                (b)

**Figure 19.** (a) Photo of the full scale Time Projection Chamber prototype. (b) Testbench setup on the beam dump platform of the COMPASS experiment at CERN. The muon beam is coming from the right side.

voltages, gas flow and temperatures a fully featured slow control system has been developed. It allows control via a graphical user interface and a web interface, and is designed to easily integrate into other systems, thus allowing for a simple integration of the TPC into existing experiments [13].

### 5.2 TPC Energy Calibration

The preferred method to estimate the energy resolution and perform a relative channel-to-channel calibration for a GEM-TPC in order to improve the overall dE/dx resolution is to feed radioactive $^{83m}$Kr into the detector. It originates from EC decay of $^{83}$Rb, mostly populating the isometric state at 41.6 keV which has a half-life of 1.83 h. By this decay different conversion electrons and photons with an energy between 9.4 keV and 41.6 keV are emitted which are detected inside the TPC volume. The calculated energy spectrum from $^{83m}$Kr decays taking into account photon conversion factors (for Ar/$CO_2$ 90/10) inside a TPC is shown in Fig 20(a). A rubidium source with an activity of 5 MBq has been produced at the HISKP cyclotron in Bonn with a $^{81}$Br($\alpha$,2n)$^{83}$Rb reaction (at 54 MeV with $\sigma_{\mathrm{max}}$ = 1200 mb) and has been integrated into the existing TPC setup with a bypass valve to switch between normal operation and calibration mode. Decay rates of the order of $10^6$ per second have been observed inside the GEM-TPC, also showing a very stable operation in a high rate environment.

Charge clusters inside the TPC from krypton decays differ significantly from clusters originating from particle tracks in size and overall charge. Nevertheless, inside a magnetic field of 600 mT, the clusters in the x-y plane are almost point-like or have a circular structure. Therefore the standard peak-finding clustering can be used. A larger cut on cluster length was used, as depending on the effective range inside the drift gas, higher energetic clusters might have a range corresponding to 10-15 samples for standard settings.

To perform a relative channel-wise calibration, a different method was also tested and compared to the leader-pad method that has been used before. Channel-wise spectra containing the cluster-charge are generated and rela-

DOI: 10.1051/epjconf/201713405003

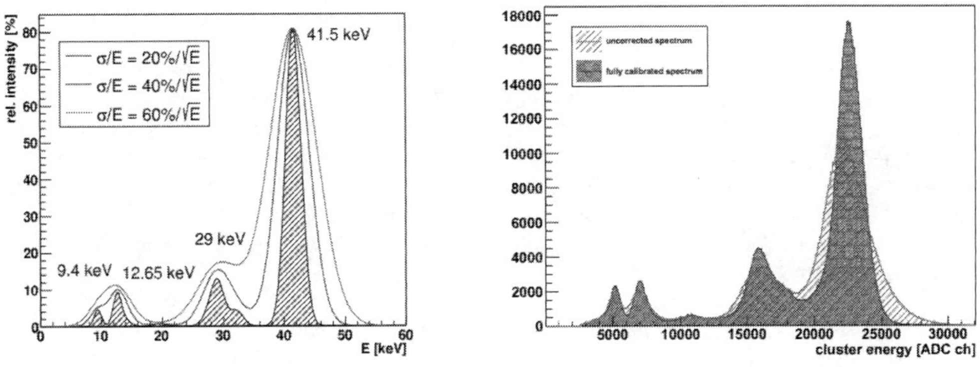


(a) (b)

**Figure 20.** (a) Theoretical $^{83m}$Kr spectrum for different energy resolutions (b) Measured energy distribution before and after calibration.

**Table 1.** The relative resolutions of the peaks in the krypton spectrum, with and without relative gain calibration.

| peak [keV] | uncorrected | corrected |
|---|---|---|
| 9.4 | 11.0 % | 6.6 % |
| 12.65 | 8.5 % | 6.0 % |
| 28.95 | 6.6 % | 4.0 % |
| 41.6 | 6.4 % | 3.6 % |

tive gain factors normalized to the median value of all 10254 channels are calculated by comparing the peak positions for each channel. Using this method, a much faster and more stable convergence was achieved than with the leader-pad method while delivering comparable results. It also simplifies identification of the main peaks as no hard cuts to preselect clusters have to be applied. Relative calibration is done iteratively rerunning the whole reconstruction and recursively correcting the gain factors. In that way a gain uniformity of less than 1% after 3 iterations can be achieved. [4]

For a first calculation of the energy resolution using the krypton spectrum, the relative width of the peaks was estimated from a fit to the clearly separated peaks, omitting the double-peak structure at 29.0/32.3 keV, with and without applying relative gain calibration. The spectra after correction show a significant improvement already in the main peak from 6.4% to 3.6%. The relative resolutions for the single peaks are given in Tab. 1. Taking into account noise and non-uniformities the energy resolution curve follows a $\sigma_E/E \sim 0.25/\sqrt{E}$ function, while after correction $\sigma_E/E \sim 0.18/\sqrt{E}$ was achieved. [5]

The work reported here would not have been possible without the enduring enthusiasm of many bachelor, master, and PhD students and the continuous technical support from staff members of the participating Universities. The support of the ELSA staff members and the polarized target group is gratefully recognized. We would like to thank C. Honisch, J. Müllers and M. Urban for contributions to this report. We appreciate the support of the Schweizerischer Nationalfonds, the US National Science Foundation, and the Russian Foundation for Basic Research, in particular the generous support from the Deutsche Forschungsgemeinschaft (SFB/TR16).

## References

[1] I. Jaegle *et al.* [CBELSA/TAPS Collaboration], *"Quasi-free photoproduction of eta-mesons off the deuteron"*, Eur. Phys. J. A **47** (2011) 89

[2] M. Ball *et al.* [GEM-TPC collaboration], *"GEM-TPC Technical design Report"*, arXiv:1207.0013v1 [physics.ins-det] (2012).

[3] A. Winnebeck, *"Design studies for a tracking upgrade of the Crystal Barrel experiment and installation of a tracking testbench"*, PhD. Thesis HISKP University Bonn (2009).

[4] R. Schmitz [GEM-TPC Collaboration], *"Energy calibration of a GEM-TPC Prototype with Kr-83m"*, EXA 2011 Conference Record, DOI: 10.1007/s10751-011-0555-6 (2012).

[5] R. Schmitz, *"Auflösungsstudien und Energiekalibrierung für eine GEM-TPC im Crystal-Barrel-Experiment"*, PhD. Thesis HISKP University Bonn (2017).

[6] A. R. Gabler *et al.*, *"Response of TAPS to monochromatic photons with energies between 45 and 790 MeV"*, Nucl. Instr. and Meth. in Phys. Res. A 346 (1994) 168.

[7] S. Diehl, *"New Readout Concepts for Electromagnetic Calorimeters"*, Bachelor Thesis University Gießen (2010).

[8] F.N. Afzal, *"Measurement of the double polarization observable E in $\pi^0$-, $\eta$- and $\eta'$- photoproduction"*, PhD. Thesis HISKP University Bonn (2017).

[9] H.v. Pee, private communication (2009).

[10] S. Undorf, *"Untersuchungen zur Lichtsammlung von Caesiumiodid-Kristallen"*, Bachelor Thesis HISKP University Bonn (2010).

DOI: 10.1051/epjconf/201713405003

[11] E. Aker et al., Nucl. Instr. and Meth. in Phys. Res., A 321, 69-109 (1992).

[12] Saint-Gobain Crystals, CsI(Tl), CsI(Na) Cesium Iodide Scintillation Material, Datenblatt, 2007.

[13] D. Kaiser, *"Steuerung und Überwachung einer Zeitprojektionskammer mit GEM-Auslese"*, PhD. Thesis HISKP University Bonn (2014).

[14] M. Urban, *"Aufbau eines Lichtpulsersystems zur Messung des Verstärkungsfaktors der APDs in der neuen CB-Ausleseelektronik"*, PhD. Thesis HISKP University Bonn (2017).

[15] J. Müllers, *"Design and Test of a Slowcontrol for the APD Upgrade of the Crystal Barrel Calorimeter"*, Master Thesis HISKP University Bonn (2013).

[16] P. Klassen, *"Design and Test of a new Cluster Finder for the Crystal Barrel Calorimeter"*, PhD. Thesis HISKP University Bonn (2017).

[17] J. Bieling, github.com/jobisoft/jTDC.

[18] G. Grutzeck, *"Implementierung eines Carry-Chain TDCs auf dem neuen CBDiskriminator Modul"*, Bachelor Thesis HISKP University Bonn (2016).

[19] P. Pauli, *"Studies on an Automatic Gain Stabilisation for the new APD Readout of the Crystal Barrel Calorimeter"*, Bachelor Thesis HISKP University Bonn (2015).

[20] C. Honisch *"Design, Aufbau und Test einer neuen Ausleseelektronik für das Crystal-Barrel-Kalorimeter"*, PhD Thesis, HISKP University Bonn (2014).

[21] J. Müllers *"An FPGA-based Sampling-ADC Readout for the Crystal Barrel Calorimeter"*, PhD. Thesis HISKP University Bonn (2017).

[22] T. Poller, *"Simulation, Aufbau und Test einer Schaltung zur Baseline-Verschiebung der Crystal Barrel Kalorimetersignale für den Sampling-ADC"*, Bachelor Thesis HISKP University Bonn (2016).

[23] P. Marciniewski, Angströmlaboratoriet, Uppsala, private communication (2014).

[24] J. Schultes, *"Test and Improvement of Feature-Extraction Methods for the new SADC-Readout of the Crystal Barrel Calorimeter"*, Master Thesis HISKP University Bonn (2016).

# AUTHOR INDEX